Lecture Notes in Computer Science 11825

More information about this series at http://www.springer.com/series/7408

Graham Hutton (Ed.)

Mathematics of Program Construction

13th International Conference, MPC 2019
Porto, Portugal, October 7–9, 2019
Proceedings

Springer

Editor
Graham Hutton
University of Nottingham
Nottingham, UK

ISSN 0302-9743 ISSN 1611-3349 (electronic)
Lecture Notes in Computer Science
ISBN 978-3-030-33635-6 ISBN 978-3-030-33636-3 (eBook)
https://doi.org/10.1007/978-3-030-33636-3

LNCS Sublibrary: SL2 – Programming and Software Engineering

This Springer imprint is published by the registered company Springer Nature Switzerland AG
The registered company address is: Gewerbestrasse 11, 6330 Cham, Switzerland

Preface

This volume contains the proceedings of MPC 2019, the 13th International Conference on Mathematics of Program Construction.

This conference series aims to promote the development of mathematical principles and techniques that are demonstrably practical and effective in the process of constructing computer programs. Topics of interest range from algorithmics to support for program construction in programming languages and systems. Typical areas include type systems, program analysis and transformation, programming language semantics, security, and program logics. The notion of a 'program' is interpreted broadly, ranging from algorithms to hardware.

The conference was held in Porto, Portugal, from October 7–9, 2019, and was co-located with FM 2019, the Third World Congress on Formal Methods. The previous 12 conferences were held in 2015 in Königswinter, Germany (with proceedings published as LNCS 9129); in 2012 in Madrid, Spain (LNCS 7342); in 2010 in Québec City, Canada (LNCS 6120); in 2008 in Marseille, France (LNCS 5133); in 2006 in Kuressaare, Estonia (LNCS 4014); in 2004 in Stirling, UK (LNCS 3125); in 2002 in Dagstuhl, Germany (LNCS 2386); in 2000 in Ponte de Lima, Portugal (LNCS 1837); in 1998 in Marstrand, Sweden (LNCS 1422); in 1995 in Kloster Irsee, Germany (LNCS 947); in 1992 in Oxford, UK (LNCS 669); and in 1989 in Twente, The Netherlands (LNCS 375).

This volume contains one invited paper and 15 papers selected for presentation by the Program Committee from 22 submissions. Each paper was refereed by three reviewers, and the review process was conducted online using the EasyChair system. I would like to thank the Program Committee and the external reviewers for their care and diligence in reviewing the submissions, José Nuno Oliveira and his team for their excellent local arrangements, and Assia Mahboubi and Annabelle McIver for their inspiring invited talks.

October 2019

Graham Hutton

Organization

Program Chair

Graham Hutton	University of Nottingham, UK

Program Committee

Patrick Bahr	IT University of Copenhagen, Denmark
Richard Bird	University of Oxford, UK
Corina Cîrstea	University of Southampton, UK
Brijesh Dongol	University of Surrey, UK
João F. Ferreira	University of Lisbon, Portugal
Jennifer Hackett	University of Nottingham, UK
William Harrison	University of Missouri, USA
Ralf Hinze	University of Kaiserslautern, Germany
Zhenjiang Hu	Peking University, China
Cezar Ionescu	University of Oxford, UK
Mauro Jaskelioff	National University of Rosario, Argentina
Ranjit Jhala	University of California, USA
Gabriele Keller	Utrecht University, The Netherlands
Ekaterina Komendantskaya	Heriot-Watt University, UK
Chris Martens	North Carolina State University, USA
Bernhard Möller	University of Augsburg, Germany
Shin-Cheng Mu	Academia Sinica, Taiwan
Mary Sheeran	Chalmers University of Technology, Sweden
Alexandra Silva	University College London, UK
Georg Struth	University of Sheffield, UK

External Reviewers

Robin Hirsch	Hans Leiß	Maciej Piróg
Christian Johansen	Alexandra Mendes	Kanae Tsushima
Hsiang-Shang Ko	Louis Parlant	

Local Organizer

José Nuno Oliveira	University of Minho, Portugal

Contents

Experiments in Information Flow Analysis

Annabelle McIver(✉)

Department of Computing, Macquarie University, Sydney, Australia
annabelle.mciver@mq.edu.au

Abstract. Designing programs that do not leak confidential information continues to be a challenge. Part of the difficulty arises when partial information leaks are inevitable, implying that design interventions can only limit rather than eliminate their impact.

We show, by example, how to gain a better understanding of the consequences of information leaks by modelling what adversaries might be able to do with any leaked information.

Our presentation is based on the theory of Quantitative Information Flow, but takes an experimental approach to explore potential vulnerabilities in program designs. We make use of the tool Kuifje [12] to interpret a small programming language in a probabilistic semantics that supports quantitative information flow analysis.

Keywords: Quantitative Information Flow · Probabilistic program semantics · Security · Confidentiality

1 Introduction

Consider the following scenarios:

(I) Paris needs a secretary and intends to pick the best candidate from a list of N applicants. His father King Priam knows Paris to be extremely impulsive making it highly likely that he will hire someone immediately after interviewing them, thereby running the risk that he might miss the best candidate. Knowing that he will not be able to suppress entirely Paris' impetuous nature, Priam persuades Paris to interview first a certain number n, and *only then* to select the next best from the remaining candidates. In that way Priam hopes to improve Paris' chance of hiring the best secretary. "But father, what value of n should I choose so that this strategy increases the prospect of my finding the best candidate overall? Wouldn't I be better off just relying on my instinct?"

This research was supported by the Australian Research Council Grant DP140101119.

G. Hutton (Ed.): MPC 2019, LNCS 11825, pp. 1–17, 2019.
https://doi.org/10.1007/978-3-030-33636-3_1

(II) Remus wants to hack into Romulus' account. Romulus knows this and wonders whether he should upgrade his system to include the latest password checking software. He has lately been reading about a worrying "side channel" that could reveal prefixes of his real password. However he is not sure whether the new software (which sounds quite complicated, and is very expensive) is worth it or not. How does the claimed defence against the side channel actually protect his password? What will be the impact on useability?

Both these scenarios share the same basic ingredients: there is a player/adversary who wants to make a decision within a context of partial information. In order to answer the questions posed by Paris and Romulus, we need to determine whether the information available is actually useful given their respective objectives. The result of such a determination for Paris might be that he would be able to form some judgement about the circumstances under which his proposed strategy could yield the best outcome for him. For Romulus he might be better able to decide whether his brother could marshal the resources required to breach the security defences in whichever password checking software he decides to install.

The aim of this paper is to illustrate, by example, how to analyse the consequences of any information leaks in program designs. The analysis is supported by the theory of Quantitative Information Flow (QIF). In particular QIF formalises how adversaries stand to benefit from any information leaks, allowing alternative designs to be compared in terms of specific adversarial scenarios. The presentation takes an experimental approach with the goal of understanding the impact and extent of partial information leaks. In so doing we acknowledge that this form analysis is almost certain to be incomplete, however its strength is to highlight relative strengths and weaknesses of program designs that handle confidential information.

In Sect. 2 we summarise the main ideas of QIF and show in Sects. 3 and 4 how they can be applied to the scenarios outlined above. We also touch on some logical features of modelling information flow in Sect. 5. In Sect. 6 we describe briefly the features of a recent QIF analysis tool Kuifje [12], and sketch how it can be used to describe the scenarios above and to carry out experiments to assess their information leaks.

2 Review of Quantitative Information Flow

The informal idea of a secret is that it is something about which there is some uncertainty, and the greater the uncertainty the more difficult it is to discover exactly what the secret is. For example, one's mother's maiden name might not be generally known, but if the nationality of one's mother is leaked, then it might rule out some possible names and make others more likely. Similarly, when some information about a secret becomes available to an observer (often referred to as an adversary) the uncertainty is reduced, and it becomes easier to guess its value. If that happens, we say that information (about the secret) has leaked.

Quantitative Information Flow (QIF) makes this intuition mathematically precise. Given a range of possible secret values of (finite) type $\mathcal{X}$, we model a secret as a probability distribution of type $\mathbb{D}\mathcal{X}$, because it ascribes "probabilistic uncertainty" to the secret's exact value. Given $\pi{:}\,\mathbb{D}\mathcal{X}$, we write π_x for the probability that π assigns to $x{:}\,\mathcal{X}$, with the idea that the more likely it is that the real value is some specific x, then the closer π_x will be to 1. Normally the uniform distribution over $\mathcal{X}$ models a secret which could equally take any one of the possible values drawn from its type and we might say that, beyond the existence of the secret, nothing else is known. There could, of course, be many reasons for using some other distribution, for example if the secret was the height of an individual then a normal distribution might be more realistic. In any case, once we have a secret, we are interested in analysing whether an algorithm, or protocol, that uses it might leak some information about it. To do this we define a measure for uncertainty, and use it to compare the uncertainty of the secret before and after executing the algorithm. If we find that the two measurements are different then we can say that there has been an information leak.

The original QIF analyses of information leaks in computer systems [3, 4] used Shannon entropy [17] to measure uncertainty because it captures the idea that more uncertainty implies "more secrecy", and indeed the uniform distribution corresponds to maximum Shannon entropy (corresponding to maximum "Shannon uncertainty"). More recent treatments have shown that Shannon entropy is not the best way to measure uncertainty in security contexts because it does not model scenarios relevant to the goals of the adversary. In particular there are some circumstances where a Shannon analysis actually gives a more favourable assessment of security than is actually warranted if the adversary's motivation is taken into account [18].

Alvim et al. [2] proposed a more general notion of uncertainty based on "gain functions". This is the notion we will use. A *gain function* measures a secret's uncertainty according to how it affects an adversary's actions within a given scenario. We write $\mathcal{W}$ for a (usually finite) set of actions available to an adversary corresponding to an "attack scenario" where the adversary tries to guess something (e.g. some property) about the secret. For a given secret $x{:}\,\mathcal{X}$, an adversary's choice of $w{:}\,\mathcal{W}$ results in the adversary gaining something beneficial to his objective. This gain can vary depending on the adversary's choice (w) and the exact value of the secret (x). The more effective is the adversary's choice in how to act, the more he is able to overcome any uncertainty concerning the secret's value.

Definition 1. *Given a type $\mathcal{X}$ of secrets, a gain function $g{:}\,\mathcal{W}{\times}\mathcal{X} \rightarrow \mathbb{R}$ is a real-valued function such that $g(w, x)$ determines the gain to an adversary if he chooses w and the secret is x.*

A simple example of a gain function is bv, where $\mathcal{W}{:=}\,\mathcal{X}$, and

$$\mathsf{bv}(x, x') \quad := \quad 1 \;\; \textit{if} \;\; x = x' \;\; \textit{else} \;\; 0. \tag{1}$$

For this scenario, the adversary's goal is to determine the exact value of the secret, so he receives a gain of 1 if he correctly guesses the value of a secret, and zero otherwise. Assuming that he knows the range of possible secrets $\mathcal{X}$, he therefore has $\mathcal{W}{:=}\mathcal{X}$ for his set of possible guesses.

Elsewhere the utility and expressivity of gain functions for measuring various attack scenarios relevant to security have been explored [1,2]. Given a gain function we define the *vulnerability* of a secret in $\mathbb{D}\mathcal{X}$ relative to the scenario it describes: it is the maximum average gain to an adversary. More explicitly, for each guess w, the adversary's average gain relative to π is $\sum_{x\in\mathcal{X}} g(w,x){\times}\pi_x$; thus his maximum average gain is the guess that yields the greatest average gain.

Definition 2. *Let $g{:}\,\mathcal{W}{\times}\mathcal{X}\rightarrow\mathbb{R}$ be a gain function, and $\pi{:}\,\mathbb{D}\mathcal{X}$ be a secret. The* vulnerability *$V_g[\pi]$ of the secret wrt. g is:*

$$V_g[\pi] \quad := \quad \max_{w\in\mathcal{W}} \sum_{x\in\mathcal{X}} g(w,x){\times}\pi_x .$$

For a secret $\pi{:}\,\mathbb{D}\mathcal{X}$, the vulnerability wrt. bv is $V_{\mathsf{bv}}[\pi]{:=}\max_{x{:}\,\mathcal{X}}\pi_x$, i.e. the maximum probability assigned by π to possible values of x. The adversary's best strategy for optimising his gain would therefore be to choose the value x that corresponds to the maximum probability under π. This vulnerability V_{bv} is called *Bayes Vulnerability*.

A *mechanism* is an abstract model of a protocol or algorithm that uses secrets. As the mechanism executes we assume that there are a number of observables that can depend on the actual value of the secret. We define $\mathcal{Y}$ to be the type for observables. The model of a mechanism now assigns a probability that $y{:}\,\mathcal{Y}$ can be observed given that the secret is x. Such observables could be sample timings in a timing analysis in cryptography, for example.

Definition 3. *A* mechanism *is a stochastic channel*[1] $C{:}\,\mathcal{X}{\times}\mathcal{Y}{\rightarrow}[0,1]$. *The value C_{xy} is the probability that y is observed given that the secret is x.*

Given a (prior) secret $\pi{:}\,\mathbb{D}\mathcal{X}$ and mechanism C we write $\pi\rangle C$ for the joint distribution *in $\mathcal{X}{\times}\mathcal{Y}$ defined*

$$(\pi\rangle C)_{xy} \quad := \quad \pi_x{\times}C_{xy} .$$

For each $y{:}\,\mathcal{Y}$, the marginal probability *that y is observed is $p_y{:=}\sum_{x{:}\,\mathcal{X}}(\pi\rangle C)_{xy}$. For each observable y, the corresponding* posterior probability *of the secret is the conditional $\pi|^y{:}\,\mathbb{D}\mathcal{X}$ defined $(\pi|^y)_x{:=}(\pi\rangle C)_{xy}/p_y$.*[2]

Intuitively, given a prior secret π and mechanism C, the entry $\pi_x{\times}C_{xy}$ of the joint distribution $\pi\rangle C$ is the probability that the actual secret value is x and the observation is y. This joint distribution contains two pieces of information: the probability p_y of observing y and the corresponding posterior $\pi|^y$ which represents the adversary's updated view about the uncertainty of the secret's value.

---

[1] Stochastic means that the rows sum to 1, i.e. $\sum_{y\in\mathcal{Y}} C_{xy} = 1$.

[2] We assume for convenience that when we write p_y the terms C, π and y are understood from the context. Notation suited for formal calculation would need to incorporate C and π explicitly.

If the vulnerability of the posterior increases, then information about the secret has leaked and the adversary can use it to increase his gain by changing how he chooses to act. The adversary's average overall gain, taking the observations into account, is defined to be the average posterior vulnerability (i.e. the average gain of each posterior distribution, weighted according to their respective marginals):

$$V_g[\pi\rangle C] := \sum_{y\in\mathcal{Y}} p_y \times V_g[\pi|^y] \;, \qquad \textit{where } p_y,\ \pi|^y \textit{ are defined at Definition 3.} \quad (2)$$

Now that we have Definitions 2 and 3 we can start to investigate whether the information leaked through observations $\mathcal{Y}$ actually have an impact in terms of whether it is useful to an adversary. It is easy to see that for any gain function g, prior π and mechanism C we have that $V_g[\pi] \leq V_g[\pi\rangle C]$. In fact the greater the difference between the prior and posterior vulnerability, the more the adversary is able to use the leaked information within the scenario defined by g. In our investigations of problems (I) and (II) we use the following implications of this idea.

(A) The adversary is able to use information leaked through mechanism C *only* when $V_g[\pi] < V_g[\pi\rangle C]$. In the case that the prior and posterior vulnerability are the same then, although information has leaked, the adversary acting within a scenario defined by g is not able to use the information in any way that benefits him. Of course there could be other scenarios where the information could prove to be advantageous, but those scenarios would correspond to a different gain function. If $V_g[\pi] = V_g[\pi\rangle C]$ for all gain functions g then in fact the channel has leaked no information at all that can be used by any adversary.

(B) For each observation y there is an action w that an adversary can pick to optimise his overall gain. This corresponds roughly to a strategy that an adversary might follow in "attacking" a mechanism C. Where here "attack" is a passive attack in the sense that the adversary observes the outputs of the mechanism C and then revises his opinion about the uncertainty of the secret to be that of the relevant posterior. This might mean that he can benefit by changing his mind about how to act, after he observes the mechanism.

(C) If P, Q are two different mechanisms corresponding to different designs of say a system, we can compare the information leaks as follows. If $V_g[\pi\rangle P] > V_g[\pi\rangle Q]$ then we can say that P leaks more than Q for the scenario defined by g in the context of prior π. If it turns out that $V_g[\pi\rangle P] \geq V_g[\pi\rangle Q]$ for all g and π then we can say that Q is more secure than P, written $P \sqsubseteq Q$, because in every scenario Q leaks less useful information than does P. Indeed any mechanism that only has a single observable leaks nothing at all and so it is maximal wrt. $\sqsubseteq$. We explore some aspects of $\sqsubseteq$ in Sect. 5.

In the remainder of the paper we illustrate these ideas by showing how adversaries can reason about leaked information, and how defenders can understand potential vulnerabilities of programs that partially leak information. We note that some of our examples are described using a small programming language, and therefore a fully formal treatment needs a model that can account for both state updates and information leaks. This can be done using a generalisation of Hidden Markov Models [15]; however the principal focus in all our examples is information flows concerning secrets that are initialised and never subsequently updated, and therefore any variable updates are incidental. In these cases a simple channel model, as summarised here, is sufficient to understand the information flow properties. We do however touch on some features of the programming language which, although illustrated only on very simple programs, nevertheless apply more generally in the more detailed Hidden Markov Model.

3 Experiment (I): When Is a Leak Not a (Useful) Leak?

The secretary problem and its variations [6] have a long history having been originally posed around 1955 although the earliest published accounts seem to date back to the 1960's [7–9]. They are classic problems in reasoning with incomplete information. Here we revisit the basic problem in order to illustrate the relationship between gain functions and adversarial scenarios and how gain functions can be used to model what an adversary can usefully do with partial information.

We assume that there are N candidates who have applied to be Paris' secretary and that they are numbered $1, \ldots, N$ in order of their interview schedule. However their *ability rank* in terms of their fitness for the position is modelled as a permutation $\sigma : \{1, \ldots, N\} \rightarrow \{1, \ldots, N\}$, and we assume that the precise σ is unknown to Paris, who is therefore the adversary. The best candidate scheduled in time slot i will have ability rank $\sigma[i] = N$, and the worst, interviewed at slot j will have $\sigma[j] = 1$. We write Σ for the set of all such permutations. We model Paris' prior knowledge of σ as the uniform probability distribution $u{:}\,\mathbb{D}\Sigma$—i.e. $u_\sigma = 1/N!$ for each $\sigma{:}\,\Sigma$ reflecting the idea that initially Paris cannot distinguish the candidates on suitability for the position.

When Paris interviews candidate i, he does not learn $\sigma[i]$, but he does learn how that candidate ranks in comparison to the others he has already seen. This implies that after the first interview all he can do is establish an initial baseline for subsequent comparison. For instance, after the second interview he will learn either that $\sigma[1]<\sigma[2]$ or that $\sigma[2]<\sigma[1]$. We write C_n for the channel that has as observables the relative rankings of the first n candidates. This means that C_1 has one observation (and so leaks nothing), C_2 has two observations (as above), C_3 has six observations, and in general C_n has $n!$ observations. Thus when $n = N$, C_N leaks the precise rankings of all candidates (and in so doing identifies σ). In Fig. 1 we illustrate the channel C_2 in the situation where there are only 3 candidates.

But what exactly can Paris do with these observations to help him select the best candidate, i.e. to enable him to offer the position to the candidate i that satisfies $\sigma[i] = N$? Since he makes an offer directly to candidate i at the end of the i'th interview, the risk is that he offers the job too early (he hasn't yet interviewed the best candidate) or too late (the best candidate has already been let go).

C_2	$(\sigma[1]<\sigma[2])$	$(\sigma[2]<\sigma[1])$
σ_1	1	0
σ_2	0	1
σ_3	1	0
σ_4	1	0
σ_5	0	1
σ_6	0	1

There are six possible permutations $\sigma_1, \ldots \sigma_6$. These are secret and therefore label the rows of C_2. Next, C_2's columns are labelled by the possible observations after the first two interviews. Paris is able to observe only the relative ordering of those two candidates. Either the first candidate is not as good as the second $(\sigma[1]<\sigma[2])$ or vice versa $(\sigma[2]<\sigma[1])$.
If we let σ_1 be the permutation with relative ordering $\sigma_1[1]<\sigma_1[2]<\sigma_1[3]$, and σ_2 be the permutation with relative ordering $\sigma_2[2]<\sigma_2[1]<\sigma_2[3]$ then Paris would record that "$(\sigma[1]<\sigma[2])$" for σ_1; this is denoted by a 1 in the first column. For σ_2 Paris would record the opposite, and therefore a 1 appears in the second column for σ_2.

Fig. 1. Information flow channel C_2 after interviewing 2 candidates from a total of 3

In any information flow problem, there is little point in measuring how much information is released if that measurement does not pertain to the actual decision making mechanism that an adversary uses. For example suppose that $N = 6$. A traditional information flow analysis might concentrate on measuring Shannon entropy as a way to get a handle on Paris' uncertainty and how it changes as he interviews more candidates. At the beginning, Paris knows nothing, and so his uncertainty as measured by Shannon entropy over the uniform distribution of 6! orderings is $\log_2(6!) \approx 9.5$. He doesn't know about 9 bits of hidden information. After two interviews, his uncertainty has dropped to $\log_2(360) \approx 8.5$, and after three interviews it becomes $\log_2(120) \approx 7$. The question is, how do these numbers help Paris make his decision? If candidate 2 is better than candidate 1 should he make the appointment? If he waits, should he make the appointment if candidate 3 is the best so far? With as much as 7 bits from a total of 9 still uncertain, maybe it would be better to wait?

Of course a detailed mathematical analysis [6] shows that if Paris' objective is to *maximise his probability* of selecting the best candidate using *only* the actions he has available (i.e. "appoint now" or "wait") then it would indeed be best for him **not** to wait but rather to appoint candidate 3 immediately if they present as the best so far. In fact the numbers taken from a Shannon analysis do not appear to help Paris at all to decide what he should do because Shannon entropy is not designed for analysing his objective when faced with his particular scenario and the set of actions he has available.

A more helpful analysis is to use a gain function designed to describe exactly Paris' possible actions so that his ability to use any information leaks –by enabling the capability to favour one action over another– can be evaluated. Recall that after interviewing each candidate Paris can choose to perform two actions: either to "appoint immediately" (a) or to "wait" (w) until the next best.

We let $\mathcal{W} := \{a, w\}$ represent the set of those actions after the n'th interview. Next we need to define the associated gains. For action a, the gain is 1 if he manages to appoint the overall maximum, i.e. $g_n(a, \sigma) = 1$ exactly if $\sigma[n] = N$. For w Paris does not appoint the n'th candidate but instead continues to interview candidates and will appoint the next best candidate he has seen thus far. His gain is 1 if that appointment is best overall, thus $g_n(w, \sigma) = 1$ if and only if the first $k > n$ which is strictly better than all candidates seen so far turns out to be the maximum overall.[3] We write $\sigma[i, j]$ for the subsequence of σ consisting of the interval between i and j inclusive, and $\sigma(i, j)$ for the subsequence consisting of the interval strictly between i and j (i.e. exclusive of $\sigma[i]$ and $\sigma[j]$). Furthermore we write $\sqcup\rho$ for the maximum item in a given subsequence ρ. We can now define Paris' gain function as follows:

$$
\begin{aligned}
g_n(a, \sigma) \quad &:= \quad 1 \;\; \mathit{iff} \;\; \sigma[n] = N & (3)\\
g_n(w, \sigma) \quad &:= \quad 1 \;\; \mathit{iff} \;\; \exists (N \geq k > n) \;\; s.t. \sqcup \sigma(n, k) < \sqcup \sigma[1, n] < N = \sigma[k]. & (4)
\end{aligned}
$$

Given uniform prior u (over the possible $N!$ permutations), we can approximate Paris' residual uncertainty after interviewing n candidates by considering the posterior vulnerability $V_{g_n}[u \rangle C_n]$. Using (A) from Sect. 2 we can evaluate whether the information leaked is useful by comparing $V_{g_n}[u \rangle C_n]$ with the prior vulnerability $V_{g_n}[u]$; next we can use (B) from Sect. 2 to examine, for each posterior which of a or w is Paris' best action relative to each observation. In the case that N is 6, we can show that:

$$
V_{g_2}[u] = V_{g_2}[u \rangle C_2] \;, \quad \mathit{but} \;\; V_{g_3}[u] < V_{g_3}[u \rangle C_3],
$$

i.e. there is no leakage wrt. the question "should the candidate just interviewed be selected?" after two interviews because the prior and posterior vulnerabilities wrt. g_2 are the same. On the other hand the information leaked after interviewing *three* candidates is sufficient to help him decide whether to appoint the third candidate—he should do so if that third candidate is the best so far.

To see how this reasoning works, consider the situation after interviewing two candidates as described above. Note that both outcomes occur with probability $1/2$ (since the prior is uniform). If we look at the two possible actions available in g_2 we can compare whether the probability of the second candidate being best overall actually changes from its initial value of $1/6$ with this new information. In fact it does not—the probability that the second candidate is the best overall remains at $1/6$, and the probability that the next best candidate is best overall also does not change, remaining at $0.427 = V_{g_2}[u \rangle C_2]$. This tells us that although some uncertainty has been resolved, it can't be used by Paris to help him with his objective—he will still not appoint candidate 2 even when they rank more highly than candidate 1.

[3] In the traditional analysis Paris only employs this strategy after interviewing approximately N/e candidates. Here we are interested in studying this strategy at each stage of the interviewing procedure.

On the other hand if we compare the case where Paris interviews three candidates, and observes that the best candidate so far is $\sigma[3]$, then there is a good chance that this is the best candidate overall. Indeed his best strategy is to appoint that candidate immediately—and he should only continue interviewing if $\sigma[3]$ rates lower than either the other two. Overall this gives him the greatest chance of appointing the best candidate which is $0.428 = V_{g_3}[u \rangle C_3]$. [4]

A similar analysis applies to other C_n's but when $N = 6$ his overall best strategy is to start appointing after $n = 3$.

We note that the analysis presented above relates only to the original formulation of the Secretary problem; the aim is to explain the well-known solution (to the Secretary problem) in terms of how an adversary acts in a context of partial information. More realistic variations of the Secretary problem have been proposed. For example the original formulation only speaks of hiring the best, whereas in reality hiring the second best might not be so bad. Analysing such alternative adversarial goals (such as choosing either the best or the second best) would require designing a gain function that specifically matches those alternative requirements.

4 Experiment (II): Comparing Designs

Consider now the scenario of Romulus who does not trust his brother Remus. Romulus decides to password protect his computer but has read that the password checker, depicted in Fig. 2, suffers from a security flaw which allows an adversary to identify prefixes of his real password. Observe that the program in Fig. 2 iteratively checks each character of an input password with the real password, so that the time it takes to finish checking depends on how many characters have been matched. Romulus wonders why anyone would design a password checker with such an obvious flaw since an easy fix would be to report whether the guess succeeds only after waiting until all the characters have been checked. But then Romulus remembers that he often mistypes his password and would not want to wait a long time between retries.

In Fig. 2 we assume that `pw` is Romulus' secret password, and that `gs` is an adversary's input when prompted by the system for an input. To carry out a QIF analysis of a timing attack, as described in Sect. 2, we would first create an information flow channel with observations describing the possible outcomes after executing the loop to termination, and then we would use it to determine a joint distribution wrt. a prior distribution. In this case the observations in the channel would be: "does not enter the loop", "iterates once", "iterates twice"

[4] Interestingly the two $V_{g_2}[u \rangle C_2] = V_{g_3}[u \rangle C_3]$—although the actions wrt. g_2 and g_3 are different, in the case they are interpreted over the scenario of 6 candidates they dictate the same behaviour. Under the g_2 strategy which becomes operational after interviewing 2 candidates, it is never to appoint candidate 2, but take the very next best. Under the g_3 case (which becomes operational after interviewing three candidates) it is to take candidate 3 if she is the best so far, or if not select the next best.

```
// Assume pw has been initialised as the secret password,
// and that gs is a input password that needs to be checked
i := 0;
ans := true;
while (ans && i<N) {
        Print (ans && i<N);        // Side Channel
        if (pw[i] != gs[i])
                then ans := false;
        else skip;
        i++;
}
Print ans        // Information leak
```

All assignments and updates are considered unobservable; information leaks are indicated at "Print" statements. The "timing" attack is modelled as the Remus' ability to observe the number of executions of the loop.

Fig. 2. Basic password checker, with early exit

etc. There is however an alternative approach which leads to the same (logical) outcome, but offers an opportunity for QIF tool support summarised in Sect. 6. Instead of creating the channel in one hit, we interpret each program fragment as a mini-channel with its own observations. At each stage of the computation the joint distribution described at Definition 3 is determined "on the go", by amalgamating all subsequent information leaks from new mini-channels with those already accounted for. The situation is a little more complicated because state updates also have to be included. But in the examples described here, we can concentrate on the program fragments that explicitly release information about the secret `pw`, and assume that the additional state updates (to the counter i for example) have no impact on the information leaks.

To enable this approach we introduce a "Print" statement into the programming language which is interpreted as a mini-channel—"Print" statements have no effect on the computations except to transmit information to an observer during program execution. In Fig. 2 the statement `Print(ans && i<N)` occurring each time the loop body is entered transmits that fact to the adversary by emitting the observation "condition `(ans && i<N)` is true". Similarly a second statement `Print ans` transmits whether the loop terminated with `ans` true or false. To determine the overall information flow of the loop, we interpret `Print(ans && i<N)` and `Print ans` separately as channels, and use "channel sequential composition" [14] to rationalise the various sequences of observations with what the adversary already knows. It is easy to see that this use of "Print" statements logically (at least) accounts for the timing attack because it simulates an adversary's ability to determine the number of iterations of the loop, and therefore to deduce the prefix of the `pw` of length the number of iterations, because it must be the same as the corresponding prefix of the known `gs`.

```
// Assume pw has been initialised as the secret password,
// and that gs is an input password that needs to be checked

list := [0,...,n-1];                    // Indices for checking
ans := true;
      while (ans && list!=[]) {
         Print (ans && list!=[]);       // "Smudged" leak
         i := uniform(list);            // Uniform choice
         ans := ans && (pw[i] = gs[i]);
         list := list-{i};              // Remove checked index
      }
Print ans
```

Fig. 3. Password checker, with early exit and random checking order

With this facility, Romulus can now compare various designs for password checking programs. Figure 3 models the system upgrade that Romulus is considering. Just as in Fig. 2, an early exit is permitted, but the information flow is now different because rather than checking the characters in order of index, the checking order is randomised uniformly over all possible orders. For example if Fig. 2 terminates after checking a single character of input password `gs`, then the adversary knows to begin subsequent guesses with `gs[0]`. But if Fig. 3 similarly terminates after trying a single character, all the adversary knows is that one of `gs[0]`, ... `gs[n-1]` is correct, but he does not know which one.

At first sight, this appears to offer a good trade-off between Romulus' aim of making it difficult for Remus to guess his password and his desire not to wait too long before he can try again in case he mistypes his own password. To gain better insight however we can look at some specific instances of information flow with respect to these password checking algorithms.

We assume a prior uniform distribution over all permutations of a three character password `"abc"`, i.e. we assume the adversary knows that `pw` is one of those permutations, but nothing else. Next we consider the information leaks when the various algorithms are run with input `gs = "abc"`. Since the adversary's intention is to figure out the exact value of `pw`, we use Bayes Vulnerability V_{bv} (recall 1 for definition of bv) to measure the uncertainty, where $\mathcal{X}$ is the set of permutations of `"abc"` and $\mathcal{W} := \mathcal{X}$ is the set of possible guesses. [5]

Consider first the basic password checker depicted at Fig. 2. There are three amalgamated observations: the loop can terminate after 1, 2 or 3 iterations. For each observation, we compute the marginal probability and the posterior probabilities associated with the residual uncertainty (recall Definition 3). For this example the marginal, corresponding posterior and observations are as follows:

[5] In this case, we imagine that an adversary uses an input `gs` in order to extract some information. His actions described by $\mathcal{W}$ are related to what he is able to do in the face of his current uncertainty.

```
marginal  {posterior}
1 ÷ 6        {1 ÷ 1    "abc"}   ← termination after 3 iterations
1 ÷ 6        {1 ÷ 1    "acb"}   ← termination after 2 iterations
2 ÷ 3        {1 ÷ 4    "bac"    ← termination after 1 iteration
              1 ÷ 4    "bca"
              1 ÷ 4    "cab"
              1 ÷ 4    "cba"}
```

The first observation with marginal probability $1/6$, is the case where gs and pw are the same: only then will the loop run to completion. The associated posterior probability is the point distribution over "abc". For the second observation –termination after 2 iterations– the marginal is also $1/6$ and the associated posterior is also a point distribution over "acb". That is because if gs[0] matches pw[0] but gs[1] does not match pw[1] it must be that the second and third characters in gs are swapped.

The third observation is the case where the loop terminates after one iteration and it occurs with marginal probability $2/3$. This can only happen when gs[0] is not equal to pw[0] (2 possibilities), and the second and third characters can be in either order (2 more possibilities), giving $1/2{\times}2$ for each posterior probability.

Overall the (posterior) Bayes Vulnerability has increased (from $1/6$ initially relative to the uniform prior) to $1/2$ because the chance of guessing the password, taking the observations into account, is now $1/2$.

Comparing now with the obfuscation of the checking order Fig. 3, we again have the situation of three possible observations, but now the uncertainty is completely reduced only in the case of 3 iterations. The resulting marginal, corresponding posterior and observations are as follows:

```
marginal   {posterior}
1 ÷ 6         {1 ÷ 1    "abc"}   ← termination after 3 iterations
2 ÷ 3         {1 ÷ 6    "acb"    ← termination after 2 iterations
               1 ÷ 6    "bac"
               1 ÷ 4    "bca"
               1 ÷ 4    "cab"
               1 ÷ 6    "cba"}
1 ÷ 6         {1 ÷ 3    "acb"    ← termination after 1 iteration
               1 ÷ 3    "bac"
               1 ÷ 3    "cba"}
```

In spite of the residual uncertainty in the case of an incorrect input, the overall posterior vulnerability of $7/18$ is only slightly lower than the $1/2$ for the original early exit Fig. 2.

For longer passwords, displaying the list of marginals and posteriors is not so informative; but still we can give the posterior Bayes vulnerability. We find for example that for Fig. 3, even with its early exit, only reduces the probability of guessing the password by half when compared with Fig. 2. For a 5 character password, Fig. 2 gives a posterior vulnerability of $1/24$, but Fig. 3 reduces it only to about half as much at 13/600. Perhaps Romulus' upgrade is not really worth it after all.

5 Experiment (III): Properties of the Modelling Language

```
// h, k are secret bits , already initialised

hif (h==1) then Print (h XOR k) // Print the exclusive or of h, k
           else Print 1         // Print 1
fih
```

Fig. 4. A hidden if ... then ... else

Thus far we have considered information flows in specific scenarios. In this section we look more generally at the process of formalisation, in particular the information flow properties of the programming language itself. In standard program semantics, the idea of refinement is important because it provides a sound technique for simplifying a verification exercise. If an "abstract" description of a program can be shown to satisfy some property, then refinement can be used to show that if the actual program only exhibits consistent behaviours relative to its abstract counterpart, then it too must satisfy the property. To make all of this work in practice, the abstraction/refinement technique normally demands that program constructs are "monotonic" in the sense that applying abstraction to a component leads to an abstraction of the whole program.

In Sect. 2(C) we described an information flow refinement relative to the QIF semantics so that if $P \sqsubseteq P'$ then program P leaks more information than does program P' in all possible scenarios. In the examples we have seen above, the sequence operator used to input the results of one program P into another Q is written $P;Q$. It is an important fact that sequence is monotonic, i.e. if $P \sqsubseteq P'$ then $P;Q \sqsubseteq P';Q$. Elsewhere [14,16] define a QIF semantics for a programming language in which all operators are "monotonic" wrt. information flow—this means that improving the security of any component in a program implies that the security of the overall program is also improved. The experiment in this section investigates monotonicity for conditional branching in a context of information flow.

Consider programs at Figs. 4 and 5 which assume two secrets `h, k`. Each program is the same except for the information transmitted in the case that the secret `h` is 1—in this circumstance Fig. 4 Prints the exclusive OR of `h` and `k` whereas Fig. 5 always Prints 1. In this experiment we imagine that the adversary can only observe information transmitted via the "Print" statements, and cannot (for example) observe the branching itself. This is a "hidden if statement" denoted by "`hif ... fih`", and we investigate whether it can be monotonic as a (binary) program construct.

```
// h, k are secret bits , already initialised

hif (h==1) then Print 0 // Print 0
            else Print 1 // Print 1
fih
```

Since `Print 1` is strictly more secure than `Print (h XOR k)`, shouldn't it be the case that this program is at least as secure as Fig. 4 ?

Fig. 5. A program which emits 0's and 1's

To study this question, we first observe that the statement `Print 1` is more secure than `Print (h XOR k)`; that is because `Print 1` transmits no information at all because its observable does not depend on the value of the secret, and so is an example of the most secure program (recall (C) in Sect. 2). This means if "`hif` ... `fih`" has a semantics which is monotonic, it must be the case that Fig. 5 is more secure than Fig. 4.

In these simple cases, we can give a QIF semantics for the two programs directly, taking into account the assumption that only the "Print" statements transmit observations. Assume that the variables `h`, `k` are initialised uniformly and independently. In the case of Fig. 4 the adversary reasons that if 1 is printed then this is either because the lower branch was executed (`h == 0`) or the upper branch is executed (`h == 1`) and `h XOR k` evaluates to 1. When he observes 0 however that can only happen if (`h == 1`). We can summarise the relative marginal and posterior probabilities for these cases as follows. We write the values of the variables as a pair, with `h` as the left component and `k` the right component. For example the pair `(1, 0)` means that `h` has value 1 and `k` has value 0. In the scenario just described, there is a $1/4$ marginal probability that the pair is `(1, 1)`, and in this case the adversary knows the value of `h` and `k`. But with probability $3/4$ a 1 is printed, and in this case the adversary does not know which of the remaining pairs it can be. The following marginal and posterior probabilities summarise this behaviour:

marginal	{posterior}		
$1 \div 4$	$\{1 \div 1$	"(1,1)"}	← *Prints 0*
$3 \div 4$	$\{1 \div 3$	"(0,0)"	← *Prints 1*
	$1 \div 3$	"(0,1)"	
	$1 \div 3$	"(1,0)"}	

For Fig. 5 the situation is a little simpler—when `h` is 0 a 1 is printed and vice versa. This leads to the following result.

marginal	{posterior}		
$1 \div 2$	$\{1 \div 2$	"(1,1)"	← *Prints 0*
	$1 \div 2$	"(1,0)"}	
$1 \div 2$	$\{1 \div 2$	"(0,0)"	← *Prints 1*
	$1 \div 2$	"(0,1)"}	

Comparing these two probabilistic results we see now that "`hif ...fih`" cannot be monotonic, because if we replace `Print(h XOR k)` by the more secure `Print 1` more information (not less) is leaked. For instance, under Bayes Vulnerability for `h`, the adversary can guess the value of `h` exactly under Fig. 5 but only with probability 3/4 under Fig. 4.

This experiment tells us that a hidden "`hif ...fih`" construct cannot be defined in a way that respects monotonicity of information flow, and could therefore be too risky to use.

6 Experiments with Kuifje

The experiments described above were carried out using the tool Kuifje [12] which interprets a small programming language in terms of the QIF semantics, but extended to the Hidden Markov Model alluded to in Sect. 2. It supports the usual programming constructs (assignment, sequencing, conditionals and loops) but crucially it takes into account information flows consistent with the channel model outlined in Sect. 2. In particular the "Print" statements used in our examples correspond exactly to the observations that an adversary could make during program execution. This allows a direct model for eg. known side channels that potentially expose partially computation traces during program execution.

The basic assumption built into the semantics of Kuifje is that all variables cannot be observed unless revealed fully or partially through a "Print" statement. For example `Print x` would print the value of variable `x` and so reveal it completely at that point of execution, but `Print(x>0)` would only reveal whether `x` is strictly positive or not. As usual, we also assume that the adversary knows the program code.

Kuifje is implemented in Haskell and makes extensive use of the Giry monad [10] for managing the prior, posterior and marginal probabilities in the form of "hyperdistributions". A hyperdistribution is a distribution of distributions and is based on the principle that the names of observations are redundant in terms of the analysis of information flow. Hyperdistributions therefore only summarise the posterior and marginal probabilities, because these quantities are the only ones that play a role in computing posterior vulnerabilities (recall 2). Hyperdistributions satisfy a number of useful properties relevant to QIF [15], and provide the basis for algebraic reasoning of source level code.

7 Related Work

Classical approaches to measuring insecurities in programs are based on determining a "change in uncertainty" of some "prior" value of the secret—although how to measure the uncertainty differs in each approach. For example Clark et al. [3] use Shannon entropy to estimate the number of bits being leaked; and Clarkson et al. [5] model a change in belief. Smith [18] demonstrated the importance of using measures that have some operational significance, and the idea was developed further [2] by introducing the notion of g-leakage to express such significance in a very general way. The partial order used here on programs is the same as the g-leakage order introduced by Alvim et al. [2], but it appeared also in even earlier work [14]. Its properties have been studied extensively [1].

Jacobs and Zanasi [11] use ideas based on the Giry Monad to present an abstract state transformer framework for Bayesian reasoning. Our use of hyper-distributions means that the conditioning needed for the Bayesian update has taken place in the construction of the posteriors.

8 Conclusions and Discussion

Understanding the impact of information flow is hard. The conceptual tools presented here summarise the ideas of capturing the adversary's ability to use the information released, together with a modelling language that enables the study of risks associated with information leaks in complicated algorithms and protocols. The language itself is based on the Probability Monad which has enabled its interpretation in Kuifje.

It is hoped that the ability to describe scenarios in terms of adversarial gains/losses together with Kuifje that enables detailed numerical calculation of the impact of flows will lead to a better understanding of security vulnerabilities in programs.

Acknowledgements. I thank Tom Schrijvers for having the idea of embedding these ideas in Haskell, based on Carroll Morgan's talk at IFIP WG2.1 in Vermont, and for carrying it out to produce the tool Kuifje. Together with Jeremy Gibbons all four of us wrote the first paper devoted to it [12]. (It was Jeremy who suggested the name "Kuifje", the Dutch name for TinTin—and hence his "QIF".) Carroll Morgan translated the example programs into Kuifje.

References

1. Alvim, M.S., Chatzikokolakis, K., McIver, A., Morgan, C., Palamidessi, C., Smith, G.: Additive and multiplicative notions of leakage, and their capacities. In: IEEE 27th Computer Security Foundations Symposium, CSF 2014, Vienna, Austria, 19–22 July 2014, pp. 308–322. IEEE (2014)
2. Alvim, M.S., Chatzikokolakis, K., Palamidessi, C., Smith, G.: Measuring information leakage using generalized gain functions. In: Proceedings 25th IEEE Computer Security Foundations Symposium (CSF 2012), pp. 265–279, June 2012
3. Clark, D., Hunt, S., Malacaria, P.: Quantitative analysis of the leakage of confidential data. Electr. Notes Theor. Comput. Sci. **59**(3), 238–251 (2001)
4. Clark, D., Hunt, S., Malacaria, P.: Quantified interference for a while language. Electr. Notes Theor. Comput. Sci. **112**, 149–166 (2005)
5. Clarkson, M.R., Myers, A.C., Schneider, F.B.: Belief in information flow. In: 18th IEEE Computer Security Foundations Workshop, (CSFW-18 2005), Aix-en-Provence, France, 20–22 June 2005, pp. 31–45 (2005)
6. Freeman, P.R.: The secretary problem and its extensions: a review. Int. Stat. Rev. **51**, 189–206 (1983)

7. Gardner, M.: Mathematical games. Sci. Am. **202**(2), 178–179 (1960)
8. Gardner, M.: Mathematical games. Sci. Am. **202**(3), 152 (1960)
9. Gilbert, J., Mosteller, F.: Recognising the maximum of a sequence. J. Am. Stat. Assoc. **61**, 35–73 (1966)
10. Giry, M.: A categorical approach to probability theory. In: Banaschewski, B. (ed.) Categorical Aspects of Topology and Analysis. LNM, vol. 915, pp. 68–85. Springer, Heidelberg (1982). https://doi.org/10.1007/BFb0092872
11. Jacobs, B., Zanasi, F.: A predicate/state transformer semantics for Bayesian learning. Electr. Notes Theor. Comput. Sci. **325**, 185–200 (2016)
12. Morgan, C., Gibbons, J., Mciver, A., Schrijvers, T.: Quantitative information flow with monads in haskell. In: Foundations of Probabilistic Programming. CUP (2019, to appear)
13. Mardziel, P., Alvim, M.S., Hicks, M.W., Clarkson, M.R.: Quantifying information flow for dynamic secrets. In: 2014 IEEE Symposium on Security and Privacy, SP 2014, Berkeley, CA, USA, 18–21 May 2014, pp. 540–555 (2014)
14. McIver, A., Meinicke, L., Morgan, C.: Compositional closure for Bayes risk in probabilistic noninterference. In: Abramsky, S., Gavoille, C., Kirchner, C., Meyer auf der Heide, F., Spirakis, P.G. (eds.) ICALP 2010. LNCS, vol. 6199, pp. 223–235. Springer, Heidelberg (2010). https://doi.org/10.1007/978-3-642-14162-1_19
15. McIver, A., Morgan, C., Rabehaja, T.: Abstract hidden markov models: a monadic account of quantitative information flow. In Proceedings LiCS 2015 (2015)
16. Morgan, C.: The shadow knows: refinement of ignorance in sequential programs. Sci. Comput. Program. **74**(8), 629–653 (2009). Treats Oblivious Transfer
17. Shannon, C.E.: A mathematical theory of communication. Bell Syst. Tech. J. **27**(379–423), 623–656 (1948)
18. Smith, G.: On the foundations of quantitative information flow. In: de Alfaro, L. (ed.) FoSSaCS 2009. LNCS, vol. 5504, pp. 288–302. Springer, Heidelberg (2009). https://doi.org/10.1007/978-3-642-00596-1_21

Handling Local State with Global State

Koen Pauwels[1(✉)], Tom Schrijvers[1], and Shin-Cheng Mu[2]

[1] Department of Computer Science, KU Leuven, Leuven, Belgium
{koen.pauwels,tom.schrijvers}@cs.kuleuven.be
[2] Institute of Information Science, Academia Sinica, Taipei, Taiwan
scm@iis.sinica.edu.tw

Abstract. Equational reasoning is one of the most important tools of functional programming. To facilitate its application to monadic programs, Gibbons and Hinze have proposed a simple axiomatic approach using laws that characterise the computational effects without exposing their implementation details. At the same time Plotkin and Pretnar have proposed algebraic effects and handlers, a mechanism of layered abstractions by which effects can be implemented in terms of other effects.

This paper performs a case study that connects these two strands of research. We consider two ways in which the nondeterminism and state effects can interact: the high-level semantics where every nondeterministic branch has a local copy of the state, and the low-level semantics where a single sequentially threaded state is global to all branches.

We give a monadic account of the folklore technique of handling local state in terms of global state, provide a novel axiomatic characterisation of global state and prove that the handler satisfies Gibbons and Hinze's local state axioms by means of a novel combination of free monads and contextual equivalence. We also provide a model for global state that is necessarily non-monadic.

Keywords: Monads · Effect handlers · Equational reasoning · Nondeterminism · State · Contextual equivalence

1 Introduction

Monads have been introduced to functional programming to support side effects in a rigorous, mathematically manageable manner [11,14]. Over time they have become the main framework in which effects are modelled. Various monads were developed for different effects, from general ones such as IO, state, nondeterminism, exception, continuation, environment passing, to specific purposes such as parsing. Much research was also devoted to producing practical monadic programs.

Equational reasoning about pure functional programs is particularly simple and powerful. Yet, Hutton and Fulger [7] noted that a lot less attention has been paid to reasoning about monadic programs in that style. Gibbons and Hinze [4] argue that equational reasoning about monadic programs becomes particularly

G. Hutton (Ed.): MPC 2019, LNCS 11825, pp. 18–44, 2019.
https://doi.org/10.1007/978-3-030-33636-3_2

convenient and elegant when one respects the abstraction boundaries of the monad. This is possible by reasoning in terms of axioms or laws that characterise the monad's behavior without fixing its implementation.

This paper is a case study of equational reasoning with monadic programs. Following the approach of algebraic effects and handlers [12], we consider how one monad can be implemented in terms of another—or, in other words, how one can be simulated in by another using a careful discipline. Our core contribution is a novel approach for proving the correctness of such a simulation. The proof approach is a convenient hybrid between equational reasoning based on axioms and inductive reasoning on the structure of programs. To capture the simulation we apply the algebraic effects technique of *handling* a free monad representation [15]. The latter provides a syntax tree on which to perform induction. To capture the careful discipline of the simulation we use contextual equivalence and perform inductive reasoning about program contexts. This allows us to deal with a heterogeneous axiom set where different axioms may make use of different notions of equality for programs.

We apply this proof technique to a situation where each "monad" (both the simulating monad and the simulated monad) is in fact a combination of two monads, with differing laws on how these effects interact: non-determinism and state.

In the monad we want to implement, each non-deterministic branch has its own 'local' copy of the state. This is a convenient effect interaction which is provided by many systems that solve search problems, including Prolog. A characterisation of this 'local state' monad was given by Gibbons and Hinze [4].

We realise this local state semantics in terms of a more primitive monad where a single state is sequentially threaded through the non-deterministic branches. Because this state is shared among the branches, we call this the 'global state' semantics. The appearance of local state is obtained by following a discipline of undoing changes to the state when backtracking to the next branch. This folklore backtracking technique is implemented by most sequential search systems because of its relative efficiency: remembering what to undo often requires less memory than creating multiple copies of the state, and undoing changes often takes less time than recomputing the state from scratch. To the best of our knowledge, our axiomatic characterisation of the global state monad is novel.

In brief, our contributions can be summarized as follows:

- We provide an axiomatic characterisation for the interaction between the monadic effects of non-determinism and state where the state is persistent (i.e., does not backtrack), together with a model that satisfies this characterisation.
- We prove that—with a careful discipline—our characterisation of persistent state can correctly simulate Gibbons and Hinze's monadic characterisation of backtrackable state [4]. We use our novel proof approach (the core contribution of this paper) to do so.
- Our proof also comes with a mechanization in Coq.[1]

[1] The proof can be found at https://github.com/KoenP/LocalAsGlobal.

The rest of the paper is structured as follows. First, Sect. 2 gives an overview of the main concepts used in the paper and defines our terminology. Then, Sect. 3 informally explores the differences between local and global state semantics. Next, Sect. 4 explains how to handle local state in terms of global state. Section 5 formalizes this approach and proves it correct. Finally, Sects. 6 and 7 respectively discuss related work and conclude.

2 Background

This section briefly summarises the main concepts we need for equational reasoning about effects. For a more extensive treatment we refer to the work of Gibbons and Hinze [4].

2.1 Monads, Nondeterminism and State

Monads. A monad consists of a type constructor $\mathsf{M} :: * \to *$ and two operators $\mathit{return} :: a \to \mathsf{M}\ a$ and "bind" $(\gg\!=) :: \mathsf{M}\ a \to (a \to \mathsf{M}\ b) \to \mathsf{M}\ b$ that satisfy the following *monad laws*:

$$\mathit{return}\ x \gg\!= f = f\ x \quad , \tag{1}$$

$$m \gg\!= \mathit{return} = m \quad , \tag{2}$$

$$(m \gg\!= f) \gg\!= g = m \gg\!= (\lambda x \to f\ x \gg\!= g) \quad . \tag{3}$$

Nondeterminism. The first effect we introduce is nondeterminism. Following the trail of Hutton and Fulger [7] and Gibbons and Hinze, we introduce effects based on an axiomatic characterisation rather than a specific implementation. We define a type class to capture this interface as follows:

$$\begin{array}{l}\textbf{class}\ \mathsf{Monad}\ m \Rightarrow \mathsf{MNondet}\ m\ \textbf{where}\\ \quad \emptyset \quad :: m\ a\\ \quad ([\!]) :: m\ a \to m\ a \to m\ a \quad .\end{array}$$

In this interface, $\emptyset$ denotes failure, while $m \mathbin{[\!]} n$ denotes that the computation may yield either m or n. Precisely what laws these operators should satisfy, however, can be a tricky issue. As discussed by Kiselyov [8], it eventually comes down to what we use the monad for.

It is usually expected that $([\!])$ and $\emptyset$ form a monoid. That is, $([\!])$ is associative, with $\emptyset$ as its zero:

$$(m \mathbin{[\!]} n) \mathbin{[\!]} k \;=\; m \mathbin{[\!]} (n \mathbin{[\!]} k) \quad , \tag{4}$$

$$\emptyset \mathbin{[\!]} m \;=\; m \;=\; m \mathbin{[\!]} \emptyset \quad . \tag{5}$$

It is also assumed that monadic bind distributes into ($[\!]$) from the end, while $\emptyset$ is a left zero for ($\gg\!=$):

$$\textbf{left-distributivity:} \quad (m_1 \;[\!]\; m_2) \gg\!= f \;=\; (m_1 \gg\!= f) \;[\!]\; (m_2 \gg\!= f) \;, \tag{6}$$

$$\textbf{left-zero:} \quad \emptyset \gg\!= f \;=\; \emptyset \;. \tag{7}$$

We will refer to the laws (4), (5), (6), (7) collectively as the *nondeterminism laws.*

One might intuitively expect some additional laws from a set of nondeterminism operators, such as idempotence ($p \;[\!]\; p = p$) or commutativity ($p \;[\!]\; q = q \;[\!]\; p$). However, our primary interest lies in the study of combinations of effects and – as we shall see very soon – in particular the combination of nondeterminism with *state.* One of our characterisations of this interaction would be incompatible with both idempotence and commutativity, at least if they are stated as strongly as we have done here. We will eventually introduce a weaker notion of commutativity, but it would not be instructive to do so here (as its properties would be difficult to motivate at this point).

State. The state effect provides two operators:

```
class Monad m ⇒ MState s m | m → s where
  get :: m s
  put :: s → m ()  .
```

The *get* operator retrieves the state, while *put* overwrites the state by the given value. They satisfy the *state laws*:

$$\textbf{put-put:} \quad put\ st \gg put\ st' = put\ st' \;, \tag{8}$$

$$\textbf{put-get:} \quad put\ st \gg get = put\ st \gg return\ st \;, \tag{9}$$

$$\textbf{get-put:} \quad get \gg\!= put = return\ () \;, \tag{10}$$

$$\textbf{get-get}: \quad get \gg\!= (\lambda st \rightarrow get \gg\!= k\ st) = get \gg\!= (\lambda st \rightarrow k\ st\ st) \;, \tag{11}$$

where $m_1 \gg m_2 = m_1 \gg\!= \lambda\_ \rightarrow m_2$, which has type $(\gg) :: m\ a \rightarrow m\ b \rightarrow m\ b$.

2.2 Combining Effects

As Gibbons and Hinze already noted, an advantage of defining our effects axiomatically, rather than by providing some concrete implementation, is that it is straightforward to reason about combinations of effects. In this paper, we are interested in the interaction between nondeterminism and state, specifically.

```
class (MState s m, MNondet m) ⇒ MStateNondet s m | m → s  .
```

The type class MStateNondet s m simply inherits the operators of its superclasses MState s m and MNondet m without adding new operators, and implementations of this class should comply with all laws of both superclasses.

However, this is not the entire story. Without additional 'interaction laws', the design space for implementations of this type class is left wide-open with respect to questions about how these effects interact. In particular, it seems hard to imagine that one could write nontrivial programs which are agnostic towards the question of what happens to the state of the program when the program backtracks. We discuss two possible approaches.

Local State Semantics. One is what Gibbons and Hinze call "backtrackable state", that is, when a branch of the nondeterministic computation runs into a dead end and the continuation of the computation is picked up at the most recent branching point, any alterations made to the state by our terminated branch are invisible to the continuation. Because in this scheme state is local to a branch, we will refer to these semantics as *local state semantics*. We characterise local state semantics with the following laws:

$$\textbf{right-zero:} \quad m \gg \emptyset \;=\; \emptyset \quad , \tag{12}$$
$$\textbf{right-distributivity:} \quad m \gg\!= (\lambda x \to f_1\ x \mathbin{[\!]} f_2\ x) = (m \gg\!= f_1) \mathbin{[\!]} (m \gg\!= f_2). \tag{13}$$

With some implementations of the monad, it is likely that in the lefthand side of (13), the effect of m happens once, while in the righthand side it happens twice. In (12), the m on the lefthand side may incur some effects that do not happen in the righthand side.

Having (12) and (13) leads to profound consequences on the semantics and implementation of monadic programs. To begin with, (13) implies that for $([\!])$ we have some limited notion of commutativity. For instance, both the left and right distributivity rules can be applied to the term $(\mathit{return}\ x \mathbin{[\!]} \mathit{return}\ y) \gg\!= \lambda z \to \mathit{return}\ z \mathbin{[\!]} \mathit{return}\ z$. It is then easy to show that this term must be equal to both $\mathit{return}\ x \mathbin{[\!]} \mathit{return}\ x \mathbin{[\!]} \mathit{return}\ y \mathbin{[\!]} \mathit{return}\ y$ and $\mathit{return}\ x \mathbin{[\!]} \mathit{return}\ y \mathbin{[\!]} \mathit{return}\ x \mathbin{[\!]} \mathit{return}\ y$.[2]

In fact, having (12) and (13) gives us very strong and useful commutative properties. To be clear what we mean, we give a formal definition:

Definition 1. *Let m and n be two monadic programs such that x does not occur free in m, and y does not occur free in n. We say m and n commute if*

$$\begin{aligned} m \gg\!= \lambda x \to n \gg\!= \lambda y \to f\ x\ y \;=\; \\ n \gg\!= \lambda y \to m \gg\!= \lambda x \to f\ x\ y \;\;. \end{aligned} \tag{14}$$

We say that effects ϵ and δ commute if any m and n commute as long as their only effects are respectively ϵ and δ.

One important result is that, in local state semantics, non-determinism commutes with any effect:

[2] Gibbons and Hinze [4] were mistaken in their claim that the type $s \to [(a, s)]$ constitutes a model of their backtrackable state laws; it is not a model because its $([\!])$ does not commute with itself. One could consider a relaxed semantics that admits $s \to [(a, s)]$, but that is not the focus of this paper.

Theorem 1. *If right-zero* (12) *and right-distributivity* (13) *hold in addition to the other laws, non-determinism commutes with any effect.*

Implementation-wise, (12) and (13) imply that each nondeterministic branch has its own copy of the state. To see that, let $m = put\ 1$, $f_1\ () = put\ 2$, and $f_2\ () = get$ in (13)—the state we *get* in the second branch does not change, despite the *put* 2 in the first branch. One implementation satisfying the laws is $\mathsf{M}\ s\ a = s \rightarrow \mathsf{Bag}\ (a, s)$, where $\mathsf{Bag}\ a$ is an implementation of a multiset or "bag" data structure. If we ignore the unordered nature of the Bag type, this implementation is similar to $\mathsf{StateT}\ s\ (\mathsf{ListT\ Identity})$ in the Monad Transformer Library [5]. With effect handling [9,15], the monad behaves similarly (except for the limited commutativity implied by law (13)) if we run the handler for state before that for list.

Global State Semantics. Alternatively, we can choose a semantics where state reigns over nondeterminism. In this case of non-backtrackable state, alterations to the state persist over backtracks. Because only a single state is shared over all the branches of the nondeterministic computation, we call this semantics *global state semantics.* We will return later to the question of how to define laws that capture our intuition for this kind of semantics, because (to the best of our knowledge) this constitutes a novel contribution.

Even just figuring out an implementation of a global state monad that matches our intuition is already tricky. One might believe that $\mathsf{M}\ s\ a = s \rightarrow ([a], s)$ is a natural implementation of such a monad. The usual, naive implementation of $(\gg\!=)$ using this representation, however, does not satisfy left-distributivity (6), violates monad laws, and is therefore not even a monad. The type $\mathsf{ListT}\ (\mathsf{State}\ s)$ generated using the Monad Transformer Library [5] expands to essentially the same implementation, and is flawed in the same way. More careful implementations of ListT, which do satisfy (6) and the monad laws, have been proposed [3,13]. Effect handlers (e.g. Wu [15] and Kiselyov and Ishii [9]) do produce implementations which match our intuition of a non-backtracking computation if we run the handler for non-determinism before that of state.

We provide a direct implementation to aid the intuition of the reader. Essentially the same implementation is obtained by using the type $\mathsf{ListT}\ (\mathsf{State}\ s)$ where ListT is implemented as a correct monad transformer. This implementation has a non-commutative $(\|)$.

$$\mathsf{M}\ s\ a = s \rightarrow (\mathsf{Maybe}\ (a, \mathsf{M}\ s\ a), s)\ .$$

The Maybe in this type indicates that a computation might fail to produce a result. But note that the s is outside of the Maybe: even if the computation fails to produce any result, a modified state may be returned (this is different from local state semantics). $\emptyset$, of course, returns an empty continuation (Nothing) and an unmodified state. $(\|)$ first exhausts the left branch (always collecting any state modifications it performs), before switching to the right branch.

$$
\begin{aligned}
\emptyset &= \lambda s \rightarrow (\mathsf{Nothing}, s) \ , \\
p \mathbin{[\!]} q &= \lambda s \rightarrow \mathbf{case}\ p\ s\ \mathbf{of}\ (\mathsf{Nothing}, t) \quad\ \ \rightarrow q\ t \\
&\qquad\qquad\qquad\qquad\qquad (\mathsf{Just}\ (x, r), t) \rightarrow (\mathsf{Just}\ (x, r \mathbin{[\!]} q), t) \ .
\end{aligned}
$$

The state operators are implemented in a straightforward manner.

$$
\begin{aligned}
\mathit{get} \quad &= \lambda s \rightarrow (\mathsf{Just}\ (s\ , \emptyset), s) \ , \\
\mathit{put}\ s &= \lambda t \rightarrow (\mathsf{Just}\ ((), \emptyset), s) \ .
\end{aligned}
$$

And this implementation is also a monad. The implementation of $p \gg\!= k$ extends every branch within p with k, threading the state through this entire process.

$$
\begin{aligned}
\mathit{return}\ x &= \lambda s \rightarrow (\mathsf{Just}\ (x, \emptyset), s) \ , \\
p \gg\!= k \quad &= \lambda s \rightarrow \mathbf{case}\ p\ s\ \mathbf{of}\ (\mathsf{Nothing}, t) \quad\ \ \rightarrow (\mathsf{Nothing}, t) \\
&\qquad\qquad\qquad\qquad\qquad (\mathsf{Just}\ (x, q), t) \rightarrow (k\ x \mathbin{[\!]} (q \gg\!= k))\ t \ .
\end{aligned}
$$

	0	1	2	3	4	5	6	7
0	.	.	.	.	.	Q	.	.
1	.	.	.	Q	.	.	.	.
2	.	.	.	.	.	.	Q	.
3	Q	.	.	.	.	.	.	.
4	.	.	.	.	.	.	.	Q
5	.	Q	.	.	.	.	.	.
6	.	.	.	.	Q	.	.	.
7	.	.	Q	.	.	.	.	.

(a)

	0	1	2	3	4	5	6	7
0	0	1	2	3	4	.	.	.
1	1	2	3	4	.	.	.	.
2	2	3	4	.	.	.	.	.
3	3	4	.	.	.	.	.	.
4	4	.	.	.	.	.	.	.
5	.	.	.	.	.	.	.	12
6	.	.	.	.	.	.	12	13
7	.	.	.	.	.	12	13	14

(b)

	0	1	2	3	4	5	6	7
0	0	−1	.	.	.	−5	−6	−7
1	.	0	−1	.	.	.	−5	−6
2	.	.	0	−1	.	.	.	−5
3	3	.	.	0	.	.	.	.
4	4	3	.	.	0	.	.	.
5	5	4	3	.	.	0	.	.
6	6	5	4	3	.	.	0	.
7	7	6	5	4	3	.	.	0

(c)

Fig. 1. (a) This placement can be represented by $[3, 5, 7, 1, 6, 0, 2, 4]$. (b) Up diagonals. (c) Down diagonals.

3 Motivation

In the previous section we discussed two possible semantics for the interaction of state and nondeterminism: global and local state semantics. In this section, we will further explore the differences between these two interpretations. Using the classic n-queens puzzle as an example, we show that sometimes we end up in a situation where we want to write our program according to local state semantics (which is generally speaking easier to reason about), but desire the space usage characteristics of global state semantics.

3.1 Example: The n-Queens Problem

The n-queens puzzle presented in this section is adapted and simplified from that of Gibbons and Hinze [4]. The aim of the puzzle is to place n queens on a

n by n chess board such that no two queens can attack each other. Given n, we number the rows and columns by $[0\,..\,n-1]$. Since all queens should be placed on distinct rows and distinct columns, a potential solution can be represented by a permutation xs of the list $[0\,..\,n-1]$, such that $xs\,!!\,i = j$ denotes that the queen on the ith column is placed on the jth row (see Fig. 1(a)). The specification can be written as a non-deterministic program:

$$\begin{aligned}&\mathit{queens} :: \mathsf{MNondet}\; m \Rightarrow \mathsf{Int} \rightarrow m\;[\mathsf{Int}]\\&\mathit{queens}\; n = \mathit{perm}\;[0\,..\,n-1] \ggg\!= \mathit{filt}\;\mathit{safe}\quad,\end{aligned}$$

where *perm* non-deterministically computes a permutation of its input, and the pure function $\mathit{safe} :: [\mathsf{Int}] \rightarrow \mathsf{Bool}$, to be defined later, determines whether a solution is valid. The function $\mathit{filt}\; p\; x$ returns x if $p\; x$ holds, and fails otherwise. It can be defined in terms of a standard monadic function *guard*:

$$\begin{aligned}&\mathit{filt} :: \mathsf{MNondet}\; m \Rightarrow (a \rightarrow \mathsf{Bool}) \rightarrow a \rightarrow m\; a\\&\mathit{filt}\; p\; x = \mathit{guard}\;(p\; x) \gg \mathit{return}\; x\quad,\\&\mathit{guard} :: \mathsf{MNondet}\; m \Rightarrow \mathsf{Bool} \rightarrow m\;()\\&\mathit{guard}\; b = \mathbf{if}\; b\; \mathbf{then}\; \mathit{return}\;()\; \mathbf{else}\; \emptyset\quad.\end{aligned}$$

The function *perm* can be written either as a fold or an unfold. For this problem we choose the latter, using a function *select*, which non-deterministically splits a list into a pair containing one chosen element and the rest. For example, $\mathit{select}\;[1,2,3]$ yields one of $(1,[2,3])$, $(2,[1,3])$ and $(3,[1,2])$.

$$\begin{aligned}&\mathit{select} :: \mathsf{MNondet}\; m \Rightarrow [a] \rightarrow m\;(a,[a])\\&\mathit{select}\;[\,] = \emptyset\\&\mathit{select}\;(x:xs) = \mathit{return}\;(x,xs)\;[\!]\;((\mathit{id} \times (x{:}))\;\langle\$\rangle\;\mathit{select}\;xs)\quad,\\&\mathit{perm} :: \mathsf{MNondet}\; m \Rightarrow [a] \rightarrow m\;[a]\\&\mathit{perm}\;[\,] = \mathit{return}\;[\,]\\&\mathit{perm}\;xs = \mathit{select}\;xs \ggg\!= \lambda(x,ys) \rightarrow (x{:})\;\langle\$\rangle\;\mathit{perm}\;ys\quad,\end{aligned}$$

where $f\;\langle\$\rangle\;m = m \gg\!= (\mathit{return} \cdot f)$ which applies a pure function to a monadic value, and $(f \times g)\;(x,y) = (f\;x, g\;y)$.

This specification of *queens* generates all the permutations, before checking them one by one, in two separate phases. We wish to fuse the two phases, which allows branches generates a non-safe placement to be pruned earlier, and thus produce a faster implementation.

A Backtracking Algorithm. In our representation, queens cannot be put on the same row or column. Therefore, *safe* only needs to make sure that no two queens are put on the same diagonal. An 8 by 8 chess board has 15 *up diagonals* (those running between bottom-left and top-right). Let them be indexed by $[0\,..\,14]$ (see Fig. 1(b)). Similarly, there are 15 *down diagonals* (running between top-left and bottom right, indexed by $[-7\,..\,7]$ in Fig. 1(c)). We can show, by routine program calculation, that whether a placement is safe can be checked in one left-to-right traversal—define $\mathit{safe}\;xs = \mathit{safeAcc}\;(0,[\,],[\,])\;xs$, where

$$
\begin{array}{l}
\mathit{safeAcc} :: (\mathsf{Int}, [\mathsf{Int}], [\mathsf{Int}]) \rightarrow [\mathsf{Int}] \rightarrow \mathsf{Bool} \\
\mathit{safeAcc}\ (i, \mathit{us}, \mathit{ds})\ [\,] \qquad\quad = \mathsf{True} \\
\mathit{safeAcc}\ (i, \mathit{us}, \mathit{ds})\ (x : \mathit{xs}) = \mathit{ok}\ (i', \mathit{us}', \mathit{ds}') \wedge \mathit{safeAcc}\ (i', \mathit{us}', \mathit{ds}')\ \mathit{xs} \quad , \\
\quad \textbf{where}\ (i', \mathit{us}', \mathit{ds}') = (i+1, (i+x : \mathit{us}), (i-x : \mathit{ds})) \quad , \\
\mathit{ok}\ (i, (x : \mathit{us}), (y : \mathit{ds})) = x \notin \mathit{us} \wedge y \notin \mathit{ds} \quad .
\end{array}
$$

Operationally, $(i, \mathit{us}, \mathit{ds})$ is a "state" kept by *safeAcc*, where i is the current column, while *us* and *ds* are respectively the up and down diagonals encountered so far. Function *safeAcc* behaves like a fold-left. Indeed, it can be defined using *scanl* and *all* (where $\mathit{all}\ p = \mathit{foldr}\ (\wedge)\ \mathsf{True} \cdot \mathit{map}\ p$):

$$
\begin{array}{l}
\mathit{safeAcc}\ (i, \mathit{us}, \mathit{ds}) = \mathit{all}\ \mathit{ok} \cdot \mathit{tail} \cdot \mathit{scanl}\ (\oplus)\ (i, \mathit{us}, \mathit{ds}) \quad , \\
\quad \textbf{where}\ (i, \mathit{us}, \mathit{ds}) \oplus x = (i+1, (i+x : \mathit{us}), (i-x : \mathit{ds})) \quad .
\end{array}
$$

One might wonder whether the "state" can be implemented using an actual state monad. Indeed, the following is among the theorems we have proved:

Theorem 2. *If state and non-determinism commute, we have that for all xs, st, $(\oplus)$, and ok,*

$$
\begin{array}{l}
\mathit{filt}\ (\mathit{all}\ \mathit{ok} \cdot \mathit{tail} \cdot \mathit{scanl}\ (\oplus)\ \mathit{st})\ \mathit{xs} = \\
\quad \mathit{protect}\ (\mathit{put}\ \mathit{st} \gg \mathit{foldr}\ (\odot)\ (\mathit{return}\ [\,])\ \mathit{xs}) \quad , \\
\quad \textbf{where}\ x \odot m = \mathit{get} \gg\!= \lambda \mathit{st} \rightarrow \mathit{guard}\ (\mathit{ok}\ (\mathit{st} \oplus x)) \gg \\
\qquad\qquad\qquad\qquad \mathit{put}\ (\mathit{st} \oplus x) \gg ((x{:})\ \langle\$\rangle\ m) \quad .
\end{array}
$$

The function $\mathit{protect}\ m = \mathit{get} \gg\!= \lambda \mathit{ini} \rightarrow m \gg\!= \lambda x \rightarrow \mathit{put}\ \mathit{ini} \gg \mathit{return}\ x$ saves the initial state and restores it after the computation. As for $(\odot)$, it assumes that the "state" passed around by *scanl* is stored in a monadic state, checks whether the new state $\mathit{st} \oplus x$ satisfies *ok*, and updates the state with the new value.

For Theorem 2 to hold, however, we need state and non-determinism to commute. It is so in the local state semantics, which can be proved using the non-determinism laws, (12), and (13).

Now that the safety check can be performed in a *foldr*, recalling that *perm* is an unfold, it is natural to try to fuse them into one. Indeed, it can be proved that, with $(\oplus)$, *ok*, and $(\odot)$ defined above, we have $\mathit{perm}\ \mathit{xs} \gg\!= \mathit{foldr}\ (\odot)\ (\mathit{return}\ [\,]) = \mathit{qBody}\ \mathit{xs}$, where

$$
\begin{array}{l}
\mathit{qBody} :: \mathsf{MStateNondet}\ (\mathsf{Int}, [\mathsf{Int}], [\mathsf{Int}])\ m \Rightarrow [\mathsf{Int}] \rightarrow m\ [\mathsf{Int}] \\
\mathit{qBody}\ [\,] \ = \mathit{return}\ [\,] \\
\mathit{qBody}\ \mathit{xs} = \mathit{select}\ \mathit{xs} \gg\!= \lambda(x, \mathit{ys}) \rightarrow \\
\qquad\qquad \mathit{get} \gg\!= \lambda \mathit{st} \rightarrow \mathit{guard}\ (\mathit{ok}\ (\mathit{st} \oplus x)) \gg \\
\qquad\qquad \mathit{put}\ (\mathit{st} \oplus x) \gg ((x{:})\ \langle\$\rangle\ \mathit{qBody}\ \mathit{ys}) \quad .
\end{array}
$$

The proof also heavily relies on the commutativity between non-determinism and state.

To wrap up, having fused *perm* and safety checking into one phase, we may compute *queens* by:

$$queens :: \mathsf{MStateNondet}\ (\mathsf{Int}, [\mathsf{Int}], [\mathsf{Int}])\ m \Rightarrow \mathsf{Int} \rightarrow m\ [\mathsf{Int}]$$
$$queens\ n = protect\ (put\ (0, [\,], [\,]) \gg qBody\ [0\,..\,n-1])\ \ .$$

This is a backtracking algorithm that attempts to place queens column-by-column, proceeds to the next column if *ok* holds, and backtracks otherwise. The derivation from the specification to this program relies on a number of properties that hold in the local state semantics.

3.2 Transforming Between Local State and Global State

For a monad with both non-determinism and state, the local state laws imply that each non-deterministic branch has its own state. This is not costly for states consisting of linked data structures, for example the state (Int, [Int], [Int]) in the n-queens problem. In some applications, however, the state might be represented by data structures, e.g. arrays, that are costly to duplicate: When each new state is only slightly different from the previous (say, the array is updated in one place each time), we have a wasteful duplication of information. Although this is not expected to be an issue for realistic sizes of the n-queens problem due to the relatively small state, one can imagine that for some problems where the state is very large, this can be a problem.

Global state semantics, on the other hand, has a more "low-level" feel to it. Because a single state is threaded through the entire computation without making any implicit copies, it is easier to reason about resource usage in this setting. So we might write our programs directly in the global state style. However, if we do this to a program that would be more naturally expressed in the local state style (such as our n-queens example), this will come at a great loss of clarity. Furthermore, as we shall see, although it is easier to reason about resource usage of programs in the global state setting, it is significantly more difficult to reason about their semantics. We could also write our program first in a local state style and then translate it to global state style. Doing this manually is a tedious and error-prone process that leaves us with code that is hard to maintain. A more attractive proposition is to design a systematic program transformation that takes a program written for local state semantics as input, and outputs a program that, when interpreted under global state semantics, behaves exactly the same as the original program interpreted under local state semantics.

In the remainder of the paper we define this program transformation and prove it correct. We believe that, in particular, the proof *technique* is of interest.

4 Non-determinism with Global State

So far, we have evaded giving a precise axiomatic characterisation of global state semantics: although in Sect. 2 we provided an example implementation that matches our intuition of global state semantics, we haven't provided a clear formulation of that intuition. We begin this section by finally stating the "global state law", which characterises exactly the property that sets apart non-backtrackable state from backtrackable state.

In the rest of the section, we appeal to intuition and see what happens when we work with a global state monad: what pitfalls we may encounter, and what programming pattern we may use, to motivate the more formal treatment in Sect. 5.

4.1 The Global State Law

We have already discussed general laws for nondeterministic monads (laws (4) through (7)), as well as laws which govern the interaction between state and nondeterminism in a local state setting (laws (13) and (12)). For global state semantics, an alternative law is required to govern the interactions between nondeterminism and state. We call this the *global state law*.

$$\textbf{put-or:} \qquad (put\ s \gg m)\ [\!]\ n = put\ s \gg (m\ [\!]\ n) \quad , \qquad (15)$$

This law allows the lifting of a *put* operation from the left branch of a nondeterministic choice, an operation which does not preserve meaning under local state semantics. Suppose for example that $m = \emptyset$, then by (12) and (5), the left-hand side of the equation is equal to n, whereas by (5), the right-hand side of the equation is equal to $put\ s \gg n$.

By itself, this law leaves us free to choose from a large space of implementations with different properties. For example, in any given implementation, the programs $return\ x\ [\!]\ return\ y$ and $return\ y\ [\!]\ return\ x$ may be considered semantically identical, or they may be considered semantically distinct. The same goes for the programs $return\ x\ [\!]\ return\ x$ and $return\ x$, or the programs $(put\ s \gg return\ x)\ [\!]\ m$ and $(put\ s \gg return\ x)\ [\!]\ (put\ s \gg m)$. Additional axioms will be introduced as needed to cover these properties in Sect. 5.2.

4.2 Chaining Using Non-deterministic Choice

In backtracking algorithms that keep a global state, it is a common pattern to 1. update the current state to its next step, 2. recursively search for solutions, and 3. roll back the state to the previous step (regardless of whether a solution is found). To implement such pattern as a monadic program, one might come up with something like the code below:

$$modify\ next \gg search \gg\!= modReturn\ prev \quad ,$$

where $next$ advances the state, $prev$ undoes the modification of $next$ ($prev \cdot next = id$), and $modify$ and $modReturn$ are defined by:

$$\begin{aligned} &modify\ f &&= get \gg\!= (put \cdot f) \quad , \\ &modReturn\ f\ v &&= modify\ f \gg return\ v \quad . \end{aligned}$$

Let the initial state be st and assume that *search* found three choices $m_1 \mathbin{[\!]} m_2 \mathbin{[\!]} m_3$. We wish that m_1, m_2, and m_3 all start running with state $\mathit{next}\ st$, and the state is restored to $\mathit{prev}\ (\mathit{next}\ st) = st$ afterwards. Due to (6), however, it expands to

$$\begin{aligned}
&\mathit{modify}\ \mathit{next} \gg (m_1 \mathbin{[\!]} m_2 \mathbin{[\!]} m_3) \gg\!= \mathit{modReturn}\ \mathit{prev} = \\
&\quad \mathit{modify}\ \mathit{next} \gg ((m_1 \gg\!= \mathit{modReturn}\ \mathit{prev}) \mathbin{[\!]} \\
&\qquad\qquad\qquad\qquad (m_2 \gg\!= \mathit{modReturn}\ \mathit{prev}) \mathbin{[\!]} \\
&\qquad\qquad\qquad\qquad (m_3 \gg\!= \mathit{modReturn}\ \mathit{prev}))\ .
\end{aligned}$$

With a global state, it means that m_2 starts with state st, after which the state is rolled back further to $\mathit{prev}\ st$. The computation m_3 starts with $\mathit{prev}\ st$, after which the state is rolled too far to $\mathit{prev}\ (\mathit{prev}\ st)$. In fact, one cannot guarantee that *modReturn prev* is always executed—if *search* fails and reduces to $\emptyset$, *modReturn prev* is not run at all, due to (7).

We need a way to say that "*modify next* and *modReturn prev* are run exactly once, respectively before and after all non-deterministic branches in *solve*." Fortunately, we have discovered a curious technique. Define

$$\begin{aligned}
&\mathit{side} :: \mathsf{MNondet}\ m \Rightarrow m\ a \rightarrow m\ b \\
&\mathit{side}\ m = m \gg \emptyset\ .
\end{aligned}$$

Since non-deterministic branches are executed sequentially, the program

$$\mathit{side}\ (\mathit{modify}\ \mathit{next}) \mathbin{[\!]} m_1 \mathbin{[\!]} m_2 \mathbin{[\!]} m_3 \mathbin{[\!]} \mathit{side}\ (\mathit{modify}\ \mathit{prev})$$

executes *modify next* and *modify prev* once, respectively before and after all the non-deterministic branches, even if they fail. Note that $\mathit{side}\ m$ does not generate a result. Its presence is merely for the side-effect of m, hence the name.

The reader might wonder: now that we are using $(\mathbin{[\!]})$ as a sequencing operator, does it simply coincide with $(\gg)$? Recall that we still have left-distributivity (6) and, therefore, $(m_1 \mathbin{[\!]} m_2) \gg n$ equals $(m_1 \gg n) \mathbin{[\!]} (m_2 \gg n)$. That is, $(\mathbin{[\!]})$ acts as "insertion points", where future code followed by $(\gg)$ can be inserted into! This is certainly a dangerous feature, whose undisciplined use can lead to chaos. However, we will exploit this feature and develop a safer programming pattern in the next section.

4.3 State-Restoring Operations

The discussion above suggests that one can implement backtracking, in a global-state setting, by using $(\mathbin{[\!]})$ and *side* appropriately. We can even go a bit further by defining the following variation of *put*, which restores the original state when it is backtracked over:

$$\begin{aligned}
&\mathit{put}_{\mathsf{R}} :: \mathsf{MStateNondet}\ s\ m \Rightarrow s \rightarrow m\ () \\
&\mathit{put}_{\mathsf{R}}\ s = \mathit{get} \gg\!= \lambda s_0 \rightarrow \mathit{put}\ s \mathbin{[\!]} \mathit{side}\ (\mathit{put}\ s_0)\ .
\end{aligned}$$

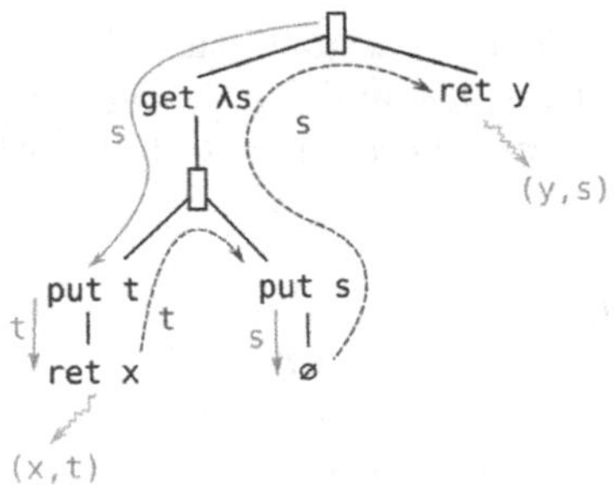

Fig. 2. Illustration of state-restoring put

To help build understanding for put_{R}, Fig. 2 shows the flow of execution for the expression $(put_{\mathsf{R}}\ t \gg ret\ x) \mathbin{[]} ret\ y$. Initially, the state is s; it gets modified to t at the $put\ t$ node after which the value x is output with the working state t. Then, because we found a result, we backtrack (since we're using global-state semantics, the state modification caused by $put\ t$ is not reversed), arriving in the *side* operation branch. The $put\ s$ operation is executed, which resets the state to s, and then the branch immediately fails, so we backtrack to the right branch of the topmost $([])$. There the value y is output with working state s.

For some further intuition about put_{R}, consider $put_{\mathsf{R}}\ s \gg comp$ where $comp$ is some arbitrary computation:

$$\begin{aligned}
& put_{\mathsf{R}}\ s \gg comp \\
= {} & (get \mathbin{>\!\!>\!\!=} \lambda s_0 \to put\ s \mathbin{[]} side\ (put\ s_0)) \gg comp \\
= {} & \quad \{ \text{ monad law, left-distributivity (6) } \} \\
& get \mathbin{>\!\!>\!\!=} \lambda s_0 \to (put\ s \gg comp) \mathbin{[]} (side\ (put\ s_0) \gg comp) \\
= {} & \quad \{ \text{ by (7) } \emptyset \gg comp = \emptyset, \text{ monad laws } \} \\
& get \mathbin{>\!\!>\!\!=} \lambda s_0 \to (put\ s \gg comp) \mathbin{[]} side\ (put\ s_0) \quad .
\end{aligned}$$

Thanks to left-distributivity (6), $(\gg comp)$ is promoted into $([])$. Furthermore, the $(\gg comp)$ after $side\ (put\ s_0)$ is discarded by (7). In words, $put_{\mathsf{R}}\ s \gg comp$ saves the current state, computes $comp$ using state s, and restores the saved state! The subscript R stands for "restore." Note also that $(put_{\mathsf{R}}\ s \gg m_1) \gg m_2 = put_{\mathsf{R}}\ s \gg (m_1 \gg m_2)$—the state restoration happens in the end.

The behaviour of put_{R} is rather tricky. It is instructive comparing

(a) $return\ x$,
(b) $put\ s \gg return\ x$,
(c) $put_{\mathsf{R}}\ s \gg return\ x$.

When run in initial state s_0, they all yield x as the result. The final states after running (a), (b) and (c) are s_0, s and s_0, respectively. However, (c) does *not* behave identically to (a) in all contexts! For example, in the context $(\gg get)$, we can tell them apart: $return\ x \gg get$ returns s_0, while $put_{\mathsf{R}}\ s \gg return\ x \gg get$ returns s, even though the program yields final state s_0.

We wish that put_{R}, when run with a global state, satisfies laws (8) through (13)—the state laws and the *local* state laws. If so, one could take a program written for a local state monad, replace all occurrences of *put* by put_{R}, and run the program with a global state. Unfortunately this is not the case: put_{R} does satisfy **put-put** (8) and **put-get** (10), but **get-put** (9) fails—$get \gg\!= put_{\mathsf{R}}$ and *return* () can be differentiated by some contexts, for example $(\gg put\ t)$. To see that, we calculate:

$$
\begin{aligned}
& (get \gg\!= put_{\mathsf{R}}) \gg put\ t \\
= {} & (get \gg\!= \lambda s \rightarrow get \gg\!= \lambda s_0 \rightarrow put\ s \,[\!]\, side\ (put\ s_0)) \gg put\ t \\
= {} & \quad \{\ \textbf{get-get}\ \} \\
& (get \gg\!= \lambda s \rightarrow put\ s \,[\!]\, side\ (put\ s)) \gg put\ t \\
= {} & \quad \{\ \text{monad laws, left-distributivity}\ \} \\
& get \gg\!= \lambda s \rightarrow (put\ s \gg put\ t) \,[\!]\, side\ (put\ s) \\
= {} & \quad \{\ \textbf{put-put}\ \} \\
& get \gg\!= \lambda s \rightarrow put\ t \,[\!]\, side\ (put\ s) \quad .
\end{aligned}
$$

Meanwhile, $return\ () \gg put\ t = put\ t$, which does not behave in the same way as $get \gg\!= \lambda s \rightarrow put\ t \,[\!]\, side\ (put\ s)$ when $s \neq t$.

In a global-state setting, the left-distributivity law (6) makes it tricky to reason about combinations of $([\!])$ and $(\gg\!=)$ operators. Suppose we have a program $(m \,[\!]\, n)$, and we construct an extended program by binding a continuation f to it such that we get $(m \,[\!]\, n) \gg\!= f$ (where f might modify the state). Under global-state semantics, the evaluation of the right branch is influenced by the state modifications performed by evaluating the left branch. So by (6), this means that when we get to evaluating the n subprogram in the extended program, it will do so with a different initial state (the one obtained after running $m \gg\!= f$) compared to the initial state in the original program (the one obtained by running m). In other words, placing our program in a different context changed the meaning of one of its subprograms. So it is difficult to reason about programs compositionally in this setting—some properties hold only when we take the entire program into consideration.

It turns out that all properties we need do hold, provided that *all* occurrences of *put* are replaced by put_{R}—problematic contexts such as *put t* above are thus ruled out. However, that "all *put* are replaced by put_{R}" is a global property, and to properly talk about it we have to formally define contexts, which is what we will do in Sect. 5. Notice though, that simulation of local state semantics by judicious use of put_{R} does not avoid the unnecessary copying mentioned in Sect. 3.2, it merely makes it explicit in the program. We will address this shortcoming in Sect. 5.6.

5 Laws and Translation for Global State Monad

In this section we give a more formal treatment of the non-deterministic global state monad. Not every implementation of the global state law allows us to accurately simulate local state semantics though, so we propose additional laws that the implementation must respect. These laws turn out to be rather intricate. To make sure that there exists a model, an implementation is proposed, and it is verified in Coq that the laws and some additional theorems are satisfied.

The ultimate goal, however, is to show the following property: given a program written for a local-state monad, if we replace all occurrences of *put* by put_{R}, the resulting program yields the same result when run with a global-state monad. This allows us to painlessly port our previous algorithm to work with a global state. To show this we first introduce a syntax for nondeterministic and stateful monadic programs and contexts. Then we imbue these programs with global-state semantics. Finally we define the function that performs the translation just described, and prove that this translation is correct.

5.1 Programs and Contexts

```
data Prog a where
  Return :: a → Prog a
  ∅      :: Prog a
  ([])   :: Prog a → Prog a → Prog a
  Get    :: (S → Prog a) → Prog a
  Put    :: S → Prog a → Prog a
```

(a)

$$\begin{array}{l}
run :: \mathsf{Prog}\ a \to \mathsf{Dom}\ a \\
\overline{ret} :: a \to \mathsf{Dom}\ a \\
\overline{\varnothing} :: \mathsf{Dom}\ a \\
(\overline{[\!]}) :: \mathsf{Dom}\ a \to \mathsf{Dom}\ a \to \mathsf{Dom}\ a \\
\overline{get} :: (\mathsf{S} \to \mathsf{Dom}\ a) \to \mathsf{Dom}\ a \\
\overline{put} :: \mathsf{S} \to \mathsf{Dom}\ a \to \mathsf{Dom}\ a
\end{array}$$

(b)

Fig. 3. (a) Syntax for programs. (b) Semantic domain.

In the previous sections we have been mixing syntax and semantics, which we avoid in this section by defining the program syntax as a free monad. This way we avoid the need for a type-level distinction between programs with local-state semantics and programs with global-state semantics. Figure 3(a) defines a syntax for nondeterministic, stateful, closed programs Prog, where the Get and Put constructors take continuations as arguments, and the $(\gg\!=)$ operator is defined as follows:

$$\begin{array}{ll}
\multicolumn{2}{l}{(\gg\!=) :: \mathsf{Prog}\ a \to (a \to \mathsf{Prog}\ b) \to \mathsf{Prog}\ b} \\
\mathsf{Return}\ x & \gg\!= f = f\ x \\
\emptyset & \gg\!= f = \emptyset \\
(m \mathbin{[\!]} n) & \gg\!= f = (m \gg\!= f) \mathbin{[\!]} (n \gg\!= f) \\
\mathsf{Get}\ k & \gg\!= f = \mathsf{Get}\ (\lambda s \to k\ s \gg\!= f) \\
\mathsf{Put}\ s\ m & \gg\!= f = \mathsf{Put}\ s\ (m \gg\!= k)\ .
\end{array}$$

One can see that $(\gg\!=)$ is defined as a purely syntactical manipulation, and its definition has laws (6) and (7) built-in.

The meaning of such a monadic program is determined by a semantic domain of our choosing, which we denote with Dom, and its corresponding domain operators $\overline{ret}$, $\overline{\varnothing}$, $\overline{get}$, $\overline{put}$ and $(\overline{[]})$ (see Fig. 3(b)). The $run :: \mathsf{Prog}\ a \rightarrow \mathsf{Dom}\ a$ function "runs" a program Prog a into a value in the semantic domain Dom a:

$$
\begin{aligned}
&run\ (\mathsf{Return}\ x) &&= \overline{ret}\ x\\
&run\ \emptyset &&= \overline{\varnothing}\\
&run\ (m_1 \mathbin{[]} m_2) &&= run\ m_1 \mathbin{\overline{[]}} run\ m_2\\
&run\ (\mathsf{Get}\ k) &&= \overline{get}\ (\lambda s \rightarrow run\ (k\ s))\\
&run\ (\mathsf{Put}\ s\ m) &&= \overline{put}\ s\ (run\ m) \ .
\end{aligned}
$$

Note that no $\gg\!=$ operator is required to define run; in other words, Dom need not be a monad. In fact, as we will see later, we will choose our implementation in such a way that there does not exist a bind operator for run.

5.2 Laws for Global State Semantics

We impose the laws upon Dom and the domain operators to ensure the semantics of a non-backtracking (global-state), nondeterministic, stateful computation for our programs. Naturally, we need laws analogous to the state laws and nondeterminism laws to hold for our semantic domain. As it is not required that a bind operator $((\gg\!=) :: \mathsf{Dom}\ a \rightarrow (a \rightarrow \mathsf{Dom}\ b) \rightarrow \mathsf{Dom}\ b)$ be defined for the semantic domain (and we will later argue that it *cannot* be defined for the domain, given the laws we impose on it), the state laws ((8) through (11)) must be reformulated to fit the continuation-passing style of the semantic domain operators.

$$\overline{put}\ s\ (\overline{put}\ t\ p) = \overline{put}\ t\ p \ , \tag{16}$$

$$\overline{put}\ s\ (\overline{get}\ k) = \overline{put}\ s\ (k\ s) \ , \tag{17}$$

$$\overline{get}\ (\lambda s \rightarrow \overline{put}\ s\ m) = m \ , \tag{18}$$

$$\overline{get}\ (\lambda s \rightarrow \overline{get}\ (\lambda t \rightarrow k\ s\ t)) = \overline{get}\ (\lambda s \rightarrow k\ s\ s) \ . \tag{19}$$

Two of the nondeterminism laws—(6) and (7)—also mention the bind operator. As we have seen earlier, they are trivially implied by the definition of $(\gg\!=)$ for Prog. Therefore, we need not impose equivalent laws for the semantic domain (and in fact, we cannot formulate them given the representation we have chosen). Only the two remaining nondeterminism laws—(4) and (5)—need to be stated:

$$(m \mathbin{\overline{[]}} n) \mathbin{\overline{[]}} p = m \mathbin{\overline{[]}} (n \mathbin{\overline{[]}} p) \ , \tag{20}$$

$$\overline{\varnothing} \mathbin{\overline{[]}} m = m \mathbin{\overline{[]}} \overline{\varnothing} = m \ . \tag{21}$$

We also reformulate the global-state law (15):

$$\overline{put}\ s\ p \mathbin{\overline{[]}} q = \overline{put}\ s\ (p \mathbin{\overline{[]}} q) \ . \tag{22}$$

It turns out that, apart from the **put-or** law, our proofs require certain additional properties regarding commutativity and distributivity which we introduce here:

$$
\begin{aligned}
&\overline{get}\ (\lambda s \rightarrow \overline{put}\ (t\ s)\ p \mathbin{\overline{[]}} \overline{put}\ (u\ s)\ q \mathbin{\overline{[]}} \overline{put}\ s\ \overline{\varnothing}) =\\
&\overline{get}\ (\lambda s \rightarrow \overline{put}\ (u\ s)\ q \mathbin{\overline{[]}} \overline{put}\ (t\ s)\ p \mathbin{\overline{[]}} \overline{put}\ s\ \overline{\varnothing}) \ ,
\end{aligned}
\tag{23}
$$

$$\overline{put}\ s\ (\overline{ret}\ x\ \overline{[\![}\ p) = \overline{put}\ s\ (\overline{ret}\ x)\ \overline{[\![}\ \overline{put}\ s\ p\ . \tag{24}$$

These laws are not considered general "global state" laws, because it is possible to define reasonable implementations of global state semantics that violate these laws, and because they are not exclusive to global state semantics.

The ($\overline{[\![}$) operator is not, in general, commutative in a global state setting. However, we will require that the order in which results are computed does not matter. This might seem contradictory at first glance. To be more precise, we do not require the property $p\ \overline{[\![}\ q = q\ \overline{[\![}\ p$ because the subprograms p and q might perform non-commuting edits to the global state. But we do expect that programs without side-effects commute freely; for instance, we expect $\mathit{return}\ x\ \overline{[\![}\ \mathit{return}\ y = \mathit{return}\ y\ \overline{[\![}\ \mathit{return}\ x$. In other words, collecting all the results of a nondeterministic computation is done with a set-based semantics in mind rather than a list-based semantics, but this does not imply that the order of state effects does not matter.

In fact, the notion of commutativity we wish to impose is still somewhat stronger than just the fact that results commute: we want the ($\overline{[\![}$) operator to commute with respect to any pair of subprograms whose modifications of the state are ignored—that is, immediately overwritten—by the surrounding program. This property is expressed by law (23). An example of an implementation which does not respect this law is one that records the history of state changes.

In global-state semantics, $\overline{put}$ operations cannot, in general, distribute over ($\overline{[\![}$). However, an implementation may permit distributivity if certain conditions are met. Law (24) states that a $\overline{put}$ operation distributes over a nondeterministic choice if the left branch of that choice simply returns a value. This law has a particularly striking implication: it disqualifies any implementation for which a bind operator $(\ \overline{\gg\!=}\ ) :: \mathsf{Dom}\ a \to (a \to \mathsf{Dom}\ b) \to \mathsf{Dom}\ b$ can be defined! Consider for instance the following program:

$$\overline{put}\ x\ (\overline{ret}\ w\ \overline{[\![}\ \overline{get}\ \overline{ret})\ \overline{\gg\!=}\ \lambda z \to \overline{put}\ y\ (\overline{ret}\ z)\ .$$

If (24) holds, this program should be equal to

$$(\overline{put}\ x\ (\overline{ret}\ w)\ \overline{[\![}\ \overline{put}\ x\ (\overline{get}\ \overline{ret}))\ \overline{\gg\!=}\ \lambda z \to \overline{put}\ y\ (\overline{ret}\ z)\ .$$

However, it is proved in Fig. 4 that the first program can be reduced to $\overline{put}\ y\ (\overline{ret}\ w\ \overline{[\![}\ \overline{ret}\ y)$, whereas the second program is equal to $\overline{put}\ y\ (\overline{ret}\ w\ \overline{[\![}\ \overline{ret}\ x)$, which clearly does not always have the same result.

To gain some intuition about why law (24) prohibits a bind operator, consider that the presence or absence of a bind operator influences what equality of programs means. Our first intuition might be that we consider two programs equal if they produce the same outputs given the same inputs. But this is too narrow a view: for two programs to be considered equal, they must also behave the same under *composition*; that is, we must be able to replace one for the other within a larger whole, without changing the meaning of the whole. The bind operator allows us to compose programs *sequentially*, and therefore its existence implies that, for two programs to be considered equal, they must also behave identically under sequential composition. Under local-state semantics, this additional requirement coincides with other notions of equality: we can't come up with a pair of programs which both produce the same outputs given the same inputs, but behave differently under sequential composition. But under global-state semantics, we can come up with such counterexamples: consider the subprograms of our previous example $\overline{put}\ x\ (\overline{ret}\ w \mathbin{\overline{[\!]}} \overline{get}\ \overline{ret})$ and $(\overline{put}\ x\ (\overline{ret}\ w) \mathbin{\overline{[\!]}} \overline{put}\ x\ (\overline{get}\ \overline{ret}))$. Clearly we expect these two programs to produce the exact same results in isolation, yet when they are sequentially composed with the program $\lambda z \to \overline{put}\ y\ (\overline{ret}\ z)$, their different nature is revealed (by (6)).

It is worth remarking that introducing either one of these additional laws disqualify the example implementation given in Sect. 2.2 (even if it is adapted for the continuation-passing style of these laws). As the given implementation records the order in which results are yielded by the computation, law (23) cannot be satisfied. And the example implementation also forms a monad, which means it is incompatible with law (24).

Machine-Verified Proofs. From this point forward, we provide proofs mechanized in Coq for many theorems. When we do, we mark the proven statement with a check mark (✓).

$$
\begin{array}{l}
\overline{put}\ x\ (\overline{ret}\ w \mathbin{\overline{[\!]}} \overline{get}\ \overline{ret}) \mathbin{\overline{\gg\!=}} \lambda z \to \overline{put}\ y\ (\overline{ret}\ z) \\
=\quad \{\ \text{definition of } (\overline{\gg\!=})\ \} \\
\overline{put}\ x\ (\overline{put}\ y\ (\overline{ret}\ w) \mathbin{\overline{[\!]}} \overline{get}\ (\lambda s \to \overline{put}\ y\ (\overline{ret}\ s))) \\
=\quad \{\ \text{by (22) and (16)}\ \} \\
\overline{put}\ y\ (\overline{ret}\ w \mathbin{\overline{[\!]}} \overline{get}\ (\lambda s \to \overline{put}\ y\ (\overline{ret}\ s))) \\
=\quad \{\ \text{by (24)}\ \} \\
\overline{put}\ y\ (\overline{ret}\ w) \mathbin{\overline{[\!]}} \overline{put}\ y\ (\overline{get}\ (\lambda s \to \overline{put}\ y\ (\overline{ret}\ s))) \\
=\quad \{\ \text{by (17) and (16)}\ \} \\
\overline{put}\ y\ (\overline{ret}\ w) \mathbin{\overline{[\!]}} \overline{put}\ y\ (\overline{ret}\ y) \\
=\quad \{\ \text{by (24)}\ \} \\
\overline{put}\ y\ (\overline{ret}\ w \mathbin{\overline{[\!]}} \overline{ret}\ y)
\end{array}
$$

(a)

$$
\begin{array}{l}
(\overline{put}\ x\ (\overline{ret}\ w) \mathbin{\overline{[\!]}} \overline{put}\ x\ (\overline{get}\ \overline{ret})) \\
\quad \mathbin{\overline{\gg\!=}} \lambda z \to \overline{put}\ y\ (\overline{ret}\ z) \\
=\quad \{\ \text{definition of } (\overline{\gg\!=})\ \} \\
\overline{put}\ x\ (\overline{put}\ y\ (\overline{ret}\ w)) \\
\quad \mathbin{\overline{[\!]}} \overline{put}\ x\ (\overline{get}\ (\lambda s \to \overline{put}\ y\ (\overline{ret}\ s))) \\
=\quad \{\ \text{by (16) and (17)}\ \} \\
\overline{put}\ y\ (\overline{ret}\ w) \mathbin{\overline{[\!]}} \overline{put}\ x\ (\overline{put}\ y\ (\overline{ret}\ x)) \\
=\quad \{\ \text{by (16)}\ \} \\
\overline{put}\ y\ (\overline{ret}\ w) \mathbin{\overline{[\!]}} \overline{put}\ y\ (\overline{ret}\ x) \\
=\quad \{\ \text{by (24)}\ \} \\
\overline{put}\ y\ (\overline{ret}\ w \mathbin{\overline{[\!]}} \overline{ret}\ x)
\end{array}
$$

(b)

Fig. 4. Proof that law (24) implies that a bind operator cannot exist for the semantic domain.

5.3 An Implementation of the Semantic Domain

We present an implementation of Dom that satisfies the laws of Sect. 5.2, and we provide machine-verified proofs to that effect. In the following implementation, we let Dom be the union of $\mathsf{M}\ s\ a$ for all a and for a given s.

The implementation is based on a multiset or Bag data structure. In the mechanization, we implement $\mathsf{Bag}\ a$ as a function $a \to \mathsf{Nat}$.

$$\begin{aligned}
&\mathbf{type}\ \mathsf{Bag}\ a\\
&\mathit{singleton}\ \ :: a \rightarrow \mathsf{Bag}\ a\\
&\mathit{emptyBag} :: \mathsf{Bag}\ a\\
&\mathit{sum}\qquad\ \ :: \mathsf{Bag}\ a \rightarrow \mathsf{Bag}\ a \rightarrow \mathsf{Bag}\ a
\end{aligned}$$

We model a stateful, nondeterministic computation with global state semantics as a function that maps an initial state onto a bag of results, and a final state. Each result is a pair of the value returned, as well as the state at that point in the computation. The use of an unordered data structure to return the results of the computation is needed to comply with law (23).

In Sect. 5.2 we mentioned that, as a consequence of law (24) we must design the implementation of our semantic domain in such a way that it is impossible to define a bind operator $\overline{\gg\!=}$ :: $\mathsf{Dom}\ a \rightarrow (a \rightarrow \mathsf{Dom}\ b) \rightarrow \mathsf{Dom}\ b$ for it. This is the case for our implementation: we only retain the final result of the branch without any information on how to continue the branch, which makes it impossible to define the bind operator.

$$\mathbf{type}\ \mathsf{M}\ s\ a = s \rightarrow (\mathsf{Bag}\ (a, s), s)$$

$\overline{\varnothing}$ does not modify the state and produces no results. $\overline{\mathit{ret}}$ does not modify the state and produces a single result.

$$\begin{aligned}
&\overline{\varnothing} :: \mathsf{M}\ s\ a\\
&\overline{\varnothing} = \lambda s \rightarrow (\mathit{emptyBag}, s)\\
&\overline{\mathit{ret}} :: a \rightarrow \mathsf{M}\ s\ a\\
&\overline{\mathit{ret}}\ x = \lambda s \rightarrow (\mathit{singleton}\ (x, s), s)
\end{aligned}$$

$\overline{\mathit{get}}$ simply passes along the initial state to its continuation. $\overline{\mathit{put}}$ ignores the initial state and calls its continuation with the given parameter instead.

$$\begin{aligned}
&\overline{\mathit{get}} :: (s \rightarrow \mathsf{M}\ s\ a) \rightarrow \mathsf{M}\ s\ a\\
&\overline{\mathit{get}}\ k = \lambda s \rightarrow k\ s\ s\\
&\overline{\mathit{put}} :: s \rightarrow \mathsf{M}\ s\ a \rightarrow \mathsf{M}\ s\ a\\
&\overline{\mathit{put}}\ s\ k = \lambda_ \rightarrow k\ s
\end{aligned}$$

The $\overline{[\!]}$ operator runs the left computation with the initial state, then runs the right computation with the final state of the left computation, and obtains the final result by merging the two bags of results.

$$\begin{aligned}
&(\overline{[\!]}) :: \mathsf{M}\ s\ a \rightarrow \mathsf{M}\ s\ a \rightarrow \mathsf{M}\ s\ a\\
&(\mathit{xs}\ \overline{[\!]}\ \mathit{ys})\ s = \mathbf{let}\ (\mathit{ansx}, s')\ \ = \mathit{xs}\ s\\
&\qquad\qquad\qquad\qquad (\mathit{ansy}, s'') = \mathit{ys}\ s'\\
&\qquad\qquad\quad\ \ \mathbf{in}\ (\mathit{sum}\ \mathit{ansx}\ \mathit{ansy}, s'')
\end{aligned}$$

Lemma 1. *This implementation conforms to every law introduced in Sect. 5.2.* ✓

5.4 Contextual Equivalence

With our semantic domain sufficiently specified, we can prove analogous properties for programs interpreted through this domain. We must take care in how we reformulate these properties however. It is certainly not sufficient to merely copy the laws as formulated for the semantic domain, substituting Prog data constructors for semantic domain operators as needed; we must keep in mind that a term in Prog a describes a syntactical structure without ascribing meaning to it. For example, one cannot simply assert that Put x (Put y p) is *equal* to Put y p, because although these two programs have the same semantics, they are not structurally identical. It is clear that we must define a notion of "semantic equivalence" between programs. We can map the syntactical structures in Prog a onto the semantic domain Dom a using *run* to achieve that. Yet wrapping both sides of an equation in *run* applications is not enough as such statements only apply at the top-level of a program. For instance, while *run* (Put x (Put y p)) = *run* (Put y p) is a correct statement, we cannot prove *run* (Return w [] Put x (Put y p)) = *run* (Return w [] Put y p) from such a law.

```
data Ctx e1 a e2 b where
  □      :: Ctx e a e a
  COr1   :: Ctx e1 a e2 b → OProg e2 b
         → Ctx e1 a e2 b
  COr2   :: OProg e2 b → Ctx e1 a e2 b
         → Ctx e1 a e2 b
  CPut   :: (Env e2 → S) → Ctx e1 a e2 b
         → Ctx e1 a e2 b
  CGet   :: (S → Bool) → Ctx e1 a (S : e2) b
         → (S → OProg e2 b) → Ctx e1 a e2 b
  CBind1 :: Ctx e1 a e2 b → (b → OProg e2 c)
         → Ctx e1 a e2 c
  CBind2 :: OProg e2 a → Ctx e1 b (a : e2) c
         → Ctx e1 b e2 c
```

(a)

```
data Env (l :: [*]) where
  Nil :: Env '[]
  Cons :: a → Env l → Env (a : l)
type OProg e a = Env e → Prog a
```

(b)

Fig. 5. (a) Environments and open programs. (b) Syntax for contexts.

So the concept of semantic equivalence in itself is not sufficient; we require a notion of "contextual semantic equivalence" of programs which allows us to formulate properties about semantic equivalence which hold in any surrounding context. Figure 5(a) provides the definition for single-hole contexts Ctx. A context C of type Ctx e_1 a e_2 b can be interpreted as a function that, given a program that returns a value of type a under environment e_1 (in other words: the type and environment of the hole), produces a program that returns a value of type b under environment e_2 (the type and environment of the whole program). Filling the hole with p is denoted by C[p]. The type of environments, Env is defined using heterogeneous lists (Fig. 5(b)). When we consider the notion of programs in contexts, we must take into account that these contexts may introduce variables which are referenced by the program. The Prog datatype however represents only closed programs. Figure 5(b) introduces the OProg type to represent "open" programs, and the Env type to represent environments. OProg e a is defined as the type of functions that construct a *closed* program of type Prog a, given an environment of type Env e. Environments, in turn, are defined as heterogeneous lists. We also define a function for mapping open programs onto the semantic domain.

$$\begin{aligned}&\mathit{orun} :: \mathsf{OProg}\ e\ a \rightarrow \mathsf{Env}\ e \rightarrow \mathsf{Dom}\ a\\ &\mathit{orun}\ p\ \mathit{env} = \mathit{run}\ (p\ \mathit{env})\quad.\end{aligned}$$

We can then assert that two programs are contextually equivalent if, for *any* context, running both programs wrapped in that context will yield the same result:

$$m_1 =_{\mathsf{GS}} m_2 \triangleq \forall \mathsf{C}.\mathit{orun}\ (\mathsf{C}[m_1]) = \mathit{orun}\ (\mathsf{C}[m_2])\quad.$$

We can then straightforwardly formulate variants of the state laws, the nondeterminism laws and the *put-or* law for this global state monad as lemmas. For example, we reformulate law (16) as

$$\mathsf{Put}\ s\ (\mathsf{Put}\ t\ p) =_{\mathsf{GS}} \mathsf{Put}\ t\ p\quad.$$

Proofs for the state laws, the nondeterminism laws and the *put-or* law then easily follow from the analogous semantic domain laws.

More care is required when we want to adapt law (24) into the Prog setting. At the end of Sect. 5.2, we saw that this law precludes the existence of a bind operator in the semantic domain. Since a bind operator for Progs exists, it might seem we're out of luck when we want to adapt law (24) to Progs. But because Progs are merely syntax, we have much more fine-grained control over what equality of programs means. When talking about the semantic domain, we only had one notion of equality: two programs are equal only when one can be substituted for the other in any context. So if, in that setting, we want to introduce a law that does not hold under a particular composition (in this case, sequential composition), the only way we can express that is to say that composition is impossible in the domain. But the fact that Prog is defined purely syntactically opens up the possibility of defining multiple notions of equality which may exist at the same time. In fact, we have already introduced syntactic equality, semantic equality, and contextual equality. It is precisely this choice of granularity that allows us to introduce laws which only hold in programs of a certain form (non-contextual laws), while other laws are much more general (contextual laws). A direct adaptation of law (24) would look something like this:

$$\begin{aligned}\forall \mathsf{C}.\ \mathsf{C}\text{ is bind-free} \Rightarrow\ &\mathit{orun}\ (\mathsf{C}[\mathsf{Put}\ s\ (\mathsf{Return}\ x \mathbin{[\,]} p)])\\ =\ &\mathit{orun}\ (\mathsf{C}[\mathsf{Put}\ s\ (\mathsf{Return}\ x) \mathbin{[\,]} \mathsf{Put}\ s\ p])\quad.\end{aligned}$$

In other words, the two sides of the equation can be substituted for one another, but only in contexts which do not contain any binds. However, in our mechanization, we only prove a more restricted version where the context must be empty (in other words, what we called semantic equivalence), which turns out to be enough for our purposes.

$$run\ (\mathsf{Put}\ s\ (\mathsf{Return}\ x \mathbin{[\!]} p)) = run\ (\mathsf{Put}\ s\ (\mathsf{Return}\ x) \mathbin{[\!]} \mathsf{Put}\ s\ p)\ .\checkmark \tag{25}$$

5.5 Simulating Local-State Semantics

We simulate local-state semantics by replacing each occurrence of Put by a variant that restores the state, as described in Sect. 4.3. This transformation is implemented by the function *trans* for closed programs, and *otrans* for open programs:

$$\begin{aligned}
&trans :: \mathsf{Prog}\ a \to \mathsf{Prog}\ a\\
&trans\ (\mathsf{Return}\ x) = \mathsf{Return}\ x\\
&trans\ (p \mathbin{[\!]} q) \quad\ = trans\ p \mathbin{[\!]} trans\ q\\
&trans\ \emptyset \qquad\quad\ \ = \emptyset\\
&trans\ (\mathsf{Get}\ p) \quad\ = \mathsf{Get}\ (\lambda s \to trans\ (p\ s))\\
&trans\ (\mathsf{Put}\ s\ p) \ = \mathsf{Get}\ (\lambda s' \to \mathsf{Put}\ s\ (trans\ p) \mathbin{[\!]} \mathsf{Put}\ s'\ \emptyset)\ ,\\
&otrans :: \mathsf{OProg}\ e\ a \to \mathsf{OProg}\ e\ a\\
&otrans\ p \qquad\quad\ = \lambda env \to trans\ (p\ env)\ .
\end{aligned}$$

We then define the function *eval*, which runs a transformed program (in other words, it runs a program with local-state semantics).

$$\begin{aligned}
&eval :: \mathsf{Prog}\ a \to \mathsf{Dom}\ a\\
&eval = run \cdot trans\ .
\end{aligned}$$

We show that the transformation works by proving that our free monad equipped with *eval* is a correct implementation for a nondeterministic, stateful monad with local-state semantics. We introduce notation for "contextual equivalence under simulated backtracking semantics":

$$m_1 =_{\mathsf{LS}} m_2 \triangleq \forall \mathsf{C}.eval\ (\mathsf{C}[m_1]) = eval\ (\mathsf{C}[m_2])\ .$$

For example, we formulate the statement that the *put-put* law (16) holds for our monad as interpreted by *eval* as

$$\mathsf{Put}\ s\ (\mathsf{Put}\ t\ p) =_{\mathsf{LS}} \mathsf{Put}\ t\ p\ .\checkmark$$

Proofs for the nondeterminism laws follow trivially from the nondeterminism laws for global state. The state laws are proven by promoting *trans* inside, then applying global-state laws. For the proof of the *get-put* law, we require the property that in global-state semantics, Put distributes over ($[\!]$) if the left branch has been transformed (in which case the left branch leaves the state unmodified). This property only holds at the top-level.

$$run\ (\mathsf{Put}\ x\ (trans\ m_1\ [\!]\ m_2)) = run\ (\mathsf{Put}\ x\ (trans\ m_1)\ [\!]\ \mathsf{Put}\ x\ m_2)\ .\checkmark \tag{26}$$

Proof of this lemma depends on law (24).

Finally, we arrive at the core of our proof: to show that the interaction of state and nondeterminism in this implementation produces backtracking semantics. To this end we prove laws analogous to the local state laws (13) and (12)

$$m \gg \emptyset =_{\mathrm{LS}} \emptyset\ ,\checkmark \tag{27}$$

$$m \gg\!= (\lambda x \to f_1\ x\ [\!]\ f_2\ x) =_{\mathrm{LS}} (m \gg\!= f_1)\ [\!]\ (m \gg\!= f_2)\ .\checkmark \tag{28}$$

We provide machine-verified proofs for these theorems. The proof for (27) follows by straightforward induction. The inductive proof (with induction on m) of law (28) requires some additional lemmas.

For the case $m = m_1\ [\!]\ m_2$, we require the property that, at the top-level of a global-state program, $([\!])$ is commutative if both its operands are state-restoring. Formally:

$$run\ (trans\ p\ [\!]\ trans\ q) = run\ (trans\ q\ [\!]\ trans\ p)\ .\checkmark \tag{29}$$

The proof of this property motivated the introduction of law (23).

The proof for both the $m = \mathsf{Get}\ k$ and $m = \mathsf{Put}\ s\ m'$ cases requires that Get distributes over $([\!])$ at the top-level of a global-state program if the left branch is state restoring.

$$run\ (\mathsf{Get}\ (\lambda s \to trans\ (m_1\ s)\ [\!]\ (m_2\ s))) = \tag{30}$$

$$run\ (\mathsf{Get}\ (\lambda s \to trans\ (m_1\ s))\ [\!]\ \mathsf{Get}\ m_2)\ .\checkmark \tag{31}$$

And finally, we require that the *trans* function is, semantically speaking, idempotent, to prove the case $m = \mathsf{Put}\ s\ m'$.

$$run\ (trans\ (trans\ p)) = run\ (trans\ p)\ .\checkmark \tag{32}$$

5.6 Backtracking with a Global State Monad

Although we can now interpret a local state program through translation to a global state program, we have not quite yet delivered on our promise to address the space usage issue of local state semantics. From the definition of put_{R} it is clear that we simply make the implicit copying of the local state semantics explicit in the global state semantics. As mentioned in Sect. 4.2, rather than using *put*, some algorithms typically use a pair of commands *modify next* and *modify prev*, with $prev \cdot next = id$, to respectively update and roll back the state. This is especially true when the state is implemented using an array or other data structure that is usually not overwritten in its entirety. Following a style similar to put_R, this can be modelled by:

$$modify_\mathsf{R} :: \mathsf{MStateNondet}\ s\ m \Rightarrow (s \to s) \to (s \to s) \to m\ ()$$
$$modify_\mathsf{R}\ next\ prev = modify\ next\ [\!]\ side\ (modify\ prev)\ .$$

Unlike put_R, $modify_\mathsf{R}$ does not keep any copies of the old state alive, as it does not introduce a branching point where the right branch refers to a variable introduced outside the branching point. Is it safe to use an alternative translation, where the pattern $get \gg\!= (\lambda s \to put\ (next\ s) \gg m)$ is not translated into $get \gg\!= (\lambda s \to put_\mathsf{R}\ (next\ s) \gg trans\ m)$, but rather into $modify_\mathsf{R}\ next\ prev \gg trans\ m$? We explore this question by extending our Prog syntax with an additional $\mathsf{Modify_R}$ construct, thus obtaining a new $\mathsf{Prog_M}$ syntax:

$$\textbf{data}\ \mathsf{Prog_M}\ a\ \textbf{where}$$
$$\ldots$$
$$\mathsf{Modify_R} :: (\mathsf{S} \to \mathsf{S}) \to (\mathsf{S} \to \mathsf{S}) \to \mathsf{Prog_M}\ a \to \mathsf{Prog_M}\ a$$

We assume that $prev \cdot next = id$ for every $\mathsf{Modify_R}\ next\ prev\ p$ in a $\mathsf{Prog_M}\ a$ program.

We then define two translation functions from $\mathsf{Prog_M}\ a$ to $\mathsf{Prog}\ a$, which both replace Puts with put_Rs along the way, like the regular *trans* function. The first replaces each $\mathsf{Modify_R}$ in the program by a direct analogue of the definition given above, while the second replaces it by $\mathsf{Get}\ (\lambda s \to \mathsf{Put}\ (next\ s)\ (trans_2\ p))$:

$$trans_1 :: \mathsf{Prog_M}\ a \to \mathsf{Prog}\ a$$
$$\ldots$$
$$trans_1\ (\mathsf{Modify_R}\ next\ prev\ p) = \mathsf{Get}\ (\lambda s \to \mathsf{Put}\ (next\ s)\ (trans_1\ p)$$
$$[\!]\ \mathsf{Get}\ (\lambda t \to \mathsf{Put}\ (prev\ t)\ \emptyset))$$
$$trans_2 :: \mathsf{Prog_M}\ a \to \mathsf{Prog}\ a$$
$$\ldots$$
$$trans_2\ (\mathsf{Modify_R}\ next\ prev\ p) = \mathsf{Get}\ (\lambda s \to put_\mathsf{R}\ (next\ s)\ (trans_2\ p))$$
$$\textbf{where}\ put_\mathsf{R}\ s\ p = \mathsf{Get}\ (\lambda t \to \mathsf{Put}\ s\ p\ [\!]\ \mathsf{Put}\ t\ \emptyset)$$

It is clear that $trans_2\ p$ is the exact same program as $trans\ p'$, where p' is p but with each $\mathsf{Modify_R}\ next\ prev\ p$ replaced by $\mathsf{Get}\ (\lambda s \to \mathsf{Put}\ (next\ s)\ p)$.

We then prove that these two transformations lead to semantically identical instances of $\mathsf{Prog}\ a$.

Lemma 2. $run\ (trans_1\ p) = run\ (trans_2\ p)$. ✓

This means that, if we make some effort to rewrite parts of our program to use the $\mathsf{Modify_R}$ construct rather than Put, we can use the more efficient translation scheme $trans_1$ to avoid introducing unnecessary copies.

n-Queens using a global state. To wrap up, we revisit the n-queens puzzle. Recall that, for the puzzle, the operator that alters the state (to check whether a chess placement is safe) is defined by

$$(i, us, ds) \oplus x = (1 + i, (i + x) : us, (i - x) : ds) \ .$$

By defining $(i, us, ds) \ominus x = (i - 1, \mathit{tail}\ us, \mathit{tail}\ ds)$, we have $(\ominus x) \cdot (\oplus x) = \mathit{id}$. One may thus compute all solutions to the puzzle, in a scenario with a shared global state, by $\mathit{run}\ (\mathit{queens}_{\mathsf{R}}\ n)$, where

$$\begin{aligned}
&\mathit{queens}_{\mathsf{R}}\ n = \mathit{put}\ (0, [\,], [\,]) \gg \mathit{qBody}\ [0\,..\,n-1] \ , \\
&\mathit{qBody}\ [\,] \ \ = \mathit{return}\ [\,] \\
&\mathit{qBody}\ xs = \mathit{select}\ xs \gg\!= \lambda(x, ys) \rightarrow \\
&\qquad\qquad (\mathit{get} \gg\!= (\mathit{guard} \cdot \mathit{ok} \cdot (\oplus x))) \gg \\
&\qquad\qquad \mathit{modify}_{\mathsf{R}}\ (\oplus x)\ (\ominus x) \gg ((x{:}) \mathbin{\langle\$\rangle} \mathit{qBody}\ ys) \ , \\
&\quad \mathbf{where}\ (i, us, ds) \oplus x = (1 + i, (i + x) : us, (i - x) : ds) \\
&\qquad\qquad (i, us, ds) \ominus x = (i - 1, \mathit{tail}\ us, \ \ \mathit{tail}\ ds) \\
&\qquad\qquad \mathit{ok}\ (_, u : us, d : ds) = (u \notin us) \wedge (d \notin ds) \ .
\end{aligned}$$

6 Related Work

6.1 Prolog

Prolog is a prominent example of a system that exposes nondeterminism with local state to the user, but is itself implemented in terms of a single global state.

Warren Abstract Machine. The folklore idea of undoing modifications upon backtracking is a key feature of many Prolog implementations, in particular those based on the Warren Abstract Machine (WAM) [1]. The WAM's global state is the program heap and Prolog programs modify this heap during unification only in a very specific manner: following the union-find algorithm, they overwrite cells that contain self-references with pointers to other cells. Undoing these modifications only requires knowledge of the modified cell's address, which can be written back in that cell during backtracking. The WAM has a special stack, called the trail stack, for storing theses addresses, and the process of restoring those cells is called *untrailing*.

The 4-Port Box Model. While trailing happens under the hood, there is a folklore Prolog programming pattern for observing and intervening at different points in the control flow of a procedure call, known as the *4-port box model*. In this model, upon the first entrance of a Prolog procedure it is *called*; it may yield a result and *exits*; when the subsequent procedure fails and backtracks, it is asked to *redo* its computation, possibly yielding the next result; finally it may *fail*. Given a Prolog procedure p implemented in Haskell, the following program prints debugging messages when each of the four ports are used:

$$\begin{aligned}
&(\mathit{putStr}\ \texttt{"call"} \mathbin{[\,]} \mathit{side}\ (\mathit{putStr}\ \texttt{"fail"})) \gg \\
&p \gg\!= \lambda x \rightarrow \\
&(\mathit{putStr}\ \texttt{"exit"} \mathbin{[\,]} \mathit{side}\ (\mathit{putStr}\ \texttt{"redo"})) \gg \mathit{return}\ x \ .
\end{aligned}$$

This technique was applied in the monadic setting by Hinze [6], and it has been our inspiration for expressing the state restoration with global state.

6.2 Reasoning About Side Effects

There are many works on reasoning and modelling side effects. Here we cover those that have most directly inspired this paper.

Axiomatic Reasoning. Our work was directly inspired by Gibbons and Hinze's proposal to reason axiomatically about programs with side effects, and their axiomatic characterisation of local state in particular [4]. We have extended their work with an axiomatic characterisation of global state and on handling the former in terms of the latter. We also provide models that satisfy the axioms, whereas their paper mistakenly claims that one model satisfies the local state axioms and that another model is monadic.

Algebraic Effects. Our formulation of implementing local state with global state is directly inspired by the effect handlers approach of Plotkin and Pretnar [12]. By making the free monad explicit our proofs benefit directly from the induction principle that Bauer and Pretnar establish for effect handler programs [2].

While Lawvere theories were originally Plotkin's inspiration for studying algebraic effects, the effect handlers community has for a long time paid little attention to them. Yet, recently Lukšič and Pretnar [10] have investigated a framework for encoding axioms (or "effect theories") in the type system: the type of an effectful function declares the operators used in the function, as well as the equalities that handlers for these operators should comply with. The type of a handler indicates which operators it handles and which equations it complies with. This type system would allow us to express at the type-level that our handler interprets local state in terms of global state.

7 Conclusions

Starting from Gibbons and Hinze's observation [4] that the interaction between state and nondeterminism can be characterized axiomatically in multiple ways, we explored the differences between local state semantics (as characterised by Gibbons and Hinze) and global state semantics (for which we gave our own non-monadic characterisation).

In global state semantics, we find that we may use $(\|)$ to simulate sequencing, and that the idea can be elegantly packaged into commands like put_{R} and $modify_{\mathsf{R}}$. The interaction between global state and non-determinism turns out to be rather tricky. For a more rigorous treatment, we enforce a more precise separation between syntax and semantics and, as a side contribution of this paper, propose a *global state law*, plus some additional laws, which the semantics should satisfy. We verified (with the help of the Coq proof assistant) that there is an implementation satisfying these laws.

Using the n-queens puzzle as an example, we showed that one can end up in a situation where a problem is naturally expressed with local state semantics, but the greater degree of control over resources that global state semantics offers is desired. We then describe a technique to systematically transform a monadic program written against the local state laws into one that, when interpreted under global state laws, produces the same results as the original program. This transformation can be viewed as a handler (in the algebraic effects sense): it implements the interface of one effect in terms of the interface of another. We also verified the correctness of this transformation in Coq.

Acknowledgements. We would like to thank Matija Pretnar, the members of IFIP WG 2.1, the participants of Shonan meeting 146 and the MPC reviewers for their insightful comments. We would also like to thank the Flemish Fund for Scientific Research (FWO) for their financial support.

References

1. Aït-Kaci, H.: Warren's Abstract Machine: A Tutorial Reconstruction (1991)
2. Bauer, A., Pretnar, M.: An effect system for algebraic effects and handlers. Logical Methods Comput. Sci. **10**(4) (2014). https://doi.org/10.2168/LMCS-10(4:9)2014
3. Gale, Y.: ListT done right alternative (2007). https://wiki.haskell.org/ListT_done_right_alternative
4. Gibbons, J., Hinze, R.: Just do it: simple monadic equational reasoning. In: Danvy, O. (ed.) International Conference on Functional Programming, pp. 2–14. ACM Press (2011)
5. Gill, A., Kmett, E.: The monad transformer library (2014). https://hackage.haskell.org/package/mtl
6. Hinze, R.: Prolog's control constructs in a functional setting - axioms and implementation. Int. J. Found. Comput. Sci. **12**(2), 125–170 (2001). https://doi.org/10.1142/S0129054101000436
7. Hutton, G., Fulger, D.: Reasoning about effects: seeing the wood through the trees. In: Symposium on Trends in Functional Programming (2007)
8. Kiselyov, O.: Laws of monadplus (2015). http://okmij.org/ftp/Computation/monads.html#monadplus
9. Kiselyov, O., Ishii, H.: Freer monads, more extensible effects. In: Reppy, J.H. (ed.) Symposium on Haskell, pp. 94–105. ACM Press (2015)
10. Lukšič, V., Pretnar, M.: Local algebraic effect theories (2019, submitted)
11. Moggi, E.: Computational lambda-calculus and monads. In: Parikh, R. (ed.) Logic in Computer Science, pp. 14–23. IEEE Computer Society Press (1989)
12. Plotkin, G., Pretnar, M.: Handlers of algebraic effects. In: Castagna, G. (ed.) ESOP 2009. LNCS, vol. 5502, pp. 80–94. Springer, Heidelberg (2009). https://doi.org/10.1007/978-3-642-00590-9_7
13. Volkov, N.: The list-t package (2014). http://hackage.haskell.org/package/list-t
14. Wadler, P.: Monads for functional programming. In: Broy, M. (ed.) Program Design Calculi: Marktoberdorf Summer School, pp. 233–264. Springer, Heidelberg (1992). https://doi.org/10.1007/978-3-662-02880-3_8
15. Wu, N., Schrijvers, T., Hinze, R.: Effect handlers in scope. In: Voigtländer, J. (ed.) Symposium on Haskell, pp. 1–12. ACM Press (2012)

Certification of Breadth-First Algorithms by Extraction

Dominique Larchey-Wendling[1(✉)] and Ralph Matthes[2]

[1] Université de Lorraine, CNRS, LORIA, Vandœuvre-lès-Nancy, France
dominique.larchey-wendling@loria.fr

[2] Institut de Recherche en Informatique de Toulouse (IRIT), CNRS and University of Toulouse, Toulouse, France
Ralph.Matthes@irit.fr

Abstract. By using pointers, breadth-first algorithms are very easy to implement efficiently in imperative languages. Implementing them with the same bounds on execution time in purely functional style can be challenging, as explained in Okasaki's paper at ICFP 2000 that even restricts the problem to binary trees but considers numbering instead of just traversal. Okasaki's solution is modular and factors out the problem of implementing queues (FIFOs) with worst-case constant time operations. We certify those FIFO-based breadth-first algorithms on binary trees by extracting them from fully specified Coq terms, given an axiomatic description of FIFOs. In addition, we axiomatically characterize the strict and total order on branches that captures the nature of breadth-first traversal and propose alternative characterizations of breadth-first traversal of forests. We also propose efficient certified implementations of FIFOs by extraction, either with pairs of lists (with amortized constant time operations) or triples of lazy lists (with worst-case constant time operations), thus getting from extraction certified breadth-first algorithms with the optimal bounds on execution time.

Keywords: Breadth-first algorithms · Queues in functional programming · Correctness by extraction · Coq

1 Introduction

Breadth-first algorithms form an important class of algorithms with many applications. The distinguishing feature is that the recursive process tries to be "equitable" in the sense that all nodes in the graph with "distance" k from the starting point are treated before those at distance $k + 1$. In particular, with infinite (but finitely branching) structures, this ensures "fairness" in that all possible branches are eventually pursued to arbitrary depth, in other words, the

The first author got financial support by the TICAMORE joint ANR-FWF project. The second author got financial support by the COST action CA15123 EUTYPES.

G. Hutton (Ed.): MPC 2019, LNCS 11825, pp. 45–75, 2019.
https://doi.org/10.1007/978-3-030-33636-3_3

recursion does not get "trapped" in an infinite branch.[1] This phenomenon that breadth-first algorithms avoid is different from a computation that gets "stuck": even when "trapped", there may be still steady "progress" in the sense of producing more and more units of output in finite time. In this paper, we will not specify or certify algorithms to work on infinite structures although we expect the lazy reading of our extracted programs (e. g., if we choose to extract towards the Haskell language) to work properly for infinite input as well and thus be fair—however without any guarantees from the extraction process. Anyway, as mentioned above, breadth-first algorithms impose a stronger, quantitative notion of "equity" than just abstract fairness to address the order of traversal of the structure.

When looking out for problems to solve with breadth-first algorithms, plain traversal of a given structure is the easiest task; to make this traversal "observable," one simply prints the visited node labels (following the imperative programming style), or one collects them in a list, in the functional programming paradigm, as we will do in this paper (leaving filtering of search "hits" aside). However, in the interest of efficiency, it is important to use a first-in, first-out queue (FIFO) to organize the waiting sub-problems (see, e. g., Paulson's book [15] on the ML language). Here, we are also concerned with functional languages, and for them, constant-time (in the worst case) implementations of the FIFO operations were a scientific challenge, solved very elegantly by Okasaki [13].

Breadth-first traversal is commonly [4] used to identify all nodes that are reachable in a graph from a given start node, and this allows creating a tree that captures the subgraph of reachable nodes, the "breadth-first tree" (for a given start node). In Okasaki's landmark paper at ICFP 2000 [14], the author proposes to revisit the problem of *breadth-first numbering*: the traversal task is further simplified to start at the root of a (binary) tree, but the new challenge is to rebuild the tree in a functional programming language, where the labels of the nodes have been replaced by the value n if the node has been visited as the n-th one in the traversal. The progress achieved by Okasaki consists in separating the concerns of implementing breadth-first numbering from an efficient implementation of FIFOs: his algorithm works for any given FIFO and inherits optimal bounds on execution time from the given FIFO implementation. Thus, breadth-first numbering can be solved as efficiently with FIFOs as traversal,[2] and Okasaki reports in his paper that quite some of his colleagues did not come up with a FIFO-based solution when asked to find any solution.

In the present paper, using the Coq proof assistant,[3] we formalize and solve the breadth-first traversal problem for finite binary trees with a rich specification in the sense of certified programming, i. e., when the output of an algorithm is a

[1] Meaning that recursion would be pursued solely in that branch.

[2] Jones and Gibbons [7] solved a variant of the problem with a "hard-wired" FIFO implementation—the one we review in Sect. 7.1—and thus do not profit from the theoretically most pleasing FIFO implementation by Okasaki [13].

[3] https://coq.inria.fr, we have been using the current version 8.9.0, and the authoritative reference on Coq is https://coq.inria.fr/distrib/current/refman/.

dependent pair (v, C_v) composed of a value v together with a proof C_v certifying that v satisfies some (partial correctness) property.[4] The key ingredient for its functional correctness are two equations already studied by Okasaki [14], but there as definitional device. Using the very same equations, we formalize and solve the breadth-first numbering problem, out of which the Coq extraction[5] mechanism [9] can extract the same algorithm as Okasaki's (however, we extract code in the OCaml[6] language), but here with all guarantees concerning the non-functional property of termination and the functional/partial correctness, both aspects together constituting its *total correctness*, i. e., the algorithm indeed provides such a breadth-first numbering of the input tree.

As did Okasaki (and Gibbons and Jones for their solution [7]), we motivate the solution by first considering the natural extension to the problem of traversing or numbering a *forest*, i. e., a finite list of trees. The forests are subsequently replaced by an abstract FIFO structure, which is finally instantiated for several implementations, including the worst-case constant-time implementation following Okasaki's paper [13] that is based on three lazy lists (to be extracted from coinductive lists in Coq for which we require as invariant that they are finite).

The breadth-first numbering problem can be slightly generalized to *breadth-first reconstruction*: the labels of the output tree are not necessarily natural numbers but come from a list of length equal to the number of labels of the input tree, and the n-th list element replaces the n-th node in the traversal, i. e., a minor variant of "breadth-first labelling" considered by Jones and Gibbons [7]. A slight advantage of this problem formulation is that a FIFO-based solution is possible by structural recursion on that list argument while the other algorithms were obtained by recursion over a measure (this is not too surprising since the length of that list coincides with the previously used measure).

In Sect. 2, we give background type theoretical material, mostly on list notation. Then we review existing tools for reasoning or defining terms by recursion/induction on a decreasing measure which combines more than one argument. We proceed to the short example of a simple interleaving algorithm that swaps its arguments when recurring on itself. In particular, we focus on how existing tools like `Program Fixpoint` or `Equations` behave in the context of extraction. Then we describe an alternative of our own—a tailored `induction-on` tactic—and we argue that it is more transparent to extraction than, e. g., `Equations`. Section 3 concentrates on background material concerning breadth-first traversal (specification, naive algorithm), including original material on a characterization of the breadth-first order, while Sect. 4 revolves around the mathematics of breadth-first traversal of forests that motivates the FIFO-based breadth-first algorithms. In Sect. 5, we use an abstractly specified datatype of FIFOs and deal with the different problems mentioned above: traversal,

4 For Coq, this method of rich specifications has been very much advocated in the Coq'Art, the first book on Coq [3].

5 The authors of the current version are Filliâtre and Letouzey, see https://coq.inria.fr/refman/addendum/extraction.html.

6 http://ocaml.org.

numbering, reconstruction. In Sect. 6, we elaborate on a level-based approach to numbering that thus is in the spirit of the naive traversal algorithm. Section 7 reports on the instantiation of the FIFO-based algorithms to two particular efficient FIFO implementations. Section 8 concludes.

The full Coq development, that also formalizes the theoretical results (the characterizations) in addition to the certification of the extracted algorithms in the OCaml language, is available on `GitHub`:

https://github.com/DmxLarchey/BFE

In Appendix A, we give a brief presentation of these Coq vernacular[7] files.

2 Preliminaries

We introduce some compact notations to represent the language of constructive type theory used in the proof assistant Coq. We describe the method called "certification by extraction" and illustrate how it challenges existing tools like `Program Fixpoint` or even `Equations` on the example of a simple interleaving algorithm. Then we introduce a tailored method to justify termination of fixpoints (or inductive proofs) using measures over, e. g., two arguments. This method is encapsulated into an `induction-on` tactic that is more transparent to extraction than the above-mentioned tools.

The type of propositions is denoted `Prop` while the type (family) of types is denoted `Type`. We use the inductive types of Booleans ($b : \mathbb{B} := 0 \mid 1$), of natural numbers ($n : \mathbb{N} := 0 \mid \mathsf{S}\, n$), of lists ($l : \mathbb{L}\, X := [] \mid x :: l$ with $x : X$) over a type parameter X. When $l = x_1 :: \cdots :: x_n :: []$, we define $|l| := n$ as the length of l and we may write $l = [x_1; \ldots; x_n]$ as well. We use $+\!\!+$ for the concatenation of two lists (called the "append" function). The function $\mathsf{rev} : \forall X, \mathbb{L}\, X \to \mathbb{L}\, X$ implements list reversal and satisfies $\mathsf{rev}\ [] = []$ and $\mathsf{rev}(x :: l) = \mathsf{rev}\ l +\!\!+ x :: []$.

For a (heterogeneous) binary relation $R : X \to Y \to \texttt{Prop}$, we define the lifting of R over lists $\forall^2 R : \mathbb{L}\, X \to \mathbb{L}\, Y \to \texttt{Prop}$ by the following inductive rules, corresponding to the Coq standard library `Forall2` predicate:

$$\frac{}{\forall^2 R\ []\ []} \qquad \frac{R\ x\ y \qquad \forall^2 R\ l\ m}{\forall^2 R\ (x :: l)\ (y :: m)}$$

Thus $\forall^2 R$ is the smallest predicate closed under the two above rules.[8] Intuitively, when $l = [x_1; \ldots; x_n]$ and $m = [y_1; \ldots; y_p]$, the predicate $\forall^2 R\ l\ m$ means $n = p \wedge R\ x_1\ y_1 \wedge \cdots \wedge R\ x_n\ y_n$.

[7] Vernacular is a Coq idiom for *syntactic sugar*, i. e., a human-friendly syntax for type theoretical notation—the "Gallina" language.

[8] This is implemented in Coq by two constructors for the two rules together with an induction (or elimination) principle that ensures its smallestness.

2.1 Certification by Extraction

Certification by extraction is a particular way of establishing the correctness of a *given implementation* of an algorithm. In this paper, algorithms are given as programs in the OCaml language.

Hence, let us consider an OCaml program t of type $X \to Y$. The way to certify such a program by extraction is to implement a Coq term

$$f_t : \forall x : X_t,\ \mathsf{pre}_t\ x \to \{y : Y_t \mid \mathsf{post}_t\ x\ y\}$$

where $\mathsf{pre}_t : X_t \to \texttt{Prop}$ is the *precondition* (i. e., the domain of use of the function) and $\mathsf{post}_t : X_t \to Y_t \to \texttt{Prop}$ is the (dependent) *postcondition* which (possibly) relates the input with the output. The precondition could be tighter than the actual domain of the program t and the postcondition may characterize some aspects of the functional behavior of t, up to its full correctness, i. e., when $\mathsf{pre}_t/\mathsf{post}_t$ satisfy the following: for any given x such that $\mathsf{pre}_t\ x$, the value $f_t\ x$ is the unique y such that $\mathsf{post}_t\ x\ y$ holds.

We consider that t is certified when the postcondition faithfully represents the intended behavior of t *and* when f_t extracts to t: the extracted term $\texttt{extract}(f_t)$ but also the extracted types $\texttt{extract}(X_t)$ and $\texttt{extract}(Y_t)$ have to match "exactly" their respectively given OCaml definitions, i. e., we want $t \equiv \texttt{extract}(f_t)$, $X \equiv \texttt{extract}(X_t)$ and $Y \equiv \texttt{extract}(Y_t)$. The above identity sign "≡" should be read as *syntactic identity*. This does not mean character by character equality between source codes but between the abstract syntax representations. Hence some slight differences are allowed, typically the name of bound variables which cannot always be controlled during extraction. Notice that $\mathsf{pre}_t\ x$ and $\mathsf{post}_t\ x\ y$ being of sort $\texttt{Prop}$, they carry only logical content and no computational content. Thus they are erased by extraction.

As a method towards certified development, extraction can also be used without having a particular output program in mind, in which case it becomes a tool for writing programs that are correct by construction. Of course, getting a clean output might also be a goal and thus, the ability to finely control the computational content is important. But when we proceed from an already written program t, this fine control becomes critical and this has a significant impact on the tools which we can use to implement f_t (see upcoming Sect. 2.3).

2.2 Verification, Certification and the Trusted Computing Base

"Certification" encompasses "verification" in the following way: it aims at producing a certificate that can be checked independently of the software which is used to do the certification—at least in theory. While verification implies trusting the verifying software, certification implies only trusting the software that is used to verify the certificate, hence in the case of Coq, possibly an alternative type-checker. Notice that one of the goals of the MetaCoq[9] project [1] is precisely to

[9] https://github.com/MetaCoq/metacoq.

produce a type-checker for (a significant fragment of) Coq, a type-checker which will itself be certified by Coq.

Extraction in Coq is very straightforward once the term has been fully specified and type-checked. Calling the command

```
Recursive Extraction some_coq_term
```

outputs the extracted OCaml program and this part is fully automatic, although it is very likely that the output is not of the intended shape on the first attempt. So there may be a back-and-forth process to fine-tune the computational content of Coq terms until their extraction is satisfactory. The method can scale to larger developments because of the principle of *compositionality*. Indeed, provided a development can be divided into manageable pieces, Coq contains the tools that help at composing small certified bricks into bigger ones. Verifying or certifying large monolithic projects is generally hard—whatever tools are involved—because guessing the proper invariants becomes humanly unfeasible.

Considering the Trusted Computing Base (TCB) of certified extraction in Coq, besides trusting a type-checker, it also requires trusting the extraction process. In his thesis [10], P. Letouzey gave a mathematical proof of correctness (w. r. t. syntactic and semantic desiderata) of the extraction principles that guide the currently implemented `Extraction` command. Arguably, there is a difference between principles and an actual implementation. The above-cited MetaCoq project also aims at producing a certified "extraction procedure to untyped lambda-calculus accomplishing the same as the Extraction plugin of Coq." Hence, for the moment, our work includes `Extraction` in its TCB but so do many other projects such as, e. g., the CompCert compiler.[10] Still concerning verifying Coq extraction in Coq, we also mention the Œuf[11] project [12], but there is work of similar nature also in the Isabelle community [6]. Notice that we expect no or little change in the resulting extracted OCaml programs (once MetaCoq or one of its competitors reaches its goal) since these must respect the computational content of Coq terms. As the principle of "certification by extraction" is to obtain programs which are correct by construction directly from Coq code, we consider certifying extraction itself to be largely orthogonal to the goal pursued here.

2.3 Extraction of Simple Interleaving with Existing Tools

This section explains some shortcomings of the standard tools that can be used to implement recursive schemes in Coq when the scheme is more complicated than just structural recursion. We specifically study the `Function`, `Program Fixpoint`, and `Equations` commands. After discussing their versatility, we focus on how they interact with the `Extraction` mechanism of Coq.

---

[10] http://compcert.inria.fr.

[11] http://oeuf.uwplse.org.

As a way to get a glimpse of the difficulty of the method, we begin with the case of the following simple interleaving function of type $\mathbb{L}X \to \mathbb{L}X \to \mathbb{L}X$ that merges two lists into one

$$[l_1; \ldots; l_n], [m_1; \ldots; m_n] \mapsto [l_1; m_1; l_2; m_2; \ldots; l_n; m_n]$$

by alternating the elements of both input lists. The algorithm we want to certify is the following one:

$$\texttt{let rec itl}\, l\, m = \texttt{match}\ l\ \texttt{with}\ [] \to m \mid x :: l \to x :: \mathsf{itl}\, m\, l \qquad (1)$$

where l and m switch roles in the second equation/match case for itl. Hence neither of these two arguments decreases structurally[12] in the recursive call and so, this definition cannot be used as such in a Coq `Fixpoint` definition. Notice that this algorithm has been identified as a challenge for program extraction in work by McCarthy *et al.* [11, p. 58], where they discuss an OCaml program of a function cinterleave that corresponds to our itl.

We insist on that specific algorithm—it is however trivial to define a function with the same functional behavior in Coq by the following equations:

$$\mathsf{itl}_{\mathsf{triv}}\ []\ m = m \qquad \mathsf{itl}_{\mathsf{triv}}\ l\ [] = l \qquad \mathsf{itl}_{\mathsf{triv}}\ (x :: l)\ (y :: m) = x :: y :: \mathsf{itl}_{\mathsf{triv}}\ l\ m$$

Indeed, these equations correspond to a *structurally decreasing* `Fixpoint` which is accepted nearly as is by Coq.[13] However, the algorithm we want to certify proceeds through the two following equations $\mathsf{itl}\ []\, m = m$ and $\mathsf{itl}\ (x :: l)\ m = x :: \mathsf{itl}\ m\ l$. While it is not difficult to show that $\mathsf{itl}_{\mathsf{triv}}$ satisfies this specification, in particular by showing

$$\texttt{Fact}\ \mathsf{itl\_triv\_fix\_1} : \forall x\, l\, m,\ \mathsf{itl}_{\mathsf{triv}}\,(x :: l)\, m = x :: \mathsf{itl}_{\mathsf{triv}}\, m\, l$$

by nested induction on l and then m, extraction of $\mathsf{itl}_{\mathsf{triv}}$, however, does not respect the expected code of itl, see Eq. (1).

While there is no structural decrease in Eq. (1), there is nevertheless an obvious decreasing measure in the recursive call of itl, i.e., $l, m \mapsto |l| + |m|$. We investigate several ways to proceed using that measure and discuss their respective advantages and drawbacks. The comments below are backed up by the file interleave.v corresponding to the following attempts. The use of Coq 8.9 is required for a recent version of the `Equations` package described below.

---

12 Structural recursion is the built-in mechanism for recursion in Coq. It means that `Fixpoints` are type-checked only when Coq can determine that at least one of the arguments of the defined fixpoint (e. g., the first, the second...) decreases structurally on each recursive call, i. e., it must be a strict sub-term in an inductive type; and this must be the same argument that decreases for each recursive sub-call. This is called the *guard condition* for recursion and it ensures termination. On the other hand, it is a very restrictive form a recursion and we study more powerful alternatives here.

13 As a general rule in this paper, when equations can be straightforwardly implemented by structural induction on one of the arguments, we expect the reader to be able to implement the corresponding `Fixpoint` and so we do not further comment on this.

- Let us first consider the case of the Function command which extends the primitive `Fixpoint` command. It allows the definition of fixpoints on decreasing arguments that may not be structurally decreasing. However the `Function` method fails very quickly because it only allows for the decrease of *one of the arguments*, and with itl, only the decrease of a combination of both arguments can be used to show termination. We could of course pack the two arguments in a pair but then, this will modify the code of the given algorithm, a modification which will undoubtedly impact the extracted code;
- Let us now consider the case of the Program Fixpoint command [17]. We can define $\mathsf{itl}_{\mathsf{pfix}}$ via a fully specified term:

$$\texttt{Program Fixpoint}\ \mathsf{itl}^{\mathrm{full}}_{\mathsf{pfix}}\ l\ m\ \{\texttt{measure}\ (|l|+|m|)\} : \{r \mid r = \mathsf{itl}_{\mathsf{triv}}\ l\ m\} := \ldots$$

  basically by giving the right-hand side of Eq. (1) in Coq syntax, and then apply first and second projections to get the result as $\mathsf{itl}_{\mathsf{pfix}}\ l\ m := \pi_1(\mathsf{itl}^{\mathrm{full}}_{\mathsf{pfix}}\ l\ m)$ and the proof that it meets its specification $\mathsf{itl}_{\mathsf{pfix}}\ l\ m = \mathsf{itl}_{\mathsf{triv}}\ l\ m$ as $\pi_2(\mathsf{itl}^{\mathrm{full}}_{\mathsf{pfix}}\ l\ m)$. The `Program Fixpoint` command generates proof obligations that are easy to solve in this case. Notice that defining $\mathsf{itl}_{\mathsf{pfix}}$ without going through the fully specified term is possible but then it is not possible to establish the postcondition $\mathsf{itl}_{\mathsf{pfix}}\ l\ m = \mathsf{itl}_{\mathsf{triv}}\ l\ m$;
- Alternatively, we can use the Equations package [18] which could be viewed as an extension/generalization of `Program Fixpoint`. In that case, we can proceed with a weakly specified term:

$$\texttt{Equations}\ \ \mathsf{itl}_{\mathsf{eqs}}\ l\ m : \mathbb{L}\,X\ \texttt{by wf}\ (|l| + |m|) < := \ldots$$

  then derive the two equations that are used to define $\mathsf{itl}_{\mathsf{eqs}}$, i. e., $\mathsf{itl}_{\mathsf{eqs}}\ []\ m = m$ and $\mathsf{itl}_{\mathsf{eqs}}\ (x :: l)\ m := x :: \mathsf{itl}_{\mathsf{eqs}}\ m\ l$. These equations are directly obtained by making use of the handy `simp` tactic provided with `Equations`. With these two equations, it is then easy to show the identity $\mathsf{itl}_{\mathsf{eqs}}\ l\ m = \mathsf{itl}_{\mathsf{triv}}\ l\ m$.

To sum up, considering the Coq implementation side only: `Function` fails because no single argument decreases; `Program Fixpoint` succeeds but via a fully specified term to get the postcondition; and `Equations` succeeds directly and generates the defining equations that are accessible through the `simp` tactic.

However, when we consider the extraction side of the problem, both `Program Fixpoint` and `Equations` do not give us the intended OCaml term of Eq. (1). On the left side of Fig. 1, we display the extracted code for $\mathsf{itl}_{\mathsf{pfix}}$ that was developed using `Program Fixpoint`, and on the right side the extracted code for $\mathsf{itl}_{\mathsf{eqs}}$ implemented using `Equations`.[14] Both present two drawbacks in our eyes. First, because "full" (resp. "eqs_0") are not made globally visible, it is impossible to inline their definitions with the `Extraction Inline` directive although this would be an obvious optimization. But then, also more problematic from an algorithmic point of view, there is the packing of the two arguments into a pair

[14] The actual extracted codes for both $\mathsf{itl}_{\mathsf{pfix}}$ and $\mathsf{itl}_{\mathsf{eqs}}$ are not that clean but we simplified them a bit to single out two specific problems.

```
let itl_pfix l m =
  let rec loop p =
    let l0    = fst p in
    let m0    = snd p in
    let full  = fun l2 m2
                → loop (l2, m2)
    in match l0 with
      | []      → m0
      | x :: l1 → x :: full m0 l1
  in loop (l, m)
```

```
let itl_eqs l m =
  let rec loop p =
    let m0    = snd p in
    let eqs0  = fun l2 m2 _
                → loop (l2, m2)
    in match fst p with
      | []      → m0
      | x :: l1 → x :: eqs0 m0 l1 __
  in loop (l, m)
```

Fig. 1. Extraction of $\mathsf{itl}_{\mathsf{pfix}}$ (`Program Fixpoint`) and $\mathsf{itl}_{\mathsf{eqs}}$ (`Equations`).

(l, m) prior to the call to the "loop" function that implements $\mathsf{itl}_{\mathsf{pfix}}$ (resp. $\mathsf{itl}_{\mathsf{eqs}}$). As a last remark, the two extracted codes look similar except that for $\mathsf{itl}_{\mathsf{eqs}}$, there is the slight complication of an extra *dummy* parameter _ added to the above definition of "eqs_0" that is then instantiated with a dummy argument __.

To sum up our above attempts, while the `Function` commands fails to handle itl because of the swap between arguments in Eq. (1), both `Program Fixpoint` and `Equations` succeed when considering the specification side, i. e., defining the function (which implicitly ensures termination) and proving its postcondition. However, extraction-wise, both generate artifacts, e. g., the pairing of arguments in this case, and which are difficult or impossible to control making the "certification by extraction" approach defined in Sect. 2.1 fail.

In the previously cited work by McCarthy *et al.* [11], the authors report that they successfully avoided "verification residues" as far as their generation of obligations for the certification of running time was concerned. We are heading for the absence of such residues from the extracted code. Let us hence now consider a last approach based on a finely tuned tactic which is both user-friendly and inlines inner fixpoints making it transparent to extraction. This is the approach we will be using for the breadth-first algorithms considered in here.

2.4 Recursion on Measures

We describe how to implement definitions of terms by induction on a measure of, e. g., two arguments. Let us consider two types $X, Y : \mathtt{Type}$, a doubly indexed family $P : X \to Y \to \mathtt{Type}$ of types and a measure $\|\cdot,\cdot\| : X \to Y \to \mathbb{N}$. We explain how to build terms or proofs of type $t : P\,x\,y$ by induction on the measure $\|x, y\|$. Hence, to build the term t, we are allowed to use instances of the induction hypothesis:

$$IH : \forall\, x'\, y',\ \|x', y'\| < \|x, y\| \to P\, x'\, y'$$

i. e., the types $P\,x'\,y'$ with smaller x'/y' arguments are (recursively) inhabited. The measures we use here are limited to $\mathbb{N}$ viewed as well-founded under the "strictly less" order. Any other type with a well-founded order would work as well.

To go beyond measures and implement a substantial fragment of general recursion, we invite the reader to consult work by the first author and Monin [8].

In the file wf_utils.v, we prove the following theorem while carefully crafting the computational content of the proof term, so that extraction yields an algorithm that is clean of spurious elements:

$$\texttt{Theorem}\ \mathsf{measure\_double\_rect}\ (P : X \to Y \to \mathsf{Type}) :$$
$$(\forall x\, y, (\forall x'\, y',\ \|x', y'\| < \|x, y\| \to P\, x'\, y') \to P\, x\, y) \to \forall x\, y,\ P\, x\, y.$$

It allows building a term of type $P\,x\,y$ by simultaneous induction on x and y using the decreasing measure $\|x, y\|$ to ensure termination. To ease the use of theorem measure_double_rect we define a *tactic notation* that is deployed as follows:

$$\texttt{induction on}\ x\ y\ \texttt{as}\ \mathit{IH}\ \texttt{with measure}\ \|x, y\|$$

However, if the `induction-on` tactic was just about `apply`ing an instance of the term measure_double_rect, it would still leave artifacts in the extracted code much like the _ (resp. __) dummy parameter (resp. argument) in the $\mathsf{itl}_{\mathsf{eqs}}$ example of Fig. 1. So, to be precise in our description, the `induction-on` tactic actually builds the needed instance of the proof term measure_double_rect on a per-use basis, i. e., the proof term is inlined by the tactic itself. This ensures perfect extraction in the sense of the result being free of any artifacts. We refer to the file wf_utils.v for detailed explanations on how the inlining works.

From the point of view of specification, our `induction-on` tactic gives the same level of flexibility when compared to `Program Fixpoint` or `Equations`, at least when termination is grounded on a decreasing measure. However, when considering extraction, we think that it offers a finer control over the computational content of Coq terms, a feature which can be critical when doing certification by extraction. We now illustrate how to use the `induction-on` tactic from the user's point of view, on the example of the simple interleaving algorithm.

2.5 Back to Simple Interleaving

To define and specify $\mathsf{itl}_{\mathsf{on}}$ using the `induction-on` tactic, we first implement a fully specified version of $\mathsf{itl}_{\mathsf{on}}$ as the *functional* equivalent of $\mathsf{itl}_{\mathsf{triv}}$.[15] We proceed with the following script, whose first line reads as: we want to define the list r together with a proof that r is equal to $\mathsf{itl}_{\mathsf{triv}}\ l\ m$—a typical rich specification.

[15] Of course, it is not operationally equivalent.

```
Definition itl_on^full l m : {r : 𝕃 X | r = itl_triv l m}.
Proof.
  induction on l m as loop with measure (|l| + |m|).
  revert loop; refine (match l with
    | nil     ↦ fun _      ↦ exist _ m 𝕆?1
    | x :: l' ↦ fun loop ↦ let (r, H_r) := loop m l' 𝕆?2
                           in exist _ (x :: r) 𝕆?3 end).
  ⋆ trivial.                                         (* proof of 𝕆?1 *)
  ⋆ simpl; omega.                                    (* proof of 𝕆?2 *)
  ⋆ subst; rewrite itl_triv_fix_1; trivial.          (* proof of 𝕆?3 *)
Defined.
```

The code inside the **refine**(...) tactic outputs terms like **exist** _ s $\mathbb{O}^?_s$ where _ is recovered by unification, and $\mathbb{O}^?_s$ is left open to be solved later by the user.[16] The constructor **exist** packs the pair $(s, \mathbb{O}^?_s)$ as a term of dependent type $\{r : \mathbb{L}\,X \mid r = \mathsf{itl}_{\mathrm{triv}}\ l\ m\}$ and thus $\mathbb{O}^?_s$ remains to be realized in the type $s = \mathsf{itl}_{\mathrm{triv}}\ l\ m$. This particular use of **refine** generates three *proof obligations* (also denoted PO)

$$\begin{array}{ll} \mathbb{O}^?_1 & //\ \ldots \vdash m = \mathsf{itl}_{\mathrm{triv}}\ []\ m \\ \mathbb{O}^?_2 & //\ \ldots \vdash |m| + |l'| < |x :: l'| + |m| \\ \mathbb{O}^?_3 & //\ \ldots, H_r : r = \mathsf{itl}_{\mathrm{triv}}\, m\, l' \vdash x :: r = \mathsf{itl}_{\mathrm{triv}}\ (x :: l')\ m \end{array}$$

later proved with their respective short proof scripts. Notice that our newly introduced **induction-on** tactic could alternatively be used to give another proof of itl_triv_fix_1 by measure induction on $|l| + |m|$. We remark that PO $\mathbb{O}^?_2$ is of different nature than $\mathbb{O}^?_1$ and $\mathbb{O}^?_3$. Indeed, $\mathbb{O}^?_2$ is a precondition, in this case a *termination certificate* ensuring that the recursive call occurs on smaller inputs. On the other hand, $\mathbb{O}^?_1$ and $\mathbb{O}^?_3$ are postconditions ensuring the functional correctness by type-checking (of the logical part of the rich specification). As a general rule in this paper, providing termination certificates will always be relatively easy because they reduce to proofs of strict inequations between arithmetic expressions, usually solved by the **omega** tactic.[17]

Considering POs $\mathbb{O}^?_2$ and $\mathbb{O}^?_3$, they contain the following hypothesis (hidden in the dots) which witnesses the induction on the measure $|l| + |m|$. It is called loop as indicated in the **induction-on** tactic used at the beginning of the script:

$$\mathsf{loop} : \forall\, l_0\, m_0,\ |l_0| + |m_0| < |x :: l'| + |m| \to \{r \mid r = \mathsf{itl}_{\mathrm{triv}}\ l_0\ m_0\}$$

It could also appear in $\mathbb{O}^?_1$ but since we do not need it to prove $\mathbb{O}^?_1$, we intentionally cleared it using **fun** _ $\mapsto$ ... Actually, loop is not used in the proofs

[16] In the Coq code, $\mathbb{O}^?_s$ is simply another _ hole left for the **refine** tactic to either fill it by unification or postpone it. Because $\mathbb{O}^?_1$, $\mathbb{O}^?_2$ and $\mathbb{O}^?_3$ are postponed in this example, we give them names for better explanations.

[17] Because termination certificates have a purely logical content, we do not care whether **omega** produces an "optimal" proof (b. t. w., it never does), but we appreciate its efficiency in solving those kinds of goals which we do not want to spend much time on, thus the gain is in time spent on developing the certified code. Efficiency of code execution is not touched since these proofs leave no traces in the extracted code.

of $\mathbb{O}_2^?$ or $\mathbb{O}_3^?$ either but it is necessary for the recursive call implemented in the `let` ... := loop ... `in` ... construct.

This peculiar way of writing terms as a combination of *programming style* and *combination of proof tactics* is possible in Coq by the use of the "swiss-army knife" tactic `refine`[18] that, via unification, allows the user to specify only parts of a term leaving holes to be filled later if unification fails to solve them. It is a major tool to finely control computational content while allowing great tactic flexibility on purely logical content.

We continue the definition of $\mathsf{itl}_{\mathrm{on}}$ as the first projection

$$\texttt{Definition}\ \mathsf{itl}_{\mathrm{on}}\ l\ m := \pi_1(\mathsf{itl}_{\mathrm{on}}^{\mathrm{full}}\ l\ m).$$

and its specification using the projection π_2 on the dependent pair $\mathsf{itl}_{\mathrm{on}}^{\mathrm{full}}\ l\ m$.

$$\texttt{Fact}\ \mathsf{itl\_on\_spec}\ l\ m : \mathsf{itl}_{\mathrm{on}}\ l\ m = \mathsf{itl}_{\mathrm{triv}}\ l\ m.$$

Notice that by asking the extraction mechanism to inline the definition of $\mathsf{itl}_{\mathrm{on}}^{\mathrm{full}}$, we get the following extracted OCaml code for $\mathsf{itl}_{\mathrm{on}}$, the one that optimally reflects the original specification of Eq. (1):

$$\texttt{let rec}\ \mathsf{itl}_{\mathrm{on}}\ l\ m = \texttt{match}\ l\ \texttt{with}\ [] \to m \mid x :: l \to x :: \mathsf{itl}_{\mathrm{on}}\ m\ l$$

Of course, this outcome of the (automatic) code extraction had to be targeted when doing the proof of $\mathsf{itl}_{\mathrm{on}}^{\mathrm{full}}$: a trivial proof would have been to choose as r just $\mathsf{itl}_{\mathrm{triv}}\ l\ m$, and the extracted code would have been just that. Instead, we did pattern-matching on l and chose in the first case m and in the second case $x :: r$, with r obtained as first component of the recursive call $\mathsf{loop}\ m\ l'$. The outer induction took care that loop stood for the function we were defining there.

Henceforth, we will take induction or recursion on measures for granted, assuming they correspond to the use of the tactic `induction-on`.

3 Traversal of Binary Trees

We present mostly standard material on traversal of binary trees that will lay the ground for the original contributions in the later sections of the paper. After defining binary trees and basic notions for them including their branches (Sect. 3.1), we describe mathematically what constitutes breadth-first traversal by considering an order on the branches (Sect. 3.2) that we characterize axiomatically (our Theorem 1). We then look at breadth-first traversal in its most elementary form (Sect. 3.3), by paying attention that it indeed meets its specification in terms of the order of the visited branches.

[18] The `refine` tactic was originally implemented by J.-C. Filliâtre.

3.1 Binary Trees

We use the type of binary trees $(a, b : \mathbb{T}X := \langle x \rangle \mid \langle a, x, b \rangle)$ where x is of the argument type X.[19] We define the $\mathsf{root} : \mathbb{T}X \to X$ of a tree, the subtrees $\mathsf{subt} : \mathbb{T}X \to \mathbb{L}(\mathbb{T}X)$ of a tree, and the size $\|\cdot\| : \mathbb{T}X \to \mathbb{N}$ of a tree by:

$$\begin{array}{rclcrclcrcl} \mathsf{root}\ \langle x \rangle & := & x & \quad & \mathsf{subt}\ \langle x \rangle & := & [] & \quad & \|\langle x \rangle\| & := & 1 \\ \mathsf{root}\ \langle \_, x, \_ \rangle & := & x & & \mathsf{subt}\ \langle a, \_, b \rangle & := & [a; b] & & \|\langle a, x, b \rangle\| & := & 1 + \|a\| + \|b\| \end{array}$$

and we extend the measure to lists of trees by $\|[t_1; \ldots; t_n]\| := \|t_1\| + \cdots + \|t_n\|$, hence $\|[]\| = 0$, $\|[t]\| = \|t\|$ and $\|l +\!\!+ m\| = \|l\| + \|m\|$. We will not need more complicated measures than $\|\cdot\|$ to justify the termination of the breadth-first algorithms to come.

A branch in a binary tree is described as a list of Booleans in $\mathbb{L}\,\mathbb{B}$ representing a list of left/right choices (0 for left and 1 for right).[20] We define the predicate $\mathsf{btb} : \mathbb{T}X \to \mathbb{L}\,\mathbb{B} \to \mathtt{Prop}$ inductively by the rules below where $t \mathbin{/\!/} l \downarrow$ denotes $\mathsf{btb}\ t\ l$ and tells whether *l is a branch in t:*

$$\frac{}{t \mathbin{/\!/} [] \downarrow} \qquad \frac{a \mathbin{/\!/} l \downarrow}{\langle a, x, b \rangle \mathbin{/\!/} 0 :: l \downarrow} \qquad \frac{b \mathbin{/\!/} l \downarrow}{\langle a, x, b \rangle \mathbin{/\!/} 1 :: l \downarrow}$$

We define the predicate $\mathsf{bpn} : \mathbb{T}X \to \mathbb{L}\,\mathbb{B} \to X \to \mathtt{Prop}$ inductively as well. The term $\mathsf{bpn}\ t\ l\ x$, also denoted $t \mathbin{/\!/} l \leadsto x$, tells whether *the node identified by l in t is decorated with x:*

$$\frac{}{t \mathbin{/\!/} [] \leadsto \mathsf{root}\ t} \qquad \frac{a \mathbin{/\!/} l \leadsto r}{\langle a, x, b \rangle \mathbin{/\!/} 0 :: l \leadsto r} \qquad \frac{b \mathbin{/\!/} l \leadsto r}{\langle a, x, b \rangle \mathbin{/\!/} 1 :: l \leadsto r}$$

We show the result that l is a branch of t if and only if it is decorated in t:

Fact $\mathsf{btb\_spec}$ $(t : \mathbb{T}X)$ $(l : \mathbb{L}\mathbb{B}) : t \mathbin{/\!/} l \downarrow \leftrightarrow \exists x, t \mathbin{/\!/} l \leadsto x$.

By two inductive rules, we define $\sim_{\mathbb{T}} : \mathbb{T}X \to \mathbb{T}Y \to \mathtt{Prop}$, *structural equivalence of binary trees*, and we lift it to structural equivalence of lists of binary trees $\sim_{\mathbb{LT}} : \mathbb{L}(\mathbb{T}X) \to \mathbb{L}(\mathbb{T}Y) \to \mathtt{Prop}$

$$\frac{}{\langle x \rangle \sim_{\mathbb{T}} \langle y \rangle} \qquad \frac{a \sim_{\mathbb{T}} a' \qquad b \sim_{\mathbb{T}} b'}{\langle a, x, b \rangle \sim_{\mathbb{T}} \langle a', y, b' \rangle} \qquad l \sim_{\mathbb{LT}} m := \forall^2(\sim_{\mathbb{T}})\ l\ m$$

i. e., when $l = [a_1; \ldots; a_n]$ and $m = [b_1; \ldots; b_p]$, the equivalence $l \sim_{\mathbb{LT}} m$ means $n = p \wedge a_1 \sim_{\mathbb{T}} b_1 \wedge \cdots \wedge a_n \sim_{\mathbb{T}} b_n$ (see Sect. 2).

Both $\sim_{\mathbb{T}}$ and $\sim_{\mathbb{LT}}$ are equivalence relations, and if $a \sim_{\mathbb{T}} b$ then $\|a\| = \|b\|$, i. e., structurally equivalent trees have the same size, and this holds for structurally equivalent lists of trees as well.

---

[19] In his paper [14], Okasaki considered unlabeled leaves. When we compare with his findings, we always tacitly adapt his definitions to cope with leaf labels, which adds only a small notational overhead.

[20] We here intend to model branches from the root to nodes, rather than from nodes to leaves. It might have been better to call the concept paths instead of branches.

3.2 Ordering Branches of Trees

A *strict order* $<_R$ is an irreflexive and transitive (binary) relation. We say it is *total* if the associated partial order $<_R \cup =$ is total in the usual sense of being connex, or equivalently, if $\forall x\, y,\ \{x <_R y\} \vee \{x = y\} \vee \{y <_R x\}$. It is *decidable and total* if:

$$\forall x\, y,\ \{x <_R y\} + \{x = y\} + \{y <_R x\}$$

where, as usual in type theory, a proof of the sum $A + B + C$ requires either a proof of A, of B or of C together with the information whether the first, second or third summand has been proven.

The *dictionary order* of type $\prec_{\text{dic}} : \mathbb{L}\,\mathbb{B} \to \mathbb{L}\,\mathbb{B} \to \texttt{Prop}$ is the lexicographic product on lists of 0's or 1's defined by

$$\frac{}{[] \prec_{\text{dic}} b :: l} \qquad \frac{}{0 :: l \prec_{\text{dic}} 1 :: m} \qquad \frac{l \prec_{\text{dic}} m}{b :: l \prec_{\text{dic}} b :: m}$$

The *breadth-first order* on $\mathbb{L}\,\mathbb{B}$ of type $\prec_{\text{bf}} : \mathbb{L}\,\mathbb{B} \to \mathbb{L}\,\mathbb{B} \to \texttt{Prop}$ is defined by

$$\frac{|l| < |m|}{l \prec_{\text{bf}} m} \qquad \frac{|l| = |m| \qquad l \prec_{\text{dic}} m}{l \prec_{\text{bf}} m}$$

i. e., the lexicographic product of *shorter* and *dictionary order* (if equal length).

Lemma 1. *$\prec_{\text{dic}}$ and $\prec_{\text{bf}}$ are decidable and total strict orders.*

We characterize $\prec_{\text{bf}}$ with the following four axioms:

Theorem 1. *Let $<_R : \mathbb{L}\,\mathbb{B} \to \mathbb{L}\,\mathbb{B} \to \texttt{Prop}$ be a relation s. t.*

(A_1) $<_R$ is a strict order (irreflexive and transitive)
(A_2) $\forall x\, l\, m,\ l <_R m \leftrightarrow x :: l <_R x :: m$
(A_3) $\forall l\, m,\ |l| < |m| \to l <_R m$
(A_4) $\forall l\, m,\ |l| = |m| \to 0 :: l <_R 1 :: m$

Then $<_R$ is equivalent to $\prec_{\text{bf}}$, i. e., $\forall l\, m,\ l <_R m \leftrightarrow l \prec_{\text{bf}} m$. Moreover the relation $\prec_{\text{bf}}$ satisfies (A_1)–(A_4).

3.3 Breadth-First Traversal, a Naive Approach

The *zipping* function $\mathsf{zip} : \mathbb{L}\,(\mathbb{L}\,X) \to \mathbb{L}\,(\mathbb{L}\,X) \to \mathbb{L}\,(\mathbb{L}\,X)$ is defined by

$$\mathsf{zip}\ []\ m := m \qquad \mathsf{zip}\ l\ []:= l \qquad \mathsf{zip}\ (x :: l)\ (y :: m) := (x +\!\!+\, y) :: \mathsf{zip}\ l\ m$$

The level-wise function $\mathsf{niv} : \mathbb{T}\,X \to \mathbb{L}\,(\mathbb{L}\,X)$ — niv refers to the French word "niveaux" — is defined by[21]

$$\mathsf{niv}\,\langle x\rangle := [x] :: [] \qquad \mathsf{niv}\,\langle a, x, b\rangle := [x] :: \mathsf{zip}\,(\mathsf{niv}\,a)\,(\mathsf{niv}\,b)$$

[21] The function niv is called "levelorder traversal" by Jones and Gibbons [7].

The $(n+1)$-th element of $\mathsf{niv}\,t$ contains the labels of t in left-to-right order that have distance n to the root. We can then define $\mathsf{bft_{std}}\,t := \mathsf{concat}(\mathsf{niv}\,t)$.[22] We lift breadth-first traversal to branches instead of decorations, defining $\mathsf{niv_{br}} : \mathbb{T}\,X \to \mathbb{L}\,(\mathbb{L}\,(\mathbb{L}\,\mathbb{B}))$ by

$$\begin{aligned} \mathsf{niv_{br}}\ \langle x\rangle &:= [[]] :: [] \\ \mathsf{niv_{br}}\ \langle a, x, b\rangle &:= [[]] :: \mathsf{zip}\,(\mathsf{map}\,(\mathsf{map}\,(0 :: \cdot))\,(\mathsf{niv_{br}}\,a))\,(\mathsf{map}\,(\mathsf{map}\,(1 :: \cdot))\,(\mathsf{niv_{br}}\,b)) \end{aligned}$$

and then $\mathsf{bft_{br}}\ t := \mathsf{concat}(\mathsf{niv_{br}}\ t)$, and we show the two following results:

`Theorem` niveaux_br_niveaux $t : \forall^2\,(\forall^2\,(\mathsf{bpn}\ t))\,(\mathsf{niv_{br}}\ t)\,(\mathsf{niv}\ t)$.

`Theorem` bft_br_std $t : \forall^2(\mathsf{bpn}\ t)\ (\mathsf{bft_{br}}\ t)\ (\mathsf{bft_{std}}\ t)$.

Hence $\mathsf{bft_{br}}\,t$ and $\mathsf{bft_{std}}\,t$ traverse the tree t in the same order, except that $\mathsf{bft_{std}}$ outputs decorating values and $\mathsf{bft_{br}}$ outputs branches.[23] We moreover show that $\mathsf{bft_{br}}\,t$ lists the branches of t in $\prec_{\mathsf{bf}}$-ascending order.

Theorem 2. *The list* $\mathsf{bft_{br}}\ t$ *is strictly sorted w. r. t.* $\prec_{\mathsf{bf}}$.

4 Breadth-First Traversal of a Forest

We lift root and subt to lists of trees and define $\mathsf{roots} : \mathbb{L}\,(\mathbb{T}\,X) \to \mathbb{L}\,X$ and $\mathsf{subtrees} : \mathbb{L}\,(\mathbb{T}\,X) \to \mathbb{L}\,(\mathbb{T}\,X)$ by

$$\mathsf{roots} := \mathsf{map\ root} \qquad\qquad \mathsf{subtrees} := \mathsf{flat\_map\ subt}$$

where flat_map is the standard list operation given by $\mathsf{flat\_map}\ f\ [x_1; \ldots; x_n] := f\ x_1 +\!\!+ \cdots +\!\!+ f\ x_n$. To justify the upcoming fixpoints/inductive proofs where recursive sub-calls occur on subtrees l, we show the following

Lemma subtrees_dec $l :\ l = [] \lor \|\mathsf{subtrees}\,l\| < \|l\|$.

Hence we can justify termination of a recursive algorithm $f\ l$ with formal argument $l : \mathbb{L}\,(\mathbb{T}\,X)$ that is calling itself on $f(\mathsf{subtrees}\ l)$, as soon as the case $f\,[]$ is computed without using recursive calls (to f). For this, we use, for instance, recursion on the measure $l \mapsto \|l\|$ but we may also use the binary measure $l, m \mapsto \|l +\!\!+ m\|$ when f has two arguments instead of just one.

4.1 Equational Characterization of Breadth-First Traversal

We first characterize $\mathsf{bft_f}$—breadth-first traversal of a forest—with four equivalent equations:

[22] Where $\mathsf{concat} := \mathsf{fold}\ (\cdot +\!\!+ \cdot)\ []$, i. e., $\mathsf{concat}\ [l_1; \ldots; l_n] = l_1 +\!\!+ \cdots +\!\!+ l_n$.
[23] The $\mathsf{bpn}\ t$ relation, also denoted $l, x \mapsto t\ /\!\!/\ l \rightsquigarrow x$, relates branches and decorations.

Theorem 3 (Characterization of breadth-first traversal of forests – recursive part). *Let* $\mathsf{bft_f}$ *be any term of type* $\mathbb{L}\,(\mathbb{T}\,X) \to \mathbb{L}\,X$ *and consider the following equations:*

(1) $\forall\, l,\ \mathsf{bft_f}\, l = \mathsf{roots}\, l +\!\!+\, \mathsf{bft_f}\,(\mathsf{subtrees}\, l)$;
(2) $\forall\, l\, m,\ \mathsf{bft_f}\,(l +\!\!+\, m) = \mathsf{roots}\, l +\!\!+\, \mathsf{bft_f}\,(m +\!\!+\, \mathsf{subtrees}\, l)$;
(3) $\forall\, t\, l,\ \mathsf{bft_f}\,(t :: l) = \mathsf{root}\, t :: \mathsf{bft_f}\,(l +\!\!+\, \mathsf{subt}\, t)$;
(Oka1) $\forall\, x\, l,\ \mathsf{bft_f}\,(\langle x\rangle :: l) = x :: \mathsf{bft_f}\, l$;
(Oka2) $\forall\, a\, b\, x\, l,\ \mathsf{bft_f}\,(\langle a, x, b\rangle :: l) = x :: \mathsf{bft_f}\,(l +\!\!+ [a; b])$.

We have the equivalence: (1) $\leftrightarrow$ (2) $\leftrightarrow$ (3) $\leftrightarrow$ (Oka1 $\wedge$ Oka2).

Proof. Equations (1) and (3) are clear instances of Eq. (2). Then (Oka1 $\wedge$ Oka2) is equivalent to (3) because they just represent a case analysis on t. So the only difficulty is to show (1) $\to$ (2) and (3) $\to$ (2). Both inductive proofs alternate the roles of l and m. So proving (2) from e.g. (1) by induction on either l or m is not possible. Following the example of the simple interleaving algorithm of Sect. 2.5, we proceed by induction on the measure $\|l +\!\!+\, m\|$. □

Equations (Oka1) and (Oka2) are used by Okasaki [14] as defining equations, while (3) is calculated from the specification by Jones and Gibbons [7]. We single out Eq. (2) above as a smooth gateway between subtrees-based breadth-first algorithms and FIFO-based breadth-first algorithms. Unlocking that "latch bolt" enabled us to show correctness properties of refined breadth-first algorithms.

Theorem 4 (Full characterization of breadth-first traversal of forests). *Adding equation* $\mathsf{bft_f}\,[\,] = [\,]$ *to any one of the equations of Theorem 3 determines the function* $\mathsf{bft_f}$ *uniquely.*

Proof. For any bft_1 and bft_2 satisfying both $\mathsf{bft}_1\,[\,] = [\,]$, $\mathsf{bft}_2\,[\,] = [\,]$ and e.g. $\mathsf{bft}_1\, l = \mathsf{roots}\, l +\!\!+\, \mathsf{bft}_1\,(\mathsf{subtrees}\, l)$ and $\mathsf{bft}_2\, l = \mathsf{roots}\, l +\!\!+\, \mathsf{bft}_2\,(\mathsf{subtrees}\, l)$, we show $\mathsf{bft}_1\, l = \mathsf{bft}_2\, l$ by induction on $\|l\|$. □

Notice that one should not confuse the uniqueness of the function—which is an extensional notion—with the uniqueness of an algorithm implementing such a function, because there are hopefully numerous possibilities.[24]

4.2 Direct Implementation of Breadth-First Traversal

We give a definition of $\mathsf{forest_{dec}} : \mathbb{L}\,(\mathbb{T}\,X) \to \mathbb{L}\,X \times \mathbb{L}\,(\mathbb{T}\,X)$ such that, provably, $\mathsf{forest_{dec}}\ l = (\mathsf{roots}\ l, \mathsf{subtrees}\ l)$, but using a simultaneous computation:

$$\begin{array}{ll} \mathsf{forest_{dec}}\,[\,] := ([\,],[\,]) \quad \mathsf{forest_{dec}}\,(\langle x\rangle :: l) := (x :: \alpha, \beta) & \text{where } (\alpha,\beta) := \mathsf{forest_{dec}}\ l. \\ \quad \mathsf{forest_{dec}}\,(\langle a, x, b\rangle :: l) := (x :: \alpha, a :: b :: \beta) & \end{array}$$

[24] Let us stress that extensionality is not very meaningful for algorithms anyway.

Then we show one way to realize the equations of Theorem 3 into a Coq term:

Theorem 5 (Existence of breadth-first traversal of forests). *One can define a Coq term* $\mathsf{bft_f}$ *of type* $\mathbb{L}(\mathbb{T}\,X) \to \mathbb{L}\,X$ *s. t.*

1. $\mathsf{bft_f}\ [\,] = [\,]$
2. $\forall l,\ \mathsf{bft_f}\ l = \mathsf{roots}\ l \mathbin{+\!\!+} \mathsf{bft_f}\,(\mathsf{subtrees}\ l)$

and s. t. $\mathsf{bft_f}$ *extracts to the following OCaml code:*[25]

$\texttt{let rec}\ \mathsf{bft_f}\ l = \texttt{match}\ l\ \texttt{with}\ [\,] \to [\,] \mid \_ \to \texttt{let}\ \alpha, \beta = \mathsf{forest_{dec}}\ l\ \texttt{in}\ \alpha\ \texttt{@}\ \mathsf{bft_f}\ \beta.$

Proof. We define the graph $\leadsto_{\mathrm{bft}}$ of the algorithm $\mathsf{bft_f}$ as binary relation of type $\mathbb{L}(\mathbb{T}\,X) \to \mathbb{L}\,X \to \texttt{Prop}$ with the two following inductive rules:

$$\frac{}{[\,] \leadsto_{\mathrm{bft}} [\,]} \qquad \frac{l \neq [\,] \qquad \mathsf{subtrees}\ l \leadsto_{\mathrm{bft}} r}{l \leadsto_{\mathrm{bft}} \mathsf{roots}\ l \mathbin{+\!\!+} r}$$

These rules follow the intended algorithm. We show that the graph $\leadsto_{\mathrm{bft}}$ is functional/deterministic, i. e.

$\texttt{Fact}\ \mathsf{bft\_f\_fun} : \forall l\, r_1\, r_2,\ l \leadsto_{\mathrm{bft}} r_1 \to l \leadsto_{\mathrm{bft}} r_2 \to r_1 = r_2.$

By induction on the measure $\|l\|$ we define a term $\mathsf{bft\_f\_full}\ l : \{r \mid l \leadsto_{\mathrm{bft}} r\}$ where we proceed as in Sect. 2.5. We get $\mathsf{bft_f}$ by the first projection $\mathsf{bft_f}\ l := \pi_1(\mathsf{bft\_f\_full}\ l)$ and derive the specification

$\texttt{Fact}\ \mathsf{bft\_f\_spec} : \forall l,\ l \leadsto_{\mathrm{bft}} \mathsf{bft_f}\ l.$

with the second projection π_2. Equations 1 and 2 follow straightforwardly from bft_f_fun and bft_f_spec. □

Hence we see that we can use the specifying Eqs. 1 and 2 of Theorem 5 to define the term $\mathsf{bft_f}$. In the case of breadth-first algorithms, termination is not very complicated because one can use induction on a measure to ensure it. In the proof, we just need to check that recursive calls occur on smaller arguments according to the given measure, and this follows from Lemma subtrees_dec.

Theorem 6 (Correctness of breadth-first traversal of forests). *For all* $t : \mathbb{T}\,X$*, we have* $\mathsf{bft_f}\ [t] = \mathsf{bft_{std}}\ t$.

Proof. We define $\leadsto_{\mathrm{niv}} : \mathbb{L}(\mathbb{T}\,X) \to \mathbb{L}(\mathbb{L}\,X) \to \texttt{Prop}$ by

$$\frac{}{[\,] \leadsto_{\mathrm{niv}} [\,]} \qquad \frac{l \neq [\,] \qquad \mathsf{subtrees}\ l \leadsto_{\mathrm{niv}} ll}{l \leadsto_{\mathrm{niv}} \mathsf{roots}\ l :: ll}$$

and we show that $l \leadsto_{\mathrm{niv}} ll \to m \leadsto_{\mathrm{niv}} mm \to l \mathbin{+\!\!+} m \leadsto_{\mathrm{niv}} \mathsf{zip}\ ll\ mm$ holds, a property from which we deduce $[t] \leadsto_{\mathrm{niv}} \mathsf{niv}\ t$. We show $l \leadsto_{\mathrm{niv}} ll \to l \leadsto_{\mathrm{bft}} \mathsf{concat}\ ll$ and we deduce $[t] \leadsto_{\mathrm{bft}} \mathsf{concat}\,(\mathsf{niv}\,t)$ hence $[t] \leadsto_{\mathrm{bft}} \mathsf{bft_{std}}\ t$. By bft_f_spec we have $[t] \leadsto_{\mathrm{bft}} \mathsf{bft_f}\ [t]$ and we conclude with bft_f_fun. □

[25] Notice that list append is denoted `@` in OCaml and $\mathbin{+\!\!+}$ in Coq.

4.3 Properties of Breadth-First Traversal

The *shape* of a tree (resp. forest) is the structure that remains when removing the values on the nodes and leaves, e. g., by mapping the base type X to the singleton type `unit`. Alternatively, one can use the $\sim_{\mathbb{T}}$ (resp. $\sim_{\mathbb{LT}}$) equivalence relation (introduced in Sect. 3.1) to characterize trees (resp. forests) which have identical shapes. We show that on a given forest shape, breadth-first traversal is an injective map:

Lemma bft_f_inj $l\ m : l \sim_{\mathbb{LT}} m \to \mathsf{bft_f}\ l = \mathsf{bft_f}\ m \to l = m.$

Proof. By induction on the measure $l, m \mapsto \|l +\!\!+ m\|$ and then case analysis on l and m. We use (Oka1&Oka2) from Theorem 3 to rewrite $\mathsf{bft_f}$ terms. The shape constraint $l \sim_{\mathbb{LT}} m$ ensures that the same equation is used for l and m. □

Hence, on a given tree shape, $\mathsf{bft_{std}}$ is also injective:

Corollary 1. *If $t_1 \sim_{\mathbb{T}} t_2$ are two trees of type $\mathbb{T}\,X$ (of the same shape) and* $\mathsf{bft_{std}}\ t_1 = \mathsf{bft_{std}}\ t_2$ *then* $t_1 = t_2$.

Proof. From Lemma bft_f_inj and Theorem 6. □

4.4 Discussion

The algorithm described in Theorem 5 can be used to compute breadth-first traversal as a replacement for the naive $\mathsf{bft_{std}}$ algorithm. We could also use other equations of Theorem 3, for instance using $\mathsf{bft_f}\ [] = []$, (Oka1) and (Oka2) to define another algorithm. The problem with equation

$$\text{(Oka2)}\quad \mathsf{bft_f}\ (\langle a, x, b\rangle :: l) = x :: \mathsf{bft_f}\ (l +\!\!+ [a; b])$$

is that it implies the use of $+\!\!+$ to append two elements at the tail of l, which is a well-known culprit that transforms an otherwise linear-time algorithm into a quadratic one. But Equation (Oka2) hints at replacing the list data-structure with a first-in, first-out queue (FIFO) for the argument of $\mathsf{bft_f}$ which brings us to the central section of this paper.

5 FIFO-Based Breadth-First Algorithms

Here come the certified algorithms in the spirit of Okasaki's paper [14]. They have the potential to be efficient, but this depends on the later implementation of the axiomatic datatype of FIFOs considered here (Sect. 5.1). We deal with traversal, numbering, reconstruction—in breadth-first order.

$$\begin{array}{lll}
\mathsf{fifo} & : & \mathsf{Type} \to \mathsf{Type} \\
\mathsf{f2l} & : & \forall X,\ \mathsf{fifo}\ X \to \mathbb{L}\, X \\
\mathsf{emp} & : & \forall X,\ \{q \mid \mathsf{f2l}\ q = []\} \\
\mathsf{enq} & : & \forall X\, q\, x,\ \{q' \mid \mathsf{f2l}\ q' = \mathsf{f2l}\ q +\!\!+ [x]\} \\
\mathsf{deq} & : & \forall X\, q,\ \mathsf{f2l}\ q \neq [] \to \{(x, q') \mid \mathsf{f2l}\ q = x :: \mathsf{f2l}\ q'\} \\
\mathsf{void} & : & \forall X\, q,\ \{b : \mathbb{B} \mid b = 1 \leftrightarrow \mathsf{f2l}\ q = []\}
\end{array}$$

Fig. 2. An axiomatization of first-in first-out queues.

5.1 Axiomatization of FIFOs

In Fig. 2, we give an axiomatic description of polymorphic first-in, first-out queues (a. k. a. FIFOs) by projecting them to lists with $\mathsf{f2l}\ \{X\} : \mathsf{fifo}\, X \to \mathbb{L}\, X$ where the notation $\{X\}$ marks X as an *implicit argument*[26] of f2l. Each axiom is fully specified using f2l: emp is the empty queue, enq the queuing function, deq the dequeuing function which assumes a non-empty queue as input, and void a Boolean test of emptiness. Notice that when q is non-empty, deq q returns a pair (x, q') where x is the dequeued value and q' the remaining queue.

A clean way of introducing such an abstraction in Coq that generates little overhead for program extraction towards OCaml is the use of a *module type* that collects the data of Fig. 2. Coq developments based on a hypothetical implementation of the module type are then organized as *functors* (i. e., modules depending on typed module parameters). Thus, all the Coq developments described in this section are such functors, and the extracted OCaml code is again a functor, now for the module system of OCaml, and with the module parameter that consists of the operations of Fig. 2 after stripping off the logical part. In other words, the parameter is nothing but a hypothetical implementation of a FIFO signature, viewed as a module type of OCaml.

Of course, the FIFO axioms have several realizations (or refinements), including a trivial and inefficient one where f2l is the identity function (and the inefficiency comes from appending the new elements at the end of the list with enq). In Sect. 7, we refine these axioms with more efficient implementations following Okasaki's insights [13] that worst-case $\mathcal{O}(1)$ FIFO operations are even possible in an elegant way with functional programming languages.

5.2 Breadth-First Traversal

We use the equations which come from Theorem 3 and Theorem 5:

$$\mathsf{bft_f}\ [] = [] \qquad \mathsf{bft_f}\ (\langle x \rangle :: l) = x :: \mathsf{bft_f}\ l \qquad \mathsf{bft_f}\ (\langle a, x, b \rangle :: l) = x :: \mathsf{bft_f}\ (l +\!\!+ [a; b])$$

[26] When a parameter X is marked implicit using the $\{X\}$ notation, it usually means that Coq is going to be able to infer the value of the argument by unification from the constraints in the context. In the case of f2l l, it means X will be deduced from the type of l which should unify with fifo X. While not strictly necessary, the mechanism of implicit arguments greatly simplifies the readability of Coq terms. In particular, it avoids an excessive use of dummy arguments _.

They suggest the definition of an algorithm for breadth-first traversal where lists are replaced with queues (FIFOs) so that (linear-time) append at the end $(\ldots +\!\!+ [a; b])$ is turned into two primitive queue operations $(\mathsf{enq}\ (\mathsf{enq}\ \ldots\ a)\ b)$. Hence, we implement FIFO-based breadth-first traversal.

Theorem 7. *There exists a fully specified Coq term*

$$\mathsf{bft\_fifo_f} : \forall q : \mathsf{fifo}\,(\mathbb{T}\,X), \{l : \mathbb{L}\,X \mid l = \mathsf{bft_f}\ (\mathsf{f2l}\ q)\}$$

s. t. $\mathsf{bft\_fifo_f}$ *extracts to the following OCaml code:*

```
let rec bft_fifo_f q =
  if void q then []
  else let t, q' = deq q
       in match t with
         | ⟨x⟩       → x :: bft_fifo_f q'
         | ⟨a, x, b⟩ → x :: bft_fifo_f (enq (enq q' a) b).
```

Proof. We proceed by induction on the measure $q \mapsto \|\mathsf{f2l}\ q\|$ following the method exposed in the interleave example of Sect. 2.5. The proof is structured around the computational content of the above OCaml code. Termination POs are easily solved by `omega`. Postconditions for correctness are proved using the above equations. □

Corollary 2. *There is a Coq term* $\mathsf{bft_{fifo}} : \mathbb{T}\,X \to \mathbb{L}\,X$ *s.t.* $\mathsf{bft_{fifo}}\ t = \mathsf{bft_{std}}\ t$ *holds for any* $t : \mathbb{T}\,X$*. Moreover,* $\mathsf{bft_{fifo}}$ *extracts to the following OCaml code:*

```
let bft_fifo t = bft_fifo_f (enq emp t).
```

Proof. From a tree t, we instantiate $\mathsf{bft\_fifo_f}$ on the one-element FIFO $(\mathsf{enq}\ \mathsf{emp}\ t)$ and thus derive the term $\mathsf{bft\_fifo\_full}\ (t : \mathbb{T}\,X) : \{l : \mathbb{L}\,X \mid l = \mathsf{bft_f}\ [t]\}$. The first projection $\mathsf{bft_{fifo}}\ t := \pi_1(\mathsf{bft\_fifo\_full}\ t)$ gives us $\mathsf{bft_{fifo}}$ and we derive $\mathsf{bft_{fifo}}\ t = \mathsf{bft_{std}}\ t$ from the combination of the second projection $\mathsf{bft_{fifo}}\ t = \mathsf{bft_f}\ [t]$ with Theorem 6. □

5.3 Breadth-First Numbering

Breadth-first numbering was the challenge proposed by Okasaki to the community and which led him to write his paper [14]. It consists in redecorating a tree with numbers in breadth-first order. The difficulty was writing an efficient algorithm in purely functional style. We choose the easy way to specify the result of breadth-first numbering of a tree: the output of the algorithm should be a tree $t : \mathbb{T}\,\mathbb{N}$ preserving the input shape and of which the breadth-first traversal of $\mathsf{bft_{std}}\ t$ is of the form $[1; 2; 3 \ldots]$.

As usual with those breadth-first algorithms, we generalize the notions to lists of trees.

```
Definition is_bfn n l := ∃k, bft_f l = [n; … ; n + k[.
```

Lemma 2. *Given a fixed shape, the breadth-first numbering of a forest is unique, i. e., for any* $n : \mathbb{N}$ *and any* $l, m : \mathbb{L}\,(\mathbb{T}\,\mathbb{N})$,

$$l \sim_{\mathbb{LT}} m \to \mathsf{is\_bfn}\ n\ l \to \mathsf{is\_bfn}\ n\ m \to l = m.$$

Proof. By Lemma bft_f_inj in Sect. 4.3. □

We give an equational characterization of breadth-first numbering of forests combined with list reversal. In the equations below, we intentionally consider a $\mathsf{bfn_f}$ function that outputs the *reverse* of the numbering of the input forest, so that, when viewed as FIFOs of trees (instead of lists of trees), both the input FIFO over $\mathbb{T}\,X$ and the output FIFO over type $\mathbb{T}\,\mathbb{N}$ correspond to left dequeuing and right enqueuing.[27] That said, Eqs. (E2) and (E3) correspond to (Oka1) and (Oka2) respectively, augmented with an extra argument used for keeping track of the numbering.

Lemma 3. *Let* $\mathsf{bfn_f} : \mathbb{N} \to \mathbb{L}\,(\mathbb{T}\,X) \to \mathbb{L}\,(\mathbb{T}\,\mathbb{N})$ *be a term. Considering the following conditions:*

(E1) $\forall n,\ \mathsf{bfn_f}\ n\ [] = []$;
(E2) $\forall n\,x\,l,\ \mathsf{bfn_f}\ n\ (\langle x\rangle :: l) = \mathsf{bfn_f}\ (1+n)\ l +\!\!+ [\langle n\rangle]$;
(E3) $\forall n\,a\,x\,b\,l,\ \exists a'\,b'\,l'$,
$\mathsf{bfn_f}\ (1+n)\ (l +\!\!+ [a;b]) = b' :: a' :: l' \wedge \mathsf{bfn_f}\ n\ (\langle a,x,b\rangle :: l) = l' +\!\!+ [\langle a',n,b'\rangle]$;
(Bfn1) $\forall n\,l,\ l \sim_{\mathbb{LT}} \mathsf{rev}(\mathsf{bfn_f}\ n\ l)$;
(Bfn2) $\forall n\,l,\ \mathsf{is\_bfn}\ n\ (\mathsf{rev}(\mathsf{bfn_f}\ n\ l))$.

We have the equivalence: (E1 ∧ E2 ∧ E3) ↔ (Bfn1 ∧ Bfn2).

Proof. From right to left, we essentially use Lemma 2. For the reverse direction, we proceed by induction on the measure $i, l \mapsto \|l\|$ in combination with Theorem 5 and Theorem 3—Equations (Oka1) and (Oka2). □

Although not explicitly written in Okasaki's paper [14], these equations hint at the use of FIFOs as a replacement for lists for both the input and output of $\mathsf{bfn_f}$. Let's see this informally for (E3): $\mathsf{bfn_f}\ n$ is to be computed on a non-empty list viewed as a FIFO, and left dequeuing gives a composite tree $\langle a, x, b\rangle$ and the remaining list/FIFO l. The subtrees a and b are enqueued to the right of l and $\mathsf{bfn_f}\ (1+n)$ called on the resulting list/FIFO. (E3) guarantees that a' and b' can be dequeued to the left from the output list/FIFO. Finally, $\langle a', n, b'\rangle$ is enqueued to the right to give the correct result, thanks to (E3). This construction will be formalized in Theorem 8.

We define the specification bfn_fifo_f_spec corresponding to breadth-first numbering of FIFOs of trees

$$\texttt{Definition}\ \mathsf{bfn\_fifo\_f\_spec}\ n\ q\ q' := \mathsf{f2l}\ q \sim_{\mathbb{LT}} \mathsf{rev}\,(\mathsf{f2l}\ q') \wedge \mathsf{is\_bfn}\ n\ (\mathsf{rev}(\mathsf{f2l}\ q'))$$

and we show the inhabitation of this specification.

[27] The fact that input and output FIFOs operate in mirror to each other was already pointed out by Okasaki in [14]. Using reversal avoids defining two types of FIFOs or bi-directional FIFOs to solve the issue.

Theorem 8. *There exists a fully specified Coq term*

$$\mathsf{bfn\_fifo_f} : \forall (n : \mathbb{N})\ (q : \mathsf{fifo}\ X),\ \{q' \mid \mathsf{bfn\_fifo\_f\_spec}\ n\ q\ q'\}$$

which extracts to the following OCaml code:

```
let rec bfn_fifo_f n q =
  if void q then emp
  else let t, q0 = deq q in match t with
    | ⟨_⟩      → enq (bfn_fifo_f (1 + n) q0) ⟨n⟩
    | ⟨a, _, b⟩ → let b', q1 = deq (bfn_fifo_f (1 + n) (enq (enq q0 a) b)) in
                 let a', q2 = deq q1
                 in enq q2 ⟨a', n, b'⟩.
```

Proof. To define $\mathsf{bfn\_fifo_f}$, we proceed by induction on the measure $n, q \mapsto \|\mathsf{f2l}\ q\|$ where the first parameter does not participate in the measure. As in Sect. 2.5, we implement a proof script which mixes tactics and programming style using the `refine` tactic. We strictly follow the above algorithm to design $\mathsf{bfn\_fifo_f}$. Of course, proof obligations like termination certificates or postconditions are generated by Coq and need to be addressed. As usual for these breadth-first algorithms, termination certificates are easy to solve thanks to the `omega` tactic. The only difficulties lie in the postcondition POs but these correspond to the (proofs of the) equations of Lemma 3. □

Corollary 3. *There is a Coq term* $\mathsf{bfn_{fifo}} : \mathbb{T}X \to \mathbb{T}\mathbb{N}$ *s. t. for any tree* $t : \mathbb{T}X$*:*

$$t \sim_{\mathbb{T}} \mathsf{bfn_{fifo}}\ t \qquad \textit{and} \qquad \mathsf{bft_{std}}\,(\mathsf{bfn_{fifo}}\ t) = [1; \ldots; \|t\|]$$

and $\mathsf{bfn_{fifo}}$ *extracts to the following OCaml code:*

```
let bfn_fifo t = let t', _ = deq (bfn_fifo_f 1 (enq emp t)) in t'.
```

Proof. Obvious consequence of Theorem 8 in conjunction with Theorem 6. □

5.4 Breadth-First Reconstruction

Breadth-first reconstruction is a generalization of breadth-first numbering—see the introduction for its description. For simplicity (since all our structures are finite), we ask that the list of labels that serves as extra argument has to be of the right length, i. e., has as many elements as there are labels in the input data-structure, while the "breadth-first labelling" function considered by Jones and Gibbons [7] just required it to be long enough so that the algorithm does not get stuck.

We define the specification of the breadth-first reconstruction of a FIFO q of trees in $\mathbb{T}X$ using a list $l : \mathbb{L}Y$ of labels

$$\texttt{Definition}\ \mathsf{bfr\_fifo\_f\_spec}\ q\ l\ q' := \mathsf{f2l}\ q \sim_{\mathbb{LT}} \mathsf{rev}\,(\mathsf{f2l}\ q') \wedge \mathsf{bft_f}\,(\mathsf{rev}\,(\mathsf{f2l}\ q')) = l.$$

We can then define breadth-first reconstruction by structural induction on the list l of labels:

$$\texttt{Fixpoint}\ \mathsf{bfr\_fifo_f}\ q\ l\ \{\texttt{struct}\ l\} : \|\mathsf{f2l}\ q\| = |l| \to \{q' \mid \mathsf{bfr\_fifo\_f\_spec}\ q\ l\ q'\}.$$

Notice the precondition $\|\mathsf{f2l}\ q\| = |l|$ stating that l contains as many labels as the total number of nodes in the FIFO q.[28] Since we use structural induction, there are no termination POs. There is however a precondition PO (easily proved) and postcondition POs are similar to those of the proof of Theorem 8. Extraction to OCaml outputs the following:

```
let rec bfr_fifo_f q = function
  | []      → emp
  | y :: l →
  let t, q0 = deq q in match t with
    | ⟨_⟩       → enq (bfr_fifo_f q0 l) ⟨y⟩
    | ⟨a, _, b⟩ → let b′, q1 = deq (bfr_fifo_f (enq (enq q0 a) b) l) in
                  let a′, q2 = deq q1
                  in enq q2 ⟨a′, y, b′⟩.
```

Notice the similarity with the code of $\mathsf{bfn\_fifo_f}$ of Theorem 8.

Theorem 9. *There is a Coq term* $\mathsf{bfr_{fifo}} : \forall (t : \mathbb{T}X)\,(l : \mathbb{L}Y),\ \|t\| = |l| \to \mathbb{T}Y$ *such that for any tree* $t : \mathbb{T}X$, $l : \mathbb{L}Y$ *and* $H : \|t\| = |l|$ *we have:*

$$t \sim_{\mathbb{T}} \mathsf{bfr_{fifo}}\ t\ l\ H \qquad \textit{and} \qquad \mathsf{bft_{std}}\,(\mathsf{bfr_{fifo}}\ t\ l\ H) = l.$$

Moreover, $\mathsf{bfr_{fifo}}$ *extracts to the following OCaml code:*

```
let bfr_fifo t l = let t′, _ = deq (bfr_fifo_f (enq emp t) l) in t′.
```

Proof. Direct application of $\mathsf{bfr\_fifo_f}\ (\mathsf{enq}\ \mathsf{emp}\ t)\ l$. □

6 Numbering by Levels

Okasaki reports in his paper [14] on his colleagues' attempts to solve the breadth-first numbering problem and mentions that most of them were level-oriented, as is the original traversal function $\mathsf{bft_{std}}$. In his Sect. 4, he describes the "cleanest" of all those solutions, and this small section is devoted to get it by extraction (and thus with certification through the method followed in this paper).

We define $\mathsf{children_f} : \forall\{K\}, \mathbb{L}\,(\mathbb{T}K) \to \mathbb{N} \times \mathbb{L}\,(\mathbb{T}K)$ such that, provably, $\mathsf{children_f}\ l = (|l|, \mathsf{subtrees}\ l)$ but using a more efficient simultaneous computation:

$$\begin{aligned} \mathsf{children_f}\ [\,] &:= (0, [\,]) \\ \mathsf{children_f}\ (\langle\_\rangle :: l) &:= (1+n, m) \qquad \text{where } (n,m) := \mathsf{children_f}\ l \\ \mathsf{children_f}\ (\langle a, \_, b\rangle :: l) &:= (1+n, a :: b :: m) \end{aligned}$$

[28] This condition could easily be weakened to $\|\mathsf{f2l}\ q\| \le |l|$ but in that case, the specification $\mathsf{bfr\_fifo\_f\_spec}$ should be changed as well.

and $\mathsf{rebuild_f} : \forall\{K\}, \mathbb{N} \to \mathbb{L}\,(\mathbb{T}\,K) \to \mathbb{L}\,(\mathbb{T}\,\mathbb{N}) \to \mathbb{L}\,(\mathbb{T}\,\mathbb{N})$

$\mathsf{rebuild_f}\ n\ []\ \_ := []$
$\mathsf{rebuild_f}\ n\ (\langle\_\rangle :: t_s)\ c_s := \langle n\rangle :: \mathsf{rebuild_f}\ (1+n)\ t_s\ c_s$
$\mathsf{rebuild_f}\ n\ (\langle\_,\_,\_\rangle :: t_s)\ (a :: b :: c_s) := \langle a, n, b\rangle :: \mathsf{rebuild_f}\ (1+n)\ t_s\ c_s$
and otherwise $\mathsf{rebuild_f}\ \_\ \_\ \_ := []$.

Since we will need to use both $\mathsf{children_f}$ and $\mathsf{rebuild_f}$ for varying values of the type parameter K, we define them as fully polymorphic here.[29] We then fix $X : \mathsf{Type}$ for the remainder of this section.

The algorithms $\mathsf{children_f}$ and $\mathsf{rebuild_f}$ are (nearly) those defined in [14, Figure 5] but that paper does not provide a specification for $\mathsf{rebuild_f}$ and thus cannot show the correctness result which follows. Instead of a specification, Okasaki offers an intuitive explanation of $\mathsf{rebuild_f}$ [14, p. 134]. Here, we will first rephrase the following lemma in natural language: the task is to obtain the breadth-first numbering of list t_s, starting with index n. We consider the list $\mathsf{subtrees}\ t_s$ of all immediate subtrees, hence of all that is "at the next level" (speaking in Okasaki's terms), and assume that c_s is the result of breadth-first numbering of those, but starting with index $|t_s| + n$, so as to skip all the roots in t_s whose number is $|t_s|$. Then, $\mathsf{rebuild_f}\ n\ t_s\ c_s$ is the breadth-first numbering of t_s. In view of this description, the first three definition clauses of $\mathsf{rebuild_f}$ are unavoidable, and the last one gives a dummy result for a case that never occurs when running algorithm $\mathsf{bfn\_level_f}$ in Theorem 10 below.

Lemma 4. *The function* $\mathsf{rebuild_f}$ *satisfies the following specification: for any* n, $t_s : \mathbb{L}\,(\mathbb{T}\,X)$ *and* $c_s : \mathbb{L}\,(\mathbb{T}\,\mathbb{N})$, *if both* $\mathsf{subtrees}\ t_s \sim_{\mathbb{LT}} c_s$ *and* $\mathsf{is\_bfn}\ (|t_s| + n)\ c_s$ *hold then* $t_s \sim_{\mathbb{LT}} \mathsf{rebuild_f}\ n\ t_s\ c_s$ *and* $\mathsf{is\_bfn}\ n\ (\mathsf{rebuild_f}\ n\ t_s\ c_s)$.

Proof. First we show by structural induction on t_s that for any $n : \mathbb{N}$ and $t_s : \mathbb{L}\,(\mathbb{T}\,\mathbb{N})$, if $\mathsf{roots}\ t_s = [n; n+1; \ldots]$ then $\mathsf{rebuild_f}\ n\ t_s\ (\mathsf{subtrees}\ t_s) = t_s$. Then we show that for any $Y, Z : \mathsf{Type}$, $n : \mathbb{N}$, $t_s : \mathbb{L}\,(\mathbb{T}\,Y)$, $t'_s : \mathbb{L}\,(\mathbb{T}\,Z)$ and any $c_s : \mathbb{L}\,(\mathbb{T}\,\mathbb{N})$, if $t_s \sim_{\mathbb{LT}} t'_s$ then $\mathsf{rebuild_f}\ n\ t_s\ c_s = \mathsf{rebuild_f}\ n\ t'_s\ c_s$. This second proof is by structural induction on proofs of the $t_s \sim_{\mathbb{LT}} t'_s$ predicate. The result follows using Lemma 2. □

The lemma suggests the recursive algorithm contained in the following theorem: in place of c_s as argument to $\mathsf{rebuild_f}$, it uses the result of the recursive call on $\mathsf{bfn\_level_f}$ with the index shifted by the number of terms in t_s (hence the number of roots in t_s—which is different from Okasaki's setting with labels only at inner nodes) and the second component of $\mathsf{children_f}\ t_s$.

Theorem 10. *There is a fully specified Coq term*

$$\mathsf{bfn\_level_f} : \forall\,(i : \mathbb{N})\,(l : \mathbb{L}\,(\mathbb{T}\,X)), \{m \mid l \sim_{\mathbb{LT}} m \wedge \mathsf{is\_bfn}\ i\ m\}$$

[29] Hence the $\forall\{K\}$ where K is declared as implicit.

which extracts to the following OCaml code:

$$\texttt{let rec}\ \mathsf{bfn\_level_f}\ i\ t_s = \texttt{match}\ t_s\ \texttt{with}$$
$$\mid [] \rightarrow []$$
$$\mid \_ \rightarrow \texttt{let}\ n, s_s = \mathsf{children_f}\ t_s\ \texttt{in}\ \mathsf{rebuild_f}\ i\ t_s\ (\mathsf{bfn\_level_f}\ (n+i)\ s_s).$$

Proof. By induction on the measure $i, l \mapsto \|l\|$. The non-trivial correctness PO is a consequence of Lemma 4. □

Corollary 4. *There is a Coq term* $\mathsf{bfn_{level}} : \mathbb{T}X \rightarrow \mathbb{T}\mathbb{N}$ *s. t. for any tree* $t : \mathbb{T}X$*:*

$$t \sim_{\mathbb{T}} \mathsf{bfn_{level}}\ t \qquad \text{and} \qquad \mathsf{bft_{std}}\,(\mathsf{bfn_{level}}\ t) = [1; \ldots ; \|t\|]$$

and $\mathsf{bfn_{level}}$ *extracts to the following OCaml code:*

$$\texttt{let}\ \mathsf{bfn_{level}}\ t = \texttt{match}\ \mathsf{bfn\_level_f}\ 1\ [t]\ \texttt{with}\ t' :: \_ \rightarrow t'.$$

Proof. Direct application of Theorems 6 and 10. □

7 Efficient Functional FIFOs

We discuss the use of our breadth-first algorithms that are parameterized over the abstract FIFO datatype with the implementations of efficient and purely functional FIFOs. As described in Sect. 5.1, the Coq developments of Sects. 5.2, 5.3 and 5.4 take the form of (Coq module) functors and their extracted code is structured as (OCaml module) functors. Using these functors means instantiating them with an implementation of the parameter, i. e., a module of the given module type. Formally, this is just application of the functor to the argument module. In our case, we implement Coq modules of the module type corresponding to Fig. 2 and then apply our (Coq module) functors to those modules. The extraction process yields the application of the extracted (OCaml module) functors to the extracted FIFO implementations. This implies a certification that the application is justified logically, i. e., that the FIFO implementation indeed satisfies the axioms of Fig. 2.

7.1 FIFOs Based on Two Lists

It is an easy exercise to implement our abstract interface for FIFOs based on pairs (l, r) of lists, with list representation $\mathsf{f2l}\ (l, r) := l +\!\!+ \mathsf{rev}\ r$. The enq operation adds the new element to the *front* of r (seen as the tail of the second part), while the deq operation "prefers" to take the elements from the *front* of l, but if l is empty, then r has to be carried over to l, which requires reversal. It is well-known that this implementation still guarantees amortized constant-time operations if list reversal is done efficiently (in linear time). As before, we obtain the implementation by automatic code extraction from constructions with the rich specifications that use measure induction for deq.

We can then instantiate our algorithms to this specific implementation, while Jones and Gibbons [7] calculated a dedicated breadth-first traversal algorithm for this implementation from the specification.

We have the advantage of a more modular approach and tool support for the code generation (once the mathematical argument in form of the rich specification is formalized in Coq). Moreover, we can benefit from a theoretically yet more efficient and still elegant implementation of FIFOs, the one devised by Okasaki [13], to be discussed next.

7.2 FIFOs Based on Three Lazy Lists

While amortized constant-time operations for FIFOs seem acceptable—although imperative programming languages can do better—Okasaki showed that also functional programming languages allow an elegant implementation of worst-case constant time FIFO operations [13].

The technique he describes relies on lazy evaluation. To access those data structures in terms extracted from Coq code, we use coinductive types, in particular finite or infinite streams (also called "colists"):

$$\texttt{CoInductive}\ \mathbb{S}\,X := \langle\rangle : \mathbb{S}\,X \mid \_\#\_ : X \to \mathbb{S}\,X \to \mathbb{S}\,X.$$

However, this type is problematic just because of the infinite streams it contains: since our inductive arguments are based on measures, we cannot afford that such infinite streams occur in the course of execution of our algorithms. Hence we need to guard our lazy lists with a purely logical *finiteness predicate* which is erased by the extraction process.

$$\begin{array}{l}\texttt{Inductive}\ \mathsf{lfin} : \mathbb{S}\,X \to \texttt{Prop} := \\ \quad \mid \mathsf{lfin\_nil}\ \ : \mathsf{lfin}\ \langle\rangle \\ \quad \mid \mathsf{lfin\_cons} : \forall x\, s,\ \mathsf{lfin}\ s \to \mathsf{lfin}\,(x \# s).\end{array}$$

We can then define the type of (finite) lazy lists as:

$$\texttt{Definition}\ \mathbb{L}_l\,X := \{s : \mathbb{S}\,X \mid \mathsf{lfin}\ s\}.$$

Compared to regular lists $\mathbb{L}\,X$, for a lazy list $(s, Hs) : \mathbb{L}_l\,X$, on the one hand, we can also do pattern-matching on s but on the other hand, we cannot define `Fixpoints` by structural induction on s. We replace it with structural induction on the proof of the predicate $H_s : \mathsf{lfin}\ s$. Although a bit cumbersome, this allows working with such lazy lists as if they were regular lists, and this practice is fully compatible with extraction because the guards of type $\mathsf{lfin}\ s$: `Prop`, being purely logical, are erased at extraction time. Here is the extraction of the type $\mathbb{L}_l$ in OCaml:

```
type α llist   = α __llist Lazy.t
and  α __llist = Lnil | Lcons of α * α llist
```

Here, we see the outcome of the effort: the (automatic) extraction process instructs OCaml to use lazy lists instead of standard lists (a distinction that does not even exist in the lazy language Haskell).

Okasaki [13] found a simple way to implement simple FIFOs[30] efficiently using triples of lazy lists. By efficiently, he means where enqueue and dequeue operations take *constant time* (in the worst case). We follow the proposed implementation using our own lazy lists $\mathbb{L}_l\, X$. Of course, his proposed code, being of beautiful simplicity, does not provide the full specifications for a correctness proof. Some important invariants are present, though. The main difficulty we faced was to give a correct specification for his intermediate rotate and make functions.

We do not enter more into the details of this implementation which is completely orthogonal to breadth-first algorithms. We invite the reader to check that we do get the exact intended extraction of FIFOs as triples of lazy lists; see the files `llists.v`, `fifo_3llists.v` and `extraction.v` described in Appendix A.

7.3 Some Remarks About Practical Complexity

However, our experience with the extracted algorithms for breadth-first numbering in OCaml indicate for smaller (with size below 2k nodes) randomly generated input trees that the FIFOs based on three lazy lists are responsible for a factor of approximately 5 in execution time in comparison with the 2-list-based implementation. For large trees (with size over 64k nodes), garbage collection severely hampers the theoretic linear-time behaviour. A precise analysis is out of scope for this paper.

8 Final Remarks

This paper shows that, despite their simplicity, breadth-first algorithms for finite binary trees present an interesting case for algorithm certification, in particular when it comes to obtain certified versions of efficient implementations in functional programming languages, as those considered by Okasaki [14].

Contributions. For this, we used the automatic extraction mechanism of the Coq system, whence we call this "breadth-first extraction." Fine-tuning the proof constructions so that the extraction process could generate the desired code—without any need for subsequent code polishing—was an engineering challenge, and the format of our proof scripts based on a hand-crafted measure induction tactic (expressed in the Ltac language for Coq [5] that is fully usable as user of Coq—as opposed to Coq developers only) should be reusable for a wide range of algorithmic problems and thus allow their solution with formal certification by program extraction.

30 He also found a simple solution for double-ended FIFOs.

We also considered variations on the algorithms that are not optimized for efficiency but illustrate the design space and also motivate the FIFO-based solutions. And we used the Coq system as theorem prover to formally verify some more abstract properties in relation with our breadth-first algorithms. This comprises as original contributions an axiomatic characterization of relations on paths in binary trees to be the breadth-first order (Theorem 1) in which the paths are visited by the breadth-first traversal algorithm. (Sect. 3.3). This also includes the identification of four different but logically equivalent ways to express the recursive behaviour of breadth-first traversal on forests (Theorem 3) and an equational characterization of breadth-first numbering of forests (Lemma 3). Among the variations, we mention that breadth-first reconstruction (Sect. 5.4) is amenable to a proof by structural recursion on the list of labels that is used for the relabeling while all the other proofs needed induction w. r. t. measures.

Perspectives. As mentioned in the introduction, code extraction of our constructions towards lazy languages such as Haskell would yield algorithms that we expect to work properly on infinite binary trees (the forests and FIFOs would still contain only finitely many elements, but those could then be infinite). The breadth-first nature of the algorithms would ensure fairness (hinted at also in the introduction). However, our present method does not certify in any way that use outside the specified domain of application (in particular, the non-functional correctness criterion of productivity is not guaranteed). We would have to give coinductive specifications and corecursively create their proofs, which would be a major challenge in Coq (cf. the experience of the second author with coinductive rose trees in Coq [16] where the restrictive guardedness criterion of Coq had to be circumvented in particular for corecursive constructions).

As further related work, we mention a yet different linear-time breadth-first traversal algorithm by Jones and Gibbons [7] that, as the other algorithms of their paper, is calculated from the specification, hence falls under the "algebra of programming" paradigm. Our methods should apply for that algorithm, too. And there is also their breadth-first reconstruction algorithm that relies on lazy evaluation of streams—a version of it which is reduced to breadth-first numbering has been discussed by Okasaki [14] to relate his work to theirs. To obtain such kind of algorithms would be a major challenge for the research in certified program extraction.

Another Related Research Direction. We mentioned directly related work throughout the paper, and we discussed certification of the program extraction procedure in Sect. 2.2. Let's briefly indicate a complementary approach. We state in our Theorems 5, 7, 8, 9 and 10 the OCaml code we wanted to obtain by extraction (and that we then got), but there is no tool support to start with that code and to work towards the (fully specified) Coq terms. The `hs-to-coq`[31] tool [19] transforms Haskell code (in place of OCaml that we chose to use) into Coq code and provides means to subsequent verification provided the Haskell code does not exploit non-termination.

[31] https://github.com/antalsz/hs-to-coq.

Acknowledgments. We are most grateful to the anonymous reviewers for their thoughtful feedback that included numerous detailed suggestions for improvement of the presentation.

A Code Correspondence

Here, we briefly describe the Coq vernacular files behind our paper that is hosted at https://github.com/DmxLarchey/BFE. Besides giving formal evidence for the more theoretical characterizations, it directly allows doing program extraction, see the README section on the given web page.
We are here presenting 24 Coq vernacular files in useful order:

- `list_utils.v`: One of the biggest files, all concerning list operations, list permutations, the lifting of relations to lists (Sect. 2) and segments of the natural numbers – auxiliary material with use at many places.
- `wf_utils.v`: The subtle tactics for measure recursion in one or two arguments with a ℕ-valued measure function (Sect. 2.4) – this is crucial for smooth extraction throughout the paper.
- `llist.v`: Some general material on coinductive lists, in particular proven finite ones (including append for those), but also the rotate operation of Okasaki [13], relevant in Sect. 7.2.
- `interleave.v`: The example of interleaving with three different methods in Sects. 2.3 (with existing tools—needs Coq v8.9 with package `Equations`) and Sect. 2.5 (with our method).
- `zip.v`: Zipping with a rich specification and relations with concatenation – just auxiliary material.
- `sorted.v`: Consequences of a list being sorted, in particular absence of duplicates in case of strict orders – auxiliary material for Sect. 3.2.
- `increase.v`: Small auxiliary file for full specification of breadth-first traversal (Sect. 3.3).
- `bt.v`: The largest file in this library, describing binary trees (Sect. 3.1), their branches and orders on those (Sect. 3.2) in relation with breadth-first traversal and structural relations on trees and forests (again Sect. 3.1).
- `fifo.v`: the module type for abstract FIFOs (Sect. 5.1).
- `fifo_triv.v`: The trivial implementation of FIFOs through lists, mentioned in Sect. 5.1.
- `fifo_2lists.v`: An efficient implementation that has amortized $\mathcal{O}(1)$ operations (see, e. g., the paper by Okasaki [13]), described in Sect. 7.1.
- `fifo_3llists.v`: The much more complicated FIFO implementation that is slower but has worst-case $\mathcal{O}(1)$ operations, invented by Okasaki [13]; see Sect. 7.2.
- `bft_std.v`: Breadth-first traversal naively with levels (specified with the traversal of branches in suitable order), presented in Sect. 3.3.
- `bft_forest.v`: Breadth-first traversal for forests of trees, paying much attention to the recursive equations that can guide the definition and/or verification (Sect. 4.1).

- `bft_inj.v`: Structurally equal forests with the same outcome of breadth-first traversal are equal, shown in Sect. 4.3.
- `bft_fifo.v`: Breadth-first traversal given an abstract FIFO, described in Sect. 5.2.
- `bfn_spec_rev.v`: Characterization of breadth-first numbering, see Lemma 3.
- `bfn_fifo.v`: The certified analogue of Okasaki's algorithm for breadth-first numbering [14], in Sect. 5.3.
- `bfn_trivial.v`: Just the instance of the previous with the trivial implementation of FIFOs.
- `bfn_level.v`: A certified reconstruction of `bfnum` on page 134 (Sect. 4 and Fig. 5) of Okasaki's article [14]. For its full specification, we allow ourselves to use breadth-first numbering obtained in `bfn_trivial.v`.
- `bfr_fifo.v`: Breadth-first reconstruction, a slightly more general task (see next file) than breadth-first numbering, presented in Sect. 5.4.
- `bfr_bfn_fifo.v`: Shows the claim that breadth-first numbering is an instance of breadth-first reconstruction (although they have been obtained with different induction principles).
- `extraction.v`: This operates extraction on-the-fly.
- `benchmarks.v`: Extraction towards `.ml` files.

References

1. Anand, A., Boulier, S., Cohen, C., Sozeau, M., Tabareau, N.: Towards certified meta-programming with typed Template-Coq. In: Avigad, J., Mahboubi, A. (eds.) ITP 2018. LNCS, vol. 10895, pp. 20–39. Springer, Cham (2018). https://doi.org/10.1007/978-3-319-94821-8_2
2. Andronick, J., Felty, A.P. (eds.): Proceedings of the 7th ACM SIGPLAN International Conference on Certified Programs and Proofs, CPP 2018, Los Angeles, CA, USA, 8–9 January 2018. ACM (2018). http://dl.acm.org/citation.cfm?id=3176245
3. Bertot, Y., Castéran, P.: Interactive Theorem Proving and Program Development. Coq'Art: The Calculus of Inductive Constructions. Texts in Theoretical Computer Science. Springer, Heidelberg (2004). https://doi.org/10.1007/978-3-662-07964-5
4. Cormen, T.H., Leiserson, C.E., Rivest, R.L.: Introduction to Algorithms. The MIT Press and McGraw-Hill Book Company (1989)
5. Delahaye, D.: A proof dedicated meta-language. Electr. Notes Theor. Comput. Sci. **70**(2), 96–109 (2002). https://doi.org/10.1016/S1571-0661(04)80508-5
6. Hupel, L., Nipkow, T.: A verified compiler from Isabelle/HOL to CakeML. In: Ahmed, A. (ed.) ESOP 2018. LNCS, vol. 10801, pp. 999–1026. Springer, Cham (2018). https://doi.org/10.1007/978-3-319-89884-1_35
7. Jones, G., Gibbons, J.: Linear-time breadth-first tree algorithms: an exercise in the arithmetic of folds and zips. Technical report, No. 71, Department of Computer Science, University of Auckland, May 1993
8. Larchey-Wendling, D., Monin, J.F.: Simulating induction-recursion for partial algorithms. In: Espírito Santo, J., Pinto, L. (eds.) 24th International Conference on Types for Proofs and Programs, TYPES 2018, Abstracts. University of Minho, Braga (2018). http://www.loria.fr/~larchey/papers/TYPES_2018_paper_19.pdf

9. Letouzey, P.: A new extraction for Coq. In: Geuvers, H., Wiedijk, F. (eds.) TYPES 2002. LNCS, vol. 2646, pp. 200–219. Springer, Heidelberg (2003). https://doi.org/10.1007/3-540-39185-1_12
10. Letouzey, P.: Programmation fonctionnelle certifiée - L'extraction de programmes dans l'assistant Coq. Ph.D. thesis, Université Paris-Sud, July 2004. https://www.irif.fr/~letouzey/download/these_letouzey_English.pdf
11. McCarthy, J.A., Fetscher, B., New, M.S., Feltey, D., Findler, R.B.: A Coq library for internal verification of running-times. Sci. Comput. Program. **164**, 49–65 (2018). https://doi.org/10.1016/j.scico.2017.05.001
12. Mullen, E., Pernsteiner, S., Wilcox, J.R., Tatlock, Z., Grossman, D.: Œuf: minimizing the Coq extraction TCB. In: Andronick and Felty [2], pp. 172–185. https://doi.org/10.1145/3167089
13. Okasaki, C.: Simple and efficient purely functional queues and deques. J. Funct. Program. **5**(4), 583–592 (1995)
14. Okasaki, C.: Breadth-first numbering: lessons from a small exercise in algorithm design. In: Odersky, M., Wadler, P. (eds.) Proceedings of the Fifth ACM SIGPLAN International Conference on Functional Programming (ICFP 2000), pp. 131–136. ACM (2000)
15. Paulson, L.C.: ML for the Working Programmer. Cambridge University Press, Cambridge (1991)
16. Picard, C., Matthes, R.: Permutations in coinductive graph representation. In: Pattinson, D., Schröder, L. (eds.) CMCS 2012. LNCS, vol. 7399, pp. 218–237. Springer, Heidelberg (2012). https://doi.org/10.1007/978-3-642-32784-1_12
17. Sozeau, M.: Subset coercions in Coq. In: Altenkirch, T., McBride, C. (eds.) TYPES 2006. LNCS, vol. 4502, pp. 237–252. Springer, Heidelberg (2007). https://doi.org/10.1007/978-3-540-74464-1_16
18. Sozeau, M.: Equations: a dependent pattern-matching compiler. In: Kaufmann, M., Paulson, L.C. (eds.) ITP 2010. LNCS, vol. 6172, pp. 419–434. Springer, Heidelberg (2010). https://doi.org/10.1007/978-3-642-14052-5_29
19. Spector-Zabusky, A., Breitner, J., Rizkallah, C., Weirich, S.: Total Haskell is reasonable Coq. In: Andronick and Felty [2], pp. 14–27. https://doi.org/10.1145/3167092

Verified Self-Explaining Computation

Jan Stolarek[1,2(✉)] and James Cheney[1,3]

[1] University of Edinburgh, Edinburgh, UK
jan.stolarek@ed.ac.uk, jcheney@inf.ed.ac.uk
[2] Lodz University of Technology, Łódź, Poland
[3] The Alan Turing Institute, London, UK

Abstract. Common programming tools, like compilers, debuggers, and IDEs, crucially rely on the ability to analyse program code to reason about its behaviour and properties. There has been a great deal of work on verifying compilers and static analyses, but far less on verifying dynamic analyses such as program slicing. Recently, a new mathematical framework for slicing was introduced in which forward and backward slicing are dual in the sense that they constitute a Galois connection. This paper formalises forward and backward dynamic slicing algorithms for a simple imperative programming language, and formally verifies their duality using the Coq proof assistant.

1 Introduction

The aim of mathematical program construction is to proceed from (formal) specifications to (correct) implementations. For example, critical components such as compilers, and various static analyses they perform, have been investigated extensively in a formal setting [10]. However, we unfortunately do not yet live in a world where all programs are constructed in this way; indeed, since some aspects of programming (e.g. exploratory data analysis) appear resistant to *a priori* specification, one could debate whether such a world is even possible. In any case, today programs "in the wild" are not always mathematically constructed. What do we do then?

One answer is provided by a class of techniques aimed at *explanation*, *comprehension* or *debugging*, often based on run-time monitoring, and sometimes with a pragmatic or even ad hoc flavour. In our view, the mathematics of constructing well-founded (meta)programs for explanation are wanting [4]. For example, dynamic analyses such as *program slicing* have many applications in comprehending and restructuring programs, but their mathematical basis and construction are far less explored compared to compiler verification [2,3].

Dynamic program slicing is a runtime analysis that identifies fragments of a program's input and source code – known together as a *program slice* – that were relevant to producing a chosen fragment of the output (a *slicing criterion*) [8,21]. Slicing has a very large literature, and there are a wide variety of dynamic slicing algorithms. Most work on slicing has focused on imperative or object-oriented programs.

G. Hutton (Ed.): MPC 2019, LNCS 11825, pp. 76–102, 2019.
https://doi.org/10.1007/978-3-030-33636-3_4

One common application of dynamic slicing is program debugging. Assume we have a program with variables x, y, and z and a programmer expects that after the program has finished running these variables will have respective values 1, 2, and 3. If a programmer unfamiliar with the program finds that after execution, variable y contains 1 where she was expecting another value, she may designate y as a slicing criterion, and dynamic slicing will highlight fragments of the source code that could have contributed to producing the incorrect result. This narrows down the amount of code that the programmer needs to inspect to correct a program. In this tiny example, of course, there is not much to throw away and the programmer can just inspect the program—the real benefit of slicing is for understanding larger programs with multiple authors. Slicing can also be used for program comprehension, i.e. to understand the behaviour of an already existing program in order to re-engineer its specification, possibly non-existent or not up-to-date.

In recent years a new semantic basis for dynamic slicing has been proposed [15,17]. It is based on the concept of Galois connections as known from order and lattice theory. Given lattices X and Y, a Galois connection is a pair of functions $g : Y \to X$ and $f : X \to Y$ such that $g(y) \leq x \iff y \leq f(x)$; then g is the *lower adjoint* and f is the *upper adjoint*. Galois connections have been advocated as a basis for mathematical program construction already, for example by Backhouse [1] and Mu and Oliveira [12]. They showed that if one can specify a problem space and show that it is one of the component functions of a Galois connection (the "easy" part), then *optimal solutions* to the problem (the "hard" part) are uniquely determined by the dual adjoint. A simple example arises from the duality between integer multiplication and division: the Galois connection $x \cdot y \leq z \iff x \leq z/y$ expresses that z/y is the greatest integer such that $(z/y) \cdot y \leq z$.

Whereas Galois connections have been used previously for constructing programs (as well as other applications such as program analysis), here we consider using Galois connections to construct programs for program slicing. In our setting, we consider lattices of *partial inputs* and *partial outputs* of a computation corresponding to possible input and output slicing criteria, as well as *partial programs* corresponding to possible slices—these are regarded as part of the input. We then define a forward semantics (corresponding to *forward slicing*) that expresses how much of the output of a program can be computed from a given partial input. Provided the forward semantics is monotone and preserves greatest lower bounds, it is the upper adjoint of a Galois connection, whose lower adjoint computes for each partial output the *smallest* partial input needed to compute it—which we consider an *explanation*. In other words, forward and backward slicing are *dual* in the sense that forward slicing computes "as much as possible" of the output given a partial input, while backward slicing computes "as little as needed" of the input to recover a partial output.

Figure 1 illustrates the idea for a small example where the "program" is an expression $(x + y, 2x)$, the input is an initial store containing $[x = 1, y = 2]$ and the output is the pair $(3, 2)$. Partial inputs, outputs, and programs are obtained

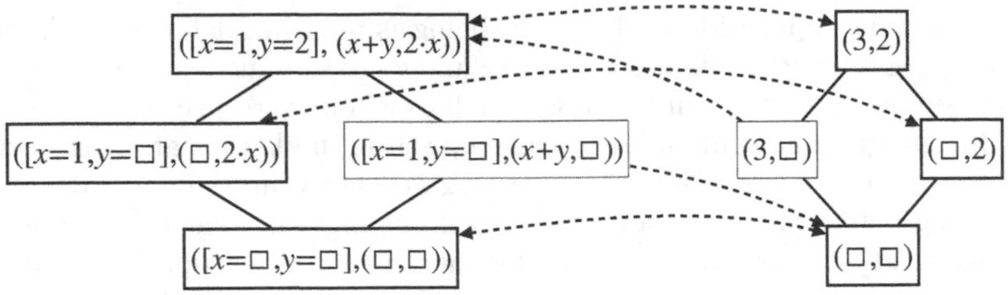

Fig. 1. Input and output lattices and Galois connection corresponding to expression $(x + y, 2 \cdot x)$ evaluated with input $[x = 1, y = 2]$ and output $(3, 2)$. Dotted lines with arrows pointing left calculate the lower adjoint, those pointing right calculate the upper adjoint, and lines with both arrows correspond to the induced isomorphism between the minimal inputs and maximal outputs. Several additional elements of the input lattice are omitted.

by replacing subexpressions by "holes" (□), as illustrated via (partial) lattice diagrams to the left and right. In the forward direction, computing the first component succeeds if both x and y are available while the second component succeeds if x is available; the backward direction computes the least input and program slice required for each partial output. Any Galois connection induces two isomorphic sublattices of the input and output, and in Fig. 1 the elements of these sublattices are enclosed in boxes with thicker lines. In the input, these elements correspond to *minimal* explanations: partial inputs and slices in which every part is needed to explain the corresponding output. The corresponding outputs are *maximal* in a sense that their minimal explanations do not explain anything else in the output.

The Galois connection approach to slicing has been originally developed for functional programming languages [15] and then extended to functional languages with imperative features [17]. So far it has not been applied to conventional imperative languages, so it is hard to compare directly with conventional slicing techniques. Also, the properties of the Galois connection framework in [15,17] have only been studied in pen-and-paper fashion. Such proofs are notoriously tricky in general and these are no exception; therefore, fully validating the metatheory of slicing based on Galois connections appears to be an open problem.

In this paper we present forward and backward slicing algorithms for a simple imperative language Imp and formally verify the correctness of these algorithms in Coq. Although Imp seems like a small and simple language, there are nontrivial technical challenges associated with slicing in the presence of mutable state, so our formalisation provides strong assurance of the correctness of our solution. To the best of our knowledge, this paper presents the first formalisation of a Galois connection slicing algorithm for an imperative language. Compared with Ricciotti et al. [17], Imp is a much simpler language than they consider, but their results are not formalised; compared with Léchenet et al. [9], we formalise dynamic rather than static slicing.

In Sect. 2 we begin by reviewing the syntax of Imp, giving illustrative examples of slicing, and then reviewing the theory of slicing using the Galois connection framework, including the properties of minimality and consistency. In Sect. 3 we introduce an instrumented, tracing semantics for Imp and present the forward and backward slicing algorithms. We formalise all of the theory from Sect. 3 in Coq and prove their duality. Section 4 highlights key elements of our Coq development, with full code available online [19]. Section 5 provides pointers to other related work.

2 Overview

2.1 Imp Slicing by Example

Arithmetic expressions	$a ::= n \mid x \mid a_1 + a_2$
Boolean expressions	$b ::= \texttt{true} \mid \texttt{false} \mid a_1 = a_2 \mid \neg b \mid b_1 \wedge b_2$
Imperative commands	$c ::= \texttt{skip} \mid x := a \mid c_1 \,;\, c_2$
	$\mid\ \texttt{while}\ b\ \texttt{do}\ \{\ c\ \} \mid \texttt{if}\ b\ \texttt{then}\ \{\ c_1\ \}\ \texttt{else}\ \{\ c_2\ \}$
Values	$v ::= v_a \mid v_b$
State	$\mu ::= \varnothing \mid \mu, x \mapsto v_a\ (x \text{ fresh})$

Fig. 2. Imp syntax

For the purpose of our analysis we use a minimal imperative programming language Imp used in some textbooks on programming languages, e.g. [13,16,22][1]. Imp contains arithmetic and logical expressions, mutable state (a list of mappings from variable names to numeric values) and a small core of imperative commands: empty instruction (`skip`), variable assignments, instruction sequencing, conditional `if` instructions, and `while` loops. An Imp program is a series of commands combined using a sequencing operator. Imp lacks more advanced features, such as functions or pointers.

Figure 2 shows Imp syntax. We use letter n to denote natural number constants and x to denote program variables. In this presentation we have omitted subtraction ($-$) and multiplication ($\cdot$) from the list of arithmetic operators, and less-or-equal comparison ($\leq$) from the list of boolean operators. All these operators are present in our Coq development and we omit them here only to make the presentation in the paper more compact. Otherwise the treatment of subtraction and multiplication is analogous to the treatment of addition, and treatment of $\leq$ is analogous to $=$.

Dynamic slicing is a program simplification technique that determines which parts of the source code contributed to a particular program output. For example, a programmer might write this simple program in Imp:

[1] In the literature Imp is also referred to as WHILE.

```
if (y = 1) then { y := x + 1 }
           else { y := y + 1 } ;
z := z + 1
```

and after running it with an input state $[\mathtt{x} \mapsto 1, \mathtt{y} \mapsto 0, \mathtt{z} \mapsto 2]$ might (wrongly) expect to obtain output state $[\mathtt{x} \mapsto 1, \mathtt{y} \mapsto 2, \mathtt{z} \mapsto 3]$. However, after running the program the actual output state will be $[\mathtt{x} \mapsto 1, \mathtt{y} \mapsto 1, \mathtt{z} \mapsto 3]$ with value of y differing from the expectation.

We can use dynamic slicing to debug this program by asking for an explanation which parts of the source code and the initial input state contributed to incorrect output value of y. We do this by formulating a *slicing criterion*, where we replace all values that we consider irrelevant in the output state (i.e. don't need an explanation for them) with *holes* (denoted with $\square$):

$$[\mathtt{x} \mapsto \square, \mathtt{y} \mapsto 1, \mathtt{z} \mapsto \square]$$

and a slicing algorithm might produce a program slice:

```
if (y = 1) then { □ }
           else { y := y + 1 } ; □
```

with a sliced input state $[\mathtt{x} \mapsto \square, \mathtt{y} \mapsto 0, \mathtt{z} \mapsto \square]$. This result indicates which parts of the original source code and input state could be ignored when looking for a fault (indicated by replacing them with holes), and which ones were relevant in producing the output indicated in the slicing criterion. The result of slicing narrows down the amount of code a programmer has to inspect to locate a bug. Here we can see that only the input variable y was relevant in producing the result; x and z are replaced with a $\square$ in the input state, indicating their irrelevance. We can also see that the first branch of the conditional was not taken (unexpectedly!) and that in the second branch y was incremented to become 1. With irrelevant parts of the program hidden away it is now easier to spot that the problem comes from a mistake in the initial state. The initial value of y should be changed to 1 so that the first branch of the conditional is taken and then y obtains an output value of 2 as expected.

Consider the same example, but a different output slicing criterion $[\mathtt{x} \mapsto \square, \mathtt{y} \mapsto \square, \mathtt{z} \mapsto 3]$. In this case, a correctly sliced program is as follows:

```
□ ; z := z + 1
```

with the corresponding sliced input $[\mathtt{x} \mapsto \square, \mathtt{y} \mapsto \square, \mathtt{z} \mapsto 2]$, illustrating that the entire first conditional statement is irrelevant. However, while the conclusion seems intuitively obvious, actually calculating this correctly takes some work: we need to ensure that none of the assignments inside the taken conditional branch affected z, and conclude from this that the value of the conditional test y = 1 is also irrelevant to the final value of z.

2.2 A Galois Connection Approach to Program Slicing

Example in Sect. 2.1 relies on an intuitive feel of how backward slicing should behave. We now address the question of how to make that intuition precise and show that slicing using the Galois connection framework introduced by Perera et al. [15] offers an answer.

Consider these two extreme cases of backward slicing behaviour:

1. For any slicing criterion backward slicing always returns a full program with no holes inserted;
2. For any slicing criterion backward slicing always returns a $\square$, i.e. it discards all the program code.

Neither of these two specifications is practically useful since they don't fulfil our intuitive expectation of "discarding program fragments irrelevant to producing a given fragment of program output". The first specification does not discard anything, bringing us no closer to understanding which code fragments are irrelevant. The second specification discards everything, including the code necessary to produce the output we want to understand. We thus want a backward slicing algorithm to have two properties:

- **Consistency**: backward slicing retains code required to produce output we are interested in.
- **Minimality**: backward slicing produces the smallest partial program and partial input state that suffice to achieve consistency.

Our first specification above does not have the minimality property; the second one does not have the consistency property. To achieve these properties we turn to order and lattice theory.

We begin with defining *partial Imp programs* (Fig. 3) by extending Imp syntax presented in Fig. 2 with *holes* (denoted using $\square$ in the semantic rules). A hole can appear in place of any arithmetic expression, boolean expression, or command. In the same way we allow values stored inside a state to be mapped to holes. For example:

$$\mu = [\mathtt{x} \mapsto 1, \mathtt{y} \mapsto \square]$$

is a *partial state* that maps variable x to 1 and variable y to a hole. We also introduce operation $\varnothing_\mu$ that takes a state μ and creates a partial state with the same domain as μ but all variables mapped to $\square$. For example if $\mu = [\mathtt{x} \mapsto 1, \mathtt{y} \mapsto 2]$ then $\varnothing_\mu = [\mathtt{x} \mapsto \square, \mathtt{y} \mapsto \square]$. A partial state that maps all its variables to holes is referred to as an *empty partial state.*

Having extended Imp syntax with holes, we define partial ordering relations on partial programs and partial states that consider holes to be syntactically smaller than any other subexpression. Figure 4 shows the partial ordering relation for arithmetic expressions. Definitions for ordering of partial boolean expressions and partial commands are analogous. Ordering for partial states is defined

Partial arithmetic expr.	$a ::= \ldots \mid \square$
Partial boolean expr.	$b ::= \ldots \mid \square$
Partial commands	$c ::= \ldots \mid \square$
Partial state	$\mu ::= \varnothing \mid \mu, x \mapsto v_a \mid \mu, x \mapsto \square$

Fig. 3. Partial Imp syntax. All elements of syntax from Fig. 2 remain unchanged, only $\square$ are added.

$$\frac{}{\square \sqsubseteq a} \qquad \frac{}{n \sqsubseteq n} \qquad \frac{}{x \sqsubseteq x} \qquad \frac{a_1 \sqsubseteq a_1' \quad a_2 \sqsubseteq a_2'}{a_1 + a_2 \sqsubseteq a_1' + a_2'}$$

Fig. 4. Ordering relation for partial arithmetic expressions.

element-wise, thus requiring that two states in the ordering relation have identical domains, i.e. store the same variables in the same order.

For every Imp program p, a set of all partial programs smaller than p forms a complete finite lattice, written $\downarrow p$ with p being the top and $\square$ the bottom element of this lattice. Partial states, arithmetic expressions, and boolean expressions form lattices in the same way. Moreover, a pair of lattices forms a (product) lattice, with the ordering relation defined component-wise:

$$(a_1, b_1) \sqsubseteq (a_2, b_2) \iff a_1 \sqsubseteq a_2 \wedge b_1 \sqsubseteq b_2$$

Figure 5 shows definition of the *join* (*least upper bound*, $\sqcup$) operation for arithmetic expressions. Definitions for boolean expressions and imperative commands are analogous. A join exists for every two elements from a complete lattice formed by a program p or state μ [6, Theorem 2.31].

Assume we have a program p paired with an input state μ that evaluates to an output state μ'. We can now formulate slicing as a pair of functions between lattices:

- **Forward slicing**: Forward slicing can be thought of as evaluation of partial programs. A function $\mathsf{fwd}_{(p,\mu)}$ takes as its input a partial program and a partial state from a lattice formed by pairing a program p and state μ. $\mathsf{fwd}_{(p,\mu)}$ outputs a partial state belonging to a lattice formed by μ'. The input to the forward slicing function is referred to as a *forward slicing criterion* and output as a *forward slice*.
- **Backward slicing**: Backward slicing can be thought of as "rewinding" a program's execution. A function $\mathsf{bwd}_{\mu'}$ takes as its input a partial state from the lattice formed by the output state μ'. $\mathsf{bwd}_{\mu'}$ outputs a pair consisting of a partial program and a partial state, both belonging to a lattice formed by program p and state μ. Input to a backward slicing function is referred to as a *backward slicing criterion* and output as a *backward slice*.

A key point above (discussed in detail elsewhere [17]) is that for imperative programs, both $\mathsf{fwd}_{(p,\mu)}$ and $\mathsf{bwd}_{\mu'}$ depend not only on p, μ, μ' but also on the

$$a \sqcup \Box = a \qquad \Box \sqcup a = a \qquad n \sqcup n = n \qquad x \sqcup x = x$$
$$(a_1 + a_2) \sqcup (a_1' + a_2') = (a_1 \sqcup a_1') + (a_2 \sqcup a_2')$$

Fig. 5. Join operation for arithmetic expressions.

particular execution path taken while evaluating p on μ. (In earlier work on slicing pure functional programs [15], traces are helpful for *implementing* slicing efficiently but not required for *defining* it.) We make this information explicit in Sect. 3 by introducing *traces* T that capture the choices made during execution. We will define the slicing algorithms inductively as relations indexed by T, but in our Coq formalisation $\mathsf{fwd}^T_{(p,\mu)}$ and $\mathsf{bwd}^T_{\mu'}$ are represented as dependently-typed functions where T is a proof term witnessing an operational derivation.

A pair of forward and backward slicing functions is guaranteed to have both the minimality and consistency properties when they form a Galois connection [6, Lemmas 7.26 and 7.33].

Definition 1 (Galois connection). *Given lattices P, Q and two functions $f : P \to Q$, $g : Q \to P$, we say f and g form a Galois connection (written $f \dashv g$) when $\forall_{p \in P, q \in Q} f(p) \sqsubseteq_Q q \iff p \sqsubseteq_P g(q)$. We call f a* lower adjoint *and g an* upper adjoint.

Importantly, for a given Galois connection $f \dashv g$, function f uniquely determines g and vice versa [6, Lemma 7.33]. This means that our choice of fwd (i.e. definition of how to evaluate partial programs on partial inputs) uniquely determines the backward slicing function bwd that will be minimal and consistent with respect to fwd, provided we can show that fwd and bwd form a Galois connection. There are many strategies to show that two functions $f : P \to Q$ and $g : Q \to P$ form a Galois connection, or to show that f or g in isolation has an upper or respectively lower adjoint. One attractive approach is to show that f preserves least upper bounds, or dually that g preserves greatest lower bounds (in either case, monotonicity follows as well). This approach is indeed attractive because it allows us to analyse just one of f or g and know that its dual adjoint exists, without even having to write it down. Indeed, in previous studies of Galois slicing [15,17], this characterisation was the one used: fwd was shown to preserve greatest lower bounds to establish the existence of its lower adjoint bwd, and then efficient versions of bwd were defined and proved correct.

For our constructive formalisation, however, we really want to give computable definitions for both fwd and bwd and prove they form a Galois connection, so while preservation of greatest lower bounds by fwd is a useful design constraint, proving it does not really save us any work. Instead, we will use the following equivalent characterisation of Galois connections [6, Lemma 7.26]:

1. f and g are monotone
2. *deflation* property holds:

$$\forall_{q \in Q}\ f(g(q)) \sqsubseteq_Q q$$

$$
\begin{array}{lrcl}
\text{Arithmetic traces} & T_a & ::= & n \mid x(v_a) \mid T_{a1} + T_{a2} \\
\text{Boolean traces} & T_b & ::= & \texttt{true} \mid \texttt{false} \mid T_{a1} = T_{a2} \mid \neg T_b \mid T_{b1} \wedge T_{b2} \\
\text{Command traces} & T_c & ::= & \texttt{skip} \mid x := T_a \mid T_1 \,;\, T_2 \\
 & & \mid & \texttt{if}_{\textsf{true}}\ T_b\ \texttt{then}\ \{\ T_1\ \} \mid \texttt{if}_{\textsf{false}}\ T_b\ \texttt{else}\ \{\ T_2\ \} \\
 & & \mid & \texttt{while}_{\textsf{false}}\ T_b \mid \texttt{while}_{\textsf{true}}\ T_b\ \texttt{do}\ \{\ T_c\ \};\ T_w
\end{array}
$$

Fig. 6. Trace syntax

3. *inflation* property holds:

$$\forall_{p \in P}\ p \sqsubseteq_P g(f(p))$$

We use this approach in our Coq mechanisation. We will first prove a general theorem that any pair of functions that fulfils properties (1)–(3) above forms a Galois connection. We will then define forward and backward slicing functions for Imp programs and prove that they are monotone, deflationary, and inflationary. Once this is done we will instantiate the general theorem with our specific definitions of forward and backward slicing to arrive at the proof that our slicing functions form a Galois connection. This is the crucial correctness property that we aim to prove. We also prove that existence of a Galois connection between forward and backward slicing functions implies consistency and minimality properties. Note that consistency is equivalent to the inflation property.

3 Dynamic Program Slicing

3.1 Tracing Semantics

Following previous work [15,17], we employ a *tracing semantics* to define the slicing algorithms. Since dynamic slicing takes account of the actual execution path followed by a run of a program, we represent the execution path taken using an explicit trace data structure. Traces are then traversed as part of both the forward and backward slicing algorithms. That is, unlike tracing evaluation, forward and backward slicing follow the structure of traces, rather than the program. Note that we are not really inventing anything new here: in our formalisation, the trace is simply a *proof term* witnessing the derivability of the operational semantics judgement. The syntax of traces is shown in Fig. 6. The structure of traces follows the structure of language syntax with the following exceptions:

- The expression trace $x(v_a)$ records both the variable name x and a value v_a that was read from program state μ;
- For conditional instructions, traces record which branch was actually taken. When the `if` condition evaluates to **true** we store traces of evaluating the condition and the `then` branch; if it evaluates to **false** we store traces of evaluating the condition and the `else` branch.

- For `while` loops, if the condition evaluates to **false** (i.e. loop body does not execute) we record only a trace for the condition. If the condition evaluates to **true** we record traces for the condition (T_b), a single execution of the loop body (T_c) and the remaining iterations of the loop (T_w).

$$\frac{}{\mu, n \Rightarrow n :: v_n} \qquad \frac{\mu(x) = v_a}{\mu, x \Rightarrow x(v_a) :: v_a} \qquad \frac{\mu, a_1 \Rightarrow T_1 :: v_1 \quad \mu, a_2 \Rightarrow T_2 :: v_2}{\mu, a_1 + a_2 \Rightarrow T_1 + T_2 :: v_1 +_{\mathbb{N}} v_2}$$

Fig. 7. Imp arithmetic expressions evaluation

$$\frac{}{\mu, \texttt{true} \Rightarrow \texttt{true} :: \textbf{true}} \qquad \frac{}{\mu, \texttt{false} \Rightarrow \texttt{false} :: \textbf{false}}$$

$$\frac{\mu, a_1 \Rightarrow T_1 :: v_1 \quad \mu, a_2 \Rightarrow T_2 :: v_2}{\mu, a_1 = a_2 \Rightarrow T_1 = T_2 :: v_1 =_{\mathbb{B}} v_2} \qquad \frac{\mu, b \Rightarrow T :: v}{\mu, \neg b \Rightarrow \neg T :: \neg_{\mathbb{B}} v}$$

$$\frac{\mu, b_1 \Rightarrow T_1 :: v_1 \quad \mu, b_2 \Rightarrow T_2 :: v_2}{\mu, b_1 \wedge b_2 \Rightarrow T_1 \wedge T_2 :: v_1 \wedge_{\mathbb{B}} v_2}$$

Fig. 8. Imp boolean expressions evaluation

Figures 7, 8, and 9 show evaluation rules for arithmetic expressions, boolean expressions, and imperative commands, respectively[2]. Traces are written in grey colour and separated with a double colon (::) from the evaluation result. Arithmetic expressions evaluate to numbers (denoted v_a). Boolean expressions evaluate to either **true** or **false** (jointly denoted as v_b). Operators with $_{\mathbb{N}}$ or $_{\mathbb{B}}$ subscripts should be evaluated as mathematical and logical operators respectively to arrive at an actual value; this is to distinguish them from the language syntax. Commands evaluate by side-effecting on the input state, producing a new state as output (Fig. 9). Only arithmetic values can be assigned to variables and stored inside a state. Assignments to variables absent from the program state are treated as no-ops. This means all variables that we want to write and read must be included (initialised) in the initial program state. We explain reasons behind this decision later in Sect. 4.3.

3.2 Forward Slicing

In this and the next section we present concrete definitions of forward and backward slicing for Imp programs. Readers may prefer to skip ahead to Sect. 3.4 for an extended example of these systems at work first. Our slicing algorithms are based on the ideas first presented in [17]. Presentation in Sect. 2.2 views the

[2] We overload the $\Rightarrow$ notation to mean one of three evaluation relations. It is always clear from the arguments which relation we are referring to.

$$\frac{}{\mu, \texttt{skip} \Rightarrow \texttt{skip} :: \mu} \qquad \frac{\mu, a \Rightarrow_a T_a :: v_a}{\mu, x := a \Rightarrow x := T_a :: \mu[x \mapsto v_a]}$$

$$\frac{\mu, c_1 \Rightarrow T_1 :: \mu' \qquad \mu', c_2 \Rightarrow T_2 :: \mu''}{\mu, c_1 \,;\, c_2 \Rightarrow T_1 \,;\, T_2 :: \mu''}$$

$$\frac{\mu, b \Rightarrow T_b :: \textsf{true} \qquad \mu, c_1 \Rightarrow T_1 :: \mu'}{\mu, \texttt{if}\ b\ \texttt{then}\ \{\ c_1\ \}\ \texttt{else}\ \{\ c_2\ \} \Rightarrow \texttt{if}_{\textsf{true}}\ T_b\ \texttt{then}\ \{\ T_1\ \} :: \mu'}$$

$$\frac{\mu, b \Rightarrow T_b :: \textsf{false} \qquad \mu, c_2 \Rightarrow T_2 :: \mu'}{\mu, \texttt{if}\ b\ \texttt{then}\ \{\ c_1\ \}\ \texttt{else}\ \{\ c_2\ \} \Rightarrow \texttt{if}_{\textsf{false}}\ T_b\ \texttt{else}\ \{\ T_2\ \} :: \mu'}$$

$$\frac{\mu, b \Rightarrow T_b :: \textsf{false}}{\mu, \texttt{while}\ b\ \texttt{do}\ \{\ c\ \} \Rightarrow \texttt{while}_{\textsf{false}}\ T_b :: \mu}$$

$$\frac{\mu, b \Rightarrow T_b :: \textsf{true} \qquad \mu, c \Rightarrow T_c :: \mu' \qquad \mu', \texttt{while}\ b\ \texttt{do}\ \{\ c\ \} \Rightarrow T_w :: \mu''}{\mu, \texttt{while}\ b\ \texttt{do}\ \{\ c\ \} \Rightarrow \texttt{while}_{\textsf{true}}\ T_b\ \texttt{do}\ \{\ T_c\ \};\ T_w :: \mu''}$$

Fig. 9. Imp command evaluation.

$$\frac{}{T :: \mu, \Box \nearrow \Box} \qquad \frac{}{n :: \mu, n \nearrow n} \qquad \frac{}{x(v_a) :: \mu, x \nearrow \mu(x)}$$

$$\frac{T_1 :: \mu, a_1 \nearrow \Box}{T_1 + T_2 :: \mu, a_1 + a_2 \nearrow \Box} \qquad \frac{T_2 :: \mu, a_2 \nearrow \Box}{T_1 + T_2 :: \mu, a_1 + a_2 \nearrow \Box}$$

$$\frac{T_1 :: \mu, a_1 \nearrow v_1 \qquad T_2 :: \mu, a_2 \nearrow v_2}{T_1 + T_2 :: \mu, a_1 + a_2 \nearrow v_1 +_{\mathbb{N}} v_2} \quad v_1, v_2 \neq \Box$$

Fig. 10. Forward slicing rules for Imp arithmetic expressions.

slicing algorithms as computable functions and we will implement them in code as such. However for the purpose of writing down the formal definitions of our algorithms we will use a relational notation. It is more concise and allows easier comparisons with previous work.

Figures 10, 11 and 12 present forward slicing rules for the Imp language[3]. As mentioned in Sect. 2.2, forward slicing can be thought of as evaluation of partial programs. Thus the forward slicing relations $\nearrow$ take a partial program, a partial state, and an execution trace as an input and return a partial value, either a partial number (for partial arithmetic expressions), a partial boolean (for partial logical expressions) or a partial state (for partial commands). For example, we can read $T :: \mu, c \nearrow \mu'$ as "Given trace T, in partial environment μ the partial command c forward slices to partial output μ'."

A general principle in the forward slicing rules for arithmetic expressions (Fig. 10) and logical expressions (Fig. 11) is that "holes propagate". This means

[3] We again overload $\nearrow$ and $\searrow$ arrows in the notation to denote one of three forward/backward slicing relations. This is important in the rules for boolean slicing, whose premises refer to the slicing relation for arithmetic expressions, and command slicing, whose premises refer to slicing relation for boolean expressions.

$$\frac{}{T :: \mu, \square \nearrow \square}$$

$$\frac{}{\mathtt{true} :: \mu, \mathtt{true} \nearrow \mathsf{true}} \qquad \frac{}{\mathtt{false} :: \mu, \mathtt{false} \nearrow \mathsf{false}}$$

$$\frac{T_1 :: a_1 \nearrow \square}{T_1 = T_2 :: \mu, a_1 = a_2 \nearrow \square} \qquad \frac{T_2 :: a_2 \nearrow \square}{T_1 = T_2 :: \mu, a_1 = a_2 \nearrow \square}$$

$$\frac{T_1 :: a_1 \nearrow v_1 \quad T_2 :: a_2 \nearrow v_2}{T_1 = T_2 :: \mu, a_1 = a_2 \nearrow v_1 =_{\mathbb{B}} v_2} \; v_1, v_2 \neq \square$$

$$\frac{T :: b \nearrow \square}{\neg T :: \mu, \neg b \nearrow \square} \qquad \frac{T :: b \nearrow v_b}{\neg T :: \mu, \neg b \nearrow \neg_{\mathbb{B}} v} \; v_b \neq \square$$

$$\frac{T_1 :: b_1 \nearrow \square}{T_1 \wedge T_2 :: \mu, b_1 \wedge b_2 \nearrow \square} \qquad \frac{T_2 :: b_2 \nearrow \square}{T_1 \wedge T_2 :: \mu, b_1 \wedge b_2 \nearrow \square}$$

$$\frac{T_1 :: b_1 \nearrow v_1 \quad T_2 :: b_2 \nearrow v_2}{T_1 \wedge T_2 :: \mu, b_1 \wedge b_2 \nearrow v_1 \wedge_{\mathbb{B}} v_2} \; v_1, v_2 \neq \square$$

Fig. 11. Forward slicing rules for Imp boolean expressions.

that whenever $\square$ appears as an argument of an operator, application of that operator forward slices to a $\square$. For example, $1 + \square$ forward slices to a $\square$ and so does $\neg\square$. In other words, if an arithmetic or logical expression contains at least one hole it will reduce to a $\square$; if it contains no holes it will reduce to a proper value. This is not the case for commands though. For example, command **if true then** 1 **else** $\square$ forward slices to 1, even though it contains a hole in the (not taken) **else** branch.

A rule worth attention is one for forward slicing of variable reads:

$$\frac{}{x(v_a) :: \mu, x \nearrow \mu(x)}$$

It is important here that we read the return value from μ and not v_a recorded in a trace. This is because μ is a partial state and also part of a forward slicing criterion. It might be that μ maps x to $\square$, in which case we must forward slice to $\square$ and not to v_a. Otherwise minimality would not hold.

Forward slicing rules for arithmetic and logical expressions both have a universal rule for forward slicing of holes that applies regardless of what the exact trace value is:

$$\frac{}{T :: \mu, \square \nearrow \square}$$

There is no such rule for forward slicing of commands (Fig. 12). There we have separate rules for forward slicing of holes for each possible trace. This is due to the side-effecting nature of commands, which can mutate the state through variable assignments. Consider this rule for forward slicing of assignments w.r.t.

$$\frac{}{\texttt{skip} :: \mu, \Box \nearrow \mu} \qquad \frac{}{\texttt{skip} :: \mu, \textbf{skip} \nearrow \mu} \qquad \frac{}{x := T_a :: \mu, \Box \nearrow \mu[x \mapsto \Box]}$$

$$\frac{T_a :: \mu, a \nearrow v_a}{x := T_a :: \mu, x := a \nearrow \mu[x \mapsto v_a]} \qquad \frac{T_1 :: \mu, \Box \nearrow \mu' \quad T_2 :: \mu', \Box \nearrow \mu''}{T_1 \,;\, T_2 :: \mu, \Box \nearrow \mu''}$$

$$\frac{T_1 :: \mu, c_1 \nearrow \mu' \quad T_2 :: \mu', c_2 \nearrow \mu''}{T_1 \,;\, T_2 :: \mu, c_1 \,;\, c_2 \nearrow \mu''} \qquad \frac{T_1 :: \mu, \Box \nearrow \mu'}{\texttt{if}_{\mathsf{true}}\ T_b\ \texttt{then}\ \{\ T_1\ \} :: \mu, \Box \nearrow \mu'}$$

$$\frac{T_b :: \mu, b \nearrow \Box \quad T_1 :: \mu, \Box \nearrow \mu'}{\texttt{if}_{\mathsf{true}}\ T_b\ \texttt{then}\ \{\ T_1\ \} :: \mu, \textbf{if}\ b\ \textbf{then}\ \{\ c_1\ \}\ \textbf{else}\ \{\ c_2\ \} \nearrow \mu'}$$

$$\frac{T_b :: \mu, b \nearrow v_b \quad T_1 :: \mu, c_1 \nearrow \mu'}{\texttt{if}_{\mathsf{true}}\ T_b\ \texttt{then}\ \{\ T_1\ \} :: \mu, \textbf{if}\ b\ \textbf{then}\ \{\ c_1\ \}\ \textbf{else}\ \{\ c_2\ \} \nearrow \mu'}\ v_b \neq \Box$$

$$\frac{T_2 :: \mu, \Box \nearrow \mu'}{\texttt{if}_{\mathsf{false}}\ T_b\ \texttt{else}\ \{\ T_2\ \} :: \mu, \Box \nearrow \mu'}$$

$$\frac{T_b :: \mu, b \nearrow \Box \quad T_2 :: \mu, \Box \nearrow \mu'}{\texttt{if}_{\mathsf{false}}\ T_b\ \texttt{else}\ \{\ T_2\ \} :: \mu, \textbf{if}\ b\ \textbf{then}\ \{\ c_1\ \}\ \textbf{else}\ \{\ c_2\ \} \nearrow \mu'}$$

$$\frac{T_b :: \mu, b \nearrow v_b \quad T_2 :: \mu, c_2 \nearrow \mu'}{\texttt{if}_{\mathsf{false}}\ T_b\ \texttt{else}\ \{\ T_2\ \} :: \mu, \textbf{if}\ b\ \textbf{then}\ \{\ c_1\ \}\ \textbf{else}\ \{\ c_2\ \} \nearrow \mu'}\ v_b \neq \Box$$

$$\frac{}{\texttt{while}_{\mathsf{false}}\ T_b :: \mu, \Box \nearrow \mu} \qquad \frac{}{\texttt{while}_{\mathsf{false}}\ T_b :: \mu, \textbf{while}\ b\ \textbf{do}\ \{\ c\ \} \nearrow \mu}$$

$$\frac{T_c :: \mu, \Box \nearrow \mu_c \quad T_w :: \mu_c, \Box \nearrow \mu_w}{\texttt{while}_{\mathsf{true}}\ T_b\ \texttt{do}\ \{\ T_c\ \};\ T_w :: \mu, \Box \nearrow \mu_w}$$

$$\frac{T_b :: \mu, b \nearrow \Box \quad T_c :: \mu, \Box \nearrow \mu_c \quad T_w :: \mu_c, \Box \nearrow \mu_w}{\texttt{while}_{\mathsf{true}}\ T_b\ \texttt{do}\ \{\ T_c\ \};\ T_w :: \mu, \textbf{while}\ b\ \textbf{do}\ \{\ c\ \} \nearrow \mu_w}$$

$$\frac{T_b :: \mu, b \nearrow v_b \quad T_c :: \mu, c \nearrow \mu_c \quad T_w :: \mu_c, \textbf{while}\ b\ \textbf{do}\ \{\ c\ \} \nearrow \mu_w}{\texttt{while}_{\mathsf{true}}\ T_b\ \texttt{do}\ \{\ T_c\ \};\ T_w :: \mu, \textbf{while}\ b\ \textbf{do}\ \{\ c\ \} \nearrow \mu_w}\ v_b \neq \Box$$

Fig. 12. Forward slicing rules for Imp commands.

a $\Box$ as a slicing criterion[4]:

$$\frac{}{x := T_a\ ::\ \mu, \Box \nearrow \mu[x \mapsto \Box]}$$

When forward slicing an assignment w.r.t. a $\Box$ we need to erase (i.e. change to a $\Box$) variable x in the state μ, which follows the principle of "propagating holes". Here having a trace is crucial to know which variable was actually assigned during the execution. Rules for forward slicing of other commands w.r.t. a $\Box$ traverse the trace recursively to make sure that all variable assignments within a trace are reached. For example:

$$\frac{T_1\ ::\ \mu, \Box \nearrow \mu'}{\texttt{if}_{\mathsf{true}}\ T_b\ \texttt{then}\ \{\ T_1\ \}\ ::\ \mu, \Box \nearrow \mu'}$$

[4] When some partial value v is used as a slicing criterion we say that we "slice w.r.t. v".

In this rule, trace T_1 is traversed recursively to arrive at a state μ' that is then returned as the final product of the rule. Notice how trace T_b is *not* traversed. This is because boolean expressions (and arithmetic ones as well) do not have side effects on the state and so there is no need to traverse them.

The problem of traversing the trace recursively to handle side-effects to the state can be approached differently. Authors of [17] have formulated a single rule, which we could adapt to our setting like this:

$$\frac{\mathcal{L} = \mathsf{writes}(T)}{T :: \mu, \square \nearrow \mu \lhd \mathcal{L}}$$

In this rule $\mathsf{writes}(T)$ means "all state locations written to inside trace T" and $\mu \lhd \mathcal{L}$ means erasing (i.e. mapping to a $\square$) all locations in μ that are mentioned in $\mathcal{L}$. Semantically this is equivalent to our rules – both approaches achieve the same effect. However, we have found having separate rules easier to formalise in a proof assistant.

3.3 Backward Slicing

Backward slicing rules are given in Figs. 13, 14 and 15. These judgements should be read left-to-right, for example, $T :: \mu \searrow \mu', c$ should be read as "Given trace T and partial output state μ, backward slicing yields partial input μ' and partial command c." Their intent is to reconstruct the smallest program code and initial state that suffice to produce, after forward slicing, a result that is at least as large as the backward slicing criterion. To this end, backward slicing crucially relies on execution traces as part of input, since slicing effectively runs a program backwards (from final result to source code).

Figures 13 and 14 share a universal rule for backward slicing w.r.t. a hole.

$$\frac{}{T :: \mu, \square \searrow \varnothing_\mu, \square}$$

This rule means that to obtain an empty result it always suffices to have an empty state and no program code. This rule always applies preferentially over other rules, which means that whenever a value, such as v_a or v_b, appears as a backward slicing criterion we know it is not a $\square$. Similarly, Fig. 15 has a rule:

$$\frac{}{T :: \varnothing \searrow \varnothing_\varnothing, \square}$$

$$\frac{}{T :: \mu, \square \searrow \varnothing_\mu, \square} \qquad \frac{}{n :: \mu, v_a \searrow \varnothing_\mu, n}$$

$$\frac{}{x(v_a) :: \mu, v_a \searrow \varnothing_\mu[x \mapsto v_a], x}$$

$$\frac{T_1 :: \mu, v_1 \searrow \mu_1, a_1 \qquad T_2 :: \mu, v_2 \searrow \mu_2, a_2}{T_1 + T_2 :: \mu, v_a \searrow \mu_1 \sqcup \mu_2, a_1 + a_2}$$

Fig. 13. Backward slicing rules for Imp arithmetic expressions.

$$\frac{}{T :: \mu, \square \searrow \varnothing_\mu, \square}$$

$$\frac{}{\mathtt{true} :: \mu, \textbf{true} \searrow \varnothing_\mu, \mathtt{true}} \qquad \frac{}{\mathtt{false} :: \mu, \textbf{false} \searrow \varnothing_\mu, \mathtt{false}}$$

$$\frac{T_1 :: \mu, v_1 \searrow \mu_1, a_1 \qquad T_2 :: \mu, v_2 \searrow \mu_2, a_2}{T_1 = T_2 :: \mu, v_b \searrow \mu_1 \sqcup \mu_2, a_1 = a_2}$$

$$\frac{T :: \mu, v_b \searrow \mu', b}{\neg T :: \mu, v_b \searrow \mu', \neg b}$$

$$\frac{T_1 :: \mu, v_1 \searrow \mu_1, b_1 \qquad T_2 :: \mu, v_2 \searrow \mu_2, b_2}{T_1 \wedge T_2 :: \mu, v_b \searrow \mu_1 \sqcup \mu_2, b_1 \wedge b_2}$$

Fig. 14. Backward slicing rules for Imp boolean expressions.

It means that backward slicing w.r.t. a state with an empty domain (i.e. containing no variables) returns an empty partial state and an empty program. Of course having a program operating over a state with no variables would be completely useless – since a state cannot be extended with new variables during execution we wouldn't observe any effects of such a program. However, in the Coq formalisation, it is necessary to handle this case because otherwise Coq will not accept that the definition of backward slicing is a total function.

In the rule for backward slicing of variable reads (third rule in Fig. 13) it might seem that v_a stored inside a trace is redundant because we know what v_a is from the slicing criterion. This is a way of showing that variables can only be sliced w.r.t. values they have evaluated to during execution. So for example if x evaluated to 17 it is not valid to backward slice it w.r.t. 42.

The rule for backward slicing of addition in Fig. 13 might be a bit surprising. Each of the subexpressions is sliced w.r.t. a value that this expression has evaluated to (v_1, v_2), and not w.r.t. v_a. It might seem we are getting v_1 and v_2 out of thin air, since they are not directly recorded in a trace. Note however that knowing T_1 and T_2 allows to recover v_1 and v_2 at the expense of additional computations. In the actual implementation we perform induction on the structure of evaluation derivations, which record values of v_1 and v_2, thus allowing us to avoid extra computations. We show v_1 and v_2 in our rules but avoid showing the evaluation relation as part of slicing notation. This is elaborated further in Sect. 4.4.

Recursive backward slicing rules also rely crucially on the join ($\sqcup$) operation, which combines smaller slices from slicing subexpressions into one slice for the whole expression.

There are two separate rules for backward slicing of variable assignments (rules 3 and 4 in Fig. 15). If a variable is mapped to a $\square$ it means it is irrelevant. We therefore maintain mapping to a $\square$ and do not reconstruct variable assignment instructions. If a variable is relevant though, i.e. it is mapped to a concrete value in a slicing criterion, we reconstruct the assignment instruction together with an arithmetic expression in the RHS. We also join state μ_a required to

$$\frac{}{T :: \varnothing \searrow \varnothing_\varnothing, \square} \qquad \frac{}{\texttt{skip} :: \mu \searrow \mu, \square} \qquad \frac{}{x := T_a :: \mu[x \mapsto \square] \searrow \mu[x \mapsto \square], \square}$$

$$\frac{T_a :: v_a \searrow \mu_a, a}{x := T_a :: \mu[x \mapsto v_a] \searrow \mu_a \sqcup \mu[x \mapsto \square], x := a} \; v_a \neq \square$$

$$\frac{T_2 :: \mu \searrow \mu', \square \quad T_1 :: \mu' \searrow \mu'', \square}{T_1 \,;\, T_2 :: \mu \searrow \mu'', \square} \qquad \frac{T_2 :: \mu \searrow \mu', \square \quad T_1 :: \mu' \searrow \mu'', c_1}{T_1 \,;\, T_2 :: \mu \searrow \mu'', c_1 \,;\, \square} \; c_1 \neq \square$$

$$\frac{T_2 :: \mu \searrow \mu', c_2 \quad T_1 :: \mu' \searrow \mu'', c_1}{T_1 \,;\, T_2 :: \mu \searrow \mu'', c_1 \,;\, c_2} \; c_2 \neq \square \qquad \frac{T_1 :: \mu \searrow \mu', \square}{\texttt{if}_{\textsf{true}}\, T_b \;\texttt{then}\; \{\, T_1 \,\} :: \mu \searrow \mu', \square}$$

$$\frac{T_1 :: \mu \searrow \mu', c_1 \quad T_b :: \texttt{true} \searrow \mu_b, b}{\texttt{if}_{\textsf{true}}\, T_b \;\texttt{then}\; \{\, T_1 \,\} :: \mu \searrow \mu' \sqcup \mu_b, \texttt{if}\; b \;\texttt{then}\; \{\, c_1 \,\} \;\texttt{else}\; \{\, \square \,\}} \; c_1 \neq \square$$

$$\frac{T_2 :: \mu \searrow \mu', \square}{\texttt{if}_{\textsf{false}}\, T_b \;\texttt{else}\; \{\, T_2 \,\} :: \mu \searrow \mu', \square}$$

$$\frac{T_2 :: \mu \searrow \mu', c_2 \quad T_b :: \texttt{false} \searrow \mu_b, b}{\texttt{if}_{\textsf{false}}\, T_b \;\texttt{else}\; \{\, T_2 \,\} :: \mu \searrow \mu' \sqcup \mu_b, \texttt{if}\; b \;\texttt{then}\; \{\, \square \,\} \;\texttt{else}\; \{\, c_2 \,\}} \; c_2 \neq \square$$

$$\frac{}{\texttt{while}_{\textsf{false}}\, T_b :: \mu \searrow \mu, \square} \qquad \frac{T_w :: \mu \searrow \mu_w, \square \quad T_c :: \mu_w \searrow \mu_c, \square}{\texttt{while}_{\textsf{true}}\, T_b \;\texttt{do}\; \{\, T_c \,\};\, T_w :: \mu \searrow \mu_c, \square}$$

$$\frac{T_w :: \mu \searrow \mu_w, \square \quad T_c :: \mu_w \searrow \mu_c, c \quad T_b :: \texttt{true} \searrow \mu_b, b}{\texttt{while}_{\textsf{true}}\, T_b \;\texttt{do}\; \{\, T_c \,\};\, T_w :: \mu \searrow \mu_c \sqcup \mu_b, \texttt{while}\; b \;\texttt{do}\; \{\, c \,\}} \; c \neq \square$$

$$\frac{T_w :: \mu \searrow \mu_w, c_w \quad T_c :: \mu_w \searrow \mu_c, c \quad T_b :: \texttt{true} \searrow \mu_b, b}{\texttt{while}_{\textsf{true}}\, T_b \;\texttt{do}\; \{\, T_c \,\};\, T_w :: \mu \searrow \mu_c \sqcup \mu_b, c_w \sqcup \texttt{while}\; b \;\texttt{do}\; \{\, c \,\}} \; c_w \neq \square$$

Fig. 15. Backward slicing rules for Imp commands.

evaluate the RHS with $\mu[\texttt{x} \mapsto \square]$. It is crucial that we erase x in μ prior to joining. Firstly, if `x` is assigned, its value becomes irrelevant prior to the assignment, unless `x` is read during evaluation of the RHS (e.g. we are slicing an assignment `x := x + 1`). In this case `x` will be included in μ_a but its value can be different than the one in μ. It is thus necessary to erase `x` in μ to make a join operation possible.

At this point, it may be helpful to review the forward rules for assignment and compare with the backward rules, illustrated via a small example. Suppose we have an assignment `z := x + y`, initially evaluated on $[\texttt{w} \mapsto 0, \texttt{x} \mapsto 1, \texttt{y} \mapsto 2, \texttt{z} \mapsto 42]$, and yielding result state $[\texttt{w} \mapsto 0, \texttt{x} \mapsto 1, \texttt{y} \mapsto 2, \texttt{z} \mapsto 3]$. The induced lattice of minimal inputs and maximal outputs consists of the following pairs:

$$[\texttt{w} \mapsto v, \texttt{x} \mapsto 1, \texttt{y} \mapsto 2, \texttt{z} \mapsto \square] \longleftrightarrow [\texttt{w} \mapsto v, \texttt{x} \mapsto 1, \texttt{y} \mapsto 2, \texttt{z} \mapsto 3]$$
$$[\texttt{w} \mapsto v, \texttt{x} \mapsto 1, \texttt{y} \mapsto \square, \texttt{z} \mapsto \square] \longleftrightarrow [\texttt{w} \mapsto v, \texttt{x} \mapsto 1, \texttt{y} \mapsto \square, \texttt{z} \mapsto \square]$$
$$[\texttt{w} \mapsto v, \texttt{x} \mapsto \square, \texttt{y} \mapsto 2, \texttt{z} \mapsto \square] \longleftrightarrow [\texttt{w} \mapsto v, \texttt{x} \mapsto \square, \texttt{y} \mapsto 2, \texttt{z} \mapsto \square]$$
$$[\texttt{w} \mapsto v, \texttt{x} \mapsto \square, \texttt{y} \mapsto \square, \texttt{z} \mapsto \square] \longleftrightarrow [\texttt{w} \mapsto v, \texttt{x} \mapsto \square, \texttt{y} \mapsto \square, \texttt{z} \mapsto \square]$$

where $\mathtt{v} \in \{\square, 0\}$ so that each line above abbreviates two concrete relationships; the lattice has the shape of a cube. Because w is not read or written by `z := x + y`, it is preserved if present in the forward direction or if required in the backward direction. Because `z` is written but not read, its initial value is always irrelevant. To obtain the backward slice of any other partial output, such as $[\mathtt{w} \mapsto \square, \mathtt{x} \mapsto 1, \mathtt{y} \mapsto \square, \mathtt{z} \mapsto 3]$, find the smallest maximal partial output containing it, and take its backward slice, e.g. $[\mathtt{w} \mapsto \square, \mathtt{x} \mapsto 1, \mathtt{y} \mapsto 1, \mathtt{z} \mapsto \square]$.

In the backward slicing rules for `if` instructions, we only reconstruct a branch of the conditional that was actually taken during execution, leaving a second branch as a $\square$. Importantly in these rules state μ_b is a minimal state sufficient for an `if` condition to evaluate to a `true` or `false` value. That state is joined with state μ', which is a state sufficient to evaluate the reconstructed branch of an `if`.

Rules for `while` slicing follow a similar approach. It might seem that the second rule for slicing $\mathtt{while}_{\mathsf{true}}$ is redundant because it is a special case of the third $\mathtt{while}_{\mathsf{true}}$ rule if we allowed $c_w = \square$. Indeed, that is the case on paper. However, for the purpose of a mechanised formalisation we require that these two rules are separate. This shows that formalising systems designed on paper can indeed be tricky and require modifications tailored to solve mechanisation-specific issues.

Readers might have noticed that whenever a backward slicing rule from Fig. 15 returns $\square$ as an output program, the state returned by the rule will be identical to the input state. One could then argue that we should reflect this in our rules by explicitly denoting that input and output states are the same, e.g.

$$\frac{T_1 :: \mu \searrow \mu, \square}{\mathtt{if}_{\mathsf{true}}\ T_b\ \mathtt{then}\ \{\ T_1\ \} :: \mu \searrow \mu, \square}$$

While it is true that in such a case states will be equal, this approach would not be directly reflected in the implementation, where slicing is implemented as a function and a result is always assigned to a new variable. However, it would be possible to prove a lemma about equality of input and output states for $\square$ output programs, should we need this fact.

3.4 An Extended Example of Backward Slicing

We now turn to an extended example that combines all the programming constructs of Imp[5]: assignments, sequencing, conditionals, and loops. Figure 16 shows a program that divides integer `a` by `b`, and produces a quotient `q`, remainder `r`, and result `res` that is set to 1 if `b` divides `a` and to 0 otherwise.

To test whether 2 divides 4 we set $\mathtt{a} \mapsto 4$, $\mathtt{b} \mapsto 2$ in the input state. The remaining variables `q`, `r` and `res` are initialised to 0 (Fig. 16a). The `while` loop body is executed twice; the loop condition is evaluated three times. Once the loop has stopped, variable `q` is set to 2 and variable `r` to 0. Since the `if` condition is

[5] This example is adapted from [9].

$[\mathtt{q} \mapsto 0, \mathtt{r} \mapsto 0, \mathtt{res} \mapsto 0, \mathtt{a} \mapsto 4, \mathtt{b} \mapsto 2]$

```
r := a;
while ( b <= r ) do {
  q := q + 1;
  r := r - b
};
if ( ! (r = 0) )
then { res := 0 }
else { res := 1 }
```

$[\mathtt{q} \mapsto 2, \mathtt{r} \mapsto 0, \mathtt{res} \mapsto 1, \mathtt{a} \mapsto 4, \mathtt{b} \mapsto 2]$

(a) Original program.

$[\mathtt{q} \mapsto \square, \mathtt{r} \mapsto \square, \mathtt{res} \mapsto \square, \mathtt{a} \mapsto 4, \mathtt{b} \mapsto 2]$

```
r := a;
while ( b <= r ) do {
  □;
  r := r - b
};
if ( ! (r = 0) )
then { □ }
else { res := 1 }
```

$[\mathtt{q} \mapsto \square, \mathtt{r} \mapsto \square, \mathtt{res} \mapsto 1, \mathtt{a} \mapsto \square, \mathtt{b} \mapsto \square]$

(b) Backward slice w.r.t. `res` $\mapsto$ 1.

Fig. 16. Slicing a program that computes whether b divides a.

false we execute the `else` branch and set `res` to 1. Figure 17 shows the execution trace.

(1) $\mathtt{r} := \mathtt{a}(4);$
(2) $\mathtt{while}_{\mathsf{true}}\ (\mathtt{b}(2) <= \mathtt{r}(4))\ \mathtt{do}\ \{$
(3) $\quad \mathtt{q} := \mathtt{q}(0) + 1;\ \mathtt{r} := \mathtt{r}(4) - \mathtt{b}(2)$
(4) $\};$
(5) $\mathtt{while}_{\mathsf{true}}\ (\mathtt{b}(2) <= \mathtt{r}(2))\ \mathtt{do}\ \{$
(6) $\quad \mathtt{q} := \mathtt{q}(1) + 1;\ \mathtt{r} := \mathtt{r}(2) - \mathtt{b}(2)$
(7) $\};$
(8) $\mathtt{while}_{\mathsf{false}}\ (\mathtt{b}(2) <= \mathtt{r}(0));$
(9) $\mathtt{if}_{\mathsf{false}}\ (\neg(\mathtt{r}(0) = 0))\ \mathtt{else}\ \{$
(10) $\quad \mathtt{res} := 1$
(11) $\}$

Fig. 17. Trace of executing an example program for $\mathtt{a} \mapsto 4$ and $\mathtt{b} \mapsto 2$.

We now want to obtain an explanation of `res`. We form a slicing criterion by setting `res` $\mapsto$ 1 (this is the value at the end of execution); all other variables are set to $\square$.

We begin by reconstructing the `if` conditional. We apply the second rule for $\mathtt{if}_{\mathsf{false}}$ slicing (Fig. 16b). This is because c_2, i.e. the body of this branch, backward slices to an assignment `res := 1`, and not to a $\square$ (in which case the first rule for $\mathtt{if}_{\mathsf{false}}$ slicing would apply). Assignment in the `else` branch is reconstructed by applying the second rule for assignment slicing. Since the value assigned to `res` is a constant it does not require presence of any variables in the state. Therefore state μ_a is empty. Moreover, variable `res` is erased in state μ; joining of μ_a and $\mu[\mathtt{res} \mapsto \square]$ results in an empty state, which indicates that the code inside the `else` branch does not rely on the program state. However, to reconstruct the condition of the `if` we need a state μ_b that contains variable `r`. From the trace we read that $\mathtt{r} \mapsto 0$, and so after reconstructing the conditional we have a state where $\mathtt{r} \mapsto 0$ and all other variables, including `res`, map to $\square$.

We now apply the third rule for sequence slicing and proceed with reconstruction of the `while` loop. First we apply a trivial $\texttt{while}_{\textsf{false}}$ rule. The rule basically says that there is no need to reconstruct a `while` loop that does not execute – it might as well not be in a program. Since the final iteration of the `while` loop was reconstructed as a $\Box$, we reconstruct the second iteration using the second $\texttt{while}_{\textsf{true}}$ backward slicing rule, i.e. the one where we have $T_w :: \mu \searrow \mu_w, \Box$ as the first premise. We begin reconstruction of the body with the second assignment $\texttt{r} := \texttt{r}(2) - \texttt{b}(2)$. Recall that the current state assigns 0 to `r`. The RHS is reconstructed using the second rule for backward slicing of assignments we have already applied when reconstructing `else` branch of the conditional. An important difference here is that `r` appears both in the LHS and RHS. Reconstruction of RHS yields a state where $\texttt{r} \mapsto 2$ and $\texttt{b} \mapsto 2$ (both values read from a trace), whereas the current state contains $\texttt{r} \mapsto 0$. Here it is crucial that we erase `r` in the current state before joining. We apply third rule of sequence slicing and proceed to reconstruct the assignment to `q` using the first rule for assignment slicing (since $\texttt{q} \mapsto \Box$ in the slicing criterion). This reconstructs the assignment as a $\Box$. We then reconstruct the first iteration of the loop using the third $\texttt{while}_{\textsf{true}}$ slicing rule, since it is the case that $c_w \neq \Box$. Assignments inside the first iteration are reconstructed following the same logic as in the second iteration, yielding a state where $\texttt{r} \mapsto 4$, $\texttt{b} \mapsto 2$, and other variables map to $\Box$.

Finally, we reconstruct the initial assignment `r := a`. Since `r` is present in the slicing criterion, we yet again apply the second rule for assignment slicing, arriving at a partial input state $[\texttt{q} \mapsto 0, \texttt{r} \mapsto 0, \texttt{res} \mapsto 0, \texttt{a} \mapsto 4, \texttt{b} \mapsto 2]$ and a partial program shown in Fig. 16b.

4 Formalisation

In the previous sections we defined the syntax and semantics of the Imp language, and provided definitions of slicing in a Galois connection framework. We have implemented all these definitions in the Coq proof assistant [20] and proved their correctness as formal theorems. The following subsections outline the structure of our Coq development. We provide references to the source code by providing the name of file and theorem or definition as (`filename.v: theorem_name, definition_name`). We will use `*` in abbreviations like `*_monotone` to point to several functions ending with `_monotone` suffix. The whole formalisation is around 5.2k lines of Coq code (not counting the comments). Full code is available online [19].

4.1 Lattices and Galois Connections

Our formalisation is built around a core set of definitions and theorems about lattices and Galois connections. Most importantly we define:

– That a relation that is reflexive, antisymmetric and transitive is a partial order (`Lattice.v: order`). When we implement concrete definitions of ordering relations we require a proof that these implementations indeed have these three properties, e.g. (`ImpPartial.v: order_aexpPO`).

- What it means for a function $f : P \to Q$ to be monotone (Lattice.v: monotone):

$$\forall_{x,y}\ x \sqsubseteq_P y \implies f(x) \sqsubseteq_Q f(y)$$

- Consistency properties as given in Sect. 2.2 (Lattice.v: inflation, deflation).
- A Galois connection of two functions between lattices P and Q (see Definition 1 in Sect. 2.2) (Lattice.v: galoisConnection).

We then prove that:

- Existence of a Galois connection between two functions implies their consistency and minimality (Lattice.v: gc_implies_consistency, gc_implies_minimality).
- Two monotone functions with deflation and inflation properties form a Galois connection (Lattice.v: cons_mono__gc).

Throughout the formalisation we operate on elements inside lattices of partial expressions ($\downarrow a$, $\downarrow b$, commands ($\downarrow c$) or states ($\downarrow \mu$). We represent values in a lattice with an inductive data type prefix[6] (PrefixSets.v: prefix) indexed by the top element of the lattice and the ordering relation[7]. Values of prefix data type store an element from a lattice together with the evidence that it is in the ordering relation with the top element. Similarly we define an inductive data type prefixO (PrefixSets.v: prefixO) for representing ordering of two elements from the same lattice. This data type stores the said two elements together with proofs that one is smaller than another and that both are smaller than the top element of a lattice.

4.2 Imp Syntax and Semantics

All definitions given in Figs. 2, 3, 4 and 5 are directly implemented in our Coq formalisation.

Syntax trees for Imp (Fig. 2), traces (Fig. 6) and partial Imp (Fig. 3) are defined as ordinary inductive data types in (Imp.v: aexp, bexp, cmd), (Imp.v: aexpT, bexpT, cmdT) and (ImpPartial.v: aexpP, bexpP, cmdP), respectively. We also define functions to convert Imp expressions to partial Imp expressions by rewriting from normal syntax tree to a partial one (ImpPartial.v: aexpPartialize, bexpPartialize, cmdPartialize).

Evaluation relations for Imp (Figs. 7, 8 and 9) and ordering relations for partial Imp (Fig. 4) are defined as inductive data types with constructors indexed by elements in the relation (Imp.v: aevalR, bevalR, cevalR and ImpPartial.v:

---

[6] Name comes from a term "prefix set" introduced in [17] to refer to a set of all partial values smaller than a given value. So a prefix set of a top element of a lattice denotes a set of all elements in that (complete) lattice.

[7] In order to make the code snippets in the paper easier to read we omit the ordering relation when indexing prefix.

aexpPO, bexpPO, comPO), respectively. For each ordering relation we construct a proof of its reflexivity, transitivity, and antisymmetry, which together proves that a given relation is a partial order (ImpPartial.v: order_aexpPO, order_bexpPO, order_comPO).

Join operations (Fig. 5) are implemented as functions (ImpPartial.v: aexpLUB, bexpLUB, comLUB). Their implementation is particularly tricky. Coq requires that all functions are total. We know that for two elements from the same lattice a join always exists, and so a join function is a total one. However, we must guarantee that a join function only takes as arguments elements from the same lattice. To this end a function takes three arguments: top element e of a lattice and two prefix values e_1, e_2 indexed by the top element e. So for example if e is a variable name x, we know that each of e_1 and e_2 is either also a variable name x or a $\square$. However, Coq does not have a built-in dependent pattern match and this leads to complications. In our example above, even if we know that e is a variable name x we still have to consider cases for e_1 and e_2 being a constant or an arithmetic operator. These cases are of course impossible, but it is the programmer's responsibility to dismiss them explicitly. This causes further complications when we prove lemmas about properties of join, e.g.:

$$e_1 \sqsubseteq e \wedge e_2 \sqsubseteq e \implies (e_1 \sqcup e_2) \sqsubseteq e$$

This proof is done by induction on the top element of a lattice, where e, e_1, and e_2 are all smaller than that element. The top element limits the possible values of e, e_1, and e_2 but we still have to consider the impossible cases and dismiss them explicitly.

4.3 Program State

Imp programs operate by side-effecting on a program state. Handling of the state was one of the most tedious parts of the formalisation.

State is defined as a data type isomorphic to an association list that maps variables to natural number values (ImpState.v: state). Partial state is defined in the same way, except that it permits partial values, i.e. variables can be mapped to a hole or a numeric value (ImpState.v: stateP). We assume that no key appears more than once in a partial state. This is not enforced in the definition of stateP itself, but rather defined as a separate inductive predicate (ImpState.v: statePWellFormed) that is explicitly passed as an assumption to any theorem that needs it. We also have a statePartialize function that turns a state into a partial state. This only changes representation from one data type to another, with no change in the state contents.

For partial states we define an ordering relation as a component-wise ordering of partial values inside a state (ImpState.v: statePO). This assumes that domains of states in ordering relation are identical (same elements in the same order), which allows us to formulate lemmas such as:

$$[\,] \leq \mu \implies \mu = [\,]$$

This lemma says that if a partial state μ is larger than an state with empty domain then μ itself must have an empty domain.

We also define a join operation on partial states, which operates element-wise on two partial states from the same lattice (`ImpState.v: stateLUB`).

As already mentioned in Sect. 3.1, the domain of the state is fixed throughout the execution. This means that updating a variable that does not exist in a state is a no-op, i.e. it returns the original state without any modifications. This behaviour is required to allow comparison of states before and after an update. Consider this lemma:

$$(\mu_1 \leq \mu_2) \wedge (v_1 \leq \mu_2(k)) \implies \mu_1[k \mapsto v_1] \leq \mu_2$$

It says that if a partial state μ_1 is smaller than μ_2 and the value stored in state μ_2 under key k is greater than v_1 then we can assign v_1 to key k in μ_1 and the ordering between states will be maintained. A corner-case for this theorem is when the key k is absent from the states μ_1 and μ_2. Looking up a non-existing key in a partial state returns a $\Box$. If k did not exist in μ_2 (and thus μ_1 as well) then $\mu_2(k)$ would return $\Box$ and so v_1 could only be a $\Box$ (per second assumption of the theorem). However, if we defined semantics of update to insert a non-existing key into the state, rather than be a no-op, the conclusion of the theorem would not hold because domain of $\mu_1[k \mapsto v_1]$ would contain k and domain of μ_2 would not, thus making it impossible to define the ordering between the two states.

The approach described above is one of several possible design choices. One alternative approach would be to require evidence that the key being updated exists in the state, making it impossible to attempt update of non-existent keys. We have experimented with this approach but found explicit handling of evidence that a key is present in a state very tedious and seriously complicating many of the proofs. In the end we decided for the approach outlined above, as it allowed us to prove all the required lemmas, with only some of them relying on an explicit assumption that a key is present in a state. An example of such a lemma is:

$$\mu[k \mapsto v](k) = v$$

which says that if we update key k in a partial state μ with value v and then immediately lookup k in the updated state we will get back the v value we just wrote to μ. However, this statement only holds if k is present in μ. If it was absent the update would return μ without any changes and then lookup would return $\Box$, which makes the theorem statement false. Thus this theorem requires passing explicit evidence that $k \in \mathsf{dom}(\mu)$ in order for the conclusion to hold.

Formalising program state was a tedious task, requiring us to prove over sixty lemmas about the properties of operations on state, totalling over 800 lines of code.

4.4 Slicing Functions

Slicing functions implement rules given in Figs. 10, 11, 12, 13, 14 and 15. We have three separate forward slicing functions, one for arithmetic expressions, one for logical expressions and one for imperative commands (`ImpSlicing.v`: `aexpFwd, bexpFwd, comFwd`). Similarly for backward slicing (`ImpSlicing.v`: `aexpBwd, bexpBwd, comBwd`).

In Sect. 2.2 we said that forward and backward slicing functions operate between two lattices. If we have an input p (an arithmetic or logical expression or a command) with initial state μ that evaluates to output p' (a number, a boolean, a state) and records a trace T then $\mathsf{fwd}^T_{(p,\mu)}$ is a forward slicing function parametrized by T that takes values from lattice generated by (p, μ) to values in lattice generated by p'. Similarly $\mathsf{bwd}^T_{p'}$ is a backward slicing function parametrized by T that takes values from lattice generated by p' to values in lattice generated by (p, μ). Therefore our implementation of forward and backward slicing functions has to enforce that:

1. (p, μ) evaluates to p' and records trace T
2. Forward slicing function takes values from lattice generated by (p, μ) to lattice generated by p'
3. Fackward slicing function takes values from lattice generated by p' to lattice generated by (p, μ)

To enforce the first condition we require that each slicing function is parametrized by inductive evidence that a given input (p, μ) evaluates to p' and records trace T. We then define input and output types of such slicing functions as belonging to relevant lattices, which is achieved using the `prefix` data type described in Sect. 4.1. This enforces the conditions above. For example, the type signature of the forward slicing function for arithmetic expressions looks like this:

```
Fixpoint aexpFwd {st : state} {a : aexp}
   {v : nat} {t : aexpT}
   (ev : t :: a, st \\ v):
   (prefix a * prefix st) -> prefix v.
```

Here `ev` is evidence that arithmetic expression `a` with input state `st` evaluates to a natural number `v` and records an execution trace `t`. The `t :: a, st \\ v` syntax is a notation for the evaluation relation. The first four arguments to `aexpFwd` are in curly braces, denoting they are implicit and can be inferred from the type of `ev`. The function then takes values from the lattice generated by `(a, st)` and returns values in the lattice generated by `v`.

In the body of a slicing function we first decompose the evaluation evidence with pattern matching. In each branch we implement logic corresponding to relevant slicing rules defined in Figs. 10, 11, 12, 13, 14 and 15. Premises appearing in the rules are turned into recursive calls. If necessary, results of these calls are analysed to decide which rule should apply. For example, when backward slicing sequences we analyse whether the recursive calls return holes or expressions to decide which of the rules should apply.

The implementation of the slicing functions faces similar problems as the implementation of joins described in Sect. 4.2. When we pattern match on the evaluation evidence, in each branch we are restricted to concrete values of the expression being evaluated. For example, if the last step in the evaluation was an addition, then we know the slicing criterion is a partial expression from a lattice formed by expression $a_1 + a_2$. Yet we have to consider the impossible cases, e.g. having an expression that is a constant, and dismiss them explicitly. Moreover, operating inside a lattice requires us not to simply return a result, but also provide a proof that this result is inside the required lattice. We rely on Coq's `refine` tactic to construct the required proof terms. All of this makes the definitions of slicing functions very verbose. For example, forward slicing of arithmetic expressions requires over 80 lines of code with over 60 lines of additional boilerplate lemmas to dismiss the impossible cases.

For each slicing function we state and prove a theorem that it is monotone (`ImpSlicing.v: *_monotone`). For each pair of forward and backward slicing functions we state theorems that these pairs of functions have deflation and inflation properties (`ImpSlicing.v: *_deflating, *_inflating`), as defined in Sect. 2.2. Once these theorems are proven we create instances of a general theorem `cons_mono__gc`, described in Sect. 4.1, which proves that our definitions form a Galois connection and are thus a correctly defined pair of slicing functions. We also create instances of the `gc_implies_minimality` theorem, one instance for each slicing function. This provides us with a formalisation of all the correctness properties, proving our main result:

Theorem 1. *Suppose* $\mu_1, c \Rightarrow T :: \mu_2$. *Then there exist total, monotone functions* $\mathsf{fwd}^T_{(c,\mu_1)} : \downarrow c \times \downarrow \mu_1 \rightarrow \downarrow \mu_2$ *and* $\mathsf{bwd}^T_{\mu_2} : \downarrow \mu_2 \rightarrow \downarrow c \times \downarrow \mu_1$. *Moreover,* $\mathsf{bwd}^T_{\mu_2} \dashv \mathsf{fwd}^T_{(c,\mu_1)}$ *form a Galois connection and in particular satisfy the minimality, inflation (consistency), and deflation properties.*

Here, the forward and backward slicing judgements are implemented as functions $\mathsf{fwd}^T_{(c,\mu_1)}$ and $\mathsf{bwd}^T_{\mu_2}$.

5 Related and Future Work

During the past few decades a plethora of slicing algorithms has been presented in the literature. See [18] for a good, although now slightly out of date, survey. Most of these algorithms have been analysed in a formal setting of some sort using pen and paper. However, work on formalising slicing in a machine checked way has been scarce. One example of such a development is [14], which formalises dynamic slicing for π-calculus in Agda using a Galois connection framework identical to the one used in this paper. The high-level outline of the formalisation is thus similar to ours. However, details differ substantially, since [14] formalises a completely different slicing algorithm for concurrent processes using a different proof assistant. Another example of formalising slicing in a proof assistant is [3], where Coq is used to perform an *a posteriori* validation of a slice obtained using an unverified program slicer. This differs from our approach of verifying correctness of a slicing algorithm itself. We see our approach of verifying correctness of the whole algorithm as a significant improvement over the validation approach. In a more recent work Léchenet et al. [9] introduce a variant of static slicing known as relaxed slicing and use Coq to formalise the slicing algorithm. Their work is identical in spirit to ours and focuses on the Imp language[8] with an extra `assert` statement.

Galois connections have been investigated previously as a tool in the mathematics of program construction, for example by Backhouse [1] and more recently by Mu and Oliveira [12]. As discussed in Sect. 1, Galois connections capture a common pattern in which one first specifies a space of possible solutions to a problem, the "easy" part, via one adjoint, and defines the mapping from problem instances to optimal solutions, the "hard" part, as the Galois dual. In the case of slicing, we have used the goal of obtaining a verifiable Galois connection, along with intuition, to motivate choices in the design of the forward semantics, and it has turned out to be easier for our correctness proof to define both directions directly.

Mechanised proofs of correctness of calculational reasoning has been considered in the Algebra of Programming in Agda (AOPA) system [11], and subsequently extended to include derivation of greedy algorithms using Galois connections [5]. Another interesting, complementary approach to program comprehension is Gibbons' *program fission* [7], in which the fusion law is applied "in reverse" to an optimized, existing program in order to attempt to discover a rationale for its behavior: for example by decomposing an optimized word-counting program into a "reforested" version that decomposes its behavior into "construct a list of all the words" and "take the length of the list". We conjecture that the traces that seem to arise as a natural intermediate structure in program slicing might be viewed as an extreme example of fission.

An important line of work on slicing theory focuses on formalising different slicing algorithms within a unified theoretical framework of *program projection* [2]. Authors of that approach develop a precise definition of what it means that one form of slicing is weaker than another. However, our dynamic slicing algorithm does not fit the framework as presented in [2]. We believe that it should be possible to extend the program projection framework so that it can encompass slicing based on Galois connections but this is left as future work.

6 Summary

Program slicing is an important tool for aiding software development. It is useful when creating new programs as well as maintaining existing ones. In this paper we have developed and formalised an algorithm for dynamic slicing of imperative programs. Our work extends the line of research on slicing based on the Galois connection framework. In the presented approach slicing consists of two components: forward slicing, that allows to execute partial programs, and backward slicing, that allows to "rewind" program execution to explain the output.

Studying slicing in a formal setting ensures the reliability of this technique. We have formalised all of the theory presented in this paper using the Coq proof

[8] Authors of [9] use the name WHILE, but the language is the same.

assistant. Most importantly, we have shown that our slicing algorithms form a Galois connection, and thus have the crucial properties of consistency and minimality. One interesting challenge in our mechanisation of the proofs was the need to modify some of the theoretical developments so that they are easier to formalise in a proof assistant – c.f. overlapping rules for backward slicing of while loops described in Sect. 3.3.

Our focus in this paper was on a simple programming language Imp. This work should be seen as a stepping stone towards more complicated formalisations of languages with features like (higher-order) functions, arrays, and pointers. Though previous work [17] has investigated slicing based on Galois connections for functional programs with imperative features, our experience formalising slicing for the much simpler Imp language suggests that formalising a full-scale language would be a considerable effort. We leave this as future work.

Acknowledgements. We gratefully acknowledge help received from Wilmer Ricciotti during our work on the Coq formalisation, and Jeremy Gibbons for comments on a draft. This work was supported by ERC Consolidator Grant Skye (grant number 682315).

References

1. Backhouse, R.: Galois connections and fixed point calculus. In: Backhouse, R., Crole, R., Gibbons, J. (eds.) Algebraic and Coalgebraic Methods in the Mathematics of Program Construction. LNCS, vol. 2297, pp. 89–150. Springer, Heidelberg (2002). https://doi.org/10.1007/3-540-47797-7_4
2. Binkley, D., Danicic, S., Gyimóthy, T., Harman, M., Kiss, A., Korel, B.: A formalisation of the relationship between forms of program slicing. Sci. Comput. Program. **62**(3), 228–252 (2006). Special issue on Source code analysis and manipulation (SCAM 2005)
3. Blazy, S., Maroneze, A., Pichardie, D.: Verified validation of program slicing. In: Proceedings of the 2015 Conference on Certified Programs and Proofs, CPP 2015, pp. 109–117. ACM, New York (2015)
4. Cheney, J., Acar, U.A., Perera, R.: Toward a theory of self-explaining computation. In: Tannen, V., Wong, L., Libkin, L., Fan, W., Tan, W.-C., Fourman, M. (eds.) In Search of Elegance in the Theory and Practice of Computation. LNCS, vol. 8000, pp. 193–216. Springer, Heidelberg (2013). https://doi.org/10.1007/978-3-642-41660-6_9
5. Chiang, Y., Mu, S.: Formal derivation of greedy algorithms from relational specifications: a tutorial. J. Log. Algebr. Meth. Program. **85**(5), 879–905 (2016). https://doi.org/10.1016/j.jlamp.2015.12.003
6. Davey, B.A., Priestley, H.A.: Introduction to Lattices and Order. Cambridge University Press, Cambridge (2002)
7. Gibbons, J.: Fission for program comprehension. In: Proceedings of Mathematics of Program Construction, 8th International Conference, MPC 2006, Kuressaare, Estonia, 3–5 July 2006, pp. 162–179 (2006). https://doi.org/10.1007/11783596_12
8. Korel, B., Laski, J.: Dynamic program slicing. Inf. Process. Lett. **29**(3), 155–163 (1988)

9. Léchenet, J., Kosmatov, N., Gall, P.L.: Cut branches before looking for bugs: certifiably sound verification on relaxed slices. Formal Aspects Comput. **30**(1), 107–131 (2018)
10. Leroy, X., Blazy, S., Kästner, D., Schommer, B., Pister, M., Ferdinand, C.: Compcert - a formally verified optimizing compiler. In: ERTS 2016: Embedded Real Time Software and Systems. SEE (2016)
11. Mu, S., Ko, H., Jansson, P.: Algebra of programming in Agda: dependent types for relational program derivation. J. Funct. Program. **19**(5), 545–579 (2009). https://doi.org/10.1017/S0956796809007345
12. Mu, S., Oliveira, J.N.: Programming from Galois connections. J. Log. Algebr. Program. **81**(6), 680–704 (2012). https://doi.org/10.1016/j.jlap.2012.05.003
13. Nielson, F., Nielson, H.R., Hankin, C.: Principles of Program Analysis. Springer, Heidelberg (1999). https://doi.org/10.1007/978-3-662-03811-6
14. Perera, R., Garg, D., Cheney, J.: Causally consistent dynamic slicing. In: CONCUR, pp. 18:1–18:15 (2016)
15. Perera, R., Acar, U.A., Cheney, J., Levy, P.B.: Functional programs that explain their work. In: ICFP, pp. 365–376. ACM (2012)
16. Pierce, B.C., et al.: Software Foundations. Electronic textbook (2017), version 5.0. http://www.cis.upenn.edu/~bcpierce/sf
17. Ricciotti, W., Stolarek, J., Perera, R., Cheney, J.: Imperative functional programs that explain their work. In: Proceedings of the ACM on Programming Languages (PACMPL) 1(ICFP), September 2017
18. Silva, J.: An analysis of the current program slicing and algorithmic debugging based techniques. Technical University of Valencia, Tech. rep. (2008)
19. Stolarek, J.: Verified self-explaining computation, May 2019. https://bitbucket.org/jstolarek/gc_imp_slicing/src/mpc_2019_submission/
20. The Coq Development Team: The Coq proof assistant, version 8.7.0, October 2017. https://doi.org/10.5281/zenodo.1028037
21. Weiser, M.: Program slicing. In: ICSE, pp. 439–449. IEEE Press, Piscataway (1981)
22. Winskel, G.: The Formal Semantics of Programming Languages: An Introduction. MIT Press, Cambridge (1993)

Self-certifying Railroad Diagrams

Or: How to Teach Nondeterministic Finite Automata

Ralf Hinze(✉)

Technische Universität Kaiserslautern, 67653 Kaiserslautern, Germany
ralf-hinze@cs.uni-kl.de

Abstract. Regular expressions can be visualized using railroad or syntax diagrams. The construction does not depend on fancy artistic skills. Rather, a diagram can be systematically constructed through simple, local transformations due to Manna. We argue that the result can be seen as a nondeterministic finite automaton with ϵ-transitions. Despite its simplicity, the construction has a number of pleasing characteristics: the number of states and the number of edges is linear in the size of the regular expression; due to sharing of sub-automata and auto-merging of states the resulting automaton is often surprisingly small. The proof of correctness relies on the notion of a subfactor. In fact, Antimirov's subfactors (partial derivatives) appear as target states of non-ϵ-transitions, suggesting a smooth path to nondeterministic finite automata without ϵ-transitions. Antimirov's subfactors, in turn, provide a fine-grained analysis of Brzozowski's factors (derivatives), suggesting a smooth path to deterministic finite automata. We believe that this makes a good story line for introducing regular expressions and automata.

1 Introduction

Everything should be as simple as it can be, but not simpler.

Albert Einstein

Regular expressions and finite automata have a long and interesting history, see Fig. 1. Kleene introduced regular expressions in the 1950's to represent events in nerve nets [10]. In the early 60's, Brzozowski [6] presented an elegant method for translating a regular expression to a deterministic finite automaton (DFA). His algorithm is based on two general properties of languages, sets of words. A language can be partitioned into a set of non-empty words and the remainder (ϵ is the set containing the empty word, $+$ is set union).

$$L = (L - \epsilon) + (L \cap \epsilon)$$

If L is given by a regular expression, the remainder $L \cap \epsilon$ can be readily computed (is L "nullable"?). A non-nullable language can be further factored:

$$L - \epsilon = a_1 \cdot (a_1 \backslash L) + \cdots + a_n \cdot (a_n \backslash L) \qquad (1)$$

G. Hutton (Ed.): MPC 2019, LNCS 11825, pp. 103–137, 2019.
https://doi.org/10.1007/978-3-030-33636-3_5

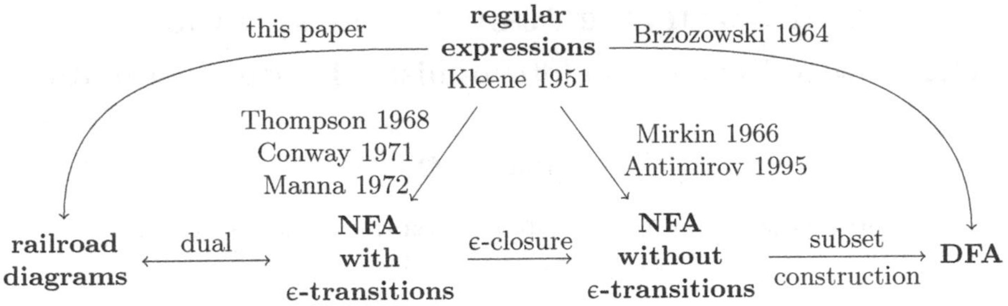

Fig. 1. A brief history of regular expressions and finite automata.

where $a_1, \ldots, a_n$ are *distinct* symbols of the alphabet and $a_1\backslash L, \ldots, a_n\backslash L$ are so-called *right factors* of L, see Sect. 3. The right factor of a regular language is again a regular language; iterating the two steps above yields a DFA. However, some care has to be exercised when implementing the algorithm: while there is only a finite number of *semantically* different right factors, they have infinitely many syntactic representations.

Thirty years later, Antimirov [3] ameliorated the problem, providing a fine-grained analysis of Brzozowski's approach. The central idea is to loosen the requirement that the symbols in (1) are distinct. In his linear form

$$L - \epsilon = a_1 \cdot L_1 + \cdots + a_n \cdot L_n \tag{2}$$

the $a_1, \ldots, a_n$ are arbitrary symbols of the alphabet, *not necessarily distinct.* Each of the languages L_i is a *right subfactor* of the left-hand side: $L - \epsilon \supseteq a_i \cdot L_i$. Because the requirement of distinctness is dropped, Antimirov's approach yields a nondeterministic finite automaton without ϵ-transitions. (Brzozowski's automaton can be recovered via the subset construction, basically grouping the subfactors by symbol.) Antimirov's representation of subfactors ensures that there is only a finite number of *syntactically* different right subfactors. Unfortunately, the argument is still somewhat involved.

An alternative advocated in this paper is to approach the problem from the other side, see Fig. 1. Regular expressions can be visualized using railroad or syntax diagrams. (I have first encountered syntax diagrams in Jensen and Wirth's "Pascal: User Manual and Report" [9].) Adapting a construction due to Manna [12], we show that a diagram can be systematically constructed through simple, local transformations. We argue that the result can be seen as a nondeterministic finite automaton with ϵ-transitions. Moreover, the automaton contains all Antimirov subfactors as target states of non-ϵ-transitions.

The remainder of the paper is organized in two strands, which are only loosely coupled:

– Sections 2, 4, and 6 introduce railroad diagrams and prove Manna's construction correct. The material is targeted at first-year students; an attempt is made to explain the construction in basic terms without, however, compromising on precision and concision. Central to the undertaking is the consistent use of inequalities and reasoning using Galois connections.

- Sections 3 and 5 highlight the theoretical background, which is based on regular algebra. The "student material" is revised providing shorter, but slightly more advanced accounts. Section 7 links railroad diagrams to Antimirov's subfactors.

The paper makes the following contributions:

- We suggest a story line for introducing regular expressions and automata, using a visual representation of regular expressions as the starting point.
- We show that diagrams form a regular algebra and we identify a sub-algebra, self-certifying diagrams, which supports simple correctness proofs.
- We show that Manna's construction can be implemented in eight lines of Haskell and highlight some of its salient features: sharing of sub-automata and auto-merging of states.
- We prove that a Manna automaton contains the subfactors of an Antimirov automaton as target states of non-ϵ-transitions.

2 Railroad Diagrams

Regular expressions can be nicely visualized using so-called railroad or syntax diagrams. Consider the regular expression $(a \mid b)^* \cdot a \cdot (a \mid b)$ over the alphabet $\Sigma = \{a, b\}$, which captures the language of all words whose penultimate symbol is an a. Its railroad diagram is shown below.

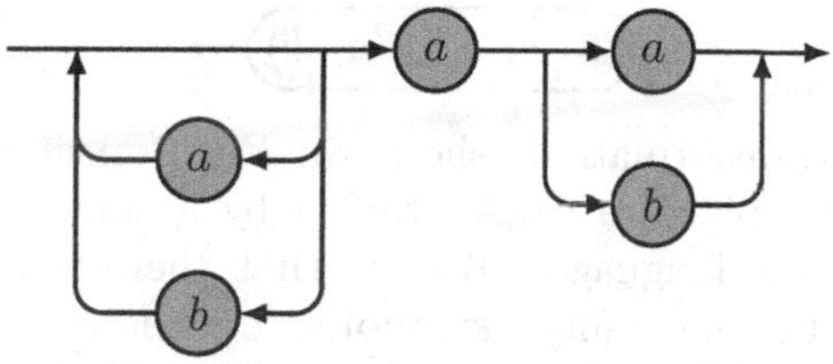

The diagram has one entry on the left, the starting point, and one exit on the right, the rail destination. Each train journey from the starting point to the destination generates a word of the language, obtained by concatenating the symbols encountered on the trip.

In the example above, the "stations" contain symbols of the alphabet. We also admit arbitrary languages (or representations of languages) as labels, so that we can visualize a regular expression or, more generally, a language at an appropriate level of detail. In the diagram below, which captures the lexical syntax of identifiers in some programming language, we have chosen not to expand the languages of letters and digits.

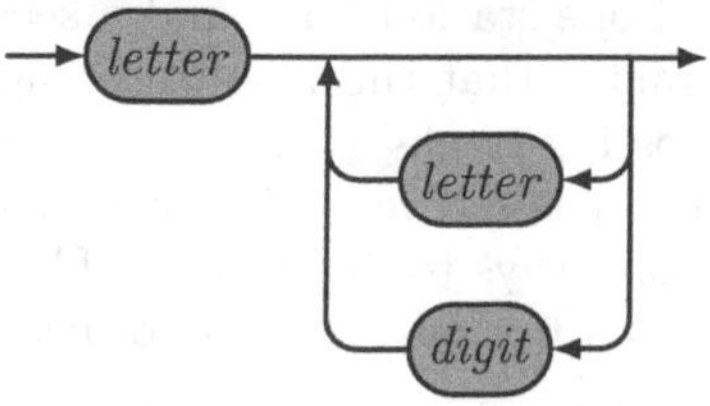

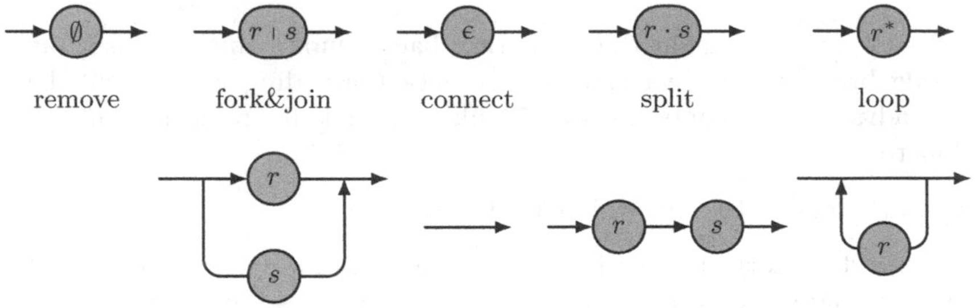

Fig. 2. Manna's rules for constructing a railroad diagram (the diagrams in the first row are translated into the corresponding diagrams in the second row).

In this more general setting, a single train journey visiting the stations

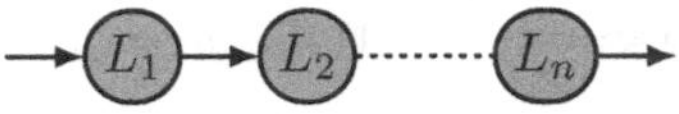

generates the language $L_1 \cdot L_2 \cdot \ldots \cdot L_n$.

The generalization to languages has the added benefit that we can construct a railroad diagram for a given regular expression in a piecemeal fashion using simple *local* transformations. The point of departure is a diagram that consists of a single station, containing the given expression. To illustrate, for our running example we have:

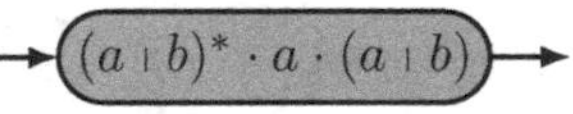

Then we apply the transformations shown in Fig. 2 until we are happy with the result. Each rule replaces a single station by a smallish railroad diagram that generates the same language. Observe that there is no rule for a single symbol—stations containing a single symbol cannot be refined. We refer to the transformation rules as *Manna's construction.* The diagrams below show two intermediate stages of the construction for our running example.

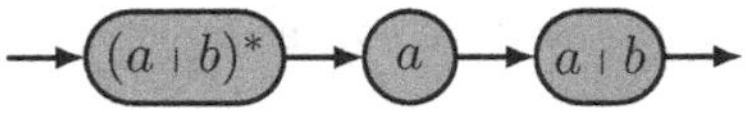

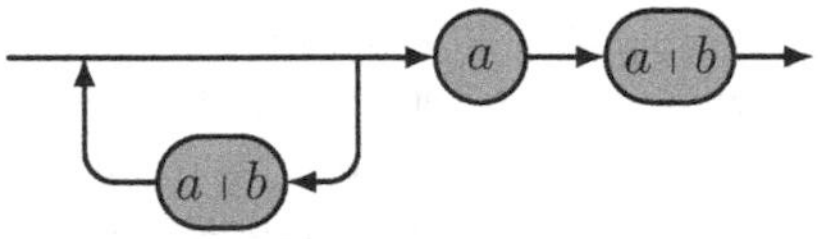

The transformations emphasize the close correspondence between regular expressions and railroad diagrams: composition is visualized by chaining two tracks; choice by forking and joining a track; iteration by constructing a loop. Each rule in Fig. 2 is compositional, for example, a station for $r \mid s$ is transformed into a diagram composed of one station for r and a second station for s. Quite attractively, the rules guarantee that the resulting diagram is planar so it can be easily drawn without crossing tracks.

When drawing diagrams by hand we typically exercise some artistic licence. For example, there are various ways to draw loops. The diagram for r^* in Fig. 2 has the undesirable feature that the sub-diagram for r is drawn from right to left.

Alternative drawings that avoid this unfortunate change of direction include:

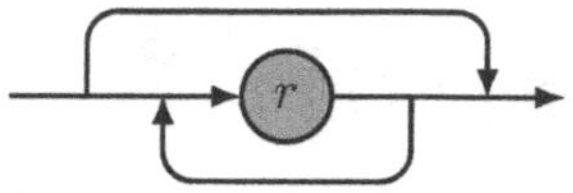

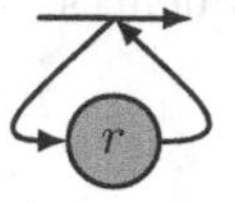

Railroad diagrams are visually appealing, but there is more to them. If we apply the transformations of Fig. 2 until all stations either contain ϵ or a single symbol, then we obtain an NFA, a nondeterministic finite automaton with ϵ-transitions. Before we make this observation precise in Sect. 4, we first introduce some background.

Exercise 1. Devise diagrams for non-zero repetitions $r^+ = r \cdot r^*$ and for optional occurrences $\epsilon \mid r$.

3 Interlude: Regular Algebras and Regular Expressions

This section details the syntax and semantics of regular expressions. It is heavily based on Backhouse's axiomatization of regular algebras [4], which, while equivalent to Conway's "Standard Kleene Algebra" or **S**-algebra [8], stresses the importance of factors.[1] I suggest that you skip the section on first reading, except perhaps for notation.

Regular Algebras. A regular algebra $(R, \leqslant, \sum, 1, \cdot)$ is a blend of a complete lattice and a monoid, interfaced by two Galois connections:

1. $(R, \leqslant)$ is a complete lattice with join $\sum$,
2. $(R, 1, \cdot)$ is a monoid,
3. for all $p \in R$, the partially applied functions $(p\,\cdot\,)$ and $(\,\cdot\, p)$ are both left adjoints in Galois connections between $(R, \leqslant)$ and itself.

We use $0 = \sum \emptyset$ and $a_1 + a_2 = \sum \{a_1, a_2\}$ as short cuts for the least element (nullary join) and binary joins. The right adjoints of $(p\,\cdot\,)$ and $(\,\cdot\, p)$ are written $(p \setminus\,)$ and $(\,/\, p)$.

Calculationally speaking, it is most useful to capture the first and the third requirement as equivalences. Join or least upper bound is defined by the following equivalence, which incidentally also establishes a Galois connection.

$$\sum A \leqslant b \iff \forall a \in A \,.\, a \leqslant b \tag{3a}$$

Instantiated to nullary and binary joins, (3a) specializes to $0 \leqslant b \iff$ *true* (that is, 0 is indeed the least element) and $a_1 + a_2 \leqslant b \iff a_1 \leqslant b \wedge a_2 \leqslant b$.

[1] Actually, Conway's axiomatization is incomplete as pointed out by Abramsky and Vickers [1]: only if the axiom $\sum\{a\} = a$ is added, his **S**-algebras are equivalent to regular algebras.

The interface between the lattice and the monoid structure is given by the following two equivalences.

$$p \cdot a \leqslant b \iff a \leqslant p \backslash b \tag{3b}$$

$$a \cdot p \leqslant b \iff a \leqslant b / p \tag{3c}$$

The element $p \backslash b$ is called a *right factor of* b (or left quotient or left derivative).

The axioms have a wealth of consequences. Adjoint functions are order-preserving; left adjoints preserve joins; right adjoints preserve meets. Consequently, the axioms imply:

$$\sum A \cdot b = \sum \{ a \cdot b \mid a \in A \} \tag{4a}$$

$$a \cdot \sum B = \sum \{ a \cdot b \mid b \in B \} \tag{4b}$$

Instantiated to nullary and binary joins, property (4a) specializes to $0 \cdot b = 0$ and $(a_1 + a_2) \cdot b = (a_1 \cdot b) + (a_2 \cdot b)$. So, composition distributes over choice.

Truth values ordered by implication form a regular algebra: $(\mathbb{B}, \Rightarrow, \exists, \mathit{true}, \wedge)$, where $\exists B = (\mathit{true} \in B)$. Note that the right adjoint of $(p \wedge\ )$ is $(p \Rightarrow\ )$.

Languages, sets of words over some alphabet Σ, are probably the most prominent example of a regular algebra. The example is actually an instance of a more general construction. Let $(R, 1, \cdot)$ be some monoid. We can lift the monoid to a monoid on sets, setting $1 = \{1\}$ and $A \cdot B = \{ a \cdot b \mid a \in A, b \in B \}$. (Note that here and elsewhere we happily overload symbols: in $1 = \{1\}$, the first occurrence of 1 denotes the unit of the lifted monoid, whereas the second occurrence is the unit of the underlying monoid.) It remains to show that $(P \cdot\ )$ and $(\ \cdot P)$ are left adjoints. We show the former; the calculation for the latter proceeds completely analogously.

$$\begin{aligned}
& P \cdot A \subseteq B \\
\iff & \quad \{ \text{ definition of composition } \} \\
& \{ p \cdot a \mid p \in P, a \in A \} \subseteq B \\
\iff & \quad \{ \text{ set inclusion } \} \\
& \forall p \in R, a \in R \,.\, p \in P \wedge a \in A \Longrightarrow p \cdot a \in B \\
\iff & \quad \{ \ (x \wedge\ ) \text{ is a left adjoint, } (x \Rightarrow\ ) \text{ preserves universal quantification } \} \\
& \forall a \in R \,.\, a \in A \Longrightarrow (\forall p \in R \,.\, p \in P \Longrightarrow p \cdot a \in B) \\
\iff & \quad \{ \text{ set inclusion } \} \\
& A \subseteq \{ a \in R \mid \forall p \in R \,.\, p \in P \Longrightarrow p \cdot a \in B \}
\end{aligned}$$

The right adjoint $P \backslash B$ is given by the formula on the right. Consequently, $(\mathcal{P}(R), \subseteq, \bigcup, 1, \cdot)$ is a regular algebra. In the case of languages, the underlying monoid is the free monoid $(\Sigma^*, \epsilon, \cdot)$, where ϵ is the empty word and composition is concatenation of words.

Occasionally, it is useful to make stronger assumptions, for example, to require that the underlying lattice is Boolean. In this case, $(\ + p)$ has a left adjoint, which we write $(\ - p)$.

Iteration. The axioms of regular algebra explicitly introduce choice and composition. Iteration is a derived concept: a^* is defined as the least solution of the inequality $1 + a \cdot x \leqslant x$ in the unknown x (which is guaranteed to exist because of completeness).

$$1 + a \cdot a^* \leqslant a^* \tag{5a}$$

$$\forall x \,.\, a^* \leqslant x \iff 1 + a \cdot x \leqslant x \tag{5b}$$

Property (5a) states that a^* is indeed a solution; the so-called fixed-point induction principle (5b) captures that a^* is the least among all solutions. The two formulas have a number of important consequences,

$$1 \leqslant a^* \qquad a^* \cdot a^* \leqslant a^* \qquad a \leqslant a^* \qquad (a^*)^* = a^*$$

which suggest why a^* is sometimes called the reflexive, transitive closure of a.

Exercise 2. Show that iteration is order-preserving: $a \leqslant b \Longrightarrow a^* \leqslant b^*$.

Exercise 3. Assuming a Boolean lattice, show $a^* = (a - 1)^*$.

Regular Expressions. Regular expressions introduce syntax for choice, composition, and iteration. Expressed as a Haskell datatype, they read:

```
type Alphabet = Char
data Reg
  = Empty              -- the empty language
  | Alt  Reg Reg       -- choice
  | Eps                -- the empty word
  | Sym Alphabet       -- a single symbol
  | Cat  Reg Reg       -- composition
  | Rep  Reg           -- iteration
```

We use the term *basic symbol* to refer to the empty word or a single symbol of the alphabet.

A regular expression denotes a language. However, there is only one constructor specific to languages: *Sym*, which turns an element of the underlying alphabet into a regular expression. The other constructors can be interpreted generically in any regular algebra.

$$\begin{aligned}
[\![\, \mathit{Empty} \,]\!] &= [\![\, \emptyset \,]\!] &&= 0 \\
[\![\, \mathit{Alt}\ r\ s \,]\!] &= [\![\, r \mid s \,]\!] &&= [\![\, r \,]\!] + [\![\, s \,]\!] \\
[\![\, \mathit{Eps} \,]\!] &= [\![\, \epsilon \,]\!] &&= 1 \\
[\![\, \mathit{Sym}\ a \,]\!] &= [\![\, a \,]\!] &&= \{a\} \\
[\![\, \mathit{Cat}\ r\ s \,]\!] &= [\![\, r \cdot s \,]\!] &&= [\![\, r \,]\!] \cdot [\![\, s \,]\!] \\
[\![\, \mathit{Rep}\ r \,]\!] &= [\![\, r^* \,]\!] &&= [\![\, r \,]\!]^*
\end{aligned}$$

The second column of the semantic equations introduces alternative notation for the Haskell constructors. They serve the sole purpose of improving the readability of examples. (Occasionally, we also omit the semantic brackets, mixing syntax and semantics.)

We additionally introduce "smart" versions of the constructors that incorporate basic algebraic identities.

$$
\begin{array}{lll}
\begin{aligned}
&\mathit{cat} :: \mathit{Reg} \to \mathit{Reg} \to \mathit{Reg} \\
&\mathit{cat}\ \mathit{Empty}\ s = \mathit{Empty} \\
&\mathit{cat}\ r\ \mathit{Empty} = \mathit{Empty} \\
&\mathit{cat}\ \mathit{Eps}\ s \quad = s \\
&\mathit{cat}\ r\ \mathit{Eps} \quad = r \\
&\mathit{cat}\ r\ s \qquad = \mathit{Cat}\ r\ s
\end{aligned}
&
\begin{aligned}
&\mathit{alt} :: \mathit{Reg} \to \mathit{Reg} \to \mathit{Reg} \\
&\mathit{alt}\ \mathit{Empty}\ s = s \\
&\mathit{alt}\ r\ \mathit{Empty} = r \\
&\mathit{alt}\ r\ s \qquad = \mathit{Alt}\ r\ s \\
& \\
&
\end{aligned}
&
\begin{aligned}
&\mathit{rep} :: \mathit{Reg} \to \mathit{Reg} \\
&\mathit{rep}\ \mathit{Empty} = \mathit{Eps} \\
&\mathit{rep}\ \mathit{Eps} \quad = \mathit{Eps} \\
&\mathit{rep}\ r \qquad = \mathit{Rep}\ r \\
& \\
&
\end{aligned}
\end{array}
$$

They ensure, in particular, that *Empty* never occurs as a sub-expression. More sophisticated identities such as $(r^*)^* = r^*$ are not captured, however, to guarantee constant running time.

Regular Homomorphisms. Whenever a new class of structures is introduced, the definition of structure-preserving maps follows hard on its heels. Regular algebras are no exception. A regular homomorphism is a monoid homomorphism that preserves joins.[2] (Or equivalently, a monoid homomorphism that is a left adjoint). For example, the test whether a language contains the empty word ("is nullable") is a regular homomorphism from languages $(\mathcal{P}(\Sigma^*), \subseteq, \bigcup, \{\epsilon\}, \cdot)$ to Booleans $(\mathbb{B}, \Rightarrow, \exists, \mathit{true}, \wedge)$. Since membership $(x \in\ )$ is a left adjoint, it remains to check that $(\epsilon \in\ )$ preserves composition and its unit,

$$
\begin{aligned}
\epsilon \in \{\epsilon\} &= \mathit{true} \\
\epsilon \in A \cdot B &= \epsilon \in A \ \wedge \ \epsilon \in B
\end{aligned}
$$

which is indeed the case.

The nullability check can be readily implemented for regular expressions.

$$
\begin{aligned}
&\mathit{nullable} :: \mathit{Reg} \to \mathit{Bool} \\
&\mathit{nullable}\ (\mathit{Empty}) = \mathit{False} \\
&\mathit{nullable}\ (\mathit{Alt}\ r\ s) = \mathit{nullable}\ r \vee \mathit{nullable}\ s \\
&\mathit{nullable}\ (\mathit{Eps}) = \mathit{True} \\
&\mathit{nullable}\ (\mathit{Sym}\ a) = \mathit{False} \\
&\mathit{nullable}\ (\mathit{Cat}\ r\ s) = \mathit{nullable}\ r \wedge \mathit{nullable}\ s \\
&\mathit{nullable}\ (\mathit{Rep}\ r) = \mathit{True}
\end{aligned}
$$

The call *nullable r* yields *True* if and only if $\epsilon \in [\![r]\!]$.

Exercise 4. Let $(R, \leqslant, \sum, 1, \cdot)$ be a regular algebra. Prove that $\sum : \mathcal{P}(R) \to R$ is a regular homomorphism from the algebra of lifted monoids to R.

[2] A regular algebra is really a blend of a *complete join-semilattice* and a monoid. It is a standard result of lattice theory that a complete join-semilattice is a complete lattice. However, a complete join-semilattice homomorphism is not necessarily a complete lattice homomorphism, as there is no guarantee that it also preserves meets.

4 Finite Automata with ϵ-Transitions

To be able to give railroad diagrams a formal treatment, we make fork and join points of tracks explicit. Our running example, the diagram for $(a \mid b)^* \cdot a \cdot (a \mid b)$, consists of six tracks, where a track is given by a source point, a labelled station, and a target point, see Fig. 3. In the example, points are natural numbers; labels are either the empty word, omitted in the pictorial representation, or symbols of the underlying alphabet.

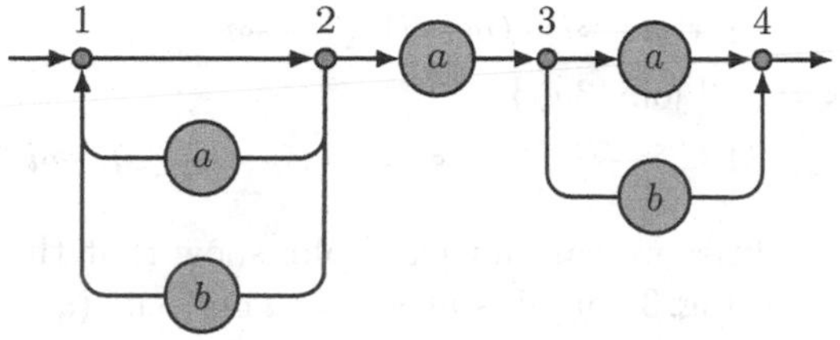

$$\{(1, \epsilon, 2), (2, a, 1), (2, b, 1), (2, a, 3), (3, a, 4), (3, b, 4)\}$$

Fig. 3. Railroad diagram for $(a \mid b)^* \cdot a \cdot (a \mid b)$.

In general, the points of a diagram are drawn from some fixed set V; its labels are elements of some monoid $(R, 1, \cdot)$. A diagram G is a set of arrows: $G \subseteq V \times R \times V$. For the application at hand, labels are given by languages: $R := \mathcal{P}(\Sigma^*)$ for some fixed alphabet Σ. An arrow (q, a, z) is sometimes written $a : q \to z$ for clarity.

Diagrams as Generators. A diagram generates a language for each pair of points q and z: we concatenate the labels along each path from q to z; the union of these languages is the language generated, in symbols $q \circ\!\!-\!\!\circ z$.

The diagram in Fig. 3 is now called $1 \circ\!\!-\!\!\circ 4$ (from entry to exit), and we have just said that it represents our running example, so we now want to formally prove that $1 \circ\!\!-\!\!\circ 4 = (a \mid b)^* \cdot a \cdot (a \mid b)$—and by the way, the sub-diagrams also equal sub-languages, so, for example, we claim that $1 \circ\!\!-\!\!\circ 2 = (a \mid b)^*$.

A path is a finite, possibly empty sequence of arrows with matching endpoints. To reason about languages generated by a diagram, we make use of the following three properties.

$$1 \subseteq i \circ\!\!-\!\!\circ i \tag{6a}$$

$$L \subseteq i \circ\!\!-\!\!\circ j \iff (i, L, j) \in G \tag{6b}$$

$$(i \circ\!\!-\!\!\circ j) \cdot (j \circ\!\!-\!\!\circ k) \subseteq i \circ\!\!-\!\!\circ k \tag{6c}$$

Properties (6a) and (6c) imply that we can go around in a loop a finite number of times.

$$(i{\circ\!\!-\!\!\circ}i)^* \subseteq i{\circ\!\!-\!\!\circ}i \tag{6d}$$

The proof is a straightforward application of the fixed-point induction principle.

$$\begin{aligned}
& (i{\circ\!\!-\!\!\circ}i)^* \subseteq i{\circ\!\!-\!\!\circ}i \\
\Longleftarrow \quad & \{\text{ fixed-point induction (5b) }\} \\
& 1 + (i{\circ\!\!-\!\!\circ}i) \cdot (i{\circ\!\!-\!\!\circ}i) \subseteq i{\circ\!\!-\!\!\circ}i \\
\Longleftrightarrow \quad & \{\text{ join } (3a) \} \\
& 1 \subseteq i{\circ\!\!-\!\!\circ}i \quad \wedge \quad (i{\circ\!\!-\!\!\circ}i) \cdot (i{\circ\!\!-\!\!\circ}i) \subseteq i{\circ\!\!-\!\!\circ}i
\end{aligned}$$

Using these inequalities we can, for example, show that the language generated by the diagram in Fig. 3 contains at least $(a \mid b)^* \cdot a \cdot (a \mid b)$.

$$\begin{aligned}
& (a \mid b)^* \cdot a \cdot (a \mid b) \\
= \quad & \{\text{ unit of composition }\} \\
& \epsilon \cdot ((a \mid b) \cdot \epsilon)^* \cdot a \cdot (a \mid b) \\
\subseteq \quad & \{\text{ arrows (6b) and monotonicity of operators }\} \\
& (1{\circ\!\!-\!\!\circ}2) \cdot ((2{\circ\!\!-\!\!\circ}1) \cdot (1{\circ\!\!-\!\!\circ}2))^* \cdot (2{\circ\!\!-\!\!\circ}3) \cdot (3{\circ\!\!-\!\!\circ}4) \\
\subseteq \quad & \{\text{ composition (6c) and iteration (6d) }\} \\
& (1{\circ\!\!-\!\!\circ}2) \cdot (2{\circ\!\!-\!\!\circ}2) \cdot (2{\circ\!\!-\!\!\circ}3) \cdot (3{\circ\!\!-\!\!\circ}4) \\
\subseteq \quad & \{\text{ composition (6c) }\} \\
& 1{\circ\!\!-\!\!\circ}4
\end{aligned}$$

Of course, we would actually like to show that $1{\circ\!\!-\!\!\circ}4$ is equal to $(a \mid b)^* \cdot a \cdot (a \mid b)$. We could argue that there are no other paths that contribute to the language generated. While this is certainly true, the statement does not seem to lend itself to a nice calculational argument. Fortunately, there is an attractive alternative, which we dub self-certifying diagrams.

Self-certifying Diagrams. The central idea is to record our expectations about the languages generated in the diagram itself, using languages as points: $V := \mathcal{P}(\Sigma^*)$. Continuing our running example, we replace the natural numbers of Fig. 3 by languages, see Fig. 4. The arrow $(r, a, a \mid b)$, for example, records that we can generate the language r if we start at the source point of the arrow; if we start at its end point, we can only generate a or b. In general, a point is identified with the language of all paths from the position to the unique exit. (As an aside, note that the point r is drawn twice in Fig. 4. Our mathematical model does not distinguish between these two visual copies. So the arrow (r, ϵ, r) is actually a self-loop, which could be safely removed.)

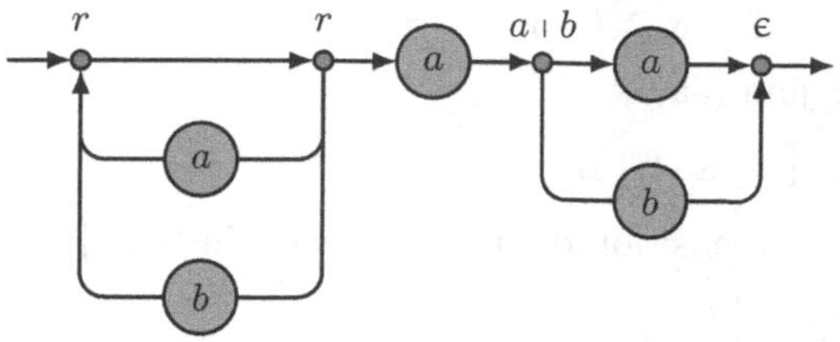

$$\{(r, \epsilon, r), (r, a, r), (r, b, r), (r, a, a \mid b), (a \mid b, a, \epsilon), (a \mid b, b, \epsilon)\}$$

Fig. 4. Annotated railroad diagram for $r = (a \mid b)^* \cdot a \cdot (a \mid b)$.

Of course, not every combination of languages makes sense. We require that the target is a *subfactor* of the source:

$$\alpha \circ\!\!\longrightarrow\!\! \textcircled{L} \!\!\longrightarrow\!\!\circ\ \beta \quad \text{is admissible} \quad :\Longleftrightarrow \quad \alpha \supseteq L \cdot \beta \tag{7}$$

Observe the use of inequalities: we require $\alpha \supseteq L \cdot \beta$, not $\alpha = L \cdot \beta$. Equality is too strong as there may be further arrows with the same source. The source point only provides an upper bound on the language generated from this point onward.

Theorem 1. *If all arrows of a diagram are admissible, then*

$$\alpha \supseteq (\alpha \circ\!\!-\!\!\circ \beta) \cdot \beta \tag{8}$$

for all points $\alpha \in V$ *and* $\beta \in V$.

Returning briefly to our running example, it is not too hard to check that all arrows in Fig. 4 are admissible. We can therefore conclude that $r \supseteq (r \circ\!\!-\!\!\circ \epsilon) \cdot \epsilon$. Since we already know that the diagram generates at least r, we have $r \circ\!\!-\!\!\circ \epsilon = r$, as desired.

Proof (Theorem 1). First, we show that a single path is admissible if all of its arrows are. The empty path is admissible as $\alpha \supseteq \epsilon \cdot \alpha$. The concatenation of two admissible paths again gives an admissible path:

$$\begin{aligned}
& \alpha \supseteq K \cdot \beta \ \wedge \ \beta \supseteq L \cdot \gamma \\
\Longrightarrow \quad & \{ \text{ monotonicity of composition } \} \\
& \alpha \supseteq K \cdot (L \cdot \gamma) \\
\Longleftrightarrow \quad & \{ \text{ associativity of composition } \} \\
& \alpha \supseteq (K \cdot L) \cdot \gamma
\end{aligned}$$

The language $Q \circ\!\!-\!\!\circ Z$ is the join of all languages generated by paths from Q to Z. To establish the theorem it suffices to show that admissibility is closed under arbitrary joins.

$$
\begin{aligned}
& \forall L \in \mathcal{L} \;.\; \alpha \supseteq L \cdot \beta \\
\iff & \quad \{ \text{ join (3a) } \} \\
& \alpha \supseteq \bigcup \{ L \cdot \beta \mid L \in \mathcal{L} \} \\
\iff & \quad \{ \text{ composition distributes over join (4a) } \} \\
& \alpha \supseteq (\bigcup \mathcal{L}) \cdot \beta
\end{aligned}
$$

Diagrams as Acceptors. So far we have emphasized the generative nature of diagrams. However, if a diagram is simple enough, it can also be seen as an acceptor. Indeed, if all the stations contain a basic symbol, then the diagram amounts to an NFA. The difference is merely one of terminology and presentation. In automata theory, points are called states and arrows are called transitions. Furthermore, transitions are typically drawn as labelled edges, as on the right below.

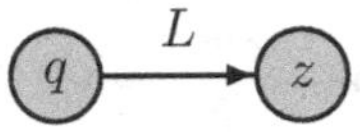

(I prefer railroad diagrams over standard drawings of automata, as the latter seem to put the visual emphasis on the wrong entity but, perhaps, this is a matter of personal taste.)

You may want to skim through the next section on first reading and proceed swiftly to Sect. 6, which justifies Manna's construction.

5 Interlude: The Regular Algebra of Diagrams

The purpose of this section is to investigate the algebraic structure of diagrams. We show that diagrams also form a regular algebra, provided the labels are drawn from a monoid, and we identify an important sub-algebra: self-certifying diagrams.

If R is a monoid, then diagrams $(\mathcal{P}(V \times R \times V), \subseteq, \bigcup, 1, \cdot)$ form a regular algebra where

$$1 = \{ (i, 1, i) \mid i \in V \} \tag{9a}$$

$$F \cdot G = \{ (i, a \cdot b, k) \mid \exists j \in V \;.\; (i, a, j) \in F, (j, b, k) \in G \} \tag{9b}$$

The unit is the diagram of self-loops $1 : i \to i$; composition takes an arrow $a : i \to j$ in F and an arrow $b : j \to k$ in G to form an arrow $a \cdot b : i \to k$ in $F \cdot G$. Diagrams are a blend of lifted monoids and relations: for $V := \mathbf{1}$ we obtain the regular algebra of lifted monoids; for $R := \mathbf{1}$ we obtain the regular algebra of (untyped) relations. (Here $\mathbf{1}$ is a singleton set.)

The lattice underlying diagrams is a powerset lattice and hence complete. It is not too hard to show that composition is associative with 1 as its unit. It remains to prove that $(P\cdot)$ and $(\cdot P)$ are left adjoints. As before, we only provide one calculation.

$$
\begin{array}{ll}
& P \cdot A \subseteq B \\
\iff & \{ \text{ definition of composition (9b) } \} \\
& \{ (i, p \cdot a, k) \mid \exists j \in V \,.\, (i,p,j) \in P, (j,a,k) \in A \} \subseteq B \\
\iff & \{ \text{ set inclusion } \} \\
& \forall i \in V, p \in R, a \in R, k \in V \,. \\
& \qquad (\exists j \in V \,.\, (i,p,j) \in P \wedge (j,a,k) \in A) \Longrightarrow (i, p \cdot a, k) \in B \\
\iff & \{ \text{ existential quantification is the join in the lattice of predicates } \} \\
& \forall i \in V, p \in R, j \in V, a \in R, k \in V \,. \\
& \qquad (i,p,j) \in P \wedge (j,a,k) \in A \Longrightarrow (i, p \cdot a, k) \in B \\
\iff & \{ \; (x \wedge\,) \text{ is a left adjoint } \} \\
& \forall i \in V, p \in R, j \in V, a \in R, k \in V \,. \\
& \qquad (j,a,k) \in A \Longrightarrow ((i,p,j) \in P \Longrightarrow (i, p \cdot a, k) \in B) \\
\iff & \{ \; (x \Rightarrow\,) \text{ preserves universal quantification } \} \\
& \forall j \in V, a \in R, k \in V \,. \\
& \qquad (j,a,k) \in A \Longrightarrow (\forall i \in V, p \in R \,.\, (i,p,j) \in P \Longrightarrow (i, p \cdot a, k) \in B) \\
\iff & \{ \text{ set inclusion } \} \\
& A \subseteq \{ (j,a,k) \mid \forall i \in V, p \in R \,.\, (i,p,j) \in P \Longrightarrow (i, p \cdot a, k) \in B \}
\end{array}
$$

The right adjoint $P \setminus B$ is given by the formula on the right.

For the remainder of the section we assume that R is a regular algebra. The translation of regular expressions to diagrams reduces the word problem (is a given word w an element of the language denoted by the regular expression r?) to a path-finding problem (is there a w-labelled path in the diagram corresponding to r?). Now, a directed path in the diagram G is an arrow in G^*, the reflexive, transitive closure of G. Using the closure of a diagram, we can provide a more manageable definition of $q \circ\!\!-\!\!\circ z$, the language generated by *paths* from q to z in G: we set $q \circ\!\!-\!\!\circ z := q \circ\!\!-\!(G^*)\!-\!\!\circ z$ where

$$q \circ\!\!-\!(F)\!-\!\!\circ z = \sum \{ a \in R \mid (q,a,z) \in F \} \tag{10}$$

is the language generated by all *arrows* from q to z in F.

It is appealing that we can use regular algebra to reason about the semantics of regular expressions, their visualization using diagrams, as well as their "implementation" in terms of nondeterministic finite automata. For example, in Sect. 4 we have mentioned in passing that we can safely remove ϵ-labelled self-loops from a diagram. This simple optimization is justified by $G^* = (G - 1)^*$, a general property of iteration, see Exercise 3.

Exercise 5. Let $D = \{(0, \epsilon, 1), (0, L, 0)\}$. Calculate D^* using the so-called *star decomposition* rule $(F + G)^* = F^* \cdot G \cdot F^* \cdot G \cdot \ldots \cdot G \cdot F^* = F^* \cdot (G \cdot F^*)^*$.

Admissibility Revisited. In Sect. 4 we have introduced the concept of an admissible arrow. (Section 6, which proves Manna's construction correct, relies heavily on this concept.) Let us call a diagram admissible if all its arrows are admissible. An alternative, but equivalent definition builds on $\circ\!\!-\!\!(G)\!\!-\!\!\circ$:

$$G \text{ admissible} \quad :\Longleftrightarrow \quad \forall \alpha \in V, \beta \in V \, . \, \alpha \supseteq (\alpha \circ\!\!-\!\!(G)\!\!-\!\!\circ \beta) \cdot \beta$$

The proof that the two definitions are indeed equivalent is straightforward.

$$\begin{aligned} & \forall \alpha \in V, \beta \in V \, . \, \alpha \supseteq (\alpha \circ\!\!-\!\!(G)\!\!-\!\!\circ \beta) \cdot \beta \\ \Longleftrightarrow \quad & \{ \text{ definition of } \circ\!\!-\!\!(G)\!\!-\!\!\circ \text{ (10) } \} \\ & \forall \alpha \in V, \beta \in V \, . \, \alpha \supseteq \left(\sum \{ L \mid (\alpha, L, \beta) \in G \}\right) \cdot \beta \\ \Longleftrightarrow \quad & \{ \text{ composition distributes over join (4a) } \} \\ & \forall \alpha \in V, \beta \in V \, . \, \alpha \supseteq \sum \{ L \cdot \beta \mid (\alpha, L, \beta) \in G \} \\ \Longleftrightarrow \quad & \{ \text{ join (3a) } \} \\ & \forall (\alpha, L, \beta) \in G \, . \, \alpha \supseteq L \cdot \beta \end{aligned}$$

Admissible (or self-certifying) diagrams form a sub-algebra of the algebra of diagrams $(\mathcal{P}(V \times R \times V), \subseteq, \bigcup, 1, \cdot)$: the unit 1 is admissible, $F \cdot G$ is admissible if both F and G are, $\sum \mathcal{G}$ is admissible if every diagram $G \in \mathcal{G}$ is. (The proof of Theorem 1 shows exactly that.) As a consequence,

$$G^* \text{ admissible} \quad \Longleftarrow \quad G \text{ admissible}$$

which is the import of Theorem 1.

Mathematical Models of Diagrams. There are many options for modelling diagrams or graphs. One important consideration is whether edges have an identity. A common definition of a labelled graph introduces two sets, a set of nodes and a set of edges with mappings from edges to source node, label, and target node. While our definition admits multiple directed edges between two nodes, it cannot capture multiple edges with the same label, a feature that we do not consider important for the application at hand. Backhouse's notion of a matrix [4] simplifies further by not allowing multiple edges at all. In his model a graph[3] is a function $V \times V \to R$ that maps a pair of nodes, source and target, to an element of the underlying regular algebra.[4] Matrices ordered pointwise also form a regular algebra, provided R is one.

---

[3] Backhouse uses the terms matrix and graph interchangeably.

[4] Backhouse's definition is actually more general: a matrix is given by a function $r \to R$ where $r \subseteq V \times V$ is a fixed relation, the dimension of the matrix. This allows him to distinguish between non-existent edges and edges that are labelled with 0.

We haven chosen to allow multiple edges as we wish to model Manna's transformation for choice, which replaces a single edge by two edges with the same source and target, see Fig. 2. In some sense, the operation $q \circ\!\!-\!(G)\!-\!\!\circ z$ undoes this step by joining the labels of all arrows from q to z. In fact, the operation is a regular homomorphism from diagrams to matrices, mapping a diagram G to a matrix $\circ\!\!-\!(G)\!-\!\!\circ$, where application to source and target points is written $q \circ\!\!-\!(G)\!-\!\!\circ z$. We need to establish the following three properties (the right-hand sides of the formulas implicitly define join, unit, and composition of matrices).

$$q \circ\!\!-\!(\textstyle\bigcup \mathcal{G})\!-\!\!\circ z = \sum \{\, q \circ\!\!-\!(G)\!-\!\!\circ z \mid G \in \mathcal{G} \,\} \tag{11a}$$

$$q \circ\!\!-\!(1)\!-\!\!\circ z = (1 \lhd q = z \rhd 0) \tag{11b}$$

$$q \circ\!\!-\!(F \cdot G)\!-\!\!\circ z = \sum \{\, (q \circ\!\!-\!(F)\!-\!\!\circ i) \cdot (i \circ\!\!-\!(G)\!-\!\!\circ z) \mid i \in V \,\} \tag{11c}$$

Here $a \lhd c \rhd b$ is shorthand for **if** c **then** a **else** b, known as the Hoare conditional choice operator. The proof of (11a) relies on properties of set comprehensions.

$$
\begin{aligned}
& q \circ\!\!-\!(\textstyle\bigcup \mathcal{G})\!-\!\!\circ z \\
= \quad & \{ \text{ definition (10) } \} \\
& \sum \left\{ a \in R \mid (q, a, z) \in \bigcup \mathcal{G} \right\} \\
= \quad & \{ \text{ set comprehension } \} \\
& \sum \{\, a \in R \mid G \in \mathcal{G}, (q, a, z) \in G \,\} \\
= \quad & \{ \text{ book-keeping law, see below } \} \\
& \sum \left\{ \sum \{\, a \in R \mid (q, a, z) \in G \,\} \mid G \in \mathcal{G} \right\} \\
= \quad & \{ \text{ definition (10) } \} \\
& \sum \{\, q \circ\!\!-\!(G)\!-\!\!\circ z \mid G \in \mathcal{G} \,\}
\end{aligned}
$$

The penultimate step of the proof uses the identity $\sum \{ \sum \{ a_i \mid i \in I \} \mid I \in \mathcal{I} \} = \sum \{ a_i \mid I \in \mathcal{I}, i \in I \}$. Written in a point-free style the formula is known as the book-keeping law: $\sum \circ \mathcal{P} \sum = \sum \circ \bigcup$. (Categorically speaking, the identity follows from the fact that every complete join-semilattice is an algebra for the powerset monad, see Ex. VI.2.1 in [11].)

For (11b) we reason,

$$
\begin{aligned}
& q \circ\!\!-\!(1)\!-\!\!\circ\, z \\
= \quad & \{ \text{ definition (10) } \} \\
& \textstyle\sum \{ a \in R \mid (q, a, z) \in 1 \} \\
= \quad & \{ \text{ definition of unit (9a) } \} \\
& \textstyle\sum \{ a \in R \mid (q, a, z) \in \{ (i, 1, i) \mid i \in V \} \} \\
= \quad & \{ \text{ membership } \} \\
& \textstyle\sum \{ 1 \mid q = z \} \\
= \quad & \{ \text{ set comprehension } \} \\
& 1 \lhd q = z \rhd 0
\end{aligned}
$$

The proof of (11c) relies again on the book-keeping law. The calculation is most perspicuous if read from bottom to top.

$$
\begin{aligned}
& q \circ\!\!-\!(F \cdot G)\!-\!\!\circ\, z \\
= \quad & \{ \text{ definition (10) } \} \\
& \textstyle\sum \{ x \in R \mid (q, x, z) \in F \cdot G \} \\
= \quad & \{ \text{ definition of composition (9b) } \} \\
& \textstyle\sum \{ x \in R \mid (q, x, z) \in \{ (q, a \cdot b, z) \mid \exists i \in V \,.\, (q, a, i) \in F, (i, b, z) \in G \} \} \\
= \quad & \{ \text{ membership } \} \\
& \textstyle\sum \{ a \cdot b \mid \exists i \in V \,.\, (q, a, i) \in F, (i, b, z) \in G \} \\
= \quad & \{ \text{ book-keeping law, see above } \} \\
& \textstyle\sum \Big\{ \sum \{ a \cdot b \mid (q, a, i) \in F, (i, b, z) \in G \} \mid i \in V \Big\} \\
= \quad & \{ \text{ composition preserves joins (4a) } \} \\
& \textstyle\sum \Big\{ \big(\sum \{ a \in R \mid (q, a, i) \in F \} \big) \cdot \big(\sum \{ b \in R \mid (i, b, z) \in G \} \big) \mid i \in V \Big\} \\
= \quad & \{ \text{ definition (10) } \} \\
& \textstyle\sum \{ (q \circ\!\!-\!(F)\!-\!\!\circ\, i) \cdot (i \circ\!\!-\!(G)\!-\!\!\circ\, z) \mid i \in V \}
\end{aligned}
$$

Exercise 6. Show that properties (6a)–(6c) are consequences of (11a)–(11c).

6 Construction of Finite Automata with ϵ-Transitions

We now have the necessary prerequisites in place to formalize and justify Manna's construction of diagrams. Recall that the construction works by repeatedly replacing a single arrow by a small diagram until all labels are basic symbols.

The point of departure is the diagram below that consists of a single, admissible arrow,

$G_0 :$ 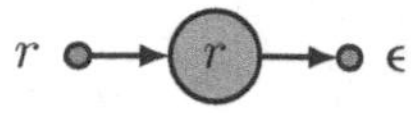

where r is the given regular expression. Clearly, we have $r \multimap\!(G_0^*)\!\multimap \epsilon = r$, where $q \multimap\!(F)\!\multimap z$ is the language generated by all arrows from q to z in F (10).

For each of the following transformations we show that (1) admissibility of arrows is preserved (*correctness*); and (2) for each path in the original diagram, there is a corresponding path in the transformed diagram (*completeness*). Correctness implies that for each graph G_i generated in the process we have $r \supseteq r \multimap\!(G_i^*)\!\multimap \epsilon$. Completeness ensures that $Q \multimap\!(G_i^*)\!\multimap Z \subseteq Q \multimap\!(G_{i+1}^*)\!\multimap Z$, which implies $r \subseteq r \multimap\!(G_i^*)\!\multimap \epsilon$.[5]

We consider each of the transformations of Fig. 2 in turn.

Case ∅: there is nothing to prove.

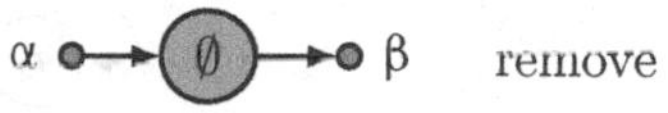

Case $r \mid s$: we replace the single arrow by two arrows sharing source and target.

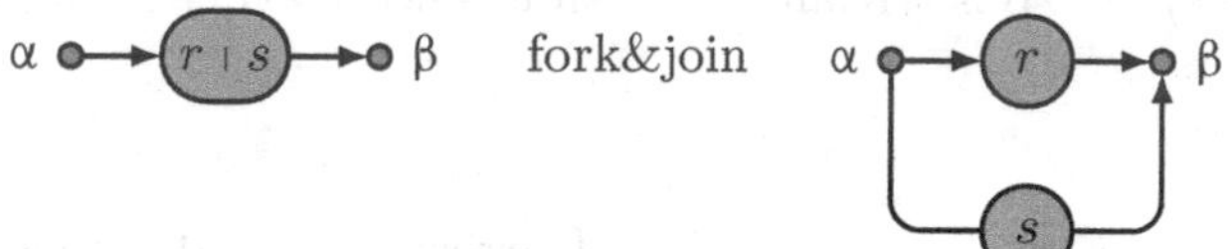

To establish correctness and completeness, we reason:

$$\begin{array}{ll} & \alpha \supseteq (r \mid s) \cdot \beta \\ \iff & \{\text{ distributivity (4a) }\} \\ & \alpha \supseteq (r \cdot \beta) \mid (s \cdot \beta) \\ \iff & \{\text{ join (3a) }\} \\ & \alpha \supseteq r \cdot \beta \ \wedge\ \alpha \supseteq s \cdot \beta \end{array} \qquad \begin{array}{ll} & r \mid s \\ \subseteq & \{\text{ arrows (6b) and monotonicity }\} \\ & (\alpha \multimap \beta) \cup (\alpha \multimap \beta) \\ = & \{\text{ idempotency }\} \\ & \alpha \multimap \beta \end{array}$$

Case ϵ: we keep ϵ-arrows (we may choose to omit ϵ-labels in diagrams though),

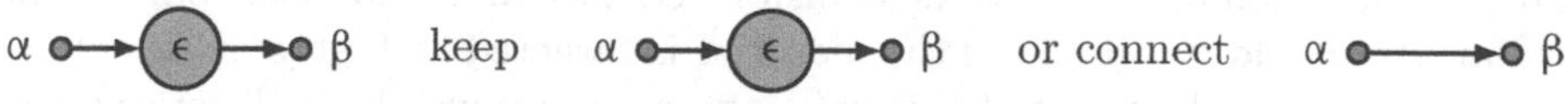

so there is nothing to prove.

Case $r \cdot s$: we split the composition introducing an intermediate point.

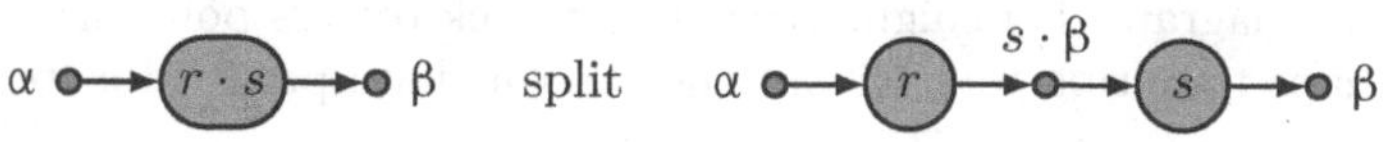

---

[5] We use "correctness" and "completeness" only in this very narrow sense.

We have to show that the two arrows on the right are admissible and that they generate $r \cdot s$.

$$
\begin{array}{ll}
& \alpha \supseteq (r \cdot s) \cdot \beta \\
\iff & \{\ \text{associativity}\ \} \\
& \alpha \supseteq r \cdot (s \cdot \beta) \\
\iff & \{\ \text{reflexivity}\ \} \\
& \alpha \supseteq r \cdot (s \cdot \beta) \ \wedge\ s \cdot \beta \supseteq s \cdot \beta
\end{array}
$$

$$
\begin{array}{ll}
& r \cdot s \\
\subseteq & \{\ \text{arrows (6b) and monotonicity}\ \} \\
& (\alpha \circ\!\!-\!\!\circ\, s \cdot \beta) \cdot (s \cdot \beta \circ\!\!-\!\!\circ\, \beta) \\
\subseteq & \{\ \text{composition (6c)}\ \} \\
& \alpha \circ\!\!-\!\!\circ\, \beta
\end{array}
$$

Case r^*: we introduce an intermediate point, $i = r^* \cdot \beta$, which serves as source and target of a looping arrow.

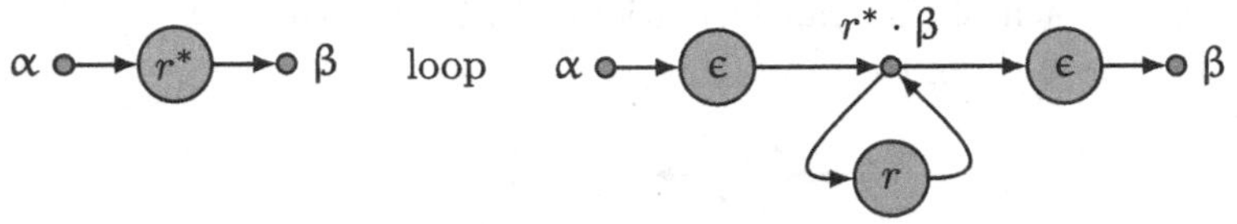

We have to show that the three arrows are admissible and that they generate r^*. The arrow $\alpha \circ\!\!-\!(\epsilon)\!-\!\!\circ\, r^* \cdot \beta$ is trivially admissible since $\alpha \circ\!\!-\!(r^*)\!-\!\!\circ\, \beta$ is. It remains to show

$$
\begin{array}{ll}
& i \supseteq i \\
\iff & \{\ r^* = \epsilon \mid r \cdot r^*\ \} \\
& i \supseteq (\epsilon \mid r \cdot r^*) \cdot \beta \\
\iff & \{\ \text{distributivity (4a)}\ \} \\
& i \supseteq (\epsilon \cdot \beta) \mid (r \cdot i) \\
\iff & \{\ \text{join (3a)}\ \} \\
& i \supseteq \epsilon \cdot \beta \ \wedge\ i \supseteq r \cdot i
\end{array}
$$

$$
\begin{array}{ll}
& \epsilon \cdot r^* \cdot \epsilon \\
\subseteq & \{\ \text{arrows (6b) and monotonicity}\ \} \\
& (\alpha \circ\!\!-\!\!\circ\, i) \cdot (i \circ\!\!-\!\!\circ\, i)^* \cdot (i \circ\!\!-\!\!\circ\, \beta) \\
\subseteq & \{\ \text{iteration (6d)}\ \} \\
& (\alpha \circ\!\!-\!\!\circ\, i) \cdot (i \circ\!\!-\!\!\circ\, i) \cdot (i \circ\!\!-\!\!\circ\, \beta) \\
\subseteq & \{\ \text{composition (6c)}\ \} \\
& \alpha \circ\!\!-\!\!\circ\, \beta
\end{array}
$$

A few remarks are in order.

The calculations are entirely straightforward. The calculations on the left rely on basic properties of regular algebra; the calculations on the right capture visual arguments—it is quite obvious that for each path in the original diagram there is a corresponding path in the transformed diagram. That's the point—the calculations should be simple as the material is targeted at first-year students.

Quite interestingly, we need not make any assumptions about disjointness of states, which is central to the McNaughton-Yamada-Thompson algorithm [2]. For example, there is no guarantee that the intermediate state for composition, $s \cdot \beta$, is not used elsewhere in the diagram. Admissibility of arrows ensures that sharing of sub-diagrams is benign. We will get back to this point shortly.

It is perhaps tempting to leave out the intermediate point i for iteration:

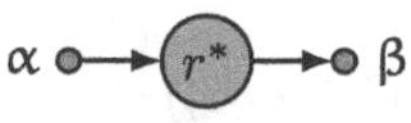

loop

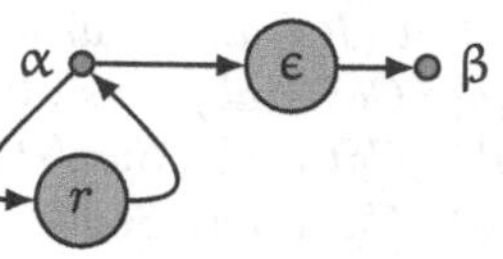

Wrong!

While the diagram on the right is complete, it is not correct: the arrow (α, r, α) is not admissible, consider, for example, $\alpha = a \mid b^*$ and $r = b^*$. Since the arrow (α, r, α) is generally part of a larger diagram, it might contribute to other edges emerging from α. A simple case analysis shows that the intermediate point i is necessary—there is no diagram involving only α and β that will do the trick.

Exercise 7. Do the transformations work in both directions?

Implementation in Haskell. Manna's transformation rules can be seen as the specification of a non-deterministic algorithm as the rules can be applied in any order. However, they serve equally well as the basis for an inductive construction. We represent arrows by triples.

```
type Arrow label = (Reg, label, Reg)
diagram :: Reg → Set (Arrow (Basic Alphabet))
diagram r = diagram′ (r, r, Eps)
```

The worker function *diagram′* maps a single arrow to a diagram, a set of arrows. (We assume the existence of a suitable library for manipulating finite sets.) Basic symbols are represented by elements of *Basic Alphabet*: *Eps′* represents the empty word ϵ, *Sym′ a* represents a singleton word, the symbol a.

```
data Basic a = Eps′ | Sym′ a
diagram′ :: Arrow Reg → Set (Arrow (Basic Alphabet))
diagram′ (α, Empty, β)  = ∅
diagram′ (α, Alt r s, β) = diagram′ (α, r, β) ∪ diagram′ (α, s, β)
diagram′ (α, Eps, β)    = {(α, Eps′, β)}
diagram′ (α, Sym a, β)  = {(α, Sym′ a, β)}
diagram′ (α, Cat r s, β) = diagram′ (α, r, i) ∪ diagram′ (i, s, β)
  where i = cat s β
diagram′ (α, Rep r, β)  = {(α, Eps′, i)} ∪ diagram′ (i, r, i) ∪ {(i, Eps′, β)}
  where i = cat (Rep r) β
```

Voilà. A regular expression compiler in eight lines.

The following session shows the algorithm in action (a is shorthand for *Sym* `'a'`, b for *Sym* `'b'`, and ab for *Alt a b*).

$\ggg$ *diagram* (*Cat* (*Rep ab*) (*Cat a ab*))
{(*Cat* (*Rep ab*) (*Cat a ab*), *Eps'*, *Cat* (*Rep ab*) (*Cat a ab*))
(*Cat* (*Rep ab*) (*Cat a ab*), *Sym'* 'a', *Cat* (*Rep ab*) (*Cat a ab*))
(*Cat* (*Rep ab*) (*Cat a ab*), *Sym'* 'b', *Cat* (*Rep ab*) (*Cat a ab*))
(*Cat* (*Rep ab*) (*Cat a ab*), *Eps'*, *Cat a ab*)
(*Cat a ab*, *Sym'* 'a', *ab*)
(*ab*, *Sym'* 'a', *Eps*),
(*ab*, *Sym'* 'b', *Eps*)}

The diagram produced for our running example is almost the same as before: compared to the one in Fig. 4 we have one additional vertex, labelled $a \cdot (a \mid b)$.

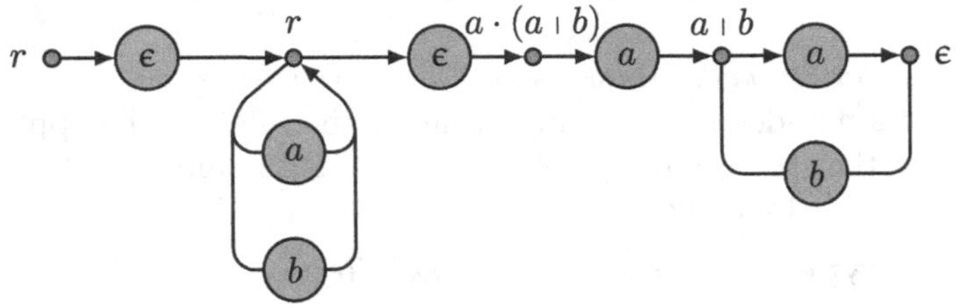

The algorithm has a number of pleasing characteristics. First of all, the number of points and arrows is linear in the size of the given regular expression: the number of intermediate points (states) is bounded by the number of compositions and iterations; the number of arrows (transitions) is bounded by the number of basic symbols plus twice the number of iterations. In practice, the actual size may be significantly smaller through sharing of sub-diagrams. Consider the regular expressions *Alt* (*Cat r t*) (*Cat s t*) and *Cat* (*Alt r s*) *t*, which are syntactically different but semantically equal by distributivity. Due to the use of sets, the generated diagrams are equal, as well. The sub-diagram for the replicated sub-expression *t* is automagically shared, as the following calculation demonstrates.

$$
\begin{aligned}
& \mathit{diagram}' \, (\alpha, \mathit{Alt}\,(\mathit{Cat}\; r\; t)\,(\mathit{Cat}\; s\; t), \beta) \\
= \;& \{ \text{ definition of } \mathit{diagram}' \} \\
& \mathit{diagram}' \, (\alpha, r, i) \cup \mathit{diagram}' \, (i, t, \beta) \\
& \qquad \cup \mathit{diagram}' \, (\alpha, s, i) \cup \mathit{diagram}' \, (i, t, \beta) \;\textbf{where}\; i = \mathit{cat}\; t\; \beta \\
= \;& \{ \text{ idempotency of union } \} \\
& \mathit{diagram}' \, (\alpha, r, i) \cup \mathit{diagram}' \, (\alpha, s, i) \\
& \qquad \cup \mathit{diagram}' \, (i, t, \beta) \;\textbf{where}\; i = \mathit{cat}\; t\; \beta \\
= \;& \{ \text{ definition of } \mathit{diagram}' \} \\
& \mathit{diagram}' \, (\alpha, \mathit{Cat}\,(\mathit{Alt}\; r\; s)\; t, \beta)
\end{aligned}
$$

As another example, *Alt r r* and *r* are mapped to the same diagram. The examples demonstrate that it is undesirable that diagrams for sub-expressions have disjoint states. Sharing of sub-diagrams is a feature, not a bug. Sharing comes at a small cost though: due to the use of sets (rather than lists), the running-time of *diagram* is $\Theta(n \log n)$ rather than linear.

In Sect. 5 we have noted that ϵ-labelled self-loops can be safely removed from a diagram. (Recall that this optimization is justified by $G^* = (G - 1)^*$.)

$$
\begin{aligned}
&\mathit{trim} :: \mathit{Set}\ (\mathit{Arrow}\ (\mathit{Basic}\ a)) \to \mathit{Set}\ (\mathit{Arrow}\ (\mathit{Basic}\ a))\\
&\mathit{trim}\ g = [a \mid a \leftarrow g, \mathit{not}\ (\mathit{self\text{-}loop}\ a)]\\
&\mathit{self\text{-}loop} :: \mathit{Arrow}\ (\mathit{Basic}\ a) \to \mathit{Bool}\\
&\mathit{self\text{-}loop}\ (i, \mathit{Eps}', \quad j) = i == j\\
&\mathit{self\text{-}loop}\ (i, \mathit{Sym}'\ a, j) = \mathit{False}
\end{aligned}
$$

Arrows of the form (r, ϵ, r) are actually not that rare. Our running example features one. More generally, expressions of the form *Cat r* (*Rep s*) generate a self-loop.

$$
\begin{aligned}
&\quad \mathit{diagram}'\ (\alpha, \mathit{Cat}\ r\ (\mathit{Rep}\ s), \beta)\\
= &\quad \{\ \text{definition of } \mathit{diagram}'\ \}\\
&\quad \mathit{diagram}'\ (\alpha, r, i) \cup \mathit{diagram}'\ (i, \mathit{Rep}\ s, \beta)\ \textbf{where}\ i = \mathit{cat}\ (\mathit{Rep}\ s)\ \beta\\
= &\quad \{\ \text{definition of } \mathit{diagram}'\ \}\\
&\quad \mathit{diagram}'\ (\alpha, r, i) \cup \{(i, \mathit{Eps}', i)\}\\
&\qquad\quad \cup\ \mathit{diagram}'\ (i, s, i) \cup \{(i, \mathit{Eps}', \beta)\}\ \textbf{where}\ i = \mathit{cat}\ (\mathit{Rep}\ s)\ \beta
\end{aligned}
$$

The intermediate states for composition and iteration are identified, i, further reducing the total number of states.

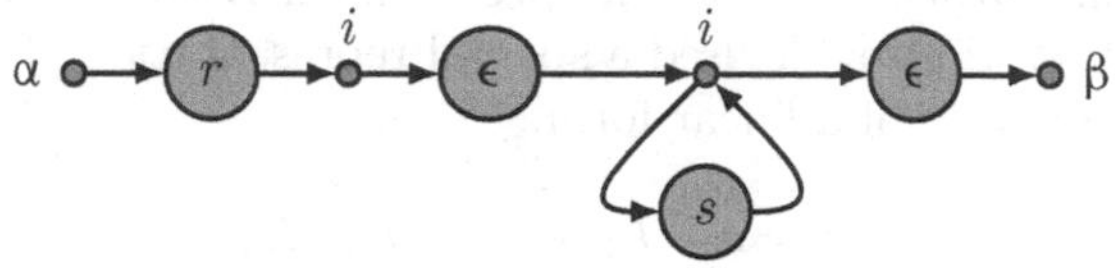

7 Finite Automata Without ϵ-Transitions

It is advisable to eliminate ϵ-transitions before translating a nondeterministic finite automaton into an executable program as ϵ-loops might cause termination problems. (Incidentally, Thompson [16] reports that his compilation scheme does not work for $(a^*)^*$ as the code goes into an infinite loop—he also proposes a fix.) There are at least two ways forward.

One option is to apply a graph transformation. Quite pleasingly, the transformation is based on a general property of iteration: $(a + b)^* = (a^* \cdot b)^* \cdot a^*$, see Exercise 5. Let $E = \{(v, \epsilon, w) \mid v, w \in V\}$ be the complete, ϵ-labelled diagram and let G be the diagram generated for r. We can massage the reflexive, transitive closure of G as follows.

$$G^* = ((G \cap E) + (G - E))^* = ((G \cap E)^* \cdot (G - E))^* \cdot (G \cap E)^* \tag{12}$$

The sub-diagram $C = (G \cap E)^*$ is the so-called ϵ-closure of G. The path-finding problem in G can be reduced to a path-finding problem in $C \cdot (G - E)$:

$$
\begin{aligned}
& w \in r \multimapinv (G^*) \multimap \epsilon \\
\Longleftrightarrow \quad & \{ \text{ see above (12) } \} \\
& w \in r \multimapinv ((C \cdot (G - E))^* \cdot C) \multimap \epsilon \\
\Longleftrightarrow \quad & \{ \text{ composition (11c) } \} \\
& w \in \sum \{ (r \multimapinv ((C \cdot (G - E))^*) \multimap f) \cdot (f \multimapinv (C) \multimap \epsilon) \mid f \in V \} \\
\Longleftrightarrow \quad & \{ \text{ membership } \} \\
& \exists f \in V \,.\, w \in (r \multimapinv ((C \cdot (G - E))^*) \multimap f) \cdot (f \multimapinv (C) \multimap \epsilon) \\
\Longleftrightarrow \quad & \{ \; q \multimapinv (C) \multimap z \subseteq \epsilon \text{ and property of free monoid } \} \\
& \exists f \in V \,.\, w \in r \multimapinv ((C \cdot (G - E))^*) \multimap f \;\; \wedge \;\; \epsilon \in f \multimapinv (C) \multimap \epsilon
\end{aligned}
$$

Let us call a point f with $\epsilon \in f \multimapinv (C) \multimap \epsilon$ an accepting or final state. The calculation demonstrates that we can reduce the word problem, $w \in [\![r]\!]$, to the problem of finding an w-labelled path from the start state r to an accepting state f in $C \cdot (G - E)$. The final formula explains in a sense why NFAs *without* ϵ-transitions need to have a *set* of final states, whereas a railroad diagram has exactly one entry and one exit.

Another option is to integrate the computation of the ϵ-closure into the construction of the automaton itself. This is exactly what Antimirov's scheme does, which we review next. (The purpose of the following is to show that a railroad diagram contains all of Antimirov's subfactors.)

Antimirov's Linear Forms. To avoid the problems of Brzozowski's construction outlined in Sect. 1, Antimirov devised a special representation of regular expressions based on the notion of a linear form.

$$L - \epsilon = a_1 \cdot L_1 + \cdots + a_n \cdot L_n$$

We represent a linear form by a *set* of pairs, consisting of a symbol and a regular expression.

$$\textbf{type}\ \mathit{Lin} = \mathit{Set}\ (\mathit{Alphabet}, \mathit{Reg})$$

Observe that the type is non-recursive; the subfactors are still given by regular expressions. Antimirov's insight was that in order to guarantee a finite number of *syntactically* different subfactors, it is sufficient to apply the ACI-properties of choice to the top-level of an expression.

Turning a regular expression into a linear form is straightforward except, perhaps, for composition and iteration. It is *not* the case that the linear form of $r \cdot s - \epsilon$ is given by the composition of the forms for $r - \epsilon$ and $s - \epsilon$. (Do you see why?) The following calculation points us into the right direction.

$$r \cdot s - \epsilon = (r - \epsilon) \cdot s + (r \cap \epsilon) \cdot s - \epsilon = (r - \epsilon) \cdot s + (r \cap \epsilon) \cdot (s - \epsilon)$$

The final formula suggests that we need to compose a linear form, the representation of $r - \epsilon$, with a standard regular expression, namely s.

$$
\begin{array}{ll}
\textbf{infixr } 7 \bullet & \\
(\bullet) :: \mathit{Lin} \to \mathit{Reg} \to \mathit{Lin} & \\
\mathit{lf} \bullet \mathit{Empty} = \emptyset & \\
\mathit{lf} \bullet \mathit{Eps} & = \mathit{lf} \\
\mathit{lf} \bullet s & = [(a, \mathit{cat}\ r\ s) \mid (a, r) \leftarrow \mathit{lf}]
\end{array}
$$

Thus, $\mathit{lf} \bullet s$ composes every subfactor of lf with s. (Just in case you wonder, $[e \mid q]$ above is a monad, not a list comprehension.) The calculation for iteration makes use of $a^* = (a - 1)^*$, see Exercise 3.

$$r^* - \epsilon = (r - \epsilon)^* - \epsilon = (r - \epsilon) \cdot (r - \epsilon)^* = (r - \epsilon) \cdot r^*$$

Again, we need to compose a linear form, the representation of $r - \epsilon$, with a standard regular expression, namely r^* itself.

Given these prerequisites, Antimirov's function lf [3], which maps r to the linear form of $r - \epsilon$, can be readily implemented in Haskell.

$$
\begin{array}{lll}
\mathit{linear\text{-}form} :: \mathit{Reg} \to \mathit{Lin} & & \\
\mathit{linear\text{-}form}\ (\mathit{Empty}) = \emptyset & & \\
\mathit{linear\text{-}form}\ (\mathit{Alt}\ r\ s) = \mathit{linear\text{-}form}\ r \cup \mathit{linear\text{-}form}\ s & & \\
\mathit{linear\text{-}form}\ (\mathit{Eps}) & = \emptyset & \\
\mathit{linear\text{-}form}\ (\mathit{Sym}\ a) & = \{(a, \mathit{Eps})\} & \\
\mathit{linear\text{-}form}\ (\mathit{Cat}\ r\ s) & & \\
\quad \mid \mathit{nullable}\ r & = \mathit{linear\text{-}form}\ r \bullet s \cup \mathit{linear\text{-}form}\ s & \text{-- } r \cap \epsilon = \epsilon \\
\quad \mid \mathit{otherwise} & = \mathit{linear\text{-}form}\ r \bullet s & \text{-- } r \cap \epsilon = \emptyset \\
\mathit{linear\text{-}form}\ (\mathit{Rep}\ r) & = \mathit{linear\text{-}form}\ r \bullet \mathit{Rep}\ r &
\end{array}
$$

For our running example, we obtain

$$
\begin{array}{l}
\rangle\!\rangle\!\rangle \quad \mathit{linear\text{-}form}\ (\mathit{Cat}\ (\mathit{Rep}\ ab)\ (\mathit{Cat}\ a\ ab)) \\
\{(\texttt{'a'}, ab), (\texttt{'a'}, \mathit{Cat}\ (\mathit{Rep}\ ab)\ (\mathit{Cat}\ a\ ab)), (\texttt{'b'}, \mathit{Cat}\ (\mathit{Rep}\ ab)\ (\mathit{Cat}\ a\ ab))\}
\end{array}
$$

Antimirov's Subfactors. Putting the automata glasses on, $\mathit{linear\text{-}form}\ r$ computes the outgoing edges of r. The successor states of r are given by the immediate subfactors.

$$\mathit{subfactors}\ r = [\beta \mid (a, \beta) \leftarrow \mathit{linear\text{-}form}\ r] \tag{13}$$

In order to easily compare Antimirov's construction to Manna's, it is useful to concentrate on subfactors. To this end, we unfold the specification (13) to obtain

$$
\begin{aligned}
&\mathit{subfactors} :: \mathit{Reg} \to \mathit{Set}\ \mathit{Reg}\\
&\mathit{subfactors}\ (\mathit{Empty}) = \emptyset\\
&\mathit{subfactors}\ (\mathit{Alt}\ r\ s) = \mathit{subfactors}\ r \cup \mathit{subfactors}\ s\\
&\mathit{subfactors}\ (\mathit{Eps}) \quad = \emptyset\\
&\mathit{subfactors}\ (\mathit{Sym}\ a) = \{\mathit{Eps}\}\\
&\mathit{subfactors}\ (\mathit{Cat}\ r\ s)\\
&\quad \mid \mathit{nullable}\ r \qquad = \mathit{subfactors}\ r \circ s \cup \mathit{subfactors}\ s\\
&\quad \mid \mathit{otherwise} \qquad = \mathit{subfactors}\ r \circ s\\
&\mathit{subfactors}\ (\mathit{Rep}\ r) \ = \mathit{subfactors}\ r \circ \mathit{Rep}\ r
\end{aligned}
$$

The operator "∘" is a version of "•" that works on sets of states, rather than sets of edges.

$$
\begin{aligned}
&\mathbf{infixr}\ 7 \circ\\
&(\circ) :: \mathit{Set}\ \mathit{Reg} \to \mathit{Reg} \to \mathit{Set}\ \mathit{Reg}\\
&rs \circ \mathit{Empty} = \emptyset\\
&rs \circ \mathit{Eps} \quad = rs\\
&rs \circ s \qquad = [\mathit{cat}\ r\ s \mid r \leftarrow rs]
\end{aligned}
$$

The reflexive, transitive closure of *subfactors* applied to the given regular expression r then yields the states of the Antimirov automaton: starting with $\{r\}$ we iterate $\mathit{subfactors}^{\dagger}$ until a fixed-point is reached, where $f^{\dagger}$ is the so-called Kleisli extension of f, defined $f^{\dagger}\ X = \bigcup\{ f\ x \mid x \in X \}$. That is easy enough—however, it is, perhaps, not immediately clear that the set of all subfactors, immediate and transitive ones, is finite.

Manna's Construction Revisited. We claim that Antimirov's subfactors appear as target states of non-ϵ-transitions in the corresponding railroad diagram. Given the specification,

$$\mathit{targets}\ r = [\beta \mid (\alpha, \mathit{Sym}'\ a, \beta) \leftarrow \mathit{diagram}\ r]$$

it is a straightforward exercise to derive $\mathit{targets}\ r = \mathit{targets}'\ (r, \mathit{Eps})$ where $\mathit{targets}'$ is defined

$$
\begin{aligned}
&\mathit{targets}' :: (\mathit{Reg}, \mathit{Reg}) \to \mathit{Set}\ \mathit{Reg}\\
&\mathit{targets}'\ (\mathit{Empty}, \ \beta) = \emptyset\\
&\mathit{targets}'\ (\mathit{Alt}\ r\ s, \ \beta) = \mathit{targets}'\ (r, \beta) \cup \mathit{targets}'\ (s, \beta)\\
&\mathit{targets}'\ (\mathit{Eps}, \beta) \quad = \emptyset\\
&\mathit{targets}'\ (\mathit{Sym}\ a, \ \beta) = \{\beta\}\\
&\mathit{targets}'\ (\mathit{Cat}\ r\ s, \beta) = \mathit{targets}'\ (r, \mathit{cat}\ s\ \beta) \cup \mathit{targets}'\ (s, \beta)\\
&\mathit{targets}'\ (\mathit{Rep}\ r, \ \beta) = \mathit{targets}'\ (r, \mathit{cat}\ (\mathit{Rep}\ r)\ \beta)
\end{aligned}
$$

This is basically the definition of *diagram'*, only that we ignore source points and labels and discard ϵ-labelled arrows.

The definition of *targets'* is tantalizingly close to *subfactors*, except that the former makes use of an accumulating parameter, whereas the latter does not. Removing the accumulating parameter is, of course, a matter of routine.

$$
\begin{aligned}
&\mathit{targets} :: \mathit{Reg} \to \mathit{Set}\ \mathit{Reg} \\
&\mathit{targets}\ (\mathit{Empty}) &&= \emptyset \\
&\mathit{targets}\ (\mathit{Alt}\ r\ s) &&= \mathit{targets}\ r \cup \mathit{targets}\ s \\
&\mathit{targets}\ (\mathit{Eps}) &&= \emptyset \\
&\mathit{targets}\ (\mathit{Sym}\ a) &&= \{\mathit{Eps}\} \\
&\mathit{targets}\ (\mathit{Cat}\ r\ s) &&= \mathit{targets}\ r \circ s \cup \mathit{targets}\ s \\
&\mathit{targets}\ (\mathit{Rep}\ r) &&= \mathit{targets}\ r \circ \mathit{Rep}\ r
\end{aligned}
$$

(The transformation is only meaning-preserving if the accumulation is based on a monoid. Alas, concatenation of regular expressions is not associative, so *targets* and *targets′* produce sets of syntactically different regular expressions that are, however, semantically equivalent: $[\![\mathit{targets}'\ (r, \beta)]\!] = [\![\mathit{targets}\ r \circ \beta]\!]$. We choose to ignore this technicality.)

Theorem 2. *Antimirov's subfactors are contained in Manna's automaton.*

$$\mathit{subfactors}\ r \subseteq \mathit{targets}\ r \tag{14a}$$

$$\mathit{targets}^{\dagger}\ (\mathit{targets}\ r) \subseteq \mathit{targets}\ r \tag{14b}$$

Recall that $f^{\dagger}$ is the Kleisli extension of f to sets.

We first establish the following properties of *targets*.

$$\mathit{targets}\ (\mathit{cat}\ r\ s) \subseteq \mathit{targets}\ r \circ s \cup \mathit{targets}\ s \tag{14c}$$

$$\mathit{targets}^{\dagger}\ (rs\ \circ\ s) \subseteq \mathit{targets}^{\dagger}\ rs \circ s \cup \mathit{targets}\ s \tag{14d}$$

The proof of (14c) is straightforward and omitted. (If r is non-null, the inequality can be strengthened to an equality.) For (14d) we reason,

$$
\begin{aligned}
& \mathit{targets}^{\dagger}\ (rs \circ s) \\
= \quad & \{ \text{ definition of } f^{\dagger} \ \} \\
& \textstyle\bigcup \{ \mathit{targets}\ (\mathit{cat}\ r\ s) \mid r \in rs \} \\
\subseteq \quad & \{ \text{ property of of } \mathit{targets} \text{ (14c) } \} \\
& \textstyle\bigcup \{ \mathit{targets}\ r \circ s \cup \mathit{targets}\ s \mid r \in rs \} \\
= \quad & \{ \text{ set union } \} \\
& \textstyle\bigcup \{ \mathit{targets}\ r \circ s \mid r \in rs \} \cup \mathit{targets}\ s \\
= \quad & \{ \text{ ``}\circ\text{'' distributes over set union } \} \\
& \textstyle\bigcup \{ \mathit{targets}\ r \mid r \in rs \} \circ s \cup \mathit{targets}\ s \\
= \quad & \{ \text{ definition of } f^{\dagger} \ \} \\
& \mathit{targets}^{\dagger}\ rs \circ s \cup \mathit{targets}\ s
\end{aligned}
$$

Proof (Theorem 2). It is fairly obvious that *subfactors* r is a subset of *targets* r as only the clause for concatenation is different. The proof of (14b) proceeds by induction of the structure of r.

Case *Empty* **and** *Eps***:**

$$
\begin{aligned}
& \mathit{targets}^{\dagger}\ (\mathit{targets}\ \mathit{Empty}) \\
= \ & \{ \text{ definition of } \mathit{targets} \ \} \\
& \mathit{targets}^{\dagger}\ \emptyset \\
= \ & \{ \text{ definition of } f^{\dagger} \ \} \\
& \emptyset
\end{aligned}
$$

Case *Alt r s***:**

$$
\begin{aligned}
& \mathit{targets}^{\dagger}\ (\mathit{targets}\ (\mathit{Alt}\ r\ s)) \\
= \ & \{ \text{ definition of } \mathit{targets} \ \} \\
& \mathit{targets}^{\dagger}\ (\mathit{targets}\ r \cup \mathit{targets}\ s) \\
= \ & \{ \ f^{\dagger} \text{ distributes over set union } \} \\
& \mathit{targets}^{\dagger}\ (\mathit{targets}\ r) \cup \mathit{targets}^{\dagger}\ (\mathit{targets}\ s) \\
\subseteq \ & \{ \text{ induction assumption and monotonicity } \} \\
& \mathit{targets}\ r \cup \mathit{targets}\ s \\
= \ & \{ \text{ definition of } \mathit{targets} \ \} \\
& \mathit{targets}\ (\mathit{Alt}\ r\ s)
\end{aligned}
$$

Case *Sym a***:**

$$
\begin{aligned}
& \mathit{targets}^{\dagger}\ (\mathit{targets}\ (\mathit{Sym}\ a)) \\
= \ & \{ \text{ definition of } \mathit{targets} \ \} \\
& \mathit{targets}^{\dagger}\ \{\mathit{Eps}\} \\
= \ & \{ \text{ definition of } f^{\dagger} \ \} \\
& \bigcup \{\mathit{targets}\ \mathit{Eps}\} \\
= \ & \{ \text{ definition } \mathit{targets} \text{ and } \bigcup\{\emptyset\} = \emptyset \ \} \\
& \emptyset
\end{aligned}
$$

Case *Cat r s*:

$$
\begin{aligned}
& \mathit{targets}^{\dagger}\ (\mathit{targets}\ (\mathit{Cat}\ r\ s)) \\
= \ & \{\ \text{definition of } \mathit{targets}\ \} \\
& \mathit{targets}^{\dagger}\ (\mathit{targets}\ r \circ s \cup \mathit{targets}\ s) \\
= \ & \{\ f^{\dagger} \text{ distributes over set union }\} \\
& \mathit{targets}^{\dagger}\ (\mathit{targets}\ r \circ s) \cup \mathit{targets}^{\dagger}\ (\mathit{targets}\ s) \\
= \ & \{\ \text{(14d)}\ \} \\
& \mathit{targets}^{\dagger}\ (\mathit{targets}\ r) \circ s \cup \mathit{targets}\ s \cup \mathit{targets}^{\dagger}\ (\mathit{targets}\ s) \\
\subseteq \ & \{\ \text{induction assumption and monotonicity}\ \} \\
& \mathit{targets}\ r \circ s \cup \mathit{targets}\ s \cup \mathit{targets}\ s \\
= \ & \{\ \text{definition of } \mathit{targets} \text{ and idempotency}\ \} \\
& \mathit{targets}\ (\mathit{Cat}\ r\ s)
\end{aligned}
$$

Case *Rep r*:

$$
\begin{aligned}
& \mathit{targets}^{\dagger}\ (\mathit{targets}\ (\mathit{Rep}\ r)) \\
= \ & \{\ \text{definition of } \mathit{targets}\ \} \\
& \mathit{targets}^{\dagger}\ (\mathit{targets}\ r \circ \mathit{Rep}\ r) \\
= \ & \{\ \text{(14d)}\ \} \\
& \mathit{targets}^{\dagger}\ (\mathit{targets}\ r) \circ \mathit{Rep}\ r \cup \mathit{targets}\ (\mathit{Rep}\ r) \\
\subseteq \ & \{\ \text{induction assumption and monotonicity}\ \} \\
& \mathit{targets}\ r \circ \mathit{Rep}\ r \cup \mathit{targets}\ (\mathit{Rep}\ r) \\
= \ & \{\ \text{definition of } \mathit{targets} \text{ and idempotency}\ \} \\
& \mathit{targets}\ (\mathit{Rep}\ r)
\end{aligned}
$$

We may conclude that a regular expression r has only a finite number of *syntactically* different right subfactors, as each subfactor appears as a target state in the corresponding railroad diagram. Moreover, subfactors have a very simple structure: they are compositions of sub-expressions of r.

Just in case you wonder, the converse is not true—not every target is also an Antimirov subfactor. If the regular expression contains *Empty* as a sub-expression, then Manna's automaton may contain unreachable states. Consider, for example, *Cat* (*Cat Empty a*) *b*.

8 Related Work

Manna's Generalized Transition Graphs. Our diagrams are modelled after Manna's generalized transition graphs, directed graphs labelled with regular expressions. For the discussion, it is useful to remind us of the different ways of defining labelled graphs.

$$\begin{array}{ccccccc} \text{labelled graph} & & \text{state-transition} & & \text{adjacency list} & & \text{adjacency matrix} \\ \mathcal{P}(V \times \Sigma \times V) & \cong & V \times \Sigma \to \mathcal{P}(V) & \cong & V \to \mathcal{P}(\Sigma \times V) & \cong & V \times V \to \mathcal{P}(\Sigma) \\ (i, a, j) \in G & & \delta(i, a) \ni j & & \mathit{lf}\ i \ni (a, j) & & i \circ\!\!-\!\!\circ j \ni a \end{array}$$

The isomorphisms are based on the one-to-one correspondence between relations and set-valued functions, $\mathcal{P}(A \times B) \cong A \to \mathcal{P}(B)$. Each "view" is in use. Manna models deterministic finite automata as labelled graphs with certain restrictions on the edges to ensure determinacy. The standard definition of nondeterministic finite automata is based on state-transition functions [15]. The adjacency list representation underlies Antimirov's linear forms where $V := \mathcal{P}(\Sigma)$, see Sect. 7. The adjacency matrix representation emphasizes the generative nature of automata, see Sect. 5. (Actually, there is also a fifth alternative, $\Sigma \to \mathcal{P}(V \times V)$, which, however, does not seem to be popular.)

The design space has a further dimension: we can equip the type of labels with structure. Manna first generalizes DFAs to transition graphs by allowing words as labels and then to generalized transition graphs, where arrows are labelled with regular expressions.

Σ	symbol	NFA
$\{\epsilon\} \cup \Sigma$	basic symbol	NFA with ϵ-transitions
Σ^*	word	Manna's transition graphs
$\mathcal{P}(\Sigma^*)$	language	Manna's generalized transition graphs

Generalizing Σ^* to an arbitrary monoid, we obtain the diagrams of Sect. 5, which form a regular algebra. They are isomorphic to Backhouse's matrices [4], where the underlying regular algebra is given by a lifted monoid.

$$\begin{array}{ccc} \text{diagram} & & \text{matrix} \\ \mathcal{P}(V \times M \times V) & \cong & V \times V \to \mathcal{P}(M) \\ (i, w, j) \in G & & i \circ\!\!-\!\!\circ j \ni w \end{array}$$

The isomorphism is also a regular homomorphism.

The McNaughton-Yamada-Thompson Algorithm. One of the first algorithms for converting a regular expression to a DFA is due to McNaughton and Yamada [13]. Though not spelled out explicitly, their algorithm first constructs an NFA without ϵ-transitions, which is subsequently converted to a DFA using the subset construction. In the first phase, they annotate each symbol of the regular expression with a position, which corresponds roughly to a station and its target point in our setting. The NFA is obtained by suitably connecting "terminal" and "initial" positions of sub-expressions for composition and iteration. Nonetheless, the idea of using an NFA as an intermediary is usually attributed to Thompson [16].

A standard textbook algorithm, the McNaughton-Yamada-Thompson algorithm [2], is based on their ideas. The algorithm proceeds by induction over the structure of the regular expression, as illustrated in Fig. 5. Each automaton has one start state (with no outgoing transitions) and one accepting state (with no incoming transitions). To avoid interference, states must be suitably renamed when sub-automata are combined. (The exact nature of the states is actually somewhat unclear—in the illustration they are not even named.) The resulting automata feature quite a few ϵ-transitions, not all of which are present in the original articles.

It is instructive to scrutinize Thompson's algorithm for "regular expression search" [16]. He explains the workings of his compiler using $a \cdot (b \mid c) \cdot d$ as a running example, illustrating the steps with diagrams. Interestingly, the illustrations are quite close to railway diagrams. Consequently, his algorithm can be easily recast in our framework, see Fig. 6. Thompson uses ϵ-transitions only for choice (and iteration which involves choice). Renaming of states is necessary for sub-automata that do not end in ϵ.

In this case, β must be appended to each state. Turning to an implementation in Haskell, the renaming operation has a familiar ring.

$$
\begin{array}{l}
\textbf{infixr } 7 \odot \\
(\odot) :: \mathit{Set}\ (\mathit{Arrow}\ a) \to \mathit{Reg} \to \mathit{Set}\ (\mathit{Arrow}\ a) \\
g \odot \mathit{Empty} = \emptyset \\
g \odot \mathit{Eps} \quad\; = g \\
g \odot s \qquad\;\; = [(\mathit{cat}\ \alpha\ s, r, \mathit{cat}\ \beta\ s) \mid (\alpha, r, \beta) \leftarrow g]
\end{array}
$$

We have introduced similar operations for Antimirov's linear forms and subfactors. Observe that the diagram $G \odot \beta$ is admissible if G is.

Thompson's translation is then captured by the following Haskell program (ignoring the fact that his compiler actually produces IBM 7094 machine code).

$$
\begin{array}{ll}
\multicolumn{2}{l}{\mathit{thompson} :: \mathit{Reg} \to \mathit{Set}\ (\mathit{Reg}, \mathit{Basic\ Alphabet}, \mathit{Reg})} \\
\mathit{thompson}\ \mathit{Eps} & = \{(\mathit{Eps}, \mathit{Eps}', \mathit{Eps})\} \\
\mathit{thompson}\ (\mathit{Sym}\ a) & = \{(\mathit{Sym}\ a, \mathit{Sym}'\ a, \mathit{Eps})\} \\
\mathit{thompson}\ (\mathit{Cat}\ r\ s) & = \mathit{thompson}\ r \odot s \cup \mathit{thompson}\ s \\
\mathit{thompson}\ \mathit{Empty} & = \emptyset \\
\mathit{thompson}\ (\mathit{Alt}\ r\ s) & = \{(\mathit{Alt}\ r\ s, \mathit{Eps}', r)\} \\
& \quad \cup \{(\mathit{Alt}\ r\ s, \mathit{Eps}', s)\} \cup \mathit{thompson}\ r \cup \mathit{thompson}\ s \\
\mathit{thompson}\ (\mathit{Rep}\ r) & = \{(\mathit{Rep}\ r, \mathit{Eps}', \mathit{Eps})\} \\
& \quad \cup \{(\mathit{Rep}\ r, \mathit{Eps}', \mathit{cat}\ r\ (\mathit{Rep}\ r))\} \cup \mathit{thompson}\ r \odot \mathit{Rep}\ r
\end{array}
$$

Even though similar in appearance, Thompson's compiler is quite different from Manna's construction. Manna's algorithm is *iterative* or top-down: *diagram* iteratively replaces a single arrow by a smallish diagram. Thompson's algorithm is *recursive* or bottom-up: *thompson* recursively combines sub-automata for subexpressions. For composition (and iteration which involves composition), this requires explicit renaming of states. In Manna's construction renaming is, in some sense, implicit through the use of an accumulating parameter. (As an aside, *thompson* also enables sharing of sub-automata as renaming does not operate on anonymous states, but on subfactors, which are semantically meaningful; consider, for example, $\mathit{Alt}\ (\mathit{Cat}\ r\ t)\ (\mathit{Cat}\ s\ t)$ and $\mathit{Cat}\ (\mathit{Alt}\ r\ s)\ t$.)

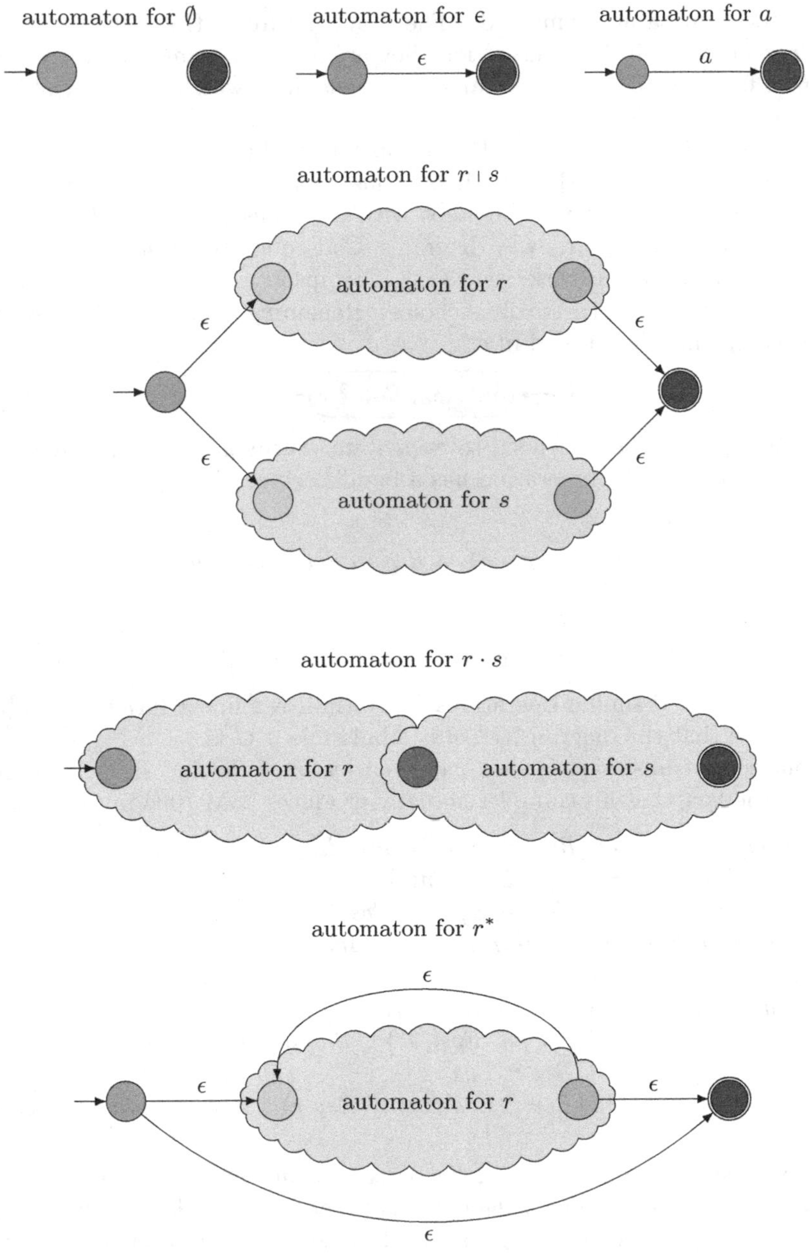

Fig. 5. McNaughton-Yamada-Thompson construction of an NFA with ϵ-transitions.

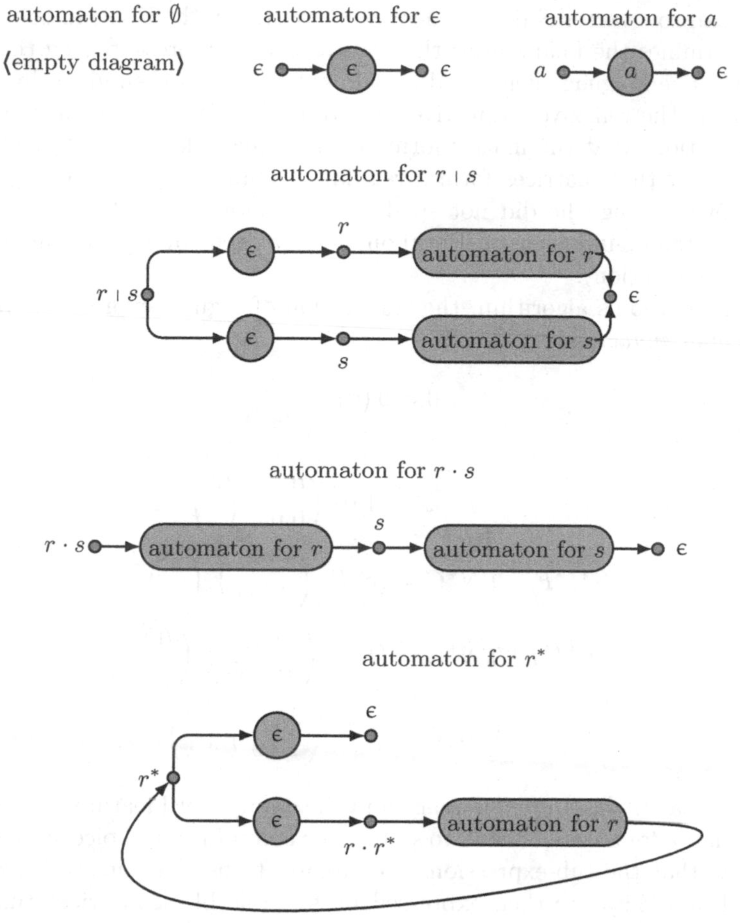

Fig. 6. Thompson's "original" construction of an NFA with ϵ-transitions.

Conways's Linear Mechanisms. A related construction is given by Conway [8]. In his seminal book on "Regular Algebra and Finite Machines" he defines a *linear mechanism*, which amounts to a nondeterministic finite automaton with ϵ-transitions. Interestingly, his automata feature both multiple final states *and* multiple start states, which allows for a very symmetric treatment.

He represents an automaton by an $n \times n$ matrix, where n is the number of vertices. To illustrate, the automaton of Fig. 3 is captured by

$$S = \begin{pmatrix}1\,0\,0\,0\end{pmatrix} \qquad M = \begin{pmatrix} 0 & 1 & 0 & 0 \\ a+b & 0 & a & 0 \\ 0 & 0 & 0 & a+b \\ 0 & 0 & 0 & 0 \end{pmatrix} \qquad F = \begin{pmatrix}0\\0\\0\\1\end{pmatrix}$$

The Boolean row vector S determines the start states; the Boolean column vector F determines the final states; the transitions are represented by the square matrix M. The language generated by the automaton is then given by SM^*F, where M^* is the reflexive, transitive closure of M. On a historical note, Conway [8] mentions that the matrix formula was already known to P. J. Cleave in 1961. The fact that matrices form a regular algebra was probably apparent to Conway, even though he did not spell out the details. Nonetheless, Conway's work had a major influence on Backhouse's treatment of regular algebra (personal communication).

Like Thompson's algorithm, the translation of regular expressions proceeds *recursively* or *inductively*.

$$\begin{aligned}
0 &= 0\,(0)^*\,0 \\
1 &= 1\,(1)^*\,1 \\
a &= (1\;0)\begin{pmatrix}0 & a\\ 0 & 0\end{pmatrix}^*\begin{pmatrix}0\\1\end{pmatrix} \\
SM^*F + TN^*G &= (S\;T)\begin{pmatrix}M & 0\\ 0 & N\end{pmatrix}^*\begin{pmatrix}F\\G\end{pmatrix} \\
SM^*F \cdot TN^*G &= (S\;0)\begin{pmatrix}M & FT\\ 0 & N\end{pmatrix}^*\begin{pmatrix}0\\G\end{pmatrix} \\
(SM^*F)^* &= (0\;1)\begin{pmatrix}M & F\\ S & 0\end{pmatrix}^*\begin{pmatrix}0\\1\end{pmatrix}
\end{aligned}$$

The automata for the empty language and the empty word feature a single state; the automaton for a symbol has two states. For composition, choice, and iteration we assume that the sub-expressions are already translated into automata. The combined automata are then expressed by suitable block matrices that detail the interaction of the component automata. For example, the automaton for iteration r^* adds one state to the automaton for r; this state is both a start and a final state; there is an ϵ-transition from the new state to each start state of r and from each final state of r to the new state.

Like the McNaughton-Yamada-Thompson construction, Conway's automata feature quite a few ϵ-transitions as there is no sharing of sub-automata. For example, the automaton for composition adds an ϵ-transition from each final state of the first to each initial state of the second automaton (represented by the matrix FT). On the positive side, the correctness of the translation can be readily established using the following characterization of iteration [8].

$$\begin{pmatrix}A & B\\ C & D\end{pmatrix}^* = \begin{pmatrix}X^* & A^*BY^*\\ D^*CX^* & Y^*\end{pmatrix} \quad \textbf{where} \quad \begin{cases} X = A + BD^*C \\ Y = D + CA^*B \end{cases}$$

The block matrix consists of two square matrices A and D, which represent sub-automata, and two rectangular matrices B and C, which record transitions between the sub-automata. The entries on the right specify the possible paths in the combined automaton, for example, A^*BY^* contains all paths from A to D: a path in A (ie A^*), followed by an edge from A to D (ie B), followed by a path from D to D (ie Y^*). Given this decomposition, the correctness of the translation can be shown using straightforward equational reasoning.

Brzozowski's Factors and Antimirov's Subfactors. Brzozowski [6] showed that a regular expression has only a finite number of *syntactically* different factors, if expressions are compared modulo associativity, commutativity, and idempotence of choice. Antimirov [3] pointed out that some care has to be exercised when computing the factors. Brzozowski uses the following formula for composition: $a \setminus (r \cdot s) = (a \setminus r) \cdot s + \delta\, r \cdot (a \setminus s)$, where $\delta\, r = 1$ if r is nullable and $\delta\, r = 0$ otherwise. To ensure finiteness, the definition of δ must actually be unfolded: $a \setminus (r \cdot s) = \textbf{if}\ \textit{nullable}\ r\ \textbf{then}\ (a \setminus r) \cdot s + (a \setminus s)\ \textbf{else}\ (a \setminus r) \cdot s$.

Antimirov further realized that it is sufficient to apply the ACI-properties of choice only to the top-level of a term, see his definition of linear form. His approach essentially derives a system of equations of the form (2) from a regular expression. Based on the same idea, Mirkin [14] gave an algorithm for constructing an NFA, predating Antimirov's work by almost two decades. Champarnaud and Ziadi [7] pointed out the similarity, attempting to show that the two approaches actually yield the same automata. Unfortunately, their proof contains an error, which was later corrected by Broda et al. [5]. In more detail, Champarnaud and Ziadi claim that the function *subfactors*, which they call π, computes the immediate *and* transitive subfactors, whereas it only determines the former. Broda et al. pointed out that *subfactors* must be replaced by *targets*. (Contrary to their claim, even then the two constructions are not identical as pointed out in Sect. 7: *targets* may include unreachable states not present in Antimirov's construction. A minor technicality, which can be fixed by excluding *Empty* as a constructor.) The corrected definition of π seems to fall out of thin air though—it is pleasing to see that it is obtained as a projection of Manna's automaton.

9 Conclusion

Regular algebra is the algebra of three operators central to programming: composition, choice, and iteration. As such, it is perhaps the most fundamental algebraic structure in computing science.

Roland Backhouse

I could not agree more. We have used regular algebra to reason both about languages and diagrams. Students of computing science should see at least a glimpse of regular algebra in their first term. Regular expressions and railroad diagrams provide an ideal starting point. Manna's construction, which ties the two concepts together, is both charming and challenging. It is charming because the transformations are local, supporting an iterative, step by step refinement of diagrams. It is challenging for the same reason: one has to ensure that different parts of the diagram do not interfere. This is where subfactors (called partial derivatives elsewhere) enter the scene. Standard textbook proofs of the equivalence of regular expressions and nondeterministic finite automata often involve verbose arguments, making implicit assumptions about disjointness of state sets ("Here, i is a new state, ..."). By contrast, subfactors facilitate simple, calculational correctness proofs, based on fundamental properties of Galois connections. What equational reasoning with factors is for DFAs, inequational reasoning with subfactors is for NFAs.

(On a personal note, I think that it is a mistake to introduce a finite automaton in this particular context as a quintuple $(\Sigma, S, s_0, F, \delta)$ where S is some anonymous, unstructured set of states. Subfactors as states serve as important scaffolding that should only be removed in a final abstraction step—once Kleene's Theorem is established or other uses for finite automata have been introduced.)

Despite its simplicity, Manna's construction has a number of pleasing characteristics: the number of states and the number of edges is linear in the size of the regular expression; due to sharing of sub-automata and auto-merging of states the resulting automaton is often surprisingly small. This demonstrates that disjointness of state sets is undesirable or, put differently, "renaming" should be semantically meaningful: "$\bullet$" and "$\odot$" can be seen as incarnations of the distributive law.

Finally, it has been satisfying to be able to relate Manna's construction to Antimirov's subfactors through simple program transformations, based on accumulating parameters.

Acknowledgements. A big thank you is due to Bernhard Möller for pointing me to Manna's textbook on the "Mathematical Theory of Computation" [12]. Thanks are also due to Clare Martin and Sebastian Schweizer for proof-reading a preliminary version of the manuscript. The TikZ diagrams for Thompson's construction, see Fig. 5, are due to Ian McLoughlin, who kindly granted permission to use them. As always, the anonymous referees of MPC provided detailed feedback, suggesting numerous improvements regarding presentation and organization of the material. In response to their comments I moved the section on "Railroad Diagrams" to the front and I clarified Antimirov's contribution. Finally, I owe a particular debt of gratitude to Roland Backhouse, who provided numerous comments, pointing out, in particular, the relevance of Conway's work. The paragraph on "linear mechanisms" was added in response to his suggestions.

References

1. Abramsky, S., Vickers, S.: Quantales, observational logic and process semantics. Math. Struct. Comput. Sci. **3**(2), 161–227 (1993). https://doi.org/10.1017/S0960129500000189
2. Aho, A.V., Lam, M.S., Sethi, R., Ullman, J.D.: Compilers: Principles, Techniques, & Tools, 2nd edn. Pearson Addison-Wesley, Boston (2007)

3. Antimirov, V.: Partial derivatives of regular expressions and finite automaton constructions. Theor. Comput. Sci. **155**(2), 291–319 (1996). https://doi.org/10.1016/0304-3975(95)00182-4. http://www.sciencedirect.com/science/article/pii/0304397595001824
4. Backhouse, R.: Regular algebra applied to language problems. J. Logic Algebraic Program. **66**(2), 71–111 (2006). https://doi.org/10.1016/j.jlap.2005.04.008. http://www.sciencedirect.com/science/article/pii/S1567832605000329
5. Broda, S., Machiavelo, A., Moreira, N., Reis, R.: On the average state complexity of partial derivative automata: an analytic combinatorics approach. Int. J. Found. Comput. Sci. **22**(07), 1593–1606 (2011). https://doi.org/10.1142/S0129054111008908
6. Brzozowski, J.A.: Derivatives of regular expressions. J. ACM **11**(4), 481–494 (1964). https://doi.org/10.1145/321239.321249. http://doi.acm.org/10.1145/321239.321249
7. Champarnaud, J.M., Ziadi, D.: From Mirkin's prebases to Antimirov's word partial derivatives. Fundam. Inf. **45**(3), 195–205 (2001)
8. Conway, J.H.: Regular Algebra and Finite Machines. Chapman and Hall, London (1971)
9. Jensen, K., Wirth, N.: Pascal: User Manual and Report, 2nd edn. Springer, Heidelberg (1978)
10. Kleene, S.C.: Representation of events in nerve nets and finite automata. Technical report, RM-704, U.S. Air Force, Project RAND, Research Memorandum, December 1951
11. Mac Lane, S.: Categories for the Working Mathematician. Graduate Texts in Mathematics, 2nd edn. Springer, Heidelberg (1998)
12. Manna, Z.: Introduction to Mathematical Theory of Computation. McGraw-Hill Book Company, New York (1974)
13. McNaughton, R., Yamada, H.: Regular expressions and state graphs for automata. IRE Trans. Electron. Comput. **EC-9**(1), 39–47 (1960). https://doi.org/10.1109/TEC.1960.5221603
14. Mirkin, B.G.: An algorithm for constructing a base in a language of regular expressions. Eng. Cybern. **5**, 110–116 (1966)
15. Rabin, M.O., Scott, D.: Finite automata and their decision problems. IBM J. Res. Dev. **3**(2), 114–125 (1959). https://doi.org/10.1147/rd.32.0114
16. Thompson, K.: Programming techniques: Regular expression search algorithm. Commun. ACM **11**(6), 419–422 (1968). https://doi.org/10.1145/363347.363387. http://doi.acm.org/10.1145/363347.363387

How to Calculate with Nondeterministic Functions

Richard Bird[1(✉)] and Florian Rabe[2]

[1] Department of Computer Science, Oxford University, Wolfson Building, Parks Road, Oxford OX1 3QD, UK
bird@cs.ox.ac.uk

[2] Laboratoire de Recherche en Informatique, University Paris Sud, Rue Noetzlin, 91405 Orsay Cedex, France

Abstract. While simple equational reasoning is adequate for the calculation of many algorithms from their functional specifications, it is not up to the task of dealing with others, particularly those specified as optimisation problems. One approach is to replace functions by relations, and equational reasoning by reasoning about relational inclusion. But such a wholesale approach means one has to adopt a new and sometimes subtle language to argue about the properties of relational expressions. A more modest proposal is to generalise our powers of specification by allowing certain nondeterministic, or multi-valued functions, and to reason about refinement instead. Such functions will not appear in any final code. Refinement calculi have been studied extensively over the years and our aim in this article is just to explore the issues in a simple setting and to justify the axioms of refinement using the semantics suggested by Morris and Bunkenburg.

1 Introduction

We set the scene by considering the following Haskell definition for an archetypal optimisation problem:

$$\begin{aligned}&\mathit{mcc} :: [\mathit{Item}] \to \mathit{Candidate}\\ &\mathit{mcc} = \mathit{minWith}\ \mathit{cost} \cdot \mathit{candidates}\end{aligned}$$

The function *mcc* computes a candidate with minimum cost. The function *minWith* can be defined by

$$\begin{aligned}&\mathit{minWith} :: \mathit{Ord}\ b \Rightarrow (a \to b) \to [a] \to a\\ &\mathit{minWith}\ f = \mathit{foldr1}\ \mathit{smaller}\\ &\qquad\qquad\mathbf{where}\ \mathit{smaller}\ x\ y = \mathbf{if}\ f\ x \leqslant f\ y\ \mathbf{then}\ x\ \mathbf{else}\ y\end{aligned}$$

Applied to a finite, nonempty list of candidates, *minWith cost* returns the first candidate with minimum cost. The function *candidates* takes a finite list of items and returns a finite, nonempty list of candidates. We will suppose that the construction uses *foldr*:

G. Hutton (Ed.): MPC 2019, LNCS 11825, pp. 138–154, 2019.
https://doi.org/10.1007/978-3-030-33636-3_6

$$\begin{aligned}&\mathit{candidates} :: [\mathit{Item}] \rightarrow [\mathit{Candidate}]\\&\mathit{candidates}\ \mathit{xs} = \mathit{foldr}\ \mathit{step}\ [c_0]\ \mathit{xs}\\&\qquad\qquad\qquad\mathbf{where}\ \mathit{step}\ x\ \mathit{cs} = \mathit{concatMap}\ (\mathit{additions}\ x)\ \mathit{cs}\end{aligned}$$

The value c_0 is some default candidate for an empty list of items. The function *concatMap* is defined by

$$\mathit{concatMap}\ f = \mathit{concat} \cdot \mathit{map}\ f$$

and $\mathit{additions} :: \mathit{Item} \rightarrow \mathit{Candidate} \rightarrow [\mathit{Candidate}]$ takes a new item and a candidate and constructs a nonempty list of extended candidates. For example, if the candidates were the permutations of a list, then c_0 would be the empty list and *additions* x would be a list of all the ways x can be inserted into a given permutation. For example,

$$\mathit{additions}\ 1\ [2,4,3] = [[1,2,4,3],[2,1,4,3],[2,4,1,3],[2,4,3,1]]$$

A greedy algorithm for *mcc* arises as the result of successfully fusing the function *minWith cost* with *candidates*. Operationally speaking, instead of building the complete list of candidates and then selecting a best one, we construct a single best candidate at each step. The usual formulation of the fusion rule for *foldr* states that

$$\mathit{foldr}\ f\ (h\ e)\ \mathit{xs} = h\ (\mathit{foldr}\ g\ e\ \mathit{xs})$$

for all finite lists xs provided the fusion condition

$$h\ (g\ x\ y) = f\ x\ (h\ y)$$

holds for all x and y. In fact the fusion condition is required to hold only for all y of the form $y = \mathit{foldr}\ g\ e\ \mathit{xs}$; this version is called *context-sensitive* fusion.

For our problem, $h = \mathit{minWith}\ \mathit{cost}$ and $g = \mathit{step}$ but f is unknown. Abbreviating *candidates xs* to *cs*, the context-sensitive fusion condition reads

$$\mathit{minWith}\ \mathit{cost}\ (\mathit{step}\ x\ \mathit{cs}) = \mathit{add}\ x\ (\mathit{minWith}\ \mathit{cost}\ \mathit{cs})$$

for some function *add*. To see if it holds, and to discover *add* in the process, we can reason:

$$\begin{aligned}&\quad \mathit{minWith}\ \mathit{cost}\ (\mathit{step}\ x\ \mathit{cs})\\=&\quad \{\ \text{definition of } \mathit{step}\ \}\\&\quad \mathit{minWith}\ \mathit{cost}\ (\mathit{concatMap}\ (\mathit{additions}\ x)\ \mathit{cs})\\=&\quad \{\ \text{distributive law (see below)}\ \}\\&\quad \mathit{minWith}\ \mathit{cost}\ (\mathit{map}\ (\mathit{minWith}\ \mathit{cost} \cdot \mathit{additions}\ x)\ \mathit{cs})\\=&\quad \{\ \text{define } \mathit{add}\ x = \mathit{minWith}\ \mathit{cost} \cdot \mathit{additions}\ x\ \}\\&\quad \mathit{minWith}\ \mathit{cost}\ (\mathit{map}\ (\mathit{add}\ x)\ \mathit{cs})\\=&\quad \{\ \text{greedy condition (see below)}\ \}\\&\quad \mathit{add}\ x\ (\mathit{minWith}\ \mathit{cost}\ \mathit{cs})\end{aligned}$$

The distributive law used in the second step is the fact that

$$\mathit{minWith}\ f\ (\mathit{concat}\ \mathit{xss}) = \mathit{minWith}\ f\ (\mathit{map}\ (\mathit{minWith}\ f)\ \mathit{xss})$$

provided *xss* is a finite list of finite, nonempty lists. Equivalently,

$$\mathit{minWith}\ f\ (\mathit{concatMap}\ g\ \mathit{xs}) = \mathit{minWith}\ f\ (\mathit{map}\ (\mathit{minWith}\ f \cdot g)\ \mathit{xs})$$

provided *xs* is a finite list and *g* returns finite, nonempty lists. The proof of the distributivity law is straightforward but we omit details.

Summarising this short calculation, we have shown that

$$\mathit{mcc} = \mathit{foldr}\ \mathit{add}\ c_0\ \textbf{where}\ \mathit{add}\ x = \mathit{minWith}\ \mathit{cost} \cdot \mathit{additions}\ x$$

provided the following *greedy condition* holds for all x and *xs*:

$$\mathit{minWith}\ \mathit{cost}\ (\mathit{map}\ (\mathit{add}\ x)\ \mathit{cs}) = \mathit{add}\ x\ (\mathit{minWith}\ \mathit{cost}\ \mathit{cs})$$

where $\mathit{cs} = \mathit{candidates}\ \mathit{xs}$.

That all seems simple enough. However, the fly in the ointment is that, in order to establish the greedy condition when there may be more than one candidate in *cs* with minimum cost, we need to prove the very strong fact that

$$\mathit{cost}\ c \leqslant \mathit{cost}\ c' \quad \Leftrightarrow \quad \mathit{cost}\ (\mathit{add}\ x\ c) \leqslant \mathit{cost}\ (\mathit{add}\ x\ c') \tag{1}$$

for all candidates c and c' in *cs*. To see why, observe that if c is the first candidate with minimum cost in a list of candidates, then $\mathit{add}\ x\ c$ has to be the first candidate with minimum cost in the list of extended candidates. This follows from our definition of *minWith* which selects the first element with minimum cost in a list of candidates. To ensure that the extension of a candidate c' earlier in the list has a larger cost we have to show that

$$\mathit{cost}\ c' > \mathit{cost}\ c \quad \Rightarrow \quad \mathit{cost}\ (\mathit{add}\ x\ c') > \mathit{cost}\ (\mathit{add}\ x\ c) \tag{2}$$

for all c and c' in *cs*. To ensure that the extension of a candidate c' later in the list does not have a smaller cost we have to show that

$$\mathit{cost}\ c \leqslant \mathit{cost}\ c' \quad \Rightarrow \quad \mathit{cost}\ (\mathit{add}\ x\ c) \leqslant \mathit{cost}\ (\mathit{add}\ x\ c') \tag{3}$$

for all c and c' in *cs*. The conjunction of (2) and (3) is (1). The problem is that (1) is so strong that it rarely holds in practice. As evidence for this assertion, the appendix briefly discusses one example. A similar condition is needed if, say, *minWith* returned the last element in a list with minimum cost, so the problem is not to do with the specific definition of *minWith*. What we really need is a form of reasoning that allows us to establish the necessary fusion condition from the simple monotonicity condition (3) alone, and the plain fact of the matter is that equational reasoning with any definition of *minWith* is simply not adequate to provide it.

It follows that we have to abandon equational reasoning. One approach is to replace our functional framework with a relational one, and to reason instead about the inclusion of one relation in another. Such an approach has been suggested in a number of places, including our own [1]. But, for the purposes of presenting a simple introduction to the subject of greedy algorithms in Haskell, this solution is way too drastic, more akin to a heart transplant than a tube of solvent for occasional use. The alternative, if it can be made to work smoothly, is to introduce nondeterministic functions, also called *multi-valued* functions in mathematics, and to reason about refinement.

The necessary intuitions and syntax are introduced in Sect. 2. Section 3 gives a formal calculus and Sect. 4 a denotational semantics for our language. The soundness of the semantics establishes the consistency of the calculus. We have formalised syntax, calculus, and semantics in the logical framework LF [2] and are in the process of also formalizing the soundness proof; the formalisation is not given in this paper but is available online[1].

2 Nondeterminism and Refinement

Suppose we introduce *MinWith* as a nondeterministic function, specified only by the condition that if x is a possible value of $MinWith\ f\ xs$, where xs is a finite nonempty list, then x is an element of xs and for all elements y of xs we have $f\ x \leqslant f\ y$. Note the initial capital letter: *MinWith* is not part of Haskell. It is not our intention to extend Haskell with nondeterministic functions; instead nondeterminism is simply there to extend our powers of specification and cannot appear in any final algorithm.

Suppose we define $y \leftarrow F\ x$ to mean that y is one possible output of the nondeterministic function F applied to a value x. In words, y is a possible *refinement* of the nondeterministic expression $F\ x$. For example, $1 \leftarrow MinWith\ (const\ 0)\ [1,2]$ and $2 \leftarrow MinWith\ (const\ 0)\ [1,2]$. More generally, if E_1 and E_2 are possibly nondeterministic expressions of the same type T, we will write $E_1 \leftarrow E_2$ to mean that for all values v of T we have

$$v \leftarrow E_1 \;\Rightarrow\; v \leftarrow E_2$$

We define two nondeterministic expressions of the same type to be equal if they both have the same set of refinements: $E_1 = E_2$ if

$$v \leftarrow E_1 \;\Leftrightarrow\; v \leftarrow E_2$$

for all v. Equivalently,

$$E_1 = E_2 \;\Leftrightarrow\; E_1 \leftarrow E_2 \wedge E_2 \leftarrow E_1$$

which just says that $\leftarrow$ is anti-symmetric. Our task is to make precise the exact rules allowed for reasoning about $\leftarrow$ and to prove that these rules do not lead to contradictions.

[1] https://github.com/florian-rabe/nondet.

To illustrate some of the pitfalls that have to be avoided, we consider three examples. First, here is the distributive law again in which *minWith* is replaced by *MinWith*:

$$\mathit{MinWith}\ f\ (\mathit{concat}\ \mathit{xss}) = \mathit{MinWith}\ f\ (\mathit{map}\ (\mathit{MinWith}\ f)\ \mathit{xss})$$

If this equation is to hold for all finite, nonempty lists *xss* of finite, nonempty lists, and we do indeed want it to, then it has to mean there is no refinement of one side that is not also a refinement of the other side. It does *not* mean that the equation should hold for all possible implementations of *MinWith*, and it cannot mean that because it is false. Suppose we define *minWith* to return the *second* best candidate in a list of candidates, or the only best candidate if there is only one. In particular,

$$\begin{array}{ll} \mathit{minWith}\ (\mathit{const}\ 0)\ (\mathit{concat}\ [[a],[b,c]]) & = b \\ \mathit{minWith}\ (\mathit{const}\ 0)\ (\mathit{map}\ (\mathit{minWith}\ (\mathit{const}\ 0))\ [[a],[b,c]]) & = c \end{array}$$

The results are different so the distributive law fails. What the distributive law has to mean is the conjunction of the following two assertions, in which M abbreviates *MinWith cost*:

$$\begin{array}{lll} x \leftarrow M\ (\mathit{concat}\ \mathit{xss}) & \Rightarrow & (\exists \mathit{xs} : \mathit{xs} \leftarrow \mathit{map}\ M\ \mathit{xss} \wedge x \leftarrow M\ \mathit{xs}) \\ (\mathit{xs} \leftarrow \mathit{map}\ M\ \mathit{xss} \wedge x \leftarrow M\ \mathit{xs}) & \Rightarrow & x \leftarrow M\ (\mathit{concat}\ \mathit{xss}) \end{array}$$

It is easy enough to show that these two assertions do hold though we omit details.

For the remaining two examples, define

$$\mathit{Choose}\ x\ y = \mathit{MinWith}\ (\mathit{const}\ 0)\ [x, y]$$

so $x \leftarrow \mathit{Choose}\ x\ y$ and $y \leftarrow \mathit{Choose}\ x\ y$. Do we have

$$\mathit{double}\ (\mathit{Choose}\ 1\ 2) = \mathit{Choose}\ 1\ 2 + \mathit{Choose}\ 1\ 2$$

where $\mathit{double}\ x = x + x$? The answer is no, because

$$\begin{array}{l} \quad x \leftarrow \mathit{double}\ (\mathit{Choose}\ 1\ 2) \\ \Leftrightarrow \exists y : y \leftarrow \mathit{Choose}\ 1\ 2 \wedge x = \mathit{double}\ y \\ \Leftrightarrow x = 2 \vee x = 4 \end{array}$$

while

$$\begin{array}{l} \quad x \leftarrow \mathit{Choose}\ 1\ 2 + \mathit{Choose}\ 1\ 2 \\ \Leftrightarrow \exists y, z : y \leftarrow \mathit{Choose}\ 1\ 2 \wedge z \leftarrow \mathit{Choose}\ 1\ 2 \wedge x = y + z \\ \Leftrightarrow x = 2 \vee x = 3 \vee x == 4 \end{array}$$

We have only that $\mathit{double}\ (\mathit{Choose}\ x\ y) \leftarrow \mathit{Choose}\ x\ y + \mathit{Choose}\ x\ y$.

For the third example, it is easy enough to show, for all f_1, f_2 and x that

$$\mathit{Choose}\ (f_1\ x)\ (f_2\ x) = \mathit{Choose}\ f_1\ f_2\ x$$

but it would be wrong to conclude by η conversion that

$$\lambda x.\ \mathit{Choose}\ (f_1\ x)\ (f_2\ x) = \mathit{Choose}\ f_1\ f_2$$

We have

$$f \leftarrow \lambda x.\ \mathit{Choose}\ (f_1\ x)\ (f_2\ x) \Leftrightarrow \forall x : f\ x = f_1\ x \vee f\ x = f_2\ x$$

However,

$$f \leftarrow \mathit{Choose}\ f_1\ f_2 \Leftrightarrow (\forall x : f\ x = f_1\ x) \vee (\forall x : f\ x = f_2\ x)$$

The results are different. The η rule, namely $f = \lambda x. f\ x$, does not hold if f is a nondeterministic function such as $\mathit{Choose}\ f_1\ f_2$.

What else do we want? Certainly, we want a refinement version of the fusion law for *foldr*, namely that over finite lists we have

$$\mathit{foldr}\ f\ e'\ xs \leftarrow H\ (\mathit{foldr}\ g\ e\ xs)$$

for all finite lists xs provided that $e' \leftarrow H\ e$ and $f\ x\ (H\ y) \leftarrow H\ (g\ x\ y)$. Here is the proof of the fusion law. The base case is immediate and the induction step is as follows:

$$\begin{array}{ll}
& \mathit{foldr}\ f\ e'\ (x : xs) \\
= & \quad \{\ \text{definition of } \mathit{foldr}\ \} \\
& f\ x\ (\mathit{foldr}\ f\ e'\ xs) \\
\leftarrow & \quad \{\ \text{induction, and monotonicity of refinement (see below)}\ \} \\
& f\ x\ (H\ (\mathit{foldr}\ g\ e\ xs)) \\
\leftarrow & \quad \{\ \text{fusion condition, and monotonicity of refinement}\ \} \\
& H\ (g\ x\ (\mathit{foldr}\ g\ e\ xs)) \\
= & \quad \{\ \text{definition of } \mathit{foldr}\ \} \\
& H\ (\mathit{foldr}\ g\ e\ (x : xs))
\end{array}$$

The appeal to the monotonicity of refinement is the assertion

$$E_1 \leftarrow E_2 \;\Rightarrow\; F\ E_1 \leftarrow F\ E_2$$

So this condition is also required to hold.

Let us see what else we might need by redoing the calculation of the greedy algorithm for *mcc*. This time we start with the specification

$$\mathit{mcc} \leftarrow \mathit{MinWith}\ \mathit{cost} \cdot \mathit{candidates}$$

For the fusion condition we reason:

$$
\begin{array}{ll}
& \mathit{MinWith}\ \mathit{cost}\ (\mathit{step}\ x\ \mathit{cs}) \\
= & \quad \{\ \text{definition of } \mathit{step}\ \} \\
& \mathit{MinWith}\ \mathit{cost}\ (\mathit{concatMap}\ (\mathit{additions}\ x)\ \mathit{cs}) \\
= & \quad \{\ \text{distributive law}\ \} \\
& \mathit{MinWith}\ \mathit{cost}\ (\mathit{map}\ (\mathit{MinWith}\ \mathit{cost} \cdot \mathit{additions}\ x)\ \mathit{cs}) \\
\rightarrow & \quad \{\ \text{suppose } \mathit{add}\ x \leftarrow \mathit{MinWith}\ \mathit{cost} \cdot \mathit{additions}\ x\ \} \\
& \mathit{MinWith}\ \mathit{cost}\ (\mathit{map}\ (\mathit{add}\ x)\ \mathit{cs}) \\
\rightarrow & \quad \{\ \text{greedy condition (see below)}\ \} \\
& \mathit{add}\ x\ (\mathit{MinWith}\ \mathit{cost}\ \mathit{cs})
\end{array}
$$

We write $E_1 \rightarrow E_2$ as an alternative to $E_2 \leftarrow E_1$. The second step makes use of the distributive law, and the third step is an instance of the monotonicity of refinement.

Let us now revisit the greedy condition. This time we only have to show

$$\mathit{add}\ x\ (\mathit{MinWith}\ \mathit{cost}\ \mathit{cs}) \leftarrow \mathit{MinWith}\ \mathit{cost}\ (\mathit{map}\ (\mathit{add}\ x)\ \mathit{cs})$$

where $\mathit{add}\ x \leftarrow \mathit{MinWith}\ \mathit{cost} \cdot \mathit{additions}\ x$. Unlike the previous version, this claim follows from the monotonicity condition (3). To spell out the details, suppose c is a candidate in cs with minimum cost. We have only to show that

$$\mathit{add}\ x\ c \leftarrow \mathit{MinWith}\ \mathit{cost}\ (\mathit{map}\ (\mathit{add}\ x)\ \mathit{cs})$$

Equivalently, that

$$\mathit{cost}\ (\mathit{add}\ x\ c) \leqslant \mathit{cost}\ (\mathit{add}\ x\ c')$$

for all candidates c' on cs. But this follows from (3) and the fact that $\mathit{cost}\ c \leqslant \mathit{cost}\ c'$.

Summarising, we can now define $\mathit{mcc} = \mathit{foldr}\ \mathit{add}\ c_0$ provided (3) holds for a suitable refinement of *add*. Unlike the previous calculation, the new one is sufficient to deal with most examples of greedy algorithms, at least when candidate generation is expressed in terms of *foldr*.

We have concentrated on greedy algorithms and the function *MinWith*, but there is another nondeterministic function *ThinBy*, which is needed in the study of thinning algorithms. Not every optimisation problem can be solved by a greedy algorithm, and between the extremes of maintaining just one candidate at each step and maintaining all possible candidates, there is the option of keeping only a subset of candidates in play. That is where *ThinBy* comes in. It is a function with type

$$\mathit{ThinBy} :: (a \rightarrow a \rightarrow \mathit{Bool}) \rightarrow [a] \rightarrow [a]$$

Thus $\mathit{ThinBy}\ (\ll)\ \mathit{xs}$ takes a comparison function $\ll$ and a list xs as arguments and returns a subsequence ys of xs such that for all x in xs there is a y in ys

with $y \ll x$. The subsequence is not specified further, so *ThinBy* is nondeterministic. We mention *ThinBy* to show that there is more than one nondeterministic function of interest in the study of deriving algorithms from specifications.

The task now before us is to find a suitable axiomatisation for a theory of refinement and to give a model to show the soundness and consistency of the axioms. Essentially, this axiomatisation is the one proposed in [3,4] but simplified by leaving out some details inessential for our purposes.

3 An Axiomatic Basis

Rather than deal with specific nondeterministic functions such as *MinWith* and *ThinBy*, we can phrase the required rules in terms of a binary choice operator ($\sqcap$). Thus,

$$E_1 \sqcap E_2 = \mathit{MinWith}\ (\mathit{const}\ 0)\ [E_1, E_2]$$

We also have

$$\mathit{MinWith}\ f\ xs = \mathit{foldr1}\ (\sqcap)\ [x \mid x \leftarrow xs, \mathit{and}\ [f\ x \leqslant f\ y \mid y \leftarrow xs]]$$

so *MinWith* can be defined in terms of ($\sqcap$). Below we write $\sqcap/$ for *foldr1* ($\sqcap$). Thus $\sqcap/$ takes a finite, nonempty list of arguments and returns an arbitrary element of the list.

To formulate the axioms we need a language of types and expressions, and we choose the simply-typed lambda calculus. Types are given by the grammar

$$T ::= B \mid T \rightarrow T$$

B consists of the base types, such as *Int* and *Bool*. We could have included pair types explicitly, as is done in [3], but for present purposes it is simpler to omit them. Expressions are given by the grammar

$$E ::= C \mid V \mid \sqcap/\ [E_1, E_2, ..., E_n] \mid E\ E \mid \lambda V : T.\, E$$

where $n > 0$ and each of $E_1, E_2, ..., E_n$ are expressions of the same type. We omit the type of the bound variable in a λ-abstraction if it can be inferred, and we write $E_1 \sqcap E_2$ for $\sqcap/\ [E_1, E_2]$. Included in the constants C are constant functions such as the addition function $+$ on integers (written infix as usual) and integer literals $0, 1, -1, ...$. The typing rules are standard; in particular, $\sqcap/\ [E_1, E_2, ..., E_n]$, has type T if all E_i do.

Boolean formulas are formed using equality $E_1 = E_2$ and refinement $E_1 \leftarrow E_2$ of expressions as well as universal and existential quantification and the propositional connectives in the usual way. We use the same type of Booleans both for programs and for formulas about them, but only some Boolean expressions are practical in programs (e.g., propositional connectives and equality at base types). Additionally, in order to state the axioms, we need a predicate $\mathit{pure}(E)$ to distinguish a subclass of expressions, called *pure* expressions. The intention is to define a semantics in which a pure expression denotes a single value, except for lambda abstractions with impure bodies, which denote a set of functions. We add rules such that $\mathit{pure}(E)$ holds if E is

- a constant C applied to any number of pure arguments (including C itself if there are no arguments),
- a lambda abstraction (independent of whether its body is pure).

Like any predicate symbol, purity is closed under equality, i.e., if E_1 is pure and we can prove $E_1 = E_2$, then so is E_2. For example, 2 and $E_1 + E_2$ for pure E_1 and E_2 are pure because 2 and $+$ are constants. Also $\lambda y.\, 1 \sqcap y$ is pure because it is a lambda abstraction, and $(\lambda x.\, \lambda y.\, x \sqcap y)\ 1$ is pure because it is equal by β-reduction (see below) to the former. Furthermore, $2 \sqcap 2$ is pure because it is equal to 2 (using the axioms given below), but $(\lambda y.\, 1 \sqcap y)\ 2$ and $1 \sqcap 2$ are both impure. In what follows we use lowercase letters for pure expressions and uppercase letters for possibly impure expressions.

The reason for introducing pure expressions is in the statement of our first two axioms, the rules of β and η conversion. The β rule is that if e is a pure expression, then

$$(\lambda x.\, E)\ e = E\ (x := e) \tag{4}$$

where $E\ (x := e)$ denotes the expression E with all free occurrences of x replaced by e. Intuitively, the purity restriction to β-reduction makes sense because the bound variable of the lambda abstraction only ranges over values and therefore may only be substituted with pure expressions.

The η rule asserts that if f is a pure function, then

$$f = \lambda x.\, f\ x \tag{5}$$

The purity restriction to η-expansion makes sense because lambda-abstractions are always pure and thus can never equal an impure function.

Our notion of purity corresponds to the *proper* expressions of [3] except that we avoid the axiom that variables are pure. Our first draft used that axiom, but we were unable to formalise the calculus until we modified that aspect. The reason why the axiom is problematic is that it forces a distinction between meta-variables (which may be impure) and object variables (which must be pure). That precludes using higher-order abstract syntax when representing and reasoning about the language, e.g., in a logical framework like [2], and highly complicates the substitution properties of the language. However, just like in [3], our binders will range only over values, which our calculus captures by adding a purity assumption for the bound variable whenever traversing into the body of a binder. For example, the ξ rule for equality reasoning under a lambda becomes:

$$\frac{\mathit{pure}(x)\ \vdash\ E = F}{\vdash\ \lambda x.E = \lambda x.F}$$

As we will see below, without the above purity restrictions we could derive a contradiction with the remaining five axioms, which are as follows:

$$E_1 \leftarrow E_2 \Leftrightarrow \forall x : x \leftarrow E_1 \Rightarrow x \leftarrow E_2 \tag{6}$$

$$E_1 = E_2 \Leftrightarrow \forall x : x \leftarrow E_1 \Leftrightarrow x \leftarrow E_2 \tag{7}$$

$$x \leftarrow \sqcap/ \, [E_1, E_2, ..., E_n] \Leftrightarrow x \leftarrow E_1 \;\vee\; x \leftarrow E_2 \;\vee ... \vee\; x \leftarrow E_n \tag{8}$$

$$x \leftarrow F\ E \Leftrightarrow \exists f, e : f \leftarrow F \wedge e \leftarrow E \wedge x \leftarrow f\ e \tag{9}$$

$$f \leftarrow \lambda x.\, E \Leftrightarrow \forall x : f\ x \leftarrow E \tag{10}$$

Recall that free lower case variables range over pure expressions only, i.e., the free variables x and f are assumed pure.

From (6) and (7) we obtain that ($\leftarrow$) is reflexive, transitive and anti-symmetric. From (8) we obtain that ($\sqcap$) is associative, commutative and idempotent. Axioms (8) and (9) are sufficient to establish

$$F\ (\sqcap/ \, [E_1, E_2, ..., E_n]) = \sqcap/ \, [F\ E_1, F\ E_2, ..., F\ E_n] \tag{11}$$

Here is the proof:

$$
\begin{array}{ll}
 & x \leftarrow F\ (\sqcap/ \, [E_1, E_2, ..., E_n]) \\
\Leftrightarrow & \{\ (9)\ \} \\
 & \exists f, e : f \leftarrow F \wedge e \leftarrow \sqcap/ \, [E_1, E_2, ..., E_n] \wedge x \leftarrow f\ e \\
\Leftrightarrow & \{\ (8)\ \} \\
 & \exists i, f, e : f \leftarrow F \wedge e \leftarrow E_i \wedge x \leftarrow f\ e \\
\Leftrightarrow & \{\ (9)\ \} \\
 & \exists i : x \leftarrow F\ E_i \\
\Leftrightarrow & \{\ (8)\ \} \\
 & x \leftarrow \sqcap/ \, [F\ E_1, F\ E_2, ..., F\ E_n]
\end{array}
$$

It follows from (11) and (4) that

$$(\lambda x.\, x + x)\ (1 \sqcap 2) = (\lambda x.\, x + x)\ 1 \sqcap (\lambda x.\, x + x)\ 2 = 2 \sqcap 4$$

If, however, (4) was allowed to hold for arbitrary expressions, then we would have

$$(\lambda x.\, x + x)\ (1 \sqcap 2) = (1 \sqcap 2) + (1 \sqcap 2) = 2 \sqcap 3 \sqcap 4$$

which is a contradiction.

We can also show, for example, that $\lambda x.\, x \sqcap 3$ and $id \sqcap const\ 3$ are different functions even though they are extensionally the same:

$$(\lambda x.\, x \sqcap 3)\ x = x \sqcap 3 = (id \sqcap const\ 3)\ x$$

Consider the function $h = \lambda f.\, f\ 1 + f\ 2$. We have by β reduction that

$$h\ (\lambda x.\, x \sqcap 3) = (\lambda x.\, x \sqcap 3)\ 1 + (\lambda x.\, x \sqcap 3)\ 2 = (1 \sqcap 3) + (2 \sqcap 3) = 3 \sqcap 4 \sqcap 5 \sqcap 6$$

while, on account of (11), we have

$$h\ (id \sqcap const\ 3) = h\ id \sqcap h\ (const\ 3) = (1+2) \sqcap (3+3) = 3 \sqcap 6$$

Thus two nondeterministic functions can be extensionally equal without being the same function. That explains the restriction of the η rule to pure functions. Finally, (9) gives us that

$$\begin{array}{l} G_1 \leftarrow G_2 \Rightarrow F \cdot G_1 \leftarrow F \cdot G_2 \\ F_1 \leftarrow F_2 \ \Rightarrow F_1 \cdot G \leftarrow F_2 \cdot G \end{array}$$

where $(\cdot) = (\lambda f.\lambda g.\lambda x.f\ (g\ x))$.

To complete the presentation of the calculus, we need to give the rules for the logical operators used in the axioms. The rule for the propositional connectives are the standard ones and are omitted. But the rules for the quantifies are subtle because we have to ensure the quantifiers range over pure expressions only. In single-conclusion natural deduction style, these are

$$\frac{pure(x) \vdash F}{\vdash \forall x{:}F} \qquad \frac{\vdash \forall x{:}F \quad \vdash pure(e)}{\vdash F(x{:=}e)}$$

$$\frac{\vdash F(x{:=}e) \quad \vdash pure(e)}{\vdash \exists x{:}F} \qquad \frac{\vdash \exists x{:}F \quad pure(x), F \vdash G}{\vdash G}$$

Here $pure(e)$ is the purity predicate, whose axioms are described above.

4 A Denotational Semantics

To establish the consistency of the axiomatisation we give a denotational semantics for nondeterministic expressions. As the target language of our semantics, we use standard set theory, with the notations $A \to B$ and $\lambda x \in A.b$ for functions (with $\in A$ omitted if clear).

Overview. The basic intuition of the interpretation function $[\![-]\!]$ is given in the following table where we write $\mathbb{P}^*\ A$ for the set of non-empty subsets of A:

Syntax	Semantics
type T	set $[\![T]\!]$
context declaring $x : T$	environment mapping $\rho : x \mapsto [\![T]\!]$
expression $E : T$	non-empty subset $[\![E]\!] \in \mathbb{P}^*[\![T]\!]$
refinement $E_1 \leftarrow E_2$	subset $[\![E_1]\!]_\rho \subseteq [\![E_2]\!]_\rho$
function type $S \to T$	set–valued functions $[\![S]\!] \to \mathbb{P}^*[\![T]\!]$
choice $E_1 \sqcap E_2$	union $[\![E_1]\!]_\rho \cup [\![E_2]\!]_\rho$
purity $pure(E)$ for $E : T$	$[\![E]\!]_\rho$ is generated by a single $v \in [\![T]\!]$

Thus, types denotes sets, and non-deterministic expressions denote sets of values. Functions are set-valued, and choice is simply union.

Additionally, for each type T, we will define the operation

$$[\![T]\!] \ni v \mapsto v^{\leftarrow} \in \mathbb{P}^*[\![T]\!],$$

which embeds the single (deterministic) values into the power set. We call it *refinement closure* because $v^{\leftarrow}$ is the set of all values that we want to allow as a refinement of v. This allows defining the refinement ordering $\leqslant_T$ on $[\![T]\!]$ by $v \leqslant_T w$ iff $v^{\leftarrow} \subseteq w^{\leftarrow}$. While we will not need it in the sequel, it is helpful to define because for every expression $E : T$, the set $[\![E]\!]$ will be downward closed with respect to $\leqslant_T$. One could add an expression $\perp$ as a value with no refinements other than itself, which denotes the empty set. But doing so would mean that $\perp$ would be a refinement of every expression, which we choose not to have. That explains the restriction to non-empty sets in our semantics. Note that $\leqslant_T$ is not the same as the usual approximation ordering on Haskell expressions of a given type with $\perp$ as the least element.

Choice and Refinement. We define

$$[\![\sqcap/[E_1, ..., E_n]]\!]_\rho = [\![E_1]\!]_\rho \cup \ldots \cup [\![E_n]\!]_\rho$$

This captures our intuition that a choice refines to any of its arguments, i.e., it denotes all values denoted by any argument. This is tied to the intuition that the refinement property corresponds to the subset condition on denotations. For example, $E_1 \leftarrow E_1 \sqcap E_2$ corresponds to $[\![E_1]\!]_\rho \subseteq [\![E_1 \sqcap E_2]\!]_\rho$.

Pure expressions $e : T$ cannot be properly refined. At base types, they are interpreted as singletons. For the general case, we have to relax this idea somewhat and require only $[\![e]\!]_\rho = v^{\leftarrow}$ for some $v \in [\![T]\!]$.

Variables. As usual, expressions with free variables are interpreted relative to an environment ρ. Analogously to variables ranging over pure expressions, the environment maps every variable $x : T$ to a value $v \in [\![T]\!]$ (but not to a subset of $[\![T]\!]$ as one might expect). Consequently, the denotation of a variable is defined by applying the refinement closure

$$[\![x]\!]_\rho = \rho(x)^{\leftarrow}$$

Base Types and Constants. The interpretation of base types is straightforward, and we define

$$\begin{aligned} [\![\mathit{Int}]\!] &= \mathbb{Z} \\ [\![\mathit{Bool}]\!] &= \mathbb{B} \end{aligned}$$

Moreover, we define $v^{\leftarrow} = \{v\}$ for $v \in [\![B]\!]$ for every base type B. In particular, we have $v \leqslant_B w$ iff $v = w$. In other words, the refinement ordering on base types is *flat*.

We would like to interpret all constants C in this straightforward way as well, but that is not as easy. In general, we assume that for every user-declared constant $C : T$, a denotation $\overline{C} \in [\![T]\!]$ is provided. Then we define

$$[\![C]\!]_\rho = \overline{C}^{\leftarrow}.$$

However, we cannot simply assume that $\overline{C}$ is the standard denotation that we would use to interpret a deterministic type theory. For example, for $+ : Int \to Int \to Int$, we cannot define $\overline{+}$ as the usual addition $+_{\mathbb{Z}} : \mathbb{Z} \to \mathbb{Z} \to \mathbb{Z}$ because we need a value $\overline{+} : \mathbb{Z} \to \mathbb{P}^*(\mathbb{Z} \to \mathbb{P}^*\mathbb{Z})$.

For first-order constants, i.e., constants $C : B_1 \to \ldots \to B_n \to B$ where B and all B_i are base types (e.g., the constant $+$), we can still lift the standard interpretation relatively easily: If $f : [\![B_1]\!] \to \ldots \to [\![B_n]\!] \to [\![B]\!]$ is the intended interpretation for C, we define

$$\overline{C} : [\![B_1]\!] \to \mathbb{P}^*([\![B_2]\!] \to \ldots \to \mathbb{P}^*([\![B_n]\!] \to \mathbb{P}^*[\![B]\!])\ldots)$$

by

$$\overline{C} = \lambda x_1.\{\lambda x_2.\ldots\{\lambda x_n.\{f\, x_1\, \ldots,\, x_n\}\}\ldots\}$$

Because all B_i are base types, this yields we have $[\![C]\!]_\rho = \overline{C}^{\leftarrow} = \{\overline{C}\}$. For $n = 0$, this includes constants $C : B$, e.g., $[\![1]\!]_\rho = \{1\}$ and accordingly for all integer literals.

But we cannot systematically lift standard interpretations of higher-order constants C accordingly. Instead, we must provide $\overline{C}$ individually for each higher-order constant. But for the purposes of program calculation, this is acceptable because we only have to do it once for the primitive constants of the language. In [3], this subtlety is handled by restricting attention to first-order constants.

Functions. We define the interpretation of function types as follows:

$$[\![S \to T]\!] = [\![S]\!] \to \mathbb{P}^*[\![T]\!]$$

and for $f \in [\![S \to T]\!]$ we define

$$f^{\leftarrow} = \{g : [\![S \to T]\!] \mid g(v) \subseteq f(v) \text{ for all } v \in [\![S]\!]\}$$

Thus, the refinement ordering on functions acts point-wise: $g \leqslant_{S \to T} f$ iff $g(v) \subseteq f(v)$ for all $v \in [\![S]\!]$.

For example, there are nine functions of type $[\![Bool \to Bool]\!]$ with $\mathbb{B} = \{0, 1\}$ whose tables are as follows:

	f_0	f_1	f_2	f_3	f_4	f_5	f_6	f_7	f_8
0	$\{0,1\}$	$\{0,1\}$	$\{0\}$	$\{1\}$	$\{0,1\}$	$\{0\}$	$\{0\}$	$\{1\}$	$\{1\}$
1	$\{0,1\}$	$\{0\}$	$\{0,1\}$	$\{0,1\}$	$\{1\}$	$\{0\}$	$\{1\}$	$\{0\}$	$\{1\}$

For example, $f_7 = \overline{\neg}$ is the lifting of the usual negation function. The ordering $\leqslant_{Bool \to Bool}$ has top element f_0 and the four bottom elements f_5, f_6, f_7 and f_8.

Finally, the clauses for the denotation of λ and application terms are

$$[\![\lambda x : S.E]\!]_\rho = (\lambda v \in [\![S]\!].[\![E]\!]_{\rho(x:=v)})^{\leftarrow} \tag{12}$$

$$[\![F\,E]\!]_\rho = \bigcup\{f(e) \mid f \in [\![F]\!]_\rho,\ e \in [\![E]\!]_\rho\} \tag{13}$$

Here the notation $\rho(x := v)$ means the environment ρ extended with the binding of v to x. Because every expression in already interpreted as a set and function expressions must be interpreted as set-valued functions, a λ-abstraction can be interpreted essentially as the corresponding semantic function. We only need to apply the refinement closure. Equivalently, we could rewrite (12) using

$$(\lambda v \in [\![S]\!].[\![E]\!]_{\rho(x:=v)})^{\leftarrow} = \{f \mid f(v) \subseteq [\![E]\!]_{\rho(x:=v)} \text{ for all } v \in [\![S]\!]\}$$

The clause for application captures our intuition of monotonicity of refinement: $F\,E$ is interpreted by applying all possible denotations f of F to all possible denotations e of E; each such application returns a set, and we take the union of all these sets.

Formulas. Because formulas are a special case of expressions, they are interpreted as non-empty subsets of $[\![Bool]\!] = \{0, 1\}$. We write $\top$ for the truth value $\{1\}$ denoting truth. The truth value $\{0, 1\}$ will never occur (unless the user wilfully interprets a constant in a way that returns it).

The denotation of all Boolean constants and expressions is as usual. The denotation of the quantifiers and the special predicates is defined by:

$$[\![E_1 \leftarrow E_2]\!]_\rho = \top \quad \text{iff} \quad [\![E_1]\!]_\rho \subseteq [\![E_2]\!]_\rho \tag{14}$$

$$[\![pure(E)]\!]_\rho = \top \quad \text{iff} \quad [\![E]\!]_\rho = v^{\leftarrow} \text{ for some } v \in [\![S]\!] \tag{15}$$

$$[\![\forall_S x : F]\!]_\rho = \top \quad \text{iff} \quad [\![F]\!]_{\rho(x:=v)} = \top \text{ for all } v \in [\![S]\!] \tag{16}$$

$$[\![\exists_S x : F]\!]_\rho = \top \quad \text{iff} \quad [\![F]\!]_{\rho(x:=v)} = \top \text{ for some } v \in [\![S]\!] \tag{17}$$

Note that the quantified variables seamlessly range only over values.

Soundness and Consistency. We can now state the soundness of our calculus as follows:

Theorem 1 (Soundness). *If F is provable, then $[\![F]\!]_\rho = \top$ for every environment ρ for the free variables of F. In particular, if $E_1 \leftarrow E_2$ is provable, then $[\![E_1]\!]_\rho \subseteq [\![E_2]\!]_\rho$ for all environments ρ.*

Proof. As usual, the proof proceeds by induction on derivations.

In particular, we must justify the axioms (4)–(10). We concentrate on (4), which requires us to show

$$[\![(\lambda x : S.E)\,e]\!]_\rho = [\![E(x := e)]\!]_\rho$$

for all expressions E, all pure expressions e and all environments ρ. The proof divides into two cases according to the two axioms for purity: either e is an application of a constant to pure arguments, in which case $[\![e]\!]_\rho$ is a singleton set, or e is a lambda abstraction. For the former we will need the fact that if e is single-valued, then $[\![E(x := e)]\!]_\rho = [\![E]\!]_{\rho(x:=![\![e]\!]_\rho)}$ where $!\{v\} = v$. This *substitution lemma* can be proved by structural induction on E. That means we can argue:

$$
\begin{array}{l}
[\![(\lambda x : S.E)\, e]\!]_\rho \\
= \{13\} \\
\bigcup\{f(v) \mid f \in [\![\lambda x.E]\!]_\rho,\ v \in [\![e]\!]_\rho\} \\
= \{12\} \\
\bigcup\{f(v) \mid f(w) \subseteq [\![E]\!]_{\rho(x:=w)} \text{ for all } w \in [\![S]\!],\ v \in [\![e]\!]_\rho\} \\
= \{\text{subsumed sets can be removed from a union}\} \\
\bigcup\{f(v) \mid f(w) = [\![E]\!]_{\rho(x:=w)} \text{ for all } w \in [\![S]\!],\ v \in [\![e]\!]_\rho\} \\
= \{[\![e]\!]_\rho \subseteq [\![S]\!]\} \\
\bigcup\{[\![E]\!]_{\rho(x:=v)} \mid v \in [\![e]\!]_\rho\} \\
= \{e \text{ is single-valued}\} \\
[\![E]\!]_{\rho(x:=![\![e]\!]_\rho)} \\
= \{\text{substitution lemma}\} \\
[\![E(x := e)]\!]_\rho
\end{array}
$$

For the second case, where e is a lambda abstraction $\lambda y : T.\, F$, we need the fact that

$$[\![(\lambda x.E)\,(\lambda y.F)]\!]_\rho = [\![E]\!]_{\rho(x:=\lambda v.[\![F]\!]_{\rho(y:=v)})}$$

This fact can be established as a corollary to the *monotonicity lemma* which asserts $[\![E]\!]_{\rho(x:=f)} \subseteq [\![E]\!]_{\rho(x:=g)}$ whenever $f(v) \subseteq g(v)$ holds for all $v \in [\![S]\!]$. for all expressions E and environments ρ. The monotonicity lemma can be proved by structural induction on E. The corollary above is now proved by reasoning

$$
\begin{array}{l}
[\![(\lambda x.E)\,(\lambda y.F)]\!]_\rho \\
= \{13\} \\
\bigcup\{h(f) \mid h \in [\![\lambda x.E]\!]_\rho,\ f \in [\![\lambda y.F]\!]_\rho\} \\
= \{\text{as in previous calculation}\} \\
\bigcup\{[\![E]\!]_{\rho(x:=f)} \mid f \in [\![\lambda y.F]\!]_\rho\} \\
= \{12\} \\
\bigcup\{[\![E]\!]_{\rho(x:=f)} \mid f(v) \subseteq [\![F]\!]_{\rho(y:=v)} \text{ for all } v \in [\![T]\!]\} \\
= \{\subseteq\text{-direction: monotonicity lemma; } \supseteq\text{-direction: } X \subseteq \bigcup Y \text{ if } X \in Y\} \\
[\![E]\!]_{\rho(x:=\lambda v.[\![F]\!]_{\rho(y:=v)})}
\end{array}
$$

It remains to show that the latter is equal to $[\![E(x := \lambda y.F)]\!]_\rho$. Here we proceed by structural induction on E. We omit the details. The other axioms are proved by similar reasoning.

As a straightforward consequence of soundness, we have

Theorem 2 (Consistency). *Our calculus is consistent, i.e., we cannot derive a contradiction.*

Proof. If we could derive a contradiction, then soundness would yield a contradiction in set theory.

Technically, our calculus is only consistent under the assumption that set theory is consistent. We can strengthen that result by using a much weaker target language than set theory for our semantics. Indeed, standard higher-order logic (using an appropriate definition of power set) is sufficient.

5 Summary

The need for nondeterministic functions arose while the first author was preparing a text on an introduction to Algorithm Design using Haskell. The book, which is co-authored by Jeremy Gibbons, will be published by Cambridge University Press next year. Two of the six parts of the book are devoted to greedy algorithms and thinning algorithms. To make the material as accessible as possible, we wanted to stay close to Haskell and that meant we did not want to make the move from functions to relations, as proposed for instance in [1]. Instead, we made use of just two nondeterministic functions, *MinWith* and *ThinBy* (or three if you count *MaxWith*), and reasoned about refinement rather than equality when the need arose. The legitimacy of the calculus, as propounded above, is not given in the book. The problems associated with reasoning about nondeterminism were discussed at the Glasgow meeting of WG2.1 in 2016, when the second author came on board. Our aim has been to write a short and hopefully sufficient introduction to the subject of nondeterminism for functional programmers rather than logicians. In this enterprise we made much use of the very readable papers by Joe Morris and Alexander Bunkenberg.

Appendix

Here is the example, known as the *paragraph* problem. Consider the task of dividing a list of words into a list of lines so that each line is subject to a maximum line width of w. Each line is a list of words and its width is the sum of the length of the words plus the number of inter-word spaces. There is an obvious greedy algorithm for this problem, namely to add the next word to the current line if it will fit, otherwise to start a newline with the word. For what cost function does the greedy algorithm produce a division with minimum cost?

The obvious answer is that such a division has the minimum possible number of lines. So it has, but we cannot calculate this algorithm from a specification involving *minWith length*. To see why, consider a list of words whose lengths are $[3,6,1,8,1,8]$ (the words are not important, only their lengths matter). Taking $w = 12$, there are four shortest possible layouts, of which two are

$$p_1 = [[3,6,1],[8],[1,8]]$$
$$p_2 = [[3,6],[1,8,1],[8]]$$

Let $add\ x\ p$ be the function that adds the next word x to the end of the last line if the result will still fit into a width of 12, or else begins a new line. In particular

$$q_1 = add\ 2\ p_1 = [[3,6,1],[8],[1,8],[2]]$$
$$q_2 = add\ 2\ p_2 = [[3,6],[1,8,1],[8,2]]$$

We have

$$length\ p_1 \leqslant length\ p_2 \wedge length\ q_1 > length\ q_2$$

so the monotonicity condition fails. The situation can be redeemed by strengthening the cost function to read

$$cost\ p = (length\ p, width\ (last\ p))$$

In words one paragraph costs less than another if its length is shorter, or if the lengths are equal and the width of the last line is shorter. Minimising *cost* will also minimise *length*. This time we do have

$$cost\ p \leqslant cost\ p' \Rightarrow cost\ (add\ x\ p) \leqslant cost\ (add\ x\ p')$$

as can be checked by considering the various cases, so the monotonicity condition holds. However, we also have

$$cost\ (add\ 5\ p_1) = cost\ (add\ 5\ p_2) = (4,5)$$

and $cost\ p_2 < cost\ p_1$, so the strong monotonicity condition (1) fails.

References

1. Bird, R.S., de Moor, O.: The Algebra of Programming. Prentice-Hall International Series in Computer Science, Hemel Hempstead (1997)
2. Harper, R., Honsell, F., Plotkin, G.: A framework for defining logics. J. Assoc. Comput. Mach. **40**(1), 143–184 (1993)
3. Morris, J.M., Bunkenburg, A.: Specificational functions. ACM Trans. Program. Lang. Syst. **21**(3), 677–701 (1999)
4. Morris, J.M., Bunkenburg, A.: Partiality and nondeterminacy in program proofs. Formal Aspects Comput. **10**, 76–96 (1998)
5. Morris, J.M., Tyrrell, M.: Dually nondeterministic functions. ACM Trans. Program. Lang. Syst. **30**(6) (2008). Article 34

Setoid Type Theory—A Syntactic Translation

Thorsten Altenkirch[1], Simon Boulier[2], Ambrus Kaposi[3(✉)], and Nicolas Tabareau[2]

[1] University of Nottingham, Nottingham, UK
Thorsten.Altenkirch@nottingham.ac.uk
[2] Inria, Nantes, France
{simon.boulier,nicolas.tabareau}@inria.fr
[3] Eötvös Loránd University, Budapest, Hungary
akaposi@inf.elte.hu

Abstract. We introduce setoid type theory, an intensional type theory with a proof-irrelevant universe of propositions and an equality type satisfying function extensionality, propositional extensionality and a definitional computation rule for transport. We justify the rules of setoid type theory by a syntactic translation into a pure type theory with a universe of propositions. We conjecture that our syntax is complete with regards to this translation.

Keywords: Type theory · Function extensionality · Proof irrelevance · Univalence

1 Introduction

Extensional type theory (ETT [23]) is a convenient setting for formalising mathematics: equality reflection allows replacing provably equal objects with each other without the need for any clutter. On paper this works well, however computer checking ETT preterms is hard because they don't contain enough information to reconstruct their derivation. From Hofmann [16] and later work [26,32] we know that any ETT derivation can be rewritten in intensional type theory (ITT) extended with two axioms: function extensionality and uniqueness of identity proofs (UIP). ITT preterms contain enough information to allow computer checking, but the extra axioms[1] introduce an inconvenience: they prevent certain computations. The axioms act like new neutral terms which even appear in the empty context: a boolean in the empty context is now either true or false or a

[1] The problem is only with the axiom of function extensionality as adding UIP using Streicher's axiom K [29] doesn't pose a challenge to normalisation.

This work was supported by the European Union, co-financed by the European Social Fund (EFOP-3.6.3-VEKOP-16-2017-00002), by COST Action EUTypes CA15123, by ERC Starting Grant CoqHoTT 637339, EPSRC grant EP/M016994/1 and by USAF, Airforce office for scientific research, award FA9550-16-1-0029 and by the National Research, Development and Innovation Office – NKFIH project 127740.

G. Hutton (Ed.): MPC 2019, LNCS 11825, pp. 155–196, 2019.
https://doi.org/10.1007/978-3-030-33636-3_7

neutral term coming from the axiom. This is a practical problem: for computer formalisation one main advantage of type theory over to set theory is that certain equalities are trivially true by computation, and the additional axioms limit this computational power.

In general, the usage of an axiom is justified by a model [18] in which the axiom holds. For example, the cubical set model [8] justifies the univalence axiom, the reflexive graph model [6] justifies parametricity, the groupoid model [19] justifies the negation of UIP, the setoid model [1, 17] justifies function extensionality. A model can help designing a new type theory in which the axiom holds and which has full computational power, i.e. normalisation. Examples are cubical type theory [12] inspired by the cubical set model [8] and observational type theory [4] inspired by the setoid model [1].

In this paper we revisit the problem of designing a type theory based on the setoid model. We derive *setoid type theory* from the setoid model using an intermediate *syntactic translation.*

Most models interpret syntactic objects by metatheoretic structures, usually the ones they are named after. In the cubical model, a context (or a closed type) is a cubical set, in the groupoid model a context is a groupoid, and so on. Syntactic models [10] are special kinds of models: they interpret syntax by the syntax of another (or the same) theory. We call the interpretation function into such a model a *syntactic translation.* Equal (convertible) terms are equal objects in a model, which means that convertible terms are translated to convertible terms in the case of a syntactic model. This restricts the number of models that can be turned into syntactic models. A sufficient (but not necessary) criterion to turn a model into a syntactic model is the *strictness* of the model which means that all the equality proofs in the model are given by reflexivity (i.e. they are definitional equalities of the metatheory). Giving the metatheory an explicit syntax and renaming it target theory, a strict model can be turned into a syntactic translation from the source theory to the target theory. We will give examples of this process later on.

The setoid model given by Altenkirch [1] is a strict model, hence it can be phrased as a syntactic translation. A closed type in the setoid model is a set together with an equivalence relation. There are several ways to turn this model into a syntactic model, but in one of these a closed type is given by (1) a type, (2) a binary relation on terms of that type and (3) terms expressing that the relation (2) is reflexive, symmetric and transitive. We will define the syntax for setoid type theory by reifying parts of this model: we add the definitions of the relation (2) and its properties (3) as new term formers to type theory. The new equality type (identity type) will be the relation (2). The equalities describing the translation will be turned into new definitional equality rules of the syntax. Thus the new equality type will satisfy function extensionality and propositional extensionality by definition.

We also extend the setoid translation with a new rule making the elimination principle of equality compute definitionally.

In this paper we do not aim to give a precise definition of the notion of syntactic model or the relationship between different kinds of models. Our main goal is to obtain a convenient syntax for setoid type theory.

Structure of the Paper. After summarising related work, in Sect. 2 we introduce $\mathsf{MLTT}_{\mathsf{Prop}}$, Martin-Löf type theory extended with a definitionally proof irrelevant universe of propositions [14]. In Sect. 3, we illustrate how to turn models into syntactic translations by the examples of the standard (set) model and the graph model. One of the syntactic translation variants of the graph model turns out to be Bernardy et al.'s parametricity translation [7]. The model corresponding to this translation is not strict, showing that strictness is not a necessary condition for a model to have a syntactic variant. In Sect. 4 we define the setoid model as a syntactic translation. We also show that this translation can be extended with a new component saying that transport (the eliminator of equality) computes definitionally. In Sect. 5, we reflect the setoid translation into the syntax of $\mathsf{MLTT}_{\mathsf{Prop}}$ obtaining a new definition of a heterogeneous equality type. We also show that the translation of Sect. 4 extends to this new equality type and we compare it to the old-style inductive definition of equality. We conclude in Sect. 6.

Contributions. Our main contribution is the new heterogeneous equality type which, as opposed to John Major equality [4], is not limited to proof-irrelevant equality and is much simpler than cubical equality types [12,28]. As opposed to [4,28] we do not need to go through extensional type theory to justify our syntax but we do this by a direct translation into a pure intensional type theory. In addition to function extensionality, our setoid type theory supports propositional extensionality and a definitional computation rule for transport, which is also a new addition to the setoid model. The results were formalised in Agda.

Formalisation. The model variant ($|-|_0$ variant in Sect. 3) of the setoid translation has been formalised [21] in Agda using the built-in definitionally proof irrelevant Prop universe of Agda. The formalisation includes the definitional computation rule for transport and does not use any axioms. In addition to what is described in this paper, we show that this model supports quotient types and universes of sets where equality is given by equality of codes.

1.1 Related Work

A general description of syntactic translations for type theory is given in [10]. In contrast with this work, our translations are defined on intrinsic (well-typed) terms. A translation inspired by [7] for deriving computation rules from univalence is given in [30]. This work does not define a new type theory but recovers some computational power lost by adding the univalence axiom. A syntactic translation for the presheaf model is given in [20].

The setoid model was first described by [17] in order to add extensionality principles to type theory such as function extensionality and equality of logically equivalent propositions. A strict variant of the setoid model was given by [1] using a definitionally proof-irrelevant universe of propositions. Recently, support for such a universe was added to Agda and Coq [14] allowing a full formalisation of Altenkirch's setoid model. Observational type theory (OTT) [4] is a syntax for the setoid model differing from our setoid type theory by using a different notion of heterogeneous equality type, McBride's John Major equality [25]. We show the consistency of our theory using the setoid translation, while OTT is translated to extensional type theory for this purpose [4]. XTT [28] is a cubical variant of observational type theory where the equality type is defined using an interval pretype. Supporting this pretype needs much more infrastructure than our new rules for setoid type theory.

A very powerful extensionality principle is Voevodsky's univalence axiom [27]. The cubical set model of type theory [8] is a constructive model justifying this axiom. A type theory extracted from this model is cubical type theory [12]. The relationship between the cubical set model and cubical type theory is similar to the relationship between the setoid model and setoid type theory.

Previously we attempted to use a heterogeneous equality type similar to the one coming from the setoid translation to define a cubical type theory [3]. This work however is unfinished: the combinatorial complexity arising from equalities between equalities so far prevents us from writing down all the computation rules for that theory. In the setoid case, this blow up is avoided by forcing the equality to be a proposition.

Compared to cubical type theories [12,28], our setoid type theory has the advantage that the equality type satisfies more definitional equalities: while in cubical type theory equality of pairs is isomorphic[2] to the pointwise equalities of the first and second components, in our case the isomorphism is replaced by a definitional equality. The situation is similar for other type formers. These additional definitional equalities are the main motivation for Herbelin's proposal for a cubical type theory [15]. As setoid type theory supports UIP (Streicher's axiom K, [29]), it is incompatible with full univalence. The universe of propositions in setoid type theory satisfies propositional extensionality, which is the version of univalence for mere propositions. However, this is not a subobject classifier in the sense of Topos Theory since it doesn't classify propositions in the sense of HoTT (it seems to be a quasi topos though).

Setoid type theory is not homotopy type theory restricted to homotopy level 0 (the level of sets, or h-sets). This is because the universe of propositions we have is static: we don't have that for any type, if any two elements of it are equal, then it is a proposition. The situation is similar for the groupoid model [19] which features a static universe of sets (h-sets).

[2] This is a definitional isomorphism: A and B are definitionally isomorphic, if there is an $f : A \to B$, a $g : B \to A$ and $\lambda x.f\,(g\,x) = \lambda x.x$ and vice versa where $=$ is definitional equality.

2 $\mathsf{MLTT}_{\mathsf{Prop}}$

$\mathsf{MLTT}_{\mathsf{Prop}}$ is an intensional Martin-Löf type theory with Π, Σ, Bool types and a static universe of strict propositions. We present $\mathsf{MLTT}_{\mathsf{Prop}}$ using an algebraic (intrinsic) syntax [2], that is, there are only well-typed terms so preterms or typing relations are never mentioned. Conversion (definitional equality) rules are given by equality constructors (using the metatheoretic equality $=$), so the whole syntax is quotiented by conversion. As a consequence, all of the constructions in this paper have to preserve typing and definitional equality. In this section we explain the syntax for this type theory listing the most important rules. The full signature of the algebraic theory $\mathsf{MLTT}_{\mathsf{Prop}}$ is given in Appendix A.

There are four sorts: contexts, types, substitutions and terms. Contexts and types are stratified into separate (predicative, cumulative) levels, as indicated by the indices i, j. In the Agda formalisation we use explicit lifting operations instead of cumulativity.

$$\frac{}{\mathsf{Con}_i : \mathsf{Set}} \qquad \frac{\Gamma : \mathsf{Con}_i}{\mathsf{Ty}_j\,\Gamma : \mathsf{Set}} \qquad \frac{\Gamma : \mathsf{Con}_i \qquad \Delta : \mathsf{Con}_j}{\mathsf{Sub}\,\Gamma\,\Delta : \mathsf{Set}} \qquad \frac{\Gamma : \mathsf{Con}_i \qquad A : \mathsf{Ty}_j\,\Gamma}{\mathsf{Tm}\,\Gamma\,A : \mathsf{Set}}$$

We use the following naming conventions for metavariables: universe levels i, j; contexts $\Gamma, \Delta, \Theta, \Omega$; types A, B, C; terms t, u, v, w, a, b, c, e; substitutions δ, ν, τ, ρ. Constructors of the syntax are written in red to help distinguish from definitions. Most constructors have implicit arguments, e.g. type substitution below $-[-]$ takes the two contexts Γ and Δ as implicit arguments.

The syntax for the substitution calculus is the following. It can also be seen as an unfolding of category with families (CwF, [13]) with the difference that we write variable names and implicit weakenings instead of De Bruijn indices.

$$\frac{}{\cdot : \mathsf{Con}_0} \qquad \frac{\Gamma : \mathsf{Con}_i \qquad A : \mathsf{Ty}_j\,\Gamma}{(\Gamma\,, x : A) : \mathsf{Con}_{i \sqcup j}} \qquad \frac{A : \mathsf{Ty}_i\,\Delta \qquad \delta : \mathsf{Sub}\,\Gamma\,\Delta}{A[\delta] : \mathsf{Ty}_i\,\Gamma}$$

$$\frac{\Gamma : \mathsf{Con}_i}{\mathsf{id}_\Gamma : \mathsf{Sub}\,\Gamma\,\Gamma} \qquad \frac{\delta : \mathsf{Sub}\,\Theta\,\Delta \qquad \nu : \mathsf{Sub}\,\Gamma\,\Theta}{\delta \circ \nu : \mathsf{Sub}\,\Gamma\,\Delta} \qquad \frac{\Gamma : \mathsf{Con}_i}{\epsilon_\Gamma : \mathsf{Sub}\,\Gamma\,\cdot}$$

$$\frac{\delta : \mathsf{Sub}\,\Gamma\,\Delta \qquad t : \mathsf{Tm}\,\Gamma\,(A[\delta])}{(\delta\,, x \mapsto t) : \mathsf{Sub}\,\Gamma\,(\Delta\,, x : A)} \qquad \frac{\delta : \mathsf{Sub}\,\Gamma\,(\Delta\,, x : A)}{\delta : \mathsf{Sub}\,\Gamma\,\Delta} \qquad \frac{\delta : \mathsf{Sub}\,\Gamma\,(\Delta\,, x : A)}{x[\delta] : \mathsf{Tm}\,\Gamma\,(A[\delta])}$$

$$\frac{t : \mathsf{Tm}\,\Delta\,A \qquad \delta : \mathsf{Sub}\,\Gamma\,\Delta}{t[\delta] : \mathsf{Tm}\,\Gamma\,(A[\delta])} \qquad [\mathsf{Id}] : A[\mathsf{id}] = A \qquad [\circ] : A[\delta \circ \nu] = A[\delta][\nu]$$

$$\mathsf{id}\circ : \mathsf{id} \circ \delta = \delta \qquad \circ\mathsf{id} : \delta \circ \mathsf{id} = \delta \qquad \circ\circ : (\delta \circ \nu) \circ \tau = \delta \circ (\nu \circ \tau)$$

$$\cdot\eta : (\delta : \mathsf{Sub}\,\Gamma\,\cdot) = \epsilon \qquad ,\beta_0 : (\delta\,, x \mapsto t) = \delta \qquad ,\beta_1 : x[(\delta\,, x \mapsto t)] = t$$

$$,\eta : (\delta\,, x \mapsto x[\delta]) = \delta \qquad ,\circ : (\delta\,, x \mapsto t) \circ \nu = (\delta \circ \nu\,, x \mapsto t[\nu])$$

There are two ways of forming a context: the empty context and context extension (or comprehension; here $i \sqcup j$ denotes the maximum of i and j). In context extension, the : after the variable x is just notation, it differs from the metatheoretic colon. A substitution $\mathsf{Sub}\,\Gamma\,\Delta$ can be thought of as a list of terms, one for each type in Δ, all given in context Γ. Such a substitution δ acts on a type $A : \mathsf{Ty}_i\,\Delta$ by $A[\delta] : \mathsf{Ty}_i\,\Gamma$. Note that $-[-]$ is a constructor, not an operation, that is, we are defining an explicit substitution calculus. There are five ways to form substitutions: identity id, composition $-\circ-$, the empty substitution ϵ, extending a substitution with a term and forgetting the last term in the substitution (this is an implicit constructor). Terms can be formed using a variable (projecting out the last component from a substitution) and by action of a substitution.

We use variable names for readability, however these should be understood formally as the well-typed De Bruijn indices of CwFs, hence we consider α-equivalent terms equal. In the formalisation we use De Bruijn indices. We denote variables by $x, y, z, f, \gamma, \alpha$.

The equalities of the substitution calculus can be summarised as follows: type substitution is functorial, contexts and substitutions form a category with a terminal object $\cdot$ and substitutions $\mathsf{Sub}\,\Gamma\,(\Delta\,, x : A)$ are in a natural one-to-one correspondence with substitutions $\delta : \mathsf{Sub}\,\Gamma\,\Delta$ and terms $\mathsf{Tm}\,\Gamma\,(A[\delta])$. Ordinary variables can be recovered by $x := x[\mathsf{id}]$. Weakenings are implicit.

In equation $\cdot\eta$, a type annotation is added on δ to show that this equation is only valid for δs with codomain $\cdot$. Implicit weakenings are present in equations $,\beta_0$ and $,\eta$. Note that equation $,\circ$ is only well-typed because of a previous equation: $t[\nu]$ has type $A[\delta][\nu]$, but it needs type $A[\delta \circ \nu]$ to be used in an extended substitution. In our informal notation, we use extensional type theory [23] as metatheory, hence we do not write such transports explicitly.[3] However all of our constructions can be translated to an intensional metatheory with function extensionality and uniqueness of identity proofs (UIP) following [16,26,32].

We sometimes omit the arguments written in subscript as e.g. we write id instead of id_Γ. We write $t[x \mapsto u]$ for $t[(\mathsf{id}\,, x \mapsto u)]$.

Dependent function space is given by the following syntax.

$$\frac{A : \mathsf{Ty}_i\,\Gamma \qquad B : \mathsf{Ty}_j\,(\Gamma\,, x : A)}{\Pi(x : A).B : \mathsf{Ty}_{i \sqcup j}\,\Gamma} \qquad \frac{t : \mathsf{Tm}\,(\Gamma\,, x : A)\,B}{\lambda x.t : \mathsf{Tm}\,\Gamma\,(\Pi(x : A).B)}$$

$$\frac{t : \mathsf{Tm}\,\Gamma\,(\Pi(x : A).B)}{t \mathbin{@} x : \mathsf{Tm}\,(\Gamma\,, x : A)\,B} \qquad \Pi\beta : (\lambda x.t) \mathbin{@} x = t \qquad \Pi\eta : \lambda x.t \mathbin{@} x = t$$

$$\Pi[] : (\Pi(x : A).B)[\nu] = \Pi(x : A[\nu]).B[\nu] \qquad \lambda[] : (\lambda x.t)[\nu] = \lambda x.t[\nu]$$

We write $A \Rightarrow B$ for $\Pi(x : A).B$ when x does not appear in B. The usual application can be recovered from the categorical application $@$ using a substitution and we use the same $@$ notation: $t \mathbin{@} u := (t \mathbin{@} x)[x \mapsto u]$. $\Pi[]$ and $\lambda[]$ are the substitution laws for Π and λ, respectively. A substitution law for $@$ can be derived using $\lambda[]$, $\Pi\beta$ and $\Pi\eta$.

[3] Note that this does not mean that when defining our setoid model we rely on extensionality of the metatheory: our models will be given as syntactic translations as described in Sect. 3.

The syntax of dependent pairs and booleans follows the same principles and is given in Appendix A for the completeness of the presentation.

We have a hierarchy of universes of strict propositions. Any two elements of a proposition are definitionally equal: this is expressed by the rule irr_a (recall that $=$ is the equality of the metatheory).

$$\frac{}{\mathsf{Prop}_i : \mathsf{Ty}_{i+1}\,\Gamma} \qquad \frac{a : \mathsf{Tm}\,\Gamma\,\mathsf{Prop}_i}{\underline{a} : \mathsf{Ty}_i\,\Gamma} \qquad \frac{u : \mathsf{Tm}\,\Gamma\,\underline{a} \qquad v : \mathsf{Tm}\,\Gamma\,\underline{a}}{\mathsf{irr}_a : u = v}$$

This universe is closed under dependent function space, dependent sum, unit and empty types. Decoding an element of Prop is written using underline instead of the usual El. We use a, b, c as metavariables of type Prop and lowercase π and σ for the proposition constructors. The domain of the function space needs not be a proposition however needs to have the same universe level. For π and σ the constructors and destructors are overloaded. We also write $\Rightarrow$ and $\times$ for the non-dependent versions of π and σ. The syntax is given below (for the substitution laws see Appendix A).

$$\frac{A : \mathsf{Ty}_i\,\Gamma \qquad b : \mathsf{Tm}\,(\Gamma\,, x : A)\,\mathsf{Prop}_j}{\pi(x : A).b : \mathsf{Tm}\,\Gamma\,\mathsf{Prop}_{i \sqcup j}} \qquad \frac{t : \mathsf{Tm}\,(\Gamma\,, x : A)\,\underline{b}}{\lambda x.t : \mathsf{Tm}\,\Gamma\,\underline{\pi(x : A).b}}$$

$$\frac{t : \mathsf{Tm}\,\Gamma\,\underline{\pi(x : A).b}}{t\,@\,x : \mathsf{Tm}\,(\Gamma\,, x : A)\,\underline{b}} \qquad \frac{a : \mathsf{Tm}\,\Gamma\,\mathsf{Prop}_i \qquad b : \mathsf{Tm}\,(\Gamma\,, x : \underline{a})\,\mathsf{Prop}_j}{\sigma(x : a).b : \mathsf{Tm}\,\Gamma\,\mathsf{Prop}_{i \sqcup j}}$$

$$\frac{u : \mathsf{Tm}\,\Gamma\,\underline{a} \qquad v : \mathsf{Tm}\,\Gamma\,\underline{b[x \mapsto u]}}{(u\,, v) : \mathsf{Tm}\,\Gamma\,\underline{\sigma(x : a).b}} \qquad \frac{t : \mathsf{Tm}\,\Gamma\,\underline{\sigma(x : a).b}}{\mathsf{pr}_0\,t : \mathsf{Tm}\,\Gamma\,\underline{a}} \qquad \frac{t : \mathsf{Tm}\,\Gamma\,\underline{\sigma(x : a).b}}{\mathsf{pr}_1\,t : \mathsf{Tm}\,\Gamma\,\underline{a[x \mapsto \mathsf{pr}_0\,t]}}$$

$$\frac{}{\top : \mathsf{Tm}\,\Gamma\,\mathsf{Prop}_0} \qquad \frac{}{\mathsf{tt} : \mathsf{Tm}\,\Gamma\,\underline{\top}} \qquad \frac{}{\bot : \mathsf{Tm}\,\Gamma\,\mathsf{Prop}_0} \qquad \frac{C : \mathsf{Ty}_i\,\Gamma \qquad t : \mathsf{Tm}\,\Gamma\,\underline{\bot}}{\mathsf{exfalso}\,t : \mathsf{Tm}\,\Gamma\,C}$$

Note that we do not need to state definitional equalities of proofs of propositions such as β for function space, as they are true by irr. Definitional proof-irrelevance also has the consequence that for any two pairs $(t\,, u)$ and $(t'\,, u')$ which both have type $\Sigma(x : A).\underline{b}$, whenever $t = t'$ we have $(t\,, u) = (t'\,, u')$. We will use this fact later.

3 From Model to Translation

In this section, as a warm-up for the setoid translation, we illustrate the differences between models and syntactic translations by defining three different syntactic translation variants of the standard model (Subsect. 3.1) and then showing what the corresponding translations for the graph model are (Subsect. 3.2). One of these will be the parametricity translation of Bernardy et al. [7].

A *model* of $\mathsf{MLTT}_{\mathsf{Prop}}$ is an algebra of the signature given in Sect. 2 and fully in Appendix A. In categorical terms, a model is a CwF with extra structure but informally expressed with named variables. The syntax is the initial model which means that for every model M there is an unique interpretation function from the syntax to M (usually called the recursor or eliminator). Below we define models by their interpretation functions: we first provide the specification of the functions (what contexts, types, substitutions and terms are mapped to), then provide the implementation of the functions by listing how they act on each constructor of the syntax. This includes the equality constructors (conversion rules) that the interpretation needs to preserve. For a formal definition of how to derive the notion of model, interpretation function and related notions from a signature, see [22].

As opposed to the notion of model, the notion *syntactic model* (or its interpretation function, *syntactic translation*) is informal. Contexts in a model are usually given by some metatheoretic structure (e.g. sets, graphs, cubical sets, setoids, etc.) and similarly for types, substitutions and terms. In a syntactic model, contexts are given by syntax of another type theory called the *target theory* (this syntax could be contexts of the target theory, terms of the target theory, or a combination of both and also some equalities, etc.). We will illustrate the possibilities with several examples below. It is usually harder to define syntactic models than models, because of the equalities (conversion rules) the model has to satisfy. In a model these equalities are propositional equalities of the metatheory, while in a syntactic model equalities are definitional equalities (conversion rules) of the target theory. A basic example to illustrate this difference is given by extensional type theory (ETT). ETT has a *model* in an intensional metatheory with function extensionality (the standard interpretation $|-|_0$ below works: equality reflection is mapped to function extensionality). However there is no *syntactic translation* from extensional type theory to intensional type theory with function extensionality (the [16,26,32] translations do not preserve definitional equalities).

In the following, $|-|_0$ is a model and $|-|_1$, $|-|_2$ and $|-|_3$ are variants which are syntactic translations.

3.1 Standard Model

The standard interpretation $|-|_0$ (aka set interpretation, or metacircular interpretation) is specified as follows.

$$\frac{\Gamma : \mathsf{Con}_i}{|\Gamma|_0 : \mathsf{Set}_i} \qquad \frac{A : \mathsf{Ty}_j\,\Gamma}{|A|_0 : |\Gamma|_0 \to \mathsf{Set}_j} \qquad \frac{\delta : \mathsf{Sub}\,\Gamma\,\Delta}{|\delta|_0 : |\Gamma|_0 \to |\Delta|_0} \qquad \frac{t : \mathsf{Tm}\,\Gamma\,A}{|t|_0 : (\gamma : |\Gamma|_0) \to |A|_0\,\gamma}$$

Contexts are mapped to metatheoretic types, types to families of types over the interpretation of the context, substitutions become functions and terms dependent functions. We illustrate the implementation by listing some components for contexts and function space.

$$\begin{aligned} |\cdot|_0 &:= \top \\ |\Gamma\,, x : A|_0 &:= (\gamma : |\Gamma|_0) \times |A|_0\,\gamma \\ |\Pi(x : A).B|_0\,\gamma &:= (\alpha : |A|_0\,\gamma) \to |B|_0\,(\gamma, \alpha) \end{aligned}$$

$$
\begin{aligned}
|\lambda x.t|_0\,\gamma\,\alpha &:= |t|_0\,(\gamma,\alpha) \\
|t\,@\,x|_0\,(\gamma,\alpha) &:= |t|_0\,\gamma\,\alpha \\
|\Pi\beta|_0 &: |(\lambda x.t)\,@\,x|_0\,(\gamma,\alpha) = |\lambda x.t|_0\,\gamma\,\alpha = |t|_0\,(\gamma,\alpha)
\end{aligned}
$$

The empty context is interpreted by the unit type (note that it is written in black, this is just the metatheoretic unit). Extended contexts are interpreted by metatheoretic Σ types, Π types are interpreted by function space, λ becomes metatheoretic abstraction, $@$ becomes function application, $\Pi\beta$ and $\Pi\eta$ hold by definition.

The Standard Interpretation $|-|_1$. If we make the metatheory explicit, the previous set interpretation can be seen as a syntactic translation from $\mathsf{MLTT}_{\mathsf{Prop}}$ to $\mathsf{MLTT}_{\mathsf{Prop}}$ extended with a hierarchy of Coquand universes. The latter is no longer called metatheory because the metatheory is now the one in which we talk about both the source and the target theory.

The syntax for Coquand universes[4] is the following.

$$
\frac{}{\mathsf{U}_i : \mathsf{Ty}_{i+1}\,\Gamma} \qquad \frac{A : \mathsf{Ty}_i\,\Gamma}{\mathsf{c}\,A : \mathsf{Tm}\,\Gamma\,\mathsf{U}_i} \qquad \frac{a : \mathsf{Tm}\,\Gamma\,\mathsf{U}_i}{\mathsf{El}\,a : \mathsf{Ty}_i\,\Gamma} \qquad \mathsf{El}\,(\mathsf{c}\,A) = A \qquad \mathsf{c}\,(\mathsf{El}\,a) = a
$$

The specification of this interpretation is as follows (we don't distinguish the source and the target theory in our notation).

$$
\frac{\Gamma : \mathsf{Con}_i}{|\Gamma|_1 : \mathsf{Tm}\cdot\mathsf{U}_i} \qquad \frac{A : \mathsf{Ty}_j\,\Gamma}{|A|_1 : \mathsf{Tm}\cdot(\mathsf{El}\,|\Gamma|_1 \Rightarrow \mathsf{U}_j)}
$$

$$
\frac{\delta : \mathsf{Sub}\,\Gamma\,\Delta}{|\delta|_1 : \mathsf{Tm}\cdot(\mathsf{El}\,|\Gamma|_1 \Rightarrow \mathsf{El}\,|\Delta|_1)} \qquad \frac{t : \mathsf{Tm}\,\Gamma\,A}{|t|_1 : \mathsf{Tm}\cdot(\Pi(\gamma : \mathsf{El}\,|\Gamma|_1).\mathsf{El}\,(|A|_1\,@\,\gamma))}
$$

A context becomes a term of type U in the empty target context. A type becomes a term of a function type with codomain U. Note the difference between the arrows $\rightarrow$ and $\Rightarrow$. A substitution becomes a term of function type and a term becomes a term of a dependent function type where we use target theory application $@$ in the codomain.

The difference between $|-|_0$ and $|-|_1$ is that the latter uses an explicit syntax, otherwise the constructions are the same. They both interpret contexts, types, substitutions and terms all as terms. Type dependency is modelled by Π types and comprehension is modelled by Σ types. The interpretation of $\Pi\beta$ illustrates this nicely: apart from the target theory $\Pi\beta$, $\Pi\eta$ is needed for dealing with type dependency and $\Sigma\eta$ for a comprehension law.

$$
\begin{aligned}
|\cdot|_1 &:= \mathsf{c}\,\top \\
|\Gamma, x : A|_1 &:= \mathsf{c}\,(\Sigma(\gamma : \mathsf{El}\,|\Gamma|_1).\mathsf{El}\,(|A|_1\,@\,\gamma)) \\
|\Pi(x : A).B|_1 &:= \lambda\gamma.\mathsf{c}\,(\Pi(\alpha : \mathsf{El}\,(|A|_1\,@\,\gamma)).\mathsf{El}\,(|B|_1\,@\,(\gamma\,,\alpha)))
\end{aligned}
$$

[4] We learnt this representation of Russell universes from Thierry Coquand.

$$
\begin{aligned}
|\lambda x.t|_1 &:= \lambda\gamma.\lambda\alpha.|t|_1 \mathbin{@} (\gamma \mathbin{,} \alpha) \\
|t \mathbin{@} x|_1 &:= \lambda\gamma.|t|_1 \mathbin{@} \mathsf{pr}_0\,\gamma \mathbin{@} \mathsf{pr}_1\,\gamma \\
|\Pi\beta|_1 &: |(\lambda x.t) \mathbin{@} x|_1 = \lambda\gamma.|\lambda x.t|_1 \mathbin{@} \mathsf{pr}_0\,\gamma \mathbin{@} \mathsf{pr}_1\,\gamma = \\
&\quad \lambda\gamma.(\lambda\gamma.\lambda\alpha.|t|_1 \mathbin{@} (\gamma \mathbin{,} \alpha)) \mathbin{@} \mathsf{pr}_0\,\gamma \mathbin{@} \mathsf{pr}_1\,\gamma \overset{\Pi\beta}{=} \\
&\quad \lambda\gamma.|t|_1 \mathbin{@} (\mathsf{pr}_0\,\gamma \mathbin{,} \mathsf{pr}_1\,\gamma) \overset{\Sigma\eta}{=} \lambda\gamma.|t|_1 \mathbin{@} \gamma \overset{\Pi\eta}{=} |t|_1
\end{aligned}
$$

$|{-}|_1$ can be seen as $|{-}|_0$ composed with a quoting operation returning the syntax of a metatheoretic term (see [5,31]).

Strict vs Non-strict Models. It is easy to implement the $|{-}|_0$ interpretation in Agda as it supports all the constructors and equalities of $\mathsf{MLTT}_{\mathsf{Prop}}$, it has Π and Σ types with definitional η laws, and has the required hierarchy of universes. Because all the equalities hold as definitional equalities in Agda, the proofs of these equalities are just reflexivity. We call such a model a *strict model.* A strict model can be always turned into a syntactic translation (to the metatheory as target theory) the same way as we turned $|{-}|_0$ into $|{-}|_1$. A non-strict model is one where some of the interpretations of equalities need a proof, that is, they cannot be given by reflexivity.

Note that the notion of strict model is relative to the metatheory. The same model can be strict in one metatheory and not in another one. For example, all models are strict in a metatheory with extensional equality. The standard model $|{-}|_0$ is strict in Agda, however if we turn off definitional η for Σ types using the pragma `--no-eta`[5], it becomes non-strict as the definitional η law is needed to interpret the syntactic equality $(\delta \mathbin{,} x \mapsto x[\delta]) = \delta$. The model can be still defined because a propositional η law can be proven (the equalities of the model are given by propositional equalities of the metatheory). However this model cannot be turned into a syntactic translation into a target theory with no definitional η for Σ types.

There are models which are not strict, but can be still turned into a syntactic translation. An example is the $_{0a}$ variant of the graph model defined in Subsect. 3.2.

We can define two more variants of the standard model by changing what models type dependency and comprehension. $|{-}|_2$ models type dependency of the source theory with type dependency in the target theory, but still models comprehension using Σ types. $|{-}|_3$ models type dependency by type dependency and comprehension by comprehension.

The standard interpretation $|{-}|_2$ does not need a universe in the target theory. In general, this translation works for any source theory once the target theory has $\top$ and Σ types (in addition to all the constructors that the source theory has).

[5] Agda version 2.6.0.

$$\frac{\Gamma : \mathsf{Con}_i}{|\Gamma|_2 : \mathsf{Ty}_i \cdot} \qquad \frac{A : \mathsf{Ty}_j\,\Gamma}{|A|_2 : \mathsf{Ty}_j\,(\cdot\,,\gamma : |\Gamma|_2)}$$

$$\frac{\delta : \mathsf{Sub}\,\Gamma\,\Delta}{|\delta|_2 : \mathsf{Tm}\,(\cdot\,,\gamma : |\Gamma|_2)\,|\Delta|_2} \quad \frac{t : \mathsf{Tm}\,\Gamma\,A}{|t|_2 : \mathsf{Tm}\,(\cdot\,,\gamma : |\Gamma|_2)\,|A|_2}$$

The interpretation of $\Pi\beta$ still needs $\Sigma\eta$ because comprehension is given by Σ types, however the type dependency parts are dealt with by substitution laws. For example, we use substitution to write $|B|_2[\gamma \mapsto (\gamma, \alpha)]$ in the interpretation of Π while in the $|-|_1$ variant we used application $|B|_1 \mathbin{@} (\gamma, \alpha)$.

$$\begin{aligned}
|\cdot|_2 &:= \top \\
|\Gamma\,, x : A|_2 &:= \Sigma(\gamma : |\Gamma|_2).|A|_2 \\
|\Pi(x : A).B|_2 &:= \Pi(\alpha : |A|_2).|B|_2[\gamma \mapsto (\gamma\,, \alpha)] \\
|\lambda x.t|_2 &:= \lambda\alpha.|t|_2[\gamma \mapsto (\gamma\,, \alpha)] \\
|t \mathbin{@} x|_2 &:= \big(|t|_2[\gamma \mapsto \mathsf{pr}_0\,\gamma]\big) \mathbin{@} \mathsf{pr}_1\,\gamma \\
|\Pi\beta|_2 &: |(\lambda x.t) \mathbin{@} x|_2 = \big(|\lambda x.t|_2[\gamma \mapsto \mathsf{pr}_0\,\gamma]\big) \mathbin{@} \mathsf{pr}_1\,\gamma = \\
&\quad \big((\lambda\alpha.|t|_2[\gamma \mapsto (\gamma\,, \alpha)])[\gamma \mapsto \mathsf{pr}_0\,\gamma]\big) \mathbin{@} \mathsf{pr}_1\,\gamma = \\
&\quad \big(\lambda\alpha.|t|_2[\gamma \mapsto (\mathsf{pr}_0\,\gamma\,, \alpha)]\big) \mathbin{@} \mathsf{pr}_1\,\gamma \overset{\Pi\beta}{=} \\
&\quad |t|_2[\gamma \mapsto (\mathsf{pr}_0\,\gamma\,, \mathsf{pr}_1\,\gamma)] \overset{\Sigma\eta}{=} |t|_2[\gamma \mapsto \gamma] = |t|_2
\end{aligned}$$

The standard interpretation $|-|_3$ The last variant of the standard interpretation is simply the identity translation: everything is mapped to itself. The source and the target theory can be exactly the same.

$$\frac{\Gamma : \mathsf{Con}_i}{|\Gamma|_3 : \mathsf{Con}_i} \qquad \frac{A : \mathsf{Ty}_j\,\Gamma}{|A|_3 : \mathsf{Ty}_j\,|\Gamma|_3} \qquad \frac{\delta : \mathsf{Sub}\,\Gamma\,\Delta}{|\delta|_3 : \mathsf{Sub}\,|\Gamma|_3\,|\Delta|_3} \qquad \frac{t : \mathsf{Tm}\,\Gamma\,A}{|t|_3 : \mathsf{Tm}\,|\Gamma|_3\,|A|_3}$$

Here the interpretation of $\Pi\beta$ obviously only needs $\Pi\beta$.

$$\begin{aligned}
|\cdot|_3 &:= \cdot \\
|\Gamma\,, x : A|_3 &:= |\Gamma|_3\,, x : |A|_3 \\
|\Pi(x : A).B|_3 &:= \Pi(x : |A|_3).|B|_3 \\
|\lambda x.t|_3 &:= \lambda x.|t|_3 \\
|t \mathbin{@} x|_3 &:= |t|_3 \mathbin{@} x \\
|\Pi\beta|_3 &: |(\lambda x.t) \mathbin{@} x|_3 = (\lambda x.|t|_3) \mathbin{@} x \overset{\Pi\beta}{=} |t|_3
\end{aligned}$$

3.2 Graph Model

In this subsection we define variants of the graph model corresponding to the $_0$, $_1$, $_2$, $_3$ variants of the standard model. Here each syntactic component is mapped to two components $|-|$ and $-^\sim$. The $|-|$ components are the same as in the case of the standard model.

The metatheoretic $_0$ variant of the graph interpretation is specified as follows.

$$\frac{\Gamma : \mathsf{Con}_i}{\begin{array}{l}|\Gamma|_0 : \mathsf{Set}_i \\ \Gamma^{\sim_0} : |\Gamma|_0 \to |\Gamma|_0 \to \mathsf{Set}_i\end{array}} \qquad \frac{A : \mathsf{Ty}_j\,\Gamma}{\begin{array}{l}|A|_0 : |\Gamma|_0 \to \mathsf{Set}_j \\ A^{\sim_0} : \Gamma^{\sim_0}\,\gamma_0\,\gamma_1 \to \\ \qquad |A|_0\,\gamma_0 \to |A|_0\,\gamma_1 \to \mathsf{Set}_j\end{array}}$$

$$\frac{\delta : \mathsf{Sub}\,\Gamma\,\Delta}{\begin{array}{l}|\delta|_0 : |\Gamma|_0 \to |\Delta|_0 \\ \delta^{\sim_0} : \Gamma^{\sim_0}\,\gamma_0\,\gamma_1 \to \Delta^{\sim_0}\,(|\delta|_0\,\gamma_0)\,(|\delta|_0\,\gamma_1)\end{array}} \qquad \frac{t : \mathsf{Tm}\,\Gamma\,A}{\begin{array}{l}|t|_0 : (\gamma : |\Gamma|_0) \to |A|_0\,\gamma \\ t^{\sim_0} : (\gamma_{01} : \Gamma^{\sim_0}\,\gamma_0\,\gamma_1) \to \\ \qquad A^{\sim_0}\,\gamma_{01}\,(|t|_0\,\gamma_0)\,(|t|_0\,\gamma_1)\end{array}}$$

Models of type theory are usually named after what contexts are mapped to (a set for the set model, a setoid for the setoid model, etc.). In the graph model a context is mapped to a graph: a set of vertices and for every two vertex, a set of arrows between those. In short, a set and a proof-relevant binary relation over it. Types are interpreted as displayed graphs over a base graph: a family over each vertex and a heterogeneous binary relation over each arrow. Substitutions become graph homomorphisms and terms their displayed variants. Note that in the types of $A^{\sim_0}$, $\delta^{\sim_0}$ and $t^{\sim_0}$, we implicitly quantified over γ_0 and γ_1.

Variant $_1$ of the graph interpretation is specified as follows. Again, we need a Coquand universe U in the target.

$$\frac{\Gamma : \mathsf{Con}_i}{\begin{array}{l}|\Gamma|_1 : \mathsf{Tm}\cdot\mathsf{U}_i \\ \Gamma^{\sim_1} : \mathsf{Tm}\cdot(\mathsf{El}\,|\Gamma|_1 \Rightarrow \mathsf{El}\,|\Gamma|_1 \Rightarrow \mathsf{U}_i)\end{array}}$$

$$\frac{A : \mathsf{Ty}_i\,\Gamma}{\begin{array}{l}|A|_1 : \mathsf{Tm}\cdot(\mathsf{El}\,|\Gamma|_1 \Rightarrow \mathsf{El}\,|\Delta|_1) \\ A^{\sim_1} : \mathsf{Tm}\cdot\big(\mathsf{El}\,(\Gamma^{\sim_1} @ \gamma_0 @ \gamma_1) \Rightarrow \mathsf{El}\,(|A|_1 @ \gamma_0) \Rightarrow \mathsf{El}\,(|A|_1 @ \gamma_1) \Rightarrow \mathsf{U}_i\big)\end{array}}$$

$$\frac{\delta : \mathsf{Sub}\,\Gamma\,\Delta}{\begin{array}{l}|\delta|_1 : \mathsf{Tm}\cdot(\mathsf{El}\,|\Gamma|_1 \Rightarrow \mathsf{U}_i) \\ \delta^{\sim_1} : \mathsf{Tm}\cdot\Big(\mathsf{El}\,(\Gamma^{\sim_1} @ \gamma_0 @ \gamma_1) \Rightarrow \mathsf{El}\,\big(\Delta^{\sim_1} @ (|\delta|_1 @ \gamma_0) @ (|\delta|_1 @ \gamma_1)\big)\Big)\end{array}}$$

$$\frac{t : \mathsf{Tm}\,\Gamma\,A}{\begin{array}{l}|t|_1 : \mathsf{Tm}\cdot(\Pi(\gamma : \mathsf{El}\,|\Gamma|_1).\mathsf{El}\,(|A|_1 @ \gamma)) \\ t^{\sim_1} : \mathsf{Tm}\cdot\Big(\Pi(\gamma_{01} : \mathsf{El}\,(\Gamma^{\sim_1} @ \gamma_0 @ \gamma_1)).\mathsf{El}\,\big(A^{\sim_1} @ \gamma_{01} @ (|t|_1 @ \gamma_0) @ (|t|_1 @ \gamma_1)\big)\Big)\end{array}}$$

The relations become target theory terms which have function types with codomain U. We used implicit quantification in the target theory for ease of reading. For example, the type of $\delta^{\sim_1}$ should be understood as

$$\mathsf{Tm}\cdot\big(\Pi(\gamma_0 : |\Gamma|_1).\Pi(\gamma_0 : |\Gamma|_1).\mathsf{El}\,(\Gamma^{\sim_1} @ \gamma_0 @ \gamma_1) \Rightarrow \Delta^{\sim_1} @ (|\delta|_1 @ \gamma_0) @ (|\delta|_1 @ \gamma_1)\big).$$

Variant $_2$ *of the graph interpretation* is specified as follows.

$$\frac{\Gamma : \mathsf{Con}_i}{\begin{array}{l}|\Gamma|_2 : \mathsf{Ty}_i\,\cdot \\ \Gamma^{\sim_2} : \mathsf{Ty}_i\,(\cdot\,, \gamma_0 : |\Gamma|_2\,, \gamma_1 : |\Gamma|_2)\end{array}}$$

$$\frac{A : \mathsf{Ty}_i\,\Gamma}{\begin{array}{l}|A|_2 : \mathsf{Ty}_i\,(\cdot\,, \gamma : |\Gamma|_2) \\ A^{\sim_2} : \mathsf{Ty}_i\,\big(\cdot\,, \gamma_{01} : \Gamma^{\sim_2}\,, \alpha_0 : |A|_2[\gamma \mapsto \gamma_0]\,, \alpha_1 : |A|_2[\gamma \mapsto \gamma_1]\big)\end{array}}$$

$$\frac{\delta : \mathsf{Sub}\,\Gamma\,\Delta}{\begin{array}{l}|\delta|_2 : \mathsf{Tm}\,(\cdot\,, \gamma : |\Gamma|_2)\,|\Delta|_2 \\ \delta^{\sim_2} : \mathsf{Tm}\,(\cdot\,, \gamma_{01} : \Gamma^{\sim_2})\,\big(\Delta^{\sim_2}\big[\gamma_0 \mapsto |\delta|_2[\gamma \mapsto \gamma_0]\,, \gamma_1 \mapsto |\delta|_2[\gamma \mapsto \gamma_1]\big]\big)\end{array}}$$

$$\frac{t : \mathsf{Tm}\,\Gamma\,A}{\begin{array}{l}|t|_2 : \mathsf{Tm}\,(\cdot\,, \gamma : |\Gamma|_2)\,|A|_2 \\ t^{\sim_2} : \mathsf{Tm}\,(\cdot\,, \gamma_{01} : \Gamma^{\sim_2})\,\big(A^{\sim_2}\big[\alpha_0 \mapsto |t|_2[\gamma \mapsto \gamma_0]\,, \alpha_1 \mapsto |t|_2[\gamma \mapsto \gamma_1]\big]\big)\end{array}}$$

This variant shows that the extra $^{\sim_2}$ components in the model can be expressed without using Π or a universe: type dependency is enough to express e.g. that $\Gamma^{\sim_2}$ is indexed over two copies of $|\Gamma|_2$. Here, analogously to the usage of implicit Πs in variant $_1$, we use implicit context extensions in the target theory. For example, the type of $\delta^{\sim_2}$ should be understood as

$$\mathsf{Tm}\,(\cdot\,, \gamma_0 : |\Gamma|_2\,, \gamma_1 : |\Gamma|_2\,, \gamma_{01} : \Gamma^{\sim_2})\,\big(\Delta^{\sim_2}\big[\gamma_0 \mapsto |\delta|_2[\gamma \mapsto \gamma_0]\,, \gamma_1 \mapsto |\delta|_2[\gamma \mapsto \gamma_1]\big]\big).$$

Defining variant $_3$ of the graph interpretation is not as straightforward as the previous ones. As $|\Gamma|_3 : \mathsf{Con}$, we need a notion of binary relation over a context. One solution is going back to the $|-|_0$ model and using the equivalence between indexed families and fibrations [11, p. 221]:

$$A \to \mathsf{Set} \quad \simeq \quad (A' : \mathsf{Set}) \times (A' \to A).$$

This means that we replace $\Gamma^{\sim_0} : |\Gamma|_0 \to |\Gamma|_0 \to \mathsf{Set}$ with a set $\Gamma^{\sim_{0a}}$ and two projections 0_{0a}, 1_{0a} which give the domain and codomain of the arrow. This interpretation is specified as follows (the $|-|$ components are the same as in the $_0$ model, so they don't have the $_a$ subscript).

$$\frac{\Gamma : \mathsf{Con}_i}{\begin{array}{l}|\Gamma|_0 \quad : \mathsf{Set}_i \\ \Gamma^{\sim_{0a}} \quad : \mathsf{Set}_i \\ 0_{0a}\,\Gamma : \Gamma^{\sim_{0a}} \to |\Gamma|_0 \\ 1_{0a}\,\Gamma : \Gamma^{\sim_{0a}} \to |\Gamma|_0\end{array}}$$

$$\frac{A : \mathsf{Ty}_i\, \Gamma}{\begin{array}{l} |A|_0 \;: |\Gamma|_0 \to \mathsf{Set}_i \\ A^{\sim_{0a}} : (\gamma_{01} : \Gamma^{\sim_{0a}}) \to |A|_0\,(0_{0a}\,\Gamma\,\gamma_{01}) \to |A|_0\,(1_{0a}\,\Gamma\,\gamma_{01}) \to \mathsf{Set}_i \end{array}}$$

$$\frac{\delta : \mathsf{Sub}\,\Gamma\,\Delta}{\begin{array}{l} |\delta|_0 \quad : |\Gamma|_0 \to |\Delta|_0 \\ \delta^{\sim_{0a}} \quad : \Gamma^{\sim_{0a}} \to \Delta^{\sim_{0a}} \\ 0_{0a}\,\delta : (\gamma_{01} : \Gamma^{\sim_{0a}}) \to 0_{0a}\,\Delta\,(\delta^{\sim_{0a}}\,\gamma_{01}) = |\delta|_0\,(0_{0a}\,\Gamma\,\gamma_{01}) \\ 1_{0a}\,\delta : (\gamma_{01} : \Gamma^{\sim_{0a}}) \to 1_{0a}\,\Delta\,(\delta^{\sim_{0a}}\,\gamma_{01}) = |\delta|_0\,(1_{0a}\,\Gamma\,\gamma_{01}) \end{array}}$$

$$\frac{t : \mathsf{Tm}\,\Gamma\,A}{\begin{array}{l} |t|_0 \;: (\gamma : |\Gamma|_0) \to |A|_0\,\gamma \\ t^{\sim_{0a}} : (\gamma_{01} : \Gamma^{\sim_{0a}}) \to A^{\sim_{0a}}\,\gamma_{01}\,\big(|t|_0\,(0_{0a}\,\Gamma\,\gamma_{01})\big)\,\big(|t|_0\,(1_{0a}\,\Gamma\,\gamma_{01})\big) \end{array}}$$

The fact that contexts are given as fibrations forces substitutions to include some equalities, while types are still indexed. This is an example of a model which can be turned into a syntactic translation, but is not strict in Agda. The reason is that equalities are needed to interpret substitutions and in turn we use these equalities to interpret some conversion rules. For example, the 0_{0a} and 1_{0a} components in the interpretation of $- \circ -$ are given by transitivity of equality, so associativity of substitutions needs associativity of transitivity (or UIP). We believe however that this model would be strict in a setoid type theory (Sect. 5).

In the corresponding $_{3a}$ *variant* a context is interpreted by two contexts and two projection substitutions. This is the same as the parametricity translation of Bernardy et al. [7]. We list the $|-|_3$ part separately because we will reuse it in $^{\sim_{3b}}$.

$$\frac{\Gamma : \mathsf{Con}_i}{|\Gamma|_3 : \mathsf{Con}_i} \qquad \frac{A : \mathsf{Ty}_i\,\Gamma}{|A|_3 : \mathsf{Ty}_i\,|\Gamma|_3} \qquad \frac{\delta : \mathsf{Sub}\,\Gamma\,\Delta}{|\delta|_3 : \mathsf{Sub}\,|\Gamma|_3\,|\Delta|_3} \qquad \frac{t : \mathsf{Tm}\,\Gamma\,A}{|t|_3 : \mathsf{Tm}\,|\Gamma|_3\,|A|_3}$$

$$\frac{\Gamma : \mathsf{Con}_i}{\begin{array}{l} \Gamma^{\sim_{3a}} \quad : \mathsf{Con}_i \\ 0_{3a}\,\Gamma : \mathsf{Sub}\,\Gamma^{\sim_{3a}}\,|\Gamma|_3 \\ 1_{3a}\,\Gamma : \mathsf{Sub}\,\Gamma^{\sim_{3a}}\,|\Gamma|_3 \end{array}} \qquad \frac{\delta : \mathsf{Sub}\,\Gamma\,\Delta}{\begin{array}{l} \delta^{\sim_{3a}} \quad : \mathsf{Sub}\,\Gamma^{\sim_{3a}}\,\Delta^{\sim_{3a}} \\ 0_{3a}\,\delta : 0_{3a}\,\Delta \circ \delta^{\sim_{3a}} = |\delta|_3 \circ 0_{3a}\,\Gamma \\ 1_{3a}\,\delta : 1_{3a}\,\Delta \circ \delta^{\sim_{3a}} = |\delta|_3 \circ 1_{3a}\,\Gamma \end{array}}$$

$$\frac{A : \mathsf{Ty}_i\,\Gamma}{A^{\sim_{3a}} : \mathsf{Ty}_i\,\big(\Gamma^{\sim_{3a}}, \alpha_0 : |A|_3[0_{3a}\,\Gamma], \alpha_1 : |A|_3[1_{3a}\,\Gamma]\big)}$$

$$\frac{t : \mathsf{Tm}\,\Gamma\,A}{t^{\sim_{3a}} : \mathsf{Tm}\,\Gamma^{\sim_{3a}}\,\big(A^{\sim_{3a}}\big[\alpha_0 \mapsto |t|_3[0_{3a}\,\Gamma], \alpha_1 \mapsto |t|_3[1_{3a}\,\Gamma]\big]\big)}$$

The $_{3b}$ *Variant of the Graph Interpretation.* Another solution is to define $\Gamma^{\sim_3}$ in an indexed way by referring to substitutions into $|\Gamma|_3$. This is how we define $^{\sim_{3b}}$. The $|-|_3$ parts are the same as in $_{3a}$.

$$\frac{\Gamma : \mathsf{Con}_i \qquad \rho_0, \rho_1 : \mathsf{Sub}\,\Omega\,|\Gamma|_3}{\begin{array}{l}\Gamma^{\sim_{3b}}\,\rho_0\,\rho_1 : \mathsf{Ty}_i\,\Omega \\ (\Gamma^{\sim_{3b}}\,\rho_0\,\rho_1)[\nu] = \Gamma^{\sim_{3b}}\,(\rho_0 \circ \nu)\,(\rho_1 \circ \nu)\end{array}}$$

$$\frac{A : \mathsf{Ty}_i\,\Gamma \qquad \rho_{01} : \Gamma^{\sim_{3b}}\,\rho_0\,\rho_1 \qquad t_0 : \mathsf{Tm}\,\Omega\,(|A|_3[\rho_0]) \qquad t_1 : \mathsf{Tm}\,\Omega\,(|A|_3[\rho_1])}{\begin{array}{l}A^{\sim_{3b}}\,\rho_{01}\,t_0\,t_1 : \mathsf{Ty}_i\,\Omega \\ (A^{\sim_{3b}}\,\rho_{01}\,t_0\,t_1)[\nu] = A^{\sim_{3b}}\,(\rho_{01}[\nu])\,(t_0[\nu])\,(t_1[\nu])\end{array}}$$

$$\frac{\delta : \mathsf{Sub}\,\Gamma\,\Delta \qquad \rho_{01} : \mathsf{Tm}\,\Omega\,(\Gamma^{\sim_{3b}}\,\rho_0\,\rho_1)}{\begin{array}{l}\delta^{\sim_{3b}}\,\rho_{01} : \mathsf{Tm}\,\Omega\,(\Delta^{\sim_{3b}}\,(|\delta|_3 \circ \rho_0)\,(|\delta|_3 \circ \rho_1)) \\ (\delta^{\sim_{3b}}\,\rho_{01})[\nu] = \delta^{\sim_{3b}}\,(\rho_{01}[\nu])\end{array}}$$

$$\frac{t : \mathsf{Tm}\,\Gamma\,A \qquad \rho_{01} : \mathsf{Tm}\,\Omega\,(\Gamma^{\sim_{3b}}\,\rho_0\,\rho_1)}{\begin{array}{l}t^{\sim_{3b}}\,\rho_{01} : \mathsf{Tm}\,\Omega\,\big(A^{\sim_{3b}}\,\rho_{01}\,(|t|_3[\rho_0])\,(|t|_3[\rho_1])\big) \\ (t^{\sim_{3b}}\,\rho_{01})[\nu] = t^{\sim_{3b}}\,(\rho_{01}[\nu])\end{array}}$$

Ω, ρ_0 and ρ_1 are implicit parameters of $A^{\sim_{3b}}$, $\delta^{\sim_{3b}}$ and $t^{\sim_{3b}}$. The advantage of the $^{\sim_{3b}}$ compared to the $^{\sim_{3a}}$ is that we don't need the projection substitutions for contexts and the naturality conditions for substitutions, the disadvantage is that we need the extra equalities expressing substitution laws.

4 The Setoid Model as a Translation

In this section, after recalling the setoid model, we turn it into a syntactic translation from $\mathsf{MLTT}_{\mathsf{Prop}}$ to $\mathsf{MLTT}_{\mathsf{Prop}}$. We follow the approach of graph model variant $_{3b}$ (Sect. 3.2) and extend it into a setoid syntactic translation where a context is modelled not only by a set and a relation, but a set and an equivalence relation.

4.1 The Setoid Model

In the setoid model [1], a context is given by a set together with an equivalence relation which, in contrast with the graph model, is proof-irrelevant. We think about this relation as the explicit equality relation for the set. A type is interpreted by a displayed setoid together with a coercion and coherence operation.

Coercion transports between families at related objects and coherence says that this coercion respects the displayed relation.

$$
\begin{array}{l}
\Gamma : \mathsf{Con}_i \\
\hline
|\Gamma|_0 : \mathsf{Set}_i \\
\Gamma^{\sim_0} : |\Gamma|_0 \to |\Gamma|_0 \to \mathsf{Prop}_i \\
\mathsf{R}^0_\Gamma \ : (\gamma : |\Gamma|_0) \to \Gamma^{\sim_0}\,\gamma\,\gamma \\
\mathsf{S}^0_\Gamma \ : \Gamma^{\sim_0}\,\gamma_0\,\gamma_1 \to \Gamma^{\sim_0}\,\gamma_1\,\gamma_0 \\
\mathsf{T}^0_\Gamma \ : \Gamma^{\sim_0}\,\gamma_0\,\gamma_1 \to \\
\qquad \Gamma^{\sim_0}\,\gamma_1\,\gamma_2 \to \\
\qquad \Gamma^{\sim_0}\,\gamma_0\,\gamma_2
\end{array}
\qquad
\begin{array}{ll}
\multicolumn{2}{c}{A : \mathsf{Ty}_j\,\Gamma} \\
\hline
|A|_0 & : |\Gamma|_0 \to \mathsf{Set}_j \\
A^{\sim_0} & : \Gamma^{\sim_0}\,\gamma_0\,\gamma_1 \to |A|_0\,\gamma_0 \to |A|_0\,\gamma_1 \to \mathsf{Prop}_j \\
\mathsf{R}^0_A & : (\alpha : |A|_0\,\gamma) \to A^{\sim_0}\,(\mathsf{R}^0_\Gamma\,\gamma)\,\alpha\,\alpha \\
\mathsf{S}^0_A & : A^{\sim_0}\,\gamma_{01}\,\alpha_0\,\alpha_1 \to A^{\sim_0}\,(\mathsf{S}^0_\Gamma\,\gamma_{01})\,\alpha_1\,\alpha_0 \\
\mathsf{T}^0_A & : A^{\sim_0}\,\gamma_{01}\,\alpha_0\,\alpha_1 \to A^{\sim_0}\,\gamma_{12}\,\alpha_1\,\alpha_2 \to \\
& \quad A^{\sim_0}\,(\mathsf{T}^0_\Gamma\,\gamma_{01}\,\gamma_{12})\,\alpha_0\,\alpha_2 \\
\mathsf{coe}^0_A & : \Gamma^{\sim_0}\,\gamma_0\,\gamma_1 \to |A|_0\,\gamma_0 \to |A|_0\,\gamma_1 \\
\mathsf{coh}^0_A & : (\gamma_{01} : \Gamma^{\sim_0}\,\gamma_0\,\gamma_1) \to (\alpha_0 : |A|_0\,\gamma_0) \to \\
& \quad A^{\sim_0}\,\gamma_{01}\,\alpha_0\,(\mathsf{coe}^0_A\,\gamma_{01}\,\alpha_0)
\end{array}
$$

This notion of family of setoids is different from Altenkirch's original one [1] but is equivalent to it [9, Sect. 1.6.1]. Substitutions and terms are specified the same as in the graph model (see $|-|_0$ in Sect. 3.2). There is no need for R, S, T components because these are provable by proof irrelevance (unlike in the groupoid model [19,24]).

4.2 Specification of the Translation

In the following we turn the setoid interpretation $_0$ into a setoid syntactic translation following the $_{3b}$ variant of the graph translation. We drop the $_{3b}$ indices to ease the reading. We expect the other variants to be definable as well.

An $\mathsf{MLTT}_{\mathsf{Prop}}$ context is mapped to six components: a context, a binary propositional relation over substitutions into that context, reflexivity, symmetry and transitivity of this relation and a substitution law for $-^\sim$. Note that we use implicit arguments, e.g. $\Gamma^\sim$ takes an $\Omega : \mathsf{Con}$ implicitly and S_Γ takes $\Omega : \mathsf{Con}$, $\rho_0, \rho_1 : \mathsf{Sub}\,\Omega\,|\Gamma|$ implicitly.

$$
\begin{array}{ll}
\multicolumn{2}{c}{\Gamma : \mathsf{Con}_i} \\
\hline
|\Gamma| & : \mathsf{Con}_i \\
\Gamma^\sim & : \mathsf{Sub}\,\Omega\,|\Gamma| \to \mathsf{Sub}\,\Omega\,|\Gamma| \to \mathsf{Tm}\,\Omega\,\mathsf{Prop}_i \\
\Gamma^\sim[] & : (\Gamma^\sim\,\rho_0\,\rho_1)[\nu] = \Gamma^\sim\,(\rho_0 \circ \nu)\,(\rho_1 \circ \nu) \\
\mathsf{R}_\Gamma & : (\rho : \mathsf{Sub}\,\Omega\,|\Gamma|) \to \mathsf{Tm}\,\Omega\,\underline{\Gamma^\sim\,\rho\,\rho} \\
\mathsf{S}_\Gamma & : \mathsf{Tm}\,\Omega\,\underline{\Gamma^\sim\,\rho_0\,\rho_1} \to \mathsf{Tm}\,\Omega\,\underline{\Gamma^\sim\,\rho_1\,\rho_0} \\
\mathsf{T}_\Gamma & : \mathsf{Tm}\,\Omega\,\underline{\Gamma^\sim\,\rho_0\,\rho_1} \to \mathsf{Tm}\,\Omega\,\underline{\Gamma^\sim\,\rho_1\,\rho_2} \to \mathsf{Tm}\,\Omega\,\underline{\Gamma^\sim\,\rho_0\,\rho_2}
\end{array}
$$

A type is interpreted by a type over the interpretation of the context, a heterogeneous relation over the relation for contexts which is reflexive, symmetric and transitive (over the corresponding proofs for the contexts). Moreover, there is a coercion function which relates types substituted by related substitutions. Coherence expresses that coercion respects the relation (coh). The $^\sim$ relation and coe come with substitution laws.

$$\begin{array}{ll}
\multicolumn{2}{c}{A : \mathsf{Ty}_j\,\Gamma} \\
\hline
|A| & : \mathsf{Ty}_j\,|\Gamma| \\
A^\sim & : \mathsf{Tm}\,\Omega\,\underline{\Gamma^\sim\,\rho_0\,\rho_1} \to \mathsf{Tm}\,\Omega\,(|A|[\rho_0]) \to \mathsf{Tm}\,\Omega\,(|A|[\rho_1]) \to \mathsf{Tm}\,\Omega\,\mathsf{Prop}_j \\
A^\sim[] & : (A^\sim\,\rho_{01}\,t_0\,t_1)[\nu] = A^\sim\,(\rho_{01}[\nu])\,(t_0[\nu])\,(t_1[\nu]) \\
\mathsf{R}_A & : (t : \mathsf{Tm}\,\Omega\,(|A|[\rho])) \to \mathsf{Tm}\,\Omega\,\underline{A^\sim\,(\mathsf{R}_\Gamma\,\rho)\,t\,t} \\
\mathsf{S}_A & : \mathsf{Tm}\,\Omega\,\underline{A^\sim\,\rho_{01}\,t_0\,t_1} \to \mathsf{Tm}\,\Omega\,\underline{A^\sim\,(\mathsf{S}_\Gamma\,\rho_{01})\,t_1\,t_0} \\
\mathsf{T}_A & : \mathsf{Tm}\,\Omega\,\underline{A^\sim\,\rho_{01}\,t_0\,t_1} \to \mathsf{Tm}\,\Omega\,\underline{A^\sim\,\rho_{12}\,t_1\,t_2} \to \mathsf{Tm}\,\Omega\,\underline{A^\sim\,(\mathsf{T}_\Gamma\,\rho_{01}\,\rho_{12})\,t_0\,t_2} \\
\mathsf{coe}_A & : \mathsf{Tm}\,\Omega\,\underline{\Gamma^\sim\,\rho_0\,\rho_1} \to \mathsf{Tm}\,\Omega\,(|A|[\rho_0]) \to \mathsf{Tm}\,\Omega\,(|A|[\rho_1]) \\
\mathsf{coe}[]_A & : (\mathsf{coe}_A\,\rho_{01}\,t_0)[\nu] = \mathsf{coe}_A\,(\rho_{01}[\nu])\,(t_0[\nu]) \\
\mathsf{coh}_A & : (\rho_{01} : \mathsf{Tm}\,\Omega\,\underline{\Gamma^\sim\,\rho_0\,\rho_1})(t_0 : \mathsf{Tm}\,\Omega\,(|A|[\rho_0])) \to \\
& \quad \mathsf{Tm}\,\Omega\,\underline{A^\sim\,\rho_{01}\,t_0\,(\mathsf{coe}_A\,\rho_{01}\,t_0)}
\end{array}$$

A substitution is interpreted by a substitution which respects the relations.

$$\begin{array}{ll}
\multicolumn{2}{c}{\delta : \mathsf{Sub}\,\Gamma\,\Delta} \\
\hline
|\delta| & : \mathsf{Sub}\,|\Gamma|\,|\Delta| \\
\delta^\sim & : \mathsf{Tm}\,\Omega\,\underline{\Gamma^\sim\,\rho_0\,\rho_1} \to \mathsf{Tm}\,\Omega\,\underline{\Delta^\sim\,(|\delta| \circ \rho_0)\,(|\delta| \circ \rho_1)}
\end{array}$$

A term is interpreted by a term which respects the relations.

$$\begin{array}{ll}
\multicolumn{2}{c}{t : \mathsf{Tm}\,\Gamma\,A} \\
\hline
|t| & : \mathsf{Tm}\,|\Gamma|\,|A| \\
t^\sim & : (\rho_{01} : \mathsf{Tm}\,\Omega\,\underline{\Gamma^\sim\,\rho_0\,\rho_1}) \to \mathsf{Tm}\,\Omega\,\underline{A^\sim\,\rho_{01}\,(|t|[\rho_0])\,(|t|[\rho_1])}
\end{array}$$

Note that we do not need substitution laws for those components which don't have any parameters (the $|-|$ ones) and those which result in a term of an underlined type. The laws for the latter ones hold by proof irrelevance.

4.3 Implementation of the Translation

We implement this specification by explaining what the different components of $\mathsf{MLTT}_{\mathsf{Prop}}$ are mapped to. All details can be found in Appendix B. As in the case of the $|-|_3$ standard and graph interpretations, the $|-|$ components in the setoid translation are almost always identity functions. The only exception is the case of Π where a function is interpreted by a function which preserves equality:

$$|\Pi(x : A).B| := \Sigma(f : \Pi(x : |A|).|B|).\underline{\pi(x_0 : |A|).\pi(x_1 : |A|).}$$
$$\underline{\pi(x_{01} : \underline{A^\sim\,(\mathsf{R}_\Gamma\,\mathsf{id})\,x_0\,x_1}).B^\sim\,(\mathsf{R}_\Gamma\,\mathsf{id}\,,x_{01})\,(f\,@\,x_0)\,(f\,@\,x_1)}$$

Equality of functions is defined by saying that the first component (pr_0) of the interpretation of the function preserves equality:

$$(\Pi(x : A).B)^\sim\,\rho_{01}\,t_0\,t_1 := \pi(x_0 : |A|[\rho_0]).\pi(x_1 : |A|[\rho_1]).\pi(x_{01} : \underline{A^\sim\,\rho_{01}\,x_0\,x_1}).$$
$$B^\sim\,(\rho_{01}\,,x_{01})\,(\mathsf{pr}_0\,t_0\,@\,x_0)\,(\mathsf{pr}_0\,t_1\,@\,x_1),$$

We need the second component to implement reflexivity for the function space: $\mathsf{R}_{\Pi(x:A).B}\,t := \mathsf{pr}_1\,t$. Equality for extended contexts and Σ types is pointwise, equality of booleans is given by a decision procedure, equality of propositions is logical equivalence and equality of proofs of propositions is trivial:

$$
\begin{aligned}
(\Gamma\,, x : A)^{\sim}\,(\rho_0\,, x \mapsto t_0)\,(\rho_1\,, x \mapsto t_1) &:= \sigma(\rho_{01} : \Gamma^{\sim}\,\rho_0\,\rho_1).A^{\sim}\,\rho_{01}\,t_0\,t_1\\
(\Sigma(x:A).B)^{\sim}\,\rho_{01}\,(u_0\,,v_0)\,(u_1\,,v_1) &:= \sigma(x_{01} : A^{\sim}\,\rho_{01}\,u_0\,u_1).B^{\sim}\,(\rho_{01}\,,x_{01})\,v_0\,v_1\\
\mathsf{Bool}^{\sim}\,\rho_{01}\,t_0\,t_1 &:= \mathsf{if}\;t_0\;\mathsf{then}\;(\mathsf{if}\;t_1\;\mathsf{then}\;\top\;\mathsf{else}\;\bot)\\
&\qquad\qquad\;\; \mathsf{else}\;\;(\mathsf{if}\;t_1\;\mathsf{then}\;\bot\;\mathsf{else}\;\top)\\
\mathsf{Prop}_i{}^{\sim}\,\rho_{01}\,a_0\,a_1 &:= (\underline{a_0} \Rightarrow a_1) \times (\underline{a_1} \Rightarrow a_0)\\
\underline{a}^{\sim}\,\rho_{01}\,t_0\,t_1 &:= \top
\end{aligned}
$$

Symmetry for Π types takes as input an equality proof x_{01}, applies symmetry on it at the domain type, then applies the proof of equality, then applies symmetry at the codomain type: $\mathsf{S}_{\Pi(x:A).B}\,t_{01} := \lambda x_0\,x_1\,x_{01}.\mathsf{S}_B\,(t_{01} \mathbin{@} x_1 \mathbin{@} x_0 \mathbin{@} \mathsf{S}_A\,x_{01})$. Coercion needs to produce a function $t_0 : \mathsf{Tm}\,\Omega\,(|\Pi(x:A).B|[\rho_0])$ and has to produce one of type $|\Pi(x:A).B|[\rho_1]$. The first component is given by

$$\lambda x_1.\mathsf{coe}_B\,(\rho_{01}\,,\mathsf{coh}_A\,(\mathsf{S}_\Gamma\,\rho_{01})\,x_1)\,(\mathsf{pr}_0\,t_0 \mathbin{@} \mathsf{coe}_A\,(\mathsf{S}_\Gamma\,\rho_{01})\,x_1).$$

First the input is coerced backwards by $\mathsf{coe}^*{}_A$ (from $|A|[\rho_1]$ to $|A|[\rho_0]$), then the function is applied, then the output is coerced forwards by coe_B. Backwards coercion $\mathsf{coe}^*{}_A$ is defined using coe_A and S_Γ. Backwards coherence is defined in a similar way, see Appendix B.2 for details.

Reflexivity, symmetry and transitivity are pointwise for Σ types. For Bool, they are defined using if – then – else –, e.g. $\mathsf{R}_{\mathsf{Bool}}\,t := \mathsf{if}\;t\;\mathsf{then}\;\mathsf{tt}\;\mathsf{else}\;\mathsf{tt}$. As Bool is a closed type, coercion is the identity function and coherence is trivial.

Reflexivity for propositions is given by two identity functions: $\mathsf{R}_{\mathsf{Prop}_i}\,a := (\lambda x.x, \lambda x.x)$. Symmetry swaps the functions: $\mathsf{S}_{\mathsf{Prop}_i}\,(a_{01},a_{10}) := (a_{10},a_{01})$. Coercion is the identity, and hence coherence is given by two identity functions. For $\underline{a}$ types, reflexivity, symmetry, transitivity and coherence are all trivial (tt). Coercion is $\underline{a}$ is more interesting: it uses a function from the logical equivalence given by $a^{\sim}\,\rho_{01}$.

$$\mathsf{coe}_{\underline{a}}\,\rho_{01}\,t_0 := \mathsf{pr}_0\,(a^{\sim}\,\rho_{01}) \mathbin{@} t_0$$

The rest of the setoid translation follows that of the setoid model [1], see Appendix B for all the details.

4.4 Extensions

The Identity Type. We extend the signature of $\mathsf{MLTT}_{\mathsf{Prop}}$ given in Sect. 2 with Martin-Löf's inductive identity type with a propositional computation rule, function extensionality and propositional extensionality by the following rules.

$$\frac{A : \mathsf{Ty}_i\,\Gamma \qquad u, v : \mathsf{Tm}\,\Gamma\,A}{\mathsf{Id}_A\,u\,v : \mathsf{Tm}\,\Gamma\,\mathsf{Prop}_i}$$

$$\frac{P : \mathsf{Ty}_i\,(\Gamma\,, x : A) \qquad e : \mathsf{Tm}\,\Gamma\,\underline{\mathsf{Id}_A\,u\,v} \qquad t : \mathsf{Tm}\,\Gamma\,(P[x \mapsto u])}{\mathsf{transport}_{x.P}\,e\,t : \mathsf{Tm}\,\Gamma\,(P[x \mapsto v])}$$

$$\frac{u : \mathsf{Tm}\,\Gamma\,A}{\mathsf{refl}_u : \mathsf{Tm}\,\Gamma\,\underline{\mathsf{Id}_A\,u\,u}} \qquad \frac{P : \mathsf{Ty}_i\,(\Gamma\,, x : A) \qquad t : \mathsf{Tm}\,\Gamma\,(P[x \mapsto u])}{\mathsf{Id}\beta\,t : \mathsf{Tm}\,\Gamma\,\underline{\mathsf{Id}_{P[x \mapsto u]}\,(\mathsf{transport}_{x.P}\,\mathsf{refl}_u\,t)\,t}}$$

$$\frac{t_0, t_1 : \mathsf{Tm}\,\Gamma\,(\Pi(x : A).B) \qquad e : \mathsf{Tm}\,\Gamma\,\underline{\big(\Pi(x : A).\mathsf{Id}_B\,(t_0 \mathbin{@} x)\,(t_1 \mathbin{@} x)\big)}}{\mathsf{funext}\,e : \mathsf{Tm}\,\Gamma\,\underline{\mathsf{Id}_{\Pi(x : A).B}\,t_0\,t_1}}$$

$$\frac{a_0, a_1 : \mathsf{Tm}\,\Gamma\,\mathsf{Prop} \qquad t : \mathsf{Tm}\,\Gamma\,\underline{(\underline{a_0} \Rightarrow a_1) \times (\underline{a_1} \Rightarrow a_0)}}{\mathsf{propext}\,t : \mathsf{Tm}\,\Gamma\,\underline{\mathsf{Id}_{\mathsf{Prop}}\,a_0\,a_1}}$$

$$\begin{aligned}
&\mathsf{Id}[] &&: (\mathsf{Id}_A\,u\,v)[\nu] &&= \mathsf{Id}_{A[\nu]}\,(u[\nu])\,(v[\nu])\\
&\mathsf{transport}[] &&: (\mathsf{transport}_{x.P}\,e\,t)[\nu] &&= \mathsf{transport}_{x.P[\nu]}\,(e[\nu])\,(t[\nu])
\end{aligned}$$

Note that the dependent eliminator for Id (usually called J) can be derived from $\mathsf{transport}$ in the presence of UIP (as in our setting).

The setoid translation given in Subsects. 4.2–4.3 translates from $\mathsf{MLTT}_{\mathsf{Prop}}$ to $\mathsf{MLTT}_{\mathsf{Prop}}$. However it extends to a translation from $\mathsf{MLTT}_{\mathsf{Prop}}$ + identity type to $\mathsf{MLTT}_{\mathsf{Prop}}$. $|\mathsf{Id}_A\,u\,v|$ is defined as $A^{\sim}\,(\mathsf{R}_\Gamma\,\mathsf{id})\,|u|\,|v|$, $|\mathsf{transport}_{x.P}\,e\,t|$ is given by $\mathsf{coe}_P\,(\mathsf{R}_\Gamma\,\mathsf{id}\,, |e|)\,|t|$. Function extensionality and propositional extensionality are also justified. See Appendix B.3 for the translation of all the rules of the identity type.

Definitional Computation Rule for $\mathsf{transport}$. We can extend the setoid translation with the following new component for types:

$$\frac{A : \mathsf{Ty}_i\,\Gamma}{\mathsf{coeR}_A : (\rho : \mathsf{Sub}\,\Omega\,|\Gamma|)(t : \mathsf{Tm}\,\Omega\,(|A|[\rho])) \to \mathsf{coe}_A\,(\mathsf{R}_\Gamma\,\rho)\,t = t}$$

This expresses that coercion along reflexivity is the identity. Once we have this, the propositional computation rule of $\mathsf{transport}$ becomes definitional:

$$\begin{aligned}
&|\mathsf{transport}_{x.P}\,\mathsf{refl}_u\,t| =\\
&\mathsf{coe}_P\,(\mathsf{R}_\Gamma\,\mathsf{id}_{|\Gamma|}, |\mathsf{refl}_u|)\,|t| \overset{\mathsf{irr}}{=}\\
&\mathsf{coe}_P\,(\mathsf{R}_{\Gamma\,, x : A}\,\mathsf{id}_{|\Gamma\,, x : A|})\,|t| \overset{\mathsf{coeR}_P\,\mathsf{id}\,|t|}{=}\\
&|t| \quad .
\end{aligned}$$

Adding this rule to the setoid translation amounts to checking whether this equality holds for all type formers, we do this in Appendix B.3. Our Agda formalisation of the setoid model [21] also includes this rule and no axioms are required to justify it.

5 Setoid Type Theory

In this section we extend the signature of $\mathsf{MLTT}_{\mathsf{Prop}}$ given in Sect. 2 with a new heterogeneous equality type. This extended algebraic theory is called *setoid type theory*. The heterogeneous equality type is inspired by the setoid translation of the previous section. The idea is that we simply add the rules of the setoid translation as new term formers to $\mathsf{MLTT}_{\mathsf{Prop}}$. If we did this naively, this would mean adding the operations $|-|$, $^\sim$, R, S, T, coe, coh as new syntax and all the $:=$ definitions of Sect. 4.3 as definitional equalities to the syntax. However this would not result in a usable type theory: $A^\sim$ would not be a relation between terms of type A, but terms of type $|A|$, so we wouldn't even have a general identity type. Our solution is to *not* add $|-|$ as new syntax (as it is mostly the identity anyway), but only the other components. Moreover, we only add those equalities from the translation which are not derivable by irr.

Thus we extend $\mathsf{MLTT}_{\mathsf{Prop}}$ with the following new constructors which explain equality of contexts. This is a homogeneous equivalence relation on substitutions into the context. These new constructors follow the components $\Gamma^\sim$, R_Γ, S_Γ, T_Γ in the setoid translation (Sect. 4.2) except that they do not refer to $|-|$.

$$\frac{\Gamma : \mathsf{Con}_i \qquad \rho_0, \rho_1 : \mathsf{Sub}\,\Omega\,\Gamma}{\Gamma^\sim\,\rho_0\,\rho_1 : \mathsf{Tm}\,\Omega\,\mathsf{Prop}_i} \qquad \frac{\Gamma : \mathsf{Con}_i \qquad \rho : \mathsf{Sub}\,\Omega\,\Gamma}{\mathsf{R}_\Gamma\,\rho : \mathsf{Tm}\,\Omega\,\underline{\Gamma^\sim\,\rho\,\rho}}$$

$$\frac{\Gamma : \mathsf{Con}_i \qquad \rho_{01} : \mathsf{Tm}\,\Omega\,\underline{\Gamma^\sim\,\rho_0\,\rho_1}}{\mathsf{S}_\Gamma\,\rho_{01} : \mathsf{Tm}\,\Omega\,\underline{\Gamma^\sim\,\rho_1\,\rho_0}}$$

$$\frac{\Gamma : \mathsf{Con}_i \qquad \rho_{01} : \mathsf{Tm}\,\Omega\,\underline{\Gamma^\sim\,\rho_0\,\rho_1} \qquad \rho_{12} : \mathsf{Tm}\,\Omega\,\underline{\Gamma^\sim\,\rho_1\,\rho_2}}{\mathsf{T}_\Gamma\,\rho_{01}\,\rho_{12} : \mathsf{Tm}\,\Omega\,\underline{\Gamma^\sim\,\rho_0\,\rho_2}}$$

Note that while $^\sim$ was an operation defined by induction on the syntax, $^\sim$ is a constructor of the syntax. On types, $^\sim$ is heterogeneous: it is a relation between two terms of the same type but substituted by substitutions which are related by $\Gamma^\sim$. It is reflexive, symmetric and transitive and comes with coercion and coherence operators.

$$\frac{A : \mathsf{Ty}_j\,\Gamma \qquad \rho_{01} : \mathsf{Tm}\,\Omega\,\underline{\Gamma^\sim\,\rho_0\,\rho_1} \qquad t_0 : \mathsf{Tm}\,\Omega\,(A[\rho_0]) \qquad t_1 : \mathsf{Tm}\,\Omega\,(A[\rho_1])}{A^\sim\,\rho_{01}\,t_0\,t_1 : \mathsf{Tm}\,\Omega\,\mathsf{Prop}_j}$$

$$\frac{A : \mathsf{Ty}_j\,\Gamma \qquad t : \mathsf{Tm}\,\Omega\,(A[\rho])}{\mathsf{R}_A\,t : \mathsf{Tm}\,\Omega\,\underline{A^\sim\,(\mathsf{R}_\Gamma\,\rho)\,t\,t}} \qquad \frac{A : \mathsf{Ty}_j\,\Gamma \qquad t_{01} : \mathsf{Tm}\,\Omega\,\underline{A^\sim\,\rho_{01}\,t_0\,t_1}}{\mathsf{S}_A\,t_{01} : \mathsf{Tm}\,\Omega\,\underline{A^\sim\,(\mathsf{S}_\Gamma\,\rho_{01})\,t_0\,t_1}}$$

$$\frac{A : \mathsf{Ty}_j\,\Gamma \qquad t_{01} : \mathsf{Tm}\,\Omega\,\underline{A^\sim\,\rho_{01}\,t_0\,t_1} \qquad t_{12} : \mathsf{Tm}\,\Omega\,\underline{A^\sim\,\rho_{12}\,t_1\,t_2}}{\mathsf{T}_A\,t_{01}\,t_{12} : \mathsf{Tm}\,\Omega\,\underline{A^\sim\,(\mathsf{T}_\Gamma\,\rho_{01}\,\rho_{12})\,t_0\,t_2}}$$

$$\frac{A : \mathsf{Ty}_j\,\Gamma \qquad \rho_{01} : \mathsf{Tm}\,\Omega\,\underline{\Gamma^\sim\,\rho_0\,\rho_1} \qquad t_0 : \mathsf{Tm}\,\Omega\,A[\rho_0]}{\mathsf{coe}_A\,\rho_{01}\,t_0 : \mathsf{Tm}\,\Omega\,(A[\rho_1]) \qquad \mathsf{coh}_A\,\rho_{01}\,t_0 : \mathsf{Tm}\,\Omega\,\underline{A^\sim\,\rho_{01}\,t_0\,(\mathsf{coe}_A\,\rho_{01}\,t_0)}}$$

On substitutions and terms $^\sim$ expresses congruence.

$$\frac{\delta : \mathsf{Sub}\,\Gamma\,\Delta \qquad \rho_{01} : \mathsf{Tm}\,\Omega\,\underline{\Gamma^\sim\,\rho_0\,\rho_1}}{\delta^\sim\,\rho_{01} : \mathsf{Tm}\,\Omega\,\underline{\Delta^\sim\,(\delta\circ\rho_0)\,(\delta\circ\rho_1)}} \qquad \frac{t : \mathsf{Tm}\,\Gamma\,A \qquad \rho_{01} : \mathsf{Tm}\,\Omega\,\underline{\Gamma^\sim\,\rho_0\,\rho_1}}{t^\sim\,\rho_{01} : \mathsf{Tm}\,\Omega\,\underline{A^\sim\,\rho_{01}\,(t[\rho_0])\,(t[\rho_1])}}$$

We state the following definitional equalities on how the equality types and coercions compute.

$$\begin{aligned}
\cdot^\sim\,\epsilon\,\epsilon &= \top\\
(\Gamma\,,x:A)^\sim\,(\rho_0\,,x\mapsto t_0)\,(\rho_1\,,x\mapsto t_1) &= \sigma(\rho_{01}:\Gamma^\sim\,\rho_0\,\rho_1).A^\sim\,\rho_{01}\,t_0\,t_1\\
(A[\delta])^\sim\,\rho_{01}\,t_0\,t_1 &= A^\sim\,(\delta^\sim\,\rho_{01})\,t_0\,t_1\\
(\Pi(x:A).B)^\sim\,\rho_{01}\,t_0\,t_1 &= \pi(x_0:A[\rho_0]).\pi(x_1:A[\rho_1]).\pi(x_{01}:\underline{A^\sim\,\rho_{01}\,x_0\,x_1}).\\
&\qquad B^\sim\,(\rho_{01}\,,x_{01})\,(t_0\,@\,x_0)\,(t_1\,@\,x_1)\\
(\Sigma(x:A).B)^\sim\,\rho_{01}\,(u_0\,,v_0)\,(u_1\,,v_1) &= \sigma(u_{01}:A^\sim\,\rho_{01}\,u_0\,u_1).B^\sim\,(\rho_{01}\,,u_{01})\,v_0\,v_1\\
\mathsf{Bool}^\sim\,\rho_{01}\,t_0\,t_1 &= \mathsf{if}\,t_0\,\mathsf{then}\,(\mathsf{if}\,t_1\,\mathsf{then}\,\top\,\mathsf{else}\,\bot)\,\mathsf{else}\,(\mathsf{if}\,t_1\,\mathsf{then}\,\bot\,\mathsf{else}\,\top)\\
\mathsf{Prop}^\sim\,\rho_{01}\,a_0\,a_1 &= (\underline{a_0}\Rightarrow a_1)\times(\underline{a_1}\Rightarrow a_0)\\
\underline{a}^\sim\,\rho_{01}\,t_0\,t_1 &= \top\\
\mathsf{coe}_{A[\delta]}\,\rho_{01}\,t_0 &= \mathsf{coe}_A\,(\delta^\sim\,\rho_{01})\,t_0\\
\mathsf{coe}_{\Pi(x:A).B}\,\rho_{01}\,t_0 &= \lambda x_1.\mathsf{coe}_B\,\big(\rho_{01}\,,\mathsf{S}_A\,(\mathsf{coh}_A\,(\mathsf{S}_\Gamma\,\rho_{01})\,x_1)\big)\\
&\qquad\qquad \big(t_0\,@\,\mathsf{coe}_A\,(\mathsf{S}_\Gamma\,\rho_{01})\,x_1\big)\\
\mathsf{coe}_{\Sigma(x:A).B}\,\rho_{01}\,(u_0\,,v_0) &= \big(\mathsf{coe}_A\,\rho_{01}\,u_0\,,\mathsf{coe}_B\,(\rho_{01}\,,\mathsf{coh}_A\,\rho_{01}\,u_0)\,v_0\big)\\
\mathsf{coe}_{\mathsf{Bool}}\,\rho_{01}\,t_0 &= t_0\\
\mathsf{coe}_{\mathsf{Prop}}\,\rho_{01}\,a_0 &= a_0\\
\mathsf{coe}_{\underline{a}}\,\rho_{01}\,t_0 &= \mathsf{pr}_0\,(a^\sim\,\rho_{01})\,@\,t_0
\end{aligned}$$

In addition, we have the following substitution laws.

$$\begin{aligned}
(\Gamma^\sim\,\rho_0\,\rho_1)[\nu] &= \Gamma^\sim\,(\rho_0\circ\nu)\,(\rho_1\circ\nu)\\
(A^\sim\,\rho_{01}\,t_0\,t_1)[\nu] &= A^\sim\,(\rho_{01}[\nu])\,(t_0[\nu])\,(t_1[\nu])\\
(\mathsf{coe}_A\,\rho_{01}\,t_0)[\nu] &= \mathsf{coe}_A\,(\rho_{01}[\nu])\,(t_0[\nu])
\end{aligned}$$

We only need to state these for $\Gamma^\sim$, $A^\sim$ and coe_A as all the other rules coming from the translation are true by irr. Note that e.g. the equality for $(\Pi(x:A).B)^\sim$ is not merely a convenience, but this is the rule which adds function extensionality.

We conclude the definition of setoid type theory by adding the definitional equality for coercing along reflexivity.

$$\mathsf{coe}_A\,(\mathsf{R}_\Gamma\,\rho)\,t = t$$

Justification. The setoid translation extends to all the extra rules of setoid type theory. As all the new syntactic components are terms, we have to implement the $|-|$ and the $-^{\sim}$ operations for terms as specified in Sect. 4.2. Most components are modelled by their black counterparts because the purpose of the new rules is precisely to reflect the extra structure of the setoid translation. All the equalities are justified (T^3 is three steps transitivity, see Appendix C for all the details).

$$
\begin{aligned}
|\Gamma^{\sim}\,\rho_0\,\rho_1| &:= \Gamma^{\sim}\,|\rho_0|\,|\rho_1| \\
|A^{\sim}\,\rho_{01}\,t_0\,t_1| &:= A^{\sim}\,|\rho_{01}|\,|t_0|\,|t_1| \\
(\Gamma^{\sim}\,\rho_0\,\rho_1)^{\sim}\,\tau_{01} &:= \big(\lambda\rho_{01}.\mathsf{T}^3{}_{\Gamma}\,(\mathsf{S}_{\Gamma}\,(\rho_0{}^{\sim}\,\tau_{01}))\,\rho_{01}\,(\rho_1{}^{\sim}\,\tau_{01}), \\
&\qquad \lambda\rho_{01}.\mathsf{T}^3{}_{\Gamma}\,(\rho_0^{\sim}\,\tau_{01})\,\rho_{01}\,(\mathsf{S}_{\Gamma}\,(\rho_1^{\sim}\,\tau_{01}))\big) \\
(A^{\sim}\,\rho_{01}\,t_0\,t_1)^{\sim}\,\tau_{01} &:= \big(\lambda t_{01}.\mathsf{T}^3{}_{A}\,(\mathsf{S}_{A}\,(t_0{}^{\sim}\,\tau_{01}))\,t_{01}\,(t_1{}^{\sim}\,\tau_{01}), \\
&\qquad \lambda t_{01}.\mathsf{T}^3{}_{A}\,(t_0{}^{\sim}\,\tau_{01})\,t_{01}\,(\mathsf{S}_{A}\,(t_1{}^{\sim}\,\tau_{01}))\big)
\end{aligned}
$$

$$
\begin{aligned}
&|\mathsf{R}_{\Gamma}\,\rho| &&:= \mathsf{R}_{\Gamma}\,|\rho| &&|\mathsf{R}_{A}\,t| &&:= \mathsf{R}_{A}\,|t| \\
&|\mathsf{S}_{\Gamma}\,\rho_{01}| &&:= \mathsf{S}_{\Gamma}\,|\rho_{01}| &&|\mathsf{S}_{A}\,t_{01}| &&:= \mathsf{S}_{A}\,|t_{01}| \\
&|\mathsf{T}_{\Gamma}\,\rho_{01}\,\rho_{12}| &&:= \mathsf{T}_{\Gamma}\,|\rho_{01}|\,|\rho_{12}| &&|\mathsf{T}_{A}\,t_{01}\,t_{12}| &&:= \mathsf{T}_{A}\,|t_{01}|\,|t_{12}| \\
&|\mathsf{coe}_{A}\,\rho_{01}\,t_0| &&:= \mathsf{coe}_{A}\,|\rho_{01}|\,|t_0| &&|\delta^{\sim}\,\rho_{01}| &&:= \delta^{\sim}\,|\rho_{01}| \\
&|\mathsf{coh}_{A}\,\rho_{01}\,t_0| &&:= \mathsf{coh}_{A}\,|\rho_{01}|\,|t_0| &&|t^{\sim}\,\rho_{01}| &&:= t^{\sim}\,|\rho_{01}|
\end{aligned}
$$

Relationship to Martin-Löf's identity type (as given in Sect. 4.4). The rules of the traditional identity are admissible in setoid type theory. The translation is the following.

$$
\begin{aligned}
&\mathsf{Id}_A\,u\,v &&:= A^{\sim}\,(\mathsf{R}_{\Gamma}\,\mathsf{id})\,u\,v \\
&\mathsf{refl}_u &&:= \mathsf{R}_A\,u \\
&\mathsf{transport}_{x.P}\,e\,t &&:= \mathsf{coe}_P\,(\mathsf{R}_{\Gamma}\,\mathsf{id}\,,e)\,t \\
&\mathsf{Id}\beta\,t &&:= \mathsf{S}_{P[x\mapsto u]}\,(\mathsf{coh}_P\,(\mathsf{R}_{\Gamma}\,\mathsf{id}\,,\mathsf{R}_A\,u)\,t) \\
&\mathsf{funext}\,e &&:= \lambda x_0\,x_1\,x_{01}.\mathsf{T}_B\,(e\,@\,x_0)\,(t_1{}^{\sim}\,(\mathsf{R}_{\Gamma}\,\mathsf{id})\,@\,x_0\,@\,x_1\,@\,x_{01}) \\
&\mathsf{propext}\,t &&:= t \\
&\mathsf{Id}[] &&: \;(\mathsf{Id}_A\,u\,v)[\nu] = (A^{\sim}\,(\mathsf{R}_{\Gamma}\,\mathsf{id})\,u\,v)[\nu] = \\
& && \quad A^{\sim}\,((\mathsf{R}_{\Gamma}\,\mathsf{id})[\nu])\,(u[\nu])\,(v[\nu]) \overset{\mathsf{irr}}{=} A^{\sim}\,(\nu^{\sim}\,(\mathsf{R}_{\Theta}\,\mathsf{id}))\,(u[\nu])\,(v[\nu]) = \\
& && \quad (A[\nu])^{\sim}\,(\mathsf{R}_{\Theta}\,\mathsf{id})\,(u[\nu])\,(v[\nu]) = \mathsf{Id}_{A[\nu]}\,(u[\nu])\,(v[\nu]) \\
&\mathsf{transport}[] &&: \;(\mathsf{transport}_{x.P}\,e\,t)[\nu] = (\mathsf{coe}_P\,(\mathsf{R}_{\Gamma}\,\mathsf{id}\,,e)\,t)[\nu] \overset{\mathsf{coe}[]_P}{=} \\
& && \quad \mathsf{coe}_P\,((\mathsf{R}_{\Gamma}\,\mathsf{id}\,,e)[\nu])\,(t[\nu]) \overset{\mathsf{irr}}{=} \mathsf{coe}_P\,(\nu^{\sim}\,(\mathsf{R}_{\Theta}\,\mathsf{id})\,,e[\nu])\,(t[\nu]) = \\
& && \quad \mathsf{coe}_{P[\nu]}\,(\mathsf{R}_{\Theta}\,\mathsf{id}\,,e[\nu])\,(t[\nu]) = \mathsf{transport}_{x.P[\nu]}\,(e[\nu])\,(t[\nu])
\end{aligned}
$$

The other direction does not work. For example, the following definitional equalities do not hold in $\mathsf{MLTT_{Prop}}$ extended with Martin-Löf's identity type, however they hold in setoid type theory where $\mathsf{transport}$ is translated as above:

$$\text{“constant predicate”}: \mathsf{transport}_{x.\mathsf{Bool}}\, e\, t = t$$
$$\text{“funext computes”}: \quad \mathsf{transport}_{f.P[y \mapsto f \,@\, u]}\, (\mathsf{funext}\, e)\, t = \mathsf{transport}_{y.P}\, (e \,@\, u)\, t$$

As setoid type theory reflects the setoid translation, we conjecture that it is complete, that is, if $|t| = |t'|$ for any two terms $t, t' : \mathsf{Tm}\,\Gamma\,A$ of setoid type theory, then $t = t'$.

6 Conclusions and Further Work

We have presented a type theory which justifies both function extensionality and propositional extensionality. Compared with [1], it adds propositional extensionality and a definitional computation rule for transport, presents an equational theory and the results are checked formally. Compared with [4], it provides a translation into intensional type theory without requiring extensional type theory as a reference.

It is clear that the theory follows the setoid translation, hence we conjecture completeness with respect to this model. A corollary would be canonicity for our theory.

We expect that the translation can be extended with a universe of setoids where equality is equality of codes and quotient inductive types. Our Agda formalisation of the setoid model already supports such a universe and quotient types.

The theory is less powerful than cubical type theory [12] but the semantic justification is much more straightforward and for many practical applications, this type theory is sufficient. It also supports some definitional equalities which do not hold in cubical type theory. We believe that our programme can be extended, first of all to obtain a syntax for a groupoid type theory using our informal method to derive a theory from a translation. We also expect that we could derive an alternative explanation and implementation of homotopy type theory.

A Full Syntax of $\mathsf{MLTT_{Prop}}$

Sorts:

$$\frac{}{\mathsf{Con}_i : \mathsf{Set}} \qquad \frac{\Gamma : \mathsf{Con}_i}{\mathsf{Ty}_j\,\Gamma : \mathsf{Set}} \qquad \frac{\Gamma : \mathsf{Con}_i \qquad \Delta : \mathsf{Con}_j}{\mathsf{Sub}\,\Gamma\,\Delta : \mathsf{Set}} \qquad \frac{\Gamma : \mathsf{Con}_i \qquad A : \mathsf{Ty}_j\,\Gamma}{\mathsf{Tm}\,\Gamma\,A : \mathsf{Set}}$$

Substitution calculus:

$$\frac{}{\cdot : \mathsf{Con}_0} \qquad \frac{\Gamma : \mathsf{Con}_i \qquad A : \mathsf{Ty}_j\, \Gamma}{(\Gamma\,, x : A) : \mathsf{Con}_{i \sqcup j}} \qquad \frac{A : \mathsf{Ty}_i\, \Delta \qquad \delta : \mathsf{Sub}\, \Gamma\, \Delta}{A[\delta] : \mathsf{Ty}_i\, \Gamma}$$

$$\frac{\Gamma : \mathsf{Con}_i}{\mathsf{id}_\Gamma : \mathsf{Sub}\, \Gamma\, \Gamma} \qquad \frac{\delta : \mathsf{Sub}\, \Theta\, \Delta \qquad \nu : \mathsf{Sub}\, \Gamma\, \Theta}{\delta \circ \nu : \mathsf{Sub}\, \Gamma\, \Delta} \qquad \frac{\Gamma : \mathsf{Con}_i}{\epsilon_\Gamma : \mathsf{Sub}\, \Gamma\, \cdot}$$

$$\frac{\delta : \mathsf{Sub}\, \Gamma\, \Delta \qquad t : \mathsf{Tm}\, \Gamma\, (A[\delta])}{(\delta\,, x \mapsto t) : \mathsf{Sub}\, \Gamma\, (\Delta\,, x : A)} \qquad \frac{\delta : \mathsf{Sub}\, \Gamma\, (\Delta\,, x : A)}{\delta : \mathsf{Sub}\, \Gamma\, \Delta} \qquad \frac{\delta : \mathsf{Sub}\, \Gamma\, (\Delta\,, x : A)}{x[\delta] : \mathsf{Tm}\, \Gamma\, (A[\delta])}$$

$$\frac{t : \mathsf{Tm}\, \Delta\, A \qquad \delta : \mathsf{Sub}\, \Gamma\, \Delta}{t[\delta] : \mathsf{Tm}\, \Gamma\, (A[\delta])} \qquad [\mathsf{Id}] : A[\mathsf{id}] = A \qquad [\circ] : A[\delta \circ \nu] = A[\delta][\nu]$$

$$\mathsf{id}\circ : \mathsf{id} \circ \delta = \delta \qquad \circ\mathsf{id} : \delta \circ \mathsf{id} = \delta \qquad \circ\circ : (\delta \circ \nu) \circ \tau = \delta \circ (\nu \circ \tau)$$

$$\cdot\eta : (\delta : \mathsf{Sub}\, \Gamma\, \cdot) = \epsilon \qquad ,\beta_0 : (\delta\,, x \mapsto t) = \delta \qquad ,\beta_1 : x[(\delta\,, x \mapsto t)] = t$$

$$,\eta : (\delta\,, x \mapsto x[\delta]) = \delta \qquad ,\circ : (\delta\,, x \mapsto t) \circ \nu = (\delta \circ \nu\,, x \mapsto t[\nu])$$

Π types:

$$\frac{A : \mathsf{Ty}_i\, \Gamma \qquad B : \mathsf{Ty}_j\, (\Gamma\,, x : A)}{\Pi(x : A).B : \mathsf{Ty}_{i \sqcup j}\, \Gamma} \qquad \frac{t : \mathsf{Tm}\, (\Gamma\,, x : A)\, B}{\lambda x.t : \mathsf{Tm}\, \Gamma\, (\Pi(x : A).B)}$$

$$\frac{t : \mathsf{Tm}\, \Gamma\, (\Pi(x : A).B)}{t\, @\, x : \mathsf{Tm}\, (\Gamma\,, x : A)\, B} \qquad \Pi\beta : (\lambda x.t)\, @\, x = t \qquad \Pi\eta : \lambda x.t\, @\, x = t$$

$$\Pi[] : (\Pi(x : A).B)[\nu] = \Pi(x : A[\nu]).B[\nu] \qquad \lambda[] : (\lambda x.t)[\nu] = \lambda x.t[\nu]$$

Σ types (we write $A \times B$ for $\Sigma(x : A).B$ when x does not appear in B):

$$\frac{A : \mathsf{Ty}_i\, \Gamma \qquad B : \mathsf{Ty}_j\, (\Gamma\,, x : A)}{\Sigma(x : A).B : \mathsf{Ty}_{i \sqcup j}\, \Gamma} \qquad \frac{u : \mathsf{Tm}\, \Gamma\, A \qquad v : \mathsf{Tm}\, \Gamma\, (B[x \mapsto u])}{(u\,, v) : \mathsf{Tm}\, \Gamma\, (\Sigma(x : A).B)}$$

$$\frac{t : \mathsf{Tm}\, \Gamma\, (\Sigma(x : A).B)}{\mathsf{pr}_0\, t : \mathsf{Tm}\, \Gamma\, A} \qquad \frac{t : \mathsf{Tm}\, \Gamma\, (\Sigma(x : A).B)}{\mathsf{pr}_1\, t : \mathsf{Tm}\, \Gamma\, (B[x \mapsto \mathsf{pr}_0\, t])}$$

$$\Sigma\beta_0 : \mathsf{pr}_0\, (u\,, v) = u \qquad \Sigma\beta_1 : \mathsf{pr}_1\, (u\,, v) = v \qquad \Sigma\eta : (\mathsf{pr}_0\, t\,, \mathsf{pr}_1\, t) = t$$

$$\Sigma[] : (\Sigma(x : A).B)[\nu] = \Sigma(x : A[\nu]).B[\nu] \qquad ,[] : (u\,, v)[\nu] = (u[\nu]\,, v[\nu])$$

Booleans:

$$\frac{}{\mathsf{Bool} : \mathsf{Ty}_0\, \Gamma} \qquad \frac{}{\mathsf{true} : \mathsf{Tm}\, \Gamma\, \mathsf{Bool}} \qquad \frac{}{\mathsf{false} : \mathsf{Tm}\, \Gamma\, \mathsf{Bool}}$$

$$\frac{\begin{array}{ll} C & : \mathsf{Ty}_i\, (\Gamma\,, x : \mathsf{Bool}) \\ t & : \mathsf{Tm}\, \Gamma\, \mathsf{Bool} \\ u & : \mathsf{Tm}\, \Gamma\, (C[x \mapsto \mathsf{true}]) \\ v & : \mathsf{Tm}\, \Gamma\, (C[x \mapsto \mathsf{false}]) \end{array}}{\mathsf{if}\, t\, \mathsf{then}\, u\, \mathsf{else}\, v : \mathsf{Tm}\, \Gamma\, (C[x \mapsto t])}$$

$$\begin{array}{ll}
\mathsf{Bool}\beta_{\mathsf{true}} & : \mathsf{if}\;\mathsf{true}\;\mathsf{then}\;u\;\mathsf{else}\;v = u \\
\mathsf{Bool}\beta_{\mathsf{false}} & : \mathsf{if}\;\mathsf{false}\;\mathsf{then}\;u\;\mathsf{else}\;v = v \\
\mathsf{Bool}[] & : \mathsf{Bool}[\nu] = \mathsf{Bool} \\
\mathsf{true}[] & : \mathsf{true}[\nu] = \mathsf{true} \\
\mathsf{false}[] & : \mathsf{false}[\nu] = \mathsf{false} \\
\mathsf{if}[] & : (\mathsf{if}\;t\;\mathsf{then}\;u\;\mathsf{else}\;v)[\nu] = \mathsf{if}\;t[\nu]\;\mathsf{then}\;u[\nu]\;\mathsf{else}\;v[\nu]
\end{array}$$

Propositions:

$$\frac{}{\mathsf{Prop}_i : \mathsf{Ty}_{i+1}\,\Gamma} \qquad \frac{a : \mathsf{Tm}\,\Gamma\,\mathsf{Prop}_i}{\underline{a} : \mathsf{Ty}_i\,\Gamma} \qquad \frac{u : \mathsf{Tm}\,\Gamma\,\underline{a} \qquad v : \mathsf{Tm}\,\Gamma\,\underline{a}}{\mathsf{irr}_a : u = v}$$

$$\frac{A : \mathsf{Ty}_i\,\Gamma \qquad b : \mathsf{Tm}\,(\Gamma\,, x : A)\,\mathsf{Prop}_j}{\pi(x : A).b : \mathsf{Tm}\,\Gamma\,\mathsf{Prop}_{i \sqcup j}} \qquad \frac{t : \mathsf{Tm}\,(\Gamma\,, x : A)\,\underline{b}}{\lambda x.t : \mathsf{Tm}\,\Gamma\,\underline{\pi(x : A).b}}$$

$$\frac{t : \mathsf{Tm}\,\Gamma\,\underline{\pi(x : A).b}}{t\,@\,x : \mathsf{Tm}\,(\Gamma\,, x : A)\,\underline{b}} \qquad \frac{a : \mathsf{Tm}\,\Gamma\,\mathsf{Prop}_i \qquad b : \mathsf{Tm}\,(\Gamma\,, x : \underline{a})\,\mathsf{Prop}_j}{\sigma(x : a).b : \mathsf{Tm}\,\Gamma\,\mathsf{Prop}_{i \sqcup j}}$$

$$\frac{u : \mathsf{Tm}\,\Gamma\,\underline{a} \qquad v : \mathsf{Tm}\,\Gamma\,\underline{b[x \mapsto u]}}{(u\,, v) : \mathsf{Tm}\,\Gamma\,\underline{\sigma(x : a).b}} \qquad \frac{t : \mathsf{Tm}\,\Gamma\,\underline{\sigma(x : a).b}}{\mathsf{pr}_0\,t : \mathsf{Tm}\,\Gamma\,\underline{a}} \qquad \frac{t : \mathsf{Tm}\,\Gamma\,\underline{\sigma(x : a).b}}{\mathsf{pr}_1\,t : \mathsf{Tm}\,\Gamma\,\underline{a[x \mapsto \mathsf{pr}_0\,t]}}$$

$$\frac{}{\top : \mathsf{Tm}\,\Gamma\,\mathsf{Prop}_0} \qquad \frac{}{\mathsf{tt} : \mathsf{Tm}\,\Gamma\,\underline{\top}} \qquad \frac{}{\bot : \mathsf{Tm}\,\Gamma\,\mathsf{Prop}_0} \qquad \frac{C : \mathsf{Ty}_i\,\Gamma \qquad t : \mathsf{Tm}\,\Gamma\,\underline{\bot}}{\mathsf{exfalso}\,t : \mathsf{Tm}\,\Gamma\,C}$$

$$\begin{array}{ll}
\mathsf{Prop}[] & : \mathsf{Prop}_i[\nu] = \mathsf{Prop}_i \\
\underline{[]} & : \underline{a}[\nu] = \underline{a[\nu]} \\
\pi[] & : (\pi(x : A).b)[\nu] = \pi(x : A[\nu]).b[\nu] \\
\sigma[] & : (\sigma(x : a).b)[\nu] = \sigma(x : a[\nu]).b[\nu] \\
,[] & : (u\,, v)[\nu] = (u[\nu]\,, v[\nu]) \\
\top[] & : \top[\nu] = \top \\
\bot[] & : \bot[\nu] = \bot \\
\mathsf{exfalso}[] & : (\mathsf{exfalso}\,t)[\nu] = \mathsf{exfalso}\,(t[\nu])
\end{array}$$

B Complete Implementation of the Setoid Translation

B.1 Specification

$$\frac{\Gamma : \mathsf{Con}_i}{\begin{array}{ll} |\Gamma| & : \mathsf{Con}_i \\ \Gamma^{\sim} & : \mathsf{Sub}\,\Omega\,|\Gamma| \to \mathsf{Sub}\,\Omega\,|\Gamma| \to \mathsf{Tm}\,\Omega\,\mathsf{Prop}_i \\ \Gamma^{\sim}[] & : (\Gamma^{\sim}\,\rho_0\,\rho_1)[\nu] = \Gamma^{\sim}\,(\rho_0 \circ \nu)\,(\rho_1 \circ \nu) \\ \mathsf{R}_\Gamma & : (\rho : \mathsf{Sub}\,\Omega\,|\Gamma|) \to \mathsf{Tm}\,\Omega\,\underline{\Gamma^{\sim}\,\rho\,\rho} \\ \mathsf{S}_\Gamma & : \mathsf{Tm}\,\Omega\,\underline{\Gamma^{\sim}\,\rho_0\,\rho_1} \to \mathsf{Tm}\,\Omega\,\underline{\Gamma^{\sim}\,\rho_1\,\rho_0} \\ \mathsf{T}_\Gamma & : \mathsf{Tm}\,\Omega\,\underline{\Gamma^{\sim}\,\rho_0\,\rho_1} \to \mathsf{Tm}\,\Omega\,\underline{\Gamma^{\sim}\,\rho_1\,\rho_2} \to \mathsf{Tm}\,\Omega\,\underline{\Gamma^{\sim}\,\rho_0\,\rho_2} \end{array}}$$

$$\frac{A : \mathsf{Ty}_j\,\Gamma}{\begin{array}{ll} |A| & : \mathsf{Ty}_j\,|\Gamma| \\ A^{\sim} & : \mathsf{Tm}\,\Omega\,\underline{\Gamma^{\sim}\,\rho_0\,\rho_1} \to \mathsf{Tm}\,\Omega\,(|A|[\rho_0]) \to \mathsf{Tm}\,\Omega\,(|A|[\rho_1]) \to \mathsf{Tm}\,\Omega\,\mathsf{Prop}_j \\ A^{\sim}[] & : (A^{\sim}\,\rho_{01}\,t_0\,t_1)[\nu] = A^{\sim}\,(\rho_{01}[\nu])\,(t_0[\nu])\,(t_1[\nu]) \\ \mathsf{R}_A & : (t : \mathsf{Tm}\,\Omega\,(|A|[\rho])) \to \mathsf{Tm}\,\Omega\,\underline{A^{\sim}\,(\mathsf{R}_\Gamma\,\rho)\,t\,t} \\ \mathsf{S}_A & : \mathsf{Tm}\,\Omega\,\underline{A^{\sim}\,\rho_{01}\,t_0\,t_1} \to \mathsf{Tm}\,\Omega\,\underline{A^{\sim}\,(\mathsf{S}_\Gamma\,\rho_{01})\,t_1\,t_0} \\ \mathsf{T}_A & : \mathsf{Tm}\,\Omega\,\underline{A^{\sim}\,\rho_{01}\,t_0\,t_1} \to \mathsf{Tm}\,\Omega\,\underline{A^{\sim}\,\rho_{12}\,t_1\,t_2} \to \mathsf{Tm}\,\Omega\,\underline{A^{\sim}\,(\mathsf{T}_\Gamma\,\rho_{01}\,\rho_{12})\,t_0\,t_2} \\ \mathsf{coe}_A & : \mathsf{Tm}\,\Omega\,\underline{\Gamma^{\sim}\,\rho_0\,\rho_1} \to \mathsf{Tm}\,\Omega\,(|A|[\rho_0]) \to \mathsf{Tm}\,\Omega\,(|A|[\rho_1]) \\ \mathsf{coe}[]_A & : (\mathsf{coe}_A\,\rho_{01}\,t_0)[\nu] = \mathsf{coe}_A\,(\rho_{01}[\nu])\,(t_0[\nu]) \\ \mathsf{coh}_A & : (\rho_{01} : \mathsf{Tm}\,\Omega\,\underline{\Gamma^{\sim}\,\rho_0\,\rho_1})(t_0 : \mathsf{Tm}\,\Omega\,(|A|[\rho_0])) \to \\ & \quad \mathsf{Tm}\,\Omega\,\underline{A^{\sim}\,\rho_{01}\,t_0\,(\mathsf{coe}_A\,\rho_{01}\,t_0)} \end{array}}$$

$$\frac{\delta : \mathsf{Sub}\,\Gamma\,\Delta}{\begin{array}{ll} |\delta| & : \mathsf{Sub}\,|\Gamma|\,|\Delta| \\ \delta^{\sim} & : \mathsf{Tm}\,\Omega\,\underline{\Gamma^{\sim}\,\rho_0\,\rho_1} \to \mathsf{Tm}\,\Omega\,\underline{\Delta^{\sim}\,(|\delta| \circ \rho_0)\,(|\delta| \circ \rho_1)} \end{array}}$$

$$\frac{t : \mathsf{Tm}\,\Gamma\,A}{\begin{array}{ll} |t| & : \mathsf{Tm}\,|\Gamma|\,|A| \\ t^{\sim} & : (\rho_{01} : \mathsf{Tm}\,\Omega\,\underline{\Gamma^{\sim}\,\rho_0\,\rho_1}) \to \mathsf{Tm}\,\Omega\,\underline{A^{\sim}\,\rho_{01}\,(|t|[\rho_0])\,(|t|[\rho_1])} \end{array}}$$

Abbreviations. The operations coe^* and coh^* are the counterparts of coe^* and coh^* in the symmetric direction. The two T^3 operations are "three steps" transitivity.

$$\begin{array}{l} \mathsf{coe}^*{}_A\,(\rho_{01} : \mathsf{Tm}\,\Omega\,\underline{\Gamma^{\sim}\,\rho_0\,\rho_1})(t_1 : \mathsf{Tm}\,\Omega\,(|A|[\rho_1])) : \mathsf{Tm}\,\Omega\,(|A|[\rho_0]) \\ \qquad := \mathsf{coe}_A\,(\mathsf{S}_\Gamma\,\rho_{01})\,t_1 \\ \mathsf{coh}^*{}_A\,(\rho_{01} : \mathsf{Tm}\,\Omega\,\underline{\Gamma^{\sim}\,\rho_0\,\rho_1})(t_1 : \mathsf{Tm}\,\Omega\,(|A|[\rho_1])) : \mathsf{Tm}\,\Omega\,\underline{A^{\sim}\,\rho_{01}\,(\mathsf{coe}^*{}_A\,\rho_{01}\,t_1)\,t_1} \\ \qquad := \mathsf{S}_A\,(\mathsf{coh}_A\,(\mathsf{S}_\Gamma\,\rho_{01})\,t_1) \end{array}$$

$$\mathsf{T}^3{}_\Gamma\,(\rho_{01} : \mathsf{Tm}\,\Omega\,\underline{\Gamma^\sim\,\rho_0\,\rho_1})(\rho_{12} : \mathsf{Tm}\,\Omega\,\underline{\Gamma^\sim\,\rho_1\,\rho_2})(\rho_{12} : \mathsf{Tm}\,\Omega\,\underline{\Gamma^\sim\,\rho_2\,\rho_3})$$
$$: \mathsf{Tm}\,\Omega\,\underline{\Gamma^\sim\,\rho_0\,\rho_3} := \mathsf{T}_\Gamma\,\rho_{01}\,(\mathsf{T}_\Gamma\,\rho_{12}\,\rho_{23})$$
$$\mathsf{T}^3{}_A\,(t_{01} : \mathsf{Tm}\,\Omega\,\underline{A^\sim\,\rho_{01}\,t_0\,t_1})(t_{12} : \mathsf{Tm}\,\Omega\,\underline{A^\sim\,\rho_{12}\,t_1\,t_2})(t_{23} : \mathsf{Tm}\,\Omega\,\underline{A^\sim\,\rho_{23}\,t_2\,t_3})$$
$$: \mathsf{Tm}\,\Omega\,\underline{A^\sim\,(\mathsf{T}^3{}_\Gamma\,\rho_{01}\,\rho_{12}\,\rho_{23})\,t_0\,t_3} := \mathsf{T}_A\,t_{01}\,(\mathsf{T}_A\,t_{12}\,t_{23})$$

B.2 Implementation

We implement this specification by listing what the different components of $\mathsf{MLTT}_{\mathsf{Prop}}$ are mapped to. We follow the order of the presentation of $\mathsf{MLTT}_{\mathsf{Prop}}$ in Sect. 2.

The $|{-}|$ part of the model is almost the same as the identity translation (variant ${}_3$ in Sect. 3.1). The only difference is for Π types which are interpreted by a subset of all Π types: they need to also respect the relation.

Substitution Calculus. The $|{-}|$, ${-}^\sim$, R_-, etc. components can be given one after the other as there is no interdependency for the substitution calculus. For the substitution calculus, $|{-}|$ is the same as the set interpretation, ${-}^\sim$ is the same as in the graph model.

Set (identity) interpretation for the substitution calculus.

$$
\begin{aligned}
|\cdot| &:= \cdot \\
|\Gamma\,, x : A| &:= |\Gamma|\,, x : |A| \\
|A[\delta]| &:= |A|[|\delta|] \\
|\mathsf{id}_\Gamma| &:= \mathsf{id}_{|\Gamma|} \\
|\delta \circ \nu| &:= |\delta| \circ |\nu| \\
|\epsilon_\Gamma| &:= \epsilon_\Gamma \\
|(\delta\,, x \mapsto t)| &:= (|\delta|\,, x \mapsto |t|) \\
|\delta| &:= |\delta| \\
|x[\delta]| &:= x[|\delta|] \\
|t[\delta]| &:= |t|[|\delta|]
\end{aligned}
$$

$$
\begin{aligned}
|[\mathsf{id}]| &: \quad |A[\mathsf{id}]| = |A|[\mathsf{id}] \overset{[\mathsf{id}]}{=} |A| \\
|[\circ]| &: \quad |A[\delta \circ \nu]| = |A|[|\delta| \circ |\nu|] \overset{[\circ]}{=} |A|[|\delta|][|\nu|] = |A[\delta][\nu]| \\
|\mathsf{id}\circ| &: \quad |\mathsf{id} \circ \delta| = \mathsf{id} \circ |\delta| \overset{\mathsf{id}\circ}{=} |\delta| \\
|\circ\mathsf{id}| &: \quad |\delta \circ \mathsf{id}| = |\delta| \circ \mathsf{id} \overset{\circ\mathsf{id}}{=} |\delta| \\
|\circ\circ| &: \quad |(\delta \circ \nu) \circ \tau| = (|\delta| \circ |\nu|) \circ |\tau| \overset{\circ\circ}{=} |\delta| \circ (|\nu| \circ |\tau|) = |\delta \circ (\nu \circ \tau)| \\
|\epsilon\eta| &: \quad (|\delta| : \mathsf{Sub}\,|\Gamma|\,|\cdot|) = (|\delta| : \mathsf{Sub}\,|\Gamma|\,\cdot) \overset{\epsilon\eta}{=} \epsilon = |\epsilon|
\end{aligned}
$$

$$
\begin{aligned}
&|,\beta_0| &&: |(\delta, x \mapsto t)| = (|\delta|, x \mapsto |t|) \overset{,\beta_0}{=} |\delta| \\
&|,\beta_1| &&: |x[(\delta, x \mapsto t)]| = x[(|\delta|, x \mapsto |t|)] \overset{,\beta_1}{=} |t| \\
&|,\eta| &&: |(\delta, x \mapsto x[\delta])| = (|\delta|, x \mapsto x[|\delta|]) \overset{,\eta}{=} |\delta| \\
&|,\circ| &&: |(\delta, x \mapsto t) \circ \nu| = (|\delta|, x \mapsto |t|) \circ |\nu| \overset{,\circ}{=} (|\delta| \circ |\nu|, x \mapsto |t|[|\nu|]) = \\
& && \quad |(\delta \circ \nu, x \mapsto t[\nu])|
\end{aligned}
$$

Note that $|\delta| := |\delta|$ means that implicit weakening inside $|-|$ was interpreted by implicit weakening outside the $|-|$ operation.

Logical predicates.

$$
\begin{aligned}
&\cdot^{\sim}\, \epsilon\, \epsilon && := \top \\
&(\Gamma, x : A)^{\sim}\, (\rho_0, x \mapsto t_0)\, (\rho_1, x \mapsto t_1) := \sigma(\rho_{01} : \Gamma^{\sim}\, \rho_0\, \rho_1).A^{\sim}\, \rho_{01}\, t_0\, t_1 \\
&(A[\delta])^{\sim}\, \rho_{01}\, t_0\, t_1 := A^{\sim}\, (\delta^{\sim}\, \rho_{01})\, t_0\, t_1 \\
&\mathsf{id}_{\Gamma}{}^{\sim}\, \rho_{01} && := \rho_{01} \\
&(\delta \circ \nu)^{\sim}\, \rho_{01} && := \delta^{\sim}\, (\nu^{\sim}\, \rho_{01}) \\
&\epsilon^{\sim}\, \rho_{01} && := \mathsf{tt} \\
&(\delta, x \mapsto t)^{\sim}\, \rho_{01} && := (\delta^{\sim}\, \rho_{01}, t^{\sim}\, \rho_{01}) \\
&\delta^{\sim}\, \rho_{01} && := \mathsf{pr}_0\, (\delta^{\sim}\, \rho_{01}) \\
&(x[\delta])^{\sim}\, \rho_{01} && := \mathsf{pr}_1\, (\delta^{\sim}\, \rho_{01}) \\
&(t[\delta])^{\sim}\, \rho_{01} && := t^{\sim}\, (\delta^{\sim}\, \rho_{01}) \\
&[\mathsf{id}]^{\sim} && : (A[\mathsf{id}])^{\sim}\, \rho_{01}\, t_0\, t_1 = A^{\sim}\, (\mathsf{id}^{\sim}\, \rho_{01})\, t_0\, t_1 = A^{\sim}\, \rho_{01}\, t_0\, t_1 \\
&[\circ]^{\sim} && : (A[\delta \circ \nu])^{\sim}\, \rho_{01}\, t_0\, t_1 = A^{\sim}\, ((\delta \circ \nu)^{\sim}\, \rho_{01})\, t_0\, t_1 = \\
& && \quad A^{\sim}\, (\delta^{\sim}\, (\nu^{\sim}\, \rho_{01}))\, t_0\, t_1 = (A[\delta][\nu])^{\sim}\, \rho_{01}\, t_0\, t_1 \\
&\mathsf{id}\circ^{\sim} && : (\mathsf{id} \circ \delta)^{\sim}\, \rho_{01} \overset{\mathsf{irr}}{=} \delta^{\sim}\, \rho_{01} \\
&\circ\mathsf{id}^{\sim} && : (\delta \circ \mathsf{id})^{\sim}\, \rho_{01} \overset{\mathsf{irr}}{=} \delta^{\sim}\, \rho_{01} \\
&\circ\circ^{\sim} && : ((\delta \circ \nu) \circ \tau)^{\sim}\, \rho_{01} \overset{\mathsf{irr}}{=} (\delta \circ (\nu \circ \tau))^{\sim}\, \rho_{01} \\
&\epsilon\eta^{\sim} && : \delta^{\sim}\, \rho_{01} \overset{\mathsf{irr}}{=} \epsilon^{\sim}\, \rho_{01} \\
&,\beta_0{}^{\sim} && : (\delta, x \mapsto t)^{\sim}\, \rho_{01} \overset{\mathsf{irr}}{=} \delta^{\sim}\, \rho_{01} \\
&,\beta_1{}^{\sim} && : (x[(\delta, x \mapsto t)])^{\sim}\, \rho_{01} \overset{\mathsf{irr}}{=} t^{\sim}\, \rho_{01} \\
&,\eta^{\sim} && : (\delta, x \mapsto x[\delta])^{\sim}\, \rho_{01} \overset{\mathsf{irr}}{=} \delta^{\sim}\, \rho_{01} \\
&,\circ^{\sim} && : ((\delta, x \mapsto t) \circ \nu)^{\sim}\, \rho_{01} \overset{\mathsf{irr}}{=} (\delta \circ \nu, x \mapsto t[\nu])^{\sim}\, \rho_{01}
\end{aligned}
$$

Substitution laws for logical predicates.

$$
\begin{aligned}
&\cdot^{\sim}[] && : (\cdot^{\sim}\, \epsilon\, \epsilon)[\nu] = \top[\nu] \overset{\top[]}{=} \top = \cdot^{\sim}\, (\epsilon[\nu])\, (\epsilon[\nu]) \\
&(\Gamma, x : A)^{\sim}[] && : ((\Gamma, x : A)^{\sim}\, (\rho_0, x \mapsto t_0)\, (\rho_1, x \mapsto t_1))[\nu] =
\end{aligned}
$$

$$\sigma(\rho_{01} : (\Gamma^\sim\,\rho_0\,\rho_1)[\nu]).(A^\sim\,\rho_{01}\,t_0\,t_1)[\nu] \overset{\Gamma^\sim[],A^\sim[]}{=}$$
$$\sigma(\rho_{01} : \Gamma^\sim\,(\rho_0 \circ \nu)\,(\rho_1 \circ \nu)).A^\sim\,(\rho_{01}[\nu])\,(t_0[\nu])\,(t_1[\nu]) =$$
$$(\Gamma\,, x : A)^\sim\,(\rho_0 \circ \nu\,, x \mapsto t_0[\nu])\,(\rho_1 \circ \nu\,, x \mapsto t_1[\nu]) \overset{,\circ}{=}$$
$$(\Gamma\,, x : A)^\sim\,((\rho_0\,, x \mapsto t_0) \circ \nu)\,((\rho_1\,, x \mapsto t_1) \circ \nu)$$

$$(A[\delta])^\sim[] \quad : ((A[\delta])^\sim\,\rho_{01}\,t_0\,t_1)[\nu] = (A^\sim\,(\delta^\sim\,\rho_{01})\,t_0\,t_1)[\nu] \overset{A^\sim[]}{=}$$
$$A^\sim\,((\delta^\sim\,\rho_{01})[\nu])\,(t_0[\nu])\,(t_1[\nu]) \overset{\mathsf{irr}}{=}$$
$$A^\sim\,(\delta^\sim\,(\rho_{01}[\nu]))\,(t_0[\nu])\,(t_1[\nu]) =$$
$$(A[\delta])^\sim\,(\rho_{01}[\nu])\,(t_0[\nu])\,(t_1[\nu])$$

Reflexivity, symmetry and transitivity.

$$\begin{aligned}
&\mathsf{R}_\cdot\,\epsilon &&:= \mathsf{tt}\\
&\mathsf{R}_{\Gamma\,, x : A}\,(\rho\,, x \mapsto t) &&:= (\mathsf{R}_\Gamma\,\rho\,, \mathsf{R}_A\,t)\\
&\mathsf{R}_{A[\delta]}\,t &&:= \mathsf{R}_A\,t\\
&\mathsf{R}_{[\mathsf{id}]} &&: \mathsf{R}_{A[\mathsf{id}]}\,t = \mathsf{R}_A\,t\\
&\mathsf{R}_{[\circ]} &&: \mathsf{R}_{A[\delta\circ\nu]}\,t = \mathsf{R}_A\,t = \mathsf{R}_{A[\delta][\nu]}\,t\\
&\mathsf{S}_\cdot\,\mathsf{tt} &&:= \mathsf{tt}\\
&\mathsf{S}_{\Gamma\,, x : A}\,(\rho_{01}\,, t_{01}) &&:= (\mathsf{S}_\Gamma\,\rho_{01}\,, \mathsf{S}_A\,t_{01})\\
&\mathsf{S}_{A[\delta]}\,t_{01} &&:= \mathsf{S}_A\,t_{01}\\
&\mathsf{S}_{[\mathsf{id}]} &&: \mathsf{S}_{A[\mathsf{id}]}\,t_{01} = \mathsf{S}_A\,t_{01}\\
&\mathsf{S}_{[\circ]} &&: \mathsf{S}_{A[\delta\circ\nu]}\,t_{01} = \mathsf{S}_A\,t_{01} = \mathsf{S}_{A[\delta][\nu]}\,t_{01}\\
&\mathsf{T}_\cdot\,\mathsf{tt}\,\mathsf{tt} &&:= \mathsf{tt}\\
&\mathsf{T}_{\Gamma\,, x : A}\,(\rho_{01}\,, t_{01})\,(\rho_{12}\,, t_{12}) &&:= (\mathsf{T}_\Gamma\,\rho_{01}\,\rho_{12}\,, \mathsf{T}_A\,t_{01}\,t_{12})\\
&\mathsf{T}_{A[\delta]}\,t_{01}\,t_{12} &&:= \mathsf{T}_A\,t_{01}\,t_{12}\\
&\mathsf{T}_{[\mathsf{id}]} &&: \mathsf{T}_{A[\mathsf{id}]}\,t_{01}\,t_{12} = \mathsf{T}_A\,t_{01}\,t_{12}\\
&\mathsf{T}_{[\circ]} &&: \mathsf{T}_{A[\delta\circ\nu]}\,t_{01}\,t_{12} = \mathsf{T}_A\,t_{01}\,t_{12} = \mathsf{T}_{A[\delta][\nu]}\,t_{01}\,t_{12}
\end{aligned}$$

Coercion and coherence.

$$\begin{aligned}
&\mathsf{coe}_{A[\delta]}\,\rho_{01}\,t_0 := \mathsf{coe}_A\,(\delta^\sim\,\rho_{01})\,t_0\\
&\mathsf{coe}_{[\mathsf{id}]} \quad : \mathsf{coe}_{A[\mathsf{id}]}\,\rho_{01}\,t_0 = \mathsf{coe}_A\,(\mathsf{id}^\sim\,\rho_{01})\,t_0 = \mathsf{coe}_A\,\rho_{01}\,t_0\\
&\mathsf{coe}_{[\circ]} \quad : \mathsf{coe}_{A[\delta\circ\nu]}\,\rho_{01}\,t_0 = \mathsf{coe}_A\,(\delta^\sim\,(\nu^\sim\,\rho_{01}))\,t_0 = \mathsf{coe}_{A[\delta][\nu]}\,\rho_{01}\,t_0\\
&\mathsf{coe}[]_{A[\delta]} \quad : (\mathsf{coe}_{A[\delta]}\,\rho_{01}\,t_0)[\nu] = (\mathsf{coe}_A\,(\delta^\sim\,\rho_{01})\,t_0)[\nu] \overset{\mathsf{coe}[]_A}{=}\\
&\qquad \mathsf{coe}_A\,((\delta^\sim\,\rho_{01})[\nu])\,(t_0[\nu]) \overset{\mathsf{irr}}{=} \mathsf{coe}_A\,(\delta^\sim\,(\rho_{01}[\nu]))\,(t_0[\nu]) = \mathsf{coe}_{A[]}\\
&\mathsf{coh}_{A[\delta]}\,\rho_{01}\,t_0 := \mathsf{coh}_A\,(\delta^\sim\,\rho_{01})\,t_0\\
&\mathsf{coh}_{[\mathsf{id}]} \quad : \mathsf{coh}_{A[\mathsf{id}]}\,\rho_{01}\,t_0 \overset{\mathsf{irr}}{=} \mathsf{coh}_A\,\rho_{01}\,t_0
\end{aligned}$$

$$\mathsf{coh}_{[\circ]} \quad : \quad \mathsf{coh}_{A[\delta\circ\nu]}\,\rho_{01}\,t_0 \overset{\mathsf{irr}}{=} \mathsf{coh}_{A[\delta][\nu]}\,\rho_{01}\,t_0$$

Π

$$|\Pi(x:A).B| \quad := \Sigma(f:\Pi(x:|A|).|B|).\underline{\pi(x_0:|A|).\pi(x_1:|A|).}$$
$$\underline{\pi(x_{01}:\underline{A^{\sim}\,(\mathsf{R}_\Gamma\,\mathsf{id})\,x_0\,x_1}).B^{\sim}\,(\mathsf{R}_\Gamma\,\mathsf{id}\,,x_{01})\,(f \mathbin{@} x_0)\,(f \mathbin{@} x_1)}$$

$$(\Pi(x:A).B)^{\sim}\,\rho_{01}\,t_0\,t_1 := \pi(x_0:|A|[\rho_0]).\pi(x_1:|A|[\rho_1]).\pi(x_{01}:\underline{A^{\sim}\,\rho_{01}\,x_0\,x_1}).$$
$$B^{\sim}\,(\rho_{01}\,,x_{01})\,(\mathsf{pr}_0\,t_0 \mathbin{@} x_0)\,(\mathsf{pr}_0\,t_1 \mathbin{@} x_1)$$

$$(\Pi(x:A).B)^{\sim}[] \quad : \quad ((\Pi(x:A).B)^{\sim}\,\rho_{01}\,t_0\,t_1)[\nu] =$$
$$\pi(x_0:A[\rho_0\circ\nu]).\pi(x_1:A[\rho_1\circ\nu]).\pi(x_{01}:\underline{A^{\sim}\,(\rho_{01}[\nu])\,x_0\,x_1}).$$
$$B^{\sim}\,(\rho_{01}[\nu]\,,x_{01})\,(\mathsf{pr}_0\,(t_0[\nu]) \mathbin{@} x_0)\,(\mathsf{pr}_0\,(t_1[\nu]) \mathbin{@} x_1) =$$
$$(\Pi(x:A).B)^{\sim}\,(\rho_{01}[\nu])\,(t_0[\nu])\,(t_1[\nu])$$

$$\mathsf{R}_{\Pi(x:A).B}\,t \quad := \mathsf{pr}_1\,t$$

$$\mathsf{S}_{\Pi(x:A).B}\,t_{01} \quad := \lambda x_0\,x_1\,x_{01}.\mathsf{S}_B\,(t_{01} \mathbin{@} x_1 \mathbin{@} x_0 \mathbin{@} \mathsf{S}_A\,x_{01})$$

$$\mathsf{T}_{\Pi(x:A).B}\,t_{01}\,t_{12} \quad :=$$
$$\lambda x_0\,x_2\,x_{02}.\mathsf{T}_B\,(t_{01} \mathbin{@} x_0 \mathbin{@} \mathsf{coe}_A\,\rho_{01}\,x_0 \mathbin{@} \mathsf{coh}_A\,\rho_{01}\,x_0)$$
$$(t_{12} \mathbin{@} \mathsf{coe}_A\,\rho_{01}\,x_0 \mathbin{@} x_2 \mathbin{@} \mathsf{T}_A\,(\mathsf{S}_A\,(\mathsf{coh}_A\,\rho_{01}\,x_0))\,x_{02})$$

$$\mathsf{coe}_{\Pi(x:A).B}\,\rho_{01}\,t_0 := \Big(\lambda x_1.\mathsf{coe}_B\,(\rho_{01}\,,\mathsf{coh}^*{}_A\,\rho_{01}\,x_1)\,(\mathsf{pr}_0\,t_0 \mathbin{@} \mathsf{coe}^*{}_A\,\rho_{01}\,x_1)\,,$$
$$\lambda x_1\,x_2\,x_{12}.\Big(\mathsf{T}^3{}_B\,\big(\mathsf{S}_B\,(\mathsf{coh}_B\,(\rho_{01}\,,x_{10})\,(\mathsf{pr}_0\,t_0 \mathbin{@} x_0))\big)$$
$$\big(\mathsf{pr}_1\,t_0 \mathbin{@} x_0 \mathbin{@} x_3 \mathbin{@} \mathsf{T}^3{}_A\,(\mathsf{S}_A\,x_{10})\,x_{12}\,x_{23}\big)$$
$$\big(\mathsf{coh}_B\,(\rho_{01}\,,x_{23})\,(\mathsf{pr}_0\,t_0 \mathbin{@} x_3)\big)\Big)$$
$$[x_0 \mapsto \mathsf{coe}^*{}_A\,\rho_{01}\,x_1\,,x_{10} \mapsto \mathsf{coh}^*{}_A\,\rho_{01}\,x_1,$$
$$x_3 \mapsto \mathsf{coe}^*{}_A\,\rho_{01}\,x_2\,,x_{23} \mapsto \mathsf{coh}^*{}_A\,\rho_{01}\,x_2]\Big)$$

$$\mathsf{coe}[]_{\Pi(x:A).B} \quad :$$
$$(\mathsf{coe}_{\Pi(x:A).B}\,\rho_{01}\,t_0)[\nu] =$$
$$(\lambda x_1.(\mathsf{coe}_B\,(\rho_{01}\,,\mathsf{coh}^*{}_A\,\rho_{01}\,x_1)\,(\mathsf{pr}_0\,t_0 \mathbin{@} \mathsf{coe}^*{}_A\,\rho_{01}\,x_1))[\nu]\,,\dots) \overset{\mathsf{coe}[]_B}{=}$$
$$(\lambda x_1.\mathsf{coe}_B\,(\rho_{01}[\nu]\,,(\mathsf{coh}^*{}_A\,\rho_{01}\,x_1)[\nu])\,(\mathsf{pr}_0\,(t_0[\nu]) \mathbin{@} (\mathsf{coe}^*{}_A\,\rho_{01}\,x_1)[\nu])\,,\dots) \overset{\mathsf{coe}[]_A}{=}$$
$$(\lambda x_1.\mathsf{coe}_B\,(\rho_{01}[\nu]\,,(\mathsf{coh}^*{}_A\,(\rho_{01}[\nu])\,x_1))\,(\mathsf{pr}_0\,(t_0[\nu]) \mathbin{@} \mathsf{coe}^*{}_A\,(\rho_{01}[\nu])\,x_1),\dots) \overset{\mathsf{irr}}{=}$$
$$\mathsf{coe}_{\Pi(x:A).B}\,(\rho_{01}[\nu])\,(t_0[\nu])$$

$$\mathsf{coh}_{\Pi(x:A).B}\,\rho_{01}\,t_0 :=$$
$$\lambda x_0\,x_1\,x_{01}.\mathsf{T}_B\,(\mathsf{pr}_1\,t_0 \mathbin{@} x_0 \mathbin{@} \mathsf{coe}^*{}_A\,\rho_{01}\,x_1 \mathbin{@} \mathsf{T}_A\,x_{01}\,(\mathsf{S}_A\,(\mathsf{coh}^*{}_A\,\rho_{01}\,x_1)))$$
$$(\mathsf{coh}_B\,(\rho_{01}\,,\mathsf{coh}^*{}_A\,\rho_{01}\,x_1)\,(\mathsf{pr}_0\,t_0 \mathbin{@} \mathsf{coe}^*{}_A\,\rho_{01}\,x_1))$$

$$|\lambda x.t| \quad := (\lambda x.|t|, \lambda x_0\,x_1\,x_{01}.t^{\sim}\,(\mathsf{R}_\Gamma\,\mathsf{id}\,,x_{01}))$$

$$
\begin{aligned}
(\lambda x.t)^{\sim}\,\rho_{01} &:= \lambda x_0\,x_1\,x_{01}.t^{\sim}\,(\rho_{01}\,,x_{01})\\
|t \mathbin{@} x| &:= \mathsf{pr}_0\,|t| \mathbin{@} x\\
(t \mathbin{@} x)^{\sim}\,(\rho_{01}\,,x_{01}) &:= t^{\sim}\,\rho_{01} \mathbin{@} x_0 \mathbin{@} x_1 \mathbin{@} x_{01}
\end{aligned}
$$

$$
\begin{aligned}
|\Pi\beta| \;&:\; |(\lambda x.t) \mathbin{@} x| = \mathsf{pr}_0\,|\lambda x.t| \mathbin{@} x = (\lambda x.|t|) \mathbin{@} x \overset{\Pi\beta}{=} |t|\\
|\Pi\eta| \;&:\; |\lambda x.t \mathbin{@} x| = \big(\lambda x.|t \mathbin{@} x|\,,\lambda x_0\,x_1\,x_{01}.(t \mathbin{@} x)^{\sim}\,(\mathsf{R}_\Gamma\,\mathsf{id}\,,x_{01})\big) =\\
&\quad \big(\lambda x.\mathsf{pr}_0\,|t| \mathbin{@} x\,,\lambda x_0\,x_1\,x_{01}.t^{\sim}\,(\mathsf{R}_\Gamma\,\mathsf{id}) \mathbin{@} x_0 \mathbin{@} x_1 \mathbin{@} x_{01}\big) \overset{\Pi\eta}{=}\\
&\quad (\mathsf{pr}_0\,|t|\,,t^{\sim}\,(\mathsf{R}_\Gamma\,\mathsf{id})) \overset{\mathsf{irr}}{=} (\mathsf{pr}_0\,|t|\,,\mathsf{pr}_1\,|t|) \overset{\Sigma\eta}{=} |t|\\
\Pi\beta^{\sim} \;&:\; ((\lambda x.t) \mathbin{@} x)^{\sim}\,\rho_{01} \overset{\mathsf{irr}}{=} t^{\sim}\,\rho_{01}\\
\Pi\eta^{\sim} \;&:\; (\lambda x.t \mathbin{@} x)^{\sim}\,\rho_{01} \overset{\mathsf{irr}}{=} t^{\sim}\,\rho_{01}\\
|\Pi[]| \;&:
\end{aligned}
$$

$$
\begin{aligned}
&|(\Pi(x:A).B)[\nu]| = |\Pi(x:A).B|[|\nu|] \overset{\Pi[],\pi[],A^{\sim}[],B^{\sim}[]}{=}\\
&\Sigma(f:\Pi(x:|A|[|\nu|]).|B|[|\nu|]).\underline{\pi(x_0:|A|[|\nu|]).\pi(x_1:|A|[|\nu|]).}\\
&\underline{\pi(x_{01}:\underline{A^{\sim}\,((\mathsf{R}_\Gamma\,\mathsf{id})[|\nu|])\,x_0\,x_1}).B^{\sim}\,((\mathsf{R}_\Gamma\,\mathsf{id})[|\nu|]\,,x_{01})\,(f \mathbin{@} x_0)\,(f \mathbin{@} x_1)} \overset{\mathsf{irr}}{=}\\
&|\Pi(x:A[\nu]).B[\nu]|
\end{aligned}
$$

$\Pi[]^{\sim}$:

$$
\begin{aligned}
&((\Pi(x:A).B)[\nu])^{\sim}\,\rho_{01}\,t_0\,t_1 = (\Pi(x:A).B)^{\sim}\,(\nu^{\sim}\,\rho_{01})\,t_0\,t_1 =\\
&\pi(x_0:A[\nu\circ\rho_0]).\pi(x_1:A[\nu\circ\rho_1]).\pi(x_{01}:\underline{A^{\sim}\,(\nu^{\sim}\,\rho_{01})\,x_0\,x_1}).\\
&B^{\sim}\,(\nu^{\sim}\,\rho_{01}\,,x_{01})\,(\mathsf{pr}_0\,t_0 \mathbin{@} x_0)\,(\mathsf{pr}_0\,t_1 \mathbin{@} x_1) = (\Pi(x:A[\nu]).B[\nu])^{\sim}\,\rho_{01}\,t_0\,t_1
\end{aligned}
$$

$$
\begin{aligned}
\mathsf{R}_{\Pi[]} \;&:\; \mathsf{R}_{(\Pi(x:A).B)[\nu]}\,t \overset{\mathsf{irr}}{=} \mathsf{R}_{\Pi(x:A[\nu]).B[\nu]}\,t\\
\mathsf{S}_{\Pi[]} \;&:\; \mathsf{S}_{(\Pi(x:A).B)[\nu]}\,t_{01} \overset{\mathsf{irr}}{=} \mathsf{S}_{\Pi(x:A[\nu]).B[\nu]}\,t_{01}\\
\mathsf{T}_{\Pi[]} \;&:\; \mathsf{T}_{(\Pi(x:A).B)[\nu]}\,t_{01}\,t_{12} \overset{\mathsf{irr}}{=} \mathsf{T}_{\Pi(x:A[\nu]).B[\nu]}\,t_{01}\,t_{12}\\
\mathsf{coe}_{\Pi[]} \;&:
\end{aligned}
$$

$$
\begin{aligned}
&\mathsf{coe}_{(\Pi(x:A).B)[\nu]}\,\rho_{01}\,t_0 = \mathsf{coe}_{\Pi(x:A).B}\,(\nu^{\sim}\,\rho_{01})\,t_0 =\\
&\big(\lambda x_1.\mathsf{coe}_B\,(\nu^{\sim}\,\rho_{01}\,,\mathsf{coh^*}_A\,(\nu^{\sim}\,\rho_{01})\,x_1)\,(\mathsf{pr}_0\,t_0 \mathbin{@} \mathsf{coe^*}_A\,(\nu^{\sim}\,\rho_{01})\,x_1)\,,\dots\big) \overset{\mathsf{irr}}{=}\\
&\big(\lambda x_1.\mathsf{coe}_{B[\nu]}\,(\rho_{01}\,,\mathsf{coh^*}_{A[\nu]}\,\rho_{01}\,x_1)\,(\mathsf{pr}_0\,t_0 \mathbin{@} \mathsf{coe^*}_{A[\nu]}\,\rho_{01}\,x_1)\,,\dots\big) =\\
&\mathsf{coe}_{\Pi(x:A[\nu]).B[\nu]}\,\rho_{01}\,t_0
\end{aligned}
$$

$$
\begin{aligned}
\mathsf{coh}_{\Pi[]} \;&:\; \mathsf{coh}_{(\Pi(x:A).B)[\nu]}\,\rho_{01}\,t_0 \overset{\mathsf{irr}}{=} \mathsf{coh}_{\Pi(x:A[\nu]).B[\nu]}\,\rho_{01}\,t_0\\
|\lambda[]| \;&:\; |(\lambda x.t)[\nu]| = |\lambda x.t|[|\nu|] \overset{\lambda[]}{=} \lambda x.|t|[|\nu|] = |\lambda x.t[\nu]|\\
\lambda[]^{\sim} \;&:\; ((\lambda x.t)[\nu])^{\sim}\,\rho_{01} \overset{\mathsf{irr}}{=} (\lambda x.t[\nu])^{\sim}\,\rho_{01}
\end{aligned}
$$

Σ

$$
\begin{aligned}
&|\Sigma(x:A).B| && := \Sigma(x:|A|).|B| \\
&(\Sigma(x:A).B)^\sim\,\rho_{01}\,(u_0\,,v_0)\,(u_1\,,v_1) := \sigma(x_{01}:A^\sim\,\rho_{01}\,u_0\,u_1).B^\sim\,(\rho_{01}\,,x_{01})\,v_0\,v_1 \\
&(\Sigma(x:A).B)^\sim[] && : \\
&\quad ((\Sigma(x:A).B)^\sim\,\rho_{01}\,(u_0\,,v_0)\,(u_1\,,v_1))[\nu] = \\
&\quad \sigma(x_{01}:(A^\sim\,\rho_{01}\,u_0\,u_1)[\nu]).(B^\sim\,(\rho_{01}\,,x_{01})\,v_0\,v_1)[\nu] \overset{A^\sim[],B^\sim[]}{=} \\
&\quad \sigma(x_{01}:A^\sim\,(\rho_{01}[\nu])\,(u_0[\nu])\,(u_1[\nu])).B^\sim\,(\rho_{01}[\nu]\,,x_{01})\,(v_0[\nu])\,(v_1[\nu]) = \\
&\quad (\Sigma(x:A).B)^\sim\,(\rho_{01}[\nu])\,((u_0\,,v_0)[\nu])\,((u_1\,,v_1)[\nu]) \\
&\mathsf{R}_{\Sigma(x:A).B}\,(u\,,v) && := (\mathsf{R}_A\,u\,,\mathsf{R}_B\,v) \\
&\mathsf{S}_{\Sigma(x:A).B}\,(u_{01}\,,v_{01}) && := (\mathsf{S}_A\,u_{01}\,,\mathsf{S}_B\,v_{01}) \\
&\mathsf{T}_{\Sigma(x:A).B}\,(u_{01}\,,v_{01})\,(u_{12}\,,v_{12}) := (\mathsf{T}_A\,u_{01}\,u_{12}\,,\mathsf{Tm}_B\,v_{01}\,v_{12}) \\
&\mathsf{coe}_{\Sigma(x:A).B}\,\rho_{01}\,(u_0\,,v_0) := (\mathsf{coe}_A\,\rho_{01}\,u_0\,,\mathsf{coe}_B\,(\rho_{01}\,,\mathsf{coh}_A\,\rho_{01}\,u_0)\,v_0) \\
&\mathsf{coe}[]_{\Sigma(x:A).B} && : \\
&\quad (\mathsf{coe}_{\Sigma(x:A).B}\,\rho_{01}\,(u_0\,,v_0))[\nu] = \\
&\quad ((\mathsf{coe}_A\,\rho_{01}\,u_0)[\nu]\,,(\mathsf{coe}_B\,(\rho_{01}\,,\mathsf{coh}_A\,\rho_{01}\,u_0)\,v_0)[\nu]) \overset{\mathsf{coe}[]_A,\mathsf{coe}[]_B}{=} \\
&\quad (\mathsf{coe}_A\,(\rho_{01}[\nu])\,(u_0[\nu])\,,\mathsf{coe}_B\,(\rho_{01}[\nu]\,,\mathsf{coh}_A\,(\rho_{01}[\nu])\,(u_0[\nu]))\,(v_0[\nu])) = \\
&\quad \mathsf{coe}_{\Sigma(x:A).B}\,(\rho_{01}[\nu])\,((u_0\,,v_0)[\nu]) \\
&\mathsf{coh}_{\Sigma(x:A).B}\,\rho_{01}\,(u_0\,,v_0) := (\mathsf{coh}_A\,\rho_{01}\,u_0\,,\mathsf{coh}_B\,(\rho_{01}\,,\mathsf{coh}_A\,\rho_{01}\,u_0)\,v_0) \\
&|(u\,,v)| && := (|u|\,,|v|) \\
&(u\,,v)^\sim\,\rho_{01} && := (u^\sim\,\rho_{01}\,,v^\sim\,\rho_{01}) \\
&|\mathsf{pr}_0\,t| && := \mathsf{pr}_0\,|t| \\
&(\mathsf{pr}_0\,t)^\sim\,\rho_{01} && := \mathsf{pr}_0\,(t^\sim\,\rho_{01}) \\
&|\mathsf{pr}_1\,t| && := \mathsf{pr}_1\,|t| \\
&(\mathsf{pr}_1\,t)^\sim\,\rho_{01} && := \mathsf{pr}_1\,(t^\sim\,\rho_{01}) \\
&|\Sigma\beta_0| && : |\mathsf{pr}_0\,(u\,,v)| = \mathsf{pr}_0\,(|u|\,,|v|) \overset{\Sigma\beta_0}{=} |u| \\
&\Sigma\beta_0{}^\sim && : (\mathsf{pr}_0\,(u\,,v))^\sim\,\rho_{01} \overset{\mathsf{irr}}{=} u^\sim\,\rho_{01} \\
&|\Sigma\beta_1| && : |\mathsf{pr}_1\,(u\,,v)| = \mathsf{pr}_1\,(|u|\,,|v|) \overset{\Sigma\beta_1}{=} |v| \\
&\Sigma\beta_1{}^\sim && : (\mathsf{pr}_1\,(u\,,v))^\sim\,\rho_{01} \overset{\mathsf{irr}}{=} v^\sim\,\rho_{01} \\
&|\Sigma\eta| && : |(\mathsf{pr}_0\,t\,,\mathsf{pr}_1\,t)| = (\mathsf{pr}_0\,|t|\,,\mathsf{pr}_1\,|t|) \overset{\Sigma\eta}{=} |t| \\
&\Sigma\eta^\sim && : (\mathsf{pr}_0\,t\,,\mathsf{pr}_1\,t)^\sim\,\rho_{01} \overset{\mathsf{irr}}{=} t^\sim\,\rho_{01} \\
&|\Sigma[]| && : |(\Sigma(x:A).B)[\nu]| = |\Sigma(x:A).B|[|\nu|] \overset{\Sigma[]}{=} \\
& && \quad \Sigma(x:|A|[|\nu|]).|B|[|\nu|] = |\Sigma(x:A[\nu]).B[\nu]| \\
&\Sigma[]^\sim && : ((\Sigma(x:A).B)[\nu])^\sim\,\rho_{01}\,(u_0\,,v_0)\,(u_1\,,v_1) =
\end{aligned}
$$

$$(\Sigma(x:A).B)^{\sim}\,(\nu^{\sim}\,\rho_{01})\,(u_0\,,v_0)\,(u_1\,,v_1) =$$
$$\sigma(x_{01}:A^{\sim}\,(\nu^{\sim}\,\rho_{01})\,u_0\,u_1).B^{\sim}\,(\nu^{\sim}\,\rho_{01}\,,x_{01})\,v_0\,v_1 =$$
$$(\Sigma(x:A[\nu]).B[\nu])^{\sim}\,\rho_{01}\,(u_0\,,v_0)\,(u_1\,,v_1)$$

$\mathsf{R}_{\Sigma[]}$: $\mathsf{R}_{(\Sigma(x:A).B)[\nu]}\,t \overset{\mathsf{irr}}{=} \mathsf{R}_{\Sigma(x:A[\nu]).B[\nu]}\,t$

$\mathsf{S}_{\Sigma[]}$: $\mathsf{S}_{(\Sigma(x:A).B)[\nu]}\,t_{01} \overset{\mathsf{irr}}{=} \mathsf{S}_{\Sigma(x:A[\nu]).B[\nu]}\,t_{01}$

$\mathsf{T}_{\Sigma[]}$: $\mathsf{T}_{(\Sigma(x:A).B)[\nu]}\,t_{01}\,t_{12} \overset{\mathsf{irr}}{=} \mathsf{T}_{\Sigma(x:A[\nu]).B[\nu]}\,t_{01}\,t_{12}$

$\mathsf{coe}_{\Sigma[]}$: $\mathsf{coe}_{(\Sigma(x:A).B)[\nu]}\,\rho_{01}\,(u_0\,,v_0) =$
$\mathsf{coe}_{\Sigma(x:A).B}\,(\nu^{\sim}\,\rho_{01})\,(u_0\,,v_0) =$
$\big(\mathsf{coe}_A\,(\nu^{\sim}\,\rho_{01})\,u_0\,,\mathsf{coe}_B\,(\nu^{\sim}\,\rho_{01}\,,\mathsf{coh}_A\,(\nu^{\sim}\,\rho_{01})\,u_0)\,v_0\big) =$
$\big(\mathsf{coe}_{A[\nu]}\,\rho_{01}\,u_0\,,\mathsf{coe}_{B[\nu]}\,(\rho_{01}\,,\mathsf{coh}_{A[\nu]}\,\rho_{01}\,u_0)\,v_0\big) =$
$\mathsf{coe}_{\Sigma(x:A[\nu]).B[\nu]}\,\rho_{01}\,(u_0\,,v_0)$

$\mathsf{coh}_{\Sigma[]}$: $\mathsf{coh}_{(\Sigma(x:A).B)[\nu]}\,\rho_{01}\,t_0 \overset{\mathsf{irr}}{=} \mathsf{coh}_{\Sigma(x:A[\nu]).B[\nu]}\,\rho_{01}\,t_0$

$|,[]|$: $|(u\,,v)[\nu]| = |(u\,,v)|[|\nu|] = (|u|\,,|v|)[|\nu|] \overset{,[]}{=}$
$(|u|[|\nu|]\,,|v|[|\nu|]) = |(u[\nu]\,,v[\nu])|$

$,[]^{\sim}$: $((u\,,v)[\nu])^{\sim}\,\rho_{01} \overset{\mathsf{irr}}{=} (u[\nu]\,,v[\nu])^{\sim}\,\rho_{01}$

Bool

$|\mathsf{Bool}|$:= Bool

$\mathsf{Bool}^{\sim}\,\rho_{01}\,t_0\,t_1$:= $\mathsf{if}\ t_0\ \mathsf{then}\ (\mathsf{if}\ t_1\ \mathsf{then}\ \top\ \mathsf{else}\ \bot)\ \mathsf{else}\ (\mathsf{if}\ t_1\ \mathsf{then}\ \bot\ \mathsf{else}\ \top)$

$\mathsf{Bool}^{\sim}[]$:

$(\mathsf{Bool}^{\sim}\,\rho_{01}\,t_0\,t_1)[\nu] =$
$\mathsf{if}\ t_0[\nu]\ \mathsf{then}\ (\mathsf{if}\ t_1[\nu]\ \mathsf{then}\ \top\ \mathsf{else}\ \bot)\ \mathsf{else}\ (\mathsf{if}\ t_1[\nu]\ \mathsf{then}\ \bot\ \mathsf{else}\ \top) =$
$\mathsf{Bool}^{\sim}\,(\rho_{01}[\nu])\,(t_0[\nu])\,(t_1[\nu])$

$\mathsf{R}_{\mathsf{Bool}}\,t$:= $\mathsf{if}\ t\ \mathsf{then}\ \mathsf{tt}\ \mathsf{else}\ \mathsf{tt}$

$\mathsf{S}_{\mathsf{Bool}}\,t_{01}$:= $\mathsf{if}\ t_0\ \mathsf{then}\ (\mathsf{if}\ t_1\ \mathsf{then}\ \mathsf{tt}\ \mathsf{else}\ \mathsf{exfalso}\,t_{01})$
$\mathsf{else}\ (\mathsf{if}\ t_1\ \mathsf{then}\ \mathsf{exfalso}\,t_{01}\ \mathsf{else}\ \mathsf{tt})$

$\mathsf{T}_{\mathsf{Bool}}\,t_{01}\,t_{12}$:= $\mathsf{if}\ t_0\ \mathsf{then}\ \big(\mathsf{if}\ t_1\ \mathsf{then}\ (\mathsf{if}\ t_2\ \mathsf{then}\ \mathsf{tt}\ \mathsf{else}\ \mathsf{exfalso}\,t_{12})\ \mathsf{else}\ \mathsf{exfalso}\,t_{01}\big)$
$\mathsf{else}\ \big(\mathsf{if}\ t_1\ \mathsf{then}\ \mathsf{exfalso}\,t_{01}\ \mathsf{else}\ (\mathsf{if}\ t_2\ \mathsf{then}\ \mathsf{exfalso}\,t_{12}\ \mathsf{else}\ \mathsf{tt})\big)$

$\mathsf{coe}_{\mathsf{Bool}}\,\rho_{01}\,t_0$:= t_0

$\mathsf{coe}[]_{\mathsf{Bool}}$: $(\mathsf{coe}_{\mathsf{Bool}}\,\rho_{01}\,t_0)[\nu] = t_0[\nu] = \mathsf{coe}_{\mathsf{Bool}}\,\rho_{01}\,(t_0[\nu])$

$\mathsf{coh}_{\mathsf{Bool}}\,\rho_{01}\,t_0$:= $\mathsf{if}\ t_0\ \mathsf{then}\ \mathsf{tt}\ \mathsf{else}\ \mathsf{tt}$

$|\mathsf{true}|$:= true

$\mathsf{true}^{\sim}\,\rho_{01}$:= tt

$|\mathsf{false}|$:= false

$$\begin{aligned}
&\mathsf{false}^\sim\,\rho_{01} &&:= \mathsf{tt}\\
&|\mathsf{if}\,t\,\mathsf{then}\,u\,\mathsf{else}\,v| &&:= \mathsf{if}\,|t|\,\mathsf{then}\,|u|\,\mathsf{else}\,|v|\\
&(\mathsf{if}\,t\,\mathsf{then}\,u\,\mathsf{else}\,v)^\sim\,\rho_{01} &&:= \mathsf{if}\,t[\rho_0]\,\mathsf{then}\,(\mathsf{if}\,t[\rho_1]\,\mathsf{then}\,u^\sim\,\rho_{01}\,\mathsf{else}\,\mathsf{exfalso}\,(u^\sim\,\rho_{01}))\\
& && \qquad\qquad\mathsf{else}\,(\mathsf{if}\,t[\rho_1]\,\mathsf{then}\,\mathsf{exfalso}\,(v^\sim\,\rho_{01})\,\mathsf{else}\,v^\sim\,\rho_{01})
\end{aligned}$$

$$\begin{aligned}
&|\mathsf{Bool}\beta_{\mathsf{true}}| &&: |\mathsf{if}\,\mathsf{true}\,\mathsf{then}\,u\,\mathsf{else}\,v| = \mathsf{if}\,\mathsf{true}\,\mathsf{then}\,|u|\,\mathsf{else}\,|v| \overset{\mathsf{Bool}\beta_{\mathsf{true}}}{=} |u|\\
&|\mathsf{Bool}\beta_{\mathsf{false}}| &&: |\mathsf{if}\,\mathsf{false}\,\mathsf{then}\,u\,\mathsf{else}\,v| = \mathsf{if}\,\mathsf{false}\,\mathsf{then}\,|u|\,\mathsf{else}\,|v| \overset{\mathsf{Bool}\beta_{\mathsf{false}}}{=} |v|\\
&|\mathsf{Bool}[]| &&: |\mathsf{Bool}[\nu]| = |\mathsf{Bool}|[|\nu|] \overset{\mathsf{Bool}[]}{=} \mathsf{Bool} = |\mathsf{Bool}|\\
&\mathsf{Bool}[]^\sim &&: (\mathsf{Bool}[\nu])^\sim\,\rho_{01}\,t_0\,t_1 = \mathsf{Bool}^\sim\,(\nu^\sim\,\rho_{01})\,t_0\,t_1 =\\
& && \quad \mathsf{if}\,t_0\,\mathsf{then}\,(\mathsf{if}\,t_1\,\mathsf{then}\,\top\,\mathsf{else}\,\bot)\,\mathsf{else}\,(\mathsf{if}\,t_1\,\mathsf{then}\,\bot\,\mathsf{else}\,\top) =\\
& && \quad \mathsf{Bool}^\sim\,\rho_{01}\,t_0\,t_1\\
&\mathsf{R}_{\mathsf{Bool}[]} &&: \mathsf{R}_{\mathsf{Bool}[\nu]}\,t \overset{\mathsf{irr}}{=} \mathsf{R}_{\mathsf{Bool}}\,t\\
&\mathsf{S}_{\mathsf{Bool}[]} &&: \mathsf{S}_{\mathsf{Bool}[\nu]}\,t_{01} \overset{\mathsf{irr}}{=} \mathsf{S}_{\mathsf{Bool}}\,t_{01}\\
&\mathsf{T}_{\mathsf{Bool}[]} &&: \mathsf{T}_{\mathsf{Bool}[\nu]}\,t_{01}\,t_{12} \overset{\mathsf{irr}}{=} \mathsf{T}_{\mathsf{Bool}}\,t_{01}\,t_{12}\\
&\mathsf{coe}_{\mathsf{Bool}[]} &&: \mathsf{coe}_{\mathsf{Bool}[\nu]}\,\rho_{01}\,t_0 = \mathsf{coe}_{\mathsf{Bool}}\,(\nu^\sim\,\rho_{01})\,t_0 = t_0 = \mathsf{coe}_{\mathsf{Bool}}\,\rho_{01}\,t_0\\
&\mathsf{coh}_{\mathsf{Bool}[]} &&: \mathsf{coh}_{\mathsf{Bool}[\nu]}\,\rho_{01}\,t_0 \overset{\mathsf{irr}}{=} \mathsf{coh}_{\mathsf{Bool}}\,\rho_{01}\,t_0\\
&|\mathsf{true}[]| &&: |\mathsf{true}[\nu]| = \mathsf{true}[|\nu|] \overset{\mathsf{true}[]}{=} \mathsf{true} = |\mathsf{true}|\\
&\mathsf{true}[]^\sim &&: (\mathsf{true}[\nu])^\sim\,\rho_{01} \overset{\mathsf{irr}}{=} \mathsf{true}^\sim\,\rho_{01}\\
&|\mathsf{false}[]| &&: |\mathsf{false}[\nu]| = \mathsf{false}[|\nu|] \overset{\mathsf{false}[]}{=} \mathsf{false} = |\mathsf{false}|\\
&\mathsf{false}[]^\sim &&: (\mathsf{false}[\nu])^\sim\,\rho_{01} \overset{\mathsf{irr}}{=} \mathsf{false}^\sim\,\rho_{01}\\
&|\mathsf{if}[]| &&: |(\mathsf{if}\,t\,\mathsf{then}\,u\,\mathsf{else}\,v)[\nu]| = (\mathsf{if}\,|t|\,\mathsf{then}\,|u|\,\mathsf{else}\,|v|)[|\nu|] \overset{\mathsf{if}[]}{=}\\
& && \quad \mathsf{if}\,|t|[|\nu|]\,\mathsf{then}\,|u|[|\nu|]\,\mathsf{else}\,|v|[|\nu|] = |\mathsf{if}\,t[\nu]\,\mathsf{then}\,u[\nu]\,\mathsf{else}\,v[\nu]|\\
&\mathsf{if}[]^\sim &&: ((\mathsf{if}\,t\,\mathsf{then}\,u\,\mathsf{else}\,v)[\nu])^\sim\,\rho_{01} \overset{\mathsf{irr}}{=} (\mathsf{if}\,t[\nu]\,\mathsf{then}\,u[\nu]\,\mathsf{else}\,v[\nu])^\sim\,\rho_{01}
\end{aligned}$$

Prop

$$\begin{aligned}
&|\mathsf{Prop}_i| &&:= \mathsf{Prop}_i\\
&\mathsf{Prop}_i{}^\sim\,\rho_{01}\,a_0\,a_1 &&:= (\underline{a_0} \Rightarrow a_1) \times (\underline{a_1} \Rightarrow a_0)\\
&\mathsf{Prop}_i{}^\sim[] &&: (\mathsf{Prop}_i{}^\sim\,\rho_{01}\,a_0\,a_1)[\nu] \overset{\pi[],\sigma[],\underline{[]}}{=}\\
& && \quad (\underline{a_0[\nu]} \Rightarrow a_1[\nu]) \times (\underline{a_1[\nu]} \Rightarrow a_0[\nu]) =\\
& && \quad \mathsf{Prop}_i{}^\sim\,(\rho_{01}[\nu])\,(a_0[\nu])\,(a_1[\nu])\\
&\mathsf{R}_{\mathsf{Prop}_i}\,a &&:= (\lambda x.x\,,\lambda x.x)\\
&\mathsf{S}_{\mathsf{Prop}_i}\,(a_{01}\,,a_{10}) &&:= (a_{10}\,,a_{01})\\
&\mathsf{T}_{\mathsf{Prop}_i}\,(a_{01}\,,a_{10})\,(a_{12}\,,a_{21}) &&:= (\lambda x_0.a_{12} \mathbin{@} (a_{01} \mathbin{@} x_0)\,,\lambda x_2.a_{10} \mathbin{@} (a_{21} \mathbin{@} x_2))
\end{aligned}$$

$$
\begin{aligned}
&\mathsf{coe}_{\mathsf{Prop}_i}\,\rho_{01}\,a_0 && := a_0\\
&\mathsf{coe}[]_{\mathsf{Prop}_i} && : \quad (\mathsf{coe}_{\mathsf{Prop}_i}\,\rho_{01}\,a_0)[\nu] = a_0[\nu] = \mathsf{coe}_{\mathsf{Prop}_i}\,\rho_{01}\,(a_0[\nu])\\
&\mathsf{coh}_{\mathsf{Prop}_i}\,\rho_{01}\,a_0 && := (\lambda x.x\,,\lambda x.x)\\
&|\underline{a}| && := \underline{|a|}\\
&\underline{a}^{\sim}\,\rho_{01}\,t_0\,t_1 && := \top\\
&\underline{a}^{\sim}[] && : \quad (\underline{a}^{\sim}\,\rho_{01}\,t_0\,t_1)[\nu] = \top[\nu] \overset{\top[]}{=} \top = \underline{a}^{\sim}\,(\rho_{01}[\nu])\,(t_0[\nu])\,(t_1[\nu])\\
&\mathsf{R}_{\underline{a}}\,t && := \mathsf{tt}\\
&\mathsf{S}_{\underline{a}}\,t_{01} && := \mathsf{tt}\\
&\mathsf{T}_{\underline{a}}\,t_{01}\,t_{12} && := \mathsf{tt}\\
&\mathsf{coe}_{\underline{a}}\,\rho_{01}\,t_0 && := \mathsf{pr}_0\,(a^{\sim}\,\rho_{01}) @ t_0\\
&\mathsf{coe}[]_{\underline{a}} && : \quad (\mathsf{coe}_{\underline{a}}\,\rho_{01}\,t_0)[\nu] = \mathsf{pr}_0\,((a^{\sim}\,\rho_{01})[\nu]) @ t_0[\nu] \overset{\mathsf{irr}}{=}\\
& && \qquad \mathsf{pr}_0\,(a^{\sim}\,(\rho_{01}[\nu])) @ t_0[\nu] = \mathsf{coe}_{\underline{a}}\,(\rho_{01}[\nu])\,(t_0[\nu])\\
&\mathsf{coh}_{\underline{a}}\,\rho_{01}\,t_0 && := \mathsf{tt}\\
&|\mathsf{irr}_a| && : \quad |u : \mathsf{Tm}\,\Gamma\,\underline{a}| = (|u| : \mathsf{Tm}\,|\Gamma|\,\underline{|a|}) \overset{\mathsf{irr}_{|a|}}{=} (|v| : \mathsf{Tm}\,|\Gamma|\,\underline{|a|}) =\\
& && \qquad |v : \mathsf{Tm}\,\Gamma\,\underline{a}|\\
&|\pi(x:A).b| && := \pi(x:|A|).|b|\\
&(\pi(x:A).b)^{\sim}\,\rho_{01} && := \big(\lambda f_0\,x_1.\mathsf{pr}_0\,(b^{\sim}\,(\rho_{01}\,,\mathsf{coh}^*{}_A\,\rho_{01}\,x_1)) @ (f_0 @ \mathsf{coe}^*{}_A\,\rho_{01}\,x_1)\,,\\
& && \qquad \lambda f_1\,x_0.\mathsf{pr}_1\,(b^{\sim}\,(\rho_{01}\,,\mathsf{coh}_A\,\rho_{01}\,x_0)) @ (f_1 @ \mathsf{coe}_A\,\rho_{01}\,x_0)\big)\\
&|\lambda x.t| && := \lambda x.|t|\\
&(\lambda x.t)^{\sim}\,\rho_{01} && := \mathsf{tt}\\
&|t @ x| && := |t| @ x\\
&(t @ x)^{\sim}\,\rho_{01} && := \mathsf{tt}\\
&|\sigma(x:a).b| && := \sigma(x:|a|).|b|\\
&(\sigma(x:a).b)^{\sim}\,\rho_{01} && := \big(\lambda z_0.(\mathsf{pr}_0\,(a^{\sim}\,\rho_{01}) @ \mathsf{pr}_0\,z_0\,,\mathsf{pr}_0\,(b^{\sim}\,(\rho_{01}\,,\mathsf{tt})) @ \mathsf{pr}_1\,z_0),\\
& && \qquad \lambda z_1.(\mathsf{pr}_1\,(a^{\sim}\,\rho_{01}) @ \mathsf{pr}_0\,z_1\,,\mathsf{pr}_1\,(b^{\sim}\,(\rho_{01}\,,\mathsf{tt})) @ \mathsf{pr}_1\,z_1)\big)
\end{aligned}
$$

$$
\begin{aligned}
&|(u\,,v)| && :=(|u|\,,|v|)\\
&(u\,,v)^{\sim}\,\rho_{01} && :=\mathsf{tt}\\
&|\mathsf{pr}_0\,t| && :=\mathsf{pr}_0\,|t|\\
&(\mathsf{pr}_0\,t)^{\sim}\,\rho_{01} && :=\mathsf{tt}\\
&|\mathsf{pr}_1\,t| && :=\mathsf{pr}_1\,|t|\\
&(\mathsf{pr}_1\,t)^{\sim}\,\rho_{01} && :=\mathsf{tt}\\
&|\top| && :=\top\\
&\top^{\sim}\,\rho_{01} && :=(\lambda x.x\,,\lambda x.x)\\
&|\mathsf{tt}| && :=\mathsf{tt}
\end{aligned}
$$

$$
\begin{aligned}
&\mathsf{tt}^{\sim}\,\rho_{01} &&:= \mathsf{tt}\\
&|\bot| &&:= \bot\\
&\bot^{\sim}\,\rho_{01} &&:= (\lambda x.x\,,\lambda x.x)\\
&|\mathsf{exfalso}\,t| &&:= \mathsf{exfalso}\,|t|\\
&(\mathsf{exfalso}\,t)^{\sim}\,\rho_{01} &&:= \mathsf{exfalso}\,(|t|[\rho_0])
\end{aligned}
$$

$$
\begin{aligned}
&|\mathsf{Prop}[]| &&: |\mathsf{Prop}[\nu]| = |\mathsf{Prop}|[|\nu|] \overset{\mathsf{Prop}[]}{=} \mathsf{Prop} = |\mathsf{Prop}|\\
&\mathsf{Prop}[]^{\sim} &&: (\mathsf{Prop}[\nu])^{\sim}\,\rho_{01}\,a_0\,a_1 = \mathsf{Prop}^{\sim}\,(\nu^{\sim}\,\rho_{01})\,a_0\,a_1 =\\
& && \quad (\underline{a_0} \Rightarrow a_1) \times (\underline{a_1} \Rightarrow a_0) = \mathsf{Prop}^{\sim}\,\rho_{01}\,a_0\,a_1\\
&\mathsf{R}_{\mathsf{Prop}[]} &&: \mathsf{R}_{\mathsf{Prop}[\nu]}\,a \overset{\mathsf{irr}}{=} \mathsf{R}_{\mathsf{Prop}}\,a\\
&\mathsf{S}_{\mathsf{Prop}[]} &&: \mathsf{S}_{\mathsf{Prop}[\nu]}\,a_{01} \overset{\mathsf{irr}}{=} \mathsf{S}_{\mathsf{Prop}}\,a_{01}\\
&\mathsf{T}_{\mathsf{Prop}[]} &&: \mathsf{T}_{\mathsf{Prop}[\nu]}\,a_{01}\,a_{12} \overset{\mathsf{irr}}{=} \mathsf{T}_{\mathsf{Prop}}\,a_{01}\,a_{12}\\
&\mathsf{coe}_{\mathsf{Prop}[]} &&: \mathsf{coe}_{\mathsf{Prop}[\nu]}\,\rho_{01}\,a_0 = \mathsf{coe}_{\mathsf{Prop}}\,(\nu^{\sim}\,\rho_{01})\,a_0 = a_0 = \mathsf{coe}_{\mathsf{Prop}}\,\rho_{01}\,a_0\\
&\mathsf{coh}_{\mathsf{Prop}[]} &&: \mathsf{coh}_{\mathsf{Prop}[\nu]}\,\rho_{01}\,a_0 \overset{\mathsf{irr}}{=} \mathsf{coh}_{\mathsf{Prop}}\,\rho_{01}\,a_0\\
&|\underline{[]}| &&: |\underline{a}[\nu]| = \underline{|a|}[|\nu|] \overset{\underline{a}[]}{=} \underline{|a|[|\nu|]} = |\underline{a[\nu]}|\\
&\underline{[]}^{\sim} &&: (\underline{a}[\nu])^{\sim}\,\rho_{01}\,a_0\,a_1 = \underline{a}^{\sim}\,(\nu^{\sim}\,\rho_{01})\,a_0\,a_1 = \top = \underline{a[\nu]}^{\sim}\,\rho_{01}\,a_0\,a_1\\
&\mathsf{R}_{\underline{[]}} &&: \mathsf{R}_{\underline{a}[\nu]}\,t \overset{\mathsf{irr}}{=} \mathsf{R}_{\underline{a[\nu]}}\,t\\
&\mathsf{S}_{\underline{[]}} &&: \mathsf{S}_{\underline{a}[\nu]}\,t_{01} \overset{\mathsf{irr}}{=} \mathsf{S}_{\underline{a[\nu]}}\,t_{01}\\
&\mathsf{T}_{\underline{[]}} &&: \mathsf{T}_{\underline{a}[\nu]}\,t_{01}\,t_{12} \overset{\mathsf{irr}}{=} \mathsf{T}_{\underline{a[\nu]}}\,t_{01}\,t_{12}\\
&\mathsf{coe}_{\underline{[]}} &&: \mathsf{coe}_{\underline{a}[\nu]}\,\rho_{01}\,t_0 = \mathsf{coe}_{\underline{a}}\,(\nu^{\sim}\,\rho_{01})\,t_0 = \mathsf{pr}_0\,(a^{\sim}\,(\nu^{\sim}\,\rho_{01}))\,@\,t_0 =\\
& && \quad \mathsf{pr}_0\,((a[\nu])^{\sim}\,\rho_{01})\,@\,t_0 = \mathsf{coe}_{\underline{a[\nu]}}\,\rho_{01}\,t_0\\
&\mathsf{coh}_{\underline{[]}} &&: \mathsf{coh}_{\underline{a}[\nu]}\,\rho_{01}\,t_0 \overset{\mathsf{irr}}{=} \mathsf{coh}_{\underline{a[\nu]}}\,\rho_{01}\,t_0\\
&|\pi[]| &&: |(\pi(x:A).b)[\nu]| = (\pi(x:|A|).|b|)[|\nu|] \overset{\pi[]}{=}\\
& && \quad \pi(x:|A|[|\nu|]).|b|[|\nu|] = |\pi(x:A[\nu]).b[\nu]|\\
&\pi[]^{\sim} &&: ((\pi(x:A).b)[\nu])^{\sim}\,\rho_{01} \overset{\mathsf{irr}}{=} (\pi(x:A[\nu]).b[\nu])^{\sim}\,\rho_{01}\\
&|\sigma[]| &&: |(\sigma(x:a).b)[\nu]| = (\sigma(x:|a|).|b|)[|\nu|] \overset{\sigma[]}{=}\\
& && \quad \sigma(x:|a|[|\nu|]).|b|[|\nu|] = |\sigma(x:a[\nu]).b[\nu]|\\
&\sigma[]^{\sim} &&: ((\sigma(x:a).b)[\nu])^{\sim}\,\rho_{01} \overset{\mathsf{irr}}{=} (\sigma(x:a[\nu]).b[\nu])^{\sim}\,\rho_{01}\\
&|\top[]| &&: |\top[\nu]| = |\top|[|\nu|] \overset{\top[]}{=} \top = |\top|\\
&\top[]^{\sim} &&: (\top[\nu])^{\sim}\,\rho_{01} = \top^{\sim}\,(\nu^{\sim}\,\rho_{01}) = (\lambda x.x, \lambda x.x) = \top^{\sim}\,\rho_{01}\\
&|\bot[]| &&: |\bot[\nu]| = |\bot|[|\nu|] \overset{\bot[]}{=} \bot = |\bot|\\
&\bot[]^{\sim} &&: (\bot[\nu])^{\sim}\,\rho_{01} = \bot^{\sim}\,(\nu^{\sim}\,\rho_{01}) = (\lambda x.x\,,\lambda x.x) = \bot^{\sim}\,\rho_{01}
\end{aligned}
$$

$|\mathsf{exfalso}[]|$: $|(\mathsf{exfalso}\,t)[\nu]| = (\mathsf{exfalso}\,|t|)[|\nu|] \overset{\mathsf{exfalso}[]}{=} \mathsf{exfalso}\,(|t|[|\nu|]) = |\mathsf{exfalso}\,(t[\nu])|$

$\mathsf{exfalso}[]^{\sim}$: $((\mathsf{exfalso}\,t)[\nu])^{\sim}\,\rho_{01} = (\mathsf{exfalso}\,t)^{\sim}\,(\nu^{\sim}\,\rho_{01}) = \mathsf{exfalso}\,(|t|[|\nu|\circ\rho_0]) = \mathsf{exfalso}\,(|t[\nu]|[\rho_0]) = (\mathsf{exfalso}\,(t[\nu])^{\sim}\,\rho_{01}$

B.3 Extensions

Identity Type

$$|\mathsf{Id}_A\,u\,v| := A^{\sim}\,(\mathsf{R}_\Gamma\,\mathsf{id})\,|u|\,|v|$$

$$(\mathsf{Id}_A\,t_0\,t_1)^{\sim}\,\rho_{01} := \big(\lambda x_{01}.\mathsf{T}^3{}_A\,(\mathsf{S}_A\,(t_0{}^{\sim}\,\rho_{01}))\,x_{01}\,(t_1{}^{\sim}\,\rho_{01}),\ \lambda x_{01}.\mathsf{T}^3{}_A\,(t_0{}^{\sim}\,\rho_{01})\,x_{01}\,(\mathsf{S}_A\,(t_1{}^{\sim}\,\rho_{01}))\big)$$

$$|\mathsf{refl}_u| := \mathsf{R}_A\,|u|$$

$$\mathsf{refl}_t{}^{\sim}\,\rho_{01} := \mathsf{tt}$$

$$|\mathsf{transport}_{x.P}\,e\,t| := \mathsf{coe}_P\,(\mathsf{R}_\Gamma\,\mathsf{id}\,,|e|)\,|t|$$

$$(\mathsf{transport}_{x.P}\,e\,t)^{\sim}\,\rho_{01} := \mathsf{T}^3{}_P\,\big(\mathsf{S}_P\,(\mathsf{coh}_P\,((\mathsf{R}_\Gamma\,\mathsf{Id}\,,|e|)[\rho_0])\,(|t|[\rho_0]))\big)\,(t^{\sim}\,\rho_{01})\,\big(\mathsf{coh}_P\,((\mathsf{R}_\Gamma\,\mathsf{Id}\,,|e|)[\rho_1])\,(|t|[\rho_1])\big)$$

$$|\mathsf{Id}\beta| := \mathsf{S}_{P[x\mapsto|u|]}\,\big(\mathsf{coh}_P\,(\mathsf{R}_\Gamma\,\mathsf{id}\,,\mathsf{R}_A\,|u|)\,|t|\big)$$

$$(\mathsf{Id}\beta\,t)^{\sim}\,\rho_{01} := \mathsf{tt}$$

$$|\mathsf{funext}\,e| := \lambda x_0\,x_1\,x_{01}.\mathsf{T}_B\,(|e| \mathbin{@} x_0)\,(t_1{}^{\sim}\,(\mathsf{R}_\Gamma\,\mathsf{id}) \mathbin{@} x_0 \mathbin{@} x_1 \mathbin{@} x_{01})$$

$$(\mathsf{funext}\,e)^{\sim}\,\rho_{01} := \mathsf{tt}$$

$$|\mathsf{propext}\,t| := |t|$$

$$(\mathsf{propext}\,t)^{\sim}\,\rho_{01} := \mathsf{tt}$$

$|\mathsf{Id}[]|$: $|(\mathsf{Id}_A\,u\,v)[\nu]| = (A^{\sim}\,(\mathsf{R}_\Gamma\,\mathsf{id})\,|u|\,|v|)[|\nu|] \overset{A^{\sim}[]}{=} A^{\sim}\,((\mathsf{R}_\Gamma\,\mathsf{id})[|\nu|])\,(|u|[|\nu|])\,(|v|[|\nu|]) \overset{\mathsf{irr}}{=} A^{\sim}\,(\nu^{\sim}\,(\mathsf{R}_\Theta\,\mathsf{id}))\,(|u|[|\nu|])\,(|v|[|\nu|]) = (A[\nu])^{\sim}\,(\mathsf{R}_\Theta\,\mathsf{id})\,(|u|[|\nu|])\,(|v|[|\nu|]) = |\mathsf{Id}_{A[\nu]}\,(u[\nu])\,(v[\nu])|$

$\mathsf{Id}[]^{\sim}$: $((\mathsf{Id}_A\,u\,v)[\nu])^{\sim}\,\rho_{01} \overset{\mathsf{irr}}{=} (\mathsf{Id}_{A[\nu]}\,(u[\nu])\,(v[\nu]))^{\sim}\,\rho_{01}$

$|\mathsf{transport}[]|$: $|(\mathsf{transport}_{x.P}\,e\,t)[\nu]| = (\mathsf{coe}_P\,(\mathsf{R}_\Gamma\,\mathsf{id}\,,|e|)\,|t|)[|\nu|] \overset{\mathsf{coe}[]_P}{=} \mathsf{coe}_P\,((\mathsf{R}_\Gamma\,\mathsf{id}\,,|e|)[|\nu|])\,(|t|[|\nu|]) \overset{\mathsf{irr}}{=} \mathsf{coe}_P\,(\nu^{\sim}\,(\mathsf{R}_\Theta\,\mathsf{id})\,,|e|[|\nu|])\,(|t|[|\nu|]) = \mathsf{coe}_{P[\nu]}\,(\mathsf{R}_\Theta\,\mathsf{id}\,,|e|[|\nu|])\,(|t|[|\nu|]) = |\mathsf{transport}_{x.P[\nu]}\,(e[\nu])\,(t[\nu])|$

$\mathsf{transport}[]^{\sim}$: $((\mathsf{transport}_{x.P}\,e\,t)[\nu])^{\sim}\,\rho_{01}$

Definitional Computation Rule.

$$\begin{aligned}
&\mathsf{coeR}_{A[\delta]}\,\rho\,t && : \mathsf{coe}_{A[\delta]}\,(\mathsf{R}_\Gamma\,\rho)\,t = \mathsf{coe}_A\,(\delta^\sim\,(\mathsf{R}_\Gamma\,\rho))\,t \overset{\mathsf{irr}}{=} \\
& && \quad \mathsf{coe}_A\,(\mathsf{R}_\Delta\,(|\delta| \circ \rho))\,t \overset{\mathsf{coeR}_A\,(|\delta|\circ\rho)\,t}{=} t \\
&\mathsf{coeR}_{\Pi(x\,:\,A).B}\,\rho\,t && : \\
& \quad \mathsf{coe}_{\Pi(x\,:\,A).B}\,(\mathsf{R}_\Gamma\,\rho)\,t = \\
& \quad \big(\lambda x.\mathsf{coe}_B\,(\mathsf{R}_\Gamma\,\rho\,,\mathsf{coh}^*{}_A\,(\mathsf{R}_\Gamma\,\rho)\,x)\,(\mathsf{pr}_0\,t\,@\,\mathsf{coe}^*{}_A\,(\mathsf{R}_\Gamma\,\rho)\,x)\,,\dots\big) = \\
& \quad \big(\lambda x.\mathsf{coe}_B\,(\mathsf{R}_\Gamma\,\rho\,,\mathsf{S}_A\,(\mathsf{coh}_A\,(\mathsf{S}_\Gamma\,(\mathsf{R}_\Gamma\,\rho))\,x))\,(\mathsf{pr}_0\,t\,@\,\mathsf{coe}_A\,(\mathsf{S}_\Gamma\,(\mathsf{R}_\Gamma\,\rho))\,x)\,,\dots\big) \overset{\mathsf{irr}}{=} \\
& \quad \big(\lambda x.\mathsf{coe}_B\,(\mathsf{R}_\Gamma\,\rho\,,\mathsf{S}_A\,(\mathsf{coh}_A\,(\mathsf{R}_\Gamma\,\rho)\,x))\,(\mathsf{pr}_0\,t\,@\,\mathsf{coe}_A\,(\mathsf{R}_\Gamma\,\rho)\,x)\,,\dots\big) \overset{\mathsf{coeR}_A\,\rho\,x}{=} \\
& \quad \big(\lambda x.\mathsf{coe}_B\,(\mathsf{R}_\Gamma\,\rho\,,\mathsf{S}_A\,(\mathsf{coh}_A\,(\mathsf{R}_\Gamma\,\rho)\,x))\,(\mathsf{pr}_0\,t\,@\,x)\,,\dots\big) \overset{\mathsf{irr}}{=} \\
& \quad \big(\lambda x.\mathsf{coe}_B\,(\mathsf{R}_{\Gamma\,,\,x\,:\,A}\,(\rho\,,x \mapsto x))\,(\mathsf{pr}_0\,t\,@\,x)\,,\dots\big) \overset{\mathsf{coeR}_B\,(\rho\,,\,x\mapsto x)\,(\mathsf{pr}_0\,t\,@\,x)}{=} \\
& \quad \big(\lambda x.\mathsf{pr}_0\,t\,@\,x\,,\dots\big) \overset{\Pi\eta}{=} (\mathsf{pr}_0\,t\,,\dots) \overset{\mathsf{irr}}{=} (\mathsf{pr}_0\,t\,,\dots) = t \\
&\mathsf{coeR}_{\Sigma(x\,:\,A).B}\,\rho\,(u\,,v) && : \mathsf{coe}_{\Sigma(x\,:\,A).B}\,(\mathsf{R}_\Gamma\,\rho)\,(u\,,v) = \\
& && \quad (\mathsf{coe}_A\,(\mathsf{R}_\Gamma\,\rho)\,u\,,\mathsf{coe}_B\,(\mathsf{R}_\Gamma\,\rho\,,\mathsf{coh}_A\,(\mathsf{R}_\Gamma\,\rho)\,u)\,v) \overset{\mathsf{coeR}_A\,\rho\,u}{=} \\
& && \quad (u\,,\mathsf{coe}_B\,(\mathsf{R}_\Gamma\,\rho\,,\mathsf{coh}_A\,(\mathsf{R}_\Gamma\,\rho)\,u)\,v) \overset{\mathsf{irr}}{=} \\
& && \quad (u\,,\mathsf{coe}_B\,(\mathsf{R}_{\Gamma\,,\,x\,:\,A}\,(\rho\,,x\mapsto u))\,v) \overset{\mathsf{coeR}_B\,(\rho\,,\,x\mapsto u)\,v}{=} (u\,,v) \\
&\mathsf{coeR}_{\mathsf{Bool}}\,\rho\,t && : \mathsf{coe}_{\mathsf{Bool}}\,(\mathsf{R}_\Gamma\,\rho)\,t = t \\
&\mathsf{coeR}_{\mathsf{Prop}}\,\rho\,a && : \mathsf{coe}_{\mathsf{Prop}}\,(\mathsf{R}_\Gamma\,\rho)\,a = a \\
&\mathsf{coeR}_{\underline{a}}\,\rho\,t && : \mathsf{coe}_{\underline{a}}\,(\mathsf{R}_\Gamma\,\rho)\,t = \mathsf{pr}_0\,(a^\sim\,(\mathsf{R}_\Gamma\,\rho))\,@\,t \overset{\mathsf{irr}_{a[\rho]}}{=} t
\end{aligned}$$

C Justification of the Rules of Setoid Type Theory

The setoid model justifies all the extra rules of setoid type theory. As all the new syntactic components are terms, we have to implement the $|-|$ and the $-^\sim$ operations for terms as specified in Sect. 4.2. Most components are modelled by their black counterparts.

$$\begin{aligned}
&|\Gamma^\sim\,\rho_0\,\rho_1| && := \Gamma^\sim\,|\rho_0|\,|\rho_1| \\
&(\Gamma^\sim\,\rho_0\,\rho_1)^\sim\,\tau_{01} && := \big(\lambda\rho_{01}.\mathsf{T}^3{}_\Gamma\,(\mathsf{S}_\Gamma\,(\rho_0{}^\sim\,\tau_{01}))\,\rho_{01}\,(\rho_1{}^\sim\,\tau_{01})\,, \\
& && \qquad \lambda\rho_{01}.\mathsf{T}^3{}_\Gamma\,(\rho_0^\sim\,\tau_{01})\,\rho_{01}\,(\mathsf{S}_\Gamma\,(\rho_1^\sim\,\tau_{01}))\big) \\
&|\mathsf{R}_\Gamma\,\rho| && := \mathsf{R}_\Gamma\,|\rho| \\
&(\mathsf{R}_\Gamma\,\rho)^\sim\,\tau_{01} && := \mathsf{tt} \\
&|\mathsf{S}_\Gamma\,\rho_{01}| && := \mathsf{S}_\Gamma\,|\rho_{01}| \\
&(\mathsf{S}_\Gamma\,\rho_{01})^\sim\,\tau_{01} && := \mathsf{tt}
\end{aligned}$$

$$
\begin{aligned}
&|\mathsf{T}_\Gamma\,\rho_{01}\,\rho_{12}| &&:= \mathsf{T}_\Gamma\,|\rho_{01}|\,|\rho_{12}|\\
&(\mathsf{T}_\Gamma\,\rho_{01}\,\rho_{12})^\sim\,\gamma_{01} &&:= \mathsf{tt}\\
&|A^\sim\,\rho_{01}\,t_0\,t_1| &&:= A^\sim\,|\rho_{01}|\,|t_0|\,|t_1|\\
&(A^\sim\,\rho_{01}\,t_0\,t_1)^\sim\,\tau_{01} &&:= \big(\lambda t_{01}.\mathsf{T}^3{}_A\,(\mathsf{S}_A\,(t_0{}^\sim\,\tau_{01}))\,t_{01}\,(t_1{}^\sim\,\tau_{01})\,,\\
& &&\qquad \lambda t_{01}.\mathsf{T}^3{}_A\,(t_0{}^\sim\,\tau_{01})\,t_{01}\,(\mathsf{S}_A\,(t_1{}^\sim\,\tau_{01}))\big)\\
&|\mathsf{R}_A\,t| &&:= \mathsf{R}_A\,|t|\\
&(\mathsf{R}_A\,t)^\sim\,\tau_{01} &&:= \mathsf{tt}\\
&|\mathsf{S}_A\,t_{01}| &&:= \mathsf{S}_A\,|t_{01}|\\
&(\mathsf{S}_A\,t_{01})^\sim\,\tau_{01} &&:= \mathsf{tt}\\
&|\mathsf{T}_A\,t_{01}\,t_{12}| &&:= \mathsf{T}_A\,|t_{01}|\,|t_{12}|\\
&(\mathsf{T}_A\,t_{01}\,t_{12})^\sim\,\tau_{01} &&:= \mathsf{tt}\\
&|\mathsf{coe}_A\,\rho_{01}\,t_0| &&:= \mathsf{coe}_A\,|\rho_{01}|\,|t_0|\\
&(\mathsf{coe}_A\,\rho_{01}\,t_0)^\sim\,\tau_{01} &&:= \mathsf{T}^3{}_A\,\big(\mathsf{S}_A\,(\mathsf{coh}_A\,(|\rho_{01}|\,\tau_0)\,(|t_0|\,\tau_0))\big)\,(t_0{}^\sim\,\tau_{01})\\
& &&\qquad \big(\mathsf{coh}_A\,(|\rho_{01}|\,\tau_1)\,(|t_0|\,\tau_1)\big)\\
&|\mathsf{coh}_A\,\rho_{01}\,t_0| &&:= \mathsf{coh}_A\,|\rho_{01}|\,|t_0|\\
&(\mathsf{coh}_A\,\rho_{01}\,t_0)^\sim\,\tau_{01} &&:= \mathsf{tt}\\
&|\delta^\sim\,\rho_{01}| &&:= \delta^\sim\,|\rho_{01}|\\
&(\delta^\sim\,\rho_{01})^\sim\,\tau_{01} &&:= \mathsf{tt}\\
&|t^\sim\,\rho_{01}| &&:= t^\sim\,|\rho_{01}|\\
&(t^\sim\,\rho_{01})^\sim\,\tau_{01} &&:= \mathsf{tt}
\end{aligned}
$$

All the equalities are justified. Here we only list how the $|-|$ part of the translation justifies the equalities, $-^\sim$ justifies everything automatically by irr, as all the new syntax are terms and $-^\sim$ on a term returns a proof of a proposition.

- $|\cdot^\sim\,\epsilon\,\epsilon| = \cdot^\sim\,|\epsilon|\,|\epsilon| = \top = |\top|$
- $|(\Gamma\,,x\!:\!A)^\sim\,(\rho_0\,,x\mapsto t_0)\,(\rho_1\,,x\mapsto t_1)| =$
  $(\Gamma\,,x\!:\!A)^\sim\,|(\rho_0\,,x\mapsto t_0)|\,|(\rho_1\,,x\mapsto t_1)| =$
  $(\Gamma\,,x\!:\!A)^\sim\,(|\rho_0|\,,x\mapsto|t_0|)\,(|\rho_1|\,,x\mapsto|t_1|) =$
  $\sigma(\rho_{01}\!:\!\Gamma^\sim\,|\rho_0|\,|\rho_1|).A^\sim\,\rho_{01}\,|t_0|\,|t_1| =$
  $|\sigma(\rho_{01}\!:\!\Gamma^\sim\,\rho_0\,\rho_1).A^\sim\,\rho_{01}\,t_0\,t_1|$
- $|(A[\delta])^\sim\,\rho_{01}\,t_0\,t_1| = (A[\delta])^\sim\,|\rho_{01}|\,|t_0|\,|t_1| = A^\sim\,(\delta^\sim\,|\rho_{01}|)\,|t_0|\,|t_1| =$
  $|A^\sim\,(\delta^\sim\,\rho_{01})\,t_0\,t_1|$
- $|(\Pi(x\!:\!A).B)^\sim\,\rho_{01}\,t_0\,t_1| = (\Pi(x\!:\!A).B)^\sim\,|\rho_{01}|\,|t_0|\,|t_1| =$
  $\pi(x_0\!:\!|A|[|\rho_0|]).\pi(x_1\!:\!|A|[|\rho_1|]).\pi(x_{01}\!:\!\underline{A^\sim\,|\rho_{01}|\,x_0\,x_1}).$
  $B^\sim\,(|\rho_{01}|\,,x_{01})\,(\mathsf{pr}_0\,|t_0|\,@\,x_0)\,(\mathsf{pr}_0\,|t_1|\,@\,x_1) =$
  $\pi(x_0\!:\!|A|[|\rho_0|]).\pi(x_1\!:\!|A|[|\rho_1|]).\pi(x_{01}\!:\!\underline{A^\sim\,|\rho_{01}|\,x_0\,x_1}).$
  $B^\sim\,(|\rho_{01}|\,,x_{01})\,|t_0\,@\,x_0|\,|t_1\,@\,x_1| =$

$|\pi(x_0 : A[\rho_0]).\pi(x_1 : A[\rho_1]).\pi(x_{01} : \underline{A^{\sim}\,\rho_{01}\,x_0\,x_1}).$
$B^{\sim}\,(\rho_{01}\,,x_{01})\,(t_0 \,@\, x_0)\,(t_1 \,@\, x_1)|$

- $|(\Sigma(x : A).B)^{\sim}\,\rho_{01}\,(u_0\,,v_0)\,(u_1\,,v_1)| =$
  $(\Sigma(x : A).B)^{\sim}\,|\rho_{01}|\,(|u_0|\,,|v_0|)\,(|u_1|\,,|v_1|) =$
  $\sigma(x_{01} : A^{\sim}\,|\rho_{01}|\,|u_0|\,|u_1|).B^{\sim}\,(|\rho_{01}|\,,x_{01})\,|v_0|\,|v_1| =$
  $|\sigma(u_{01} : A^{\sim}\,\rho_{01}\,u_0\,u_1).B^{\sim}\,(\rho_{01}\,,u_{01})\,v_0\,v_1|$
- $|\mathsf{Bool}^{\sim}\,\rho_{01}\,t_0\,t_1| = \mathsf{Bool}^{\sim}\,|\rho_{01}|\,|t_0|\,|t_1| =$
  $\mathsf{if}\ |t_0|\ \mathsf{then}\ (\mathsf{if}\ |t_1|\ \mathsf{then}\ \top\ \mathsf{else}\ \bot)\ \mathsf{else}\ (\mathsf{if}\ |t_1|\ \mathsf{then}\ \bot\ \mathsf{else}\ \top) =$
  $|\mathsf{if}\ t_0\ \mathsf{then}\ (\mathsf{if}\ t_1\ \mathsf{then}\ \top\ \mathsf{else}\ \bot)\ \mathsf{else}\ (\mathsf{if}\ t_1\ \mathsf{then}\ \bot\ \mathsf{else}\ \top)|$
- $|\mathsf{Prop}^{\sim}\,\rho_{01}\,a_0\,a_1| = \mathsf{Prop}^{\sim}\,|\rho_{01}|\,|a_0|\,|a_1| = (\underline{|a_0|} \Rightarrow |a_1|) \times (\underline{|a_1|} \Rightarrow |a_0|) =$
  $|(\underline{a_0} \Rightarrow a_1) \times (\underline{a_1} \Rightarrow a_0)|$
- $|\underline{a}^{\sim}\,\rho_{01}\,t_0\,t_1| = \underline{a}^{\sim}\,|\rho_{01}|\,|t_0|\,|t_1| = \top = |\top|$
- $|\mathsf{coe}_{A[\delta]}\,\rho_{01}\,t_0| = \mathsf{coe}_{A[\delta]}\,|\rho_{01}|\,|t_0| = \mathsf{coe}_A\,(\delta^{\sim}\,|\rho_{01}|)\,|t_0| = |\mathsf{coe}_A\,(\delta^{\sim}\,\rho_{01})\,t_0|$
- $|\mathsf{coe}_{\Pi(x : A).B}\,\rho_{01}\,t_0| = \mathsf{coe}_{\Pi(x : A).B}\,|\rho_{01}|\,|t_0| =$
  $\big(\lambda x_1.\mathsf{coe}_B\,(|\rho_{01}|\,,\mathsf{coh}^*{}_A\,|\rho_{01}|\,x_1)\,(\mathsf{pr}_0\,|t_0| \,@\, \mathsf{coe}^*{}_A\,|\rho_{01}|\,x_1)\,,\dots\big) =$
  $\big(\lambda x_1.\mathsf{coe}_B\,(|\rho_{01}|\,,\mathsf{coh}^*{}_A\,|\rho_{01}|\,x_1)\,(|t_0 \,@\, \mathsf{coe}^*{}_A\,\rho_{01}\,x_1|)\,,\dots\big) =$
  $|\lambda x_1.\mathsf{coe}_B\,(\rho_{01}\,,\mathsf{coh}^*{}_A\,\rho_{01}\,x_1)\,(t_0 \,@\, \mathsf{coe}^*{}_A\,\rho_{01}\,x_1)|$
- $|\mathsf{coe}_{\Sigma(x : A).B}\,\rho_{01}\,(u_0\,,v_0)| = \mathsf{coe}_{\Sigma(x : A).B}\,|\rho_{01}|\,(|u_0|\,,|v_0|) =$
  $\big(\mathsf{coe}_A\,|\rho_{01}|\,|u_0|\,,\mathsf{coe}_B\,(|\rho_{01}|\,,\mathsf{coh}_A\,|\rho_{01}|\,|u_0|)\,|v_0|\big) =$
  $\big|\big(\mathsf{coe}_A\,\rho_{01}\,u_0\,,\mathsf{coe}_B\,(\rho_{01}\,,\mathsf{coh}_A\,\rho_{01}\,u_0)\,v_0\big)\big|$
- $|\mathsf{coe}_{\mathsf{Bool}}\,\rho_{01}\,t_0| = \mathsf{coe}_{\mathsf{Bool}}\,|\rho_{01}|\,|t_0| = |t_0|$
- $|\mathsf{coe}_{\mathsf{Prop}_i}\,\rho_{01}\,a_0| = \mathsf{coe}_{\mathsf{Prop}_i}\,|\rho_{01}|\,|a_0| = |a_0|$
- $|\mathsf{coe}_{\underline{a}}\,\rho_{01}\,t_0| = \mathsf{coe}_{\underline{a}}\,|\rho_{01}|\,|t_0| = \mathsf{pr}_0\,(a^{\sim}\,|\rho_{01}|) \,@\, |t_0| = |\mathsf{pr}_0\,(a^{\sim}\,\rho_{01}) \,@\, t_0|$
- $|(\Gamma^{\sim}\,\rho_0\,\rho_1)[\nu]| = (\Gamma^{\sim}\,|\rho_0|\,|\rho_1|)[|\nu|] \overset{\Gamma^{\sim}[]}{=} \Gamma^{\sim}\,(|\rho_0| \circ |\nu|)\,(|\rho_1| \circ |\nu|) =$
  $|\Gamma^{\sim}\,(\rho_0 \circ \nu)\,(\rho_1 \circ \nu)|$
- $|(A^{\sim}\,\rho_{01}\,t_0\,t_1)[\nu]| = (A^{\sim}\,|\rho_{01}|\,|t_0|\,|t_1|)[|\nu|] \overset{A^{\sim}[]}{=}$
  $A^{\sim}\,(|\rho_{01}|[|\nu|])\,(|t_0|[|\nu|])\,(|t_1|[|\nu|]) = |A^{\sim}\,(\rho_{01}[\nu])\,(t_0[\nu])\,(t_1[\nu])|$
- $|(\mathsf{coe}_A\,\rho_{01}\,t_0)[\nu]| = (\mathsf{coe}_A\,|\rho_{01}|\,|t_0|)[|\nu|] \overset{\mathsf{coe}[]_A}{=} \mathsf{coe}_A\,(|\rho_{01}|[|\nu|])\,(|t_0|[|\nu|]) =$
  $|\mathsf{coe}_A\,(\rho_{01}[\nu])\,(t_0[\nu])|$
- $|\mathsf{coe}_A\,(\mathsf{R}_\Gamma\,\rho)\,t| = \mathsf{coe}_A\,(\mathsf{R}_\Gamma\,|\rho|)\,|t| \overset{\mathsf{coeR}_A}{=} |t|$

References

1. Altenkirch, T.: Extensional equality in intensional type theory. In: 14th Symposium on Logic in Computer Science, pp. 412–420 (1999)
2. Altenkirch, T., Kaposi, A.: Type theory in type theory using quotient inductive types. In: Bodik, R., Majumdar, R. (eds.) Proceedings of POPL 2016, St. Petersburg, FL, USA, January 2016, pp. 18–29. ACM (2016). https://doi.org/10.1145/2837614.2837638
3. Altenkirch, T., Kaposi, A.: Towards a cubical type theory without an interval. In: Uustalu, T. (ed.) TYPES 2015. Leibniz International Proceedings in Informatics (LIPIcs), vol. 69, pp. 3:1–3:27. Schloss Dagstuhl-Leibniz-Zentrum fuer Informatik, Dagstuhl, Germany (2018). https://doi.org/10.4230/LIPIcs.TYPES.2015.3
4. Altenkirch, T., Mcbride, C., Swierstra, W.: Observational equality, now! In: PLPV 2007: Proceedings of the 2007 Workshop on Programming Languages Meets Program Verification, pp. 57–58. ACM (2007)
5. Anand, A., Boulier, S., Cohen, C., Sozeau, M., Tabareau, N.: Towards certified meta-programming with typed TEMPLATE-COQ. In: Avigad, J., Mahboubi, A. (eds.) ITP 2018. LNCS, vol. 10895, pp. 20–39. Springer, Cham (2018). https://doi.org/10.1007/978-3-319-94821-8_2
6. Atkey, R., Ghani, N., Johann, P.: A relationally parametric model of dependent type theory. In: Jagannathan, S., Sewell, P. (eds.) Proceedings of POPL 2014, San Diego, CA, USA, 20–21 January 2014, pp. 503–516. ACM (2014)
7. Bernardy, J.P., Jansson, P., Paterson, R.: Proofs for free – parametricity for dependent types. J. Funct. Program. **22**(02), 107–152 (2012). https://doi.org/10.1017/S0956796812000056
8. Bezem, M., Coquand, T., Huber, S.: A model of type theory in cubical sets. In: 19th International Conference on Types for Proofs and Programs (TYPES 2013), vol. 26, pp. 107–128 (2014)
9. Boulier, S.: Extending type theory with syntactic models. Ph.D. thesis, École des Mines de Nantes (2018)
10. Boulier, S., Pédrot, P.M., Tabareau, N.: The next 700 syntactical models of type theory. In: Proceedings of the 6th ACM SIGPLAN Conference on Certified Programs and Proofs, CPP 2017, pp. 182–194. ACM, New York (2017). https://doi.org/10.1145/3018610.3018620
11. Cartmell, J.: Generalised algebraic theories and contextual categories. Ann. Pure Appl. Logic **32**, 209–243 (1986)
12. Cohen, C., Coquand, T., Huber, S., Mörtberg, A.: Cubical type theory: a constructive interpretation of the univalence axiom. In: Uustalu, T. (ed.) TYPES 2015. Leibniz International Proceedings in Informatics (LIPIcs), vol. 69, pp. 5:1–5:34. Schloss Dagstuhl-Leibniz-Zentrum fuer Informatik, Dagstuhl, Germany (2018)
13. Dybjer, P.: Internal type theory. In: Berardi, S., Coppo, M. (eds.) TYPES 1995. LNCS, vol. 1158, pp. 120–134. Springer, Heidelberg (1996). https://doi.org/10.1007/3-540-61780-9_66
14. Gilbert, G., Cockx, J., Sozeau, M., Tabareau, N.: Definitional proof-irrelevance without K. In: Proceedings of POPL 2019, Lisbon, Portugal, January 2019
15. Herbelin, H.: Syntactic investigations into cubical type theory. In: Espírito Santo, J., Pinto, L. (eds.) TYPES 2018. University of Minho (2018)
16. Hofmann, M.: Conservativity of equality reflection over intensional type theory. In: TYPES 1995, pp. 153–164 (1995)

17. Hofmann, M.: Extensional concepts in intensional type theory. Thesis, University of Edinburgh, Department of Computer Science (1995)
18. Hofmann, M.: Syntax and semantics of dependent types. In: Semantics and Logics of Computation, pp. 79–130. Cambridge University Press (1997)
19. Hofmann, M., Streicher, T.: The groupoid interpretation of type theory. In: In Venice Festschrift, pp. 83–111. Oxford University Press (1996)
20. Jaber, G., Lewertowski, G., Pédrot, P., Sozeau, M., Tabareau, N.: The definitional side of the forcing. In: Grohe, M., Koskinen, E., Shankar, N. (eds.) Proceedings of LICS 2016, New York, NY, USA, 5–8 July 2016, pp. 367–376. ACM (2016). https://doi.org/10.1145/2933575.2935320
21. Kaposi, A.: The setoid model formalized in Agda, May 2019. https://bitbucket.org/akaposi/setoid/src/master/agda
22. Kaposi, A., Kovács, A., Altenkirch, T.: Constructing quotient inductive-inductive types. Proc. ACM Program. Lang. **3**(POPL), 2:1–2:24 (2019). https://doi.org/10.1145/3290315
23. Martin-Löf, P.: Intuitionistic type theory, Studies in Proof Theory, vol. 1. Bibliopolis (1984)
24. Matthieu Sozeau, N.T.: Univalence for free (2013). http://hal.inria.fr/hal-00786589/en
25. McBride, C.: Elimination with a motive. In: Callaghan, P., Luo, Z., McKinna, J., Pollack, R., Pollack, R. (eds.) TYPES 2000. LNCS, vol. 2277, pp. 197–216. Springer, Heidelberg (2002). https://doi.org/10.1007/3-540-45842-5_13
26. Oury, N.: Extensionality in the calculus of constructions. In: Hurd, J., Melham, T. (eds.) TPHOLs 2005. LNCS, vol. 3603, pp. 278–293. Springer, Heidelberg (2005). https://doi.org/10.1007/11541868_18
27. Program, T.U.F.: Homotopy type theory: univalent foundations of mathematics. Technical report, Institute for Advanced Study (2013)
28. Sterling, J., Angiuli, C., Gratzer, D.: Cubical syntax for reflection-free extensional equality. In: Geuvers, H. (ed.) Proceedings of the 4th International Conference on Formal Structures for Computation and Deduction (FSCD 2019), vol. 131 (2019). https://doi.org/10.4230/LIPIcs.FSCD.2019.32
29. Streicher, T.: Investigations into intensional type theory. Habilitation thesis (1993)
30. Tabareau, N., Tanter, É., Sozeau, M.: Equivalences for free. In: ICFP 2018 - International Conference on Functional Programming, Saint-Louis, United States, September 2018. https://doi.org/10.1145/3234615
31. van der Walt, P., Swierstra, W.: Engineering proof by reflection in agda. In: Hinze, R. (ed.) IFL 2012. LNCS, vol. 8241, pp. 157–173. Springer, Heidelberg (2013). https://doi.org/10.1007/978-3-642-41582-1_10
32. Winterhalter, T., Sozeau, M., Tabareau, N.: Using reflection to eliminate reflection. In: Santo, J.E., Pinto, L. (eds.) 24th International Conference on Types for Proofs and Programs, TYPES 2018. University of Minho (2018)

Cylindric Kleene Lattices for Program Construction

Brijesh Dongol[1(✉)], Ian Hayes[2], Larissa Meinicke[2], and Georg Struth[3]

[1] University of Surrey, Guildford, UK
b.dongol@surrey.ac.uk
[2] University of Queensland, Brisbane, Australia
[3] University of Sheffield, Sheffield, UK

Abstract. Cylindric algebras have been developed as an algebraisation of equational first order logic. We adapt them to cylindric Kleene lattices and their variants and present relational and relational fault models for these. This allows us to encode frames and local variable blocks, and to derive Morgan's refinement calculus as well as an algebraic Hoare logic for while programs with assignment laws. Our approach thus opens the door for algebraic calculations with program and logical variables instead of domain-specific reasoning over concrete models of the program store. A refinement proof for a small program is presented as an example.

1 Introduction

Kleene algebras and similar formalisms have found their place in program construction and verification. Kleene algebras with tests [19] have been used for calculating complex program equivalences; the rules of propositional Hoare logic—Hoare logic without assignments laws—can be derived from their axioms [20]. Demonic refinement algebras [29] have been applied to non-trivial program transformations in the refinement calculus [3]. Modal Kleene algebras [7,8] have been linked with predicate transformer semantics and found applications in program correctness. More recently, links between Kleene algebras and Morgan-style refinement calculi [23] have been established; program construction and verification components based on Kleene algebras have been formalised in proof assistants such as Coq [9,25] or Isabelle/HOL [2,15,28].

The Isabelle components are based on shallow embeddings of while programs, Hoare logic and refinement calculi. Programs, assertions and correctness specifications are modelled as semantic objects directly within Isabelle's higher-order logic. Explicit data types for the syntax of programs, assertions or logics of programs, and explicit semantic maps for their interpretation can thus be avoided. Kleene algebras, as abstract semantics for while programs, propositional Hoare logics or propositional refinement calculi, fit very naturally into this approach. Yet assignments and their laws are currently formalised in concrete program

Dongol and Struth are supported by EPSRC Grant EP/R032556/2; Hayes, Meinicke and Dongol are supported by ARC Discovery Grant DP190102142.

G. Hutton (Ed.): MPC 2019, LNCS 11825, pp. 197–225, 2019.
https://doi.org/10.1007/978-3-030-33636-3_8

store semantics that form models of the algebras. With a shallow embedding, program construction and verification is thus performed in these concrete semantics. Other familiar features of refinement calculi, such as variable frames or local variable blocks, cannot be expressed in Kleene algebras either. How algebra could handle such important features remains open.

Yet algebra *can* deal with bindings, scopes and variables. Nominal Kleene algebras [13] can model the first two features, and cylindric algebras of Henkin, Monk and Tarski [17] the third, albeit in the setting of boolean algebras, where notions of variables and quantification are added in an algebratisation of first-order equational logic. They introduce a family of cylindrification operators $c_\kappa\, x$ that abstract existential quantification $\exists_\kappa x$ of first-order formulas.

Henkin, Monk and Tarski give a standard interpretation of c_κ in *cylindric set algebras* [17, p. 166]. In this setting, cylindrification is defined over $\mathcal{P}\, X^\alpha$ for some set X and ordinal α.[1] Elements of a cylindric set algebra are therefore functions of type $\alpha \to X$, or sequences $x = (x_0, x_1, \dots)$ of "length" α. In logic, if α is a set of logical variables and X the carrier set of a structure, these correspond to valuations. Geometrically, X^α corresponds to an α-dimensional Cartesian space with base X where x_κ represents the κth coordinate. Apart from the usual boolean operations on sets, cylindric set algebras use a family of cylindrification operators $C^c_\kappa : \mathcal{P}\, X^\alpha \to \mathcal{P}\, X^\alpha$ for $\kappa < \alpha$—the superscript c stands for "classical". For each $A \subseteq X^\alpha$,

$$C^c_\kappa A = \{y \in X^\alpha \mid \exists x \in A.\ x \approx_\kappa y\},$$

where $x \approx_\kappa y$ if x and y are equal, except at κ, (i.e. $\forall \lambda \neq \kappa.\ x_\lambda = y_\lambda$). Geometrically, $C^c_\kappa A$ thus translates A along the κ-axis and constructs a cylinder in some hyperspace.

Our main idea is to generalise cylindrification from boolean algebras to Kleene lattices (thus foregoing the complement operator of boolean algebra, while adding a monoidal composition and a star). We explain it through *relational cylindrification* C_κ, which acts on programs modelled by relations in $\mathcal{P}\,(X^\alpha \times X^\alpha)$, where X^α represents program stores as functions from variables in α to values in X. Cylindrifying relation R in variable κ by $C_\kappa\, R$ means adding any combination of values for κ to elements of R. We therefore say that C_κ *liberates* variable κ in program R,

$$C_\kappa R = \{(a, b) \in X^\alpha \times X^\alpha \mid \exists (c, d) \in R.\ a \approx_\kappa c \wedge b \approx_\kappa d\}.$$

Note that κ is liberated (can take on any value) independently in both the first and second coordinates of R.[2]

[1] In applying Henkin, Monk and Tarski's work to program algebra we do not rely much on the use of ordinals; sets usually suffice.

[2] Expressing the relation as a predicate in the Z style [16,27], i.e. representing before values of a variable by x and after values by x', relational cylindrification corresponds to the predicate $\exists_{x,x'} R$.

The cylindrification of the identity relation, $C_\kappa \, Id_{X^\alpha}$, in particular, liberates variable κ while constraining all other variables to satisfy the identity relation on X^α. Henkin, Monk and Tarski [17, §1.7] have generalised cylindrification to finite sets of variables so that $c_{(\{\kappa_0,\dots,\kappa_{n-1}\})}x = c_{\kappa_0} \dots c_{\kappa_{n-1}} x$, where the parentheses on the left are part of their syntax. For a set of variables Γ, a program R may be restricted to only change variables in Γ by conjoining it with $C_{(\Gamma)} Id_{X^\alpha}$, i.e. $R \cap C_{(\Gamma)} Id_{X^\alpha}$, which we abbreviate to $\Gamma : x$ to match the syntax of frames in Morgan's refinement calculus [24]. A local variable κ with a scope over some program R is obtained by first liberating the local κ over the program and then constraining any non-local κ to not change, i.e. $(C_\kappa R) \cap C_{(\overline{\{\kappa\}})} Id_{X^\alpha}$, which we abbreviate as $(\mathbf{var}\, \kappa . R)$. Finally, assignment statements are encoded by framed specification statements, where tests are used to abstract from expressions, and variable substitutions are handled using another concept from cylindric algebras, namely diagonal elements.

Our main contribution lies in the formal development of this new extension and application of cylindrification. This opens the door to algebraic calculations with variables in imperative programs where set-theoretic reasoning in concrete store semantics is so far required. Our technical contributions are as follows.

- We extend Kleene algebras (Sect. 2) to cylindric Kleene lattices (Sect. 4), explore their basic properties and prove their soundness with respect to a relational (fault) semantics for imperative programs (Sects. 3 and 5).
- Generalised cylindrification liberates a set of variables, rather than a single variable (Sect. 6). It is used to show that the frames of Morgan's refinement calculus (Sect. 8) and local variable blocks (Sect. 9) can be expressed in cylindric Kleene lattices. Based on these encodings we derive the laws of Morgan's refinement calculus with frames and those of Hoare logic (Sect. 7), both with assignment laws.
- Synchronous cylindrification (Sect. 10) supports the cylindrification of tests in the relational model. It is used in combination with diagonal elements (representing equality in equational logic) to define substitutions algebraically (Sect. 11). These are then used to define variable assignments (Sect. 12).
- We explain how simple refinement proofs can be performed in our framework by purely algebraic and symbolic reasoning.
- We propose liberation Kleene lattices (Sect. 13) as a conceptually simpler and more fine-grained variant, and prove that the axioms of cylindric Kleene lattices are derivable from those of liberation Kleene lattices.

Many of our results have been verified with Isabelle/HOL, but verification and refinement components based on cylindric Kleene algebras remain work in progress. All Isabelle proofs are accessible online[3].

Overall, many of the concepts needed for our development could be readily adapted from cylindric algebra. Henkin, Monk and Tarski's textbook [17] has been a surprising source of insights from a seemingly unrelated area. We follow their notational conventions closely.

[3] https://github.com/gstruth/liberation.

2 L-Monoids and Kleene Lattices

This section briefly recalls the basic algebraic structures used in this article. Cylindric variants are presented in Sect. 4, liberation algebras are introduced in Sect. 13. We work with l-monoids instead of dioids and Kleene lattices instead of Kleene algebras because a meet operation is crucial for defining the concepts we care about: frames, local variables and variable assignments.

Definition 1 (l-monoid). *A* lattice-ordered monoid *(*l-monoid*) [4] is a structure* $(L, +, \cdot, ;, 0, 1)$ *such that* $(L, +, \cdot, 0)$ *is a lattice with join operation* $+$*, meet operation* $\cdot$*, and least element* 0*;* $(L, ;, 1)$ *is a monoid and the distributivity axioms* $x ; (y + z) = x ; y + x ; z$ *and* $(x + y) ; z = x ; y + x ; z$ *and annihilation axioms* $0 ; x = 0$ *and* $x ; 0 = 0$ *hold for all* $x, y, z \in L$*. An l-monoid is* weak *if the axiom* $x ; 0 = 0$ *is absent.*

Definition 2 (Kleene lattice). *A (weak)* Kleene lattice *[1,18] is a (weak) l-monoid,* K*, equipped with a star operation* $^* : K \to K$*, that satisfies the unfold and induction axioms*

$$1 + x ; x^* \leq x^*, \qquad z + x ; y \leq y \Rightarrow x^* ; z \leq y,$$
$$1 + x^* ; x \leq x^*, \qquad z + y ; x \leq y \Rightarrow z ; x^* \leq y.$$

The unfold and induction laws in the first line and those in the second line above are opposites: the order of composition has been swapped.

Forgetting the meet operation in l-monoids yields dioids (i.e., semirings with idempotent addition); forgetting meet in Kleene lattices yields Kleene algebras.

Definition 3 (l-monoid with tests). *A (weak)* l-monoid with tests *is a structure* $(L, B, +, \cdot, ;, 0, 1, \neg)$ *where* $B \subseteq L$*,* $\neg$ *is a partial operation defined on* B *such that* $(B, +, \cdot, 0, 1, \neg)$ *is a boolean algebra in which* $;$ *and* $\cdot$ *coincide and* $(L, +, \cdot, ;, 0, 1)$ *is a (weak) l-monoid. In addition, for all* $p \in B$ *and* $x, y \in K$*,*

$$p ; (x \cdot y) = (p ; x) \cdot (p ; y), \qquad \text{and} \qquad (x \cdot y) ; p = (x ; p) \cdot (y ; p).$$

Definition 4 (Kleene lattice with tests). *A (weak)* Kleene lattice with tests *is a (weak) l-monoid with tests that is also a (weak) Kleene lattice.*

Alternatively, Kleene lattices can be based on the operation $^+ : K \to K$ that satisfies the following unfold and induction axioms

$$x + x ; x^+ = x^+, \qquad z + x ; y \leq y \Rightarrow z + x^+ ; z \leq y,$$

and their opposites $x + x^+ ; x = x^+$ and $z + y ; x \leq y \Rightarrow z + z ; x^+ \leq y$, even when the unit 1 is absent. In the presence of this unit, the identities $x^+ = x ; x^*$ and $x^* = 1 + x^+$ make the two variants interderivable.

3 Relation Kleene Lattices

Before cylindrifying l-monoids and Kleene lattices in the next section, we sketch the relational model and the relational fault model of these algebras. First of all, these form the basis of the standard relational program semantics to which we restrict our attention. Secondly, they are used in the soundness proofs of the cylindric and liberation algebras that we axiomatise. Last, but not least, they provide valuable intuitions for the algebraic development.

A standard model of Kleene algebra with tests is formed by the algebra of binary relations over a set X. In this model, $+$ is interpreted as union, ; as relational composition ($(a,b) \in R \,;\, S \Leftrightarrow \exists c \in X.\ (a,c) \in R \wedge (c,b) \in S$), 0 as $\emptyset$, 1 as the identity relation on X, ($(a,b) \in Id_X \Leftrightarrow a = b$), and * as the reflexive-transitive closure operation ($R^* = \bigcup_{i<\omega} R^i$, for $R^0 = Id_X$ and $R^{i+1} = R \,;\, R^i$). As our basis is a lattice, $\cdot$ is interpreted as intersection. Finally, tests are *subidentities*, that is, elements of $\mathcal{P}Id_X = \{R \subseteq X \times X \mid R \subseteq Id_X\}$. These distribute over infs in both arguments with respect to sequential composition. Test complementation is defined by $Id_X - (\_)$. The test algebra $\mathcal{P}Id_X$ forms a subalgebra of any algebra $\mathcal{P}(X \times X)$ of binary relations—in fact a complete atomic boolean algebra. The following result is therefore routine.

Proposition 1. *Let X be a set. Then $(\mathcal{P}(X \times X), \mathcal{P}Id_X, \cup, \cap, ;, \emptyset, Id_X, -, ^*)$ is a Kleene lattice with tests—the* full relation Kleene lattice with tests *over X.*

Weak Kleene lattices with tests are formed by relations that model faults or nontermination over $X \times X_\perp$, where $X_\perp = X \cup \{\perp\}$ and $\perp \notin X$ is an element that represents a fault or non-termination. We refer to this model as the *relational fault model.* We partition each $R \subseteq X \times X_\perp$ into its *proper part* $R_p \subseteq X \times X$ and its *faulting part* $R_f \subseteq X \times \{\perp\}$, that is, $R = R_p \cup R_f$ and $R_p \cap R_f = \emptyset$. Redefining $R \,;\, S = R_f \cup R_p; S$ then makes faults override compositions, representing R as (R_p, R_f) and S by (S_p, S_f) yields a semidirect product, which is well known in semigroup theory:

$$(R_p, R_f) \,;\, (S_p, S_f) = (R_p \,;\, S_p, R_f \cup R_p \,;\, S_f). \tag{1}$$

With $(R_p, R_f)^0 = (Id_X, \emptyset)$ and $(R_p, R_f)^{i+1} = (R_p, R_f) \,;\, (R_p, R_f)^i$ we define

$$(R_p, R_f)^* = \bigcup_{i<\omega} (R_p, R_f)^i. \tag{2}$$

An inductive argument shows that * satisfies the Kleene algebra axioms and that

$$(R_p, R_f)^* = (R_p^*, R_p^* \,;\, R_f).$$

Proposition 2. *Let X be a set. Then $(\mathcal{P}(X \times X_\perp), \mathcal{P}Id_X, \cup, \cap, ;, \emptyset, Id_X, -, ^*)$, with composition (1) and star (2), forms a weak Kleene lattice with tests—the* full weak relation Kleene lattice with tests *over X.*

The identity of the pair representation with respect to ; is $(Id_X, \emptyset)$; its left zero is $(\emptyset, \emptyset)$. All tests are proper and test complementation is restricted to the proper part. Right annihilation fails because $(R_p, R_f) ; (\emptyset, \emptyset) = (\emptyset, R_f) \neq (\emptyset, \emptyset)$ whenever $R_f \neq \emptyset$. Algebraic proofs for this development can be found in Appendix A; it has been formalised with Isabelle.

Each subalgebra (K, B), with $K \subseteq \mathcal{P}(X \times X)$ and $B \subseteq \mathcal{P}Id_X$, of a full (weak) relation Kleene lattice with tests over X is a *(weak) relation Kleene lattice with tests* over X.

The relation algebras described in this section have of course a much richer structure. Firstly, we ignore the fact that relations have converses and can be complemented, yet this only means that we focus on the programming concepts that matter. Secondly, relational composition preserves sups in both arguments, whereas the redefined composition (1) preserves sups in its first and non-empty sups in its second argument. Non-preservation of empty sups in the second argument is of course due to the absence of right annihilation.

4 Cylindric L-Monoids and Kleene Lattices

This section extends l-monoids and Kleene lattices from Sect. 2 by a family of cylindrification operators. In other words, we generalise the classical cylindric algebras of Henkin, Monk and Tarski [17] from boolean algebras to Kleene algebras. The axiomatisations have been developed, minimised and proved to be independent using Isabelle/HOL. Apart from the axioms, we present some simple algebraic properties, all of which have been verified with Isabelle. The relational models from Sect. 3 are extended to models for cylindric l-monoids and Kleene lattices in Sect. 5. In reading the following definition a suitable intuition is that $c_\kappa x$ represents an abstraction of existential quantification $\exists_\kappa x$.

Definition 5 (cylindric l-monoid). *Let α be an ordinal. A (*weak*) *cylindric l-monoid *(*CLM*) of dimension α is a structure $(L, +, \cdot, ;, 0, 1, c_\kappa)_{\kappa<\alpha}$ such that $(L, +, \cdot, ;, 0, 1)$ is a (weak) l-monoid and each $c_\kappa : L \to L$ satisfies:*

$$c_\kappa\, 0 = 0, \tag{C1}$$

$$x \leq c_\kappa\, x, \tag{C2}$$

$$c_\kappa\,(x \cdot c_\kappa\, y) = c_\kappa\, x \cdot c_\kappa\, y, \tag{C3}$$

$$c_\kappa c_\lambda\, x = c_\lambda c_\kappa\, x, \tag{C4}$$

$$c_\kappa\,(x + y) = c_\kappa\, x + c_\kappa\, y, \tag{C5}$$

$$c_\kappa\,(x \,;\, c_\kappa\, y) = c_\kappa\, x \,;\, c_\kappa\, y, \tag{C6}$$

$$c_\kappa\,(c_\kappa\, x \,;\, y) = c_\kappa\, x \,;\, c_\kappa\, y, \tag{C7}$$

$$\kappa \neq \lambda \;\Rightarrow\; c_\kappa\, 1 \cdot c_\lambda\, 1 = 1, \tag{C8}$$

$$(c_\kappa\, 1 \,;\, c_\lambda\, 1) \cdot (c_\kappa\, 1 \,;\, c_\mu\, 1) = c_\kappa\,(c_\lambda\, 1 \cdot c_\mu\, 1), \tag{C9}$$

$$c_\kappa\,(c_\lambda\, 1) = c_\kappa\, 1 \,;\, c_\lambda\, 1. \tag{C10}$$

Classical cylindric algebra is axiomatised over a boolean algebra instead of a Kleene lattice; a monoidal structure is absent. Cylindric algebras usually consider diagonal elements $d_{\kappa\lambda}$ as well [17]. In this sense CLM is *diagonal free*. CLMs with diagonals are introduced in Sect. 11.

Axioms (C1), (C2), (C3) and (C4) are those of classical cylindric algebra [17, p.162]; (C5) is derivable in that context because it is based on a boolean algebra. The axioms (C6) and (C7) appear in a previous abelian-semiring-based approach to cylindrification by Giacobazzi, Debray and Levi [14]. Axioms (C8)–(C10) are new. In axioms (C1), (C5), (C9) and (C10), = could have been weakened to $\leq$. Isabelle's counterexample generators show that the axioms are independent. We write 1_κ instead of $c_\kappa\, 1$. Intuitively, such elements are identities of ; except for κ. The next lemmas establish basic facts about cylindrification. The properties in the first one are known from classical cylindric algebras.

Lemma 1. *[17, §1.2] In every weak* CLM*,*

1. $c_\kappa\, c_\kappa\, x = c_\kappa\, x$, *(HMT1.2.3)*
2. $c_\kappa\, x = 0 \Leftrightarrow x = 0$, *(HMT1.2.1)*
3. $c_\kappa\, x = x \Leftrightarrow \exists y.\ c_\kappa\, y = x$, *(HMT1.2.4)*
4. $x \cdot c_\kappa\, y = 0 \Leftrightarrow y \cdot c_\kappa\, x = 0$, *(HMT1.2.5)*
5. $x \leq y \Rightarrow c_\kappa\, x \leq c_\kappa\, y$, *(HMT1.2.7)*
6. $x \leq c_\kappa\, y \Leftrightarrow c_\kappa\, x \leq c_\kappa\, y$, *(HMT1.2.9)*
7. $c_\kappa\, x \cdot c_\lambda\, y = 0 \Leftrightarrow c_\lambda\, x \cdot c_\kappa\, y = 0$. *(HMT1.2.15)*

Axiom (C2) and Lemma 1(1) and (5) can be summarised as follows.

Lemma 2. *In every weak* CLM*, c_κ is a closure operator.*

The next lemma collects properties beyond classical cylindrical algebra.

Lemma 3. *In every weak* CLM*,*

1. $c_\kappa\, (x\,; y) \leq c_\kappa\, x\,; c_\kappa\, y$,
2. $1_\kappa\,; x\,; 1_\kappa \leq c_\kappa\, x$,
3. $1 \leq 1_\kappa$,
4. $1_\kappa\,; 0 = 0$,
5. $1_\kappa\,; 1_\kappa = 1_\kappa$,
6. $1_\kappa\,; 1_\lambda = 1_\lambda\,; 1_\kappa$,
7. $c_\kappa\, (1_\lambda \cdot 1_\mu) = 1_\kappa\,; (1_\lambda \cdot 1_\mu)$,
8. $1_\kappa + 1_\lambda \leq 1_\kappa\,; 1_\lambda$.

Lemma 3(2) may be strengthened to an equality in the relational model of CLM, but neither in trace models [21] nor in the algebra; the following lemma gives a counterexample.

Lemma 4. *There is a* CLM *in which $c_\kappa\, x \not\leq 1_\kappa\,; x\,; 1_\kappa$.*

Proof. Consider the CLM with $L = \{0, 1, a\}$, join and meet defined by $0 < a < 1$ and composition by $a; a = a$. It can be checked that $c_\kappa : 0 \mapsto 0, a \mapsto 1, 1 \mapsto 1$ satisfies (C1)–(C7). Yet $c_\kappa\, a = 1 \neq a = 1\,; a\,; a = 1_\kappa\,; a\,; 1_\kappa$. □

In any (weak) CLM L, let

$$L_{c_\kappa} = \{x \in L \mid c_\kappa\, x = x\}$$

denote the set of cylindrified elements in dimension κ. Similarly, we define $L^l_{1_\kappa}$, $L^r_{1_\kappa}$ and L_{1_κ} as the sets of fixpoints of $1_\kappa\,;(\_)$, $(\_)\,;1_\kappa$ and $1_\kappa\,;(\_)\,;1_\kappa$, respectively. Lemma 1(3) implies that L_{c_κ} is equal to the image of L under c_κ. Analogous facts hold for the other three functions.

Proposition 3. *Let L be a (weak)* CLM *and let $\kappa < \alpha$. Then*

1. *$(L^l_{1_\kappa}, +, \cdot, ;, 0, 1_\kappa)$ forms a (weak) sub-l-semigroup of L with left unit 1_κ,*
2. *$(L^r_{1_\kappa}, +, \cdot, ;, 0, 1_\kappa)$ forms a sub-l-semigroup of L with right unit 1_κ and if L is a strong* CLM*, then $L^l_{1_\kappa}$ and $L^r_{1_\kappa}$ are isomorphic,*
3. *$(L_{1_\kappa}, +, \cdot, ;, 0, 1_\kappa)$ forms a sub-l-monoid of $L^l_{1_\kappa}$ and $L^r_{1_\kappa}$,*
4. *$(L_{c_\kappa}, +, \cdot, ;, 0, 1_\kappa)$ forms a (weak) sub-l-monoid of L_{1_κ}.*

Proof.

1. For $L^l_{1_\kappa}$, it is well known that any principal right-ideal of an idempotent in a monoid forms a subsemigroup with the idempotent as left unit. By Lemma 3(5), 1_κ is an idempotent; $L^l_{1_\kappa}$ is the principal right-ideal generated by 1_κ by definition. Closure with respect to sups follows from the dioid axioms in L and idempotence of 1_κ; inf-closure from $1_\kappa\,;(1_\kappa\,;x \cdot 1_\kappa\,;y) = 1_\kappa\,;x \cdot 1_\kappa\,;y$, which has been checked with Isabelle.
2. The proof for $L^r_{1_\kappa}$ follows from that of $L^l_{1_\kappa}$ by opposition, using the dual identity $(x\,;1_\kappa \cdot y\,;1_\kappa)\,;1_\kappa = x\,;1_\kappa \cdot y\,;1_\kappa$ for inf-closure. Right annihilation in $L^r_{1_\kappa}$ follows from Lemma 3(4). In the strong case, the isomorphism is given by opposition.
3. The subalgebra proof for L_{1_κ} follows from (1) and (2). Checking that L_{1_κ} is a subalgebra of both $L^l_{1_\kappa}$ and $L^r_{1_\kappa}$ is straightforward: by idempotence of 1_κ, every fixpoint of L_{1_κ} is a fixpoint of $L^l_{1_\kappa}$ and $L^r_{1_\kappa}$.
4. For L_{c_κ}, closure with respect to $+$, $\cdot$ and ; is immediate from the axioms. Sup-closure, for instance, means checking that $c_\kappa\,(c_\kappa\, x + c_\kappa\, y) = c_\kappa\, x + c_\kappa\, y$. Finally, 1_κ is the unit in the subalgebra because $1_\kappa\,;c_\kappa\, x = c_\kappa\, x = c_\kappa\, x\,;1_\kappa$. This property, which also establishes that L_{c_κ} is a subalgebra of L_{1_κ}, has been confirmed by Isabelle. □

By Lemma 4, the sets of fixpoints of L_{c_κ} and L_{1_κ} need not coincide. Separating the remaining sets of fixpoints with Isabelle's counterexample generators is a simple exercise and need not be expanded.

Definition 6 (cylindric Kleene lattice). *A (*weak*)* cylindric Kleene lattice *(*CKL*) of dimension α is a (weak) cylindric l-monoid of dimension α that is also a (weak) Kleene lattice, and in which*

$$c_\kappa\, x^+ \leq (c_\kappa\, x)^+. \tag{C11}$$

Isabelle's counterexample generators show that 1 need not be in K_{c_κ} for any κ, in particular not in the relational models described in Sect. 5. Together with Proposition 3 this explains why a $+$-axiom appears in CKL, and not a $*$-axiom. Next we list properties of cylindric Kleene lattices.

Lemma 5. *In every weak* CKL*,*

1. $c_\kappa\, x^* \leq 1_\kappa \,;\, (c_\kappa\, x)^*$,
2. $c_\kappa\, (c_\kappa\, x)^+ = (c_\kappa\, x)^+$,
3. $1_\kappa \,;\, (c_\kappa\, x)^+ = (c_\kappa\, x)^+$,
4. $1_\kappa \,;\, (c_\kappa\, x)^+ = (c_\kappa\, x)^+ \,;\, 1_\kappa$,
5. $c_\kappa\, (c_\kappa\, x)^* = 1_\kappa \,;\, (c_\kappa\, x)^*$,
6. $1_\kappa \,;\, (c_\kappa\, x)^* = (c_\kappa\, x)^* \,;\, 1_\kappa$,
7. $(1_\kappa + 1_\lambda)^+ = 1_\kappa \,;\, 1_\lambda = (1_\kappa + 1_\lambda)^*$,
8. $1_\kappa^+ = 1_\kappa = 1_\kappa^*$.

Finally, Proposition 3 extends to CKL.

Proposition 4. $(K_{c_\kappa}, +, \cdot, ;, 0, 1_\kappa, (\_)^+)$ *is a (weak) sub-Kleene lattice of the (weak)* CKL K *for each* $\kappa < \alpha$.

The cases of $K^l_{1_\kappa}$, $K^r_{1_\kappa}$ and K_{1_κ} are analogous. The first two benefit from the fact that $(\_)^+$ can be used to define sub-Kleene lattices $(K^l_{1_\kappa}, +, \cdot, ;, 0, 1_\kappa, (\_)^+)$ and its opposite $(K^r_{1_\kappa}, +, \cdot, ;, 0, 1_\kappa, (\_)^+)$ that do not require 1_κ.

5 Relational Cylindrification

In constructions of cylindric algebras of formulas of predicate logic, sequences in X^α correspond to valuations [17]. They associate variables of first-order formulas with values in their models. In imperative programming languages, functions from variables in α to values in X form the standard model of program stores, and the standard denotational semantics interprets programs as relations between these. Our aim is to model cylindrifications over such relations.

Hence we consider relations $R \subseteq X^\alpha \times X^\alpha$ and *relational cylindrifications* $C_\kappa : \mathcal{P}(X^\alpha \times X^\alpha) \to \mathcal{P}(X^\alpha \times X^\alpha)$ that liberate the value of variable κ in both coordinates of ordered pairs. Formally, we therefore define

$$C_\kappa\, R = \{(a,b) \in X^\alpha \times X^\alpha \mid \exists c, d \in X^\alpha.\ (a,b) \approx_\kappa (c,d) \wedge (c,d) \in R\},$$

where $\approx_\kappa$ has been extended pointwise to an equivalence on pairs: $(a,b) \approx_\kappa (c,d)$ if and only if $a \approx_\kappa c$ and $b \approx_\kappa d$.

Operationally, therefore, $C_\kappa\, R$ is constructed from R by adding all those pairs to R that are equal to some element of R, except at κ, in both their first and their second coordinate. In particular,

$$(a,b) \in (Id_{X^\alpha})_\kappa \Leftrightarrow a \approx_\kappa b.$$

On pairs, $(a, b) \approx_\kappa (c, d) \Leftrightarrow \exists e, f.\ (a, b) = (c[\kappa \leftarrow e], d[\kappa \leftarrow f])$. Thus

$$C_\kappa\, R = \{(a, b) \mid \exists e, f.\ (a[\kappa \leftarrow e], b[\kappa \leftarrow f]) \in R\}$$

presents relational cylindrification in a way that is particularly suggestive for programming: $C_\kappa\, R$ is obtained from R by updating variable κ "asynchronously" in the pre-state and post-state of R in all possible ways.

We henceforth write Id and Id_κ when the underlying set X^α is obvious. An important property is that the relational cylindrification of Id suffices to express all other relational cylindrifications.

Lemma 6. *Let $R \subseteq X^\alpha \times X^\alpha$. Then*

$$C_\kappa\, R = Id_\kappa\, ; R\, ; Id_\kappa.$$

We have proved this fact with Isabelle. By Lemma 4, CKL is too weak to capture this property, but we expect it to fail, for instance, in trace models for which cylindrification by κ liberates κ in every state in the trace [21], not just the first and last states, i.e. $\approx_\kappa$ is lifted to apply to every state in the traces.

Some rewriting may be helpful to understand the actions of $Id_\kappa\, ; (\_)$ and $(\_)\, ; Id_\kappa$ on relations: $Id_\kappa\, ; R = \{(a, b) \mid \exists c \in X.\ (a[\kappa \leftarrow c], b) \in R\}$ and $R\, ; Id_\kappa$ acts similarly on second coordinates. Thus $Id_\kappa ; R$ models a left-handed relational cylindrification of first coordinates and $R\, ; Id_\kappa$ its right-handed opposite.

For faulting relations, $C_\kappa : \mathcal{P}\,(X^\alpha \times X^\alpha_\perp) \to \mathcal{P}\,(X^\alpha \times X^\alpha_\perp)$ is determined by Lemma 6 as $(Id_\kappa, \emptyset)\, ; (R_p, R_f); (Id_\kappa, \emptyset)$, which yields

$$C_\kappa\, R = (C_\kappa\, R_p, Id_\kappa\, ; R_f).$$

Hence we cylindrify the proper part of R and the first coordinate of its faulting part. This prevents the leakage of faults into proper parts of relations. We recall that $\mathcal{P} Id_{X^\alpha}$ is the set of subidentities over X^α.

Proposition 5. *For every ordinal α and set X,*

1. $(\mathcal{P}\,(X^\alpha \times X^\alpha), \mathcal{P}\, Id_{X^\alpha}, \cup, \cap, ;, \emptyset, Id_{X^\alpha}, -, ^*, C_\kappa)_{\kappa<\alpha}$ *is a* CKL *with tests;*
2. $(\mathcal{P}(X^\alpha \times X^\alpha_\perp), \mathcal{P}\, Id_{X^\alpha}, \cup, \cap, ;, \emptyset, Id_{X^\alpha}, -, ^*, C_\kappa)_{\kappa<\alpha}$, *with composition (1) and star (2), is a weak* CKL *with tests.*

Proof. Liberation Kleene lattices and their weak variants are introduced in Sect. 13. Proposition 12 in that section shows that every (weak) liberation Kleene lattice is a (weak) CKL. Lemma 13 in the same section shows that the liberation Kleene lattice axioms hold in $\mathcal{P}\,(X^\alpha \times X^\alpha)$ while $\mathcal{P}\,(X^\alpha \times X^\alpha_\perp)$ satisfies the weak liberation Kleene lattice axioms. □

We call $\mathcal{P}\,(X^\alpha \times X^\alpha)$ the (full) *relation* CKL *with tests* over X^α and $\mathcal{P}\,(X^\alpha \times X^\alpha_\perp)$ the (full) *weak relation* CKL *with tests* over X^α.

Henkin, Monk and Tarski show that classical cylindric algebras are closed under direct products. Yet $\mathcal{P}\, X^\alpha \times \mathcal{P}\, X^\alpha$ and $\mathcal{P}\,(X^\alpha \times X^\alpha)$ are not isomorphic and thus our axiomatisation of CKL cannot be explained in terms of a simple pair construction on classical cylindric algebras. Nevertheless, many properties, for instance in Lemmas 1 and 5, translate from their setting into ours, and relations in $\mathcal{P}(X^\alpha \times X^\alpha)$ can of course be encoded as predicates in $\mathcal{P}X^{2\alpha}$ or higher dimensions.[4] As the elementary theory of binary relations is captured by classical cylindric algebra, it can be expected that at least relation CLM can be expressed in this setting, yet rather indirectly.[5]

6 Generalised Cylindrification

Modelling frames in Morgan's refinement calculus through cylindrification requires the consideration of sets of variables, at least finite ones, and the liberation of these. Henkin, Monk and Tarski [17, §1.7] have already generalised cylindrification from single variables to finite sets. We merely need to translate their approach into CKL, and this is the purpose of this section. Once again, all properties in this section have been verified with Isabelle.

For a finite subset Γ of an ordinal α, we follow Henkin, Monk and Tarski in defining

$$c_{(\emptyset)} = id \qquad \text{and} \qquad c_{(\kappa,\Gamma)} = c_\kappa \circ c_{(\Gamma)},$$

where id is the identity function on X^α, $\circ$ is function composition, and $c_{(\kappa,\Gamma)}$ abbreviates $c_{(\{\kappa\}\cup\Gamma)}$. A simple proof by induction shows that

$$c_{(\Gamma)} \circ c_{(\Delta)} = c_{(\Gamma\cup\Delta)} \tag{HMT1.7.3}$$

holds for all finite subsets Γ and Δ of α.

Henkin, Monk and Tarski call an element x of a CKL *rectangular* if

$$c_{(\Gamma)}\, x \cdot c_{(\Delta)}\, x = c_{(\Gamma\cap\Delta)}\, x \tag{HMT1.10.6}$$

holds for all finite sets Γ and Δ. They show in the classical setting that x is rectangular if and only if $c_{(\kappa,\Gamma)}\, x \cdot c_{(\lambda,\Gamma)}\, x = c_{(\Gamma)}\, x$ holds for all κ, λ and finite Γ, such that $\kappa \neq \lambda$. By defining rectangular elements of a CKL in the same way, their proof transfers to CKL. We henceforth abbreviate $c_{(\Gamma)}\, 1$ as $1_{(\Gamma)}$. Our main interest in rectangularity lies in the following inf-closure property.

Lemma 7. *In every relation* CKL, *Id is rectangular; for all finite Γ and Δ,*

$$Id_{(\Gamma)} \cap Id_{(\Delta)} = Id_{(\Gamma\cap\Delta)}.$$

---

[4] This is similar to the predicative encoding of relations in the Z style [16,27], in which the value of a variable κ in the initial state is represented by κ and its value in the final state is represented by κ'; relational cylindrification in Z is represented by $\exists_{\kappa,\kappa'} R$, i.e. $C_\kappa C_{\kappa'} R$ in the relational model. That is, relations are encoded using a set of variables, which for each program variable κ also contains κ'.

[5] We are grateful to an anonymous referee for pointing out an encoding.

Proof. Defining the equivalence $a \approx_\Gamma b$ as $\forall \lambda \notin \Gamma.\ a_\lambda = b_\lambda$, it is easy to check that $(a, b) \in Id_{(\Gamma)} \Leftrightarrow a \approx_\Gamma b$. Hence

$$\begin{aligned}(a,b) \in Id_{(\Gamma)} \cap Id_{(\Delta)} &\Leftrightarrow a \approx_\Gamma b \wedge a \approx_\Delta b \\ &\Leftrightarrow \forall \lambda.\ (\lambda \notin \Gamma \Rightarrow a_\lambda = b_\lambda) \wedge (\lambda \notin \Delta \Rightarrow a_\lambda = b_\lambda) \\ &\Leftrightarrow \forall \lambda.\ \lambda \notin (\Gamma \cap \Delta) \Rightarrow a_\lambda = b_\lambda \\ &\Leftrightarrow a \approx_{\Gamma \cap \Delta} b \\ &\Leftrightarrow (a,b) \in Id_{(\Gamma \cup \Delta)}.\end{aligned}$$

□

At the moment, we are nevertheless neither able to derive rectangularity of 1 from the CKL axioms nor to refute its derivability.

Question 1. Do the CKL axioms imply that 1 is rectangular? Otherwise, is there any finitary extension of these axioms that implies this fact?

We henceforth indicate explicitly, whenever rectangularity of 1 is assumed.

Henkin, Monk and Tarski have also shown that the axioms of classical cylindric algebras generalise to finite sets. This fact extends to CKL as well.

Lemma 8. *In every CKL the following generalisations of axioms (C1)–(C11) hold. For all finite $\Gamma, \Delta, E \subseteq \alpha$,*

1. $c_{(\Gamma)}\, 0 = 0$,
2. $x \le c_{(\Gamma)}\, x$,
3. $c_{(\Gamma)}\, (x \cdot c_{(\Gamma)}\, y) = c_{(\Gamma)}\, x \cdot c_{(\Gamma)}\, y$,
4. $c_{(\Gamma)} c_{(\Delta)}\, x = c_{(\Delta)} c_{(\Gamma)}\, x$,
5. $c_{(\Gamma)}\, (x + y) = c_{(\Gamma)}\, x + c_{(\Gamma)}\, y$,
6. $c_{(\Gamma)}\, (x\, ; c_{(\Gamma)}\, y) = c_{(\Gamma)}\, x\, ; c_{(\Gamma)}\, y$,
7. $c_{(\Gamma)}\, (c_{(\Gamma)}\, x\, ; y) = c_{(\Gamma)}\, x\, ; c_{(\Gamma)}\, y$,
8. $\Gamma \cap \Delta = \emptyset \Rightarrow 1_{(\Gamma)} \cdot 1_{(\Delta)} = 1$, *assuming* 1 *is rectangular,*
9. $(1_{(\Gamma)}\, ; 1_{(\Delta)}) \cdot (1_{(\Gamma)}\, ; 1_{(E)}) = 1_{(\Gamma)}; (1_{(\Delta)} \cdot 1_{(E)})$, *assuming* 1 *is rectangular,*
10. $c_{(\Gamma)}\, 1_{(\Delta)} = 1_{(\Gamma)}\, ; 1_{(\Delta)}$,
11. $c_{(\Gamma)}\, x^+ \le (c_{(\Gamma)}\, x)^+$.

In addition,

12. $\Gamma \subseteq \Delta \Rightarrow c_{(\Gamma)}\, x \le c_{(\Delta)}\, x$,
13. $1_{(\Gamma)}\, ; 1_{(\Delta)} = 1_{(\Gamma \cup \Delta)}$,
14. $(1_{(\Gamma)})^* = 1_{(\Gamma)} = (1_{(\Gamma)})^+$.

These properties, plus rectangularity of 1, could be used for a set-based axiomatisation of cylindrification, in which the c_k appear as special cases.

At the end of this section we study the algebra of generalised cylindrified units $1_{(\Gamma)}$. First of all, these units need not be closed under sups.

Lemma 9. *In some (relation)* CKL*, generalised cylindrified units need not be closed under sups.*

Proof. Let $X = \{a, b\}$ and $\alpha = 2$. Then, for $\kappa < \alpha$,

$$Id_\kappa = \left\{ \left(\begin{pmatrix} x_0 \\ x_1 \end{pmatrix}, \begin{pmatrix} y_0 \\ y_1 \end{pmatrix} \right) \in X^2 \times X^2 \;\middle|\; x_{1-\kappa} = y_{1-\kappa} \right\}.$$

It is easy to check that $Id \neq Id_k \neq Id_0 \cup Id_1$. In addition,

$$\left(\begin{pmatrix} a \\ b \end{pmatrix}, \begin{pmatrix} b \\ a \end{pmatrix} \right) \notin Id_0 \cup Id_1$$

and hence $Id_0 \cup Id_1 \neq Id_0 \,;\, Id_1 = Id_{(\{0,1\})} = X^2 \times X^2$. Therefore $Id_0 \cup Id_1$ is none of the generalised cylindrified units Id, Id_0, Id_1, $Id_{(\{0,1\})}$ in $X^2 \times X^2$. □

Proposition 6. *Let K be a (weak)* CKL *and suppose that* 1 *is rectangular. Let*

$$\mathbb{1} = \{1_{(\Gamma)} \mid \Gamma \text{ is a finite subset of } \alpha\}.$$

1. *Then $(\mathbb{1}, ;, \cdot)$ forms a distributive lattice with sup* ;, *inf* $\cdot$ *and least element* 1;
2. *if α is finite, then $\mathbb{1}$ forms a finite boolean algebra with greatest element $1_{(\alpha)}$;*
3. *the map $1_{(\_)}$ from the set of finite subsets of α into $\mathbb{1}$ is a surjective lattice morphism that preserves minimal and (existing) maximal elements.*

Proof.

1. Composition in $\mathbb{1}$ is clearly associative, commutative and idempotent by Lemma 8. The distributivity laws between ; and $\cdot$ follow from Lemma 8(9), (10) and identity (HMT1.7.3). The absorption laws $1_{(\Gamma)} \,;\, (1_{(\Gamma)} \cdot 1_{(\Delta)}) = 1_{(\Gamma)}$ and $1_{(\Gamma)} \cdot (1_{(\Gamma)} \,;\, 1_{(\Delta)}) = 1_{(\Gamma)}$ have been verified with Isabelle. By rectangularity, $\mathbb{1}$ is closed under infs; by Lemma 8(13), the set is closed under composition. By definition, $1_{(0)} = 1$.
2. For finite α, Lemma 8(12) implies that $1_{(\alpha)}$ is the greatest element in $\mathbb{1}$.
3. The map $1_{(\_)}$ preserves sups by Lemma 8(13), infs by rectangularity of 1, least elements by (1) and greatest elements by (2), whenever α is finite. Surjectivity is obvious.

□

Isabelle's counterexample generators show that $1_{(\_)}$ need not be injective in CKL. Hence the lattice of these finite sets need not to be isomorphic to the lattice $\mathbb{1}$.

Lemma 10. *Let $\mathcal{P}(X^\alpha \times X^\alpha)$ by a relation* CKL *with $|X| > 1$. Then $Id_{(\_)}$ is a lattice isomorphism.*

Proof. Relative to Proposition 6, it remains to show that $Id_{(\_)}$ is injective. First we consider singleton sets. For $|X| > 1$, Id is obviously a strict subset of any Id_κ. Hence $\kappa \neq \lambda$ implies $Id_\kappa \cap Id_\lambda \neq Id_\kappa$ by (C8) and therefore $Id_\kappa \neq Id_\lambda$.

Next, suppose $\Gamma \neq \Delta = \{\lambda_1, \ldots, \lambda_n\}$ and, without loss of generality, that $\kappa \in \Gamma$, but $\kappa \notin \Delta$. Then $Id_{(\Delta)} = Id_{\lambda_1} ; \ldots ; Id_{\lambda_n}$ by Lemma 8(13) and $Id_\kappa \neq Id_{\lambda_i}$ for all $\lambda_i \in \Delta$ by injectivity on singleton sets. Thus $Id_\kappa \not\leq Id_{(\Delta)}$, because Id_κ and the Id_{λ_i} are all atoms, and therefore $1_{(\Gamma)} \neq 1_{(\Delta)}$. □

Injectivity of $1_{(\_)}$ can therefore be assumed safely for relation CKL, but other models require additional investigations. Whether this property should be turned into another CKL axiom is left for future work.

7 Propositional Refinement Calculus

Armstrong, Gomes and Struth have extended Kleene algebras with tests to refinement Kleene algebras with tests and derived the rules of a propositional variant of Morgan's refinement calculus—no frames, no local variables, no assignment laws—in this setting [2]. In the next section we show how the rules of a propositional refinement calculus with frames can be derived from the CKL axioms. Assignment laws are derived from the axioms of CKL with diagonals in Sect. 12. In this section we merely adapt the definition of refinement Kleene algebras with tests to our purposes.

Kleene algebra with tests captures propositional Hoare logic in a partial correctness setting. For a program $x \in K$ and tests $p, q \in B$, validity of the Hoare triple can be encoded as

$$\{p\}x\{q\} \Leftrightarrow p\,;x \leq x\,;q \Leftrightarrow p\,;x\,;\neg q = 0.$$

By the right-hand identity, the Hoare triple for precondition p, program x and postcondition q holds if it is impossible to execute x from states where p holds and, if the program terminates, end up in states where q does not hold. This intuition for partial correctness is easily backed up by the relational model.

In a refinement Kleene algebra [2], a specification statement $[p, q]$, where $p, q \in B$, is modelled as the largest program that satisfies $\{p\}(\_)\{q\}$. We adapt this definition to CKL.

Definition 7. *A* refinement cylindric Kleene lattice with tests *is a distributive* CKL *with tests expanded by an operation* $[\_,\_] : B \times B \to K$ *that satisfies*

$$p\,;x\,;\neg q = 0 \Leftrightarrow x \leq [p, q]. \tag{3}$$

It follows that $[p, q]$ satisfies $\{p\}[p, q]\{q\}$ and that it is indeed the greatest program that does so. It is also easy to check that in relation CKL,

$$[P, Q] = \bigcup\{R \subseteq X^\alpha \times X^\alpha \mid \{P\}R\{Q\}\},$$

which further confirms this programming intuition.

In addition, CKL with tests—like Kleene algebra with tests—provides an algebraic semantics of conditionals and while-loops that is consistent with the relational one.

$$\textbf{if } b \textbf{ then } x \textbf{ else } y = b\,;x + \neg b\,;y, \tag{4}$$
$$\textbf{while } b \textbf{ do } x = (b;x)^*;\neg b. \tag{5}$$

8 Variable Frames

Our first application to program construction shows that CKL is expressive enough to capture the variable frames of Morgan's refinement calculus [23]. For the sake of simplicity, we restrict our attention to a partial correctness setting.

In contrast to standard notations for the refinement calculus [3,23], our lattice is the dual of the refinement lattice; the standard refinement ordering $\sqsubseteq$ is the opposite of $\leq$. Hence y is a refinement of x, denoted $x \sqsubseteq y$ if and only if $x \geq y$.

In this context, we fix a CKL K with tests. We call elements of K *programs* and finite subsets of α *frames*. A frame represents the set of variables a program may modify. The program $x \cdot 1_{(\Gamma)}$ restricts x so that it may only modify variables in Γ. Using Morgan's refinement calculus notation, we define

$$\Gamma{:}x = x \cdot 1_{(\Gamma)} \tag{6}$$

for a program x restricted to frame Γ. This is consistent with relation CKL, where for a relation R and variable κ,

$$\kappa{:}R = \{(a,b) \mid (a,b) \in R \wedge \exists c.\ a = b[\kappa \leftarrow c]\}.$$

This constrains the values of all variables other than κ to remain unchanged by R, while κ is liberated and may be modified ad libitum. The generalisation to finite sets is straightforward. The following framing laws are helpful for the derivation of the laws of Morgan's refinement calculus in Proposition 7 below. They have been verified with Isabelle.

Lemma 11. *In any* CKL,

1. $\Gamma{:}x \leq x$,
2. $\Gamma \subseteq \Delta \Rightarrow \Gamma{:}x \leq \Delta{:}x$,
3. $x \leq y \Rightarrow \Gamma{:}x \leq \Gamma{:}y$,
4. $(\Gamma{:}x);(\Gamma{:}y) \leq \Gamma{:}(x;y)$,
5. $(\Gamma{:}x)^* \leq \Gamma{:}(x^*)$,
6. $\Gamma{:}x = x$, *if* $x \leq 1$.

By Lemma 11, it is a refinement to add or restrict a frame by (1) and (2). By (3), framing is isotone with respect to refinement. Equivalently to frame isotonicity, $(\Gamma : x) + (\Gamma : y) \leq \Gamma : (x + y)$. Framing distributes over sequential composition and iteration by (4) and (5). A frame has no effect on a test by (6). The distribution over sequential composition in (4) is only a refinement because the left-hand side constrains variables outside Γ to be unchanged from the initial state to the middle state and the middle state to the final state, whereas the right-hand side only has an initial-to-final constraint.

This prepares us for the main result of this section, which adapts the refinement laws derived by Armstrong, Gomes and Struth [2] to framed specifications.

Proposition 7. *The following refinement laws are derivable in any refinement* CKL *with tests.*

1. $\Gamma{:}[p,p] \geq 1$,
2. $\Gamma{:}[p,q] \geq \Gamma{:}[p',q']$ *if* $p' \geq p \wedge q \geq q'$,
3. $\Gamma{:}[0,1] \geq \Gamma{:}x$,
4. $x \geq \Gamma{:}[1,0]$,

5. $\Gamma:[p,q] \geq \Gamma:[p,r]; \Gamma:[r,q]$,
6. $\Gamma:[p,q] \geq$ **if** b **then** $\Gamma:[b \cdot p, q]$ **else** $\Gamma:[\neg b \cdot p, q]$,
7. $\Gamma:[p, \neg b \cdot p] \geq$ **while** b **do** $\Gamma:[b \cdot p, p]$.

We have verified this result with Isabelle relative to Armstrong, Gomes and Struth's proof. Assuming that the refinement laws obtained by deleting all occurrences of frames from (1)–(7) hold, we have shown that the corresponding laws with frames are derivable using a simple formalisation within CKL without tests and refinement statements. For (1), we have shown that $1 \leq x$ implies $\Gamma:1 \leq \Gamma:x$, which is an instance of Lemma 11(3). Similarly, (2) and (3) are instances of frame isotonicity. For (4), we have verified that $x \leq y$ implies $\Gamma:x \leq y$, for (5) that $x;y \leq z$ implies $\Gamma:x;\Gamma:y \leq \Gamma:z$, for (6) that $v;x+w;y \leq z$ implies $v;\Gamma:x+w;\Gamma:y \leq \Gamma:z$ whenever $v, w \leq 1$, and for (7) that $(v;x)^*;w \leq y$ implies $(v\,;\Gamma:x)^*\,;w \leq \Gamma:y$ whenever $v, w \leq 1$. All proofs use properties from Lemma 11. None of them depends on rectangularity of generalised cylindrified units.

9 Local Variable Blocks

Next we show how local variable blocks can be expressed in CKL for which 1 is rectangular. Intuitively, a local variable block introduces a variable κ having as scope a program x. The definition allows for the fact that outside the local variable block κ may (or may not) be in use as a program variable. The outer κ is unmodified by the local variable block (as represented in the definition by the conjunction of $1_{(\overline{\kappa})}$) but the body of the block is free to update the local κ as it sees fit (as represented by the cylindrification $c_\kappa\, x$). We define a local variable block (**var** κ. x) that introduces a local variable κ with scope x as

$$\mathbf{var}\ \kappa.\ x = (c_\kappa\, x) \cdot 1_{(\overline{\kappa})}. \tag{7}$$

It requires α to be a finite ordinal, so that the set $\overline{\kappa} = \alpha - \{\kappa\}$ is finite and hence $1_{(\overline{\kappa})}$ well defined. The following law allows a local variable κ to be introduced so that κ can be used to hold intermediate results of a computation.

Lemma 12. *Let K be a* CKL *for a finite ordinal α and in which* 1 *is rectangular. For all $\kappa < \alpha$ and $\Gamma \subseteq \alpha$, if $\kappa \notin \Gamma$ and $x \in K_\kappa$, that is, $c_\kappa x = x$, then*

$$\Gamma:x = \mathbf{var}\ \kappa.\ (\kappa, \Gamma):x.$$

Proof.

$$\begin{aligned}
\mathbf{var}\ \kappa.\ (\kappa,\Gamma):x &= (c_\kappa\,(x \cdot c_\kappa\, 1_{(\Gamma)})) \cdot 1_{(\overline{\kappa})} && \text{by definitions (6) and (7)}\\
&= x \cdot 1_{(\kappa,\Gamma)} \cdot 1_{(\overline{\kappa})} && \text{by (C3) and } c_\kappa x = x\\
&= x \cdot 1_{((\kappa,\Gamma)\cap\overline{\kappa})} && \text{as 1 is rectangular}\\
&= x \cdot 1_{(\Gamma)} && \text{as } \kappa \notin \Gamma,\ (\{\kappa\} \cup \Gamma) \cap \overline{\{\kappa\}} = \Gamma\\
&= \Gamma:x.
\end{aligned}$$

□

Because both cylindrification and meet are isotone so is a local variable block.

Lemma 13. *For any* $\kappa < \alpha$, *if* $x \leq y$, *then* $\mathbf{var}\ \kappa.\ x \leq \mathbf{var}\ \kappa.y$.

Introducing a local variable in a refinement is facilitated by Morgan's law (6.1) [23]. An algebraic variant of this refinement can be derived as follows.

Lemma 14. *Let* (K, B) *be a* CKL *for a finite ordinal* α *and in which* 1 *is rectangular. For all* $\kappa < \alpha$ *and* $\Gamma \subseteq \alpha$, *if* $\kappa \notin \Gamma$ *and* $p, q \in B_\kappa$, *i.e.* $c_\kappa p = p$ *and* $c_\kappa q = q$,

$$\Gamma : [p, q] = \mathbf{var}\ \kappa.\ (\kappa, \Gamma) : [p, q].$$

Proof. From Lemma 12 it suffices to show $[p, q] \in K_\kappa$ given that $p, q \in B_\kappa$, hence $c_\kappa[p, q] = [p, q]$. From (C2) it then suffices to show $c_\kappa[p, q] \leq [p, q]$.

$$\begin{aligned}
c_\kappa[p, q] \leq [p, q] &\Leftrightarrow p; c_\kappa[p, q]; \neg q = 0 && \text{by (3)}\\
&\Leftrightarrow c_\kappa p; c_\kappa[p, q]; c_\kappa \neg q = 0 && \text{as } p, q \in B_\kappa\\
&\Leftrightarrow c_\kappa(p; [p, q]; \neg q) = 0 && \text{by (C6) and (C7)}\\
&\Leftarrow p; [p, q]; \neg q = 0 && \text{by (C1)}\\
&\Leftrightarrow [p, q] \leq [p, q]. && \text{by (3)}
\end{aligned}$$

□

This law extends the refinement laws from Proposition 7 to local variable blocks.

10 Synchronous Cylindrification

Next we turn to the definition of variable assignments in CKL. This, however requires some preparation. In this section, we set up the link between CKL-style cylindrification and the classical one, which we need to apply to the tests in specification statements to model assignments. Section 11 introduces diagonal elements and substitutions as additional ingredients that are definable in CKL and needed for assignments, which are finally discussed in Sect. 12.

We have already emphasised in Sect. 5 that relational cylindrification liberates the variables in the first and second coordinates of pairs asynchronously, and this is in particular the case for subidentities, which correspond to predicates or sets. As an undesirable side effect, by Lemma 1(3), tests in CKL are not closed with respect to cylindrification: an element x of a (weak) CKL is a fixpoint of c_κ if and only if x itself has already been cylindrified by c_κ. In relational CKL, therefore, no test except $\emptyset$ is a fixpoint of any C_κ, no cylindrification of any test except $\emptyset$ is a test and $C_\kappa[\mathcal{P}Id_X] \cap \mathcal{P}Id_X = \{\emptyset\}$.

Hence if $\lceil \_ \rceil$ denotes the bijection from sets into relational subidentities, and C^c_κ denotes classical cylindrification, then $\lceil C^c_\kappa P \rceil \neq C_\kappa \lceil P \rceil$ except when predicate P is $\emptyset$.

Equality of $\lceil C^c_\kappa P \rceil$ and $C_\kappa \lceil P \rceil$ requires "synchronising" relational cylindrifications to ensure that the values of the cylindrified variable κ match in the first and the second coordinate. Synchronised relational cylindrification can be expressed in CKL as

$$\widehat{c}_\kappa x = c_\kappa x \cdot 1,$$

so that $\widehat{C}_\kappa R = C_\kappa R \cap Id_X$ and therefore, for any set P,

$$\widehat{C}_\kappa \lceil P \rceil = \{(a, a) \mid \exists c \in X.\ (a[\kappa \leftarrow c], a[\kappa \leftarrow c]) \in \lceil P \rceil\}.$$

It is then easy to check that

$$\lceil \widehat{C}_\kappa P \rceil = \widehat{C}_\kappa \lceil P \rceil$$

for any set P and hence $\lfloor \widehat{C}_\kappa P \rfloor = C^c_\kappa \lfloor P \rfloor$ for any relational subidentity P and the inverse bijection $\lfloor \_ \rfloor$.

The definition of $\widehat{c}_\kappa$ and its relational instance $\widehat{C}_\kappa$ implies that $\widehat{C}\kappa[\mathcal{P}Id_X] \subseteq \mathcal{P}Id_X$. Thus relational subidentities are closed under $\widehat{C}_\kappa$. Yet, for a general CKL with tests $B \neq 1{\downarrow} = \{x \in K \mid x \leq 1\}$ it cannot be guaranteed that $\widehat{c}_\kappa[B] \subseteq B$. Nevertheless we may require that $B = 1{\downarrow}$, which is consistent with relational models and many others. In fact, all applications of $\widehat{c}_\kappa$ in this article are restricted to tests that satisfy this property.

The current axiomatisation of the relationship between tests and the cylindrifications is not sufficient to prove some properties that we know to be true for the relational model. For example, for relations, we must add the following additional axiom relating the two notions of cylindrification for $p \in B$, where $B = 1{\downarrow}$:

$$c_\kappa\, p = 1_\kappa \,;\, \widehat{c}_\kappa\, p = \widehat{c}_\kappa\, p \,;\, 1_\kappa. \tag{8}$$

From this assumption, we have that test $\widehat{c}_\kappa\, p$ commutes over 1_κ for any test $p \in B$, i.e. $\widehat{c}_\kappa\, p \,;\, 1_\kappa = \widehat{c}_\kappa\, p \,;\, 1_\kappa \,;\, \widehat{c}_\kappa\, p$, giving us the following lemma, which is an important property used in Sect. 12 to derive properties of assignment statements.

Lemma 15. *If* $p = \widehat{c}_{(\Gamma)} p$ *then,* $\Gamma\colon [p \cdot q, r] = \Gamma\colon [p \cdot q, p \cdot r]$.

Proof. Refinement from left to right follows from Proposition 7(2). For the reverse direction we begin the proof by expanding using the definition of a frame.

$$
\begin{array}{lll}
& [p \cdot q,\ r] \cdot 1_{(\Gamma)} \leq [p \cdot q,\ p \cdot r] \cdot 1_{(\Gamma)} & \\
\Leftrightarrow & [p \cdot q,\ r] \cdot 1_{(\Gamma)} \leq [p \cdot q,\ p \cdot r] & \text{property of meet} \\
\Leftrightarrow & (p \cdot q); [p \cdot q,\ r] \cdot 1_{(\Gamma)}; (\neg p + \neg r) = 0 & \text{by (3) and De Morgan} \\
\Leftarrow & (p; 1_{(\Gamma)}; \neg p = 0) \wedge ((p \cdot q); [p \cdot q,\ r]; \neg r = 0) & \text{distributing and simplifying} \\
\Leftrightarrow & \widehat{c}_{(\Gamma)} p; 1_{(\Gamma)}; \neg p = 0 & \text{assumption } p = \widehat{c}_{(\Gamma)} p \text{ and (3)} \\
\Leftrightarrow & \widehat{c}_{(\Gamma)} p; 1_{(\Gamma)}; \widehat{c}_{(\Gamma)} p; \neg p = 0 & \text{commutativity assumption} \\
\Leftrightarrow & p; 1_{(\Gamma)}; 0 = 0 & \text{assumption and } p; \neg p = 0
\end{array}
$$

The later holds because 0 is an annihilator for tests.

Lemma 16. *If* $p = \widehat{c}_{(\Gamma)}p$ *and* $p \cdot r_2 \leq r_1$ *then,* $\Gamma : [p \cdot q, r_1] \geq \Gamma : [p \cdot q, r_2]$.

Proof.

$$\begin{aligned} \Gamma : [p \cdot q, r_2] &= \Gamma : [p \cdot q, p \cdot r_2] && \text{by Lemma 15} \\ &\leq \Gamma : [p \cdot q, r_1]. && \text{by Proposition 7(2)} \end{aligned}$$

□

11 Diagonals and Substitution

Our next step toward modelling assignments algebraically requires capturing substitutions algebraically. Once again, Henkin Monk and Tarski have paved the way for us [17, §1.5]. Yet their concept of variable substitution in classical cylindric algebra depends on another concept, which is integral to their approach, and we have so far neglected: that of diagonal elements, which abstract equality in equational logic.

In standard cylindric set algebras, *diagonal elements* [17] are defined, for each $\kappa, \lambda < \alpha$, as

$$D_{\kappa\lambda} = \{x \in X^\alpha \mid x_\kappa = x_\lambda\}.$$

Henkin, Monk and Tarski [17] give a geometric interpretation of $D_{\kappa\lambda}$ as a hyperplane in X^α that is described by the equation $x_\kappa = x_\lambda$. For instance, for $\alpha = 2$, D_{01} corresponds to the diagonal line between the coordinate axes 0 and 1; for $\alpha = 3$, D_{01} is the plane spanned by that diagonal and 3-axis.

While diagonalisation could be generalised to relational diagonalisation, we only require diagonal elements on the boolean subalgebra of tests, which is captured by the standard approach, in combination with synchronised cylindrification $\widehat{c}_\kappa$. Henkin, Monk and Tarski's axioms for classical cylindric algebra therefore lead us to the following definition.

Definition 8. *A* cylindric Kleene lattice with enriched tests *is a* CKL *equipped with a family of elements* $(d_{\kappa\lambda})_{\kappa,\lambda<\alpha} \subseteq B = 1\!\downarrow$ *that satisfy*

$$d_{\kappa\kappa} = 1, \tag{D1}$$

$$d_{\lambda\mu} = \widehat{c}_\kappa(d_{\lambda\kappa} \cdot d_{\kappa\mu}), \qquad \textit{if } \kappa \notin \{\lambda, \mu\}, \tag{D2}$$

$$\widehat{c}_\kappa(d_{\kappa\lambda} \cdot p) \cdot \widehat{c}_\kappa(d_{\kappa\lambda} \cdot \neg p) = 0, \qquad \textit{if } \kappa \neq \lambda. \tag{D3}$$

The axioms (D1)–(D3) are precisely the diagonal axioms of classical cylindric algebras [17]. They are applied to tests only and use $\widehat{c}_\kappa$ instead of c_κ. Axiom (D3) captures a notion of variable substitution. In fact, Henkin, Monk and Tarski define

$$s^\kappa_\lambda p = \begin{cases} p, & \text{if } \kappa = \lambda, \\ \widehat{c}_\kappa(d_{\kappa\lambda} \cdot p), & \text{if } \kappa \neq \lambda \end{cases} \tag{9}$$

to indicate that λ is substituted for κ in p. Axiom (D3) can then be rewritten as $s^{\kappa}_{\lambda} p \cdot s^{\kappa}_{\lambda} \neg p = 0$. The substitution operator s^{κ}_{λ} satisfies the following properties, which have been verified with Isabelle, and turn out to be useful in the following sections.

Lemma 17. *Let (K, B) be a* CKL *with enriched tests. If $p, q \in B$ and $\kappa, \lambda, \mu < \alpha$, then*

1. $s^{\kappa}_{\lambda}(p + q) = s^{\kappa}_{\lambda} p + s^{\kappa}_{\lambda} q$, *(HMT1.5.3(i))*
2. $\neg s^{\kappa}_{\lambda} p = s^{\kappa}_{\lambda} \neg p$, *(HMT1.5.3(ii))*
3. $s^{\kappa}_{\lambda} 1 = 1$,
4. $s^{\kappa}_{\lambda}(p \cdot q) = s^{\kappa}_{\lambda} p \cdot s^{\kappa}_{\lambda} q$,
5. $s^{\kappa}_{\lambda}(d_{\kappa\mu}) = d_{\lambda\mu}$ *if* $\kappa \neq \mu$, *(HMT1.5.4(i))*
6. $s^{\kappa}_{\lambda}(d_{\mu\nu}) = d_{\mu\nu}$ *if* $\kappa \notin \{\mu, \nu\}$, *(HMT1.5.4(ii))*
7. $s^{\kappa}_{\tau}(d_{\kappa\lambda} \cdot d_{\mu\nu}) = d_{\tau\lambda} \cdot d_{\mu\nu}$ *for distinct* $\kappa, \lambda, \mu, \nu, \tau$.

12 Assignments

Assignment statements are usually of the form $\kappa := e$, where e is an expression on the programming variables. Expressions are not available in CKL with enriched tests, however we can use framed specification statements to abstract the behaviour of assignments. For any $p \in B$ we write $\kappa :\in p$ to denote a non-deterministic assignment of variable κ to a value such that the final state of the command satisfies test p. It is defined as

$$\kappa :\in p = \kappa : [1, p].$$

A special case of this is the direct assignment of one variable to another, written $\kappa := \lambda$, which is defined by taking predicate p to be the diagonal $d_{\kappa\lambda}$:

$$\kappa := \lambda = \kappa : [1, d_{\kappa\lambda}]$$

For example, if κ is fresh in expression e, the assignment $\kappa := e$ can be encoded using the non-deterministic assignment command as $\kappa :\in (\kappa = e)$, where $(\kappa = e)$ is abstracted to a test in the algebra. For the more general case we can choose a variable λ different from κ that is fresh in e and write

$$(\textbf{var}\ \lambda.\ \lambda := \kappa\,;\, \kappa :\in (\kappa = e[\kappa \backslash \lambda]))$$

where, in the program model, $e[\kappa \backslash \lambda]$ is the expression e with λ substituted for κ, but in the algebra $(\kappa = e[\kappa \backslash \lambda])$ is simply abstracted as a test.

The following propositions are used to verify the algebraic equivalent of the assignment law defined by Morgan [23, p.8]. In order to more simply represent the precondition, we introduce two notations on tests: the *inner cylindrification* $c^{\partial}_{\kappa} p$ is the De Morgan dual of $\widehat{c}_{\kappa}$ and corresponds to universal quantification in first order logic [17, §1.4]; and $p \to q$ is a shorthand for implication in the boolean algebra of tests.

$$c^{\partial}_{\kappa} p = \neg \widehat{c}_{\kappa} \neg p, \tag{10}$$

$$p \to q = \neg p + q. \tag{11}$$

In the proposition below the test $c^{\partial}_{\kappa}(r \to q)$ can be interpreted as saying that for all values of κ, test r implies q, i.e. it is a test describing the states from which substituting κ for any value satisfying r will certainly result in a post-state q.

Proposition 8. *Suppose B is the test subalgebra of a CKL with enriched tests. If $q, r \in B$ and $\kappa < \alpha$ and $p \le c^{\partial}_{\kappa}(r \to q)$, then*

$$\kappa\!:\![p,\ q] \ge \kappa :\in r. \tag{12}$$

Proof. The application of Lemma 16 requires $c^{\partial}_{\kappa}(r \to q) \cdot r \le q$, which can be shown as follows.

$\widehat{c}_{\kappa}(r \cdot \neg q) \ge r \cdot \neg q$	by (C2)
$\Leftrightarrow\ \neg\widehat{c}_{\kappa}(r \cdot \neg q) \le \neg(r \cdot \neg q)$	negating both sides and reversing the order
$\Rightarrow\ c^{\partial}_{\kappa}(r \to q) \cdot r \le (\neg r + q) \cdot r$	by definition of c^{∂} and conjoin r to both sides
$\Leftrightarrow\ c^{\partial}_{\kappa}(r \to q) \cdot r \le q \cdot r$	boolean simplification
$\Rightarrow\ c^{\partial}_{\kappa}(r \to q) \cdot r \le q$	as $q \cdot r \le q$.

It also requires that $\widehat{c}_{(\Gamma)} c^{\partial}_{(\Gamma)} p = c^{\partial}_{(\Gamma)} p$, which has been shown in [17, Theorem 1.4.4(ii)].

$\kappa\!:\![p,\ q] \ge \kappa\!:\![c^{\partial}_{\kappa}(r \to q),\ q]$	by Lemma 7(2)
$\ge \kappa\!:\![c^{\partial}_{\kappa}(r \to q),\ r]$	by Lemma 16 as $c^{\partial}_{\kappa}(r \to q) \cdot r \le q$
$\ge \kappa\!:\![1,\ r]$	by Lemma 7(2)
$= \kappa :\in r$	From (12).

□

When r is $d_{\kappa\lambda}$, the test $c^{\partial}_{\kappa}(r \to q)$ simplifies to $s^{\kappa}_{\lambda} q$, the substitution of λ for κ in test q.

Proposition 9. *Suppose B is the test subalgebra of a CKL with enriched tests. If $q \in B$ and $\kappa, \lambda < \alpha$ and $p \le s^{\kappa}_{\lambda} q$, then*

$$\kappa\!:\![p,\ q] \ge \kappa := \lambda.$$

Proof. We have $c^{\partial}_{\kappa}(d_{\kappa\lambda} \to q) = \neg c_{\kappa}(d_{\kappa\lambda} \cdot \neg q) = \neg s^{\kappa}_{\lambda} \neg q = s^{\kappa}_{\lambda} q$ by Lemma 17(2).

$\kappa\!:\![p,\ q] \ge \kappa\!:\![s^{\kappa}_{\lambda} q,\ q]$	by Proposition 7(2)
$= \kappa\!:\![c^{\partial}_{\kappa}(d_{\kappa\lambda} \to q),\ q]$	by above reasoning
$\ge \kappa :\in d_{\kappa\lambda}$	taking r to be $d_{\kappa\lambda}$ in Proposition 8
$= \kappa := \lambda$	by definition (12).

Propositions 8 and 9 encode the assignment law defined by Morgan [23, p.8] for our non-deterministic assignment statement, and for the special case where we assign one variable directly to another. These propositions can be equivalently expressed in Hoare logic using the specification statement definition (3) and the Hoare logic encoding.

Proposition 10. *Suppose B is the test subalgebra of a* CKL *with enriched tests. If $p \in B$ and $\kappa, \lambda < \alpha$, then*

$$\begin{aligned} &\{p\}\ \kappa :\in r\ \{q\}, && \text{if } p \leq c_\kappa^\partial(r \to q), \\ &\{p\}\ \kappa := \lambda\ \{q\}, && \text{if } p \leq s_\lambda^\kappa\, q. \end{aligned}$$

Frames and diagonals together allow one to make use of *logical variables* and constants (e.g., natural numbers) within a specification. In Example 1, we consider a derivation of a program that swaps the values of variables λ and κ. This example is given by Morgan [23]; the difference here is that the derivation is purely algebraic. The example uses Morgan's *following assignment* law, in which a specification statement is refined to another specification statement followed by an assignment command. The next lemma derives this in the algebra.

Lemma 18. *Suppose B is the test subalgebra of a* CKL *with enriched tests. If $p, q \in B$ and $\kappa, \lambda, \mu < \alpha$, then*

$$\lambda, \kappa\colon[p, q] \geq \lambda, \kappa\colon[p, s_\mu^\kappa\, q]; \kappa := \mu.$$

Proof.

$$\begin{aligned} \lambda, \kappa\colon[p, q] &\geq \lambda, \kappa\colon[p, s_\mu^\kappa\, q]; \lambda, \kappa\colon[s_\mu^\kappa\, q, q] && \text{by Proposition 7(5)} \\ &\geq \lambda, \kappa\colon[p, s_\mu^\kappa\, q]; \kappa\colon[s_\mu^\kappa\, q, q] && \text{by Lemma 11(2)} \\ &\geq \lambda, \kappa\colon[p, s_\mu^\kappa\, q]; \kappa := \mu && \text{by Proposition 9.} \end{aligned}$$

□

Example 1. The swapping variables example can be handled entirely within the algebra. Suppose $\kappa_1, \kappa_2, \lambda_1, \lambda_2, \tau < \alpha$ are distinct. As in many refinement proofs (see [23]), λ_1 and λ_2 are logical variables used to specify the initial values of program variables κ_1 and κ_2. The first step uses Lemma 12 to introduce local variable τ that we use to temporarily store the value of κ_2:

$$\begin{aligned} &\kappa_1, \kappa_2\colon[d_{\kappa_1\lambda_1} \cdot d_{\kappa_2\lambda_2}, d_{\kappa_1\lambda_2} \cdot d_{\kappa_2\lambda_1}] \\ =&\mathbf{var}\ \tau.\ \tau, \kappa_1, \kappa_2\colon[d_{\kappa_1\lambda_1} \cdot d_{\kappa_2\lambda_2}, d_{\kappa_1\lambda_2} \cdot d_{\kappa_2\lambda_1}]. \end{aligned}$$

Using Lemma 13 this can be refined by refining the body of the local variable block as follows.

$$\begin{aligned} &\tau, \kappa_1, \kappa_2\colon[d_{\kappa_1\lambda_1} \cdot d_{\kappa_2\lambda_2}, d_{\kappa_1\lambda_2} \cdot d_{\kappa_2\lambda_1}] \\ \geq\ &\tau, \kappa_1, \kappa_2\colon[d_{\kappa_1\lambda_1} \cdot d_{\kappa_2\lambda_2}, s_\tau^{\kappa_1}\,(d_{\kappa_1\lambda_2} \cdot d_{\kappa_2\lambda_1})]; \kappa_1 := \tau && \text{by Lemma 18} \\ \geq\ &\tau, \kappa_1, \kappa_2\colon[d_{\kappa_1\lambda_1} \cdot d_{\kappa_2\lambda_2}, d_{\tau\lambda_2} \cdot d_{\kappa_2\lambda_1}]; \kappa_1 := \tau && \text{by Lemma 17(7).} \end{aligned}$$

Applying this pattern twice yields

$$\begin{aligned}
&\tau, \kappa_1, \kappa_2 \colon [d_{\kappa_1\lambda_1} \cdot d_{\kappa_2\lambda_2}, d_{\tau\lambda_2} \cdot d_{\kappa_2\lambda_1}]; \kappa_1 := \tau \\
&\geq \tau, \kappa_1, \kappa_2 \colon [d_{\kappa_1\lambda_1} \cdot d_{\kappa_2\lambda_2}, d_{\tau\lambda_2} \cdot d_{\kappa_1\lambda_1}]; \kappa_2 := \kappa_1; \kappa_1 := \tau \\
&\geq \tau, \kappa_1, \kappa_2 \colon [d_{\kappa_1\lambda_1} \cdot d_{\kappa_2\lambda_2}, d_{\kappa_2\lambda_2} \cdot d_{\kappa_1\lambda_1}]; \tau := \kappa_2; \kappa_2 := \kappa_1; \kappa_1 := \tau \\
&\geq 1; \tau := \kappa_2; \kappa_2 := \kappa_1; \kappa_1 := \tau \qquad \text{by Proposition 7(1).}
\end{aligned}$$

Eliminating the identity 1 and substituting the refined body back in the local variable block, the final code is

$$\mathbf{var}\ \tau.\ (\tau := \kappa_2; \kappa_2 := \kappa_1; \kappa_1 := \tau).$$

13 Beyond Cylindrification: Liberation Algebras

An interesting axiomatic question arises from the fact that, by Lemmas 6 and 21, the identity

$$c_\kappa\, x = 1_\kappa \,;\, x \,;\, 1_\kappa$$

holds in (weak) relational CKL, whereas, by Lemma 4, it is not derivable in CKL. On the one hand, non-derivability is desirable, because the identity fails in program trace models of CKL [21]. On the other hand, it shifts the focus from cylindrification to identities 1_κ and raises the question of axiomatising elements 1_κ for $\kappa < \alpha$ directly over (weak) Kleene lattices and defining the cylindrification operators c_κ explicitly via the identity above. This section describes the initial steps for such an approach. The elements 1_κ are now written more simply as κ for $\kappa < \alpha$.

Definition 9 (LLM). *A* (weak) liberation l-monoid *is a (weak) l-monoid L that is equipped with a family $(\kappa)_{\kappa<\alpha}$ of elements that satisfy*

$$\kappa \,;\, 0 = 0, \tag{L1}$$
$$1 \leq \kappa, \tag{L2}$$
$$\kappa \,;\, (x \cdot (\kappa \,;\, y)) = (\kappa \,;\, x) \cdot (\kappa \,;\, y), \tag{L3}$$
$$(x \cdot (y \,;\, \kappa)) \,;\, \kappa = (x \,;\, \kappa) \cdot (y \,;\, \kappa), \tag{L4}$$
$$\kappa \,;\, \lambda = \lambda \,;\, \kappa, \tag{L5}$$
$$\kappa \neq \lambda \;\Rightarrow\; \kappa \cdot \lambda = 1, \tag{L6}$$
$$(\kappa \,;\, \lambda) \cdot (\kappa \,;\, \mu) = \kappa \,;\, (\lambda \cdot \mu), \tag{L7}$$
$$(\kappa \,;\, \mu) \cdot (\lambda \,;\, \mu) = (\kappa \cdot \lambda) \,;\, \mu. \tag{L8}$$

As expected, there is a close correspondence between these axioms and axioms (C1)–(C10), although analogues of (C5)–(C7) and (C10) are derivable in this context, and therefore redundant.

Definition 10 (LKL). *A* (weak) liberation Kleene lattice *is a (weak) Kleene lattice with a family $(\kappa)_{\kappa<\alpha}$ of elements that satisfy (L1)–(L8).*

We have checked independence of these axioms in Isabelle. Extensions to (weak) liberation Kleene lattices with tests are straightforward.

Proposition 11. *Every weak* LLM *is a weak* CLM *with* $c_\kappa x = \kappa ; x ; \kappa$.

Rewriting the CLM axioms with $c_\kappa x = \kappa ; x ; \kappa$ and deriving the results from the LLM axioms is straightforward with Isabelle. Axiom (C6), for instance, becomes $\kappa ; x ; \kappa ; y ; \kappa ; \kappa = \kappa ; x ; \kappa ; \kappa ; y ; \kappa$, which is derivable because any κ can be shown to be an idempotent with respect to ; in LLM by taking x and y to both be 1 in (L3). Axiom (C3) becomes $\kappa ; (x \cdot (\kappa ; y ; \kappa)) ; \kappa = (\kappa ; x ; \kappa) \cdot (\kappa ; y ; \kappa)$, which can be obtained from (L3) and (L4).

Unlike for CKL, a special star axiom is not needed for liberation algebras. The following lemma has been obtained with Isabelle.

Lemma 19. *In every weak* LKL,

1. $x^+ ; \kappa \leq (x ; \kappa)^+$,
2. $\kappa ; x^+ \leq (\kappa ; x)^+$,
3. $\kappa ; x^+ ; \kappa \leq (\kappa ; x ; \kappa)^+$.

The proof of (1) is very simple: $x^+ ; \kappa = x^* ; x ; \kappa \leq (x ; \kappa)^* ; x ; \kappa = (x ; \kappa)^+$. Using the last identity then yields the following result.

Proposition 12. *Every weak* LKL *is a weak* CKL *with* $c_\kappa x = \kappa ; x ; \kappa$.

In addition, the LKL axioms are sound with respect to relational models.

Proposition 13. *The (weak)* LKL *axioms hold in the relational (fault) model with* κ *interpreted as* Id_κ *for all* $\kappa < \alpha$.

Proof. The relational variants of the LKL axioms have been verified with Isabelle. An algebraic proof for the weak case is given in Proposition 16, Appendix B. It has been checked with Isabelle. □

The next two facts generalise Lemma 2 and Proposition 4 from Sect. 4.

Lemma 20. *In every weak* LLM, $\kappa ; (\_)$ *and* $(\_) ; \kappa$ *are closure operators.*

Writing K^l_κ for the set of fixpoints of $\kappa ; (\_)$, K^r_κ for those of $(\_) ; \kappa$ and K_κ for those of $\kappa ; (\_) ; \kappa$ yields the following result.

Proposition 14. *Let* K *be a (weak)* LKL *and let* $\kappa < \alpha$. *Then*

1. $(K^l_\kappa, +, \cdot, ;, 0, \kappa, (\_)^+)$ *is a (weak) sub-Kleene lattice of* K *with left unit* κ,
2. $(K^r_\kappa, +, \cdot, ;, 0, \kappa, (\_)^+)$ *is a sub-Kleene lattice of* L *with right unit* κ *and if* K *is a strong* LKL, *then* K^l_κ *and* K^r_κ *are isomorphic,*
3. $(K_\kappa, +, \cdot, ;, 0, \kappa, (\_)^+)$ *is a sub-Kleene lattice of* K^l_κ *and* K^r_κ.

The proofs are very similar to those for c_κ.

The results for κ ; (_) and (_) ; κ reveal a duality in relational cylindrification without faults that is not present in the traditional approach. We already pointed out in Sect. 5 that, in the relational model, Id_κ ; R corresponds to a left-handed cylidrification of R and $R; Id_\kappa$ to its right-handed opposite. One can therefore introduce handedness via opposition to cylindrification over l-monoids by axiomatising left-handed cylindrification c^l_κ and right-handed cylindrification c^r_κ and split the axioms (C6) and (C7) accordingly. This yields a more fine-grained view on cylindrification in models with opposition duality. In addition, left-handed and right-handed cylindrifications commute (i.e. $c^l_\kappa \circ c^r_\lambda = c^r_\lambda \circ c^l_\kappa$) and $c_\kappa = c^l_\kappa \circ c^r_\kappa$ holds in the relational model but not in general. Details have been worked out in a companion article [21]. The handed cylindrifications are akin to forward and backward modal operators, yet defined over Kleene lattices instead of boolean algebras.

14 Conclusion

We have shown that cylindrification can be adapted to Kleene lattices and their relational models in such a way that variable assignments, frames and local variable blocks can be modelled. Based on this, we have derived the laws of Morgan's refinement calculus and the rules of Hoare logic, including those for assignments. The scope of algebraic approaches to program construction has therefore been extended, with the potential of fully algebraic reasoning about imperative programs.

Nevertheless, many questions about cylindric Kleene lattices and their relatives remain open and deserve further investigation. Instead of the obvious questions on completeness or decidability, we focus on conceptual ones.

First, it is easy to check that relational cylindrifications preserves arbitrary sups and hence have upper adjoints. This situation is well known from classical cylindric algebra, where the standard *outer cylindrifications* c_κ are accompanied by *inner cylindrifications* c^∂_κ that are related by De Morgan duality. Geometrically, these describe greatest cylinders with respect to κ that are contained in a given set. In cylindric algebras of formulas, inner cylindrification gives the algebra of universal quantification. In an extension of CKL, where lattices need not be complemented, dual cylindrifications can be axiomatised by adjunction. In extensions of LKL, they can be defined explicitly as $c^\partial_\kappa\ x = 1_\kappa \backslash x / 1_\kappa$, where $\backslash$ and $/$ are residuals, as they appear in action algebras [18] and action logic [26]. Our Isabelle components already contain axiomatisations for these structures, but so far we do not have any use for them.

Second, our refinement calculus and Hoare logic are restricted to partial program correctness for the sake of simplicity; yet the relational fault model is relevant to total correctness and our Isabelle components are based on weak cylindric Conway lattices and weak liberation Conway lattices, in which iteration is weak enough to be either finite, as in Kleene lattices, or possibly infinite, as in demonic refinement algebra [29]. Almost all properties presented in our paper hold in fact in this more general setting, and our relational models are a fortiori models of these generalisations. The relevance of these algebras to models with finite or possibly infinite traces, and the derivation of while rules and refinement laws for total program correctness remains to be explored.

For concurrent programs with a semantics based on a set of traces, cylindrification can be applied to liberate a variable κ in every state of each trace, in the same way that liberation of a relation liberates κ in both the initial and final states. In that setting liberation can be used in the definition of a local variable block in a similar fashion to the way it is used here [21]. A trace-based semantics for the liberation operator was given in [5, §4.6 and §5.6].[6] The generalisation of cylindric algebra presented in this paper applies directly to the trace-based model used for concurrency. That model also uses sets, binary relations, a subset of commands that form instantaneous tests (isomorphic to sets of states), subsets of commands representing program steps and environment steps (each of which is isomorphic to binary relations on states). Factoring out the cylindric algebra and applying it in each of these contexts allows one to reuse the properties of cylindric algebra in each of these contexts, thus simplifying the mechanisation of the theory.

Finally, while part of the theory and many of the proofs in this article have been formalised with Isabelle/HOL, the question whether our approach may lead to program construction and verification components that support an algebraic treatment of variable assignments requires further exploration. This seems a particularly promising avenue for future research.

Acknowledgements. We thank Simon Doherty for discussions on earlier versions of this work.

A Construction of Weak Kleene Lattices

Instead of proving Proposition 2, we show that it is a corollary to a standard semidirect product construction, which is well known from semigroup theory. All proofs in this appendix have been verified with Isabelle.

An *l-monoid module* of an l-monoid L and a semilattice S with least element 0 is an action $\circ : L \to S \to S$ that satisfies

$$\begin{aligned}
(p\,;q) \circ x &= p \circ (q \circ x),\\
(p+q) \circ x &= p \circ x + q \circ x,\\
p \circ (x+y) &= p \circ x + p \circ y,\\
1 \circ x &= x,\\
0 \circ x &= 0.
\end{aligned}$$

It follows that $p \circ 0 = 0$.

The *semidirect product* $L \ltimes S$ on $L \times S$ is defined by

$$(p, x) \ltimes (q, y) = (p\,;q, x + p \circ y).$$

[6] In that paper $c_\kappa x$ is written $x \backslash \kappa$.

The relational redefinition of composition in Sect. 3 is a simple instance of this standard algebraic concept. It is easy to check that $(1, 0)$ is the unit of $\ltimes$ and $(0, 0)$ a left annihilator. In addition, we define join and meet pointwise on pairs as $(p, x) + (q, y) = (p + q, x + y)$ and $(p, x) \cdot (q, y) = (p \cdot q, x \cdot y)$. The following fact is routine. Most axioms have already been checked elsewhere [6,11].

Proposition 15. *Let L be an l-monoid and S a semilattice with 0. Then $L \ltimes S$ forms a weak l-monoid.*

If K is a Kleene lattice, we define a *Kleene lattice module* by adding the axiom

$$x + p \circ y \leq y \Rightarrow p^* \circ x \leq y.$$

Hence the action axioms for Kleene lattice modules are essentially those for Kleene modules [12]. Finally, we define the star on products as

$$(p, x)^* = (p^*, p^* \circ x).$$

It follows that $(p, x)^+ = (p^+, p^* \circ x)$.

Proposition 15 then extends as follows.

Theorem 1. *Let K be a Kleene lattice and S a semilattice with 0. Then $K \ltimes S$ forms a weak Kleene lattice.*

Dongol, Hayes and Struth [11] present a similar result in the less general setting of quantale modules, which however captures the relational fault model in Sect. 3. A formalisation with Isabelle can be found in the Archive of Formal Proofs [10], including a verification of the properties of the relational star presented in Sect. 3. Cranch, Laurence and Struth [6] present a second proof in the more general setting of regular algebras that satisfy strictly weaker induction axioms. It gives a good impression of the manipulations needed in our present proof. Möller and Struth [22] present a third proof for total correctness in the setting of modal Kleene algebras. Instead of semidirect products, it is based on wreath products (cf. [30]).

B Construction of Weak Liberation Kleene Lattices

Instead of proving Proposition 13 for relational cylindrification we give an algebraic proof based on a new algebraic definition. This proof also supports an indirect proof of Proposition 2. All proofs in this appendix have once again been checked with Isabelle.

A *cylindric Kleene lattice module* is a Kleene lattice module over a cylindric Kleene lattice with cylindrification defined by

$$\widetilde{c}_\kappa\,(p, x) = (c_\kappa\, p, 1_\kappa \circ x).$$

By this definition, $\widetilde{c}_\kappa\,(1, 0) = (1_\kappa, 0)$ and $(p, x) \ltimes (1_\kappa, 0) = (p\,; 1_\kappa, x)$.

First we derive an algebraic variant of Lemma 6 that is suitable for the relational fault model.

Lemma 21. *Let L be a* CKL *and S a semilattice with* 0*. Then*

$$c_\kappa p = 1_\kappa \circ p \circ 1_\kappa \ \Rightarrow\ \widetilde{c}_\kappa\,(p, x) = \widetilde{c}_\kappa\,(1, 0) \ltimes (p, x) \ltimes \widetilde{c}_\kappa\,(1, 0).$$

Next we turn to the algebraic proof that subsumes Proposition 13.

A *Liberation Kleene lattice module* is a Kleene lattice module defined over a liberation Kleene lattice.

Proposition 16. *Let K be a* LKL *and S a semilattice with* 0*, such that*

$$1_\kappa \circ (x \cdot (1_\kappa \circ y)) = (1_\kappa \circ x) \cdot (1_\kappa \circ y)$$

holds for all $x, y \in S$. Then $K \ltimes S$ forms a weak LKL.

References

1. Andréka, H., Mikulás, S., Németi, I.: The equational theory of Kleene lattices. Theoret. Comput. Sci. **412**(52), 7099–7108 (2011)
2. Armstrong, A., Gomes, V.B.F., Struth, G.: Building program construction and verification tools from algebraic principles. Formal Aspects Comput. **28**(2), 265–293 (2016)
3. Back, R.-J., von Wright, J.: Refinement calculus - a systematic introduction. Springer, New York (1999). https://doi.org/10.1007/978-1-4612-1674-2
4. Birkhoff, G.: Lattice Theory. American Mathematical Society, New York (1940)
5. Colvin, R.J., Hayes, I.J., Meinicke, L.A.: Designing a semantic model for a wide-spectrum language with concurrency. Formal Aspects Comput. **29**, 853–875 (2016)
6. Cranch, J., Laurence, M.R., Struth, G.: Completeness results for omega-regular algebras. J. Logical Algebric Methods Program. **84**(3), 402–425 (2015)
7. Desharnais, J., Möller, B., Struth, G.: Kleene algebra with domain. ACM TOCL **7**(4), 798–833 (2006)
8. Desharnais, J., Struth, G.: Internal axioms for domain semirings. Sci. Comput. Program. **76**(3), 181–203 (2011)
9. The Coq development team. The Coq proof assistant reference manual. LogiCal Project. Version 8.0 (2004)
10. Dongol, B., Gomes, V.F.B., Hayes, I.J., Struth, G.: Partial semigroups and convolution algebras. Arch. Formal Proofs (2017)
11. Dongol, B., Hayes, I.J., Struth, G.: Relational convolution, generalised modalities and incidence algebras. CoRR, abs/1702.04603 (2017)
12. Ehm, T., Möller, B., Struth, G.: Kleene modules. In: Berghammer, R., Möller, B., Struth, G. (eds.) RelMiCS 2003. LNCS, vol. 3051, pp. 112–123. Springer, Heidelberg (2004). https://doi.org/10.1007/978-3-540-24771-5_10
13. Gabbay, M.J., Ciancia, V.: Freshness and name-restriction in sets of traces with names. In: Hofmann, M. (ed.) FoSSaCS 2011. LNCS, vol. 6604, pp. 365–380. Springer, Heidelberg (2011). https://doi.org/10.1007/978-3-642-19805-2_25
14. Giacobazzi, R., Debray, S.K., Levi, G.: A generalized semantics for constraint logic programs. In: FGCS, pp. 581–591 (1992)
15. Gomes, V.B.F., Struth, G.: Modal Kleene algebra applied to program correctness. In: Fitzgerald, J., Heitmeyer, C., Gnesi, S., Philippou, A. (eds.) FM 2016. LNCS, vol. 9995, pp. 310–325. Springer, Cham (2016). https://doi.org/10.1007/978-3-319-48989-6_19

16. Hayes, I. (ed.): Specification Case Studies, 2nd edn. Prentice Hall International, Englewood Cliffs (1993)
17. Henkin, L., Donald Monk, J., Tarski, A.: Cylindric Algebras, Part I., volume 64 of Studies in logic and the foundations of mathematics. North-Holland Pub. Co. (1971)
18. Kozen, D.: On action algebras. In: van Eijk, J., Visser, A. (eds.) Logic and Information Flow, pp. 78–88. MIT Press, Cambridge (1994)
19. Kozen, D.: Kleene algebra with tests. ACM Trans. Program. Lang. Syst. **19**(3), 427–443 (1997)
20. Kozen, D.: On Hoare logic and Kleene algebra with tests. ACM Trans. Comput. Log. **1**(1), 60–76 (2000)
21. Meinicke, L.A., Hayes, I.J.: Handling localisation in rely/guarantee concurrency: an algebraic approach. arXiv:1907.04005 [cs.LO] (2019)
22. Möller, B., Struth, G.: wp is wlp. In: MacCaull, W., Winter, M., Düntsch, I. (eds.) RelMiCS 2005. LNCS, vol. 3929, pp. 200–211. Springer, Heidelberg (2006). https://doi.org/10.1007/11734673_16
23. Morgan, C.: Programming From Specifications. Prentice-Hall, Upper Saddle River (1990)
24. Morgan, C.C.: Programming from Specifications, 2nd edn. Prentice Hall, Hemel Hempstead (1994)
25. Pous, D.: Kleene algebra with tests and Coq tools for while programs. In: Blazy, S., Paulin-Mohring, C., Pichardie, D. (eds.) ITP 2013. LNCS, vol. 7998, pp. 180–196. Springer, Heidelberg (2013). https://doi.org/10.1007/978-3-642-39634-2_15
26. Pratt, V.: Action logic and pure induction. In: van Eijck, J. (ed.) JELIA 1990. LNCS, vol. 478, pp. 97–120. Springer, Heidelberg (1991). https://doi.org/10.1007/BFb0018436
27. Spivey, J.M.: The Z notation: a reference manual, 2nd edn. Prentice Hall International, Englewood Cliffs (1992)
28. Struth, G.: Hoare semigroups. Math. Struct. Comput. Sci. **28**(6), 775–799 (2018)
29. von Wright, J.: Towards a refinement algebra. Sci. Comput. Program. **51**(1–2), 23–45 (2004)
30. Wells, C.: Some applications of the wreath product construction. Am. Math. Monthly **83**(5), 317–338 (1976)

A Hierarchy of Monadic Effects for Program Verification Using Equational Reasoning

Reynald Affeldt[1(✉)], David Nowak[2], and Takafumi Saikawa[3]

[1] National Institute of Advanced Industrial Science and Technology, Tsukuba, Japan
reynald.affeldt@aist.go.jp

[2] Univ. Lille, CNRS, Centrale Lille, UMR 9189 - CRIStAL - Centre de Recherche en Informatique Signal et Automatique de Lille, 59000 Lille, France

[3] Nagoya University & Peano System Inc., Nagoya, Japan

Abstract. One can perform equational reasoning about computational effects with a purely functional programming language thanks to monads. Even though equational reasoning for effectful programs is desirable, it is not yet mainstream. This is partly because it is difficult to maintain pencil-and-paper proofs of large examples. We propose a formalization of a hierarchy of effects using monads in the Coq proof assistant that makes equational reasoning practical. Our main idea is to formalize the hierarchy of effects and algebraic laws like it is done when formalizing hierarchy of traditional algebras. We can then take advantage of the sophisticated rewriting capabilities of Coq to achieve concise proofs of programs. We also show how to ensure the consistency of our hierarchy by providing rigorous models. We explain the various techniques we use to formalize a rich hierarchy of effects (with nondeterminism, state, probability, and more), to mechanize numerous examples from the literature, and we furthermore discuss extensions and new applications.

1 Introduction

Our goal is to provide a framework to produce formal proofs of semantical correctness for programs with effects. To formalize effects, we use *monads*. The notion of monad is one of the category-theoretic frameworks that are used to formalize effects in programming languages and reason about them. It is not the only available option for this purpose (for example, algebraic effects provide an alternative [36, § 5]), but monads comparatively have a longer history in proving themselves useful for the study of semantics [28] as well as for actual programming languages like Haskell as a construct to represent effects [40]. Though there exist a few formalizations of monads in proof assistants, they do not support well our interest in proving programs. Existing formalizations often focus on category theory [17,39] or on meta-theory of programming languages [9]. In contrast, proving programs raises specific practical challenges, among which the generic problem of combining monads is a central issue.

G. Hutton (Ed.): MPC 2019, LNCS 11825, pp. 226–254, 2019.
https://doi.org/10.1007/978-3-030-33636-3_9

In the practical use-cases of monads in programming, a programmer often has to combine two or more monads in order to deal with several effects in the same context. The combination of monads can be carried out in an ad-hoc way [22]. There exist more generic ways to combine monads under specific conditions [21] including the distributive law between monads, which are unfortunately not always satisfied [38] and therefore do not provide a practical solution.

In this paper, we propose a formalization of monads in the Coq proof assistant that addresses monad combination in a practical way. The main idea is to favor a good representation of the hierarchy of effects and their equational theory in terms of interfaces. In other words, monads are composed as in Haskell. We insist on interfaces but this does not preclude the formal construction of models: they just come afterwards. It happens that this corresponds to the presentation of monads as used in monadic equational reasoning [13], so that a direct consequence of our approach is that we can reproduce formally *and* faithfully pencil-and-paper proofs from the literature.

When it comes to proving properties of effectful programs, there is more than the hierarchy of effects: one also needs to provide practical tools to perform equational reasoning. With this respect, the second aspect of our approach is to leverage the rewriting capabilities of Coq by favoring a *shallow embedding*. Shallow embedding is a well-known encoding technique through which one can reuse the native language of the proof assistant at hand. This bears the promise of a reduced formalization effort and it indeed experimentally met some success [16,20] (formal verification using a shallow embedding often relies on a combination of monads and Hoare logic, e.g., [19]). However, most formal verification frameworks proceeds via a deep embedding of the target language, which requires substantial instrumentations of syntax and semantics, resulting in technical lemmas that are difficult to use, which in turn call for meta-programming. Though this paper favors shallow embedding, it does not prevent syntactical reasoning, as we will demonstrate.

Our main contribution in this paper is to demonstrate a combination of formalization techniques that make formal reasoning about effectful programs in Coq practical:

- We formalize a rich hierarchy of effects (failure, exception, nondeterminism, state, probability, and more) whose heart is the theory by Gibbons and Hinze [13] that we extend with more monads and formal models. The key technique is packed classes [11], a methodology used in the MathComp library [26] to formalize the hierarchy of mathematical structures. We do not know of another mechanization with that many monads.
- We provide many definitions and lemmas that allow for the mechanization of several examples. Because we use a shallow embedding, we can leverage Coq native rewriting capabilities, in particular SSReflect's `rewrite` tactic [15].
- The proof scripts we obtain are faithful to the original proofs. We benchmark our library against numerous examples of the literature (most examples from [12,13,30,31]) and observe that formal proofs closely match their pencil-and-paper counterparts and that they can actually be shorter thanks to the

terseness of SSReflect's tactic language. We also apply our framework to new examples such as the formalization of the semantics of an imperative language.

Outline. In Sects. 2 and 3, we show how we build a hierarchy of algebraic laws on top of the theory of monads. In Sect. 4, we illustrate its usability for mechanizing pencil-and-paper proofs. We then deal with syntactic properties in Sect. 5. In Sect. 6, we show how we can give models to our algebraic laws, thus ensuring their consistency. In Sect. 7, we discuss some technical aspects of our formalization of monads that are specific to Coq. We finally discuss related work in Sect. 8 before concluding in Sect. 9.

2 Build a Hierarchy of Algebraic Laws on Top of the Theory of Monads

The heart of our formalization is a hierarchy of effects. Each effect is represented by a monad with some additional algebraic structure that defines the effect, providing effect operators and equations that capture the properties of operators. These effects form a hierarchy in the sense that each effect is the result of a series of extensions starting from the theory of functors, each step extending an existing one in such a way that it shares operators and properties with its parents. We use the methodology of packed classes, which was originally used to formalize mathematical structures [11]. We explain how we use packed classes to formalize monads in Sect. 2.1 and to combine monads in Sect. 2.2. The next section makes a thorough presentation of the complete hierarchy (depicted in Fig. 1).

2.1 Basic Layers: Theories of Functors and Monads

Our formalization of monads starts with a formal definition of functors. This is in contrast to the hierarchy from Gibbons and Hinze [13], where the monad's functor action on morphisms (*fmap*) is defined using *bind* (hereafter, we use the infix notation $\gg=$ for *bind*); starting with functors simplifies the organization of lemmas used in monadic equational reasoning and results in a more robust hierarchy.

Functors as the Base Packed Class. The class of functors is defined in the module `Functor` below. The definition follows the usual one in category theory [25] except that the domain and codomain of functors are fixed to `Type`. In set-theoretical semantics, `Type` is interpreted as the universe of sets, thus rendering our functors to be the endofunctors on the category **Set** of sets and functions.

We use Coq modules only to get a namespace. Inside this namespace, functors are defined by the dependent record `class_of` with one field `f` satisfying the functor laws (the naming should be self-explanatory, see Table 2, Appendix B in case of doubt). The type of functors `t` is a dependent record[1] with a function `m` of type `Type -> Type`, which is the object part of the functor, that satisfies the `class_of` interface. The morphism part appears as `f` in the record. We define `Fun` to refer to it, but the purpose of the definition is essentially technical. It does not reduce (thanks to the `simpl never` declaration) and can therefore be used to provide a stable notation: `F # g` denotes the action of a functor `F` on a function `g`. Last, we provide a notation `functor` that denotes the type `Functor.t` outside of the module and a coercion so that functors can be used as if they were functions (by taking the first projection `m` of the dependent record that represents their type).

```
Module Functor.
Record class_of (m : Type -> Type) : Type := Class {
  f : forall A B, (A -> B) -> m A -> m B ;
  _ : FunctorLaws.id f ;
  _ : FunctorLaws.comp f }.
Structure t : Type := Pack { m : Type -> Type ; class : class_of m }.
Module Exports.
Definition Fun (F : t) : forall A B, (A -> B) -> m F A -> m F B :=
  let: Pack _ (Class f _ _) := F
  return forall A B, (A -> B) -> m F A -> m F B in f.
Arguments Fun _ [A] [B] : simpl never.
Notation functor := t.
Coercion m : functor >-> Funclass.
End Exports.
End Functor.
Export Functor.Exports.
Notation "F # g" := (Fun F g).
```

Monads as a Packed Class Extension. A monad in category theory is defined as an endofunctor M with two natural transformations $\eta : \mathrm{Id} \to M$ (where Id is the identity endofunctor) and $\mu : M^2 \to M$ satisfying some laws [25]. Following the above definition, our class of monads is defined as an extension of the class of functors.

Inside the module `Monad` below, the interface of monads is captured by the dependent record `mixin_of` with two fields `ret` and `join`, that correspond to η and μ respectively, satisfying the monad laws (Table 2, Appendix B). The type of monads `Monad.t` is a dependent record with a function `Monad.m` of type `Type -> Type` that satisfies a `class_of` interface; the latter extends the class of functors (its `base`) with the `mixin` of monads. Thanks to the definition `baseType`, a monad can also be seen as a functor. This fact is handled transparently by the type system of Coq thanks to the `Canonical` command.

```
Module Monad.
Record mixin_of (M : functor) : Type := Mixin {
```

[1] `Record` and `Structure` are synonymous but the latter is used to emphasize that it is to be made `Canonical`.

```
  ret : forall A, A -> M A ;
  join : forall A, M (M A) -> M A ;
  _ : JoinLaws.ret_naturality ret ;
  _ : JoinLaws.join_naturality join ;
  _ : JoinLaws.left_unit ret join ;
  _ : JoinLaws.right_unit ret join ;
  _ : JoinLaws.associativity join }.
Record class_of (M : Type -> Type) := Class {
  base : Functor.class_of M ; mixin : mixin_of (Functor.Pack base) }.
Structure t : Type := Pack { m : Type -> Type ; class : class_of m }.
Definition baseType (M : t) := Functor.Pack (base (class M)).
Module Exports.
(* intermediate definitions of Ret and Join omitted *)
Notation monad := t.
Coercion baseType : monad >-> functor.
Canonical baseType.
End Exports.
End Monad.
Export Monad.Exports.
```

The monad above is defined in terms of `ret` and `join`. In programming, the operator *bind* is more common. Using Coq notation, its type can be written `forall A B, M A -> (A -> M B) -> M B`. The second argument of type `A -> M B` is a Coq function that represents a piece of effectful program. This concretely shows that we are heading for a framework using a shallow embedding. We provide an alternative way to define monads using *ret* and *bind*. Let us assume that we are given `ret` and `bind` functions that satisfy the monad laws:

```
Variable M : Type -> Type.
Variable bind : forall A B, M A -> (A -> M B) -> M B.
Variable ret : forall A, A -> M A.
Hypothesis bindretf : BindLaws.left_neutral bind ret.
Hypothesis bindmret : BindLaws.right_neutral bind ret.
Hypothesis bindA : BindLaws.associative bind.
```

We can then define `fmap` that satisfies the functor laws:

```
Definition fmap A B (f : A -> B) (m : M A) := bind m (ret (A:=B) \o f).
Lemma fmap_id : FunctorLaws.id fmap.
Lemma fmap_o : FunctorLaws.comp fmap.
```

We can use these lemmas to build `M'` of type `functor` and use `M'` to define `join`:

```
Definition join A (pp : M' (M' A)) := bind pp id.
```

It is now an exercise to prove that `ret` and `join` satisfy the monad laws, using which we eventually build `M` of type `monad`. We call `Monad_of_ret_bind` this construction that we use in the rest of this paper.

2.2 Extensions: Specific Monads as Combined Theories

In the previous section, we explained the case of a simple extension: one structure that extends another. In this section we explain how a structure extends *two* structures. Here, we just explain how we combine theories, how we provide concrete models for combined theories is the topic of Sect. 6.

For the sake of illustration, we use the nondeterminism monad that extends both the failure monad and the choice monad. The failure monad `failMonad` extends the class of monads (Sect. 2.1) with a failure operator `fail` that is a left-zero of *bind*. Since the extension methodology is the same as in Sect. 2.1, we provide the code with little explanations[2]:

```
Module MonadFail.
Record mixin_of (M : monad) : Type := Mixin {
  fail : forall A, M A ;
  _ : BindLaws.left_zero (@Bind M) fail }.
Record class_of (m : Type -> Type) := Class {
  base : Monad.class_of m ; mixin : mixin_of (Monad.Pack base) }.
Structure t := Pack { m : Type -> Type ; class : class_of m }.
Definition baseType (M : t) := Monad.Pack (base (class M)).
Module Exports.
(* intermediate definition of Fail omitted *)
Notation failMonad := t.
Coercion baseType : failMonad >-> monad.
Canonical baseType.
End Exports.
End MonadFail.
Export MonadFail.Exports.
```

The choice monad `altMonad` extends the class of monads with a choice operator `alt` (infix notation: `[~]`; prefix: `[~p]`) that is associative and such that *bind* distributes leftwards over choice (the complete code is displayed in Appendix A).

The nondeterminism monad `nondetMonad` defined below extends both the failure monad and the choice monad. This extension is performed by first selecting the failure monad as the `base` whose base itself is further required to satisfy the mixin of the choice monad (see `base2` below). As a result, a nondeterminism monad can be regarded both as a failure monad (definition `baseType`) or as a choice monad (definition `alt_of_nondet`): both views are declared as `Canonical`.

```
Module MonadNondet.
Record mixin_of (M : failMonad) (a : forall A, M A -> M A -> M A) : Type :=
  Mixin { _ : BindLaws.left_id (@Fail M) a ;
          _ : BindLaws.right_id (@Fail M) a }.
Record class_of (m : Type -> Type) : Type := Class {
  base : MonadFail.class_of m ;
  base2 : MonadAlt.mixin_of (Monad.Pack (MonadFail.base base)) ;
  mixin : @mixin_of (MonadFail.Pack base) (MonadAlt.alt base2) }.
```

[2] Just note that the prefix @ turns off implicit arguments in Coq.

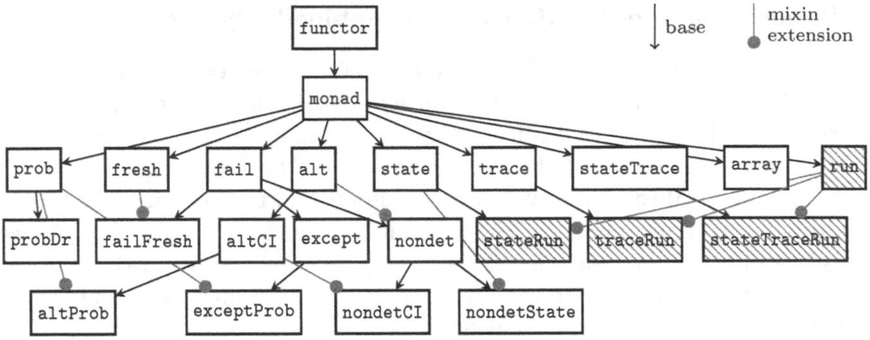

Fig. 1. Hierarchy of effects formalized. See Table 3 for the algebraic laws. In the Coq scripts [3], the monad **xyz** appears as **xyzMonad**.

```
Structure t : Type := Pack { m : Type -> Type ; class : class_of m }.
Definition baseType (M : t) := MonadFail.Pack (base (class M)).
Module Exports.
Notation nondetMonad := t.
Coercion baseType : nondetMonad >-> failMonad.
Canonical baseType.
Definition alt_of_nondet (M : nondetMonad) : altMonad :=
  MonadAlt.Pack (MonadAlt.Class (base2 (class M))).
Canonical alt_of_nondet.
End Exports.
End MonadNondet.
Export MonadNondet.Exports.
```

3 More Monads from Our Hierarchy of Effects

This section complements the previous one by explaining more monads from our hierarchy of effects (Fig. 1). We explain these monads in particular because they are used later in the paper[3] They are all obtained using the combination technique previously explained in Sect. 2.2.

3.1 The Exception Monad

The exception monad **exceptMonad** extends the failure monad (Sect. 2.2) with a **Catch** operator with monoidal properties (the **Fail** operator being the neutral) and the property that unexceptional bodies need no handler [13, §5]:

[3] The exception monad is used in the motivating example of Sect. 4.1, state-related monads are used in particular to discuss the relation with deep embedding in Sect. 5.1, the state-trace monad is used in the application of Sect. 5.2, and a model of the probability monad is provided in Sect. 6.2.

```
Record mixin_of (M : failMonad) : Type := Mixin {
  catch : forall A, M A -> M A -> M A ;
  _ : forall A, right_id Fail (@catch A) ;
  _ : forall A, left_id Fail (@catch A) ;
  _ : forall A, associative (@catch A) ;
  _ : forall A x, left_zero (Ret x) (@catch A) }.
```

The algebraic laws are given self-explanatory names; see Table 1, Appendix B in case of doubt.

3.2 The State Monad and Derived Structures

The state monad is certainly the first monad that comes to mind when speaking of effects. It denotes computations that transform a state (type `S` below). It comes with a `Get` operator to yield a copy of the state and a `Put` operator to overwrite it. These functions are constrained by four laws [13]:

```
Record mixin_of (M : monad) (S : Type) : Type := Mixin {
  get : M S ;
  put : S -> M unit ;
  _ : forall s s', put s >> put s' = put s' ;
  _ : forall s, put s >> get = put s >> Ret s ;
  _ : get >>= put = skip ;
  _ : forall k : S -> S -> M S,
      get >>= (fun s => get >>= k s) = get >>= fun s => k s s }.
```

Reification of State Monads. We introduce a `Run` operator to reify state-related monads (this topic is briefly exposed in [13, §6.2], we use reification in Sect. 3.3). First, the operator `run` defines the semantics of `Ret` and `Bind` according to the following equations:

```
Record mixin_of S (M : monad) : Type := Mixin {
 run : forall A, M A -> S -> A * S ;
 _ : forall A (a : A) s, run (Ret a) s = (a, s) ;
 _ : forall A B (m : M A) (f : A -> M B) s,
    run (do a <- m ; f a) s = let: (a', s') := run m s in run (f a') s' }.
```

The type of `run` shows that it turns a state into a pair of a value and a state. We call the monad that extends `monad` with such an operator a `runMonad`. Second, we combine `stateMonad` with `runMonad` and extend it with `Run` equations for `Get` and `Put`; this forms the `stateRunMonad`:

```
Record mixin_of S (M : runMonad S)
  (get : M S) (put : S -> M unit) : Type := Mixin {
  _ : forall s, Run get s = (s, s) ;
  _ : forall s s', Run (put s') s = (tt, s') }.
```

Monads with the `Run` operator appear shaded in Fig. 1, they can be given concrete models so as to run sample programs inside Coq (there are toy examples in [3, file `smallstep_examples.v`]).

The Backtrackable-State Monad. The monad `nondetStateMonad` combines state with nondeterminism (recall that the nondeterminism monad is itself already the result of such a combination) and extends their properties with the properties of *backtrackable-state* ([13, §6], [30, §4]):

```
Record mixin_of (M : nondetMonad) : Type := Mixin {
  _ : BindLaws.right_zero (@Bind M) (@Fail _) ;
  _ : BindLaws.right_distributive (@Bind M) [~p] }.
```

Failure is a right zero of composition to discard any accumulated stateful effects and composition distributes over choice.

3.3 The State-Trace Monad

The state-trace monad is the result of combining a state monad with a trace monad. Our trace monad extends monads with a `Mark` operator to record events:

```
Record mixin_of T (m : Type -> Type) : Type :=
  Mixin { mark : T -> m unit }.
```

We call the operators of the state-trace monad `st_get`, `st_put`, and `st_mark` (notations: `stGet`, `stPut`, `stMark`). `stGet` and `stPut` fulfill laws similar to the ones of `Get` and `Put`, but their interactions with `stMark` call for two more laws:

```
Record mixin_of S T (M : monad) : Type := Mixin {
  st_get : M S ;
  st_put : S -> M unit ;
  st_mark : T -> M unit ;
  _ : forall s s', st_put s >> st_put s' = st_put s' ;
  _ : forall s, st_put s >> st_get = st_put s >> Ret s ;
  _ : st_get >>= st_put = skip ;
  _ : forall k : S -> S -> M S,
      st_get >>= (fun s => st_get >>= k s) = st_get >>= fun s => k s s ;
  _ : forall s e, st_put s >> st_mark e = st_mark e >> st_put s ;
  _ : forall e (k : _ -> _ S),
      st_get >>= (fun v => st_mark e >> k v) = st_mark e >> st_get >>= k }
```

3.4 The Probability Monad

First, we define a type `prob` of probabilities [4] as reals of type `R` between 0 and 1:

```
(* Module Prob *)
Record t := mk { p :> R ; pO1 : 0 <= p <= 1 }.
Definition O1 (p : t) := pO1 p.
Arguments O1 : simpl never.
Notation prob := t.
Notation "'`Pr' q" := (@mk q (@O1 _)).
```

This definition is interesting because the notation makes it possible to write concrete probabilities succinctly: the proof that the real is between 0 and 1 is hidden and can be inferred automatically. For example, the probability $\frac{1}{2}$ is written `Pr /2`, the probability $\bar{p} = 1 - p$ (where p is a probability) is written `Pr p.~`, etc. This is under the condition that we equip Coq with appropriate canonical structures. For example, here follows the registration of the proof $0 \leq \frac{1}{p} \leq 1$ that makes it possible to write `Pr /2` (`IZR` injects integers into reals):

```
Lemma prob_IZR (p : positive) : 0 <= / IZR (Zpos p) <= 1.
Canonical probIZR (p : positive) := @Prob.mk _ (prob_IZR p).
```

The above type and notation for probabilities lead us to the following mixin for the probability monad [13, § 8]:

```
1  Record mixin_of (M : monad) : Type := Mixin {
2    choice : forall (p : prob) A, M A -> M A -> M A
3             where "mx <| p |> my" := (choice p mx my) ;
4    _ : forall A (mx my : M A), mx <| `Pr 0 |> my = my ;
5    _ : forall A (mx my : M A), mx <| `Pr 1 |> my = mx ;
6    _ : forall A p (mx my : M A), mx <| p |> my = my <| `Pr p.~ |> mx ;
7    _ : forall A p, idempotent (@choice p A) ;
8    _ : forall A (p q r s : prob) (mx my mz : M A),
9      p = r * s /\ s.~ = p.~ * q.~ ->
10     mx <| p |> (my <| q |> mz) = (mx <| r |> my) <| s |> mz ;
11   _ : forall p, BindLaws.left_distributive (@Bind M) (choice p)  }.
```

`mx <p> my` behaves as `mx` with probability `p` and as `my` with probability $\bar{p}$. Lines 6 and 7 are a skewed commutativity law and idempotence. Lines 8–10 is a quasi associativity law. Above laws are the same as convex spaces [18, Def 3]. Line 11 says that *bind* left-distributes over probabilistic choice.

3.5 Other Monads in the Hierarchy of Effects

Figure 1 pictures the hierarchy of effects that we have formalized; Table 3 (Appendix C) lists the corresponding algebraic laws. The starting point is the hierarchy of [13]. It needed to be adjusted to fit other papers [1,12,30,31]:

- As explained in Sect. 2.1, we put functors at the top to simplify formal proofs.
- The examples of [13] relying on nondeterministic choice use `altMonad`. However, the combination of nondeterminism and probability in `altProbMonad` requires idempotence and commutativity of nondeterministic choice [12]. Idempotence and commutativity are also required in the first part of [31]. We therefore insert the monad `altCIMonad` with those properties in the hierarchy, and also the monad `nondetCIMonad` to deal more specifically with the second part of [31].
- The probability monad `probMonad` is explained in Sect. 3.4. The probability monad `probDrMonad` is explained in [13, §8]. The main difference with [13] is that we extract `probMonad` from `probDrMonad` as an intermediate step.

probDrMonad extends probMonad with right distributivity of *bind* $(\cdot \gg= \cdot)$ over probabilistic choice $(\cdot \triangleleft \cdot \triangleright \cdot)$. The reason is that this property is not compatible with distributivity of probabilistic choice over nondeterministic choice $(\cdot \square \cdot)$ and therefore needs to be put aside to be able to form altProbMonad by combining probMonad and altMonad (the issue is explained in [1]).

There are two more monads that we have not explained. exceptProbMonad combines probability and exception [12, §7.1]. freshMonad and failFreshMonad are explained in [13, §9.1]; freshMonad provides an operator to generate fresh labels.

We have furthermore extended the hierarchy of [13] with reification (Sect. 3.2), the trace and state-trace monads (Sect. 3.3), and the array monad [35].

4 Monadic Equational Reasoning

The faithful mechanization of pencil-and-paper proofs by monadic equational reasoning is the main benefit of a hierarchy of effects built with packed classes. After a motivating example in Sect. 4.1, we explain how the Coq rewrite tactics copes with notation and lemma overloading in Sect. 4.2. Section 4.3 explains the technical issue of rewriting under function abstractions. Section 4.4 provides an overview of the existing proofs that we have mechanized.

4.1 Motivating Example: The Fast Product

This example shows the equivalence between a functional implementation of the product of integers with a monadic version (fastprod) [13]. On the left of Fig. 2 we (faithfully) reproduce the series of rewritings that constitute the original proof. On the right, we display the equivalent series of Coq goals and tactics.

The product of natural numbers is simply defined as foldr muln 1. A "faster" product can be implemented using the failure monad (Sect. 2.2) and the exception monad (Sect. 3.1):

```
Definition work (M : failMonad) s : M nat :=
  if O \in s then Fail else Ret (product s).
Definition fastprod (M : exceptMonad) s : M nat := Catch (work s) (Ret O).
```

We observe that the user can write a monadic program with one monad and use a notation from a monad below in the hierarchy. Concretely, here, work is written with failMonad but still uses the unit operator Ret of the base monad. The same can be said of fastprod. This is one consequence of packed classes. What happens is that Coq inserts appropriate calls to canonical structures so that the program type-checks. In fact, the program work and fastprod are actually *equal* to the following (more verbose) ones:

```
Let Work (M : failMonad) s := if O \in s
then @Fail M nat else @Ret (MonadFail.baseType M) nat (product s).
Let Fastprod (M : exceptMonad) s := @Catch M nat
(@work (MonadExcept.baseType M) s) (@Ret (MonadExcept.monadType M) nat O).
```

Pencil-and-paper proof [13, §5.1]	Coq intermediate goals and tactics
`fastprod xs`	`fastprod s`
=⟦ definition of `fastprod` ⟧	=⟦ `rewrite /fastprod` ⟧
`catch (work xs) (ret 0)`	`Catch (work s) (Ret 0)`
=⟦ specification of `work` ⟧	=⟦ `rewrite /work` ⟧
`catch (if 0 in xs then fail`	`Catch (if 0 \in s then Fail`
`else ret (product xs)) (ret 0)`	`else Ret (product s)) (Ret 0)`
=⟦ lift out the conditional ⟧	=⟦ `rewrite lift_if if_ext` ⟧
`if 0 in xs then catch fail (ret 0)`	`if 0 \in s then Catch Fail (Ret 0)`
`else catch (ret (product xs)) (ret 0)`	`else Catch (Ret (product s)) (Ret 0)`
=⟦ laws of catch, fail, and ret ⟧	=⟦ `rewrite catchfailm catchret` ⟧
`if 0 in xs then ret 0`	`if 0 \in s then Ret 0`
`else ret (product xs)`	`else Ret (product s)`
=⟦ arithmetic: 0 in xs ⇒ `product xs` = 0 ⟧	=⟦ `case: ifPn => // /product0` ⟧
`if 0 in xs then ret (product xs)`	(`product0` $\stackrel{def}{=} \forall s.\ 0 \in s \rightarrow$ `product` $s = 0$)
`else ret (product xs)`	`Ret 0`
=⟦ redundant conditional ⟧	=⟦ `move <-` ⟧
`ret (product xs)`	`Ret (product s)`

Fig. 2. Comparison between an existing proof and our Coq formalization

The Coq proof that `fastprod` is pure, i.e., that it never throws an unhandled exception, can be compared to its pencil-and-paper counterpart in Fig. 2. Both proofs are essentially the same, though in practice the Coq proof will be streamlined in two lines (of less than 80 characters) of script:

```
Lemma fastprodE s : fastprod s = Ret (product s).
Proof.
rewrite /fastprod /work lift_if if_ext catchfailm.
by rewrite catchret; case: ifPn => // /product0 <-.
Qed.
```

The fact that we achieve the same conciseness as the pencil-and-paper proof is not because the example is simple: the same can be said of all the examples we mechanized (see Sect. 4.4).

4.2 Basics of Equational Reasoning with Packed Classes

Packed classes not only allow sharing of notations but also sharing of lemmas: one can rewrite a monadic program with any algebraic law from structures below in the hierarchy of effects. SSReflect's advanced `rewrite` tactic[4] becomes available to faithfully reproduce monadic equational reasoning.

[4] SSReflect extends Coq's `rewrite` with contextual patterns, unfolding, etc. [15]. The main benefit is that semantically-close actions can be performed on the same line of script, instead of having to interleave with other Coq tactics such as `pattern` or `unfold`.

For illustration, let us consider a function that nondeterministically builds a subsequence of a list using the choice monad [12, §3.1]:

```
Variables (M : altMonad) (A : Type).
Fixpoint subs (s : seq A) : M (seq A) :=
  if s isn't h :: t then Ret [::]
  else let t' := subs t in fmap (cons h) t' [~] t'.
```

The mixed use of algebraic laws from various monads can be observed when proving that subsequences of concatenation are concatenations of subsequences:

```
1  Lemma subs_cat (xs ys : seq A) :
2    subs (xs ++ ys) = do us <- subs xs; do vs <- subs ys; Ret (us ++ vs).
3  Proof.
4  elim: xs ys => [ys |x xs IH ys].
5    rewrite /= bindretf. (* Ret is left neutral *)
6    by rewrite bindmret. (* Ret is right neutral *)
7  rewrite [in RHS]/=. (* beta-reduction of the rhs *)
8  rewrite alt_bindDl. (* left-distribution of Bind over Alt *)
9  rewrite bindA. (* associativity of Bind *)
10 rewrite [in RHS]/=. (* to be continued in Sect. 4.3 *)
```

The proof is by induction on the sequence `xs` (line 4). While the lemma `alt_bindDl` (line 8) belongs to the interface of the `altMonad` interface, the lemma `bindA` (line 9) comes from the `monad` interface.

4.3 Rewriting Under Function Abstractions

In pencil-and-paper proofs of monadic equational reasoning, whether rewriting occurs under a function abstraction or not does not make any difference. We need custom automation to support this feature in Coq which does not natively perform rewriting in this situation.

The proof from the previous section led us to the following subgoal:

```
subs ((x :: xs) ++ ys) =
do x0 <- subs xs; do us <- Ret (x :: x0); do vs <- subs ys; Ret (us ++ vs)
[~] (do us <- subs xs; do vs <- subs ys; Ret (us ++ vs))
```

We want to turn the first branch of the nondeterministic choice

```
do x0 <- subs xs; do us <- Ret (x :: x0); do vs <- subs ys; Ret (us ++ vs)
```

into

```
do x0 <- subs xs; do vs <- subs ys; Ret (x :: x0 ++ vs)
```

but since the occurrence of `Ret` of interest is under the binder "`do x0 <-`", `rewrite bindretf` fails. Instead, we "open" the continuation with a custom tactic `Open (X in subs xs >>= X)` to get a new subgoal

```
do us <- Ret (x :: x0); do vs <- subs ys; Ret (us ++ vs) = ?g x0
```

where `?g` is an existential variable. Now, `rewrite bindretf` succeeds:

```
do vs <- subs ys; Ret ((x :: x0) ++ vs) = ?g x0
```

Yet, the last `Ret` is still under a binder. We could again "open" the continuation but instead we use a custom "rewrite under" tactic `rewrite_ cat_cons` to get:

```
do x1 <- subs ys; Ret (x :: x0 ++ x1) = ?g x0
```

Now we can trigger unification to instantiate the existential variable and thus complete the intended rewriting.

In practice, there is little need for `Open` and most situations can be handled directly without revealing the existential variable using `rewrite_`. We chose to explain `Open` here because it shows how `rewrite_` is implemented.

4.4 Mechanization of Existing Pencil-and-Paper Proofs

We used our framework to mechanize the definitions, lemmas, and examples from [13] (except Sect. 10.2), from [12] (up to Sect. 7.2, which overlaps and complements [13]), examples from [30,31], and examples from [21] (up to Sect. 3). This includes in particular:

- Spark aggregation: Spark is a platform for distributed computing, in which the aggregation of data is therefore nondeterministic. Monadic equational reasoning can be used to sort out the conditions under which aggregation is actually deterministic [31, §4.2] as well as other properties. We have mechanized these results ([3], file `example_spark.v`), which are part of a larger specification [6].
- The n-queens puzzle: This puzzle is used to illustrate the combination of state and nondeterminism. We have mechanized the relations between functional and stateful implementations [13, §6–7] ([3], file `example_nqueens.v`), as well as the derivation of a version of the algorithm using monadic hylo-fusion [30, §5]. This example demonstrates the importance of commutativity lemmas, calling for syntax reflection (see Sect. 5).
- The Monty Hall problem: We have mechanized the probability calculations for several variants of the Monty Hall problem [12,13] using `probMonad`, `altProbMonad`, and `exceptProbMonad` ([3], file `example_monty.v`).
- The tree relabeling example: This example originally motivated monadic equational reasoning [13]. It amounts to show that the labels of a binary tree are distinct when the latter has been relabeled with fresh (see `freshMonad`) labels. We have mechanized this result ([3], file `example_relabeling.v`).
- The *swap* construction: This is an example of monad composition [21]. Strictly speaking, this is not monadic equational reasoning: formalization does not require a mechanism such as canonical structures. Yet, our framework proved adequate because it allows to mix in a single equation different *ret*'s and *join*'s without explicit mention of which monad they belong to; inference is automatic thanks to coercions.

The level of details provided by the authors using monadic equational reasoning is helpful and provides a way to check that our mechanization is faithful. Among the differences between pencil-and-paper and mechanized proofs, the main one is maybe function termination. Pencil-and-paper proofs assume Haskell and do not require particular care about function termination, whereas Coq functions must terminate, so that formalization requires an extra effort. See for example the formalization of `unfoldM` and `hyloM` [3] which are not structurally terminating. These difficulties are known [32] and can be addressed using standard techniques. Another difference is that Coq functions must be total, so that some Haskell functions cannot be formalized as such (e.g., `foldr1`).

We discovered a few problems in the work we have formalized. The main one was an error in a proof of monadic hylo-fusion for the n-queens puzzle from a draft paper [29] which has been reported to the author and fixed [30]. In short[5], the functional specification of the n-queens puzzle can be rewritten using `nondetStateMonad` as

```
Get >>= (fun ini => Put (0, [::], [::]) >>
  queensBody (map Z_of_nat (iota 0 n)) >>= overwrite ini)
```

in which `queensBody` can be rewritten as

```
hyloM (@opdot_queens M) [::] (@nilp _)
  select seed_select (@well_founded_size _)
```

The heart of this last step was a theorem [29, Thm 4.2] (now [30, Thm 5.1]) whose hypotheses did not properly match the ones available in the course of the proof. However, we were able to complete the proof with a variant of the theorem in question. Other problems were at the level of typos (they could be easily caught by type-checking): almost none in [13], a few in the appendices of [6] (whose mechanization has not been completed yet).

5 Properties Proved Using Syntax

Our formalization is a shallow embedding: a monadic program is a Coq function of return-type `M A` for some monad `M` and some type `A`. This is practical because we can use the Coq language to write, execute, and prove programs. However, it happens that some properties require an explicit syntax to be proved. In this section, we show how to handle such situations. The basic idea is to *locally* restrict programs to a subset characterized by a deep embedding. Section 5.1 is an example of property of backtrackable-states. Section 5.2 is an example of equivalence between an operational and a denotational semantics, the latter being given by a monad.

[5] We just show the main steps of the derivation, we cannot reproduce all the definitions for lack of space, see the source code [3] for all the details.

5.1 The Commutativity of State and Nondeterminism

The commutativity of state and nondeterminism is an important aspect of backtrackable-states [30]. Such a property can be proved directly on specific programs using their semantics but it can also be proved more generally using syntax.

The following predicate [30, Def 4.2] defines the commutativity of two computations `m` and `n` (in the same monad `M`):

```
Definition commute {M : monad} A B
  (m : M A) (n : M B) C (f : A -> B -> M C) : Prop :=
  m >>= (fun x => n >>= (fun y => f x y)) =
  n >>= (fun y => m >>= (fun x => f x y)).
```

In order to state a generic property of commutativity between nondeterminism and state monads, we first define a predicate that captures syntactically nondeterminism monads. They are written with the following (higher-order abstract [33]) syntax:

```
(* Module SyntaxNondet *)
Inductive t : Type -> Type :=
| ret : forall A, A -> t A
| bind : forall B A, t B -> (B -> t A) -> t A
| fail : forall A, t A
| alt : forall A, t A -> t A -> t A.
```

Let `denote` be a function that turns the above syntax into the corresponding monadic computation:

```
Fixpoint denote (M : nondetMonad) A (m : t A) : M A :=
match m with
| ret A a => Ret a
| bind A B m f => denote m >>= (fun x => denote (f x))
| fail A => Fail
| alt A m1 m2 => denote m1 [~] denote m2
end.
```

Using above definitions, we can write a predicate that captures computations in a `nondetStateMonad` that are actually just computations in a `nondetMonad`:

```
Definition nondetState_sub S (M : nondetStateMonad S) A (n : M A) :=
  {m | denote m = n}.
```

Eventually, it becomes possible to prove *by induction on the syntax* that two computations `m` and `n` using both state and choice commute when `m` actually does not use the state effects:

```
Lemma commute_nondetState S (M : nondetStateMonad S)
  A (m : M A) B (n : M B) C (f : A -> B -> M C) :
  nondetState_sub m -> commute m n f.
```

5.2 Equivalence Between Operational and Denotation Semantics

We consider a small imperative language with a state and an operator to generate events. We equip this language with a small-step semantics and a denotational semantics using `stateTraceMonad` (Sect. 3.3), and prove that both semantics are equivalent. We will see that we need an induction on the syntax to prove this equivalence.

Here follows the (higher-order abstract) syntax of our imperative language:

```
Inductive program : Type -> Type :=
| p_ret  : forall {A}, A -> program A
| p_bind : forall {A B}, program A -> (A -> program B) -> program B
| p_cond : forall {A}, bool -> program A -> program A -> program A
| p_get  : program S
| p_put  : S -> program unit
| p_mark : T -> program unit | ... (* see Appendix D *)
```

We give our language a small-step semantics specified with continuations in the style of CompCert [5]. We distinguish two kinds of continuations: `stop` for halting and `cont` (notation: ·;·) for sequencing:

```
Inductive continuation : Type :=
| stop : forall A, A -> continuation
| cont : forall A, program A -> (A -> continuation) -> continuation.
```

We can then define the ternary relation `step` that relates a state to the next one and optionally an event:

```
Definition state : Type := S * @continuation T S.
Inductive step : state -> option T -> state -> Prop :=
| s_ret  : forall s A a (k : A -> _), step (s, p_ret a `; k) None (s, k a)
| s_bind : forall s A B p (f : A -> program B) k,
    step (s, p_bind p f `; k) None (s, p `; fun a => f a `; k)
| s_cond_true : forall s A p1 p2 (k : A -> _),
    step (s, p_cond true p1 p2 `; k) None (s, p1 `; k)
| s_cond_false : forall s A p1 p2 (k : A -> _),
    step (s, p_cond false p1 p2 `; k) None (s, p2 `; k)
| s_get  : forall s k, step (s, p_get `; k) None (s, k s)
| s_put  : forall s s' k, step (s, p_put s' `; k) None (s', k tt)
| s_mark : forall s t k, step (s, p_mark t `; k) (Some t) (s, k tt)
| ... (* see Appendix D *)
```

Its reflexive and transitive closure `step_star` of type `state -> seq T -> state -> Prop` is defined as one expects. We prove that `step` is deterministic and that `step_star` is confluent and deterministic.

We also give our language a denotational semantics using the `stateTraceMonad`:

```
Variable M : stateTraceMonad S T.
Fixpoint denote A (p : program A) : M A :=
  match p with
  | p_ret _ v => Ret v
  | p_bind _ _ m f => do a <- denote m; denote (f a)
  | p_cond _ b p1 p2 => if b then denote p1 else denote p2
  | p_get => stGet
  | p_put s' => stPut s'
  | p_mark t => stMark t | ... (* see Appendix D *) end.
```

It is important to note here that the operators `stGet` and `stPut` can only read and update the state (of type `S`) but not the log of emitted events (of type `seq T`). Only the operator `stMark` has access to the list of emitted events but it can neither read nor overwrite it: it can only log a new event to the list.

We proved the correctness and completeness of the small-step semantics `step_star` w.r.t. the denotational semantics `denote` [3, file `smallstep_monad.v`]. For that we use only the equations of the run interface of the state-trace monad (Sect. 3.3). We now come to those parts of the proofs of correctness and completeness that require induction on the syntax. They take the form of two lemmas. Like in the previous section, we introduce a predicate to distinguish the monadic computations that can be written with the syntax of the programming language:

```
Definition stateTrace_sub A (m : M A) := { p | denote p = m }.
```

The first lemma states that once an event is emitted it cannot be deleted:

```
Lemma denote_prefix_preserved A (m : M A) : stateTrace_sub m ->
  forall s s' l1 l a, Run m (s, l1) = (a, (s', l)) ->
    exists l2, l = l1 ++ l2.
```

The second lemma states that the remaining execution of a program does not depend on the previously emitted events:

```
Lemma denote_prefix_independent A (m : M A) : stateTrace_sub m ->
  forall s l1 l2, Run m (s, l1 ++ l2) =
    let res := Run m (s, l2) in (res.1, (res.2.1, l1 ++ res.2.2)).
```

Those are natural properties that ought to be true for any monadic code, and not only the monadic code that results from the denotation of a `program`. But this is not the case with our monad. Indeed, the interface specifies those operators that should be implemented but does not prevent one to add other operators that might break the above properties of emitted events. This is why we restrict those properties to monadic code using the `stateTrace_sub` predicate, thus allowing us to prove the two above lemmas by induction on the syntax.

6 Models of Monads

Sections 2 and 3 explained how to build a hierarchy of effects. In this section, we complete this formalization by explaining how to provide models, i.e., concrete objects that validate the equational theories. Providing a model amounts to define a function of type `Type -> Type` for the base monad and instantiate all the interfaces up to the monad of interest. For illustration, we explain models of state monads and of the probability monad; see ([3], file `monad_model.v`) for simpler models.

6.1 Models of State Monads

State-Trace Monad. A model for `stateTraceMonad` (Sect. 3.3) is a function `fun A => S * seq T -> A * (S * seq T)`. We start by providing the *ret* and *bind* operators of the base `monad` using the constructor `Monad_of_ret_bind` (Sect. 2.1):

```
(* Module ModelMonad *)
Variables S : Type.
Let m := fun A => S -> A * S.
Definition state : monad.
refine (@Monad_of_ret_bind m
  (fun A a => fun s => (a, s)) (* ret *)
  (fun A B m f => fun s => uncurry f (m s)) (* bind *) _ _ _).
```

One needs to prove the monad laws to complete this definition. This gives a monad `ModelMonad.state` upon which we define the get, put, and mark operators:

```
(* Module ModelStateTrace *)
Variables (S T : Type).
Program Definition mk : stateTraceMonad S T :=
  let m := Monad.class (@ModelMonad.state (S * seq T)) in
  let stm := @MonadStateTrace.Class S T _ m
  (@MonadStateTrace.Mixin _ _ (Monad.Pack m)
    (fun s => (s.1, s)) (* st_get *)
    (fun s' s => (tt, (s', s.2))) (* st_put *)
    (fun t s => (tt, (s.1, rcons s.2 t))) (* st_mark *) _ _ _ _ _ _) in
  @MonadStateTrace.Pack S T _ stm.
```

The laws of the state-trace monad are proved automatically by Coq.

Backtrackable-State. A possible model for `nondetStateMonad` (Sect. 3.2) is `fun A => S -> {fset (A * S)}`, where `{fset X}` is the type of finite sets over `X` provided by the FINMAP library. This formalization of finite sets is based on list representations of finite predicates. The canonical representation is chosen uniquely among its permutations. This choice requires the base type `X` of `{fset X}` to be a `choiceType`, i.e., a type equipped with a choice function, thus satisfying a form of the axiom of choice. To be able to use the FINMAP library, we use a construct (`gen_choiceMixin`) from the MATHCOMP-ANALYSIS library that can turn any type into a `choiceType`. We use it to define a model for `nondetStateMonad` as follows:

```
Let choice_of_Type (T : Type) : choiceType :=
  Choice.Pack (Choice.Class (equality_mixin_of_Type T) gen_choiceMixin).
Definition _m : Type -> Type :=
  fun A => S -> {fset (choice_of_Type A * choice_of_Type S)}.
```

It remains to prove all the algebraic laws of the interfaces up to `nondetStateMonad`; see ([3], file `monad_model.v`) for details.

6.2 A Model of the Probability Monad

A theory of probability distributions provides a model for the probability monad (Sect. 3.4). For this paper, we propose the following definition of probability distribution [4]:

```
(* Module Dist *)
Record t := mk {
  f :> {fsfun A -> R with 0} ;
  f01 : all (fun x => 0 < f x) (finsupp f) &&
        \sum_(a <- finsupp f) f a == 1}.
```

The first field is a finitely-supported function `f`: it evaluates to `0` outside its support `finsupp f`. The second field contains proofs that (1) the probability function outputs positive reals and that (2) its outputs sum to 1. Let `Dist` be a notation for `Dist.t`. It has type `choiceType -> choiceType` and can therefore be used to build a `monad` (thanks to `choice_of_Type` from the previous section).

The *bind* operator is well-known: given `p : Dist A` and `g : A -> Dist B`, it returns a distribution with probability mass function $b \mapsto \sum_{a \in \mathtt{supp(p)}} \mathtt{p}(a) \cdot \mathtt{g}(a, b)$. This is implemented by the following combinator:

```
(* Module DistBind *)
Variables (A B : choiceType) (p : Dist A) (g : A -> Dist B).
Let D := ... (* definition of the support omitted *)
Definition f : {fsfun B -> R with 0} :=
  [fsfun b in D => \sum_(a <- finsupp p) p a * (g a) b | 0].
Definition d : Dist B := ... (* packaging of f omitted *)
```

The resulting combinator `DistBind.d` can be proved to satisfy the monad laws, for example, associativity:

```
Lemma DistBindA A B C (m : Dist A) (f : A -> Dist B) (g : B -> Dist C) :
  DistBind.d (DistBind.d m f) g =
    DistBind.d m (fun x => DistBind.d (f x) g).
```

Completing the model with a distribution for the *ret* operator and the other properties of monads is an exercise.

The last step is to provide an implementation for the interface of the probability monad. The probabilistic choice operator corresponds to the construction of a distribution `d` from two distributions `d1` and `d2` biased by a probability `p`:

```
(* Module Conv2Dist *)
Variables (A : choiceType) (d1 d2 : Dist A) (p : prob).
Definition d : Dist A := locked
  (ConvDist.d (I2Dist.d p) (fun i => if i == ord0 then d1 else d2)).
```

The combinator `ConvDist.d` is a generalization that handles the combination of any distribution of distributions: it is instantiated here with the binary distribution `I2Dist.d p` [4]. We finally prove that the probabilistic choice `d` have the expected properties, for example, skewed commutativity:

```
Notation "x <| p |> y" := (d x y p). (* probabilistic choice *)
Lemma convC (p : prob) (a b : Dist A) : a <| p |> b = b <| `Pr p.~ |> a.
```

7 Technical Aspects of Formalization in Coq

About Coq *Commands and Tactics.* There are several Coq commands and tactics that are instrumental in our formalization. Most importantly, we use Coq canonical structures (as implemented by the command `Canonical`) to implement packed classes (Sect. 2), but also to implement other theories such as probabilities (Sect. 3.4). We already mentioned that the `rewrite` tactic from SSReflect is important to obtain short proof scripts (Sect. 4). We take advantage of the reals of the Coq standard library which come with automation: the `field` and `lra` (linear real/rational arithmetic) tactics are important in practice to compute probabilities (for example in the Monty Hall problem).

About Useful Coq *Libraries.* We use the SSReflect library for lists because it is closer to the Haskell library than the Coq standard library. It provides Haskell-like notations (e.g., notation for comprehension) and more functions (e.g., `allpairs`, a.k.a. `cp` in Haskell). We use the Finmap library of MathComp for its finite sets (see Sect. 6.1). We also benefit from other libraries compatible with MathComp to formalize the model of the probability monad [4].

About the Use of Extra Axioms. We use axioms inherited from the MathComp-Analysis library (they are explained in [2, §5]). More precisely, we use functional extensionality in particular to identify the Coq functions that appear in the *bind* operator. We use `gen_choiceMixin` to turn `Types` into `choiceTypes` when constructing models (see Sect. 6). To provide a model for the probability monad (Sect. 6.2), we proposed a type of probability distributions that requires reals to also enjoy an axiom of choice. We also have a localized use of the axiom of proof irrelevance to prove properties of functors [3, file `monad.v`]. All these axioms make our Coq environment resemble classical set theory. We choose to go with these axioms because it does not restrict the applicability of our work: equational reasoning does not forbid a classical meta-theory with the axiom of choice.

8 Related Work

Formalization of Monads in COQ. Monads are widely used for modeling programming languages with effects. For instance, Delaware et al. formalize several monads and monad transformers, each one associated with a *feature theorem* [9]. When monads are combined, those feature theorems can then be combined to prove type soundness. In comparison, the work we formalize here contains more monads and focuses on equational reasoning about concrete programs instead of meta-theory about programming languages.

Monads have been used in COQ to verify low-level systems [19,20] or for their modular verification [23] based on free monads. Our motivation is similar: enable formal reasoning for effectful programs using monads.

There are more formalizations of monads in other proof assistants. To pick one example that can be easily compared to our mechanization, one can find a formalization of the Monty Hall problem in Isabelle [8] (but using the pGCL programming language).

About Monadic Equational Reasoning. Although enabling equational reasoning for reasoning about monadic programs seems to be a natural idea, there does not seem to be much related work. Gibbons and Hinze seem to be the first to synthesize monadic equational reasoning as an approach [1,12,13]. This viewpoint is also adopted by other authors [6,30,31,37].

Applicative functor is an alternative approach to represent effectful computations. It has been formalized in Isabelle/HOL together with the tree relabeling example [24]. This work focuses on the lifting of equations to allow for automation, while our approach is rather the one of *small-scale reflection* [14]: the construction of a hierarchy backed up by a rich library of definitions and lemmas to make the most out of the rewriting facilities of COQ.

We extended the hierarchy of Gibbons and Hinze with a state-trace monad with the intent of performing formal verification about programs written with the syntax and semantics of Sect. 5.2. There are actually more topics to explore about the formalization of tracing and monads [34].

About Formalization Techniques. We use packed classes [11] to formalize the hierarchy of effects. It should be possible to use other techniques. In fact, a preliminary version of our formalization was using a combination of telescopes and canonical structures. It did not suffer major problems but packed classes are more disciplined and are known to scale up to deep hierarchies. COQ's type classes have been reported to replace canonical structures in many situations, but we have not tested them here.

The problem of rewriting under function abstraction (Sect. 4.3) is not specific to monadic equational reasoning. For example, it also occurs when dealing with the big operators of the MATHCOMP library, a situation for which a forthcoming version of Coq provides automation [27].

9 Conclusions and Future Work

We reported on the formalization in the Coq proof assistant of an extensive hierarchy of effects with their algebraic laws, and its application to monadic equational reasoning. The key technique is the one of packed classes, which allows for the sharing of notations and properties of various monads, enforces modularity by insisting on interfaces, while preserving the ability to provide rigorous models. We also discussed other techniques of practical interest for monadic equational reasoning such as reasoning on the syntax despite dealing with a shallow embedding. As a benchmark, we applied our formalization to several pencil-and-paper proofs and furthermore formalized and proved properties of the semantics of an imperative programming language. Our approach is successful in the sense that our proof scripts closely match their paper-and-pencil counterparts. Our work also led us to revisit existing proofs and extend the hierarchy of effects originally proposed by Gibbons and Hinze. We believe that our experiments demonstrate that the formalization of monadic equational reasoning with packed classes and a shallow embedding provides a practical tool for formal verification of effectful programs.

Future Work. We have started the formalization of more examples of monadic equational reasoning [3, branch `experiments`]: [6] is underway, [10] proposes a sharing monad whose equations seems to call for more syntax reflection and brings to the table the issue of infinite data structures.

In its current state the `rewrite_` tactic (Sect. 4.3) is not completely satisfactory. Its main defect is practical: it cannot be chained with the standard `rewrite` tactic. We defer the design of a better solution to future work because the topic is actually more general (as discussed in Sect. 8).

The main task that we are now addressing is the formalization of the model of the monad that combines probability and nondeterminism. Though well-understood [7], its formalization requires a careful formalization of convexity, which is work in progress.

It remains to check whether we can improve the modularity of model construction (or even the extension of the hierarchy) through formalizing other generic methods for combining effects, such as algebraic effects and distributive laws between monads.

Acknowledgements. We acknowledge the support of the JSPS-CNRS bilateral program "FoRmal tools for IoT sEcurity" (PRC2199) and the JSPS KAKENHI Grant Number 18H03204, and thank all the participants of these projects for fruitful discussions. In particular, we thank Jacques Garrigue and Samuel Hym for taking the time to have extended discussions and giving us feedback on drafts of this paper. We also thank Cyril Cohen and Shinya Katsumata for comments about the formalization of monads.

A The Choice Monad

The following excerpt from the source code [3] corresponds to the choice monad first mentioned in Sect. 2.2:

```
Module MonadAlt.
Record mixin_of (M : monad) : Type := Mixin {
  alt : forall A, M A -> M A -> M A ;
  _ : forall A, associative (@alt A) ;
  _ : BindLaws.left_distributive (@Bind M) alt }.
Record class_of (m : Type -> Type) : Type := Class {
  base : Monad.class_of m ; mixin : mixin_of (Monad.Pack base) }.
Structure t := Pack { m : Type -> Type ; class : class_of m }.
Definition baseType (M : t) := Monad.Pack (base (class M)).
Module Exports.
Definition Alt M : forall A, m M A -> m M A -> m M A :=
  let: Pack _ (Class _ (Mixin x _ _)) := M
  return forall A, m M A -> m M A -> m M A in x.
Arguments Alt {M A} : simpl never.
Notation "'[~p]'" := (@Alt _). (* prefix notation *)
Notation "x '[~]' y" := (Alt x y). (* infix notation *)
Notation altMonad := t.
Coercion baseType : altMonad >-> monad.
Canonical baseType.
End Exports.
End MonadAlt.
Export MonadAlt.Exports.
```

B Generic Algebraic Laws

The algebraic laws used in this paper are instances of generic definitions with self-explanatory names. Table 1 summarizes the laws defined in SSReflect (file ssrfun.v from the standard distribution of Coq). Table 2 summarizes the laws introduced in this paper. The Coq definitions are available online [3].

Table 1. Algebraic laws defined in SSReflect

associative op	$\forall x, y, z.\ x \,\mathsf{op}\, (y \,\mathsf{op}\, z) = (x \,\mathsf{op}\, y) \,\mathsf{op}\, z$
left_id e op	$\forall x.\ e \,\mathsf{op}\, x = x$
right_id e op	$\forall x.\ x \,\mathsf{op}\, e = x$
left_zero z op	$\forall x.\ z \,\mathsf{op}\, x = z$
idempotent op	$\forall x.\ x \,\mathsf{op}\, x = x$

Table 2. Algebraic laws defined in this paper

Module FunctorLaws.	
id f	$\mathtt{f}\ id = id$
comp f	$\forall g, h.\ \mathtt{f}\ (g \circ h) = \mathtt{f}\ g \circ \mathtt{f}\ h$
Module JoinLaws.	
ret_naturality ret	$\forall h.\ \mathit{fmap}\ h \circ \mathtt{ret} = \mathtt{ret} \circ h$
join_naturality join	$\forall h.\ \mathit{fmap}\ h \circ \mathtt{join} = \mathtt{join} \circ \mathit{fmap}\ (\mathit{fmap}\ h)$
left_unit ret join	$\mathtt{join} \circ \mathtt{ret} = id$
right_unit ret join	$\mathtt{join} \circ \mathit{fmap}\ \mathtt{ret} = id$
associativity join	$\mathtt{join} \circ \mathit{fmap}\ \mathtt{join} = \mathtt{join} \circ \mathtt{join}$
Module BindLaws.	
associative bind	$\forall m, f, g.\ (m \gg= f) \gg= g = m \gg= \lambda x.(f(x) \gg= g)$
left_id op ret	$\forall m.\ \mathtt{ret}\ \mathtt{op}\ m = m$
right_id op ret	$\forall m.\ m\ \mathtt{op}\ \mathtt{ret} = m$
left_neutral bind ret	$\forall f.\ \mathtt{ret} \gg= f = f$
right_neutral bind ret	$\forall m.\ m \gg= \mathtt{ret} = m$
left_zero bind z	$\forall f.\ \mathtt{z} \gg= f = \mathtt{z}$
right_zero bind z	$\forall m.\ m \gg= \mathtt{z} = \mathtt{z}$
left_distributive bind op	$\forall m, n, f.\ m\ \mathtt{op}\ n \gg= f = (m \gg= f)\ \mathtt{op}\ (n \gg= f)$
right_distributive bind op	$\forall m, f, g.\ m \gg= \lambda x.(f\ x)\ \mathtt{op}\ (g\ x) = (m \gg= f)\ \mathtt{op}\ (m \gg= g)$

C Summary of Monads and Their Algebraic Laws

Table 3 summarizes the structures and the algebraic laws that we formalize and explain in this paper. Precise Coq definitions are available online [3].

D Details About the Imperative Language from Sect. 5.2

For the sake of completeness, we provide the definition of the syntax (`program`) and semantics (operational `step` and denotational `denote`) of the imperative language of Sect. 5.2 where we omitted looping constructs to help reading:

```
Inductive program : Type -> Type :=
| p_ret   : forall {A}, A -> program A
| p_bind  : forall {A B}, program A -> (A -> program B) -> program B
| p_cond  : forall {A}, bool -> program A -> program A -> program A
| p_get   : program S
| p_put   : S -> program unit
| p_mark  : T -> program unit.
| p_repeat : nat -> program unit -> program unit
| p_while : nat -> (S -> bool) -> program unit -> program unit
```

```
Variables T S : Type.
Definition state : Type := S * @continuation T S.
Inductive step : state -> option T -> state -> Prop :=
| s_ret   : forall s A a (k : A -> _), step (s, p_ret a `; k) None (s, k a)
```

Table 3. Monads Defined in this Paper and the Algebraic Laws They Introduce

Structure	Operators	Equations
functor (§2.1)	Fun/#	functor_id, functor_o
monad (§2.1)	Ret Join Bind/>>=/>>	ret_naturality join_naturality, joinretM (left unit), joinMret (right unit), joinA (associativity) bindretf (left neutral), bindmret (right neutral), bindA (associativity)
failMonad (§2.2)	Fail	bindfailf (fail left-zero of bind)
altMonad (§A)	Alt/[~]/[~p]	alt_bindDl (bind left-distributes over choice), altA (associativity)
nondetMonad (§2.2)		altmfail (right-id), altfailm (left-id)
exceptMonad (§3.1)	Catch	catchfailm (left-id), catchmfail (right-id), catchA (associativity), catchret (left-zero)
stateMonad (§3.2)	Get, Put	putget, getputskip, putput, getget
runMonad (§3.2)	Run	runret, runbind
stateRunMonad (§3.2)		runget, runput
nondetStateMonad (§3.2)		bindmfail (right-zero), alt_bindDr (bind right-distributes over choice)
traceMonad (§3.3)	Mark	
stateTraceMonad (§3.3)	stGet stPut stMark	st_getget st_putput, st_putget, st_getputskip st_putmark, st_getmark
traceRunMonad (§3.3)		runmark
stateTraceRunMonad (§3.3)		runstget, runstput, runstmark
probMonad (§3.4)	Choice	choicemm (idempotence), choice0, choice1 (identity laws), choiceA (quasi associativity), choiceC (skewed commutativity), prob_bindDl (bind left-distributes over choice)
altCIMonad (§3.5)		altmm (idempotence), altC (commutativity)
nondetCIMonad (§3.5)		
freshMonad (§3.5)	Fresh	
failFreshMonad (§3.5)	Distinct	failfresh_bindmfail (fail right-zero of bind) bassert (Distinct M) \o Symbols = Symbols
arrayMonad (§3.5)	aGet i, aPut i s	aputput, aputget, agetputskip, agetget, agetC, aputC, aputgetC
probDrMonad (§3.5)		prob_bindDr (bind right-distributes over choice)
altProbMonad (§3.5)		choiceDr (probabilistic choice right-distributes over nondeterministic choice)
exceptProbMonad (§3.5)		catchDl (catch left-distributes over choice)

```
| s_bind : forall s A B p (f : A -> program B) k,
   step (s, p_bind p f `; k) None (s, p `; fun a => f a `; k)
| s_cond_true : forall s A p1 p2 (k : A -> _),
   step (s, p_cond true p1 p2 `; k) None (s, p1 `; k)
| s_cond_false : forall s A p1 p2 (k : A -> _),
   step (s, p_cond false p1 p2 `; k) None (s, p2 `; k)
| s_get  : forall s k, step (s, p_get `; k) None (s, k s)
| s_put  : forall s s' k, step (s, p_put s' `; k) None (s', k tt)
```

```
| s_mark : forall s t k, step (s, p_mark t `; k) (Some t) (s, k tt).
| s_repeat_O : forall s p k, step (s, p_repeat O p `; k) None (s, k tt)
| s_repeat_S : forall s n p k,
    step (s, p_repeat n.+1 p `; k) None
         (s, p `; fun _ => p_repeat n p `; k)
| s_while_true : forall fuel s c p k, c s = true ->
    step (s, p_while fuel.+1 c p `; k) None
         (s, p `; fun _ => p_while fuel c p `; k)
| s_while_false : forall fuel s c p k, c s = false ->
    step (s, p_while fuel.+1 c p `; k) None (s, k tt)
| s_while_broke : forall s c p k,
    step (s, p_while O c p `; k) None (s, k tt)

Variables S T : Type.
Variable M : stateTraceMonad S T.
Fixpoint denote A (p : program A) : M A :=
  match p with
  | p_ret _ v => Ret v
  | p_bind _ _ m f => do a <- denote m; denote (f a)
  | p_cond _ b p1 p2 => if b then denote p1 else denote p2
  | p_get => stGet
  | p_put s' => stPut s'
  | p_mark t => stMark t
  | p_repeat n p => (fix loop m : M unit :=
    if m is m'.+1 then denote p >> loop m' else Ret tt) n
  | p_while fuel c p => (fix loop m : M unit :=
    if m is m'.+1
    then (do s <- stGet ; if c s then denote p >> loop m' else Ret tt)
    else Ret tt) fuel
  end.
```

References

1. Abou-Saleh, F., Cheung, K.H., Gibbons, J.: Reasoning about probability and nondeterminism. In: Workshop on Probabilistic Programming Semantics, St. Petersburg, FL, USA, 23 January 2016, January 2016
2. Affeldt, R., Cohen, C., Rouhling, D.: Formalization techniques for asymptotic reasoning in classical analysis. J. Formaliz. Reason. **11**(1), 43–76 (2018)
3. Affeldt, R., Garrigue, J., Nowak, D., Saikawa, T.: A Coq formalization of monadic equational reasoning (2018). https://github.com/affeldt-aist/monae
4. Affeldt, R., et al.: A Coq formalization of information theory and linear error-correcting codes (2018). https://github.com/affeldt-aist/infotheo
5. Appel, A.W., Blazy, S.: Separation logic for small-step CMINOR. In: Schneider, K., Brandt, J. (eds.) TPHOLs 2007. LNCS, vol. 4732, pp. 5–21. Springer, Heidelberg (2007). https://doi.org/10.1007/978-3-540-74591-4_3
6. Chen, Y.-F., Hong, C.-D., Lengál, O., Mu, S.-C., Sinha, N., Wang, B.-Y.: An executable sequential specification for spark aggregation. In: El Abbadi, A., Garbinato, B. (eds.) NETYS 2017. LNCS, vol. 10299, pp. 421–438. Springer, Cham (2017). https://doi.org/10.1007/978-3-319-59647-1_31

7. Cheung, K.H.: Distributive Interaction of Algebraic Effects. Ph.D. thesis, Merton College, University of Oxford (2017)
8. Cock, D.: Verifying probabilistic correctness in Isabelle with pGCL. In: 7th Systems Software Verification, Sydney, Australia, pp. 1–10, November 2012
9. Delaware, B., Keuchel, S., Schrijvers, T., d. S. Oliveira, B.C.: Modular monadic meta-theory. In: ACM SIGPLAN International Conference on Functional Programming (ICFP 2013), Boston, MA, USA, 25–27 September 2013, pp. 319–330 (2013)
10. Fischer, S., Kiselyov, O., Shan, C.: Purely functional lazy nondeterministic programming. J. Funct. Program. **21**(4–5), 413–465 (2011)
11. Garillot, F., Gonthier, G., Mahboubi, A., Rideau, L.: Packaging mathematical structures. In: Berghofer, S., Nipkow, T., Urban, C., Wenzel, M. (eds.) TPHOLs 2009. LNCS, vol. 5674, pp. 327–342. Springer, Heidelberg (2009). https://doi.org/10.1007/978-3-642-03359-9_23
12. Gibbons, J.: Unifying theories of programming with monads. In: Wolff, B., Gaudel, M.-C., Feliachi, A. (eds.) UTP 2012. LNCS, vol. 7681, pp. 23–67. Springer, Heidelberg (2013). https://doi.org/10.1007/978-3-642-35705-3_2
13. Gibbons, J., Hinze, R.: Just do it: simple monadic equational reasoning. In: 16th ACM SIGPLAN International Conference on Functional Programming (ICFP 2011), Tokyo, Japan, 19–21 September 2011, pp. 2–14. ACM (2011)
14. Gonthier, G., Mahboubi, A.: An introduction to small scale reflection in Coq. J. Formaliz. Reasoning **3**(2), 95–152 (2010)
15. Gonthier, G., Tassi, E.: A language of patterns for subterm selection. In: Beringer, L., Felty, A. (eds.) ITP 2012. LNCS, vol. 7406, pp. 361–376. Springer, Heidelberg (2012). https://doi.org/10.1007/978-3-642-32347-8_25
16. Greenaway, D.: Automated Proof-Producing Abstraction of C Code. Ph.D. thesis, University of New South Wales, Sydney, Australia, January 2015
17. Hirschowitz, A., Maggesi, M.: Modules over monads and initial semantics. Inf. Comput. **208**(5), 545–564 (2010)
18. Jacobs, B.: Convexity, duality and effects. In: Calude, C.S., Sassone, V. (eds.) TCS 2010. IFIP AICT, vol. 323, pp. 1–19. Springer, Heidelberg (2010). https://doi.org/10.1007/978-3-642-15240-5_1
19. Jomaa, N., Nowak, D., Grimaud, G., Hym, S.: Formal proof of dynamic memory isolation based on MMU. Sci. Comput. Program. **162**, 76–92 (2018)
20. Jomaa, N., Torrini, P., Nowak, D., Grimaud, G., Hym, S.: Proof-oriented design of a separation kernel with minimal trusted computing base. In: 18th International Workshop on Automated Verification of Critical Systems (AVOCS 2018), July 2018. Oxford, UK. Electronic Communications of the EASST Open Access Journal (2018)
21. Jones, M.P., Duponcheel, L.: Composing monads. Technical report YALEU/DCS/RR-1004, Yale University (Dec 1993)
22. King, D.J., Wadler, P.: Combining monads. In: Launchbury, J., Sansom, P. (eds.) Functional Programming, Glasgow 1992. Workshops in Computing, pp. 134–143. Springer, London (1992)
23. Letan, T., Régis-Gianas, Y., Chifflier, P., Hiet, G.: Modular verification of programs with effects and effect handlers in Coq. In: Havelund, K., Peleska, J., Roscoe, B., de Vink, E. (eds.) FM 2018. LNCS, vol. 10951, pp. 338–354. Springer, Cham (2018). https://doi.org/10.1007/978-3-319-95582-7_20
24. Lochbihler, A., Schneider, J.: Equational reasoning with applicative functors. In: Blanchette, J.C., Merz, S. (eds.) ITP 2016. LNCS, vol. 9807, pp. 252–273. Springer, Cham (2016). https://doi.org/10.1007/978-3-319-43144-4_16

25. Mac Lane, S. (ed.): Categories for the Working Mathematician. GTM, vol. 5. Springer, New York (1978). https://doi.org/10.1007/978-1-4757-4721-8
26. Mahboubi, A., Tassi, E.: Mathematical Components (2016). https://math-comp.github.io/mcb/, with contributions by Yves Bertot and Georges Gonthier. Version of 2018/08/11
27. Martin-Dorel, E., Tassi, E.: SSReflect in Coq 8.10. In: The Coq Workshop 2019, Portland, OR, USA, 8 September 2019, pp. 1–2, September 2019
28. Moggi, E.: Computational lambda-calculus and monads. In: LICS, pp. 14–23. IEEE Computer Society (1989)
29. Mu, S.C.: Functional pearls, reasoning and derivation of monadic programs, a case study of non-determinism and state, July 2017, draft. http://flolac.iis.sinica.edu.tw/flolac18/files/test.pdf. Accessed 10 July 2019
30. Mu, S.C.: Calculating a backtracking algorithm: an exercise in monadic program derivation. Technical report TR-IIS-19-003, Institute of Information Science, Academia Sinica, June 2019
31. Mu, S.C.: Equational reasoning for non-deterministic monad: a case study of Spark aggregation. Technical report TR-IIS-19-002, Institute of Information Science, Academia Sinica, June 2019
32. Mu, S., Ko, H., Jansson, P.: Algebra of programming in Agda: dependent types for relational program derivation. J. Funct. Program. **19**(5), 545–579 (2009)
33. Pfenning, F., Elliott, C.: Higher-order abstract syntax. In: ACM SIGPLAN Conference on Programming Language Design and Implementation (PLDI 1988), Atlanta, GA, USA, 22–24 June 1988, pp. 199–208. ACM (1988)
34. Piróg, M., Gibbons, J.: Tracing monadic computations and representing effects. In: 4th Workshop on Mathematically Structured Functional Programming (MSFP 2012). EPTCS, Tallinn, Estonia, 25 March 2012, vol. 76, pp. 90–111 (2012)
35. Plotkin, G., Power, J.: Notions of computation determine monads. In: Nielsen, M., Engberg, U. (eds.) FoSSaCS 2002. LNCS, vol. 2303, pp. 342–356. Springer, Heidelberg (2002). https://doi.org/10.1007/3-540-45931-6_24
36. Pretnar, M.: An introduction to algebraic effects and handlers (invited tutorial paper). Electr. Notes Theor. Comput. Sci. **319**, 19–35 (2015)
37. Shan, C.C.: Equational reasoning for probabilistic programming. In: POPL 2018 TutorialFest, January 2018
38. Varacca, D., Winskel, G.: Distributing probability over non-determinism. Math. Struct. Comput. Sci. **16**(1), 87–113 (2006)
39. Voevodsky, V., Ahrens, B., Grayson, D., et al.: UniMath-a computer-checked library of univalent mathematics. https://github.com/UniMath/UniMath
40. Wadler, P.: Comprehending monads. In: LISP and Functional Programming, pp. 61–78 (1990)

System *F* in Agda, for Fun and Profit

James Chapman[1(✉)], Roman Kireev[1], Chad Nester[2], and Philip Wadler[2]

[1] IOHK, Hong Kong, Hong Kong
{james.chapman,roman.kireev}@iohk.io
[2] University of Edinburgh, Edinburgh, UK
{cnester,wadler}@inf.ed.ac.uk

Abstract. System *F*, also known as the polymorphic λ-calculus, is a typed λ-calculus independently discovered by the logician Jean-Yves Girard and the computer scientist John Reynolds. We consider $F_{\omega\mu}$, which adds higher-order kinds and iso-recursive types. We present the first complete, intrinsically typed, executable, formalisation of System $F_{\omega\mu}$ that we are aware of. The work is motivated by verifying the core language of a smart contract system based on System $F_{\omega\mu}$. The paper is a literate Agda script [14].

1 Introduction

System *F*, also known as the polymorphic λ-calculus, is a typed λ-calculus independently discovered by the logician Jean-Yves Girard and the computer scientist John Reynolds. System *F* extends the simply-typed λ-calculus (STLC). Under the principle of Propositions as Types, the $\rightarrow$ type of STLC corresponds to implication; to this System F adds a $\forall$ type that corresponds to universal quantification over propositions. Formalisation of System *F* is tricky: it, when extended with subtyping, formed the basis for the POPLmark challenge [8], a set of formalisation problems widely attempted as a basis for comparing different systems.

System *F* is small but powerful. By a standard technique known as Church encoding, it can represent a wide variety of datatypes, including natural numbers, lists, and trees. However, while System *F* can encode the type "list of *A*" for any type *A* that can also be encoded, it cannot encode "list" as a function from types to types. For that one requires System *F* with higher-kinded types, known as System F_ω. Girard's original work also considered this variant, though Reynolds did not.

The basic idea of System F_ω is simple. Not only does each *term* have a *type*, but also each *type* level object has a *kind*. Notably, type *families* are classified by higher kinds. The first level, relating terms and types, includes an embedding of STLC (plus quantification); while the second level, relating types and kinds, is an isomorphic image of STLC.

Church encodings can represent any algebraic datatype recursive only in positive positions; though extracting a component of a structure, such as finding the tail of a list, takes time proportional to the size of the structure. Another standard technique, known as Scott encoding, can represent any algebraic type whatsoever; and extracting a component now takes constant time. However, Scott encoding requires a second

G. Hutton (Ed.): MPC 2019, LNCS 11825, pp. 255–297, 2019.
https://doi.org/10.1007/978-3-030-33636-3_10

extension to System F, to represent arbitrary recursive types, known as System F_μ. The system with both extensions is known as System $F_{\omega\mu}$, and will be the subject of our formalisation.

Terms in Systems F and F_ω are strongly normalising. Recursive types with recursion in a negative position permit encoding arbitrary recursive functions, so normalisation of terms in Systems F_μ and $F_{\omega\mu}$ may not terminate. However, constructs at the type level of Systems F_ω and $F_{\omega\mu}$ are also strongly normalising.

There are two approaches to recursive types, *equi-recursive* and *iso-recursive* [33]. In an equi-recursive formulation, the types $\mu\alpha.A[\alpha]$ and $A[\mu\alpha.A[\alpha]]$ are considered equal, while in an iso-recursive formulation they are considered isomorphic, with an *unfold* term to convert the former to the latter, and a *fold* term to convert the other way. Equi-recursive formulation makes coding easier, as it doesn't require extra term forms. But it makes type checking more difficult, and it is not known whether equi-recursive types for System $F_{\omega\mu}$ are decidable [11,19]. Accordingly, we use iso-recursive types, which are also used by Dreyer [18] and Brown and Palsberg [10].

There are also two approaches to formalising a typed calculus, *extrinsic* and *intrinsic* [35]. In an extrinsic formulation, terms come first and are assigned types later, while in an intrinsic formulation, types come first and a term can be formed only at a given type. The two approaches are sometimes associated with Curry and Church, respectively [23]. There is also the dichotomy between named variables and de Bruijn indices. De Bruijn indices ease formalisation, but require error-prone arithmetic to move a term underneath a lambda expression. An intrinsic formulation catches such errors, because they would lead to incorrect types. Accordingly, we use an intrinsic formulation with de Bruijn indices. The approach we follow was introduced by Altenkirch and Reus [6], and used by Chapman [13] and Allais et al. [2] among others.

1.1 For Fun and Profit

Our interest in System $F_{\omega\mu}$ is far from merely theoretical. Input Output HK Ltd. (IOHK) is developing the Cardano blockchain, which features a smart contract language known as Plutus [12]. The part of the contract that runs off-chain is written in Haskell with an appropriate library, while the part of the contract that runs on-chain is written using Template Haskell and compiled to a language called Plutus Core. Any change to the core language would require all participants of the blockchain to update their software, an event referred to as a *hard fork*. Hard forks are best avoided, so the goal with Plutus Core was to make it so simple that it is unlikely to need revision. The design settled on is System $F_{\omega\mu}$ with suitable primitives, using Scott encoding to represent data structures. Supported primitives include integers, bytestrings, and a few cryptographic and blockchain-specific operations.

The blockchain community puts a high premium on rigorous specification of smart contract languages. Simplicity, a proposed smart contract language for Bitcoin, has been formalised in Coq [31]. The smart contract language Michelson, used by Tezos, has also been formalised in Coq [30]. EVM, the virtual machine of Ethereum, has been formalised in K [32], in Isabelle/HOL [7,24], and in F^* [21]. For a more complete account of blockchain projects involving formal methods see [22].

IOHK funded the development of our formalisation of System $F_{\omega\mu}$ because of the correspondence to Plutus Core. The formal model in Agda and associated proofs give us high assurance that our specification is correct. Further, we plan to use the evaluator that falls out from our proof of progress for testing against the evaluator for Plutus Core that is used in Cardano.

1.2 Contributions

This paper represents the first complete intrinsically typed, executable, formalisation of System $F_{\omega\mu}$ that we are aware of. There are other intrinsically typed formalisations of fragments of System $F_{\omega\mu}$. But, as far as we are aware none are complete. András Kovács has formalised System F_ω [27] using hereditary substitutions [38] at the type level. Kovács' formalisation does not cover iso-recursive types and also does not have the two different presentations of the syntax and the metatheory relating them that are present here.

Intrinsically typed formalisations of arguably more challenging languages exist such as those of Chapman [13] and Danielsson [16] for dependently typed languages. However, they are not complete and do not consider features such as recursive types. This paper represents a more complete treatment of a different point in the design space which is interesting in its own right and has computation at the type level but stops short of allowing dependent types. We believe that techniques described here will be useful when scaling up to greater degrees of dependency.

A key challenge with the intrinsically typed approach for System F_ω is that due to computation at the type level, it is necessary to make use of the implementations of type level operations and even proofs of their correctness properties when defining the term level syntax and term level operations. Also, if we want to run term level programs, rather than just formalise them, it is vital that these proofs of type level operations compute, which means that we cannot assume any properties or rely on axioms in the metatheory such as functional extensionality. Achieving this level of completeness is a contribution of this paper as is the fact that this formalisation is executable. We do not need extensionality despite using higher order representations of renamings, substitutions, and (the semantics of) type functions. First order variants of these concepts are more cumbersome and long winded to work with. As the type level language is a strongly normalising extension of the simply-typed λ-calculus we were able to leverage work about renaming, substitution and normalisation from simply-typed λ-calculus. Albeit with the greater emphasis that proofs must compute. We learnt how to avoid using extensionality when reasoning about higher order/functional representations of renamings and substitutions from Conor McBride. The normalisation algorithm is derived from work by Allais et al. and McBride [3,29]. The normalisation proof also builds on their work, and in our opinion, simplifies and improves it as the uniformity property in the completeness proof becomes simply a type synonym required only at function type (kind in our case) rather than needing to be mutually defined with the logical relation at every type (kind), simplifying the construction and the proofs considerably. In addition we work with β-equality not $\beta\eta$-equality which, in the context of NBE makes things a little more challenging. The reason for this choice is that our smart contract core language Plutus Core has only β-equality.

Another challenge with the intrinsically typed approach for System F_ω, where typing derivations and syntax coincide, is that the presence of the conversion rule in the syntax makes computation problematic as it can block β-reduction. Solving/avoiding this problem is a contribution of this paper.

The approach to the term level and the notation borrow heavily from PLFA [37] where the chapters on STLC form essentially a blueprint for and a very relevant introduction to this work. The idea of deriving an evaluator from a proof of progress appears in PLFA, and appears to be not widely known [36].

1.3 Overview

This paper is a literate Agda program that is machine checked and executable either via Agda's interpreter or compiled to Haskell. The code (i.e. the source code of the paper) is available as a supporting artefact. In addition the complete formalisation of the extended system (Plutus Core) on which this paper is based is also available as a supporting artefact.

In the paper we aim to show the highlights of the formalisation: we show as much code as possible and the statements of significant lemmas and theorems. We hide many proofs and minor auxiliary lemmas.

Dealing with the computation in types and the conversion rule was the main challenge in this work for us. The approaches taken to variable binding, renaming/substitution and normalisation lifted relatively easily to this setting. In addition to the two versions of syntax where types are (1) not normalised and (2) completely normalised we also experimented with a version where types are in weak head normal form (3). In (1) the conversion rule takes an inductive witness of type equality relation as an argument. In (2) conversion is derivable as type equality is replaced by identity. In (3), the type equality relation in conversion can be replaced by a witness of a logical relation that computes, indeed it is the same logical relation as described in the completeness of type normalisation proof. We did not pursue this further in this work so far as this approach is not used in Plutus Core but this is something that we would like to investigate further in future.

In Sect. 2 we introduce intrinsically typed syntax (kinds, types and terms) and the dynamics of types (type equality). We also introduce the necessary syntactic operations for these definitions: type weakening and substitution (and their correctness properties) are necessary to define terms. In Sect. 3 we introduce an alternative version of the syntax where the types are β-normal forms. We also introduce the type level normalisation algorithm, its correctness proof and a normalising substitution operation on normal types. In Sect. 4 we reconcile the two versions of the syntax, prove soundness and completeness results and also demonstrate that normalising the types preserves the *semantics* of terms where semantics refers to corresponding untyped terms. In Sect. 5 we introduce the dynamics of the algorithmic system (type preserving small-step reduction) and we prove progress in Sect. 3. Preservation holds intrinsically. In Sect. 6 we provide a step-indexed evaluator that we can use to execute programs for a given number of reduction steps. In Sect. 7 we show examples of Church and Scott Numerals. In Sect. 8 we discuss extensions of the formalisation to higher kinded recursive types and intrinsically sized integers and bytestrings.

2 Intrinsically Typed Syntax of System $F_{\omega\mu}$

We take the view that when writing a program such as an interpreter we want to specify very precisely how the program behaves on meaningful input and we want to rule out meaningless input as early and as conclusively as possible. Many of the operations we define in this paper, including substitution, evaluation, and normalisation, are only intended to work on well-typed input. In a programming language with a less precise type system we might need to work under the informal assumption that we will only ever feed meaningful inputs to our programs and otherwise their behaviour is unspecified, and all bets are off. Working in Agda we can guarantee that our programs will only accept meaningful input by narrowing the definition of valid input. This is the motivation for using intrinsic syntax as the meaningful inputs are those that are guaranteed to be type correct and in Agda we can build this property right into the definition of the syntax.

In practice, in our setting, before receiving the input (some source code in a file) it would have been run through a lexing, parsing, scope checking and most importantly *type checking* phase before reaching our starting point in this paper: intrinsically typed syntax. Formalising the type checker is future work.

One can say that in intrinsically typed syntax, terms carry their types. But, we can go further, the terms are actually typing derivations. Hence, the definition of the syntax and the type system, as we present it, coincide: each syntactic constructor corresponds to one typing rule and vice versa. As such we dispense with presenting them separately and instead present them in one go.

There are three levels in this syntax:

1. kinds, which classify types;
2. types, which classify terms;
3. terms, the level of ordinary programs.

The kind level is needed as there are functions at the type level. Types appear in terms, but terms do not appear in types.

2.1 Kinds

The kinds consist of a base kind *, which is the kind of types, and a function kind.[1]

```
data Kind : Set where
  *     : Kind                     -- type
  _⇒_ : Kind → Kind → Kind -- function kind
```

Let K and J range over kinds.

[1] The code in this paper is typeset in colour.

2.2 Type Contexts

To manage the types of variables and their scopes we introduce contexts. Our choice of how to deal with variables is already visible in the representation of contexts. We will use de Bruijn indices to represent variables. While this makes terms harder to write, it makes the syntactic properties of the language clear and any potential off-by-one errors etc. are mitigated by working with intrinsically scoped terms and the fact that syntactic properties are proven correct. We intend to use the language as a compilation target so ease of manually writing programs in this language is not a high priority.

We refer to type contexts as Ctx⋆ and reserve the name Ctx for term level contexts. Indeed, when a concept occurs at both type and term level we often suffix the name of the type level version with ⋆.

Type contexts are essentially lists of types written in reverse. No names are required.

```
data Ctx⋆ : Set where
  ∅    : Ctx⋆                  -- empty context
  _,⋆_ : Ctx⋆ → Kind → Ctx⋆   -- context extension
```

Let Φ and Ψ range over contexts.

2.3 Type Variables

We use de Bruijn indices for variables. They are natural numbers augmented with additional kind and context information. The kind index tells us the kind of the variable and the context index ensures that the variable is in scope. It is impossible to write a variable that isn't in the context. Z refers to the last variable introduced on the right hand end of the context. Adding one to a variable via S moves one position to the left in the context. Note that there is no way to construct a variable in the empty context as it would be out of scope. Indeed, there is no way at all to construct a variable that is out of scope.

```
data _∋⋆_ : Ctx⋆ → Kind → Set where
  Z : ∀ {Φ J}              → Φ ,⋆ J ∋⋆ J
  S : ∀ {Φ J K} → Φ ∋⋆ J → Φ ,⋆ K ∋⋆ J
```

Let α and β range over type variables.

2.4 Types

Types, like type variables, are indexed by context and kind, ensuring well-scopedness and well-kindedness. The first three constructors ` ƛ and · are analogous to the terms of STLC. This is extended with the Π type to classify type abstractions at the type level, function type ⇒ to classify functions, and μ to classify recursive terms. Note that Π, ⇒, and μ are effectively base types as they live at kind ⋆.

```
data _⊢⋆_ Φ : Kind → Set where
  `    : ∀{J}   → Φ ∋⋆ J                      → Φ ⊢⋆ J        -- type variable
  ƛ    : ∀{K J} → Φ ,⋆ K ⊢⋆ J                 → Φ ⊢⋆ K ⇒ J    -- type lambda
  _·_  : ∀{K J} → Φ ⊢⋆ K ⇒ J → Φ ⊢⋆ K         → Φ ⊢⋆ J        -- type application
  _⇒_  :          Φ ⊢⋆ *      → Φ ⊢⋆ *        → Φ ⊢⋆ *        -- function type
  Π    : ∀{K}   → Φ ,⋆ K ⊢⋆ *                 → Φ ⊢⋆ *        -- Pi/forall type
  μ    :          Φ ,⋆ * ⊢⋆ *                 → Φ ⊢⋆ *        -- recursive type
```

Let A and B range over types.

2.5 Type Renaming

Types can contain functions and as such are subject to a nontrivial equality relation. To explain the computation equation (the β-rule) we need to define substitution for a single type variable in a type. Later, when we define terms that are indexed by their type we will need to be able to weaken types by an extra kind (Sect. 2.9) and also, again, substitute for a single type variable in a type (Sect. 2.10). There are various different ways to define the required weakening and substitution operations. We choose to define so-called parallel renaming and substitution i.e. renaming/substitution of several variables at once. Weakening and single variable substitution are special cases of these operations.

We follow Altenkirch and Reus [6] and implement renaming first and then substitution using renaming. In our opinion the biggest advantage of this approach is that it has a very clear mathematical theory. The necessary correctness properties of renaming are identified with the notion of a functor and the correctness properties of substitution are identified with the notion of a relative monad. For the purposes of reading this paper it is not necessary to understand relative monads in detail. The important thing is that, like ordinary monads, they have a return and bind and the rules that govern them are the same. It is only the types of the operations involved that are different. The interested reader may consult [5] for a detailed investigation of relative monads and [4] for a directly applicable investigation of substitution of STLC as a relative monad.

A type renaming is a function from type variables in one context to type variables in another. This is much more flexibility than we need. We only need the ability to introduce new variable on the right hand side of the context. The simplicity of the definition makes it easy to work with and we get some properties for free that we would have to pay for with a first order representation, such as not needing to define a lookup function, and we inherit the properties of functions provided by η-equality, such as associativity of composition, for free. Note that even though renamings are functions we do not require our metatheory (Agda's type system) to support functional extensionality. As pointed out to us by Conor McBride we only ever need to make use of an equation between renamings on a point (a variable) and therefore need only a pointwise version of equality on functions to work with equality of renamings and substitutions.

```
Ren⋆ : Ctx⋆ → Ctx⋆ → Set
Ren⋆ Φ Ψ = ∀ {J} → Φ ∋⋆ J → Ψ ∋⋆ J
```

Let ρ range over type renamings.

As we are going to push renamings through types we need to be able to push them under a binder. To do this safely the newly bound variable should remain untouched and other renamings should be shifted by one to accommodate this. This is exactly what the lift⋆ function does and it is defined by recursion on variables:

```
lift⋆ : ∀ {Φ Ψ} → Ren⋆ Φ Ψ → ∀ {K} → Ren⋆ (Φ ,⋆ K) (Ψ ,⋆ K)
lift⋆ ρ Z     = Z       -- leave newly bound variable untouched
lift⋆ ρ (S α) = S (ρ α) -- apply renaming to other variables and add 1
```

Next we define the action of renaming on types. This is defined by recursion on the type. Observe that we lift the renaming when we go under a binder and actually apply the renaming when we hit a variable:

```
ren⋆ : ∀ {Φ Ψ} → Ren⋆ Φ Ψ → ∀ {J} → Φ ⊢⋆ J → Ψ ⊢⋆ J
ren⋆ ρ (‘ α)     = ‘ (ρ α)
ren⋆ ρ (ƛ B)     = ƛ (ren⋆ (lift⋆ ρ) B)
ren⋆ ρ (A · B)   = ren⋆ ρ A · ren⋆ ρ B
ren⋆ ρ (A ⇒ B)  = ren⋆ ρ A ⇒ ren⋆ ρ B
ren⋆ ρ (Π B)     = Π (ren⋆ (lift⋆ ρ) B)
ren⋆ ρ (μ B)     = μ (ren⋆ (lift⋆ ρ) B)
```

Weakening is a special case of renaming. We apply the renaming S which does double duty as the variable constructor, if we check the type of S we see that it is a renaming.

Weakening shifts all the existing variables one place to the left in the context:

```
weaken⋆ : ∀ {Φ J K} → Φ ⊢⋆ J → Φ ,⋆ K ⊢⋆ J
weaken⋆ = ren⋆ S
```

2.6 Type Substitution

Having defined renaming we are now ready to define substitution for types. Substitutions are defined as functions from type variables to types:

```
Sub⋆ : Ctx⋆ → Ctx⋆ → Set
Sub⋆ Φ Ψ = ∀ {J} → Φ ∋⋆ J → Ψ ⊢⋆ J
```

Let σ range over substitutions.

We must be able to lift substitutions when we push them under binders. Notice that we leave the newly bound variable intact and make use of weaken⋆ to weaken a term that is substituted.

```
lifts⋆ : ∀ {Φ Ψ} → Sub⋆ Φ Ψ → ∀ {K} → Sub⋆ (Φ ,⋆ K) (Ψ ,⋆ K)
lifts⋆ σ Z     = ‘ Z -- leave newly bound variable untouched
lifts⋆ σ (S α) = weaken⋆ (σ α) -- apply substitution and weaken
```

Analogously to renaming, we define the action of substitutions on types:

```
sub⋆ : ∀ {Φ Ψ} → Sub⋆ Φ Ψ → ∀ {J} → Φ ⊢⋆ J → Ψ ⊢⋆ J
sub⋆ σ (‘ α)     = σ α
sub⋆ σ (ƛ B)     = ƛ (sub⋆ (lifts⋆ σ) B)
sub⋆ σ (A · B)   = sub⋆ σ A · sub⋆ σ B
sub⋆ σ (A ⇒ B) = sub⋆ σ A ⇒ sub⋆ σ B
sub⋆ σ (Π B)     = Π (sub⋆ (lifts⋆ σ) B)
sub⋆ σ (μ B)     = μ (sub⋆ (lifts⋆ σ) B)
```

Substitutions could be implemented as lists of types and then the *cons* constructor would extend a substitution by an additional term. Using our functional representation for substitutions it is convenient to define an operation for this. This effectively defines a new function that, if it is applied to the Z variable, returns our additional terms and otherwise invokes the original substitution.

```
extend⋆ : ∀{Φ Ψ} → Sub⋆ Φ Ψ → ∀{J}(A : Ψ ⊢⋆ J) → Sub⋆ (Φ ,⋆ J) Ψ
extend⋆ σ A Z     = A    -- project out additional term
extend⋆ σ A (S α) = σ α  -- apply original substitution
```

Substitution of a single type variable is a special case of parallel substitution sub⋆. Note we make use of extend⋆ to define the appropriate substitution by extending the substitution ‘ with the type A. Notice that the variable constructor ‘ serves double duty as the identity substitution:

```
_[_]⋆ : ∀ {Φ J K} → Φ ,⋆ K ⊢⋆ J → Φ ⊢⋆ K → Φ ⊢⋆ J
B [ A ]⋆ = sub⋆ (extend⋆ ‘ A) B
```

At this point the reader may well ask how we know that our substitution and renaming operations are the right ones. One indication that we have the right definitions is that renaming defines a functor, and that substitution forms a relative monad. Further, evaluation (eval defined in Sect. 3.2) can be seen as an algebra of this relative monad. This categorical structure results in clean proofs.

Additionally, without some sort of compositional structure to our renaming and substitution, we would be unable to define coherent type level operations. For example, we must have that performing two substitutions in sequence results in the same type as performing the composite of the two substitutions. We assert that these are necessary functional correctness properties and structure our proofs accordingly.

Back in our development we show that lifting a renaming and the action of renaming satisfy the functor laws where lift⋆ and ren⋆ are both functorial actions.

```
lift⋆-id   : ∀ {Φ J K}(α : Φ ,⋆ K ∋⋆ J) → lift⋆ id α ≡ α
lift⋆-comp : ∀{Φ Ψ Θ}{ρ : Ren⋆ Φ Ψ}{ρ' : Ren⋆ Ψ Θ}{J K}(α : Φ ,⋆ K ∋⋆ J)
  → lift⋆ (ρ' ∘ ρ) α ≡ lift⋆ ρ' (lift⋆ ρ α)
```

```
ren⋆-id   : ∀{Φ J}(A : Φ ⊢⋆ J) → ren⋆ id A ≡ A
ren⋆-comp : ∀{Φ Ψ Θ}{ρ : Ren⋆ Φ Ψ}{ρ' : Ren⋆ Ψ Θ}{J}(A : Φ ⊢⋆ J)
  → ren⋆ (ρ' ∘ ρ) A ≡ ren⋆ ρ' (ren⋆ ρ A)
```

Lifting a substitution satisfies the functor laws where lift⋆ is a functorial action:

```
lifts⋆-id   : ∀ {Φ J K}(x : Φ ,⋆ K ∋⋆ J) → lifts⋆ ` x ≡ ` x
lifts⋆-comp : ∀{Φ Ψ Θ}{σ : Sub⋆ Φ Ψ}{σ' : Sub⋆ Ψ Θ}{J K}(α : Φ ,⋆ K ∋⋆ J)
  → lifts⋆ (sub⋆ σ' ∘ σ) α ≡ sub⋆ (lifts⋆ σ') (lifts⋆ σ α)
```

The action of substitution satisfies the relative monad laws where ` is return and sub⋆ is bind:

```
sub⋆-id   : ∀ {Φ J}(A : Φ ⊢⋆ J) → sub⋆ ` A ≡ A
sub⋆-var  : ∀ {Φ Ψ}{σ : Sub⋆ Φ Ψ}{J}(α : Φ ∋⋆ J) → sub⋆ σ (` α) ≡ σ α
sub⋆-comp : ∀{Φ Ψ Θ}{σ : Sub⋆ Φ Ψ}{σ' : Sub⋆ Ψ Θ}{J}(A : Φ ⊢⋆ J)
  → sub⋆ (sub⋆ σ' ∘ σ) A ≡ sub⋆ σ' (sub⋆ σ A)
```

Note that the second law holds definitionally, it is the first line of the definition of sub⋆.

2.7 Type Equality

We define type equality as an intrinsically scoped and kinded relation. In particular, this means it is impossible to state an equation between types in different contexts, or of different kinds. The only interesting rule is the β-rule from the lambda calculus. We omit the η-rule as Plutus Core does not have it. The formalisation could be easily modified to include it and it would slightly simplify the type normalisation proof. The additional types ($\Rightarrow$, $\forall$, and μ) do not have any computational behaviour, and are essentially inert. In particular, the fixed point operator μ does not complicate the equational theory.

```
data _≡β_ {Φ} : ∀{J} → Φ ⊢⋆ J → Φ ⊢⋆ J → Set where
  β≡β : ∀{K J}(B : Φ ,⋆ J ⊢⋆ K)(A : Φ ⊢⋆ J) → ƛ B · A ≡β B [ A ]⋆
  -- remaining rules hidden
```

We omit the rules for reflexivity, symmetry, transitivity, and congruence rules for type constructors.

2.8 Term Contexts

Having dealt with the type level, we turn our attention to the term level.

Terms may contain types, and so the term level contexts must also track information about type variables in addition to term variables. We would like to avoid having the extra syntactic baggage of multiple contexts. We do so by defining term contexts which contain both (the kinds of) type variables and (the types of) term variables. Term contexts are indexed over type contexts. In an earlier version of this formalisation instead of indexing by type contexts we defined inductive term contexts simultaneously with a recursive erasure operation that converts a term level context to a type level context by dropping the term variables but keeping the type variables. Defining an inductive data type simultaneously with a recursive function is referred to as *induction recursion* [20]. This proved to be too cumbersome in later proofs as it can introduce a situation where there can be multiple provably equal ways to recover the same type context and expressions become cluttered with proofs of such equations. In addition to the difficulty of working with this version, it also made type checking the examples in our formalisation much slower. In the version presented here neither of these problems arise.

A context is either empty, or it extends an existing context by a type variable of a given kind, or by a term variable of a given type.

```
data Ctx : Ctx⋆ → Set where
  ∅ : Ctx ∅
  -- empty term context
  _,⋆_ : ∀{Φ} → Ctx Φ → ∀ J → Ctx (Φ ,⋆ J)
  -- extension by (the kind of) a type variable
  _,_ : ∀ {Φ} → Ctx Φ → Φ ⊢⋆ * → Ctx Φ
  -- extension by (the type of) a term variable
```

Let Γ, Δ, range over contexts. Note that in the last rule _,_, the type we are extending by may only refer to variables in the type context, a term that inhabits that type may refer to any variable in its context.

2.9 Term Variables

A variable is indexed by its context and type. While type variables can appear in types, and those types can appear in terms, the variables defined here are term level variables only.

Notice that there is only one base constructor Z. This gives us exactly what we want: we can only construct term variables. We have two ways to shift these variables to the left, we use S to shift over a type and T to shift over a kind in the context.

```
data _∋_ : ∀{Φ} → Ctx Φ → Φ ⊢⋆ * → Set where
  Z : ∀{Φ Γ} {A : Φ ⊢⋆ *}                    → Γ , A ∋ A
  S : ∀{Φ Γ} {A : Φ ⊢⋆ *} {B : Φ ⊢⋆ *} → Γ ∋ A → Γ , B ∋ A
  T : ∀{Φ Γ} {A : Φ ⊢⋆ *} {K}           → Γ ∋ A → Γ ,⋆ K ∋ weaken⋆ A
```

Let *x*, *y* range over variables. Notice that we need weakening of (System *F*) types in the (Agda) type of T. We must weaken *A* to shift it from context Γ to context $\Gamma ,^\star K$. Indeed, weaken⋆ is a function and it appears in a type. This is possible due to the rich support for dependent types and in particular inductive families in Agda. It is however a feature that must be used with care and while it often seems to be the most natural option it can be more trouble than it is worth. We have learnt from experience, for example, that it is easier to work with renamings (morphisms between contexts) ρ : Ren Γ Δ rather than context extensions $\Gamma + \Delta$ where the contexts are built from concatenation. The function

+, whose associativity holds only propositionally, is awkward to work with when it appears in type indices. Renamings do not suffer from this problem as no additional operations on contexts are needed as we commonly refer to a renaming into an *arbitrary* new context (e.g., Δ) rather than, precisely, an extension of an existing one (e.g., $\Gamma + \Delta$). In this formalisation we could have chosen to work with explicit renamings and substitutions turning operations like weaken★ into more benign constructors but this would have been overall more cumbersome and in this case we are able to work with executable renaming and substitution cleanly. Doing so cleanly is a contribution of this work.

2.10 Terms

A term is indexed by its context and type. A term is a variable, an abstraction, an application, a type abstraction, a type application, a wrapped term, an unwrapped term, or a term whose type is cast to another equal type.

```
data _⊢_ {Φ} Γ : Φ ⊢★ * → Set where
  `      : ∀{A}   → Γ ∋ A                  → Γ ⊢ A             -- variable
  ƛ      : ∀{A B} → Γ , A ⊢ B              → Γ ⊢ A ⇒ B         -- term λ
  _·_    : ∀{A B} → Γ ⊢ A ⇒ B              → Γ ⊢ A → Γ ⊢ B     -- term app
  Λ      : ∀{K B} → Γ ,★ K ⊢ B             → Γ ⊢ Π B           -- type λ
  _·★_   : ∀{K B} → Γ ⊢ Π B → (A : Φ ⊢★ K) → Γ ⊢ B [ A ]★      -- type app
  wrap   : ∀ A    → Γ ⊢ A [ μ A ]★         → Γ ⊢ μ A           -- wrap
  unwrap : ∀{A}   → Γ ⊢ μ A                → Γ ⊢ A [ μ A ]★    -- unwrap
  conv   : ∀{A B} → A ≡β B → Γ ⊢ A         → Γ ⊢ B             -- type cast
```

Let L, M range over terms. The last rule conv is required as we have computation in types. So, a type which has a β-redex in it is equal, via type equality, to the type where that redex is reduced. We want a term which is typed by the original unreduced type to also be typed by the reduced type. This is a standard typing rule but it looks strange as a syntactic constructor. See [17] for a discussion of syntax with explicit conversions.

We could give a dynamics for this syntax as a small-step reduction relation but the conv case is problematic. It is not enough to say that a conversion reduces if the underlying term reduces. If a conversion is in the function position (also called head position) in an application it would block β-reduction. We cannot prove progress directly for such a relation. One could try to construct a dynamics for this system where during reduction both terms and also types can make reduction steps and we could modify progress and explicitly prove preservation. We do not pursue this here. In the system we present here we have the advantage that the type level language is strongly normalising. In Sect. 3 we are able to make use of this advantage quite directly to solve the conversion problem in a different way. An additional motivation for us to choose the normalisation oriented approach is that in Plutus, contracts are stored and executed on chain with types normalised and this mode of operation is therefore needed anyway.

If we forget intrinsically typed syntax for a moment and consider these rules as a type system then we observe that it is not syntax directed, we cannot use it as the algorithmic specification of a type checker as we can apply the conversion rule at any point. This is why we refer to this version of the rules as *declarative* and the version presented in Sect. 3, which is (in this specific sense) syntax directed, as *algorithmic*.

3 Algorithmic Rules

In this section we remove the conversion rule from our system. Two promising approaches to achieving this are (1) to push traces of the conversion rule into the other rules which is difficult to prove complete [34] and (2) to normalise the types which collapses all the conversion proofs to reflexivity. In this paper we will pursue the latter.

In the pursuit of (2) we have another important design decision to make: which approach to take to normalisation. Indeed, another additional aspect to this is that we need not only a normaliser but a normal form respecting substitution operation. We choose to implement a Normalisation-by-Evaluation (NBE) style normaliser and use that to implement a substitution operation on normal forms.

We chose NBE as we are experienced with it and it has a clear mathematical structure (e.g., evaluation is a relative algebra for the relative monad given by substitution) which gave us confidence that we could construct a well structured normalisation proof that would compute. The NBE approach is also centred around a normalisation *algorithm*: something that we want to use. Other approaches would also work we expect. One option would be to try hereditary substitutions where the substitution operation is primary and use that to define a normaliser.

Sections 3.1–3.6 describe the normal types, the normalisation algorithm, its correctness proof, and a normalising substitution operation. Readers not interested in these details may skip to Sect. 3.7.

3.1 Normal Types

We define a data type of β-normal types which are either in constructor form or neutral. Neutral types, which are defined mutually with normal types, are either variables or (possibly nested) applications that are stuck on a variable in a function position, so cannot reduce. In this syntax, it is impossible to define an expression containing a β-redex.

```
data _⊢Nf⋆_ Φ : Kind → Set

data _⊢Ne⋆_ Φ J : Set where
  `   :           Φ ∋⋆ J                          → Φ ⊢Ne⋆ J -- type var
  _·_ : ∀{K} → Φ ⊢Ne⋆ (K ⇒ J) → Φ ⊢Nf⋆ K → Φ ⊢Ne⋆ J -- neutral app

data _⊢Nf⋆_ Φ where
  ƛ   : ∀{K J} → Φ ,⋆ K ⊢Nf⋆ J          → Φ ⊢Nf⋆ (K ⇒ J) -- type lambda
  ne  : ∀{K}   → Φ ⊢Ne⋆ K               → Φ ⊢Nf⋆ K       -- neutral type
  _⇒_ :          Φ ⊢Nf⋆ * → Φ ⊢Nf⋆ *    → Φ ⊢Nf⋆ *       -- function type
  Π   : ∀{K}   → Φ ,⋆ K ⊢Nf⋆ *          → Φ ⊢Nf⋆ *       -- pi/forall type
  μ   :          Φ ,⋆ * ⊢Nf⋆ *          → Φ ⊢Nf⋆ *       -- recursive type
```

Let A, B range over neutral and normal types.

As before, we need weakening at the type level in the definition of term level variables. As before, we define it as a special case of renaming whose correctness we verify by proving the functor laws.

```
renNf⋆     : ∀{Φ Ψ} → Ren⋆ Φ Ψ → ∀ {J} → Φ ⊢Nf⋆ J → Ψ ⊢Nf⋆ J
renNe⋆     : ∀{Φ Ψ} → Ren⋆ Φ Ψ → ∀ {J} → Φ ⊢Ne⋆ J → Ψ ⊢Ne⋆ J
weakenNf⋆ : ∀{Φ J K} → Φ ⊢Nf⋆ J → Φ ,⋆ K ⊢Nf⋆ J
```

Renaming of normal and neutral types satisfies the functor laws where renNf⋆ and renNe⋆ are both functorial actions:

```
renNf⋆-id     : ∀{Φ J}(A : Φ ⊢Nf⋆ J) → renNf⋆ id A ≡ A
renNf⋆-comp : ∀{Φ Ψ Θ}{ρ : Ren⋆ Φ Ψ}{ρ' : Ren⋆ Ψ Θ}{J}(A : Φ ⊢Nf⋆ J)
  → renNf⋆ (ρ' ∘ ρ) A ≡ renNf⋆ ρ' (renNf⋆ ρ A)

renNe⋆-id     : ∀{Φ J}(A : Φ ⊢Ne⋆ J) → renNe⋆ id A ≡ A
renNe⋆-comp : ∀{Φ Ψ Θ}{ρ : Ren⋆ Φ Ψ}{ρ' : Ren⋆ Ψ Θ}{J}(A : Φ ⊢Ne⋆ J)
  → renNe⋆ (ρ' ∘ ρ) A ≡ renNe⋆ ρ' (renNe⋆ ρ A)
```

3.2 Type Normalisation Algorithm

We use the NBE approach introduced by [9]. This is a two stage process, first we evaluate into a semantic domain that supports open terms, then we reify these semantic terms back into normal forms.

The semantic domain ⊨, our notion of semantic value is defined below. Like syntactic types and normal types it is indexed by context and kind. However, it is not a type defined as an inductive data type. Instead, it is function that returns a type. More precisely, it is a function that takes a context and, by recursion on kinds, defines a new type. At base kind it is defined to be the type of normal types. At function kind it is either a neutral type at function kind or a semantic function. If it is a semantic function then we are essentially interpreting object level (type) functions as meta level (Agda) functions. The additional renaming argument means we have a so-called *Kripke function space* ([25]). This is essential for our purposes as it allows us to introduce new free variables into the context and then apply functions to them. Without this feature we would not be able to reify from semantic values to normal forms.

```
_⊨_ : Ctx⋆ → Kind → Set
Φ ⊨ *          = Φ ⊢Nf⋆ *
Φ ⊨ (K ⇒ J) = Φ ⊢Ne⋆ (K ⇒ J) ⊎ ∀ {Ψ} → Ren⋆ Φ Ψ → Ψ ⊨ K → Ψ ⊨ J
```

Let V, W range over values. Let F, G range over meta-level (Agda) functions. The definition ⊨ is a Kripke Logical Predicate. It is also a so-called large elimination, as it is a function which returns a new type (a Set in Agda terminology). This definition is inspired by Allais et al. [3]. Their normalisation proof, which we also took inspiration from, is, in turn, based on the work of C. Coquand [15]. The coproduct at the function kind is present in McBride [29]. Our motivation for following these three approaches was to be careful not to perturb neutral terms where possible as we want to use our normaliser in substitution and we want the identity substitution for example not to modify variables. We also learned from [3] how to move the uniformity condition out of the definition of values into the completeness relation.

We will define an evaluator to interpret syntactic types into this semantic domain but first we need to explain how to reify from semantics to normal forms. This is needed first as, at base type, our semantic values are normal forms, so we need a way to convert from values to normal forms during evaluation. Note that usual NBE operations of reify and reflect are not mutually defined here as they commonly are in $\beta\eta$-NBE. This is a characteristic of the coproduct style definition above.

Reflection takes a neutral type and embeds it into a semantic type. How we do this depends on what kind we are at. At base kind *, semantic values are normal forms, so we embed our neutral term using the ne constructor. At function kind, semantic values are a coproduct of either a neutral term or a function, so we embed our neutral term using the inl constructor.

```
reflect : ∀{K Φ} → Φ ⊢Ne⋆ K → Φ ⊨ K
reflect {*}        A = ne A
reflect {K ⇒ J} A = inl A
```

Reification is the process of converting from a semantic type to a normal syntactic type. At base kind and for neutral functions it is trivial, either we already have a normal form or we have a neutral term which can be embedded. The last line, where we have a semantic function is where the action happens. We create a fresh variable of kind K using reflect and apply f to it making use of the Kripke function space by supplying f with the weakening renaming S. This creates a semantic value of kind J in context Φ , K which we can call reify recursively on. This, in turn, gives us a normal form in Φ , K ⊢Nf⋆ J. We can then wrap this normal form in a ƛ.

```
reify : ∀ {K Φ} → Φ ⊨ K → Φ ⊢Nf⋆ K
reify {*}        A     = A
reify {K ⇒ J} (inl A) = ne A
reify {K ⇒ J} (inr F) = ƛ (reify (F S (reflect (‘ Z))))
```

We define renaming for semantic values. In the semantic function case, the new renaming is composed with the existing one.

```
ren⊨ : ∀ {σ Φ Ψ} → Ren⋆ Φ Ψ → Φ ⊨ σ → Ψ ⊨ σ
ren⊨ {*}        ρ A     = renNf⋆ ρ A
ren⊨ {K ⇒ J} ρ (inl A) = inl (renNe⋆ ρ A)
ren⊨ {K ⇒ J} ρ (inr F) = inr (λ ρ’ → F (ρ’ ∘ ρ))
```

Weakening for semantic values is a special case of renaming:

```
weaken⊨ : ∀ {σ Φ K} → Φ ⊨ σ → (Φ ,⋆ K) ⊨ σ
weaken⊨ = ren⊨ S
```

Our evaluator will take an environment giving semantic values to syntactic variables, which we represent as a function from variables to values:

```
Env : Ctx⋆ → Ctx⋆ → Set
Env Ψ Φ = ∀{J} → Ψ ∋⋆ J → Φ ⊨ J
```

Let η, η' range over environments.

It is convenient to extend an environment with an additional semantic type:

```
extende : ∀{Ψ Φ} → (η : Env Φ Ψ) → ∀{K}(A : Ψ ⊨ K) → Env (Φ ,⋆ K) Ψ
extende η V Z     = V
extende η V (S α) = η α
```

Lifting of environments to push them under binders can be defined as follows. One could also define it analogously to the lifting of renamings and substitutions defined in Sect. 2.

```
lifte : ∀ {Φ Ψ} → Env Φ Ψ → ∀ {K} → Env (Φ ,⋆ K) (Ψ ,⋆ K)
lifte η = extende (weaken⊨ ∘ η) (reflect (` Z))
```

We define a semantic version of application called ·V which applies semantic functions to semantic arguments. As semantic values at function kind can either be neutral terms or genuine semantic functions we need to pattern match on them to see how to apply them. Notice that the identity renaming id is used in the case of a semantic function. This is because, as we can read of from the type of ·*V*, the function and the argument are in the same context.

```
_·V_ : ∀{Φ K J} → Φ ⊨ (K ⇒ J) → Φ ⊨ K → Φ ⊨ J
inl A ·V V = reflect (A · reify V)
inr F ·V V = F id V
```

Evaluation is defined by recursion on types:

```
eval : ∀{Φ Ψ K} → Ψ ⊢⋆ K → Env Ψ Φ → Φ ⊨ K
eval (` α)     η = η α
eval (ƛ B)     η = inr λ ρ v → eval B (extende (ren⊨ ρ ∘ η) v)
eval (A · B)   η = eval A η ·V eval B η
eval (A ⇒ B)   η = reify (eval A η) ⇒ reify (eval B η)
eval (Π B)     η = Π (reify (eval B (lifte η)))
eval (μ B)     η = μ (reify (eval B (lifte η)))
```

We can define the identity environment as a function that embeds variables into neutral terms with ‘ and then reflects them into values:

```
idEnv : ∀ Φ → Env Φ Φ
idEnv Φ = reflect ∘ ‘
```

We combine reify with eval in the identity environment idEnv to yield a normalisation function that takes types in a given context and kind and returns normal forms in the same context and kind:

```
nf : ∀{Φ K} → Φ ⊢⋆ K → Φ ⊢Nf⋆ K
nf A = reify (eval A (idEnv _))
```

In the next three sections we prove the three correctness properties about this normalisation algorithm: completeness; soundness; and stability.

3.3 Completeness of Type Normalisation

Completeness states that normalising two β-equal types yields the same normal form. This is an important correctness property for normalisation: it ensures that normalisation picks out unique representatives for normal forms. In a similar way to how we defined the semantic domain by recursion on kinds, we define a Kripke Logical Relation on kinds which is a sort of equality on values. At different kinds and for different semantic values it means different things: at base type and for neutral functions it means equality of normal forms; for semantic functions it means that in a new context and given a suitable renaming into that context, we take related arguments to related results. We also require an additional condition on semantic functions, which we call uniformity, following Allais et al. [3]. However, our definition is, we believe, simpler as uniformity is just a type synonym (rather than being mutually defined with the logical relation) and we do not need to prove any auxiliary lemmas about it throughout the completeness proof. Uniformity states that if we receive a renaming and related arguments in the target context of the renaming, and then a further renaming, we can apply the function at the same context as the arguments and then rename the result or rename the arguments first and then apply the function in the later context.

It should not be possible that a semantic function can become equal to a neutral term so we rule out these cases by defining them to be ⊥. This would not be necessary if we were doing $\beta\eta$-normalisation.

```
CR : ∀{Φ} K → Φ ⊨ K → Φ ⊨ K → Set
CR *        A       A'       = A ≡ A'
CR (K ⇒ J) (inl A) (inl A') = A ≡ A'
CR (K ⇒ J) (inr F) (inl A') = ⊥
CR (K ⇒ J) (inl A) (inr F') = ⊥
CR (K ⇒ J) (inr F) (inr F') = Unif F × Unif F' ×
  ∀ {Ψ}(ρ : Ren⋆ _ Ψ){V V' : Ψ ⊨ K} → CR K V V' → CR J (F ρ V) (F' ρ V')
```

```
where
  -- Uniformity
  Unif : ∀{Φ K J} → (∀ {Ψ} → Ren⋆ Φ Ψ → Ψ ⊨ K → Ψ ⊨ J) → Set
  Unif {Φ}{K}{J} F = ∀{Ψ Ψ'}(ρ : Ren⋆ Φ Ψ)(ρ' : Ren⋆ Ψ Ψ')(V V' : Ψ ⊨ K)
    → CR K V V' → CR J (ren⊨ ρ' (F ρ V)) (F (ρ' ∘ ρ) (ren⊨ ρ' V'))
```

The relation CR is not an equivalence relation, it is only a partial equivalence relation (PER) as reflexivity does not hold. However, as is always the case for PERs there is a limited version of reflexivity for elements that are related to some other element.

```
symCR  : ∀{Φ K}{V V' : Φ ⊨ K}      → CR K V V' → CR K V' V
transCR : ∀{Φ K}{V V' V'' : Φ ⊨ K} → CR K V V' → CR K V' V'' → CR K V V''
reflCR  : ∀{Φ K}{V V' : Φ ⊨ K}     → CR K V V' → CR K V V
```

We think of CR as equality of semantic values. Renaming of semantic values ren⊨ (defined in the Sect. 3.2) is a functorial action and we can prove the functor laws. The laws hold up to CR not up to propositional equality ≡:

```
ren⊨-id   : ∀{K Φ}{V V' : Φ ⊨ K} → CR K V V' → CR K (ren⊨ id V) V'
ren⊨-comp : ∀{K Φ Ψ Θ}(ρ : Ren⋆ Φ Ψ)(ρ' : Ren⋆ Ψ Θ){V V' : Φ ⊨ K}
  → CR K V V' → CR K (ren⊨ (ρ' ∘ ρ) V) (ren⊨ ρ' (ren⊨ ρ V'))
```

The completeness proof follows a similar structure as the normalisation algorithm. We define reflectCR and reifyCR analogously to the reflect and reify of the algorithm.

```
reflectCR : ∀{Φ K}{A A' : Φ ⊢Ne⋆ K} → A ≡ A'      → CR K (reflect A) (reflect A')
reifyCR   : ∀{Φ K}{V V' : Φ ⊨ K}    → CR K V V' → reify V ≡ reify V'
```

We define a pointwise partial equivalence for environments analogously to the definition of environments themselves:

```
EnvCR : ∀ {Φ Ψ} → (η η' : Env Φ Ψ) → Set
EnvCR η η' = ∀{K}(α : _ ∋⋆ K) → CR K (η α) (η' α)
```

Before defining the fundamental theorem of logical relations which is analogous to eval we define an identity extension lemma which is used to bootstrap the fundamental theorem. It states that if we evaluate a single term in related environments we get related results. Semantic renaming commutes with eval, and we prove this simultaneously with identity extension:

```
idext     : ∀{Φ Ψ K}{η η' : Env Φ Ψ} → EnvCR η η' → (A : Φ ⊢⋆ K)
  → CR K (eval A η) (eval A η')
ren⊨-eval : ∀{Φ Ψ Θ K}(A : Ψ ⊢⋆ K){η η' : Env Ψ Φ}(p : EnvCR η η')
  → (ρ : Ren⋆ Φ Θ ) → CR K (ren⊨ ρ (eval A η)) (eval A (ren⊨ ρ ∘ η'))
```

We have proved that semantic renaming commutes with evaluation. We also require that syntactic renaming commutes with evaluation: that we can either rename before evaluation or evaluate in a renamed environment:

```
ren-eval : ∀{Φ Ψ Θ K}(A : Θ ⊢⋆ K){η η' : Env Ψ Φ}(p : EnvCR η η')(ρ : Ren⋆ Θ Ψ)
  → CR K (eval (ren⋆ ρ A) η) (eval A (η' ∘ ρ))
```

As in our previous renaming lemma we require that we can either substitute and then evaluate or, equivalently, evaluate the underlying term in an environment constructed by evaluating everything in the substitution. This is the usual *substitution lemma* from denotational semantics and also one of the laws of an algebra for a relative monad (the other one holds definitionally):

```
subst-eval : ∀{Φ Ψ Θ K}(A : Θ ⊢⋆ K){η η' : Env Ψ Φ}
  → (p : EnvCR η η')(σ : Sub⋆ Θ Ψ)
  → CR K (eval (sub⋆ σ A) η) (eval A (λ α → eval (σ α) η'))
```

We can now prove the fundamental theorem of logical relations for CR. It is defined by recursion on the β-equality proof:

```
fund : ∀{Φ Ψ K}{η η' : Env Φ Ψ}{A A' : Φ ⊢⋆ K}
  → EnvCR η η' → A ≡β A' → CR K (eval A η) (eval A' η')
```

As for the ordinary identity environment, the proof that the identity environment is related to itself relies on reflection:

```
idCR : ∀{Φ} → EnvCR (idEnv Φ) (idEnv Φ)
idCR x = reflectCR refl
```

Given all these components we can prove the completeness result by running the fundamental theorem in the identity environment and then applying reification. Thus, our normalisation algorithm takes β-equal types to identical normal forms.

```
completeness : ∀ {K Φ} {A B : Φ ⊢⋆ K} → A ≡β B → nf A ≡ nf B
completeness p = reifyCR (fund idCR p)
```

Complications due to omitting the η-rule and the requirement to avoid extensionality were the main challenges in this section.

3.4 Soundness of Type Normalisation

The soundness property states that terms are β-equal to their normal forms which means that normalisation has preserved the meaning. i.e. that the unique representatives chosen by normalisation are actually in the equivalence class.

We proceed in a similar fashion to the completeness proof by defining a logical relation, reify/reflect, fundamental theorem, identity environment, and then plugging it all together to get the required result.

To state the soundness property which relates syntactic types to normal forms we need to convert normal forms back into syntactic types:

```
embNf : ∀{Γ K} → Γ ⊢Nf⋆ K → Γ ⊢⋆ K
embNe : ∀{Γ K} → Γ ⊢Ne⋆ K → Γ ⊢⋆ K
```

The soundness property is a Kripke Logical relation as before, defined as a Set-valued function by recursion on kinds. But this time it relates syntactic types and semantic values. In the first two cases the semantic values are normal or neutral forms and we can state the property we require easily. In the last case where we have a semantic function, we would like to state that sound functions take sound arguments to sound results (modulo the usual Kripke extension). Indeed, when doing this proof for a version of the system with $\beta\eta$-equality this was what we needed. Here, we have only β-equality for types and we were unable to get the proof to go through with the same definition. To solve this problem we added an additional requirement to the semantic function case: we require that our syntactic type of function kind A is β-equal to a λ-expression. Note this holds trivially if we have the η-rule.

```
SR : ∀{Φ} K → Φ ⊢⋆ K → Φ ⊨ K → Set
SR * A V              = A ≡β embNf V
SR (K ⇒ J) A (inl A') = A ≡β embNe A'
SR (K ⇒ J) A (inr F) = Σ (_ ,⋆ K ⊢⋆ J) λ A' → (A ≡β ƛ A') ×
  ∀{Ψ}(ρ : Ren⋆ _ Ψ){B V}
   → SR K B V → SR J (ren⋆ ρ (ƛ A') · B) (ren⊨ ρ (inr F) ·V V)
```

As before we have a notion of reify and reflect for soundness. Reflect takes soundness results about neutral terms to soundness results about semantic values and reify takes soundness results about semantic values to soundness results about normal forms:

```
reflectSR : ∀{K Φ}{A : Φ ⊢⋆ K}{A' : Φ ⊢Ne⋆ K}
  → A ≡β embNe A' → SR K A (reflect A')
reifySR : ∀{K Φ}{A : Φ ⊢⋆ K}{V : Φ ⊨ K}
  → SR K A V → A ≡β embNf (reify V)
```

We need a notion of environment for soundness, which will be used in the fundamental theorem. Here it is a lifting of the relation SR which relates syntactic types to semantic values to a relation which relates type substitutions to type environments:

```
SREnv : ∀{Φ Ψ} → Sub⋆ Φ Ψ → Env Φ Ψ → Set
SREnv {Φ} σ η = ∀{K}(α : Φ ∋⋆ K) → SR K (σ α) (η α)
```

The fundamental Theorem of Logical Relations for SR states that, for any type, if we have a related substitution and environment then the action of the substitution and environment on the type will also be related.

```
evalSR : ∀{Φ Ψ K}(A : Φ ⊢⋆ K){σ : Sub⋆ Φ Ψ}{η : Env Φ Ψ}
  → SREnv σ η → SR K (sub⋆ σ A) (eval A η)
```

The identity substitution is related to the identity environment:

```
idSR : ∀{Φ} → SREnv ‘ (idEnv Φ)
idSR = reflectSR ∘ refl≡β ∘ ‘
```

Soundness result: all types are β-equal to their normal forms.

```
soundness : ∀ {Φ J} → (A : Φ ⊢⋆ J) → A ≡β embNf (nf A)
soundness A = subst (_≡β embNf (nf A)) (sub⋆-id A) (reifySR (evalSR A idSR))
```

Complications in the definition of SR due to omitting the η-rule were the biggest challenge in this section.

3.5 Stability of Type Normalisation

The normalisation algorithm is stable: renormalising a normal form will not change it.

This property is often omitted from treatments of normalisation. For us it is crucial as in the substitution algorithm we define in the next section and in term level definitions we renormalise types.

Stability for normal forms is defined mutually with an auxiliary property for neutral types:

```
stability   : ∀{K Φ}(A : Φ ⊢Nf⋆ K) → nf (embNf A) ≡ A
stabilityNe : ∀{K Φ}(A : Φ ⊢Ne⋆ K) → eval (embNe A) (idEnv Φ) ≡ reflect A
```

We omit the proofs which are a simple simultaneous induction on normal forms and neutral terms. The most challenging part for us was getting the right statement of the stability property for neutral terms.

Stability is quite a strong property. It guarantees both that embNf ∘ nf is idempotent and that nf is surjective:

```
idempotent : ∀{Φ K}(A : Φ ⊢⋆ K)
  → (embNf ∘ nf ∘ embNf ∘ nf) A ≡ (embNf ∘ nf) A
idempotent A = cong embNf (stability (nf A))

surjective : ∀{Φ K}(A : Φ ⊢Nf⋆ K) → Σ (Φ ⊢⋆ K) λ B → nf B ≡ A
surjective A = embNf A ,, stability A
```

Note we use double comma ,, for Agda pairs as we used single comma for contexts.

3.6 Normality Preserving Type Substitution

In the previous subsections we defined a normaliser. In this subsection we will combine the normaliser with our syntactic substitution operation on types to yield a normality preserving substitution. This will be used in later sections to define intrinsically typed terms with normal types. We proceed by working with similar interface as we did for ordinary substitutions.

Normality preserving substitutions are functions from type variables to normal forms:

```
SubNf⋆ : Ctx⋆ → Ctx⋆ → Set
SubNf⋆ Φ Ψ = ∀ {J} → Φ ∋⋆ J → Ψ ⊢Nf⋆ J
```

We can lift a substitution over a new bound variable as before. This is needed for going under binders.

```
liftsNf⋆ : ∀ {Φ Ψ}→ SubNf⋆ Φ Ψ → ∀{K} → SubNf⋆ (Φ ,⋆ K) (Ψ ,⋆ K)
liftsNf⋆ σ Z     = ne (‘ Z)
liftsNf⋆ σ (S α) = weakenNf⋆ (σ α)
```

We can extend a substitution by an additional normal type analogously to ‘cons’ for lists:

```
extendNf⋆ : ∀{Φ Ψ} → SubNf⋆ Φ Ψ → ∀{J}(A : Ψ ⊢Nf⋆ J) → SubNf⋆ (Φ ,⋆ J) Ψ
extendNf⋆ σ A Z = A
extendNf⋆ σ A (S α) = σ α
```

We define the action of substitutions on normal types as follows: first we embed the normal type to be acted on into a syntactic type, and compose the normalising substitution with embedding into syntactic types to turn it into an ordinary substitution, and then use our syntactic substitution operation from Sect. 2.6. This gives us a syntactic type which we normalise using the normalisation algorithm from Sect. 3.2. This is not efficient. It has to traverse the normal type to convert it back to a syntactic type and it may run the normalisation algorithm on things that contain no redexes. However as this is a formalisation primarily, efficiency is not a priority, correctness is.

```
subNf⋆ : ∀{Φ Ψ} → SubNf⋆ Φ Ψ → ∀ {J} → Φ ⊢Nf⋆ J → Ψ ⊢Nf⋆ J
subNf⋆ ρ n = nf (sub⋆ (embNf ∘ ρ) (embNf n))
```

We verify the same correctness properties of normalising substitution as we did for ordinary substitution: namely the relative monad laws. Note that the second law subNf⋆-∋ doesn't hold definitionally this time.

```
subNf⋆-id   : ∀{Φ J}(A : Φ ⊢Nf⋆ J) → subNf⋆ (ne ∘ ‘) A ≡ A
subNf⋆-var  : ∀{Φ Ψ J}(σ : SubNf⋆ Φ Ψ)(α : Φ ∋⋆ J)
  → subNf⋆ σ (ne (‘ α)) ≡ σ α
```

```
subNf⋆-comp : ∀{Φ Ψ Θ}(σ : SubNf⋆ Φ Ψ)(σ' : SubNf⋆ Ψ Θ){J}(A : Φ ⊢Nf⋆ J)
  → subNf⋆ (subNf⋆ σ' ∘ σ) A ≡ subNf⋆ σ' (subNf⋆ σ A)
```

These properties and the definitions that follow rely on properties of normalisation and often corresponding properties of ordinary substitution. E.g. the first law subNf⋆-id follows from stability and sub⋆-id, the second law follows directly from stability (the corresponding property holds definitionally in the ordinary case), and the third law follows from soundness, various components of completeness and sub⋆-comp.

Finally, we define the special case for single type variable substitution that will be needed in the definition of terms in the next section:

```
_[_]Nf⋆ : ∀{Φ J K} → Φ ,⋆ K ⊢Nf⋆ J → Φ ⊢Nf⋆ K → Φ ⊢Nf⋆ J
A [ B ]Nf⋆ = subNf⋆ (extendNf⋆ (ne ∘ `) B) A
```

The development in this section was straightforward. The most significant hurdle was that we require a complete normalisation proof and correctness properties of ordinary substitution to prove correctness properties of substitution on normal forms. The substitution algorithm in this section is essentially a rather indirect implementation of hereditary substitution.

Before moving on we list special case auxiliary lemmas that we will need when defining renaming and substitution for terms with normal types in Sect. 5.

```
ren[]Nf⋆ : ∀{Φ Θ J K}(ρ : Ren⋆ Φ Θ)(A : Φ ,⋆ K ⊢Nf⋆ J) (B : Φ ⊢Nf⋆ K )
  → renNf⋆ ρ (A [ B ]Nf⋆) ≡ renNf⋆ (lift⋆ ρ) A [ renNf⋆ ρ B ]Nf⋆

weakenNf⋆-subNf⋆ : ∀{Φ Ψ}(σ : SubNf⋆ Φ Ψ){K}(A : Φ ⊢Nf⋆ *)
  → weakenNf⋆ (subNf⋆ σ A) ≡ subNf⋆ (liftsNf⋆ σ {K = K}) (weakenNf⋆ A)

subNf⋆-liftNf⋆ : ∀{Φ Ψ}(σ : SubNf⋆ Φ Ψ){K}(B : Φ ,⋆ K ⊢Nf⋆ *)
  → subNf⋆ (liftsNf⋆ σ) B
    ≡
    eval (sub⋆ (lifts⋆ (embNf ∘ σ)) (embNf B)) (lifte (idEnv Ψ))

subNf⋆-[]Nf⋆ : ∀{Φ Ψ K}(σ : SubNf⋆ Φ Ψ)(A : Φ ⊢Nf⋆ K)(B : Φ ,⋆ K ⊢Nf⋆ *)
  → subNf⋆ σ (B [ A ]Nf⋆)
    ≡
    eval (sub⋆ (lifts⋆ (embNf ∘ σ)) (embNf B)) (lifte (idEnv Ψ))
    [ subNf⋆ σ A ]Nf⋆
```

3.7 Terms with Normal Types

We are now ready to define the algorithmic syntax where terms have normal types and the problematic conversion rule is not needed.

The definition is largely identical except wherever a syntactic type appeared before, we have a normal type, wherever an operation on syntactic types appeared before we have the corresponding operation on normal types. Note that the kind level remains the same, so we reuse Ctx⋆ for example.

Term Contexts. Term level contexts are indexed by their type level contexts.

```
data CtxNf : Ctx⋆ → Set where
  ∅   :                                  CtxNf ∅
  _,⋆_ : ∀{Φ} → CtxNf Φ → ∀ J         → CtxNf (Φ ,⋆ J)
  _,_  : ∀{Φ} → CtxNf Φ → Φ ⊢Nf⋆ * → CtxNf Φ
```

Let Γ, Δ range over contexts.

Term Variables. Note that in the T case, we are required to weaken (normal) types.

```
data _∋Nf_ : ∀ {Φ} → CtxNf Φ → Φ ⊢Nf⋆ * → Set where
  Z : ∀{Φ Γ}{A : Φ ⊢Nf⋆ *}                          → Γ , A ∋Nf A
  S : ∀{Φ Γ}{A : Φ ⊢Nf⋆ *}{B : Φ ⊢Nf⋆ *} → Γ ∋Nf A → Γ , B ∋Nf A
  T : ∀{Φ Γ}{A : Φ ⊢Nf⋆ *}{K}              → Γ ∋Nf A → Γ ,⋆ K ∋Nf weakenNf⋆ A
```

Let *x*, *y* range over variables.

Terms. Note the absence of the conversion rule. The types of terms are unique so it is not possible to coerce a term into a different type.

```
data _⊢Nf_ {Φ} Γ : Φ ⊢Nf⋆ * → Set where
  `      : ∀{A}   → Γ ∋Nf A                          → Γ ⊢Nf A
  ƛ      : ∀{A B} → Γ , A ⊢Nf B                      → Γ ⊢Nf A ⇒ B
  _·_    : ∀{A B} → Γ ⊢Nf A ⇒ B → Γ ⊢Nf A            → Γ ⊢Nf B
  Λ      : ∀{K B} → Γ ,⋆ K ⊢Nf B                     → Γ ⊢Nf Π B
  _·⋆_   : ∀{K B} → Γ ⊢Nf Π B    → (A : Φ ⊢Nf⋆ K) → Γ ⊢Nf B [ A ]Nf⋆
  wrap   : ∀ A    → Γ ⊢Nf A [ μ A ]Nf⋆               → Γ ⊢Nf μ A
  unwrap : ∀{A}   → Γ ⊢Nf μ A                        → Γ ⊢Nf A [ μ A ]Nf⋆
```

Let *L*, *M* range over terms.

We now have an intrinsically typed definition of terms with types that are guaranteed to be normal. By side-stepping the conversion problem we can define an operational semantics for this syntax which we will do in Sect. 5. In the next section we will reflect on the correspondence between this syntax and the syntax with conversion presented in Sect. 2.

We define two special cases of subst which allow us to substitute the types of variables or terms by propositionally equal types. While it is the case that types are now represented uniquely we still want or need to prove that two types are equal, especially in the presence of (Agda) variables, cf., while the natural number 7 has a unique representation in Agda we still might want to prove that for any natural numbers m and n, $m + n \equiv n + m$.

```
conv∋Nf : ∀ {Φ Γ}{A A' : Φ ⊢Nf⋆ *} → A ≡ A' → (Γ ∋Nf A) → Γ ∋Nf A'
conv∋Nf refl α = α

conv⊢ : ∀ {Φ Γ}{A : Φ ⊢⋆ *}{A' : Φ ⊢⋆ *} → A ≡ A' → Γ ⊢ A → Γ ⊢ A'
conv⊢ refl α = α
```

We see these operations in use in Sect. 5.

4 Correspondence Between Declarative and Algorithmic Type Systems

We now have two versions of the syntax/typing rules. Should we just throw away the old one and use the new one? No. The first version is the standard textbook version and the second version is an algorithmic version suitable for implementation. To reconcile the two we prove the second version is sound and complete with respect to the first. This is analogous to proving that a typechecker is sound and complete with respect to the typing rules. Additionally, we prove that before and after normalising the type, terms erase to the same untyped terms. The constructions in this section became significantly simpler and easier after switching from inductive-recursive term contexts to indexed term contexts.

There is an interesting parallel here with the metatheory of Twelf[2]. In Twelf, hereditary substitution are central to the metatheory and the semantics is defined on a version of the syntax where both types and terms are canonical (i.e. they are normalised). In our setting only the types are normalised (viz. canonical). But, the situation is similar: there are two versions of the syntax, one with a semantics (the canonical system), and one without (the ordinary system). Martens and Crary [28] make the case that the ordinary version is the programmer's interface, or the external language in compiler terminology, and the canonical version is the internal language in compiler terminology. In their setting the payoff is also the same: by moving from a language with type equivalence to one where types are uniquely represented, the semantics and metatheory become much simpler.

There is also a parallel with how type checking algorithms are described in the literature: they are often presented an alternative set of typing rules and then they are proved sound and complete with respect to the original typing rules. We will draw on this analogy in the rest of this section as our syntaxes are also type systems.

4.1 Soundness of Typing

From a typing point of view, soundness states that anything typeable in the new type system is also typeable in the old one. From our syntactic point of view this corresponds to taking an algorithmic term and embedding it back into a declarative term.

We have already defined an operation to embed normal types into syntactic types. But, we need an additional operation here: term contexts contain types so we must embed term contexts with normal type into term contexts with syntactic types.

[2] We thank an anonymous reviewer for bringing this to our attention.

```
embCtx : ∀{Φ} → CtxNf Φ → Ctx Φ
embCtx ∅         = ∅
embCtx (Γ ,⋆ K) = embCtx Γ ,⋆ K
embCtx (Γ , A)  = embCtx Γ , embNf A
```

Embedding for terms takes a term with a normal type and produces a term with a syntactic type.

```
embTy : ∀{Φ Γ}{A : Φ ⊢Nf⋆ *} → Γ ⊢Nf A → embCtx Γ ⊢ embNf A
```

Soundness of typing is a direct corollary of embTy:

```
soundnessT : ∀{Φ Γ}{A : Φ ⊢Nf⋆ *} → Γ ⊢Nf A → embCtx Γ ⊢ embNf A
soundnessT = embTy
```

Soundness gives us one direction of the correspondence between systems. The other direction is given by completeness.

4.2 Completeness of Typing

Completeness of typing states that anything typeable by the original declarative system is typeable by the new system, i.e. we do not lose any well typed programs by moving to the new system. From our syntactic point of view, it states that we can take any declarative term of a given type and normalise its type to produce an algorithmic term with a type that is β-equal to the type we started with.

We have already defined normalisation for types. Again, we must provide an operation that normalises a context:

```
nfCtx : ∀{Φ} → Ctx Φ → CtxNf Φ
nfCtx ∅         = ∅
nfCtx (Γ ,⋆ K) = nfCtx Γ ,⋆ K
nfCtx (Γ , A)  = nfCtx Γ , nf A
```

We observe at this point (just before we use it) that conversion is derivable for the algorithmic syntax. It computes:

```
conv⊢Nf : ∀ {Φ Γ}{A A' : Φ ⊢Nf⋆ *} → A ≡ A' → Γ ⊢Nf A → Γ ⊢Nf A'
conv⊢Nf refl L = L
```

The operation that normalises the types of terms takes a declarative term and produces an algorithmic term. We omit the majority of the definition, but include the case for a conversion. In this case we have a term t of type $\Gamma \vdash A$ and a proof p that $A \equiv_\beta B$. We require a term of type Γ ⊢Nf nf B. By inductive hypothesis/recursive call AgdaFunction-nfType $t : \Gamma$ ⊢Nf nf A. But, via completeness of normalisation we know that if $A \equiv_\beta B$ then nf B ≡ nf A, so we invoke the conversion function conv⊢Nf with the completeness proof and the recursive call as arguments:

```
nfType : ∀{Φ Γ}{A : Φ ⊢⋆ *} → Γ ⊢ A → nfCtx Γ ⊢Nf nf A
nfType (conv p t) = conv⊢Nf (completeness p) (nfType t)
```

⋮ (remaining cases omitted)

The operation nfType is not quite the same as completeness. Additionally we need that the original type is β-equal to the new type. This follows from soundness of normalisation.

```
completenessT : ∀{Φ Γ}{A : Φ ⊢⋆ *} → Γ ⊢ A
  → nfCtx Γ ⊢Nf nf A × (A ≡β embNf (nf A))
completenessT {A = A} t = nfType t ,, soundness A
```

4.3 Erasure

We have two version of terms, and we can convert from one to the other. But, how do we know that after conversion, we still have the *same* term? One answer is to show that the term before conversion and the term after conversion both erase to the same untyped term. First, we define untyped (but intrinsically scoped) λ-terms:

```
data _⊢ : ℕ → Set where
  ‘  : ∀{n} → Fin n → n ⊢
  ƛ  : ∀{n} → suc n ⊢ → n ⊢
  _·_: ∀{n} → n ⊢ → n ⊢ → n ⊢
```

Following the pattern of the soundness and completeness proofs we deal in turn with contexts, variables, and then terms. In this case erasing a context corresponds to counting the number of term variables in the context:

```
len : ∀{Φ} → Ctx Φ → ℕ
len ∅          = 0
len (Γ ,⋆ K) = len Γ
len (Γ , A)  = suc (len Γ)
```

Erasure for variables converts them to elements of Fin:

```
eraseVar : ∀{Φ Γ}{A : Φ ⊢⋆ *} → Γ ∋ A → Fin (len Γ)
eraseVar Z     = zero
eraseVar (S α) = suc (eraseVar α)
eraseVar (T α) = eraseVar α
```

Erasure for terms is straightforward:

```
erase : ∀{Φ Γ}{A : Φ ⊢⋆ *} → Γ ⊢ A → len Γ ⊢
erase (‘ α)         = ‘ (eraseVar α)
```

```
erase (ƛ L)        = ƛ (erase L)
erase (L · M)      = erase L · erase M
erase (Λ L)        = erase L
erase (L ·⋆ A)     = erase L
erase (wrap A L) = erase L
erase (unwrap L) = erase L
erase (conv p L) = erase L
```

Note that we drop wrap and unwrap when erasing as these special type casts merely indicate at which isomorphic type we want the term to considered. Without types wrap and unwrap serve no purpose.

Erasure from algorithmic terms proceeds in the same way as declarative terms. The only difference is the that there is no case for conv:

```
lenNf       : ∀{Φ} → CtxNf Φ → ℕ
eraseVarNf : ∀{Φ Γ}{A : Φ ⊢Nf⋆ *} → Γ ∋Nf A → Fin (lenNf Γ)
eraseNf     : ∀{Φ Γ}{A : Φ ⊢Nf⋆ *} → Γ ⊢Nf A → lenNf Γ ⊢
```

Having defined erasure for both term representations we proceed with the proof that normalising types preserves meaning of terms. We deal with contexts first, then variables, and then terms. Normalising types in the context preserves the number of term variables in the context:

```
sameLen : ∀ {Φ}(Γ : Ctx Φ) → lenNf (nfCtx Γ) ≡ len Γ
```

The main complication in the proofs about variables and terms below is that sameLen appears in the types. It complicates each case as the subst prevents things from computing when its proof argument is not refl. This can be worked around using Agda's *with* feature which allows us to abstract over additional arguments such as those which are stuck. However in this case we would need to abstract over so many arguments that the proof becomes unreadable. Instead we prove a simple lemma for each case which achieves the same as using *with*. We show the simplest instance lemzero for the Z variable which abstracts over proof of sameLen and replaces it with an arbitrary proof p that we can pattern match on.

```
lemzero : ∀{n n'}(p : suc n ≡ suc n') → zero ≡ subst Fin p zero
lemzero refl = refl

sameVar : ∀{Φ Γ}{A : Φ ⊢⋆ *}(x : Γ ∋ A)
  → eraseVar x ≡ subst Fin (sameLen Γ) (eraseVarNf (nfTyVar x))
sameVar {Γ = Γ , _} Z = lemzero (cong suc (sameLen Γ))
⋮ (remaining cases omitted)

same : ∀{Φ Γ}{A : Φ ⊢⋆ *}(t : Γ ⊢ A)
  → erase t ≡ subst _⊢ (sameLen Γ) (eraseNf (nfType t))
```

This result indicates that when normalising the type of a term we preserve the meaning of the term where the *meaning* of a term is taken to be the underlying untyped term.

A similar result holds for embedding terms with normal types back into terms with ordinary type but we omit it here.

5 Operational Semantics

We will define the operational semantics on the algorithmic syntax. Indeed, this was the motivation for introducing the algorithmic syntax: to provide a straightforward way to define the semantics without having to deal with type equality coercions. The operational semantics is defined as a call-by-value small-step reduction relation. The relation is typed so it is not necessary to prove preservation as it holds intrinsically. We prove progress for this relation which shows that programs cannot get stuck. As the reduction relation contains β-rules we need to implement substitution for algorithmic terms before proceeding. As we did for types, we define renaming first and then use it to define substitution.

5.1 Renaming for Terms

We index term level renamings/substitutions by their type level counter parts.

Renamings are functions from term variables to terms. The type of the output variable is the type of the input variable renamed by the type level renaming.

```
RenNf : ∀ {Φ Ψ} Γ Δ → Ren⋆ Φ Ψ → Set
RenNf Γ Δ ρ = {A : _ ⊢Nf⋆ *} → Γ ∋Nf A → Δ ∋Nf renNf⋆ ρ A
```

We can lift a renaming both over a new term variable and over a new type variable. These operations are needed to push renamings under binders (λ and Λ respectively).

```
liftNf : ∀{Φ Ψ Γ Δ}{ρ⋆ : Ren⋆ Φ Ψ} → RenNf Γ Δ ρ⋆
  → {B : Φ ⊢Nf⋆ *} → RenNf (Γ , B) (Δ , renNf⋆ ρ⋆ B) ρ⋆
liftNf ρ Z = Z
liftNf ρ (S x) = S (ρ x)

⋆liftNf : ∀{Φ Ψ Γ Δ}{ρ⋆ : Ren⋆ Φ Ψ} → RenNf Γ Δ ρ⋆
  → (∀ {K} → RenNf (Γ ,⋆ K) (Δ ,⋆ K) (lift⋆ ρ⋆))
⋆liftNf ρ (T x) = conv∋Nf (trans (sym (renNf⋆-comp _)) (renNf⋆-comp _)) (T (ρ x))
```

Next we define the functorial action of renaming on terms. In the type instantiation, wrap, unwrap cases we need a proof as this is where substitutions appear in types.

```
renNf : ∀ {Φ Ψ Γ Δ}{ρ⋆ : Ren⋆ Φ Ψ} → RenNf Γ Δ ρ⋆
  → ({A : Φ ⊢Nf⋆ *} → Γ ⊢Nf A → Δ ⊢Nf renNf⋆ ρ⋆ A )
renNf ρ (` x)          = ` (ρ x)
renNf ρ (ƛ N)          = ƛ (renNf (liftNf ρ) N)
renNf ρ (L · M)        = renNf ρ L · renNf ρ M
```

```
renNf ρ (Λ N)                = Λ (renNf (*liftNf ρ) N)
renNf ρ (_·*_ {B = B} t A)  =
  conv⊢Nf (sym (ren[]Nf* _ B A)) (renNf ρ t ·* renNf* _ A)
renNf ρ (wrap A L)           =
  wrap _ (conv⊢Nf (ren[]Nf* _ A (μ A)) (renNf ρ L))
renNf ρ (unwrap {A = A} L) =
  conv⊢Nf (sym (ren[]Nf* _ A (μ A))) (unwrap (renNf ρ L))
```

Weakening by a type is a special case. Another proof is needed here.

```
weakenNf : ∀ {Φ Γ}{A : Φ ⊢Nf* *}{B : Φ ⊢Nf* *} → Γ ⊢Nf A → Γ , B ⊢Nf A
weakenNf {A = A} x =
  conv⊢Nf (renNf*-id A) (renNf (conv∋Nf (sym (renNf*-id _)) ∘ S) x)
```

We can also weaken by a kind:

```
*weakenNf : ∀ {Φ Γ}{A : Φ ⊢Nf* *}{K} → Γ ⊢Nf A → Γ ,* K ⊢Nf weakenNf* A
*weakenNf x = renNf T x
```

5.2 Substitution

Substitutions are defined as functions from type variables to terms. Like renamings they are indexed by their type level counterpart, which is used in the return type.

```
SubNf : ∀ {Φ Ψ} Γ Δ → SubNf* Φ Ψ → Set
SubNf Γ Δ ρ = {A : _ ⊢Nf* *} → Γ ∋Nf A → Δ ⊢Nf subNf* ρ A
```

We define lifting of a substitution over a type and a kind so that we can push substitutions under binders. Agda is not able to infer the type level normalising substitution in many cases so we include it explicitly.

```
liftsNf : ∀{Φ Ψ Γ Δ}(σ* : SubNf* Φ Ψ) → SubNf Γ Δ σ*
  → {B : _ ⊢Nf* *} → SubNf (Γ , B) (Δ , subNf* σ* B) σ*
liftsNf _ σ Z = ` Z
liftsNf _ σ (S x) = weakenNf (σ x)
```

```
*liftsNf : ∀{Φ Ψ Γ Δ}(σ* : SubNf* Φ Ψ) → SubNf Γ Δ σ*
  → ∀ {K} → SubNf (Γ ,* K) (Δ ,* K) (liftsNf* σ*)
*liftsNf σ* σ (T {A = A} x) =
  conv⊢Nf (weakenNf*-subNf* σ* A) (*weakenNf (σ x))
```

Having defined lifting we are now ready to define substitution on terms:

```
subNf : ∀{Φ Ψ Γ Δ}(σ* : SubNf* Φ Ψ) → SubNf Γ Δ σ*
  → ({A : Φ ⊢Nf* *} → Γ ⊢Nf A → Δ ⊢Nf subNf* σ* A)
subNf σ* σ (` k)              = σ k
```

```
subNf σ* σ (ƛ N)                = ƛ (subNf σ* (liftsNf σ* σ) N)
subNf σ* σ (L · M)              = subNf σ* σ L · subNf σ* σ M
subNf σ* σ (Λ {B = B} N)        =
  Λ (conv⊢Nf (subNf*-liftNf* σ* B) (subNf (liftsNf* σ*) (*liftsNf σ* σ) N))
subNf σ* σ (_·*_ {B = B} L M) =
  conv⊢Nf (sym (subNf*-[]Nf* σ* M B)) (subNf σ* σ L ·* subNf* σ* M)
subNf σ* σ (wrap A L)           =
  wrap _ (conv⊢Nf (subNf*-[]Nf* σ* (μ A) A) (subNf σ* σ L))
subNf σ* σ (unwrap {A = A} L) =
  conv⊢Nf (sym (subNf*-[]Nf* σ* (μ A) A)) (unwrap (subNf σ* σ L))
```

We define special cases for single type and term variable substitution into a term, but omit their long winded and not very informative definitions.

```
_[_]Nf  : ∀{Φ Γ}{A B : Φ ⊢Nf* *} → Γ , B ⊢Nf A → Γ ⊢Nf B → Γ ⊢Nf A
_*[_]Nf : ∀{Φ Γ K}{B : Φ ,* K ⊢Nf* *}
  → Γ ,* K ⊢Nf B → (A : Φ ⊢Nf* K) → Γ ⊢Nf B [ A ]Nf*
```

We now have all the equipment we need to specify small-step reduction.

5.3 Reduction

Before defining the reduction relation we define a value predicate on terms that captures which terms cannot be reduced any further. We do not wish to perform unnecessary computation so we do not compute under the binder in the ƛ case. However, we do want to have the property that when you erase a value it remains a value. A typed value, after erasure, should not require any further reduction to become an untyped value. This gives us a close correspondence between the typed and untyped operational semantics. So, it is essential in the Λ and wrap cases that the bodies are values as both of these constructors are removed by erasure.

```
data Value {Φ}{Γ} : {A : Φ ⊢Nf* *} → Γ ⊢Nf A → Set where
  V-ƛ    : ∀{A B}(L : Γ , A ⊢Nf B)                → Value (ƛ L)
  V-Λ    : ∀{K B}{L : Γ ,* K ⊢Nf B}     → Value L → Value (Λ L)
  V-wrap : ∀{A}{L : Γ ⊢Nf A [ μ A ]Nf*} → Value L → Value (wrap A L)
```

We give the dynamics of the term language as a small-step reduction relation. The relation is typed and terms on the left and right hand side have the same type so it is impossible to violate preservation. We have two congruence (xi) rules for application and only one for type application, types are unique so the type argument cannot reduce. Indeed, no reduction of types is either possible or needed. There are three computation (beta) rules, one for application, one for type application and one for recursive types. We allow reduction in almost any term argument in the xi rules except under a ƛ. Allowing reduction under Λ and wrap is required to ensure that their bodies become values. The value condition on the function term in rule $\xi\text{-}\cdot_2$ ensures that, in an application, we reduce the function before the argument. The value condition on the argument in rule *β-ƛ* ensures that the our semantics is call-by-value.

```
data _—→_ {Φ}{Γ} : {A : Φ ⊢Nf* *} → (Γ ⊢Nf A) → (Γ ⊢Nf A) → Set where
  ξ-·1       : ∀{A B}{L L' : Γ ⊢Nf A ⇒ B}{M : Γ ⊢Nf A}
    → L —→ L' → L · M —→ L' · M
  ξ-·2       : ∀{A B}{V : Γ ⊢Nf A ⇒ B}{M M' : Γ ⊢Nf A}
    → Value V → M —→ M' → V · M —→ V · M'
  ξ-Λ        : ∀{K B}{L L' : Γ ,* K ⊢Nf B}
    → L —→ L' → Λ L —→ Λ L'
  ξ-·*       : ∀{K B}{L L' : Γ ⊢Nf Π B}{A : Φ ⊢Nf* K}
    → L —→ L' → L ·* A —→ L' ·* A
  ξ-unwrap : ∀{A}{L L' : Γ ⊢Nf μ A}
    → L —→ L' → unwrap L —→ unwrap L'
  ξ-wrap : {A : Φ ,* * ⊢Nf* *}{L L' : Γ ⊢Nf A [ μ A ]Nf*}
    → L —→ L' → wrap A L —→ wrap A L'
  β-ƛ        : ∀{A B}{L : Γ , A ⊢Nf B}{M : Γ ⊢Nf A}
    → Value M → ƛ L · M —→ L [ M ]Nf
  β-Λ        : ∀{K B}{L : Γ ,* K ⊢Nf B}{A : Φ ⊢Nf* K}
    → Λ L ·* A —→ L *[ A ]Nf
  β-wrap     : ∀{A}{L : Γ ⊢Nf A [ μ A ]Nf*}
    → unwrap (wrap A L) —→ L
```

5.4 Progress and Preservation

The reduction relation is typed. The definition guarantees that the terms before and after reduction will have the same type. Therefore it is not necessary to prove type preservation.

Progress captures the property that reduction of terms should not get stuck, either a term is already a value or it can make a reduction step. Progress requires proof. We show the proof in complete detail. In an earlier version of this work when we did not reduce under Λ and we proved progress directly for closed terms, i.e. for terms in the empty context. Reducing under the Λ binder means that we need to reduce in non-empty contexts so our previous simple approach no longer works.

There are several approaches to solving this including: (1) modifying term syntax to ensure that the bodies of Λ-expressions are already in fully reduced form (the so-called *value restriction*). This means that we need only make progress in the empty context as no further progress is necessary when we are in a non-empty context. This has the downside of changing the language slightly but keeps progress simple; (2) defining neutral terms (terms whose reduction is blocked by a variable), proving a version of progress for open terms, observing that there are no neutral terms in the empty context and deriving progress for closed terms as a corollary. This has the disadvantage of having to introduce neutral terms only to rule them out and complicating the progress proof; (3) observe that Λ only binds type variables and not term variables and only term variables can block progress, prove progress for terms in contexts that contain no term variables and derive closed progress as a simple corollary. We choose option 3 here as the language remains the same and the progress proof is relatively unchanged, it just requires an extra condition on the context. The only cost is an additional predicate on contexts and an additional lemma.

Before starting the progress proof we need to capture the property of a context not containing any term variables. Our term contexts are indexed by type contexts, if we wanted to rule out type variables we could talk about term contexts indexed by the empty type context, but we cannot use the same trick for ruling out term variables. So, we use a recursive predicate on contexts NoVar. The empty context satisfies it, a context extended by (the kind of) a type variable does if the underlying context does, and a context containing (the type of) a term variable does not.

```
NoVar : ∀{Φ} → CtxNf Φ → Set
NoVar ∅ = ⊤
NoVar (Γ ,* J) = NoVar Γ
NoVar (Γ , A) = ⊥
```

We can prove easily that it is impossible to have term variable in a context containing no term variables. There is only one case and the property follows by induction on variables:

```
noVar : ∀{Φ Γ} → NoVar Γ → {A : Φ ⊢Nf* *}(x : Γ ∋Nf A) → ⊥
noVar p (T x) = noVar p x
```

We can now prove progress. The proof is the same as the one for closed terms, except for the extra argument p : NoVar Γ.

```
progress : ∀{Φ}{Γ} → NoVar Γ → {A : Φ ⊢Nf* *}(L : Γ ⊢Nf A)
  → Value L ⊎ Σ (Γ ⊢Nf A) λ L' → L —→ L'
```

The variable case is impossible.

```
progress p (` x) = ⊥-elim (noVar p x)
```

Any ƛ-expression is a value as we do not reduce under the binder.

```
progress p (ƛ L) = inl (V-ƛ L)
```

In the application case we first examine the result of the recursive call on the function term, if it is a value, it must be a ƛ-expression, so we examine the recursive call on the argument term. If this is a value then we perform β-reduction. Otherwise we make the appropriate ξ-step.

```
progress p (L · M)   with progress p L
progress p (L · M)   — inl V         with progress p M
progress p (ƛ L · M) — inl (V-ƛ L) — inl V         = inr (L [ M ]Nf ,, β-ƛ V)
progress p (L · M)   — inl V         — inr (M' ,, q) = inr (L · M' ,, ξ-·2 V q)
progress p (L · M)   — inr (L' ,, q) = inr (L' · M ,, ξ-·1 q)
```

As we must reduce under Λ and wrap in both cases we make a recursive call on their bodies and proceed accordingly. Notice that the argument p is unchanged in the recursive call to the body of a Λ as NoVar (Γ ,* K) = NoVar Γ.

```
progress p (Λ L) with progress p L
...              — inl V          = inl (V-Λ V)
...              — inr (L' ,, q) = inr (Λ L' ,, ξ-Λ q)
progress p (wrap A L) with progress p L
...                   — inl V          = inl (V-wrap V)
...                   — inr (L' ,, q) = inr (wrap A L' ,, ξ-wrap q)
```

In the type application case we first examine the result of recursive call on the type function argument. If it is a value it must be a Λ-expression and we perform β-reduction. Otherwise we perform a ξ-step.

```
progress p (L ·* A)     with progress p L
progress p (Λ L ·* A) — inl (V-Λ V)    = inr (L *[ A ]Nf ,, β-Λ)
progress p (L ·* A)     — inr (L' ,, q) = inr (L' ·* A ,, ξ-·* q)
```

In the unwrap case we examine the result of the recursive call on the body. If it is a value it must be a wrap and we perform β-reduction or a ξ-step otherwise. That completes the proof.

```
progress p (unwrap L)            with progress p L
progress p (unwrap (wrap A L)) — inl (V-wrap V) = inr (L ,, β-wrap)
progress p (unwrap L)            — inr (L' ,, q)  = inr (unwrap L' ,, ξ-unwrap q)
```

Progress in the empty context progress∅ is a simple corollary. The empty context trivially satisfies NoVar as NoVar ∅ = ⊤:

```
progress∅ : ∀{A}(L : ∅ ⊢Nf A) → Value L ⊎ Σ (∅ ⊢Nf A) λ L' → L —→ L'
progress∅ = progress tt
```

5.5 Erasure

We can extend our treatment of erasure from syntax to (operational) semantics. Indeed, when defining values were careful to ensure this was possible.

To define the β-rule we need to be able to perform substitution on one variable only. As for syntaxes in earlier sections we define parallel renaming and substitution first and get substitution on one variable as a special case. We omit the details here which are analogous to earlier sections.

```
_[_]U : ∀{n} → suc n ⊢ → n ⊢ → n ⊢
```

When erasing reduction steps below we will require two properties about pushing erasure through a normalising single variable substitution. These properties follow from properties of parallel renaming and substitution:

```
eraseNf-*[]Nf : ∀{Φ}{Γ : CtxNf Φ}{K B}(L : Γ ,* K ⊢Nf B)(A : Φ ⊢Nf* K)
  → eraseNf L ≡ eraseNf (L *[ A ]Nf)
eraseNf-[]Nf : ∀{Φ}{Γ : CtxNf Φ}{A B}(L : Γ , A ⊢Nf B)(M : Γ ⊢Nf A)
  → eraseNf L [ eraseNf M ]U ≡ eraseNf (L [ M ]Nf)
```

There is only one value in untyped lambda calculus: lambda.

```
data UValue {n} : n ⊢ → Set where
  V-ƛ : (t : suc n ⊢) → UValue (ƛ t)
```

We define a call-by-value small-step reduction relation that is intrinsically scoped.

```
data _U—→_ {n} : n ⊢ → n ⊢ → Set where
  ξ-·1 : {L L' M : n ⊢} → L U—→ L' → L · M U—→ L' · M
  ξ-·2 : {L M M' : n ⊢} → UValue L → M U—→ M' → L · M U—→ L · M'
  β-ƛ : {L : suc n ⊢}{M : n ⊢} → UValue M → ƛ L · M U—→ L [ M ]U
```

Erasing values is straightforward. The only tricky part is to ensure that in values the subterms of the values for wrap and Λ are also values as discussed earlier. This ensures that after when we erase a typed value we will always get an untyped value:

```
eraseVal : ∀{Φ A}{Γ : CtxNf Φ}{t : Γ ⊢Nf A} → Value t → UValue (eraseNf t)
eraseVal (V-ƛ t) = V-ƛ (eraseNf t)
eraseVal (V-Λ v) = eraseVal v
eraseVal (V-wrap v) = eraseVal v
```

Erasing a reduction step is more subtle as we may either get a typed reduction step (e.g., β-ƛ) or the step may disappear (e.g., β-wrap). In the latter case the erasure of the terms before and after reduction will be identical:

```
erase—→ : ∀{Φ A}{Γ : CtxNf Φ}{t t' : Γ ⊢Nf A}
  → t —→ t' → eraseNf t U—→ eraseNf t' ⊎ eraseNf t ≡ eraseNf t'
```

In the congruence cases for application what we need to do depends on the result of erasing the underlying reduction step. We make use of map for Sum types for this purpose, the first argument explains what to do if the underlying step corresponds to a untyped reduction step (we create an untyped congruence reducing step) and the second argument explains what to do if the underlying step disappears (we create an equality proof):

```
erase—→ (ξ-·1 {M = M} p) =
  Sum.map ξ-·1 (cong (_· eraseNf M)) (erase—→ p)
erase—→ (ξ-·2 {V = V} p q) =
  Sum.map (ξ-·2 (eraseVal p)) (cong (eraseNf V ·_)) (erase—→ q)
```

In the following cases the outer reduction step is removed:

```
erase—→ (ξ-·* p)      = erase—→ p
erase—→ (ξ-Λ p)       = erase—→ p
erase—→ (ξ-unwrap p) = erase—→ p
erase—→ (ξ-wrap p)    = erase—→ p
```

In the case of β-reduction for an ordinary application we always produce a corresponding untyped β-reduction step:

```
erase—→ (β-ƛ {L = L}{M = M} V) = inl (subst
  (ƛ (eraseNf L) · eraseNf M U—→_)
  (eraseNf-[]Nf L M)
  (_U—→_.β-ƛ {L = eraseNf L}{M = eraseNf M} (eraseVal V)))
```

In the other two β-reduction cases the step is always removed, e.g., unwrap (wrap *A L*) —→ *L* becomes *L* ≡ *L* :

```
erase—→ (β-Λ {L = L}{A = A}) = inr (eraseNf-⋆[]Nf L A)
erase—→ β-wrap               = inr refl
```

That concludes the proof: either a typed reduction step corresponds to an untyped one or no step at all.

We can combine erasure of values and reduction steps to get a progress like result for untyped terms via erasure. Via typed progress we either arrive immediately at an untyped value, or a typed reduction step must exist and it will corr respond to an untyped step, or the step disappears:

```
erase-progress∅ : ∀{A : ∅ ⊢Nf⋆ *}(L : ∅ ⊢Nf A)
  → UValue (eraseNf L)
  ⊎ Σ (∅ ⊢Nf A) λ L' → (L —→ L')
     × (eraseNf L U—→ eraseNf L' ⊎ eraseNf L ≡ eraseNf L')
erase-progress∅ L =
  Sum.map eraseVal (λ {(L' ,, p) → L' ,, p ,, (erase—→ p)}) (progress∅ L)
```

6 Execution

We can iterate progress an arbitrary number of times to run programs. First, we define the reflexive transitive closure of reduction. We will use this to represent traces of execution:

```
data _—→*_ {Φ Γ} : {A : Φ ⊢Nf⋆ *} → Γ ⊢Nf A → Γ ⊢Nf A → Set where
  refl—→ : ∀{A}{M : Γ ⊢Nf A} → M —→* M
  trans—→ : ∀{A}{M M' M'' : Γ ⊢Nf A}
    → M —→ M' → M' —→* M'' → M —→* M''
```

The run function takes a number of allowed steps and a term. It returns another term, a proof that the original term reduces to the new term in zero or more steps and possibly a proof that the new term is a value. If no value proof is returned this indicates that we did not reach a value in the allowed number of steps.

If we are allowed zero more steps we return failure immediately. If we are allowed more steps then we call progress to make one. If we get a value back we return straight away with a value. If we have not yet reached a value we call run recursively having spent a step. We then prepend our step to the sequence of steps returned by run and return:

```
run : ∀ {A : ∅ ⊢Nf* *} → ℕ → (M : ∅ ⊢Nf A)
  → Σ (∅ ⊢Nf A) λ M' → (M —→* M') × Maybe (Value M')
run zero M = M ,, refl—→ ,, nothing
run (suc n) M with progress∅ M
run (suc n) M — inl V = M ,, refl—→ ,, just V
run (suc n) M — inr (M' ,, p) with run n M'
... — M" ,, q ,, mV = M" ,, trans—→ p q ,, mV
```

6.1 Erasure

Given that the evaluator run produces a trace of reduction that (if it doesn't run out of allowed steps) leads to a value we can erase the trace and value to yield a trace of untyped execution leading to an untyped value. Note that the untyped trace may be shorter as some steps may disappear.

We define the reflexive transitive closure of untyped reduction analogously to the typed version:

```
data _U—→*_ {n} : n ⊢ → n ⊢ → Set where
  reflU—→  : {M : n ⊢} → M U—→* M
  transU—→ : {M M' M" : n ⊢}
    → M U—→ M' → M' U—→* M" → M U—→* M"
```

We can erase a typed trace to yield an untyped trace. The reflexive case is straightforwards. In the transitive case, we may have a step p that corresponds to an untyped or it may disappear. We use case [_,_] instead of map this time. It is like map but instead of producing another sum it (in the non-dependent case that we are in) produces a result of an the same type in each case (in our case erase M —→ erase M''). In the first case we get an untyped *step* and rest of the trace is handled by the recursive call. In the second case *eq* is an equation erase M ≡ erase M' which we use to coerce the recursive call of type erase M' —→ erase M'' into type erase M —→ erase M'' and the length of the trace is reduced:

```
erase—→* : ∀{Φ}{A : Φ ⊢Nf* *}{Γ : CtxNf Φ}{t t' : Γ ⊢Nf A}
  → t —→* t' → eraseNf t U—→* eraseNf t'
erase—→* refl—→ = reflU—→
```

```
erase—→* (trans—→ {M” = M”} p q) =
  [ (λ step → transU—→ step (erase—→* q))
  , (λ eq → subst (_U—→* eraseNf M”) (sym eq) (erase—→* q))
  ] (erase—→ p)
```

Finally we can use run to get an untyped trace leading to a value, allowed steps permitting.

```
erase-run : ∀ {A : ∅ ⊢Nf⋆ *} → ℕ → (M : ∅ ⊢Nf A)
  → Σ (0 ⊢) λ M’ → (eraseNf M U—→* M’) × Maybe (UValue M’)
erase-run n M with run n M
... — M’ ,, p ,, mv = eraseNf M’ ,, erase—→* p ,, Maybe.map eraseVal mv
```

7 Examples

Using only the facilities of System *F* without the extensions of type functions and recursive types we can define natural numbers as Church Numerals:

```
ℕᶜ : ∀{Φ} → Φ ⊢Nf⋆ *
ℕᶜ = Π ((ne (‘ Z)) ⇒ (ne (‘ Z) ⇒ ne (‘ Z)) ⇒ (ne (‘ Z)))

Zeroᶜ : ∀{Φ}{Γ : CtxNf Φ} → Γ ⊢Nf ℕᶜ
Zeroᶜ = Λ (ƛ (ƛ (‘ (S Z))))

Succᶜ : ∀{Φ}{Γ : CtxNf Φ} → Γ ⊢Nf ℕᶜ ⇒ ℕᶜ
Succᶜ = ƛ (Λ (ƛ (ƛ (‘ Z · ((‘ (S (S (T Z)))) ·⋆ (ne (‘ Z)) · (‘ (S Z)) · (‘ Z))))))

Twoᶜ : ∀{Φ}{Γ : CtxNf Φ} → Γ ⊢Nf ℕᶜ
Twoᶜ = Succᶜ · (Succᶜ · Zeroᶜ)

Fourᶜ : ∀{Φ}{Γ : CtxNf Φ} → Γ ⊢Nf ℕᶜ
Fourᶜ = Succᶜ · (Succᶜ · (Succᶜ · (Succᶜ · Zeroᶜ)))

TwoPlusTwoᶜ : ∀{Φ}{Γ : CtxNf Φ} → Γ ⊢Nf ℕᶜ
TwoPlusTwoᶜ = Twoᶜ ·⋆ ℕᶜ · Twoᶜ · Succᶜ
```

Using the full facilities of System $F_{\omega\mu}$ we can define natural numbers as Scott Numerals [1]. We the *Z* combinator instead of the *Y* combinator as it works for both lazy and strict languages.

```
G : ∀{Φ} → Φ ,⋆ * ⊢Nf⋆ *
G = Π (ne (‘ Z) ⇒ (ne (‘ (S Z)) ⇒ ne (‘ Z)) ⇒ ne (‘ Z))

M : ∀{Φ} → Φ ⊢Nf⋆ *
```

```
M = μ G

N : ∀{Φ} → Φ ⊢Nf⋆ *
N = G [ M ]Nf⋆

Zero : ∀{Φ}{Γ : CtxNf Φ} → Γ ⊢Nf N
Zero = Λ (ƛ (ƛ (` (S (Z )))))

Succ : ∀{Φ}{Γ : CtxNf Φ} → Γ ⊢Nf N ⇒ N
Succ = ƛ (Λ (ƛ (ƛ (` Z · wrap _ (` (S (S (T Z))))))))

Two : ∀{Φ}{Γ : CtxNf Φ} → Γ ⊢Nf N
Two = Succ · (Succ · Zero)

Four : ∀{Φ}{Γ : CtxNf Φ} → Γ ⊢Nf N
Four = Succ · (Succ · (Succ · (Succ · Zero)))

case : ∀{Φ}{Γ : CtxNf Φ}
  → Γ ⊢Nf N ⇒ (Π (ne (` Z) ⇒ (N ⇒ ne (` Z)) ⇒ ne (` Z)))
case = ƛ (Λ (ƛ (ƛ (
  (` (S (S (T Z)))) ·⋆ ne (` Z) · (` (S Z)) · (ƛ (` (S Z) · unwrap (` Z)))))))

Z-comb : ∀{Φ}{Γ : CtxNf Φ} →
  Γ ⊢Nf Π (Π (((ne (` (S Z)) ⇒ ne (` Z)) ⇒ ne (` (S Z)) ⇒ ne (` Z))
    ⇒ ne (` (S Z)) ⇒ ne (` Z)))
Z-comb = Λ (Λ (ƛ (ƛ (` (S Z) · ƛ (unwrap (` (S Z)) · ` (S Z) · ` Z))
  · wrap (ne (` Z) ⇒ ne (` (S (S Z))) ⇒ ne (` (S Z)))
        (ƛ (` (S Z) · ƛ (unwrap (` (S Z)) · ` (S Z) · ` Z))))))

Plus : ∀{Φ}{Γ : CtxNf Φ} → Γ ⊢Nf N ⇒ N ⇒ N
Plus = ƛ (ƛ ((Z-comb ·⋆ N) ·⋆ N · (ƛ (ƛ ((((case · ` Z) ·⋆ N)
  · ` (S (S (S Z)))) · (ƛ (Succ · (` (S (S Z)) · ` Z)))))) · ` (S Z)))

TwoPlusTwo : ∀{Φ}{Γ : CtxNf Φ} → Γ ⊢Nf N
TwoPlusTwo = (Plus · Two) · Two
```

8 Scaling up from System $F_{\omega\mu}$ to Plutus Core

This formalisation forms the basis of a formalisation of Plutus Core. There are two key extensions.

8.1 Higher Kinded Recursive Types

In this paper we used $\mu : (* \to *) \to *$. This is easy to understand and makes it possible to express simple examples directly. This corresponds to the version of recursive types one might use in ordinary System *F*. In System F_ω we have a greater degree of freedom. We have settled on an indexed version of $\mu : ((k \to *) \to k \to *) \to k \to *$ that supports the encoding of mutually defined datatypes. This extension is straightforward in iso-recursive types, in equi-recursive it is not. We chose to present the restricted version in this paper as it is simpler and sufficient to present our examples. See the accompanying paper [26] for a more detailed discussion of higher kinded recursive types.

8.2 Integers, Bytestrings and Cryptographic Operations

In Plutus Core we also extend System $F_{\omega\mu}$ with integers and bytestrings and some cryptographic operations such as checking signatures. Before thinking about how to add these features to our language, there is a choice to be made when modelling integers and bytestrings and cryptographic operations in Agda about whether we consider them internal or external to our model. We are modelling the Haskell implementation of Plutus Core which uses the Haskell bytestring library. We chose to model the Plutus Core implementation alone and consider bytestrings as an external black box. We assume (i.e. postulate in Agda) an interface given as a type for bytestrings and various operations such as take, drop, append etc. We can also make clear our expectations of this interface by assuming (postulating) some properties such as that append is associative. Using pragmas in Agda we can ensure that when we compile our Agda program to Haskell these opaque bytestring operations are compiled to the real operations of the Haskell bytestring library. We have taken a slightly different approach with integers as Agda and Haskell have native support for integers and Agda integers are already compiled to Haskell integers by default so we just make use of this builtin support. Arguably this brings integers inside our model. One could also treat integers as a blackbox. We treat cryptographic operations as a blackbox as we do with bytestrings.

To add integers and bytestrings to the System $F_{\omega\mu}$ we add type constants as types and term constants as terms. The type of a term constant is a type constant. This ensures that we can have term variables whose type is type constant but not term constants whose type is a type variable. To support the operations for integers and bytestrings we add a builtin constructor to the term language, signatures for each operation, and a semantics for builtins that applies the appropriate underlying function to its arguments. The underlying function is postulated in Agda and when compiled to Haskell it runs the appropriate native Haskell function or library function. Note that the cryptographic functions are operations on bytestrings.

Adding this functionality did not pose any particular formalisation challenges except for the fact it was quite a lot of work. However, compiling our implementation of builtins to Haskell did trigger several bugs in Agda's GHC backend which were rapidly diagnosed and fixed by the Agda developers.

8.3 Using Our Implementation for Testing

As we can compile our Agda Plutus Core interpreter to Haskell we can test the production Haskell Plutus Core interpreter against it. We make use of the production system's parser and pretty printer which we import as a Haskell library and use the same libraries for bytestrings and cryptographic functions. The parser produces intrinsically typed terms which we scope check and convert to a representation with de Bruijn indices. We cannot currently use the intrinsically typed implementation we describe in this paper directly as we must type check terms first and formalising a type checker is future work. Instead we have implemented a separate extrinsically typed version that we use for testing. After evaluation we convert the de Bruijn syntax back to a named syntax and pretty print the output. We have proven that for any well-typed term the intrinsic and extrinsic versions give the same results after erasure.

Acknowledgements. We thank the anonymous reviewers for their helpful comments and insightful constructive criticism. We thank IOHK for their support of this work. We thank our colleagues Marko Dimjašević, Kenneth MacKenzie, and Michael Peyton Jones for helpful comments on an multiple drafts. The first author would like to James McKinna for spending an afternoon explaining pure type systems and Guillaume Allais, Apostolis Xekoukoulotakis and Ulf Norell for help with diagnosing and fixing bugs that we encountered in Agda's GHC backend in the course of writing this paper.

References

1. Abadi, M., Cardelli, L., Plotkin, G.: Types for the Scott numerals (1993)
2. Allais, G., Chapman, J., McBride, C., McKinna, J.: Type-and-scope safe programs and their proofs. In: Bertot, Y., Vafeiadis, V. (eds.) Proceedings of the 6th ACM SIGPLAN Conference on Certified Programs and Proofs (CPP 2017), pp. 195–207. ACM, New York (2017). https://doi.org/10.1145/3018610.3018613
3. Allais, G., McBride, C., Boutillier, P.: New equations for neutral terms. In: Weirich, S. (ed.) Proceedings of the 2013 ACM SIGPLAN Workshop on Dependently-typed Programming (DTP 2013), pp. 13–24. ACM, New York (2013). https://doi.org/10.1145/2502409.2502411
4. Altenkirch, T., Chapman, J., Uustalu, T.: Relative monads formalised. J. Formalized Reasoning **7**(1), 1–43 (2014). https://doi.org/10.6092/issn.1972-5787/4389
5. Altenkirch, T., Chapman, J., Uustalu, T.: Monads need not be endofunctors. Logical Methods Comput. Sci. **11**(1), 1–40 (2015). https://doi.org/10.2168/LMCS-11(1:3)2015
6. Altenkirch, T., Reus, B.: Monadic presentations of lambda terms using generalized inductive types. In: Flum, J., Rodriguez-Artalejo, M. (eds.) CSL 1999. LNCS, vol. 1683, pp. 453–468. Springer, Heidelberg (1999). https://doi.org/10.1007/3-540-48168-0_32
7. Amani, S., Bégel, M., Bortin, M., Staples, M.: Towards verifying ethereum smart contract bytecode in isabelle/HOL. In: Andronick, J., Felty, A. (eds.) Proceedings of the 7th ACM SIGPLAN International Conference on Certified Programs and Proofs (CPP 2018), pp. 66–77. ACM, New York (2018). https://doi.org/10.1145/3167084
8. Aydemir, B.E., et al.: Mechanized metatheory for the masses: the PoplMark challenge. In: Hurd, J., Melham, T. (eds.) TPHOLs 2005. LNCS, vol. 3603, pp. 50–65. Springer, Heidelberg (2005). https://doi.org/10.1007/11541868_4
9. Berger, U., Schwichtenberg, H.: An inverse of the evaluation functional for typed lambda-calculus. In: Kahn, G. (ed.) Proceedings of the Sixth Annual Symposium on Logic in Computer Science (LICS 1991), pp. 203–211. IEEE Computer Society Press (1991). https://doi.org/10.1109/LICS.1991.151645
10. Brown, M., Palsberg, J.: Breaking through the normalization barrier: a self-interpreter for F-omega. In: Majumdar, R. (ed.) Proceedings of the 43rd Annual ACM SIGPLAN-SIGACT Symposium on Principles of Programming Languages, POPL 2016, pp. 5–17. ACM, New York (2016). https://doi.org/10.1145/2837614.2837623
11. Cai, Y., Giarrusso, P.G., Ostermann, K.: System F-omega with equirecursive types for datatype-generic programming. In: Majumdar, R. (ed.) Proceedings of the 43rd Annual ACM SIGPLAN-SIGACT Symposium on Principles of Programming Languages (POPL 2016), pp. 30–43. ACM, New York (2016). https://doi.org/10.1145/2837614.2837660
12. Chakravarty, M., et al.: Functional Blockchain Contracts. Technical report, IOHK (2019). https://iohk.io/research/papers/#KQL88VAR
13. Chapman, J.: Type checking and normalisation. Ph.D. thesis, University of Nottingham, UK (2009). http://eprints.nottingham.ac.uk/10824/

14. Chapman, J., Kireev, R., Nester, C., Wadler, P.: Literate Agda source of MPC 2019 paper (2019). https://github.com/input-output-hk/plutus/blob/f9f7aef94d9614b67c037337079ad89329889ffa/papers/system-f-in-agda/paper.lagda
15. Coquand, C.: A formalised proof of the soundness and completeness of a simply typed lambda-calculus with explicit substitutions. High. Order Symbolic Comput. **15**(1), 57–90 (2002). https://doi.org/10.1023/A:1019964114625
16. Danielsson, N.A.: A formalisation of a dependently typed language as an inductive-recursive family. In: Altenkirch, T., McBride, C. (eds.) TYPES 2006. LNCS, vol. 4502, pp. 93–109. Springer, Heidelberg (2007). https://doi.org/10.1007/978-3-540-74464-1_7
17. van Doorn, F., Geuvers, H., Wiedijk, F.: Explicit convertibility proofs in pure type systems. In: Proceedings of the Eighth ACM SIGPLAN International Workshop on Logical Frameworks & Meta-languages: Theory & Practice, (LFMTP 2013), pp. 25–36. ACM, New York (2013). https://doi.org/10.1145/2503887.2503890
18. Dreyer, D.: Understanding and Evolving the ML Module System. Ph.D. thesis, Carnegie Mellon University (2005)
19. Dreyer, D.: A type system for recursive modules. In: Ramsey, N. (ed.) Proceedings of the 12th ACM SIGPLAN International Conference on Functional Programming (ICFP 2007), pp. 289–302. ACM, New York (2007). https://doi.org/10.1145/1291220.1291196
20. Dybjer, P.: A general formulation of simultaneous inductive-recursive definitions in type theory. J. Symbolic Logic **65**(2), 525–549 (2000). http://www.jstor.org/stable/2586554
21. Grishchenko, I., Maffei, M., Schneidewind, C.: A semantic framework for the security analysis of ethereum smart contracts. In: Bauer, L., Küsters, R. (eds.) POST 2018. LNCS, vol. 10804, pp. 243–269. Springer, Cham (2018). https://doi.org/10.1007/978-3-319-89722-6_10
22. Harz, D., Knottenbelt, W.J.: Towards Safer Smart Contracts: A Survey of Languages and Verification Methods (2018). https://arxiv.org/abs/1809.09805
23. Hindley, J.R., Seldin, J.P.: Lambda-Calculus and Combinators: An Introduction. Cambridge University Press, Cambridge (2008)
24. Hirai, Y.: Defining the ethereum virtual machine for interactive theorem provers. In: Brenner, M., et al. (eds.) FC 2017. LNCS, vol. 10323, pp. 520–535. Springer, Cham (2017). https://doi.org/10.1007/978-3-319-70278-0_33
25. Jung, A., Tiuryn, J.: A new characterization of lambda definability. In: Bezem, M., Groote, J.F. (eds.) TLCA 1993. LNCS, vol. 664, pp. 245–257. Springer, Heidelberg (1993). https://doi.org/10.1007/BFb0037110
26. Peyton Jones, M., Gkoumas, V., Kireev, R., MacKenzie, K., Nester, C., Wadler, P.: Unraveling recursion: compiling an IR with recursion to system F. In: Hutton, G. (ed.) MPC 2019. LNCS, vol. 11825, pp. 414–443. Springer, Cham (2019). https://doi.org/10.1007/978-3-030-33636-3_15
27. Kovács, A.: System F Omega. https://github.com/AndrasKovacs/system-f-omega
28. Martens, C., Crary, K.: LF in LF: mechanizing the metatheories of LF in Twelf. In: Proceedings of the Seventh International Workshop on Logical Frameworks and Meta-languages, Theory and Practice (LFMTP 2012), pp. 23–32. ACM, New York (2012). https://doi.org/10.1145/2364406.2364410
29. McBride, C.: Datatypes of datatypes. In: Summer School on Generic and Effectful Programming, St Anne's College, Oxford (2015). https://www.cs.ox.ac.uk/projects/utgp/school/conor.pdf
30. Nomadic Labs: Michelson in Coq. Git Repository. https://gitlab.com/nomadic-labs/mi-cho-coq/
31. O'Connor, R.: Simplicity: a new language for blockchains. In: Bielova, N., Gaboardi, M. (eds.) Proceedings of the 2017 Workshop on Programming Languages and Analysis for Security (PLAS 2017), pp. 107–120. ACM, New York (2017). https://doi.org/10.1145/3139337.3139340

32. Park, D., Zhang, Y., Saxena, M., Daian, P., Roşu, G.: A formal verification tool for ethereum VM bytecode. In: Garcia, A., Pasareanu, C.S. (eds.) Proceedings of the 2018 26th ACM Join Meeting on European Software Engineering Conference and Symposium on the Foundations of Software Engineering (ESEC/FSE 2018), pp. 912–915. ACM, New York (2018). https://doi.org/10.1145/3236024.3264591
33. Pierce, B.C.: Types and Programming Languages. MIT Press, Cambridge (2002)
34. Pollack, R., Poll, E.: Typechecking in pure type systems. In: Informal Proceedings of Logical Frameworks 1992, pp. 271–288 (1992)
35. Reynolds, J.C.: What do types mean? - from intrinsic to extrinsic semantics. In: McIver, A., Morgan, C. (eds.) Programming Methodology. Monographs in Computer Science, pp. 309–327. Springer, New York (2003). https://doi.org/10.1007/978-0-387-21798-7_15
36. Wadler, P.: Programming language foundations in agda. In: Massoni, T., Mousavi, M.R. (eds.) SBMF 2018. LNCS, vol. 11254, pp. 56–73. Springer, Cham (2018). https://doi.org/10.1007/978-3-030-03044-5_5
37. Wadler, P., Kokke, W.: Programming Language Foundations in Agda. https://plfa.github.io/
38. Watkins, K., Cervesato, I., Pfenning, F., Walker, D.: Specifying properties of concurrent computations in CLF. In: Schürmann, C. (ed.) Proceedings of the Fourth International Workshop on Logical Frameworks and Meta-Languages (LFM 2004). ENTCS, vol. 199, pp. 67–87 (2008). https://doi.org/10.1016/j.entcs.2007.11.013

An Analysis of Repeated Graph Search

Roland Backhouse(✉)

School of Computer Science, University of Nottingham, Nottingham NG8 1BB, UK
rcb@cs.nott.ac.uk

Abstract. Graph-searching algorithms typically assume that a node is given from which the search begins but in many applications it is necessary to search a graph repeatedly until all nodes in the graph have been "visited". Sometimes a priority function is supplied to guide the choice of node when restarting the search, and sometimes not. We call the nodes from which a search of a graph is (re)started the "delegate" of the nodes found in that repetition of the search and we analyse the properties of the delegate function. We apply the analysis to establishing the correctness of the second stage of the Kosaraju-Sharir algorithm for computing strongly connected components of a graph.

Keywords: Directed graph · Depth-first search · Relation algebra · Regular algebra · Strongly connected component

1 Introduction

Graph-searching algorithms typically assume that a node is given from which the search begins but in many applications it is necessary to search a graph repeatedly until all nodes in the graph have been "visited". Sometimes a priority function is supplied to guide the choice of node when restarting the search, and sometimes not.

The determination of the strongly connected components of a (directed) graph using the two-stage algorithm attributed to R. Kosaraju and M. Sharir by Aho, Hopcroft and Ullman [1] is an example of both types of repeated graph search.

In the first stage, a repeated search of the given graph is executed until all nodes in the graph have been "visited". In this stage, the choice of node when restarting the search is arbitrary; it is required, however, that the search algorithm is depth-first. The output is a numbering of the nodes in order of completion of the individual searches.

In the second stage, a repeated search of the given graph—but with edges reversed—is executed; during this stage, the node chosen from which the search is restarted is the highest numbered node (as computed in the first stage) that has not been "visited" (during this second stage). Popular accounts of the algorithm [1,9] require a depth-first search once more but, as is clear from Sharir's original formulation of the algorithm [16], this is not necessary: an arbitrary graph-searching algorithm can be used in the second stage of the algorithm. Each

G. Hutton (Ed.): MPC 2019, LNCS 11825, pp. 298–328, 2019.
https://doi.org/10.1007/978-3-030-33636-3_11

individual search identifies a strongly connected component of the graph of which the node chosen to restart the search is a representative.

The task of constructing a complete, rigorous, calculational proof of the two-stage Kosaraju-Sharir algorithm is non-trivial. (Glück [13] calls it a "Herculean Task".) The task is made simpler by a proper separation of concerns: both stages use a repeated graph search, the first stage requires depth-first search but the second does not. So what are the properties of repeated graph search (in general), and what characterises depth-first search (in particular)?

The current paper is an analysis of repeated graph search in which we abstract from the details of the Kosaraju-Sharir algorithm. That is, we assume the existence of a "choice" function from the nodes of the graph to the natural numbers that is used to determine which node is chosen from which to restart the search. We call the nodes from which a search of a graph is (re)started the "delegate" of the nodes found in that repetition of the search and we analyse the properties of the delegate function assuming, first, that the choice function is arbitrary (thus allowing different nodes to have the same number) and, second, that it is injective (i.e. different nodes are have different numbers).

The properties we identify are true irrespective of the searching algorithm that is used, contrary to popular accounts of graph searching that suggest the properties are peculiar to depth-first search. For example, all the nodes in a strongly connected component of a graph are assigned the same delegate (irrespective of the graph searching algorithm used) whereas Cormen, Leiserson and Rivest's account [9, theorem 23.13, p. 490] suggests that this is a characteristic property of depth-first search.

The primary contribution of this paper is the subject of Sects. 3 and 4. The definition of "delegate" (a function from nodes to nodes) "according to a given choice function" is stated in Sect. 3.1; an algorithm to compute each node's delegate is presented in Sect. 3.2 and further refined in Sect. 3.3. The algorithm is generic in the sense that no ordering is specified for the choice of edges during the search. (In breadth-first search, edges are queued and the choice is first-in, first-out; in depth-first search, edges are stacked and the choice is first-in, last-out. Other orderings are, of course, possible.)

Section 3.4 explores the properties of the delegate function when the choice function is injective. Section 4 applies the analysis of the delegate function to establishing the correctness of the second stage of the Kosaraju-Sharir algorithm. The proof adds insight into the algorithm by identifying clearly and precisely which elements of the so-called "parenthesis theorem" and the classification of edges in a depth-first search [9,10] are vital to the identification of strongly connected components.

With the goal of achieving the combination of concision and precision, our development exploits so-called "point-free" relation algebra. This is briefly summarised in Sect. 2.

2 Relation Algebra

For the purposes of this paper, a (directed) graph G is a homogeneous, binary relation on a finite set of nodes. One way to reason about graphs—so-called "pointwise" reasoning—is to use predicate calculus with primitive terms the booleans expressing whether or not the relation G holds between a given pair of nodes. In other words, a graph is a set of pairs—the edge set of the graph—and a fundamental primitive is the membership relation expressing whether or not a given pair is an element of a given graph. In so-called "point-free" relation algebra, on the other hand, relations are the primitive elements and the focus is on the algebraic properties of the fundamental operations on relations: converse, composition, etc. Because our focus here is on paths in graphs—algebraically the reflexive, transitive closure of a graph—we base our calculations firmly on point-free relation algebra.

A relation algebra is a combination of three structures with interfaces connecting the structures. The first structure is a powerset: the homogeneous binary relations on a set A are elements of the powerset $2^{A\times A}$ (i.e. subsets of $A\times A$) and thus form a complete, universally distributive, complemented lattice. We use familiar notations for, and properties of, set union and set intersection without further ado. (Formally, set union is the supremum operator of the lattice and set intersection is the infimum operator.) The complement of a relation R will be denoted by $\neg R$; its properties are also assumed known. The symbols $\perp\!\!\!\perp$ and $\top\!\!\!\top$ are used for the least and greatest elements of the lattice (the empty relation and the universal relation, respectively).

The second structure is composition: composition of the homogeneous binary relations on a set A forms a monoid with unit denoted in this paper by I_A (or sometimes just I if there is no doubt about the type of the relations under consideration). The interface with the lattice structure is that their combination forms a universally distributive regular algebra (called a "standard Kleene algebra" by Conway [8, p. 27]). Although not a primitive, the star operator is, of course, a fundamental component of regular algebra. For relation R, the relation R^* is its reflexive-transitive closure; in particular, whereas graph G is interpreted as the edge relation, the graph G^* is interpreted as the path relation. The star operator can be defined in several different but equivalent ways (as a sum of powers or as a fixed-point operator). We assume familiarity with the different definitions as well as properties vital to (point-free) reasoning about paths in graphs such as the star-decomposition rule [3,5,7].

The third structure is converse. We denote the converse of relation R by $R^{\cup}$ (pronounced "R wok"). Converse is an involution (i.e. $(R^{\cup})^{\cup} = R$, for all R). Its interface with the lattice structure is that it is its own adjoint in a Galois connection; its interface with composition is the distributivity property

$$(R\circ S)^{\cup} \;=\; S^{\cup}\circ R^{\cup} .$$

Finally, the interface connecting all three structures is the so-called *modularity rule*: for all relations R, S and T,

$$R\circ S\cap T \;\subseteq\; R\circ(S\cap R^{\cup}\circ T). \tag{1}$$

The (easily derived and equivalent) converse

$$R \circ S \cap T \;\subseteq\; (R \cap T \circ S^{\cup}) \circ S \tag{2}$$

is also used later.

The axioms outlined above are applicable to homogeneous relations. They can, of course, be extended to heterogeneous relations by including type restrictions on the operators. (For example, the monoid structure becomes a category.) The structure is then sometimes known as an *allegory* [12]. We use $A {\sim\!\!\!<} B$ to denote the type of a relation. The set A is called the *target* and the set B is called the *source* of the relation. A homogeneous relation has type $A {\sim\!\!\!<} A$ for some set A.

Our use of heterogeneous relations in this paper is limited to functions (which we treat as a subclass of relations). Point-free relation algebra enables concise formulations of properties usually associated with functions. A relation R of type $A {\sim\!\!\!<} B$ is *functional* if

$$R \circ R^{\cup} \;\subseteq\; I_A,$$

it is *injective* if

$$R^{\cup} \circ R \;\subseteq\; I_B,$$

it is *total* if

$$I_B \;\subseteq\; R^{\cup} \circ R,$$

and it is *surjective* if

$$I_A \;\subseteq\; R \circ R^{\cup}.$$

We abbreviate "functional relation" to "function" and write $A \leftarrow B$ for the type.

(The arrowheads in $A {\sim\!\!\!<} B$ and $A \leftarrow B$ indicate that we interpret relations as having outputs and inputs, where outputs are on the left and inputs are on the right. Our terminology reflects the choice we have made: the words "functional" and "injective", and simultaneously "total" and "surjective", can be interchanged to reflect an interpretation in which inputs are on the left and outputs are on the right.)

An idiom that occurs frequently in point-free relation algebra has the form

$$f^{\cup} \circ R \circ g$$

where f and g are functional and—often but not necessarily—total. Pointwise this expresses the relation on the source of f and the source of g that holds of x and y when $f.x \,[\![R]\!]\, g.y$. (In words, the value of f at x is related by R to the value of g at y.) The idiom occurs frequently below. For example,

$$f^{\cup} \circ < \circ \, s$$

is used later to express a relation between nodes a and b of a graph when $f.a<s.b$. This is interpreted as "the search from a finishes before the search from b starts", f and s representing finish and start times, respectively.

The "points" in our algebra are typically the nodes of a graph. Inevitably, we do need to refer to specific nodes from time to time. Points are modelled as "proper, atomic coreflexives".

A *coreflexive* is a relation that is a subset of the identity relation. The coreflexives, viewed as a subclass of the homogeneous relations of a given type, form a complete, universally distributive, complemented lattice under the infimum and supremum operators (which, as we have said, we denote by the symbols commonly used for set intersection and set union, respectively). We use lower-case letters p, q etc. to name coreflexives. So, a coreflexive is a relation p such that $p \subseteq I$. We use $\sim p$ to denote the complement in the lattice of coreflexives of the coreflexive p. This is not the same as the complement of p in the lattice of relations: the relation between them is given by the equation $\sim p \;=\; I \cap \neg p$.

Elsewhere, with a different application area, we use the word "monotype" instead of "coreflexive". (See, for example, [4,11,14].) We now prefer "coreflexive" because it is application-neutral. Others use the word "test" (e.g. [13]).

In general, an *atom* in a lattice ordered by $\sqsubseteq$ and having least element $\perp\!\!\perp$ is an element x such that

$$\langle \forall y \;::\; y \sqsubseteq x \;\equiv\; y = x \vee y = \perp\!\!\perp \rangle .$$

Note that $\perp\!\!\perp$ is an atom according to this definition. If p is an atom that is different from $\perp\!\!\perp$ we say that it is a *proper* atom. A lattice is said to be *atomic* if

$$\langle \forall y \;::\; y \neq \perp\!\!\perp \;\equiv\; \langle \exists x : atom.x \wedge x \neq \perp\!\!\perp : x \sqsubseteq y \rangle \rangle .$$

In words, a lattice is atomic if every proper element includes a proper atom.

It is necessary to distinguish between atomic coreflexives and atomic relations. We use lower-case letters a, b to denote atomic coreflexives. Proper, atomic coreflexives model singleton sets in set theory; so, when applying the theory to graphs, proper, atomic coreflexive a models $\{u\}$ for some node u of the graph. Similarly, coreflexive p models a subset of the nodes, or, in the context of algorithm development, a predicate on nodes (which explains why they are sometimes called "tests").

A lattice with top element $\top\!\!\top$ and supremum operator $\sqcup$ is *saturated* (aka "full") if $\top\!\!\top$ is the supremum of the identity function on atoms, i.e. if

$$\top\!\!\top \;=\; \langle \sqcup x : atom.x : x \rangle .$$

A powerset is atomic and saturated; since we assume that both the lattice of coreflexives and the lattice of relations form powersets, this is the case for both. The coreflexives are postulated to satisfy the *all-or-nothing rule* [13]:

$$\langle \forall\, a,b,R \;::\; a \circ R \circ b = \perp\!\!\perp \vee a \circ R \circ b = a \circ \top\!\!\top \circ b \rangle .$$

Combined with the postulates about coreflexives, the all-or-nothing rule has the consequence that the lattice of relations is a saturated, atomic lattice; the proper atoms are elements of the form $a\circ\top\!\!\top\circ b$ where a and b are proper atoms of the lattice of coreflexives. In effect, the relation $a\circ\top\!\!\top\circ b$ models the pair (a,b) in a standard set-theoretic account of relation algebra; the boolean $a\circ R\circ b = a\circ\top\!\!\top\circ b$ plays a role equivalent to the boolean $(a,b)\in[\![R]\!]$ (where $[\![R]\!]$ denotes the interpretation of R as a set of pairs).

The "domain" operators play a central role in relation algebra, particularly in its use in algorithm development. The *right domain* of a relation R is the coreflexive $R{>}$ (read R "right") defined by $R{>} = I\cap\top\!\!\top\circ R$. The *left domain* $R{<}$ (read R "left") is defined similarly. The interpretation of $R{>}$ is $\{x \mid \langle\exists y :: (y,x)\in[\![R]\!]\rangle\}$. The complement of the right domain of R in the lattice of coreflexives is denoted by $R{>\!\bullet}$; similarly $R{\bullet\!<}$ denotes the complement of the left domain of R. The left and right domains should not be confused with the source and/or target of a relation (in an algebra of heterogeneous relations).

We assume some familiarity with relation algebra (specifically set calculus, relational composition and converse, and their interfaces) as well as fixed-point calculus and Galois connections. For example, monotonicity properties of the operators, together with transitivity and anti-symmetry of the subset relation, are frequently used without specific mention. On the other hand, because the properties of domains are likely to be unfamiliar, we state the properties we use in the hints accompanying proof steps.

3 Repeated Search and Delegates

In this section, we explore a property of repeated application of graph-searching starting with an empty set of "seen" nodes until all nodes have been seen.

The algorithm we consider is introduced in Sect. 3.2 and further refined in Sect. 3.3. Roughly speaking, the algorithm repeatedly searches a given graph starting from a node chosen from among the nodes not yet seen so as to maximise a "choice function"; at each iteration, the graph searched is the given graph but restricted to edges connecting nodes not yet seen. The algorithm records the chosen nodes in a function that we call a "delegate function", the "delegate" of a node a being the node from which the search that "sees" a is initiated.

Rather than begin with the algorithm, we prefer to begin with a specification of what a repeated search of a graph is intended to implement. The formal specification of the delegate function is given in Sect. 3.1.

Our formulation of the notion of a "delegate" is inspired by Cormen, Leiserson and Rivest's [9, p. 490] discussion of a "forefather" function as used in depth-first search to compute strongly connected components of a graph. However, our presentation is more general than theirs. In particular, Cormen, Leiserson and Rivest assume that the choice function is injective. We establish some consequences of this assumption in Sect. 3.4; this is followed in Sect. 3.5 by a comparative discussion of our account and that of Cormen, Leiserson and Rivest.

Aside on Terminology. I have chosen to use the word "delegate" rather than "forefather" because it has a similar meaning to the word "representative", as used in "a representative of an equivalence class". Tarjan [17], Sharir [16], Aho, Hopcroft and Ullman [1] and Cormen, Leiserson, Rivest and Stein [10, p. 619] call the representative of a strongly connected component of a graph the "root" of the component. This is a reference to the "forest" of "trees" that is (implicitly) constructed during any repeated graph search. In the two-stage algorithm, however, each stage is a repeated graph search and so to refer to the "root" could be confusing: which forest is meant? Using the word "representative" might also be confusing because it might (wrongly) suggest that the "representative" computed by an arbitrary repeated graph search is a representative of the equivalence class of strongly connected nodes in a graph. The introduction of novel terminology also has the advantage of forcing the reader to study its definition. **End of Aside**

3.1 Delegate Function

Suppose f is a total function of type $\mathbb{N} \leftarrow \mathsf{Node}$, where Node is a finite set of *nodes*. Suppose G is a graph with set of nodes Node. That is, G is a relation of type $\mathsf{Node} \sim \mathsf{Node}$. We call f the *choice function* (because it governs the choice of delegates).

A *delegate function on* G *according to* f is a relation φ of type $\mathsf{Node} \sim \mathsf{Node}$ with the properties that

$$\varphi \circ \varphi^{\cup} \subseteq I_{\mathsf{Node}} \subseteq \varphi^{\cup} \circ \varphi \text{, and} \tag{3}$$

$$\varphi \subseteq (G^{\cup})^* \quad \wedge \quad G^* \subseteq (f \circ \varphi)^{\cup} \circ \geq \circ f . \tag{4}$$

The property (3) states that φ is a total function. Property (4), expressed pointwise and in words, states that for all nodes a and b, node a is the delegate of node b equivales the conjunction of (i) there is a path in G from b to a and (ii) among all nodes c such that there is a path from b to c, node a maximises the value of the choice function f. (The relation "$\geq$" on the right side of the second inclusion is the usual at-least ordering on numbers.)

Note that, because our main motivation for studying repeated graph search is to apply the results to understanding the second stage of the Kosaraju-Sharir algorithm for computing strongly connected components of a graph, the definition of the delegate function is that appropriate to a search of $G^{\cup}$ rather than a search of G.

Delegate functions have a couple of additional properties that we exploit later. These are formulated and proved in the lemma below.

Lemma 5. If φ is a delegate function on G according to f,

$$I \subseteq G^* \circ \varphi \quad \wedge \quad G^* \subseteq (f \circ \varphi)^{\cup} \circ \geq \circ f \circ \varphi .$$

In words, there is a path in G from each node to its delegate, and if there is a path in G from node b to node c, the value of f at the delegate of b is at least the value of f at the delegate of c.

Proof. First,

$$
\begin{array}{ll}
 & I \subseteq G^* \circ \varphi \\
\Leftarrow & \{ \quad \varphi \text{ is total, i.e. } I \subseteq \varphi^{\cup} \circ \varphi \quad \} \\
 & \varphi^{\cup} \subseteq G^* \\
= & \{ \quad \text{converse} \quad \} \\
 & \varphi \subseteq (G^*)^{\cup} \\
= & \{ \quad (G^*)^{\cup} = (G^{\cup})^* \text{ and definition of delegate: (4)} \quad \} \\
 & \mathsf{true}.
\end{array}
$$

Second,

$$
\begin{array}{ll}
 & G^* \subseteq (f \circ \varphi)^{\cup} \circ \geq \circ f \circ \varphi \\
\Leftarrow & \{ \quad I \subseteq G^* \circ \varphi \text{ (see above)} \quad \} \\
 & G^* \circ G^* \circ \varphi \subseteq (f \circ \varphi)^{\cup} \circ \geq \circ f \circ \varphi \\
\Leftarrow & \{ \quad G^* \circ G^* = G^* \text{ and monotonicity} \quad \} \\
 & G^* \subseteq (f \circ \varphi)^{\cup} \circ \geq \circ f \\
= & \{ \quad \text{definition of delegate: (4)} \quad \} \\
 & \mathsf{true}.
\end{array}
$$

□

Lemma 6. If φ is a delegate function on G according to f,

$$\varphi \subseteq f^{\cup} \circ \geq \circ f.$$

In words, the delegate of a node has f-value that is at least that of the node.

Proof

$$
\begin{array}{ll}
 & \mathsf{true} \\
= & \{ \quad \text{definition: (3) and (4)} \quad \} \\
 & \varphi \circ \varphi^{\cup} \subseteq I \ \wedge \ G^* \subseteq (f \circ \varphi)^{\cup} \circ \geq \circ f \\
\Rightarrow & \{ \quad I \subseteq G^* \text{ and transitivity; converse} \quad \} \\
 & \varphi \circ \varphi^{\cup} \subseteq I \ \wedge \ I \subseteq \varphi^{\cup} \circ f^{\cup} \circ \geq \circ f \\
\Rightarrow & \{ \quad \varphi \circ I = \varphi \text{, monotonicity of composition and transitivity} \quad \} \\
 & \varphi \subseteq f^{\cup} \circ \geq \circ f.
\end{array}
$$

□

3.2 Assigning Delegates

The basic structure of the algorithm for computing a delegate function is shown in Fig. 1. It is a simple loop that initialises the coreflexive *seen* (representing a set of nodes) to $\perp\!\!\perp$ (representing the empty set of nodes) and then repeatedly chooses a node a that has the largest f-value among the nodes that do not have a delegate and adds to *seen* the coreflexive $\sim seen \circ (G^* \circ a)_<$; this coreflexive represents the nodes that do not have a delegate and from which there is a path to a in the graph. Simultaneous with the assignments to *seen*, the variable φ is initialised to $\perp\!\!\perp$ and subsequently updated by setting the φ-value of all newly "delegated" nodes to a.

$\{ \quad f \circ f^{\cup} \subseteq I_{\mathsf{N}} \wedge I_{\mathsf{Node}} \subseteq f^{\cup} \circ f \quad \}$

$\varphi, seen \;:=\; \perp\!\!\perp, \perp\!\!\perp$;

$\{$ **Invariant**: (7) thru (14) $\}$

while $seen \neq I_{\mathsf{Node}}$ **do**

  begin

   choose node a such that $a \circ seen = \perp\!\!\perp$ and $\sim seen \circ \top\!\!\top \circ a \subseteq f^{\cup} \circ \leq \circ f$

  ; $s \;:=\; \sim seen \circ (G^* \circ a)_<$

  ; $\varphi, seen \;:=\; \varphi \cup a \circ \top\!\!\top \circ s \;,\; seen \cup s$

  end

$\{ \quad \varphi \circ \varphi^{\cup} \subseteq I_{\mathsf{Node}} \subseteq \mathsf{equiv}.G \subseteq \varphi^{\cup} \circ \varphi$

$\wedge \quad \varphi \subseteq (G^{\cup} \cap \varphi^{\cup} \circ \varphi)^* \;\wedge\; G^* \subseteq (f \circ \varphi)^{\cup} \circ \geq \circ f$

$\wedge \quad \varphi = \varphi \circ \varphi \quad \}$

Fig. 1. Repeated Search. Outer Loop

For brevity in the calculations below, the temporary variable s (short for "seen") has been introduced. The sequence of assignments

$$s \;:=\; \sim seen \circ (G^* \circ a)_<$$
$$;\; \varphi, seen \;:=\; \varphi \cup a \circ \top\!\!\top \circ s \;,\; seen \cup s$$

is implemented by a generic graph-searching algorithm. The details of how this is done are given in Sect. 3.3.

Apart from being a total function, we impose no restrictions on f. If f is a constant function (for example, if $f.a = 0$ for all nodes a), the "choice" is completely arbitrary.

The relation $\mathsf{equiv}.G$ in the postcondition of the algorithm is the equivalence relation defined by

$$\mathsf{equiv}.G \;=\; G^* \cap (G^*)^{\cup} .$$

If G is a graph, two nodes related by $\mathsf{equiv}.G$ are said to be in the same *strongly connected component* of G. The first clause of the postcondition thus asserts that the computed delegate relation φ is not only a total function, as required by (3), but also that all nodes in any one strongly connected component are assigned the same delegate.

The property

$$\varphi \subseteq (G^{\cup} \cap \varphi^{\cup} \circ \varphi)^*$$

in the postcondition is stronger than the requirement $\varphi \subseteq (G^{\cup})^*$ in (4). It states that there is a path from each node to its delegate comprising nodes that all have the same delegate. (More precisely, it states that there is a path from each node to its delegate such that successive nodes on the path have the same delegate. The equivalence of these two informal interpretations is formulated in Lemma 25.)

Note the property $\varphi = \varphi \circ \varphi$ in the postcondition. Cormen, Leiserson and Rivest [9, p. 490] require that the function f is injective and use this to derive the property from the definition of a delegate ("forefather" in their terminology). We don't impose this requirement but show instead that $\varphi = \varphi \circ \varphi$ is a consequence of the algorithm used to calculate delegates. For completeness, we also show that the property is a consequence of the definition of delegate under the assumption that f is injective: see Lemma 18. Similarly, the property $\mathsf{equiv}.G \subseteq \varphi^{\cup} \circ \varphi$ can be derived from the definition of a delegate if f is assumed to be injective. Again for completeness, we also show that the property is a consequence of the definition of delegate under the assumption that f is injective: see Lemma 19.

Termination of the loop is obvious: the coreflexive *seen* represents a set of nodes that increases strictly in size at each iteration. (The chosen node a is added at each iteration.) The number of iterations of the loop body is thus at most the number of nodes in the graph, which is assumed to be finite. The principle task is thus to verify conditional correctness (correctness assuming termination, often called "partial" correctness).

The invariant properties of the algorithm are as follows:

$$\varphi{>} \;=\; seen, \tag{7}$$

$$\varphi \circ \varphi^{\cup} \subseteq seen, \tag{8}$$

$$\varphi \subseteq (G^{\cup} \cap \varphi^{\cup} \circ \varphi)^*, \tag{9}$$

$$\varphi = \varphi \circ \varphi, \tag{10}$$

$$seen \;=\; (G^* \circ seen){<}, \tag{11}$$

$$seen \circ \top\!\!\top \circ \sim\! seen \;\subseteq\; (f \circ \varphi)^{\cup} \circ \geq \circ f, \tag{12}$$

$$seen \circ G^* \circ seen \;\subseteq\; (f \circ \varphi)^{\cup} \circ \geq \circ f, \tag{13}$$

$$seen \circ \mathsf{equiv}.G \circ seen \;\subseteq\; \varphi^{\cup} \circ \varphi. \tag{14}$$

The postcondition

$$\varphi \circ \varphi^{\cup} \subseteq I_{\mathsf{Node}} \subseteq \varphi^{\cup} \circ \varphi$$

expresses the fact that, on termination, φ is functional and total; the claimed invariants (7) and (8) state that intermediate values of φ are total on *seen* and functional. The invariant (7) also guarantees that *seen* is a coreflexive. The invariants (9) and (10) are both conjuncts of the postcondition. The additional conjunct

$$\mathsf{equiv}.G \subseteq \varphi^{\cup} \circ \varphi$$

in the postcondition states that strongly connected nodes have the same delegate. The invariant (14) states that this is the case for nodes that have been assigned a delegate. Like (7) and (8), invariant (13) states that intermediate values of φ maximise f for those nodes for which a delegate has been assigned. It is therefore obvious that the postcondition is implied by the conjunction of the invariant and the termination condition. The additional invariants (11) and (12) are needed in order to establish the invariance of (13). It is straightforward to construct and check appropriate verification conditions. Full details are given in [6].

3.3 Incremental Computation

The algorithm shown in Fig. 1 assigns to the variable s (the coreflexive representing) all the nodes that do not yet have a delegate and can reach the node a. The variable φ is also updated so that a becomes the delegate of all the nodes in the set represented by s. The assignments are implemented by a generic graph-searching algorithm. Figure 2 shows the details.

The consecutive assignments in the body of the loop in Fig. 1 (to s, and to φ and *seen*) are implemented by an inner loop together with initialising assignments. The assertions should enable the reader to verify that the two algorithms are equivalent: the variables s, $seen_0$ and φ_0 are auxiliary variables used to express the property that the inner loop correctly implements the two assignments that they replace in the outer loop; in an actual implementation the assignments to these variables may be omitted (or, preferably, included but identified as auxiliary statements that can be ignored by the computation proper).

It is straightforward to verify the correctness of this algorithm. Because it involves no new techniques, it is omitted here. Full details are included in [6].

A concrete implementation of the above graph-searching algorithm involves choosing a suitable data structure in which to store the unexplored edges represented by $\sim seen \circ (G \circ seen)_{<}$. Breadth-first search stores the edges in a queue (so newly added edges are chosen in the order that they are added), whilst depth-first search stores the edges in a stack (so the most recently added edge is chosen first). Other variations enable the solution of more specific path-finding problems. For example, if edges are labelled by distances, shortest paths from a given source can be found by storing edges in a priority queue. Topological search is also an instance: edges from each node are grouped together and an edge from

$\{\ \ a{\circ}seen = \perp\!\!\perp \wedge$ (7) thru (14) $\}$

/* s , $seen_0$ and φ_0 are auxiliary variables */

$s, seen_0, \varphi_0 \ := \ a, seen, \varphi$

$\{\ \ {\sim}seen_0 \circ G \circ seen_0 \ = \ \perp\!\!\perp \ \}$

; $seen, \varphi \ := \ seen \cup a \ , \ \varphi \cup a{\circ}\top\!\!\top{\circ}a$

; $\{$ **Invariant**: $seen \ = \ s \cup seen_0 \ \wedge \ \varphi \ = \ \varphi_0 \cup a{\circ}\top\!\!\top{\circ}s$

Invariant: $a \subseteq s \subseteq {\sim}seen_0 \circ (G^* \circ a)_< \ \}$

while ${\sim}seen \circ G \circ seen \neq \perp\!\!\perp$ **do**

begin

choose node b such that $b \subseteq {\sim}seen \circ (G{\circ}seen)_<$

$\{\ \ b \subseteq {\sim}seen_0 \circ (G^* \circ a)_< \ \ \}$

; $s \ := \ s \cup b$

; $seen, \varphi \ := \ seen \cup b \ , \ \varphi \cup a{\circ}\top\!\!\top{\circ}b$

end

$\{\ \ s \ = \ {\sim}seen_0 \circ (G^* \circ a)_< \ \wedge \ seen \ = \ s \cup seen_0 \ \wedge \ \varphi = \varphi_0 \cup a{\circ}\top\!\!\top{\circ}s \ \ \}$

$\{\ \ seen \ = \ seen_0 \cup (G^* \circ a)_< \ \wedge \ \varphi \ = \ \varphi_0 \ \cup \ a \circ \top\!\!\top \circ {\sim}seen_0 \circ (G^* \circ a)_< \ \ \}$

Fig. 2. Repeated Search. Inner Loop.

a given node is chosen when the node has no unexplored incoming edges. We do not go into details any further.

For later discussion of the so-called “white-path theorem” [9, pp. 482], we list below some consequences of the invariant properties. Lemmas 15 and 16 relate arbitrary paths to paths that are restricted to unseen nodes; Lemma 17 similarly relates arbitrary paths to paths restricted to nodes that have been seen thus far.

Lemma 15. Assuming $seen \ = \ (G^* \circ seen)_<$ (i.e. (11)) and $a{\circ}seen = \perp\!\!\perp$, the following properties also hold:

$${\sim}seen \circ G^* \circ seen = \perp\!\!\perp \ \ \wedge \ \ {\sim}seen \circ G^* \circ a \ = \ ({\sim}seen \circ G)^* \circ a .$$

(In words, the properties state that there are no paths from an unseen node to a seen node and, for all unseen nodes b there is a path in G from b to a equivales there is a path in G comprising unseen nodes from b to a.)

Proof. First,

$${\sim}seen \circ G^* \circ seen$$

$=\quad \{$ domains: $[\ R = R_< \circ R\ ]$ with $R := G^* \circ seen$;

$seen \ = \ (G^* \circ seen)_< \ \ \}$

$$
\begin{array}{ll}
 & \sim seen \circ seen \circ G^* \circ seen \\
= & \{ \quad \sim seen \circ seen = \bot\!\!\bot \quad \} \\
 & \bot\!\!\bot .
\end{array}
$$

Second,

$$
\begin{array}{ll}
 & \sim seen \circ G^* \circ a \\
= & \{ \quad I = seen \cup \sim seen;\ \text{distributivity, and star decomposition:} \\
 & \quad [\ (R{\cup}S)^* = R^* \circ (S \circ R^*)^*\ ]\ \ \text{with } R,S := seen \circ G\ ,\ \sim seen \circ G\ \} \\
 & \sim seen \circ (seen \circ G)^* \circ (\sim seen \circ G \circ (seen \circ G)^*)^* \circ a \\
= & \{ \quad (seen{\circ}G)^* \ = \ I \cup seen{\circ}G{\circ}(seen{\circ}G)^* \\
 & \quad \text{distributivity and } \sim seen \circ seen = \bot\!\!\bot \quad \} \\
 & \sim seen \circ (\sim seen \circ G \circ (seen \circ G)^*)^* \circ a \\
= & \{ \quad (seen{\circ}G)^* \ = \ I \cup seen{\circ}G{\circ}(seen{\circ}G)^* \\
 & \quad \text{distributivity and } \sim seen \circ G^* \circ seen \ = \ \bot\!\!\bot \\
 & \quad (\text{whence } \sim seen \circ G \circ seen \ = \ \bot\!\!\bot) \quad \} \\
 & \sim seen \circ (\sim seen \circ G)^* \circ a \\
= & \{ \quad (\sim seen \circ G)^* \ = \ I \cup \sim seen \circ G \circ (\sim seen \circ G)^* \\
 & \quad \text{distributivity} \quad \} \\
 & \sim seen \circ a \cup \sim seen \circ \sim seen \circ G \circ (\sim seen \circ G)^* \circ a \\
= & \{ \quad \sim seen \circ a = a \text{ and } \sim seen \circ \sim seen = \sim seen\ , \\
 & \quad (\sim seen \circ G)^* \ = \ I \cup \sim seen \circ G \circ (\sim seen \circ G)^* \\
 & \quad \text{distributivity} \quad \} \\
 & (\sim seen \circ G)^* \circ a.
\end{array}
$$

□

The following two lemmas concern the properties of the variable s which is assigned the value $\sim seen \circ (G^* \circ a)<$ in Fig. 1.

Lemma 16. Assuming properties (7) thru (14) and $a{\circ}seen = \bot\!\!\bot$,

$$s \ = \ ((\sim seen \circ G)^* \circ a)<.$$

(In words, the coreflexive s represents the set of all nodes b such that there is a path in G comprising unseen nodes from b to a.)

Proof

$$
\begin{array}{ll}
& s \\
= & \{ \quad \text{definition (see fig. 1)} \quad \} \\
& {\sim}seen \circ (G^* \circ a)_< \\
= & \{ \quad \text{domains: for all coreflexives } p \text{ and all relations } R, \\
& \quad p \circ R_< \;=\; (p{\circ}R)_< \;\text{ with } p{,}R \;:=\; {\sim}seen\;,\; G^* \circ a \quad \} \\
& ({\sim}seen \circ G^* \circ a)_< \\
= & \{ \quad \text{lemma 15} \quad \} \\
& (({\sim}seen \circ G)^* \circ a)_< .
\end{array}
$$

□

Lemma 17. Assuming properties (7) thru (14) and $a{\circ}seen = \perp\!\!\perp$,

$$
s \;=\; ((s{\circ}G)^* \circ a)_< .
$$

(In words, the coreflexive s represents the set of all nodes b such that there is a path in G comprising nodes in s from b to a.)

Proof. Applying Lemma 16, the task is to prove that

$$
(({\sim}seen \circ G)^* \circ a)_< \;=\; ((s{\circ}G)^* \circ a)_< .
$$

Clearly, since $s \subseteq {\sim}seen$, the left side of this equation is at least the right side. So it suffices to prove the inclusion. This we do as follows.

$$
\begin{array}{ll}
& (({\sim}seen \circ G)^* \circ a)_< \;\subseteq\; ((s{\circ}G)^* \circ a)_< \\
\Leftarrow & \{ \quad \text{fixed-point fusion} \quad \} \\
& \quad a \subseteq ((s{\circ}G)^* \circ a)_< \\
& \wedge\; ({\sim}seen \circ G \circ ((s{\circ}G)^* \circ a)_<)_< \;\subseteq\; ((s{\circ}G)^* \circ a)_< \\
= & \{ \quad \text{first conjunct is clearly } \mathsf{true}; \\
& \qquad {\sim}seen \\
& \qquad = \quad \{ \quad \text{case analysis: } I \;=\; (G^* \circ a)_< \cup (G^* \circ a)_{\bullet<} \quad \} \\
& \qquad {\sim}seen \circ (G^* \circ a)_< \cup {\sim}seen \circ (G^* \circ a)_{\bullet<} \\
& \qquad = \quad \{ \quad \text{definition of } s \text{ (see fig. 1)} \quad \} \\
& \qquad s \cup {\sim}seen \circ (G^* \circ a)_{\bullet<} \quad \} \\
& ((s \cup {\sim}seen \circ (G^* \circ a)_{\bullet<}) \circ G \circ ((s{\circ}G)^* \circ a)_<)_< \;\subseteq\; ((s{\circ}G)^* \circ a)_< \\
= & \{ \quad \text{domains: } [\, (R \circ S_<)_< \;=\; (R{\circ}S)_< \,] \\
& \quad \text{with } R{,}S \;:=\; (s \cup {\sim}seen \circ (G^* \circ a)_{\bullet<}) \circ G,\; (s{\circ}G)^* \circ a;
\end{array}
$$

distributivity }

$(s\circ G\circ(s\circ G)^*\circ a)_< \subseteq ((s\circ G)^*\circ a)_<$

$\wedge\ (\sim seen\circ(G^*\circ a)_{\bullet<}\circ G\circ(s\circ G)^*\circ a)_< \subseteq ((s\circ G)^*\circ a)_<$

$\Leftarrow$ { first conjunct is true (since $[\ R\circ R^* \subseteq R^*\ ]$ with $R := s\circ G$);

second conjunct: $G\circ(s\circ G)^* \subseteq G^*$ and domains }

$(\sim seen\circ(G^*\circ a)_{\bullet<}\circ(G^*\circ a)_<)_< \subseteq ((s\circ G)^*\circ a)_<$

$=$ { complements: $(G^*\circ a)_{\bullet<}\circ(G^*\circ a)_< = \perp\!\!\perp$ }

true. □

3.4 Injective Choice

This section is a preliminary to the discussion in Sect. 3.5. Throughout the section, we assume that f has type $\mathbb{N}\leftarrow\mathsf{Node}$. Also, the symbol I denotes I_{Node}: the identity relation on nodes.

Previous sections have established the existence of a delegate function φ according to choice function f with the only proviso being that f is total and functional. Moreover, the property $\varphi\circ\varphi = \varphi$ is an invariant of the algorithm for computing delegates. Cormen, Leiserson and Rivest [9] derive it from the other requirements assuming that f is also injective. For completeness, this is the point-free rendition of their proof.

Lemma 18. If f is a total, *injective* function and φ is a delegate function according to f, then

$$\varphi\circ\varphi = \varphi.$$

Proof

$\varphi\circ\varphi = \varphi$

$\Leftarrow$ { assumption: f is total and injective, i.e. $f^\cup\circ f = I$ }

$f\circ\varphi\circ\varphi = f\circ\varphi$

$=$ { antisymmetry of $\geq$

and distributivity properties of total functions }

$I \subseteq (f\circ\varphi\circ\varphi)^\cup\circ\leq\circ f\circ\varphi\ \wedge\ I \subseteq (f\circ\varphi\circ\varphi)^\cup\circ\geq\circ f\circ\varphi.$

We establish the truth of both conjuncts as follows. First,

$(f\circ\varphi\circ\varphi)^\cup\circ\leq\circ f\circ\varphi$

$=$ { converse }

$\varphi^\cup\circ(f\circ\varphi)^\cup\circ\leq\circ f\circ\varphi$

$\supseteq$ { $G^* \subseteq (f{\circ}\varphi)^{\cup} \circ \geq \circ f \circ \varphi$ (lemma 5)
i.e. $(G^*)^{\cup} \subseteq (f{\circ}\varphi)^{\cup} \circ \leq \circ f \circ \varphi$
(distributivity properties of converse and $\geq^{\cup} = (\leq)$) }

$\varphi^{\cup} \circ (G^*)^{\cup}$

$\supseteq$ { $I \subseteq G^* \circ \varphi$ (lemma 5) and converse }

I.

Second,

$(f{\circ}\varphi{\circ}\varphi)^{\cup} \circ \geq \circ f \circ \varphi$

$=$ { converse }

$\varphi^{\cup} \circ (f{\circ}\varphi)^{\cup} \circ \geq \circ f \circ \varphi$

$\supseteq$ { definition of delegate: (4) and monotonicity }

$\varphi^{\cup} \circ G^* \circ \varphi$

$\supseteq$ { $I \subseteq G^*$ }

$\varphi^{\cup} \circ \varphi$

$\supseteq$ { φ is total (by definition: (3)) }

I. □

As also shown above, the property $\mathsf{equiv}.G \subseteq \varphi^{\cup} \circ \varphi$ is an invariant of the algorithm. However, if f is a total, injective function, the property follows from the definition of a delegate, as we show below.

Lemma 19. If f is a total, injective function and φ is a delegate function according to f, strongly connected nodes have the same delegate. That is

$$\mathsf{equiv}.G \subseteq \varphi^{\cup} \circ \varphi.$$

Proof

$\mathsf{equiv}.G$

$=$ { definition }

$G^* \cap (G^*)^{\cup}$

$\subseteq$ { lemma 5 }

$(f{\circ}\varphi)^{\cup} \circ \geq \circ f \circ \varphi \ \cap \ ((f{\circ}\varphi)^{\cup} \circ \geq \circ f \circ \varphi)^{\cup}$

$=$ { converse }

$(f{\circ}\varphi)^{\cup} \circ \geq \circ f \circ \varphi \ \cap \ (f{\circ}\varphi)^{\cup} \circ \leq \circ f \circ \varphi$

$$
\begin{array}{ll}
= & \{ \quad f \text{ and } \varphi \text{ are total functions, distributivity } \} \\
& (f \circ \varphi)^{\cup} \circ (\geq \cap \leq) \circ f \circ \varphi \\
= & \{ \quad \leq \text{ is antisymmetric, converse } \} \\
& \varphi^{\cup} \circ f^{\cup} \circ f \circ \varphi \\
= & \{ \quad f \text{ is injective and total, i.e. } f^{\cup} \circ f = I \ \} \\
& \varphi^{\cup} \circ \varphi .
\end{array}
$$

□

The relation $\varphi \circ G^{\cup} \circ \varphi^{\cup}$ is a relation on delegates. Viewed as a graph, it is a homomorphic image of the graph $G^{\cup}$ formed by coalescing all the nodes with the same delegate into one node. Excluding self-loops, this graph is acyclic and topologically ordered by f, as we now show.

Definition 20 (Topological Order). A *topological ordering* of a homogeneous relation R of type A is a total, injective function ord from A to the natural numbers with the property that

$$R^{+} \subseteq ord^{\cup} \circ < \circ\, ord.$$

□

A straightforward lemma is that the requirement on ord is equivalent to

$$R \subseteq ord^{\cup} \circ < \circ\, ord.$$

Note that the less-than ordering relation on numbers is an implicit parameter of the definition of topological ordering. Sometimes it is convenient to use the greater-than ordering instead. In this way, applying basic properties of converse, it is clearly the case that a topological ordering of R is also a topological ordering of $R^{\cup}$.

Lemma 21. If f is a total, injective function and φ is a delegate function according to f, the graph $\varphi \circ G^{\cup} \circ \varphi^{\cup} \cap \neg I$ is acyclic with f as a topological ordering.

Proof. It suffices to show that f is a topological ordering. The function f is, by assumption, a total, injective function of type $\mathbb{N} \leftarrow \mathsf{Node}$. Thus, by assumption, f satisfies the first requirement of being a topological ordering. (See Definition 20.) Establishing the second requirement is achieved by the following calculation.

$$
\begin{array}{ll}
& \varphi \circ G^{\cup} \circ \varphi^{\cup} \cap \neg I \ \subseteq \ f^{\cup} \circ < \circ f \\
= & \{ \quad \text{shunting rule } \} \\
& \varphi \circ G^{\cup} \circ \varphi^{\cup} \ \subseteq \ f^{\cup} \circ < \circ f \ \cup \ I \\
= & \{ \quad f \text{ is total and injective, i.e. } I = f^{\cup} \circ f \\
& \quad\ \ \text{distributivity and definition of } \leq \ \} \\
& \varphi \circ G^{\cup} \circ \varphi^{\cup} \ \subseteq \ f^{\cup} \circ \leq \circ f
\end{array}
$$

$$
\begin{array}{ll}
\Leftarrow & \{ \quad \varphi \text{ is functional, i.e. } \varphi \circ \varphi^{\cup} \subseteq I \\
& \quad \text{monotonicity, converse and transitivity} \quad \} \\
& G^{\cup} \subseteq (f \circ \varphi)^{\cup} \circ \leq \circ f \circ \varphi \\
= & \{ \quad \text{converse} \quad \} \\
& G \subseteq (f \circ \varphi)^{\cup} \circ \geq \circ f \circ \varphi \\
\Leftarrow & \{ \quad G \subseteq G^*, \text{ transitivity} \quad \} \\
& G^* \subseteq (f \circ \varphi)^{\cup} \circ \geq \circ f \circ \varphi \\
= & \{ \quad \text{lemma 5} \quad \} \\
& \mathsf{true}.
\end{array}
$$

□

An important corollary of Lemma 21 is that the finish timestamp of (repeated) depth-first search is a topological ordering of the strongly connected components of a graph. (See Sect. 3.5 for further discussion of depth-first-search timestamps and Lemma 21.)

The algorithm presented in Fig. 1 shows that, viewed as a specification of the function φ, the equation (4) always has at least one solution. However, the algorithm is non-deterministic, which means that there may be more than one solution. We now prove that (4) has a unique solution in unknown φ if the function f is total and injective.

Lemma 22. Suppose f of type $\mathbb{N} \leftarrow \mathsf{Node}$ is a total and injective function, and φ and ψ are both total functions of type $\mathsf{Node} \leftarrow \mathsf{Node}$. Then

$$
\begin{array}{ll}
& \varphi = \psi \\
\Leftarrow & (\varphi \subseteq (G^*)^{\cup} \ \wedge \ G^* \subseteq (f \circ \varphi)^{\cup} \circ \geq \circ f) \\
& \wedge \ (\psi \subseteq (G^*)^{\cup} \ \wedge \ G^* \subseteq (f \circ \psi)^{\cup} \circ \geq \circ f).
\end{array}
$$

Proof. Suppose ψ is a total function of type $\mathsf{Node} \leftarrow \mathsf{Node}$. Then

$$
\begin{array}{ll}
& \psi \subseteq (G^*)^{\cup} \ \wedge \ G^* \subseteq (f \circ \varphi)^{\cup} \circ \geq \circ f \\
\Rightarrow & \{ \quad \text{converse and transitivity} \quad \} \\
& \psi^{\cup} \subseteq (f \circ \varphi)^{\cup} \circ \geq \circ f \\
\Rightarrow & \{ \quad \psi \text{ is total, i.e. } I \subseteq \psi^{\cup} \circ \psi; \\
& \quad \text{monotonicity and transitivity} \quad \} \\
& I \subseteq (f \circ \varphi)^{\cup} \circ \geq \circ f \circ \psi.
\end{array}
$$

Interchanging φ and ψ, and combining the two properties thus obtained, we get that, if φ and ψ are both total functions of type $\mathsf{Node} \leftarrow \mathsf{Node}$,

$$
\begin{array}{ll}
 & (\varphi \subseteq (G^*)^{\cup} \;\wedge\; G^* \subseteq (f{\circ}\psi)^{\cup} \circ \geq \circ f) \\
 & \wedge\;\; (\psi \subseteq (G^*)^{\cup} \;\wedge\; G^* \subseteq (f{\circ}\varphi)^{\cup} \circ \geq \circ f) \\
\Rightarrow & \quad \{ \quad \text{see above} \quad \} \\
 & \quad I \subseteq (f{\circ}\psi)^{\cup} \circ \geq \circ f \circ \varphi \\
 & \wedge\;\; I \subseteq (f{\circ}\varphi)^{\cup} \circ \geq \circ f \circ \psi \\
= & \quad \{ \quad f,\ \varphi \text{ and } \psi \text{ are all total functions,} \\
 & \qquad \text{converse and distributivity} \quad \} \\
 & \quad I \subseteq (f{\circ}\psi)^{\cup} \circ ((\leq)\cap(\geq)) \circ f \circ \varphi \\
 & \wedge\;\; I \subseteq (f{\circ}\varphi)^{\cup} \circ ((\leq)\cap(\geq)) \circ f \circ \psi \\
= & \quad \{ \quad \text{anti-symmetry of } (\leq) \quad \} \\
 & I \subseteq (f{\circ}\psi)^{\cup} \circ f \circ \varphi \;\wedge\; I \subseteq (f{\circ}\varphi)^{\cup} \circ f \circ \psi \\
\Rightarrow & \quad \{ \quad f,\ \varphi \text{ and } \psi \text{ are functional;} \\
 & \qquad \text{hence } f \circ \psi \circ (f{\circ}\psi)^{\cup} \subseteq I \text{ and } f \circ \varphi \circ (f{\circ}\varphi)^{\cup} \subseteq I \quad \} \\
 & f{\circ}\psi \subseteq f{\circ}\varphi \;\wedge\; f{\circ}\varphi \subseteq f{\circ}\psi \\
= & \quad \{ \quad \text{anti-symmetry} \quad \} \\
 & f{\circ}\psi \;=\; f{\circ}\varphi \\
\Rightarrow & \quad \{ \quad f \text{ is an injective, total function, i.e. } f^{\cup} \circ f \;=\; I \quad \} \\
 & \psi = \varphi .
\end{array}
$$

The lemma follows by symmetry and associativity of conjunction. □

Earlier, we stated that (9) formulates the property that there is a path from each node to its delegate on which successive nodes have the same delegate. Combined with (10) and the transitivity of equality, this means that there is a path from each node to its delegate on which all nodes have the same delegate. We conclude this section with a point-free proof of this claim. Since the claim is not specific to the delegate function, we formulate the underlying lemmas (Lemmas 23 and 24) in general terms. The relevant property of the delegate function, Lemma 25, is then a simple instance.

For readers wishing to interpret Lemma 23 pointwise, the key is to note that, for total function h and arbitrary relation S, $h^{\cup} \circ h \cap S$ relates two points x and y if they are related by S and $h.x = h.y$. However, it is not necessary to do so: completion of the calculation in Lemma 24 demands the proof of Lemma 23 and this is best achieved by uninterpreted calculation. In turn, Lemma 24 is driven by Lemma 25 which expresses the delegate function φ as a least fixed point; crucially, this enables the use of fixed-point induction to reason about φ.

Lemma 23. If h is a total function,

$$h \cap R{\circ}(h^{\cup} \circ h \cap S) \;=\; h \cap (h{\cap}R){\circ}S$$

for all relations R and S.

Proof. By mutual inclusion:

$$
\begin{array}{ll}
 & h \cap (h \cap R) \circ S \\
\subseteq & \{ \quad \text{modularity rule: (1)} \quad \} \\
 & (h \cap R) \circ ((h \cap R)^{\cup} \circ h \ \cap\ S) \\
\subseteq & \{ \quad h \cap R \subseteq h, \text{ monotonicity} \quad \} \\
 & (h \cap R) \circ (h^{\cup} \circ h \ \cap\ S) \\
\subseteq & \{ \quad h \text{ is a total function, so } h \circ h^{\cup} \circ h = h \\
 & \quad\ \ h \cap R \subseteq h, \text{ distributivity and monotonicity} \quad \} \\
 & h \ \cap\ R \circ (h^{\cup} \circ h \ \cap\ S) \\
= & \{ \quad \text{idempotency (preparatory to next step)} \quad \} \\
 & h \cap h \cap R \circ (h^{\cup} \circ h \ \cap\ S) \\
\subseteq & \{ \quad \text{modularity rule: (2)} \quad \} \\
 & h \ \cap\ (h \circ (h^{\cup} \circ h \ \cap\ S)^{\cup} \ \cap\ R) \circ (h^{\cup} \circ h \ \cap\ S) \\
\subseteq & \{ \quad h \text{ is a total function, so } h \circ h^{\cup} \circ h = h \\
 & \quad\ \ (h^{\cup} \circ h \ \cap\ S)^{\cup} \subseteq h^{\cup} \circ h, \\
 & \quad\ \ \text{distributivity and monotonicity} \quad \} \\
 & h \cap (h \cap R) \circ S.
\end{array}
$$

□

Lemma 24. If h is a total function,

$$h \cap (h^{\cup} \circ h \cap R)^{*} \ = \ \langle \mu X \ :: \ h \cap (I \cup X \circ R) \rangle$$

for all relations R.

Proof. We derive the right side as follows.

$$
\begin{array}{ll}
 & h \cap (h^{\cup} \circ h \cap R)^{*} \ = \ \mu g \\
\Leftarrow & \{ \quad \text{fusion theorem} \quad \} \\
 & \langle \forall X :: h \cap (I \cup X \circ (h^{\cup} \circ h \cap R)) \ = \ g.(h \cap X) \rangle \\
= & \{ \quad \text{distributivity, lemma 23 with } R,S := X,R \quad \} \\
 & \langle \forall X :: (h \cap I) \cup (h \cap (h \cap X) \circ R) \ = \ g.(h \cap X) \rangle \\
\Leftarrow & \{ \quad \text{strengthening: } X := h \cap X \quad \} \\
 & \langle \forall X :: (h \cap I) \cup (h \cap X \circ R) \ = \ g.X \rangle
\end{array}
$$

$=$ { distributivity }

$\langle \forall X :: h \cap (I \cup X \circ R) \; = \; g.X \rangle$. □

Lemma 25

$$\varphi \; = \; \langle \mu X :: \varphi \cap (I \; \cup \; X \circ G^{\cup}) \rangle .$$

Proof

φ

$=$ { (9) (i.e., $\varphi \subseteq (G^{\cup} \cap \varphi^{\cup} \circ \varphi)^{*}$) }

$\varphi \cap (G^{\cup} \cap \varphi^{\cup} \circ \varphi)^{*}$

$=$ { lemma 24 }

$\langle \mu X :: \varphi \cap (I \; \cup \; X \circ G^{\cup}) \rangle$. □

The significance of the equality in Lemma 25 is the inclusion of the left side in the right side. (The converse is trivial.) Thus, in words, the lemma states that there is a path from each node to its delegate on which every node has the same delegate.

3.5 Summary and Discussion

We summarise the results of this section with the following theorem.

Theorem 26. Suppose f of type $\mathbb{N} \leftarrow \mathsf{Node}$ is a total function and G is a finite graph. Then the equation

$$\varphi \; :: \; \varphi \circ \varphi^{\cup} \subseteq I_{\mathsf{Node}} \subseteq \varphi^{\cup} \circ \varphi \;\; \wedge \;\; \varphi \subseteq (G^{*})^{\cup} \;\; \wedge \;\; G^{*} \subseteq (f \circ \varphi)^{\cup} \circ \geq \circ f$$

has a solution with the additional properties that the solution is a closure operator (i.e. a delegate is its own delegate):

$$\varphi \circ \varphi = \varphi ,$$

strongly connected nodes have the same delegate:

$$\mathsf{equiv}.G \subseteq \varphi^{\cup} \circ \varphi$$

and there is a path from each node to its delegate on which successive nodes have the same delegate:

$$\varphi \; \subseteq \; (G^{\cup} \cap \varphi^{\cup} \circ \varphi)^{*} .$$

More precisely, there is a path from each node to its delegate on which all nodes have the same delegate:

$$\varphi \; = \; \langle \mu X :: \varphi \cap (I \; \cup \; X \circ G^{\cup}) \rangle .$$

Moreover, a delegate has the largest f-value

$$\varphi \subseteq f^{\cup} \circ \geq \circ f .$$

If the function f is injective, the solution is unique; in this case, we call the unique solution *the* delegate function on G according to f. Moreover, f is a topological ordering of the nodes of the graph

$$\varphi \circ G^{\cup} \circ \varphi^{\cup} \cap \neg I$$

(the graph obtained from $G^{\cup}$ by coalescing all nodes with the same delegate and removing self-loops). This graph is therefore acyclic. □

This paper is inspired by Cormen, Leiserson and Rivest's account of the "forefather" function and its use in applying depth-first search to the computation of strongly connected components [9, pp. 488–494]. However, our presentation is more general than theirs; in particular, we do not assume that the choice function is injective.

The motivation for our more general presentation is primarily to kill two birds with one stone. As do Cormen, Leiserson and Rivest, we apply the results of this section to computing strongly connected components: see Sect. 4. This is one of the "birds". The second "bird" is represented by the case that the choice function is a constant function (for example, $f.a = 0$, for all nodes a). In this case, the choice of node a in the algorithm of Fig. 1 reduces to the one condition $a \circ seen = \perp\!\!\perp$ (in words, a has not yet been seen) and the function f plays no role whatsoever. Despite this high level of nondeterminism, the specification of a delegate (see Sect. 3.1) allows many solutions that are not computed by the algorithm. (For example, the identity function satisfies the specification.) The analysis of Sect. 3.2 is therefore about the properties of a function that records the history of repeated searches of a graph until all nodes have been seen: the delegate function computed by repeated graph search records for each node b, the node a from which the search that sees b was initiated.

This analysis reveals many properties of graph searching that other accounts may suggest are peculiar to depth-first search. Most notable is the property that strongly connected nodes are assigned the same delegate. As shown in Lemma 19, this is a necessary property when the choice function is injective; otherwise, it is not a necessary property but it is a property of the delegate function computed by repeated graph search, whatever graph-searching algorithm is used. The second notable property of repeated graph search is that there is a path from each node to its delegate on which all nodes have the same delegate. This is closely related to the property that Cormen, Leiserson and Rivest call the "white-path theorem" [9, pp. 482], which we discuss shortly. Our analysis shows that the property is a generic property of repeated graph search and not specific to depth-first search.

In order to discuss the so-called "white-path theorem", it is necessary to give a preliminary explanation. Operational descriptions of graph-searching algorithms often use the colours white, grey and black to describe nodes. A white node is a node that has not been seen, a grey node is a node that has been seen but not all edges from the node have been "processed", and a black node is a node that has been seen and all edges from the node have been "processed". The property "white", "grey" or "black" is, of course, time-dependent since initially all nodes are white and on termination all nodes are black.

Now Lemmas 16 and 17 express subtley different versions of what is called the "white-path theorem". Suppose a search from node a is initiated in the *outer* loop of a repeated graph search. The search finds nodes on paths starting from a. There are three formally different properties of the paths that are found:

(i) The final node on the path is white at the time the search from a is initiated.
(ii) All nodes on the path are white at the time the search from a is initiated.
(iii) All nodes on the path are white at the time the search from their predecessor on the path is initiated.

In general, if nodes are labelled arbitrarily white or non-white, the sets of paths described by (i), (ii) and (iii) are different. (They are ordered by the subset relation, with (i) being the largest and (iii) the smallest.) However, in a repeated graph search, the sets of paths satisfying (i) and (ii) are equal. This is the informal meaning of Lemma 15. Moreover, the right side of the assignment to s in Fig. 1 is the set of nodes reached by paths satisfying (i); Lemma 16 states that, in a repeated graph search, the nodes that are added by a search initiated from node a are the nodes that can be reached by a path satisfying (ii).

We claim—without formal proof—that it is also the case that, in a repeated graph search, all three sets of paths are equal. That is, the set of paths described by (iii) is also equal to the set of paths described by (i). We don't give a proof here because it is impossible to express formally without introducing additional auxiliary variables. Informally, it is clear from the implementation shown in Fig. 2, in particular the choice of nodes b and c. The introduction of timestamps does allow us to prove the claim formally for depth-first search. See Sect. 4.

Cormen, Leiserson and Rivest's [9, pp. 482] "white-path theorem" states that it is a property of depth-first search that paths found satisfy (ii). Characteristic of depth-first search is that the property is true for all nodes, and not just nodes from which a search is initiated in the outer loop.

Finally, let us briefly remark on Lemma 21. As we see later, not only can depth-first search be used to calculate the strongly connected components of a graph, in doing so it also computes a topological ordering of these components (more precisely a topological ordering of the homomorphic-image graph discussed in Sect. 4). Lemma 21 is more general than this. It states that, if the choice function is injective, it is a topological ordering of the converse of the graph obtained by coalescing all the nodes with the same delegate and then omitting self-loops. In fact, this is also true of the delegate function computed as above. We leave its proof to the reader: remembering that during execution of the algorithm φ is partial with right domain $\varphi>$, identify and verify an invariant that states that f is a topological ordering on a subgraph of $\varphi \circ G^{\cup} \circ \varphi^{\cup} \cap \neg I$.

4 Strongly Connected Components

Recall that if G is a relation, the relation $\mathsf{equiv}.G$ defined by

$$\mathsf{equiv}.G \quad = \quad G^* \cap (G^*)^{\cup}$$

is an equivalence relation; if G is a graph, two nodes related by $\mathsf{equiv}.G$ are said to be in the same *strongly connected component* of G.

An equivalence relation R on a set A is typically represented by a so-called *representative* function ρ of type $A \leftarrow A$ with the property that

$$R \;=\; \rho^{\cup} \circ \rho .$$

For each element a of A, the element $\rho.a$ is called the *representative* of the equivalence class containing a. In words, two values a and b are equivalent (under R) iff they have the same representative.

The calculation of (a representative-function representation of the) strongly connected components of a given graph is best formulated as a two-stage process. In the first stage, a repeated depth-first search of the graph is executed; the output of this stage is a function f from nodes to numbers that records the order in which the search from each node finishes; we call it the *finish timestamp*. In the second stage a repeated search of the converse of the graph is executed using the function f as choice function.

Aside on Sharir's Algorithm. As mentioned in the introduction, Aho, Hopcroft and Ullman [1] attribute the algorithm to an unpublished 1978 document by Kosaraju and to Sharir [16]. Sharir's formulation of the algorithm supposes that a forest of trees is computed in the first stage; the ordering of the nodes is then given by a reverse postorder traversal of the trees in the forest. This is non-deterministic since the ordering of the trees is arbitrary. However, a well-known fact is that the use of the finish timestamp is equivalent to ordering the trees according to the reverse of the order in which they are constructed in the first stage; its use is also more efficient and much simpler to implement. Also, contrary to the suggestion in [1,9,10] and apparently not well-known, Sharir's formulation of the algorithm demonstrates that is not necessary to use depth-first search in the second stage: any graph searching algorithm will do. **End of Aside**

In this section, we establish the correctness of the second stage assuming certain properties of the first stage. Formally, we prove that the delegate function on G according to the timestamp f is a representative function for the strongly connected components of G.

The properties that we need involve the use of an additional function s from nodes to numbers that records the order in which the search in the first stage from each node starts. More precisely, the combination of the functions s and f records the order in which searches start and finish; the functions s and f are thus called the start and finish *timestamps*, respectively. Unlike f, which is used as a choice function in the second stage, the role of s is purely as an auxiliary variable. That is, the process of recording the start timestamp can be omitted from the computation proper because it only serves to document the properties of depth-first search.

The properties of repeated depth-first search that we assume are four-fold. First, for all nodes a and b, if the search from a starts before the start of the search from b, and the search from a finishes after the search from b finishes there is a path from a to b:

$$s^{\cup}\circ{\leq}\circ s \;\cap\; f^{\cup}\circ{\geq}\circ f \;\subseteq\; G^{*}. \tag{27}$$

Second, for all nodes a and b, if the search from a starts strictly before the start of the search from b and finishes strictly before the finish of the search from b, the search from a finishes strictly before the search from b starts:

$$s^{\cup}\circ{<}\circ s \;\cap\; f^{\cup}\circ{<}\circ f \;=\; f^{\cup}\circ{<}\circ s. \tag{28}$$

Thirdly, if there is an edge from node a to node b in the graph, the search from node a finishes after the search from b starts:

$$G \subseteq f^{\cup}\circ{\geq}\circ s. \tag{29}$$

Finally, s and f are total, injective functions from the nodes to natural numbers.

Properties (27) and (28) are both consequences of the so-called "parenthesis theorem" [9, p. 480] and [10, p. 606]. (The "parenthesis theorem" bundles together the so-called "parenthesis structure" of the start and finish times with properties of paths in the graph.) Property (29) is a consequence of the classification of edges into tree/ancestor edges, fronds or vines [17]; see also [9, exercise 23.3-4, p. 484] (after correction to include self-loops as in [10, exercise 22.3-5, p. 611]). (Property (29) is sometimes stated in its contrapositive form: King and Launchbury [15], for example, formulate it as there being no "left-right cross edges".)

It may help to present further details of repeated depth-first search. The outer loop—the repeated call of depth-first search—takes the following form:

$f{,}s\; :=\; \perp\!\!\perp{,}\perp\!\!\perp\;;$

while $s{>} \neq I_{\mathsf{Node}}$ **do**

  begin

    choose node a such that $a\circ s{>}\;=\;\perp\!\!\perp$

  ; $\mathsf{dfs}(a)$

  end

$\{\quad (27)\;\wedge\;(28)\;\wedge\;(29)$

$\wedge\quad s\circ s^{\cup}\subseteq I_{\mathsf{N}}\;\wedge\; s^{\cup}\circ s = I_{\mathsf{Node}} = f^{\cup}\circ f\;\wedge\; f\circ f^{\cup}\subseteq I_{\mathsf{N}}\;\}$

The implementation of $\mathsf{dfs}(a)$ is as follows:

$$s := s \cup \overline{(\mathsf{MAX}.s \uparrow \mathsf{MAX}.f)+1} \circ \top\!\!\top \circ a$$

; **while** $a \circ G \circ s{\succ}\!\bullet \neq \perp\!\!\perp$ **do**

begin

choose node b such that $a \circ \top\!\!\top \circ b \subseteq a \circ G \circ s{\succ}\!\bullet$

; $\mathsf{dfs}(b)$

end

; $f := f \cup \overline{(\mathsf{MAX}.s \uparrow \mathsf{MAX}.f)+1} \circ \top\!\!\top \circ a$

In the above, the current "time" is given by $\mathsf{MAX}.s{\uparrow}\mathsf{MAX}.f$ (assuming that $\mathsf{MAX}.\perp\!\!\perp$ is 0, by definition): the maximum value of the combined functions s and f; the overbar denotes the conversion of a number into a coreflexive representing the singleton set containing that number. The assignment to s thus increments the time by 1 and assigns to the node a the new time as starting time; similarly, the assignment to f increments the time by 1 and assigns to the node a the new time as finish time. The coreflexive $s{\succ}\!\bullet$ is the complement of $s{>}$; thus, it represents the set of nodes from which a search has not yet been started. The body of the inner loop is repeatedly executed while there remain edges in G from a to a node from which a search has not been started; the chosen node b is then one such node.

It is a very substantial exercise to verify the postcondition of repeated depth-first search since, in order to do so, additional invariant properties must be identified and verified. We have identified 16 different conjuncts in the invariant of depth-first search. Given that there are 5 components in its implementation (two assignments, one test, one choice and one recursive call), this means that there are at least 64 (sixteen times four) verification conditions that must be checked in order to verify repeated depth-first search: the recursive call can be ignored "by induction" but the repeated invocation of depth-first search also incurs additional verification conditions. Although many of these verification conditions are straight-forward, and might be taken for granted in an informal account of the algorithm, there is still much to be done. For more information, including a detailed comparison with [9], see [6].)

Suppose s and f are the start and finish timestamps computed by a repeated depth-first search of the graph G as detailed above. Suppose φ is the delegate function on G according to the timestamp f. (Recall that, as remarked immediately following its definition in Sect. 3.1, the function φ is the function computed by the repeated search of $G^{\cup}$ in the second stage of the Kosaraju-Sharir algorithm.)

From Theorem 26, we know that

$$\mathsf{equiv}.G \subseteq \varphi^{\cup} \circ \varphi .$$

It remains to show that

$$\varphi^{\cup} \circ \varphi \subseteq \mathsf{equiv}.G .$$

We do this by showing that $\varphi \subseteq \mathsf{equiv}.G$. That is, we show that the delegate of a node according to f is strongly connected to the node. The key is to use induction, the main difficulty being to identify a suitable induction hypothesis. This is done in the following lemma. Its proof combines two properties of delegates: (i) for each node, there is a path to its delegate on which all nodes have the same delegate and (ii) the delegate has the largest f-value.

Lemma 30

$$\varphi \subseteq \langle \mu X :: f^{\cup} \circ \geq \circ f \cap (I \cup X \circ G^{\cup}) \rangle .$$

Proof

$$\begin{array}{cl}
 & \varphi \\
= & \{ \quad \text{lemma 25} \quad \} \\
 & \langle \mu X :: \varphi \cap (I \cup X \circ G^{\cup}) \rangle \\
\subseteq & \{ \quad \text{theorem 26 (specifically, } \varphi \subseteq f^{\cup} \circ \geq \circ f) \\
 & \quad\ \ \text{and monotonicity} \quad \} \\
 & \langle \mu X :: f^{\cup} \circ \geq \circ f \cap (I \cup X \circ G^{\cup}) \rangle .
\end{array}$$

□

Lemma 30 enables us to use fixed-point induction to establish a key lemma:

Lemma 31

$$\varphi \subseteq s^{\cup} \circ \leq \circ s \cap f^{\cup} \circ \geq \circ f .$$

Proof

$$\begin{array}{cl}
 & \varphi \subseteq s^{\cup} \circ \leq \circ s \cap f^{\cup} \circ \geq \circ f \\
\Leftarrow & \{ \quad \text{lemma 30} \quad \} \\
 & \langle \mu X :: f^{\cup} \circ \geq \circ f \cap (I \cup X \circ G^{\cup}) \rangle \subseteq s^{\cup} \circ \leq \circ s \cap f^{\cup} \circ \geq \circ f \\
\Leftarrow & \{ \quad \text{fixed-point induction} \quad \} \\
 & f^{\cup} \circ \geq \circ f \cap (I \cup (s^{\cup} \circ \leq \circ s \cap f^{\cup} \circ \geq \circ f) \circ G^{\cup}) \subseteq s^{\cup} \circ \leq \circ s \cap f^{\cup} \circ \geq \circ f \\
\Leftarrow & \{ \quad \text{distributivity and } [\, R \cup S = R \cup (\neg R \cap S) \,] \\
 & \quad\ \ \text{with } R,S := I \,,\, (s^{\cup} \circ \leq \circ s \cap f^{\cup} \circ \geq \circ f) \circ G^{\cup} \quad \} \\
 & \quad f^{\cup} \circ \geq \circ f \cap I \subseteq s^{\cup} \circ \leq \circ s \\
 & \wedge \ f^{\cup} \circ \geq \circ f \cap \neg I \cap (s^{\cup} \circ \leq \circ s \cap f^{\cup} \circ \geq \circ f) \circ G^{\cup} \subseteq s^{\cup} \circ \leq \circ s \\
= & \{ \quad \leq \text{ is reflexive and } s \text{ is total, so } I \subseteq s^{\cup} \circ \leq \circ s \\
 & \quad\ \ f \text{ is injective, so } f^{\cup} \circ \geq \circ f \cap \neg I = f^{\cup} \circ > \circ f \quad \} \\
 & f^{\cup} \circ > \circ f \cap (s^{\cup} \circ \leq \circ s \cap f^{\cup} \circ \geq \circ f) \circ G^{\cup} \subseteq s^{\cup} \circ \leq \circ s .
\end{array}$$

We continue with the left-hand side of the inclusion.

$$
\begin{array}{ll}
& f^{\cup}\circ{>}\circ f \;\cap\; (s^{\cup}\circ{\leq}\circ s \;\cap\; f^{\cup}\circ{\geq}\circ f)\circ G^{\cup} \\
\subseteq & \{ \quad \text{assumption (29) and converse: } G^{\cup} \subseteq s^{\cup}\circ{\leq}\circ f \quad \} \\
& f^{\cup}\circ{>}\circ f \;\cap\; (s^{\cup}\circ{\leq}\circ s \;\cap\; f^{\cup}\circ{\geq}\circ f)\circ(s^{\cup}\circ{\leq}\circ f) \\
\subseteq & \{ \quad [\, R\cap S\subseteq R \,] \text{ with } R,S \;:=\; s^{\cup}\circ{\leq}\circ s \,,\, f^{\cup}\circ{\geq}\circ f \\
& \quad\quad \text{and monotonicity} \quad \} \\
& f^{\cup}\circ{>}\circ f \;\cap\; s^{\cup}\circ{\leq}\circ s\circ s^{\cup}\circ{\leq}\circ f \\
\subseteq & \{ \quad s \text{ is functional, so } s\circ s^{\cup} \subseteq I,\ \leq \text{ is transitive} \quad \} \\
& f^{\cup}\circ{>}\circ f \;\cap\; s^{\cup}\circ{\leq}\circ f \\
= & \{ \quad \text{assumption : (28), i.e. (taking converse and complements)} \\
& \quad\quad s^{\cup}\circ{\leq}\circ f \;=\; s^{\cup}\circ{\leq}\circ s \,\cup\, f^{\cup}\circ{\leq}\circ f \quad \} \\
& f^{\cup}\circ{>}\circ f \;\cap\; (s^{\cup}\circ{\leq}\circ s \,\cup\, f^{\cup}\circ{\leq}\circ f) \\
= & \{ \quad f^{\cup}\circ{>}\circ f \cap f^{\cup}\circ{\leq}\circ f \;=\; \perp\!\!\perp \quad \} \\
& f^{\cup}\circ{>}\circ f \;\cap\; s^{\cup}\circ{\leq}\circ s \\
\subseteq & \{ \quad \text{monotonicity} \quad \} \\
& s^{\cup}\circ{\leq}\circ s.
\end{array}
$$

Combining the two calculations, the proof is complete. □

Now we can proceed to show that every node is strongly connected to its delegate.

Lemma 32. Suppose φ is the delegate function on G according to the timestamp f. Then

$$\varphi \subseteq \mathsf{equiv}.G.$$

Proof

$$
\begin{array}{ll}
& \varphi \subseteq \mathsf{equiv}.G \\
= & \{ \quad \text{definition of } \mathsf{equiv}.G\text{, distributivity} \quad \} \\
& \varphi \subseteq G^* \,\wedge\, \varphi \subseteq (G^*)^{\cup} \\
= & \{ \quad \text{by definition of delegate (see theorem 26), } \varphi \subseteq (G^*)^{\cup} \quad \} \\
& \varphi \subseteq G^* \\
\Leftarrow & \{ \quad \text{(27) is a postcondition of repeated depth-first search} \quad \} \\
& \varphi \;\subseteq\; s^{\cup}\circ{\leq}\circ s \cap f^{\cup}\circ{\geq}\circ f \\
\Leftarrow & \{ \quad \text{lemma 31} \quad \} \\
& \mathsf{true}.
\end{array}
$$

□

Theorem 33. Suppose f is the finish timestamp computed by a repeated depth-first search of a graph G. Then the delegate function on G according to f is a representative function for strongly connected components of G. That is, if φ denotes the delegate function,

$$\varphi^{\cup} \circ \varphi \;=\; \mathsf{equiv}.G.$$

Proof

$$\begin{array}{ll}
 & \varphi^{\cup} \circ \varphi \;=\; \mathsf{equiv}.G \\
= & \quad \{ \quad \text{anti-symmetry} \quad \} \\
 & \mathsf{equiv}.G \subseteq \varphi^{\cup} \circ \varphi \;\wedge\; \varphi^{\cup} \circ \varphi \subseteq \mathsf{equiv}.G \\
\Leftarrow & \quad \{ \quad \text{theorem 26, lemma 32} \quad \} \\
 & \mathsf{true} \;\wedge\; (\mathsf{equiv}.G)^{\cup} \circ \mathsf{equiv}.G \subseteq \mathsf{equiv}.G \\
= & \quad \{ \quad \mathsf{equiv}.G \text{ is symmetric and transitive} \quad \} \\
 & \mathsf{true}.
\end{array}$$

□

5 Conclusion

In one sense, this paper offers no new results. Graph-searching algorithms have been studied extensively for decades and have long been a standard part of the undergraduate curriculum in computing science. The driving force behind this work has been to disentangle different elements of the correctness of the two-stage algorithm for determining the strongly connected components of a graph: our goal has been to clearly distinguish properties peculiar to depth-first search that are vital to the first stage of the algorithm from properties of repeated graph search that are exploited in its second stage. This is important because an algorithm to determine strongly connected components of a graph does not operate in a vacuum: the information that is gleaned is used to inform other computations. For example, Sharir [16] shows how to combine his algorithm with an iterative algorithm for data-flow analysis.

The primary contribution of the paper is, however, to show how the choice of an appropriate algebraic framework enables precise, concise calculation of algorithmic properties of graphs. Although with respect to graph algorithms (as opposed to relation algebra in general) the distinction between "point-free" and "pointwise" calculations has only been made relatively recently, this was the driving force behind the author's work on applying regular algebra to path-finding problems [2,3].

The difference between point-free and pointwise calculations can be appreciated by noting that nowhere in our calculations is there an existential quantification or a nested universal quantification. Typical accounts of depth-first search make abundant use of such quantifications; the resulting formal statements are long and unwieldy, and calculations become (in our view) much harder to check:

compare, for example, the concision of the three assumptions (27), (28) and (29) with the three assumptions made by King and Launchbury [15].

Of course, our discussion of the two-stage algorithm is incomplete because we have not formally established the properties of the first (depth-first search) stage that we assume hold in the second stage. (The same criticism is true of [15].) This we have done in [6]. Although the calculations are long—primarily because there is a large number of verification conditions to be checked—we expect that they would be substantially shorter and easier to check than formal pointwise justifications of the properties of depth-first search.

Acknowledgements. Many thanks to the referees for their careful and detailed critique of the submitted paper. Thanks also for pointing out that explicit mention of the "forefather" function, studied in detail in [9], has been elided in [10].

References

1. Aho, A.V., Hopcroft, J.E., Ullman, J.D.: Data Structures and Algorithms. Addison-Wesley, Reading (1982)
2. Backhouse, R.C.: Closure algorithms and the star-height problem of regular languages. Ph.D. thesis, University of London (1975). https://spiral.imperial.ac.uk/bitstream/10044/1/22243/2/Backhouse-RC-1976-PhD-Thesis.pdf
3. Backhouse, R.C., Carré, B.A.: Regular algebra applied to path-finding problems. J. Inst. Math. Appl. **15**, 161–186 (1975)
4. Backhouse, R.C., van der Woude, J.: Demonic operators and monotype factors. Math. Struct. Comput. Sci. **3**(4), 417–433 (1993)
5. Backhouse, R.: Regular algebra applied to language problems. J. Logic Algebraic Program. **66**, 71–111 (2006)
6. Backhouse, R., Doornbos, H., Glück, R., van der Woude, J.: Algorithmic graph theory: an exercise in point-free reasoning (2019). http://www.cs.nott.ac.uk/~pasrb2/MPC/BasicGraphTheory.pdf, Also available online at ResearchGate
7. Backhouse, R.C., van den Eijnde, J.P.H.W., van Gasteren, A.J.M.: Calculating path algorithms. Sci. Comput. Program. **22**(1–2), 3–19 (1994)
8. Conway, J.H.: Regular Algebra and Finite Machines. Chapman and Hall, London (1971)
9. Cormen, T.H., Leiserson, C.E., Rivest, R.L.: Introduction to Algorithms. MIT Electrical Engineering and Computer Science Series. MIT Press, Cambridge (1990)
10. Cormen, T.H., Leiserson, C.E., Rivest, R.L., Stein, C.: Introduction to Algorithms. MIT Electrical Engineering and Computer Science Series, 3rd edn. MIT Press, Cambridge (2009)
11. Doornbos, H., Backhouse, R.: Reductivity. Sci. Comput. Program. **26**(1–3), 217–236 (1996)
12. Freyd, P.J., Ščedrov, A.: Categories. North-Holland, Allegories (1990)
13. Glück, R.: Algebraic investigation of connected components. In: Höfner, P., Pous, D., Struth, G. (eds.) RAMICS 2017. LNCS, vol. 10226, pp. 109–126. Springer, Cham (2017). https://doi.org/10.1007/978-3-319-57418-9_7
14. Hoogendijk, P., Backhouse, R.C.: Relational programming laws in the tree, list, bag, set hierarchy. Sci. Comput. Program. **22**(1–2), 67–105 (1994)

15. King, D.J., Launchbury, J.: Structuring depth-first search algorithms in Haskell. In: POPL 1995. Proceedings of the 22nd ACM SIGPLAN-SIGACT Symposium on Principles of Programmming Languages, pp. 344–354 (1995)
16. Sharir, M.: A strong-connectivity algorithm and its application in data flow analysis. Comput. Math. Appl. **7**(1), 67–72 (1981)
17. Tarjan, R.E.: Depth first search and linear graph algorithms. SIAM J. Comput. **1**, 146–160 (1972)

Shallow Embedding of Type Theory is Morally Correct

Ambrus Kaposi(✉), András Kovács, and Nicolai Kraus

Eötvös Loránd University, Budapest, Hungary
{akaposi,kovacsandras,nkraus}@inf.elte.hu

Abstract. There are multiple ways to formalise the metatheory of type theory. For some purposes, it is enough to consider specific *models* of a type theory, but sometimes it is necessary to refer to the *syntax*, for example in proofs of canonicity and normalisation. One option is to embed the syntax deeply, by using inductive definitions in a proof assistant. However, in this case the handling of definitional equalities becomes technically challenging. Alternatively, we can reuse conversion checking in the metatheory by *shallowly embedding* the object theory. In this paper, we consider the *standard model* of a type theoretic object theory in Agda. This model has the property that all of its equalities hold definitionally, and we can use it as a shallow embedding by building expressions from the components of this model. However, if we are to reason soundly about the syntax with this setup, we must ensure that distinguishable syntactic constructs do not become provably equal when shallowly embedded. First, we prove that shallow embedding is injective up to definitional equality, by modelling the embedding as a syntactic translation targeting the metatheory. Second, we use an implementation hiding trick to disallow illegal propositional equality proofs and constructions which do not come from the syntax. We showcase our technique with very short formalisations of canonicity and parametricity for Martin-Löf type theory. Our technique only requires features which are available in all major proof assistants based on dependent type theory.

Keywords: Type theory · Shallow embedding · Set model · Standard model · Canonicity · Parametricity · Agda

1 Introduction

Martin-Löf type theory [32] (MLTT) is a formal system which can be used for writing and verifying programs, and also for formalising mathematics. Proof assistants and dependently typed programming languages such as Agda [43], Coq [33], Idris [9], and Lean [36] are based on MLTT and its variations.

This work was supported by the Thematic Excellence Programme, Industry and Digitization Subprogramme, NRDI Office, 2019 and by the European Union, co-financed by the European Social Fund (EFOP-3.6.2-16-2017-00013 and EFOP-3.6.3-VEKOP-16-2017-00002).

G. Hutton (Ed.): MPC 2019, LNCS 11825, pp. 329–365, 2019.
https://doi.org/10.1007/978-3-030-33636-3_12

Specific versions of MLTT have many interesting properties, such as *canonicity*, *normalisation* or *parametricity*. Normalisation in particular is practically significant, since it enables decidable conversion checking and thus decidable type checking. These properties are of metatheoretic nature; in other words, they are answers to questions *about* type theory, rather than questions *inside* type theory. We wish to effectively study these questions in a formal and machine-checked setting.

1.1 Technical Challenges of Deep Embeddings

We refer to the type theory that we wish to study as the *object (type) theory*. If we want to use Agda (or another proof assistant) to study it, the most direct way is to use native inductive definitions to represent the syntax. This is called a *deep embedding*. Such an embedding could be an inductive type representing syntactic expressions Expr, with a constructor for every kind of term former. Examples for such constructors are the following:

$$\begin{aligned}
\mathsf{Pi} &: \mathsf{Expr} \to \mathsf{Expr} \to \mathsf{Expr} \\
\mathsf{lam} &: \mathsf{Expr} \to \mathsf{Expr} \\
\mathsf{app} &: \mathsf{Expr} \to \mathsf{Expr} \to \mathsf{Expr}
\end{aligned}$$

The idea is simple: Pi takes two expressions e_1, e_2 as arguments, and if these represent a type A and a type family B over A, then $\mathsf{Pi}\, e_1\, e_2$ represents the corresponding Π-type. Similarly, lam represents λ-abstraction and app application.

Of course, this inductive definition of Expr does not ensure that every expression "makes sense"; e.g. $\mathsf{Pi}\, e_1\, e_2$ will not make sense unless e_1 and e_2 are of the form described above. We need to additionally define inductive relations which express well-formedness and typing for specific syntactic constructs. This way of defining raw terms together with well-formedness relations is called an *extrinsic* approach.

Depending on the available notion of inductive types in the metatheory, we can use more abstract representations. For example, if inductive-inductive types [37] are available, then we can define a syntax which contains only well-formed terms [10]. In this case, we have an *intrinsic* definition for the syntax. We have the following signature for the type constructors of the embedded syntax, respectively for contexts, types, substitutions and terms:

$$\begin{aligned}
\mathsf{Con} &: \mathsf{Set} \\
\mathsf{Ty} &: \mathsf{Con} \to \mathsf{Set} \\
\mathsf{Sub} &: \mathsf{Con} \to \mathsf{Con} \to \mathsf{Set} \\
\mathsf{Tm} &: (\Gamma : \mathsf{Con}) \to \mathsf{Ty}\, \Gamma \to \mathsf{Set}
\end{aligned}$$

However, with the intrinsic inductive-inductive definitions we also need separate inductive relations expressing definitional equality. We can avoid these relations by using a *quotient inductive* [2,29] syntax instead. This way, definitional equality is given by *equality constructors*. For example, associativity of type substitution would be given as the following $[\circ]$ equality, where we also introduce substitution composition and type substitution first, and implicitly quantify over variables:

$$\begin{aligned}
\cdot \circ \cdot &: \mathsf{Sub}\,\Theta\,\Delta \to \mathsf{Sub}\,\Gamma\,\Theta \to \mathsf{Sub}\,\Gamma\,\Delta \\
\cdot[\cdot] &: \mathsf{Ty}\,\Delta \to \mathsf{Sub}\,\Gamma\,\Delta \to \mathsf{Ty}\,\Gamma \\
[\circ] &: (A\,[\sigma])\,[\delta] = A\,[\sigma \circ \delta]
\end{aligned}$$

The quotient inductive definition allows higher-level reasoning than the purely inductive-inductive one. In the former case, every metatheoretic construction automatically respects definitional equality in the syntax, since it is identified with meta-level propositional equality. In the latter case, object-level definitional equality is just a relation, and we need to explicitly prove preservation in many cases.

However, even with quotient induction, there are major technical challenges in formalising metatheory, and an especially painful issue is the obligation to explicitly refer to conversion rules even in very simple constructions. For example, we might want to take the zeroth de Bruijn index with type Bool in some extended $\Gamma \blacktriangleright \mathsf{Bool}$ typing context. For this, we first need a weakening substitution declared in the syntax (or admissible from the syntax):

$$\mathsf{weaken} : \mathsf{Sub}\,(\Gamma \blacktriangleright A)\,\Gamma$$

Now, we are able to give a general type for the zeroth de Bruijn index:

$$\mathsf{vzero} : \mathsf{Tm}\,(\Gamma \blacktriangleright A)\,(A[\mathsf{weaken}])$$

The weakening is necessary because A has type $\mathsf{Ty}\,\Gamma$, but we also want to mention it in the $\Gamma \blacktriangleright A$ context.

Now, we might try to use vzero to get a term with type $\mathsf{Tm}\,(\Gamma \blacktriangleright \mathsf{Bool})\,\mathsf{Bool}$. However, we only get $\mathsf{vzero} : \mathsf{Tm}\,(\Gamma \blacktriangleright \mathsf{Bool})\,(\mathsf{Bool}[\mathsf{weaken}])$. We also need to refer to the computation rule for substituting Bool which just forgets about the substitution:

$$\mathsf{Bool}[] : \mathsf{Bool}[\sigma] = \mathsf{Bool}$$

Hence, the desired term needs to involve transporting over the $\mathsf{Bool}[]$ equation:

$$\begin{aligned}
&\mathsf{vzeroBool} : \mathsf{Tm}\,(\Gamma \blacktriangleright \mathsf{Bool})\,\mathsf{Bool} \\
&\mathsf{vzeroBool} :\equiv \mathsf{transport}_{(\mathsf{Tm}\,\Gamma)}\,\mathsf{Bool}[]\,\mathsf{vzero}
\end{aligned}$$

This phenomenon arises with extrinsic and purely inductive-inductive syntaxes as well; in those cases, instead of transporting along an equation, we need to invoke a conversion rule for term typing. For extrinsic syntaxes, we additionally have a choice between implicit and explicit substitution, but this choice does not change the picture either.

Hence, all of the mentioned deeply embedded syntaxes require constructing explicit derivations of definitional equalities. In more complex examples, this is a technical burden which is often humanly impossible to handle. Also, proof assistants are often unable to check formalisations within sensible time because of the huge size of the involved proof terms.

1.2 Reflecting Definitional Equality

To eliminate explicit derivations of conversion, the most promising approach is to reflect object-level definitional equality as meta-level definitional equality. If this is achieved, then all conversion derivations can be essentially replaced by proofs of reflexivity, and the meta-level typechecker would implicitly construct all derivations for us.

How can we achieve this? We might consider extensional type theory with general equality reflection, or proof assistants with limited equality reflection. In Agda there is support for the latter using rewrite rules [12], which we have examined in detail for the previously described purposes. In Agda, we can just postulate the syntax of the object theory, and try to reflect the equations. This approach does work to some extent, but there are significant limitations:

- Type-directed equalities cannot be reflected, such as η-rules for empty substitutions and unit types, or definitional proof irrelevance for propositions. Rewrite rules must be syntax-directed and have a fixed direction of rewriting.
- Rewrite rules yield poor evaluation performance and hence poor type checking performance, because they are implemented using a general mechanism which does not know anything about the domain, unlike the meta-level conversion checker.
- In the current Agda implementation (version 2.6), rewrite rules are not flexible enough to capture all desired computational behavior. For example, the left hand side of a rewrite rule is treated as a rigid expression which is not refined during the matching of the rule. Given an $f : \mathsf{Bool} \to \mathsf{Bool} \to \mathsf{Bool}$ function, if we add the rewrite rule $\forall x.\, f\, x\, (\mathsf{not}\, x) = \mathsf{true}$, the expession $f\, \mathsf{true}\, \mathsf{false}$ will not be rewritten to true, since it does not rigidly match the $\mathsf{not}\, x$ on the left hand side. In practice, this means that an unbounded number of special-cased rules are required to reflect equalities for a type theory. Lifting all the restricting assumptions in the implementation of rewrite rules would require non-trivial research effort.

It seems to be difficult to capture the equational theory of a dependent object theory with general-purpose implementations of equality reflection. In the future, robust equality reflection for conversion rules may become available, but until then we have to devise workarounds. If the object theory is similar enough to the metatheory, we can reuse meta-level conversion checking using a *shallow embedding*.

In this paper we describe such a shallow embedding. The idea is that in the *standard model* of the object theory equations already hold definitionally, and so it would be convenient to reason about expressions built from the standard model as if they came from arbitrary models, e.g. from the syntax.

However, we should only use shallow embeddings in morally correct ways: only those equations should hold in the shallow embedding that also hold in the deeply embedded syntax. In other words, we should be able in principle to translate every formalisation which uses shallow embedding to a formalisation which uses deeply embedded syntax.

To address this, first we prove that shallow embedding is injective up to *definitional equality*: the metatheory can only believe two embedded terms definitionally equal if they are already equal in the object theory. This requires us to look at both the object theory and the metatheory from an external point of view and reason about embedded meta-level terms as pieces of syntax.

Second, we describe a method for hiding implementation details of the standard model, which prevents constructing terms which do not have syntactic counterparts and which also disallows morally incorrect *propositional equalities*. This hiding is realised with import mechanisms; we do not formally model it, but it is reasonable to believe that it achieves the intended purposes.

1.3 Contributions

In order to reason about the metatheory of type theory in a proof assistant, we present a version of shallow embedding which combines the advantage of shallow embeddings (many definitional equalities) with the advantage of deep embeddings (no unjustified equalities).
In detail:

1. We formalise in Agda the standard "Set" model (metacircular interpretation [22]) of a variant of MLTT with a predicative universe hierarchy, Π-types, Booleans, Σ-types and identity types (Sect. 3). All equalities hold definitionally in this model. A variation of this (see below) is the model we propose for metatheoretic reasoning.
2. For an arbitrary model of the object theory, we construct the *termified* model (Sect. 4), where contexts, types, substitutions and terms are all modelled by closed terms. We formalise the shallow embedding into Agda as the interpretation of the object syntax into its termified model. We prove that this translation is injective (Sect. 5), thereby showing that definitional equality of shallowly embedded terms coincides with object-theoretic definitional equality. This result holds *externally* to Agda (like parametricity): we need to step one level up and consider the syntax of Agda as well. Additionally, we show that internally to Agda, injectivity of the standard interpretation is not provable.

3. We describe a way of hiding the implementation of the standard model (Sect. 6), in order to rule out constructions and equality proofs which are not available in the object syntax.
4. Using shallowly embedded syntax, we provide a concise formalisation of canonicity for MLTT (Sect. 7.2), using a proof-relevant logical predicate model in a manner similar to [14] and [27]. We also provide a formalisation of a syntactic parametricity translation [6] of MLTT in Sect. 7.1.

The contents of Sects. 3, 4, 6 and 7 were formalised [30] in Agda. Additional documentation about technicalities is provided alongside the formalisation.

1.4 Related Work

Work on embedding the syntax of type theory in type theory spans a whole spectrum from fully deep embeddings through partly deep embeddings to fully shallow ones.

Deep embeddings give maximal flexibility but at the high price of explicit handling of definitional equality derivations. Extrinsic deep embeddings of type theory are given in Agda [1,18] and Coq [44]. Meta Coq provides an extrinsic deep embedding of the syntax of almost all of Coq inside Coq [5]. An intrinsic deep embedding with explicit conversion relations using inductive-inductive types is given in [10] and another one using inductive-recursive types is described by [16].

Quotient inductive-inductive types are used in [3,4] to formalise type theory in a bit more shallow way reusing propositional equality of the metatheory to represent conversion of the object theory.

Higher-order abstract syntax (HOAS) [23,39] uses shallow embedding for the substitution calculus part of the syntax while the rest (e.g. term formers such as λ and application) are given deeply, using the function space of the metalanguage to represent binders. It has been used to embed simpler languages in type theory [11,17,40], however, to our knowledge, not type theory itself.

McBride [34] uses a mixture of deep and shallow embeddings to embed an intrinsic syntax of type theory into Agda. In this work, inductively defined types and terms are given mutually with their standard interpretation, and while there are deep term and type codes, all *indexing* in the syntax is over the standard model. In a sense, this is an extension of inductive-recursive type codes to codes of terms as well. This gives a usability improvement compared to deep embedding as equality of indices is decided by the metatheory. However, definitional equality of terms still has to be represented deeply.

Shallow embedding has been used to formalise constructions on the syntax of type theory. [8,26,42] formalise the correctness of syntactic translations using shallow embeddings in Coq. [28,29] formalise syntactic translations and models of type theory depending on previous shallow models. Our work provides a framework in which these previous formalisations could be rewritten in a more principled way.

Reflection provides an interface between shallow and deep embeddings. Meta Coq [5] provides a mechanism to reify shallow Coq terms as deeply embedded syntax. The formalisation happens shallowly, making use of the typechecker of Coq, and deeply embedded terms are obtained after reification. The motivation is very similar to ours, but their syntax is extrinsic while we use an intrinsic syntax.

More generally, using type theory as an internal language of a model can be seen as working in a shallow embedding. Synthethic homotopy theory (e.g. [24]) can be seen as a shallow embedding in type theory, compared to a deep embedding where homotopy theory is built up from the ground analytically. [38] uses MLTT extended with some axioms to formalise arguments about a presheaf model, [15] uses MLTT as the internal language of a cubical set model, [29] uses MLTT as the internal language of a categories-with-families model.

Our wrapped shallow embedding (Sect. 6) resembles the method by Dan Licata [31] to add higher inductive types to Agda with eliminators computing definitionally on point constructors. He also uses implementation hiding to disallow pattern matching but retain definitional behaviour.

2 The Involved Theories

In this paper, we altogether need to involve three different theories. We give a quick overview below, then describe them and the used notation in this section.

1. **Agda**, which we use in two ways: as a metatheory when using shallow embedding, but also as an object theory, when we study embedding from an external point of view. In the latter case, we only talk about a small subset of Agda's syntax which is relevant to the current paper.
2. The **external metatheory**. We assume that this is a conventional extensional type theory with a universe hierarchy. However, we are largely agnostic and set theory with a suitable notion of universe hierarchy would be adequate as well. We primarily use the external metatheory to reason about Agda's syntax. However, since this metatheory is extensional, we can omit all coercions and transports when working inside it informally, and thus we also use it to obtain a readable notation.
3. The **object theory**, which we wish study by shallow embedding into Agda. We single out a particular version of MLTT as object theory, and describe it in detail. However, our shallow embedding should work for a wider range of object theories; we expand on this in Sect. 8.1.

2.1 Agda

Agda is a proof assistant based on intensional type theory. When we present definitions in Agda, we use a `monospace` font. We describe below the used features and notation.

Universes are named `Set i`. Also, we use universe polymorphism which allows us to quantify over `(i : Level)`. We use `zero` and `suc` for the zero and successor levels, and `i ⊔ j` for taking least upper bounds of levels.

Dependent functions are notated `(x : A) → B`. There is also an implicit function space `{x : A} → B`, such that any expression with this type is implicitly applied to an inferred `A` argument. In this paper, we also use implicit quantification over variables in type signatures. For example, instead of declaring a type as `f : {A : Set} → A → A`, we may write `f : A → A`. This shorthand (although supported in the latest 2.6 version of Agda) is not used in the actual formalisations.

We also use Σ types, unit types, Booleans and propositional equality. There are some names which coincide in the object theory and in agda, and we disambiguate them with a `m.` prefix (which stands for "meta"). So, we use `m.Σ A B` for dependent pairs with `(t m., u)` as constructor and `m.fst` and `m.snd` as projections. We use `m.Bool`, `m.true` and `m.false` for Booleans. We use `m.⊤` for the unit type with constructor `m.tt`, and use `t ≡ u` for propositional equality with `m.refl` and `m.J`.

2.2 The External Metatheory

This is an extensional type theory, with predicative universes Set_i, dependent functions $(x : A) \to B$, and dependent pairs as $(x : A) \times B$. Propositional equality is denoted $\cdot = \cdot$, with constructor refl. We have equality reflection, which means that if $p : t = u$ is derivable, then t and u are definitionally equal. We also have uniqueness of identity proofs, meaning that for any $p, q : t = u$ we also have $p = q$.

2.3 The Object Type Theory

We take an algebraic approach to the syntax and models of type theory. There is an *algebraic signature* for the object type theory, which can be viewed as a large record type, listing all syntactic constructions along with the equations for definitional equality. *Models* of a type theory are particular inhabitants of this large record type, and the *syntax* of a type theory is the *initial* model in the category of models, where morphisms are given by structure-preserving families of functions. The setup can be compared to groups, a more familiar algebraic structure: there is a signature for groups, models are particular groups, morphisms are group homomorphisms, and the initial group ("syntax") is the trivial group (free group over the empty set). A *displayed model* over a model $\mathcal{M}$ is a way of encoding a model together with a morphism into $\mathcal{M}$. Displayed models can be viewed as containing induction motives and methods for a theory (following the nomenclature of [35]), hence we need this notion to talk about induction over the syntax. For instance, a displayed model for the theory of natural numbers contains a family $P : \mathbb{N} \to \mathsf{Set}$ (the induction motive) together with induction methods showing that P is inhabited at zero and taking successors preserves P. A generic method for deriving the notions of model, morphism and displayed model from a signature is given in [29].

More concretely, our object type theory is given in Figs. 1a and b as a category with families (CwF) [20] extended with additional type formers. We present the signature of the object theory in an extensional notation, which allows us to omit transports along equations. We also implicitly quantify over variables occurring in types, and leave these parameters implicit when we apply functions as well. Additionally, we extend the usual notion of CwF with indexing by metatheoretic natural numbers, which stand for universe levels.

This notion of model yields a syntax with explicit substitutions. The core structural rules and the theory of substitutions are described by the components from Con to $,\circ$. Contexts (Con) and substitutions (Sub) form a category (id to idr). There is a contravariant, functorial action of substitutions on types and terms ($\cdot[\cdot]$ to $[\circ]$), thus types (of fixed level) form a presheaf on the category of contexts and terms form a presheaf on the category of elements of this presheaf. The empty context ($\bullet$) is the terminal object.

Contexts can be extended by $\cdot \rhd \cdot$. Substitutions can be viewed as abstract lists of terms, with $\cdot,\cdot$ allowing us to extend a substitution with a term. We can also take the "tail" and the "head" of an extended $\sigma : \mathsf{Sub}\,\Gamma\,(\Delta \rhd A)$ substitution; the tail is given by $\mathsf{p} \circ \sigma : \mathsf{Sub}\,\Gamma\,\Delta$, and the head is given by $\mathsf{q}[\sigma] : \mathsf{Tm}\,\Gamma\,A[\mathsf{p}]$. p is usually called a weakening substitution, and q corresponds to the zeroth de Bruijn index. We denote n-fold composition of the weakening substitution p by p^n (where $\mathsf{p}^0 = \mathsf{id}$), and we denote De Bruijn indices the following way: $\mathsf{v}^0 := \mathsf{q}, \mathsf{v}^1 := \mathsf{q}[\mathsf{p}], \ldots, \mathsf{v}^n := \mathsf{q}[\mathsf{p}^n]$. We define lifting of a substitution $\sigma : \mathsf{Sub}\,\Gamma\,\Delta$ by $\sigma^{\uparrow} : \mathsf{Sub}\,(\Gamma \rhd A[\sigma])\,(\Delta \rhd A) := (\sigma \circ \mathsf{p}, \mathsf{q})$. We observe that it has the property ${}^{\uparrow}[] : (\sigma^{\uparrow}) \circ (\delta, t) = (\sigma \circ \delta, t)$.

Π-types are characterised by a natural isomorphism between $\mathsf{Tm}\,\Gamma\,(\Pi\,A\,B)$ and $\mathsf{Tm}\,(\Gamma \rhd A)\,B$, with lam and app being the morphism components. This notion of application is different from the conventional one, but in our setting with explicit substitutions, the two applications are inter-derivable, and our app is simpler to interpret in models. We define conventional application as $t\,\$\,u := (\mathsf{app}\,t)[\mathsf{id}, u]$. $A \Rightarrow B$ abbreviates non-dependent functions, and is defined as $\Pi\,A\,(B[\mathsf{p}])$.

Σ-types are given by the constructor $\cdot,\cdot$ and projections fst and snd, and we also support the η-law. There is a unit type $\top$ with one constructor tt and an η-law. We have a hierarchy of universes, given by natural isomorphisms between $\mathsf{Ty}\,i\,\Gamma$ and $\mathsf{Tm}\,\Gamma\,(\mathsf{U}\,i)$ for every i. The isomorphism consists of a coding morphism (c) and a decoding morphism, denoted by underlining $\underline{\cdot}$. This presentation of universes is due to Thierry Coquand, and has been used before in [25] for instance. In the Agda formalisations, where we cannot underline, we write `El` for the decoding morphism.

We also have a propositional identity type Id, with usual constructor refl and elimination J with definitional β-rule.

$$
\begin{array}{lll}
\textbf{(a)}\ \mathsf{Con} & : & \mathbb{N} \to \mathsf{Set} \\
\mathsf{Ty} & : & \mathbb{N} \to \mathsf{Con}\, i \to \mathsf{Set} \\
\mathsf{Sub} & : & \mathsf{Con}\, i \to \mathsf{Con}\, j \to \mathsf{Set} \\
\mathsf{Tm} & : & (\Gamma : \mathsf{Con}\, i) \to \mathsf{Ty}\, j\, \Gamma \to \mathsf{Set} \\
\mathsf{id} & : & \mathsf{Sub}\, \Gamma\, \Gamma \\
\cdot \circ \cdot & : & \mathsf{Sub}\, \Theta\, \Delta \to \mathsf{Sub}\, \Gamma\, \Theta \to \mathsf{Sub}\, \Gamma\, \Delta \\
\mathsf{ass} & : & (\sigma \circ \delta) \circ \nu = \sigma \circ (\delta \circ \nu) \\
\mathsf{idl} & : & \mathsf{id} \circ \sigma = \sigma \\
\mathsf{idr} & : & \sigma \circ \mathsf{id} = \sigma \\
\cdot[\cdot] & : & \mathsf{Ty}\, i\, \Delta \to \mathsf{Sub}\, \Gamma\, \Delta \to \mathsf{Ty}\, i\, \Gamma \\
\cdot[\cdot] & : & \mathsf{Tm}\, \Delta\, A \to (\sigma : \mathsf{Sub}\, \Gamma\, \Delta) \to \mathsf{Tm}\, \Gamma\, (A[\sigma]) \\
[\mathsf{id}] & : & A[\mathsf{id}] = A \\
[\circ] & : & A[\sigma \circ \delta] = A[\sigma][\delta] \\
[\mathsf{id}] & : & t[\mathsf{id}] = t \\
[\circ] & : & t[\sigma \circ \delta] = t[\sigma][\delta] \\
\bullet & : & \mathsf{Con}\, 0 \\
\epsilon & : & \mathsf{Sub}\, \Gamma\, \bullet \\
\bullet\eta & : & (\sigma : \mathsf{Sub}\, \Gamma\, \bullet) = \epsilon \\
\cdot \rhd \cdot & : & (\Gamma : \mathsf{Con}\, i) \to \mathsf{Ty}\, j\, \Gamma \to \mathsf{Con}\, (i \sqcup j) \\
\cdot , \cdot & : & (\sigma : \mathsf{Sub}\, \Gamma\, \Delta) \to \mathsf{Tm}\, \Gamma\, (A[\sigma]) \to \mathsf{Sub}\, \Gamma\, (\Delta \rhd A) \\
\mathsf{p} & : & \mathsf{Sub}\, (\Gamma \rhd A)\, \Gamma \\
\mathsf{q} & : & \mathsf{Tm}\, (\Gamma \rhd A)\, (A[\mathsf{p}]) \\
\rhd\beta_1 & : & \mathsf{p} \circ (\sigma, t) = \sigma \\
\rhd\beta_2 & : & \mathsf{q}[\sigma, t] = t \\
\rhd\eta & : & (\mathsf{p}, \mathsf{q}) = \mathsf{id} \\
,\circ & : & (\sigma, t) \circ \nu = (\sigma \circ \nu, t[\nu]) \\
\Pi & : & (A : \mathsf{Ty}\, i\, \Gamma) \to \mathsf{Ty}\, j\, (\Gamma \rhd A) \to \mathsf{Ty}\, (i \sqcup j)\, \Gamma \\
\mathsf{lam} & : & \mathsf{Tm}\, (\Gamma \rhd A)\, B \to \mathsf{Tm}\, \Gamma\, (\Pi\, A\, B) \\
\mathsf{app} & : & \mathsf{Tm}\, \Gamma\, (\Pi\, A\, B) \to \mathsf{Tm}\, (\Gamma \rhd A)\, B \\
\Pi\beta & : & \mathsf{app}\, (\mathsf{lam}\, t) = t \\
\Pi\eta & : & \mathsf{lam}\, (\mathsf{app}\, t) = t \\
\Pi[] & : & (\Pi\, A\, B)[\sigma] = \Pi\, (A[\sigma])\, (B[\sigma^{\uparrow}]) \\
\mathsf{lam}[] & : & (\mathsf{lam}\, t)[\sigma] = \mathsf{lam}\, (t[\sigma^{\uparrow}]) \\
\Sigma & : & (A : \mathsf{Ty}\, i\, \Gamma) \to \mathsf{Ty}\, j\, (\Gamma \rhd A) \to \mathsf{Ty}\, (i \sqcup j)\, \Gamma \\
\cdot , \cdot & : & (u : \mathsf{Tm}\, \Gamma\, A) \to \mathsf{Tm}\, \Gamma\, (B[\mathsf{id}, u]) \to \mathsf{Tm}\, \Gamma\, (\Sigma\, A\, B) \\
\mathsf{fst} & : & \mathsf{Tm}\, \Gamma\, (\Sigma\, A\, B) \to \mathsf{Tm}\, \Gamma\, A \\
\mathsf{snd} & : & (t : \mathsf{Tm}\, \Gamma\, (\Sigma\, A\, B)) \to \mathsf{Tm}\, \Gamma\, (B[\mathsf{id}, \mathsf{fst}\, t])
\end{array}
$$

Fig. 1. The object type theory as a generalised algebraic structure. $\sigma^{\uparrow}$ abbreviates $(\sigma \circ \mathsf{p}, \mathsf{q})$.

(b)

$\Sigma\beta_1$: $\mathsf{fst}\,(u, v) = u$

$\Sigma\beta_2$: $\mathsf{snd}\,(u, v) = v$

$\Sigma\eta$: $(\mathsf{fst}\,t, \mathsf{snd}\,t) = t$

$\Sigma[]$: $(\Sigma\,A\,B)[\sigma] = \Sigma\,(A[\sigma])\,(B[\sigma^{\uparrow}])$

$,[]$: $(u, v)[\sigma] = (u[\sigma], v[\sigma])$

$\top$: $\mathsf{Ty}\,0\,\Gamma$

tt : $\mathsf{Tm}\,\Gamma\,\top$

$\top\eta$: $(t : \mathsf{Tm}\,\Gamma\,\top) = \mathsf{tt}$

$\top[]$: $\top[\sigma] = \top$

$\mathsf{tt}[]$: $\mathsf{tt}[\sigma] = \mathsf{tt}$

U : $(i : \mathbb{N}) \to \mathsf{Ty}\,(i+1)\,\Gamma$

$\underline{\cdot}$: $\mathsf{Tm}\,\Gamma\,(\mathsf{U}\,i) \to \mathsf{Ty}\,i\,\Gamma$

c : $\mathsf{Ty}\,i\,\Gamma \to \mathsf{Tm}\,\Gamma\,(\mathsf{U}\,i)$

$\mathsf{U}\beta$: $\underline{\mathsf{c}\,A} = A$

$\mathsf{U}\eta$: $\mathsf{c}\,\underline{a} = a$

$\mathsf{U}[]$: $(\mathsf{U}\,i)[\sigma] = (\mathsf{U}\,i)$

$\underline{[]}$: $\underline{a}[\sigma] = \underline{a[\sigma]}$

Bool : $\mathsf{Ty}\,0\,\Gamma$

true : $\mathsf{Tm}\,\Gamma\,\mathsf{Bool}$

false : $\mathsf{Tm}\,\Gamma\,\mathsf{Bool}$

if : $(C : \mathsf{Ty}\,i\,(\Gamma \triangleright \mathsf{Bool})) \to \mathsf{Tm}\,\Gamma\,(C[\mathsf{id}, \mathsf{true}]) \to \mathsf{Tm}\,\Gamma\,(C[\mathsf{id}, \mathsf{false}]) \to$
$(t : \mathsf{Tm}\,\Gamma\,\mathsf{Bool}) \to \mathsf{Tm}\,\Gamma\,(C[\mathsf{id}, t])$

$\mathsf{Bool}\beta_1$: $\mathsf{if}\,C\,u\,v\,\mathsf{true} = u$

$\mathsf{Bool}\beta_2$: $\mathsf{if}\,C\,u\,v\,\mathsf{false} = v$

$\mathsf{Bool}[]$: $\mathsf{Bool}[\sigma] = \mathsf{Bool}$

$\mathsf{true}[]$: $\mathsf{true}[\sigma] = \mathsf{true}$

$\mathsf{false}[]$: $\mathsf{false}[\sigma] = \mathsf{false}$

$\mathsf{if}[]$: $(\mathsf{if}\,C\,u\,v\,t)[\sigma] = \mathsf{if}\,(C[\sigma^{\uparrow}])\,(u[\sigma])\,(v[\sigma])\,(t[\sigma])$

Id : $(A : \mathsf{Ty}\,i\,\Gamma) \to \mathsf{Tm}\,\Gamma\,A \to \mathsf{Tm}\,\Gamma\,A \to \mathsf{Ty}\,i\,\Gamma$

refl : $(u : \mathsf{Tm}\,\Gamma\,A) \to \mathsf{Tm}\,\Gamma\,(\mathsf{Id}\,A\,u\,u)$

J : $\big(C : \mathsf{Ty}\,i\,(\Gamma \triangleright A \triangleright \mathsf{Id}\,(A[\mathsf{p}])\,(u[\mathsf{p}])\,0)\big) \to \mathsf{Tm}\,\Gamma\,(C[\mathsf{id}, u, \mathsf{refl}\,u]) \to$
$(e : \mathsf{Tm}\,\Gamma\,(\mathsf{Id}\,A\,u\,v)) \to \mathsf{Tm}\,\Gamma\,(C[\mathsf{id}, v, e[\mathsf{p}]])$

$\mathsf{Id}\beta$: $\mathsf{J}\,C\,w\,(\mathsf{refl}\,u) = w$

$\mathsf{Id}[]$: $(\mathsf{Id}\,A\,u\,v)[\sigma] = \mathsf{Id}\,(A[\sigma])\,(u[\sigma])\,(v[\sigma])$

$\mathsf{refl}[]$: $(\mathsf{refl}\,u)[\sigma] = \mathsf{refl}\,(u[\sigma])$

$\mathsf{J}[]$: $(\mathsf{J}\,C\,w\,e)[\sigma] = \mathsf{J}\,(C[\sigma^{\uparrow\uparrow}])\,(w[\sigma])\,(e[\sigma])$

Fig. 1. (*continued*)

Note that terms of Π-, Σ- and U-types are all characterized by natural isomorphisms, with substitution laws corresponding to naturality conditions. Hence, we only need to state naturality in one direction, and the other direction can be derived. For example, we only state the $\underline{[]}$ substitution rule, and the other law for substituting c can be derived.

Remark. It is important that we present the notion of signature in extensional type theory instead of in Agda. The reason is that many components in the signature are well-typed only up to previous equations in the signature, and hence would need to include transports in intensional settings. The simplest example for this is the $\triangleright\beta_2$ component with type $\mathsf{q}[\sigma, t] = t$. The left side of the equation has type $\mathsf{Tm}\,\Gamma\,(A[\mathsf{p}][\sigma, t])$, while the right side has type $\mathsf{Tm}\,\Gamma\,(A[\sigma])$, and the two types can be shown equal by $[\circ]$ and $\triangleright\beta_1$, so in intensional type theory we would need to transport one side.

Writing out the whole signature with explicit transports is difficult. The number of transports rapidly increases as later equations need to refer to transported previous types, and we may also need to introduce more transports just to rearrange previous transports over different equations. In fact, the current authors have not succeeded at writing out the type of the $\mathsf{J}[]$ substitution rule in intensional style. This illustrates the issue of explicit conversion derivations, which we previously explained in Sect. 1.1.

3 The Standard Model and Shallow Embedding

Previously, we described the notion of signature for the object theory, but as we remarked, merely writing down the signature in Agda is already impractical. Fortunately, we do not necessarily need the full intensional signature to be able to work with models of the object theory. The reason is that some equations can hold definitionally in specific models, thereby cutting down on the amount of transporting required. For example, if $[\circ]$ and $\triangleright\beta_1$ hold definitionally in a model, then the type of $\triangleright\beta_2$ need not include any transports.

The *standard model* of the object theory in Agda has the property that *all* of its equations hold definitionally. It was described previously by Altenkirch and Kaposi [3] similarly to the current presentation, although for a much smaller object theory.

Before presenting the model, we explain a departure from the signature described in Sect. 2.3. In the signature, we used natural numbers as universe levels, but in Agda, it is more convenient to use universe polymorphism and native universe levels instead. Hence, the types of the Con, Ty, Tm and Sub components become as follows:

```
Con : (i : Level) → Set (suc i)
Ty  : (j : Level) → Con i → Set (i ⊔ suc j)
Sub : Con i → Con j → Set (i ⊔ j)
Tm  : (Γ : Con i) → Ty j Γ → Set (i ⊔ j)
```

Instead of using level polymorphism, we could have used the types given in Fig. 1a together with an $\mathbb{N}$-indexed inductive-recursive universe hierarchy, which can be implemented inside Set_0 in Agda [19]. This choice would have added some boilerplate to the model. We choose now the more convenient version, but we note that the metatheory of universe polymorphism and universe polymorphic algebraic signatures should be investigated in future work.

3.1 The Standard Model

We present excerpts from the Agda formalisation, making some quantification implicit to improve readability. Let us first look at the interpretation of the type constructors of the object theory:

```
Con : (i : Level) → Set (suc i)
Con i = Set i

Ty : (j : Level) → Con i → Set (i ⊔ suc j)
Ty j Γ = Γ → Set j

Sub : Con i → Con j → Set (i ⊔ j)
Sub Γ Δ = Γ → Δ

Tm : (Γ : Con i) → Ty j Γ → Set (i ⊔ j)
Tm Γ A = (γ : Γ) → A γ
```

Contexts are interpreted as types, dependent types as type families, substitutions and terms as functions. Type and term substitution and substitution composition can be all implemented as (dependent) function composition.

```
_∘_ : Sub Θ Δ → Sub Γ Θ → Sub Γ Δ
σ ∘ δ = λ γ → σ (δ γ)

_[_] : Ty j Δ → Sub Γ Δ → Ty j Γ
A [ σ ] = λ γ → A (σ γ)

_[_] : Tm Δ A → (σ : Sub Γ Δ) → Tm Γ (A [ σ ])
t [ σ ] = λ γ → t (σ γ)
```

The empty context becomes the unit type, context extension and substitution extension are interpreted using the meta-level Σ-type.

```
• : Con zero
• = m.⊤

ε : Sub Γ •
ε = λ γ → m.tt

_▷_  : (Γ : Con i) → Ty j Γ → Con (i ⊔ j)
Γ ▷ A = m.Σ Γ A

_,_ : (σ : Sub Γ Δ) → Tm Γ (A [ σ ]) → Sub Γ (Δ ▷ A)
σ , t = λ γ → (σ γ m., t γ)

p : Sub (Γ ▷ A) Γ
p = m.fst

q : Tm (Γ ▷ A) (A [ p ])
q = m.snd
```

We interpret object-level universes with meta-level universes at the same level. Since Agda implements Russell-style universes, coding and decoding are trivial, and `Tm Γ (U j) ≡ Ty j Γ` holds definitionally in the model.

```
U : (j : Level) → Ty (suc j) Γ
U j = λ γ → Set j

El : Tm Γ (U j) → Ty j Γ
El a = a

c : Ty j Γ → Tm Γ (U j)
c A = A
```

For Π, Σ, Bool and Id, the interpretation likewise maps object-level constructions directly to their meta-level counterparts; see the formalisation [30] for details. We note here only the J[] component: its type and definition are trivial here thanks to the lack of transports. Below, `σ ↑ A` refers to the lifting of `σ : Sub Θ Γ` to `Sub (Θ ▷ A [ σ ]) (Γ ▷ A)`.

```
J[] : J C w t [ σ ]
    ≡ J (C [σ ↑ A ↑ Id (A [ p ]) (u [ p ]) q ]) (w [ σ ]) (t [ σ ])
J[] = m.refl
```

3.2 Shallow Embedding

Having access in Agda to the standard model of the object theory, we may now form expressions built out of model components, for example, we may define a polymorphic identity function as follows. Here, `v⁰` and `v¹` are shorthands for de Bruijn indices.

```
idfun : Tm • (Π (U zero) (Π (El v⁰) (El v¹)))
idfun = lam (lam v⁰)
```

The basic idea of shallow embedding is to view expressions such as `idfun` and its type, which are built from components of the standard model, as standing for expressions coming from an arbitrary model. This arbitrary model is often meant to be the syntax, but it does not necessarily have to be.

With `idfun`, we can enjoy the benefits of reflected equalities: we can write down `Π (El v⁰) (El v¹)` without transports, because the types of `vⁿ` de Bruijn indices compute by definition to `U zero` from `U zero [ pⁿ ]`.

A larger example for shallow embedding is presented in Sect. 7.2: there we prove canonicity by induction on the syntax, but represent the syntax shallowly, so we never have to prove anything about syntactic definitional equalities. Other examples are *syntactic models* [8]: this means that we build a model of an object theory from the syntax of another object theory. Every such model yields, by initiality of the syntax, a syntactic translation. We also present in Sect. 7.1 a formalisation of a syntactic parametricity translation in this style, using the same shallowly embedded theory for both the source and target syntaxes.

However, "pretending" that embedded expressions come from arbitrary models is only valid if we:

1. Do not construct more contexts, substitutions, terms or types than what are constructible in the syntax.
2. Do not prove more equations than what are provable about the syntax.

We will expand on the first concern in Sect. 6. With regards to the second concern, it would be addressed comprehensively with a proof that the standard model is *injective*. We define its statement as follows. Assume that we have a deeply embedded syntax for the object theory in Agda, with components named as `Con`, `Sub` and so on. By initiality of the syntax, there is a model morphism from the syntax to the standard model, which includes as components the following interpretation functions:

```
⟦_⟧ : Con i → Set i
⟦_⟧ : Ty j Γ → ⟦ Γ ⟧ → Set j
⟦_⟧ : Sub Γ Δ → ⟦ Γ ⟧ → ⟦ Δ ⟧
⟦_⟧ : Tm Γ A → (γ : ⟦ Γ ⟧) → ⟦ A ⟧ γ
```

Injectivity may refer to these functions; for example, injectivity on terms is stated as follows:

```
⟦⟧-injective : (t u : Tm Γ A) → ⟦ t ⟧ ≡ ⟦ u ⟧ → t ≡ u
```

However, we can show by reasoning external to Agda that injectivity of the standard model is not provable.

Theorem 1. *The injectivity of the standard model is not provable in Agda.*

Proof. We note that the object syntax includes functions which are definitionally inequal but equal extensionally, such as the following two functions:

```
f : Tm • (Π Bool Bool)
f = lam (if Bool true false v⁰)

g : Tm • (Π Bool Bool)
g = lam v⁰
```

If function extensionality is available in the metatheory, the ⟦ f ⟧ and ⟦ g ⟧ interpretations of these terms can be proven to be propositionally equal. Therefore, injectivity of the standard model and function extensionality are incompatible. But since we know that MLTT is consistent with function extensionality, it follows that injectivity of the standard model is not provable. □

This shows that the internal statement of injectivity is too strong. We weaken it by considering injectivity up to Agda's definitional equality. This requires us to step outside Agda and reason about its syntax.

3.3 An External View of the Standard Model

Let us consider some computation rules for the interpretation function of the standard model:

```
⟦ • ⟧         = m.⊤
⟦ Γ ▷ A ⟧     = m.Σ ⟦ Γ ⟧ ⟦ A ⟧
⟦ id ⟧        = λ γ → γ
⟦ σ ∘ δ ⟧     = λ γ → ⟦ σ ⟧ (⟦ δ ⟧ γ)
⟦ ε ⟧         = λ γ → m.tt
⟦ σ , t ⟧     = λ γ → (⟦ σ ⟧ γ m., ⟦ t ⟧ γ)
⟦ A [ σ ] ⟧ = λ γ → ⟦ A ⟧ (⟦ σ ⟧ γ)
⟦ t [ σ ] ⟧ = λ γ → ⟦ t ⟧ (⟦ σ ⟧ γ)
⟦ U j ⟧       = λ γ → Set j
...
```

If we consider the results of the interpretation function from the "outside", we see that interpreted object-theoretic terms evaluate to closed Agda terms. For example, if we have a context in the object theory:

```
Γ = • ▷ Bool ▷ Bool
```

Its ⟦ Γ ⟧ interpretation evaluates to the following closed Agda term (a left-nested Σ-type):

```
m.Σ (m.Σ m.⊤ (λ γ → m.Bool)) (λ γ → m.Bool)
```

Hence, externally, the interpretation function implements a syntactic translation which converts any object-theoretic construction to a closed Agda term. We model shallow embedding as this syntactic translation: whenever we write a shallowly embedded expression like `lam (if Bool true false v⁰)`, there is a corresponding expression in the object theory with the same shape, but in Agda this expression can be evaluated further by unfolding the definitions of the standard model.

In the next section we formalise this syntactic translation, and in Sect. 5 we additionally prove that it is injective. From this it follows that shallow embedding does not introduce new definitional equalities.

4 The Termification of a Model

For any given model $\mathcal{M} = (\mathsf{Con}, \mathsf{Ty}, \mathsf{Sub}, \mathsf{Tm}, \ldots)$ of the object type theory, we can construct a new model $\mathcal{T}^{\mathcal{M}} = (\mathsf{Con}_{\mathcal{T}}, \mathsf{Ty}_{\mathcal{T}}, \mathsf{Sub}_{\mathcal{T}}, \mathsf{Tm}_{\mathcal{T}}, \ldots)$. We call $\mathcal{T}^{\mathcal{M}}$ the *termification* of $\mathcal{M}$. The idea is that every context, type, substitution, and term can be regarded as a very specific term in the empty context; and all operations can be seen as operations on these terms.

If we take $\mathcal{M}$ to be the syntax, by initiality we get a morphism to $\mathcal{T}^{\mathcal{M}}$, which we use to model shallow embedding as a syntactic translation. Note that this translation formally goes *from the object theory to the object theory.* This means that we reuse the object theory to formalise the relevant syntactic fragment of Agda. This is a fairly strong simplifying assumption, which relies on Agda conforming to the CwF formulation of type theory. However, it is also necessary, because formalising the actual implementation of Agda is not feasible.

Although our main interest is the termification of the syntax, the construction works for arbitrary models, so we present it in this generality.

The four sorts of the new model $\mathcal{T}^{\mathcal{M}}$ are the following:

$$\begin{aligned}
\mathsf{Con}_{\mathcal{T}}\, i &:= \mathsf{Tm}\,\bullet\,(\mathsf{U}\, i)\\
\mathsf{Ty}_{\mathcal{T}}\, j\, \Gamma &:= \mathsf{Tm}\,\bullet\,(\underline{\Gamma} \Rightarrow (\mathsf{U}\, j))\\
\mathsf{Sub}_{\mathcal{T}}\, \Gamma\, \Delta &:= \mathsf{Tm}\,\bullet\,(\underline{\Gamma} \Rightarrow \underline{\Delta})\\
\mathsf{Tm}_{\mathcal{T}}\, \Gamma\, A &:= \mathsf{Tm}\,\bullet\,(\Pi\, \underline{\Gamma}\, \underline{\mathsf{app}\, A})
\end{aligned}$$

All contexts, types, substitutions, and terms of the new model $\mathcal{T}^{\mathcal{M}}$ are $\mathcal{M}$-terms in the empty $\mathcal{M}$-context. It is not hard to see that the definitions above type-check: for example, if we have $\Gamma : \mathsf{Con}_{\mathcal{T}}\, i$ and $A : \mathsf{Ty}_{\mathcal{T}}\, j\, \Gamma$, then by definition $\underline{\Gamma} : \mathsf{Ty}\, i\, \bullet$ and $\underline{\mathsf{app}\, A} : \mathsf{Ty}\, j\, (\bullet \rhd \underline{\Gamma})$, which means we can build $\Pi\, \underline{\Gamma}\, \underline{\mathsf{app}\, A}$ as in the definition of $\mathsf{Tm}_{\mathcal{T}}\, \Gamma\, A$.

The object theory, as shown in Figs. 1a and b, has 29 operators. In Fig. 2, we show how all 29 operators (together with the four sorts) of the model $\mathcal{T}^{\mathcal{M}}$ are constructed from components of $\mathcal{M}$. Finally, it is straightforward albeit tedious to check the 37 equalities that are required to hold. We have done the calculations both with pen and paper and in Agda. We do not give explicit paper proofs, but we refer to our formalisation instead: there, we state all equalities explicitly, and they are all proved using `m.refl`. This concludes the construction of the model $\mathcal{T}^{\mathcal{M}}$.

$$
\begin{array}{ll}
\mathsf{Con}_{\mathcal{T}}\, i & := \mathsf{Tm}\,\bullet\,(\mathsf{U}\, i) \\
\mathsf{Ty}_{\mathcal{T}}\, j\, \Gamma & := \mathsf{Tm}\,\bullet\,(\underline{\Gamma} \Rightarrow (\mathsf{U}\, j)) \\
\mathsf{Sub}_{\mathcal{T}}\, \Gamma\, \Delta & := \mathsf{Tm}\,\bullet\,(\underline{\Gamma} \Rightarrow \underline{\Delta}) \\
\mathsf{Tm}_{\mathcal{T}}\, \Gamma\, A & := \mathsf{Tm}\,\bullet\,(\Pi\, \underline{\Gamma}\, \underline{\mathsf{app}\, A}) \\
\mathsf{id}_{\mathcal{T}} & := \mathsf{lam}\, \mathsf{v}^0 \\
\sigma \circ_{\mathcal{T}} \delta & := \mathsf{lam}\, (\sigma[\epsilon]\, \$ (\delta[\epsilon]\, \$\, \mathsf{v}^0)) \\
A[\sigma]_{\mathcal{T}} & := \mathsf{lam}\, ((\mathsf{app}\, A)[\epsilon, \mathsf{app}\, (\sigma[\epsilon])]) \\
t[\sigma]_{\mathcal{T}} & := \mathsf{lam}\, ((\mathsf{app}\, t)[\epsilon, \mathsf{app}\, (\sigma[\epsilon])]) \\
\bullet_{\mathcal{T}} & := \mathsf{c}\, \top \\
\epsilon_{\mathcal{T}} & := \mathsf{lam}\, \mathsf{tt} \\
\Gamma \rhd_{\mathcal{T}} A & := \mathsf{c}\, (\Sigma\, \underline{\Gamma}\, \underline{\mathsf{app}\, A}) \\
\sigma ,_{\mathcal{T}} t & := \mathsf{lam}\, ((\mathsf{app}\, \sigma), (\mathsf{app}\, t)) \\
\mathsf{p}_{\mathcal{T}} & := \mathsf{lam}\, (\mathsf{fst}\, \mathsf{v}^0) \\
\mathsf{q}_{\mathcal{T}} & := \mathsf{lam}\, (\mathsf{snd}\, \mathsf{v}^0) \\
\Pi_{\mathcal{T}}\, A\, B & := \mathsf{lam}\, \big(\mathsf{c}\, (\Pi\, \underline{\mathsf{app}\, A}\, \underline{\mathsf{app}\, B[\epsilon, (\mathsf{v}^1, \mathsf{v}^0)]})\big) \\
\mathsf{lam}_{\mathcal{T}}\, t & := \mathsf{lam}\, (\mathsf{lam}\, (t[\epsilon]\, \$ (\mathsf{v}^1, \mathsf{v}^0))) \\
\mathsf{app}_{\mathcal{T}}\, t & := \mathsf{lam}\, (t[\epsilon]\, \$\, \mathsf{fst}\, \mathsf{v}^0\, \$\, \mathsf{snd}\, \mathsf{v}^0) \\
\Sigma_{\mathcal{T}}\, A\, B & := \mathsf{lam}\, \big(\mathsf{c}\, (\Sigma\, \underline{\mathsf{app}\, A}\, \underline{\mathsf{app}\, B[\epsilon, (\mathsf{v}^1, \mathsf{v}^0)]})\big) \\
u ,_{\mathcal{T}} v & := \mathsf{lam}\, (\mathsf{app}\, u, \mathsf{app}\, v) \\
\mathsf{fst}_{\mathcal{T}}\, t & := \mathsf{lam}\, (\mathsf{fst}\, (\mathsf{app}\, t)) \\
\mathsf{snd}_{\mathcal{T}}\, t & := \mathsf{lam}\, (\mathsf{snd}\, (\mathsf{app}\, t)) \\
\top_{\mathcal{T}} & := \mathsf{lam}\, (\mathsf{c}\, \top) \\
\mathsf{tt}_{\mathcal{T}} & := \mathsf{lam}\, \mathsf{tt} \\
\mathsf{U}_{\mathcal{T}} & := \mathsf{lam}\, (\mathsf{c}\, (\mathsf{U}\, i)) \\
\underline{a}_{\mathcal{T}} & := a \\
\mathsf{c}_{\mathcal{T}}\, A & := A \\
\mathsf{Bool}_{\mathcal{T}} & := \mathsf{lam}\, (\mathsf{c}\, \mathsf{Bool}) \\
\mathsf{true}_{\mathcal{T}} & := \mathsf{lam}\, \mathsf{true} \\
\mathsf{false}_{\mathcal{T}} & := \mathsf{lam}\, \mathsf{false} \\
\mathsf{if}_{\mathcal{T}}\, C\, u\, v\, t & := \mathsf{lam}\, \big(\mathsf{if}\, \underline{C[\epsilon]\, \$ (\mathsf{v}^1, \mathsf{v}^0)}\, (\mathsf{app}\, u)\, (\mathsf{app}\, v)\, (\mathsf{app}\, t)\big) \\
\mathsf{Id}_{\mathcal{T}}\, A\, u\, v & := \mathsf{lam}\, \big(\mathsf{c}\, \big(\mathsf{Id}\, \underline{\mathsf{app}\, A}\, (\mathsf{app}\, u)\, (\mathsf{app}\, v)\big)\big) \\
\mathsf{refl}_{\mathcal{T}}\, u & := \mathsf{lam}\, (\mathsf{refl}\, (\mathsf{app}\, u)) \\
\mathsf{J}_{\mathcal{T}}\, C\, w\, e & := \mathsf{lam}\, \big(\mathsf{J}\, \underline{C[\epsilon]\, \$ (\mathsf{v}^2, \mathsf{v}^1, \mathsf{v}^0)}\, (\mathsf{app}\, w)\, (\mathsf{app}\, e)\big)
\end{array}
$$

Fig. 2. The termification construction

5 The Injectivity Result

In this section, we show that we can shallowly embed the syntax without creating new definitional equalities.

If we apply the termification construction of Sect. 4 on the syntax Syn, we get a model $\mathcal{T}^{\mathsf{Syn}}$. Further, we have a morphism of models $[\![\cdot]\!] : \mathsf{Syn} \to \mathcal{T}^{\mathsf{Syn}}$ by the initiality of the syntax which maps $\bullet : \mathsf{Con}\,0$ to $[\![\bullet]\!] = \bullet_{\mathcal{T}}$, and which maps $\Gamma \triangleright A : \mathsf{Con}\,i$ to $[\![\Gamma \triangleright A]\!] = [\![\Gamma]\!] \triangleright_{\mathcal{T}} [\![A]\!]$, and so on.

An interesting property of the morphism $[\![\cdot]\!]$ is that it is *injective*. Before stating precisely what this means, we need the following definition:

Definition 1. *Given two contexts $\Gamma : \mathsf{Con}\,i$, $\Delta : \mathsf{Con}\,j$ in the object theory [or any model $\mathcal{M}$], we write $\Gamma \simeq \Delta$ for the type in the metatheory whose elements are quadruples $F = (F_1, F_2, F_{12}, F_{21})$ as follows: F_1 and F_2 are substitutions in the syntax [more generally, in $\mathcal{M}$] and F_{12}, F_{21} are equalities,*

$$\begin{aligned} F_1 &: \mathsf{Sub}\,\Gamma\,\Delta \\ F_2 &: \mathsf{Sub}\,\Delta\,\Gamma \\ F_{12} &: F_2 \circ F_1 = \mathsf{id}_\Gamma \\ F_{21} &: F_1 \circ F_2 = \mathsf{id}_\Delta. \end{aligned}$$

We call such a quadruple an isomorphism.

Theorem 2. *The morphism of models $[\![\cdot]\!] : \mathsf{Syn} \to \mathcal{T}^{\mathsf{Syn}}$ is injective, in the following sense:*

(T1) If $\Gamma : \mathsf{Con}\,i, \Delta : \mathsf{Con}\,j$ are contexts such that $[\![\Gamma]\!] = [\![\Delta]\!]$, then we have $\Gamma \simeq \Delta$.
(T2) If $A, B : \mathsf{Ty}\,i\,\Gamma$ are types such that $[\![A]\!] = [\![B]\!]$, then we have $A = B$.
(T3) If $\sigma, \tau : \mathsf{Sub}\,\Gamma\,\Delta$ are substitutions such that $[\![\sigma]\!] = [\![\tau]\!]$, then $\sigma = \tau$.
(T4) If $s, t : \mathsf{Tm}\,\Gamma\,A$ are terms such that $[\![s]\!] = [\![t]\!]$, then we have $s = t$.

Proof. We show the following metatheoretic statements:

(P1) For a context $\Gamma : \mathsf{Con}\,i$, we have an element $(\Gamma_1, \Gamma_2, \Gamma_{12}, \Gamma_{21})$ of

$$\Gamma \simeq \left(\bullet \triangleright \underline{[\![\Gamma]\!]}\right)$$

(P2) For a type $A : \mathsf{Ty}\,i\,\Gamma$, we have an equation

$$A_= \,:\, A \,=\, \underline{\mathsf{app}\,[\![A]\!]}[\Gamma_1]$$

(P3) For a substitution $\sigma : \mathsf{Sub}\,\Gamma\,\Delta$, we have an equation

$$\sigma_= \,:\, \sigma \,=\, \Delta_2 \circ (\epsilon, \mathsf{app}\,[\![\sigma]\!]) \circ \Gamma_1$$

(P4) For a term $t : \mathsf{Tm}\,\Gamma\,A$, we have an equation

$$t_{=} : t = (\mathsf{app}\,\llbracket t \rrbracket)[\Gamma_1]$$

Of course, the statement of the theorem follows easily from (P1)–(P4); for example, if we have $\llbracket s \rrbracket = \llbracket t \rrbracket$ as in (T4), we get $s = (\mathsf{app}\,\llbracket s \rrbracket)[\Gamma_1] = (\mathsf{app}\,\llbracket t \rrbracket)[\Gamma_1] = t$ from the above.

Before verifying (P1)–(P4), we can first convince ourselves that these expressions type-check in the extensional type theory which we use as metatheory. For (P1), this is clear. In (P2), the types are as follows:

$$\begin{array}{lll} & A & : \mathsf{Ty}\,i\,\Gamma \\ \textit{thus} & \llbracket A \rrbracket & : \mathsf{Tm}\,\bullet\,(\underline{\llbracket \Gamma \rrbracket} \Rightarrow \mathsf{U}\,i) \\ \textit{thus} & \mathsf{app}\,\llbracket A \rrbracket & : \mathsf{Tm}\,(\bullet \rhd \underline{\llbracket \Gamma \rrbracket})\,\mathsf{U}\,i \\ \textit{thus} & \underline{\mathsf{app}\,\llbracket A \rrbracket} & : \mathsf{Ty}\,i\,(\bullet \rhd \underline{\llbracket \Gamma \rrbracket}) \\ \textit{thus} & \underline{\mathsf{app}\,\llbracket A \rrbracket}[\Gamma_1] & : \mathsf{Ty}\,i\,\Gamma \end{array}$$

The case (P4) is almost identical to this, but needs to make use of (P2):

$$\begin{array}{lll} & t & : \mathsf{Tm}\,\Gamma\,A \\ \textit{thus} & \llbracket t \rrbracket & : \mathsf{Tm}\,\bullet\,(\Pi\,\underline{\llbracket \Gamma \rrbracket}\,\underline{\mathsf{app}\,\llbracket A \rrbracket}) \\ \textit{thus} & \mathsf{app}\,\llbracket t \rrbracket & : \mathsf{Tm}\,(\bullet \rhd \underline{\llbracket \Gamma \rrbracket})\,\underline{\mathsf{app}\,\llbracket A \rrbracket} \\ \textit{thus} & (\mathsf{app}\,\llbracket t \rrbracket)[\Gamma_1] & : \mathsf{Tm}\,\Gamma\,(\underline{\mathsf{app}\,\llbracket A \rrbracket}[\Gamma_1]) \\ \textit{by } A_{=} & (\mathsf{app}\,\llbracket t \rrbracket)[\Gamma_1] & : \mathsf{Tm}\,\Gamma\,A \end{array}$$

One checks similarly that (P3) type-checks.

We prove (P1)–(P4) by constructing a displayed model. As described in Sect. 2.3, this corresponds to "induction over the syntax".

To construct the displayed model, we need to cover the four sorts, 29 operators, and 37 equalities in Figs. 1a and b. The components for the four sorts are given by (P1)–(P4). Two of the 29 operators construct a context, namely $\bullet$ and $\rhd$; for these, we need to construct an isomorphism. For the remaining 27 operators, we need to prove an equality. The components for the 37 equalities are automatic: Since (P2)–(P4) are equalities, all equality components of the displayed model amount to equalities between equalities, which are trivial in our extensional metatheory. Note that none of the equalities in Figs. 1a and b are between contexts.

We start with the two operators that construct contexts. The case for the empty context is easy: we need to find $(\bullet_1, \bullet_2, \bullet_{12}, \bullet_{21})$ showing

$$\bullet \simeq (\bullet \rhd \underline{\llbracket \bullet \rrbracket})$$

This is simple:

$$
\begin{aligned}
\bullet_1 &: \ \mathsf{Sub}\,\bullet\,(\bullet \rhd \underline{[\![\bullet]\!]}) \\
\bullet_1 &:= (\epsilon, \mathsf{tt}) \\
\bullet_2 &: \ \mathsf{Sub}\,(\bullet \rhd \underline{[\![\bullet]\!]})\,\bullet \\
\bullet_2 &:= \epsilon
\end{aligned}
$$

The equality $\bullet_{12}$ follows from $\bullet\eta$, and the equality $\bullet_{21}$ follows from $\rhd\eta$ and $\top\eta$.

Next, we have the case $\Gamma \rhd A$, where we can already assume the property (P1) for Γ and (P2) for A. After unfolding the definition of $[\![\Gamma \rhd A]\!] = [\![\Gamma]\!] \rhd_{\mathcal{T}} [\![A]\!]$, we see that we have to construct an isomorphism

$$(\Gamma \rhd A) \simeq \left(\bullet \rhd \Sigma \, \underline{[\![\Gamma]\!]} \ \underline{\mathsf{app}\,[\![A]\!]}\right)$$

The two substitutions are:

$$
\begin{aligned}
(\Gamma \rhd A)_1 &: \ \mathsf{Sub}\,(\Gamma \rhd A)\,\left(\bullet \rhd \Sigma \, \underline{[\![\Gamma]\!]} \ \underline{\mathsf{app}\,[\![A]\!]}\right) \\
(\Gamma \rhd A)_1 &:= \left(\epsilon, (\mathsf{v}^0[\Gamma_1 \circ \mathsf{p}], \mathsf{v}^0)\right) \\
(\Gamma \rhd A)_2 &: \ \mathsf{Sub}\,\left(\bullet \rhd \Sigma \, \underline{[\![\Gamma]\!]} \ \underline{\mathsf{app}\,[\![A]\!]}\right)\,(\Gamma \rhd A) \\
(\Gamma \rhd A)_2 &:= \left(\Gamma_2 \circ (\epsilon, \mathsf{fst}\,\mathsf{v}^0), \mathsf{snd}\,\mathsf{v}^0\right)
\end{aligned}
$$

Quick calculations give us

$$
\begin{aligned}
& (\Gamma \rhd A)_1 \circ (\Gamma \rhd A)_2 \\
= \ & \left(\epsilon, (\mathsf{v}^0[\Gamma_1 \circ \mathsf{p}], \mathsf{v}^0)\right) \circ \left(\Gamma_2 \circ (\epsilon, \mathsf{fst}\,\mathsf{v}^0), \mathsf{snd}\,\mathsf{v}^0\right) \\
= \ & \left(\epsilon, (\mathsf{v}^0[\Gamma_1 \circ \Gamma_2 \circ (\epsilon, \mathsf{fst}\,\mathsf{v}^0)], \mathsf{snd}\,\mathsf{v}^0)\right) \\
= \ & \left(\epsilon, (\mathsf{fst}\,\mathsf{v}^0, \mathsf{snd}\,\mathsf{v}^0)\right) \\
= \ & \left(\epsilon, \mathsf{v}^0\right) \\
= \ & (\mathsf{p}, \mathsf{q}) \\
= \ & \mathsf{id}
\end{aligned}
$$

as well as

$$
\begin{aligned}
& (\Gamma \rhd A)_2 \circ (\Gamma \rhd A)_1 \\
= \ & \left(\Gamma_2 \circ (\epsilon, \mathsf{fst}\,\mathsf{v}^0), \mathsf{snd}\,\mathsf{v}^0\right) \circ \left(\epsilon, (\mathsf{v}^0[\Gamma_1 \circ \mathsf{p}], \mathsf{v}^0)\right) \\
= \ & \left(\Gamma_2 \circ (\epsilon, \mathsf{v}^0[\Gamma_1 \circ \mathsf{p}]), \mathsf{v}^0\right) \\
= \ & \left(\Gamma_2 \circ ((\mathsf{p}, \mathsf{q}) \circ (\Gamma_1 \circ \mathsf{p})), \mathsf{v}^0\right) \\
= \ & \left(\Gamma_2 \circ \Gamma_1 \circ \mathsf{p}, \mathsf{v}^0\right) \\
= \ & (\mathsf{p}, \mathsf{q}) \\
= \ & \mathsf{id}
\end{aligned}
$$

The first of the remaining 27 operations is the identity substitution $\mathsf{id} : \mathsf{Sub}\,\Gamma\,\Gamma$, where we can already assume property (P1) for Γ. We need to show

$$\mathsf{id}_= \;:\; \mathsf{id} \;=\; \Gamma_2 \circ (\epsilon, \mathsf{app}\,[\![\mathsf{id}]\!]) \circ \Gamma_1$$

We unfold $[\![\mathsf{id}]\!] = \mathsf{id}_{\mathcal{T}} = \mathsf{lam}\,\mathsf{v}^0$ and use $\Pi\eta$ to simplify the right-hand side of the equation to

$$\Gamma_2 \circ (\epsilon, \mathsf{v}^0) \circ \Gamma_1,$$

which by $\bullet\eta, \rhd\eta$ and Γ_{12} is equal to id as required.

The calculations for the remaining 26 operations are similar, Appendix A contains all of them in full detail. For completeness, the components discussed above are included in the figure as well. This completes the proof of the injectivity result. □

6 Wrapped Standard Model

In the previous section, we have shown that our specific version of shallow embedding does not introduce new definitional equalities. However, in practice we can only apply Theorem 2 if there actually exists an object-theoretic expression which is embedded, but there are many inhabitants in the standard model which do not arise as interpretations of object-theoretic expressions.

For example, contexts are interpreted as left-nested Σ-types, but since `Con i` is defined as `Set i` in the standard model, we can just inhabit `Con zero` with `m.Bool` or any small Agda type. This would be morally incorrect in a shallow embedding situation, since we might rely on properties that are not provable about the object syntax.

Additionally, even if we avoid extraneous inhabitants, some propositional equalities may be provable in the standard model, which are provable false in the syntax. In Proof 1 we gave such an example, where function extensionality yields additional equality proofs. In general, we want the freedom to assume function extensionality and other extensionality principles (e.g. for propositions or coinductive types) in the metatheory, so outlawing these principles in the metatheory is not acceptable as an enforcer of moral conduct.

Our proposed enforcement method is the following: wrap the interpretations of contexts, terms, substitutions and types in the standard model in unary record types, whose constructors are private and thus invisible to external modules. For contexts and types, the wrappers are as follows:

```
record Con' i : Set (suc i) where
  constructor mkC
  field
    |_|C : Set i

record Ty' (j : Level)(Γ : Con' i) : Set (i ⊔ suc j) where
  constructor mkT
  field
    |_|T : | Γ |C → Set j
```

We define Sub' and Tm' likewise, with mks, |_|s, mkt and |_|t, and put these four types in a module. In a different module, we define the "wrapped" standard model. The sorts in the model are defined using the wrapper types:

```
Con : (i : Level) → Set (suc i)
Con = Con'

Ty : (j : Level)(Γ : Con i) → Set (i ⊔ suc j)
Ty = Ty'

Sub : Con i → Con j → Set (i ⊔ j)
Sub = Sub'

Tm : (Γ : Con i) → Ty j Γ → Set (i ⊔ j)
Tm = Tm'
```

The rest of the model needs to be annotated with wrapping and unwrapping. Some examples for definitions, omitting type declarations for brevity:

```
id        = mks λ γ → γ
σ ∘ δ     = mks λ γ → | σ |s (| δ |s γ)
A [ σ ] = mkT λ γ → | A |T (| σ |s γ)
t [ σ ] = mkt λ γ → | t |t (| σ |s γ)
•         = mkC m.⊤
ε         = mks λ γ → m.tt
Γ ▷ A     = mkC (m.Σ | Γ |C | A |)
σ , t     = mks λ γ → (| σ |s γ m.,Σ | t |t γ)
p         = mks m.fst
q         = mkt m.snd
U j       = mkT λ γ → Set j
El a      = mkT | a |t
c A       = mkt | A |T
```

Importantly, the wrapped model still *supports all equations definitionally.* This is possible because the wrapper record types support η-equality, which expresses that `mkC | Γ |C` is definitionally equal to `Γ`, and likewise for the other wrappers. In short, unary records in Agda yield isomorphisms of types up to definitional equality.

The usage of the wrapped standard model for shallow embedding is simply as follows: we import the wrapped standard model, but do not import the module containing the wrapper types.

This way, there is no way to refer to the internals of the model. In fact, the only way to construct any inhabitants of the embedded syntax in this setup is to explicitly refer to the components of the wrapped model. For instance, `Con zero` cannot be anymore inhabited with `m.Bool`, since `m.Bool` has type `Set₀`, but we need a `Con' zero`, which we can only inhabit now using the empty context and context extension.

7 Case Studies

As a demonstration of using the shallowly embedded syntax, in this section we describe our formalisation of a syntactic parametricity translation and a canonicity proof for MLTT. These are formalised as displayed models over the syntax (that is, over the wrapped standard model described in Sect. 6).

7.1 Parametricity

Parametricity was introduced by Reynolds [41] in order to formalise the notion of representation independence. The unary version of his parametricity theorem states that terms preserve logical predicates: if a predicate holds for a semantic context, then it holds for the interpretation of the term at that context. Reynolds formulated parametricity as a model construction of System F. Bernardy et al. [6] noticed that type theory is powerful enough to express statements about its own parametricity and defined parametricity as a syntactic operation. This operation turns a context into a lifted context which has a witness of the logical predicate for each type in the original context. There is a projection from this lifted context back to the original context. A type A is turned into a predicate over A in the lifted context and a term is turned into a witness of the predicate for its type in the lifted context. We note that a more indexed version of this translation can be defined: This turns a context is into a type in the original context (that is, a predicate over the original context), a type into a predicate over the original context, a witness of the predicate for the original context and an element of the type. Substitutions and terms are turned into terms expressing preservation of the predicates. We define this indexed version of the translation in Agda.

The sorts are given as follows in our displayed model. We use `S.` prefixes to refer to the syntax, and use $\_^{s}$ superscripts on variables coming from the syntax.

```
Con : ∀ i → S.Con i → Set (suc i)
Con i Γˢ = S.Ty i Γˢ

Ty : ∀ i (Γ : Con j Γˢ) (Aˢ : S.Ty i Γˢ) → Set (suc i ⊔ j)
Ty i Γ Aˢ = S.Ty i (Γˢ S.▷ Γ S.▷ Aˢ S.[ S.p ])

Sub : ∀ (Γ : Con i Γˢ)(Δ : Con j Δˢ) → S.Sub Γˢ Δˢ → Set (i ⊔ j)
Sub Γ Δ σˢ = S.Tm (Γˢ S.▷ Γ) (Δ S.[ σˢ S.∘ S.p ])

Tm : ∀ (Γ : Con i Γˢ)(A : Ty j Γ Aˢ) → S.Tm Γˢ Aˢ → Set (i ⊔ j)
Tm Γ A tˢ = S.Tm (Γˢ S.▷ Γ) (A S.[ S.id S., tˢ S.[ S.p ] ])
```

A context over a syntactic context Γ^s is a syntactic type in Γ^s. A type over a syntactic type A^s is a syntactic type in the context Γ^s extended with two more components: Γ, that is the logical predicate for Γ^s and A^s itself (which has to be weakened using `S.p`). A substitution over σ^s is a term in context Γ^s `S.▷` Γ which has a type saying that the predicate Δ holds for σ^s. We have the analogous statement for terms. We refer to the formalisation [30] for the rest of the displayed model, it follows the original parametricity translation.

All equalities of the displayed model hold definitionally. Compared to a previous formalisation using a deep embedding [3], it is significantly shorter (322 vs. 1682 lines of code – we only counted the lines of code for the substitution calculus, Π and the universe because only these were treated in the previous formalisation). Note that although we implemented the displayed model, we did not implement the corresponding eliminator function which translates an S-term into its interpretation; we discuss such eliminators in Sect. 8.2.

7.2 Canonicity

Canoncity for type theory states that a term of type `Bool` in the empty context is equal to either `true` or `false`. Following [14,27] this can be proven by another logical predicate argument. We formalise this logical predicate as the following displayed model. We list the definitions for sorts and `Bool` for illustration.

```
Con : ∀ i → S.Con i → Set (suc i)
Con i Γˢ = S.Sub S.• Γˢ → Set i

Ty  : ∀ i (Γ : Con j Γˢ) (Aˢ : S.Ty i Γˢ) → Set (suc i ⊔ j)
Ty i Γ Aˢ = ∀ {ρˢ} → Γ ρˢ → S.Tm S.• (Aˢ S.[ ρˢ ]) → Set i
```

```
Sub : ∀ (Γ : Con i Γˢ)(Δ : Con j Δˢ) → S.Sub Γˢ Δˢ → Set (i ⊔ j)
Sub Γ Δ σˢ = ∀ {ρˢ} → Γ ρˢ → Δ (σˢ S.∘ ρˢ)

Tm : ∀ (Γ : Con i Γˢ)(A : Ty j Γ Aˢ) → S.Tm Γˢ Aˢ → Set (i ⊔ j)
Tm Γ A tˢ = ∀ {ρˢ}(ρ' : Γ ρˢ) → A ρ' (tˢ S.[ ρˢ ])

Bool : Ty zero Γ S.Bool
Bool ρ' tˢ = m.Σ m.Bool λ β → m.if _ S.true S.false β ≡ tˢ
```

A context over `Γˢ` is a proof-relevant predicate over closed substitutions into `Γˢ`. A type over `Aˢ` is a proof-relevant predicate over closed terms of type A where the type is substituted by a closed substitution for which the predicate holds. A substitution over `σˢ` is a function which says that if the predicate `Γ` holds for a closed substitution `ρˢ` then `Δ` holds for `σˢ` composed with `ρˢ`. A term over `tˢ` similarly states that if `Γ` holds for a `ρˢ`, then `A` holds for `tˢ S.[ ρˢ ]`.

The predicate `Bool` holds for a closed term `S.Bool` of type `S.Bool` if there is a metatheoretic boolean (`β : m.Bool`) which when converted to a syntactic boolean is equal to `tˢ`: in short, it holds if `tˢ` is either `S.true` or `S.false`. The equality is expressed as a metatheoretic equality `≡`, which we generally use for representing conversion for the object syntax.

The formalisation of canonicity consists of roughly 1000 lines of Agda code. However, out of this, 400 lines are automatically generated type signatures, which are of no mathematical interest, and are necessary only because of technical problems in Agda's inference of implicit parameters. These problems also prevented us from formalising the J[] component in the displayed model, but otherwise the formalisation is complete.

7.3 Termification and Injectivity

We also implemented termification (Sect. 4) in Agda as a model and it is also possible to implement the injectivity proof (Sect. 5) using the shallow embedding, without postulating an elimination principle of the shallow syntax (the Agda proof of injectivity is not yet completed). Injectivity is given by a displayed model over the syntax which contains both the termification model of the syntax and the (P1)–(P4) components of the injectivity proof as follows. We use `TS.` prefix to refer to components of the termified model for the syntax.

```
record Con i (Γˢ : S.Con i) : Set (suc i) where
  field
    ⟦_⟧ : TS.Con i
    _₁  : S.Sub Γˢ (S.• S.▷ S.El ⟦_⟧)
    _₂  : S.Sub (S.• S.▷ S.El ⟦_⟧) Γˢ
    _₁₂ : _₁ S.∘ _₂ ≡ S.id
    _₂₁ : _₂ S.∘ _₁ ≡ S.id
```

```
record Ty j (Γ : Con i Γˢ) (Aˢ : S.Ty j Γˢ) : Set (i ⊔ suc j) where
  field
    ⟦_⟧ : TS.Ty j ⟦ Γ ⟧
    _⁼  : Aˢ ≡ S.El (S.app ⟦_⟧ S.[ Γ ₁ ])

record Sub (Γ : Con i Γˢ)(Δ : Con j Δˢ)(σˢ : S.Sub Γˢ Δˢ) :
  Set (i ⊔ j) where
  field
    ⟦_⟧ : TS.Sub ⟦ Γ ⟧ ⟦ Δ ⟧
    _⁼  : σˢ ≡ (Δ ₂ S.∘ (S.p S., S.app ⟦_⟧)) S.∘ Γ ₁

record Tm (Γ : Con i Γˢ)(A : Ty j Γ Aˢ)(tˢ : S.Tm Γˢ Aˢ) :
  Set (i ⊔ j) where
  field
    ⟦_⟧ : TS.Tm ⟦ Γ ⟧ ⟦ A ⟧
    _⁼  : tˢ ≡ m.tr (S.Tm Γˢ) (A ⁼ m.⁻¹) (S.app ⟦_⟧ S.[ Γ ₁ ])
```

The ⟦_⟧ components are just the termification model while the rest of the record types implement (P1)–(P4). Compared to the proof presented in this paper using the extensional metatheory, in Agda the last equation contains an explicit transport `m.tr` over the equality proof `A ⁼`.

8 Discussion

8.1 Range of Embeddable Object Theories

So far, we focused on a particular object theory, which was described in Sect. 2.3 in detail. However, there is a rather wide range of object theories suitable for shallow embedding. There are some features which the object theory must possess. We discuss these in the following in an informal way.

First, object theories must support a "standard model" in the metatheory, which is injective in the external sense described in our paper. External injectivity is important: for example, for a large class of algebraic theories, *terminal models* exist (see e.g. [29]), where every type is interpreted as the unit type. The motivation of shallow embedding is to get more definitional equalities, but in terminal models we get too much of it, because all inhabitants are definitionally equal. Injectivity filters out dubious embeddings like terminal models.

The notion of standard model is itself informal. We may say that a standard model should interpret object-level constructions with essentially the same meta-level constructions. This is clearly the case when we model type theories in Agda which are essentially syntactic fragments of Agda. However, this should not be taken rigidly, as there might be externally injective shallow embeddings which do not fall into the standard case of embedding syntactic fragments. Thus far we have not investigated such theories; this could be a potential line of future work.

Some language-like theories, although widely studied, do not seem to support shallow embedding. For example, partial programming languages do not admit a standard `Set`-interpretation; they may have other models, but those are unlikely to support useful definitional equalities, when implemented in MLTT. However, a potential future proof assistant for synthetic domain theory [7] could support useful shallow embedding for partial languages. Likewise, variants of type theories such as cubical [13] or modal type theories could present further opportunities for shallow embeddings which are not available in MLTT.

On the other hand, undecidable definitional equality in the object theory does not necessarily preclude shallow embedding. For example, we could add equality reflection to the object theory considered in this paper, thereby making its definitional equality undecidable. Assuming `funext : (∀ x → f x ≡ g x) → f ≡ g`, we can interpret equality reflection as follows in the standard model:

```
reflect : (t u : Tm Γ A) → Tm Γ (Id A t u) → t ≡ u
reflect t u p = funext p
```

So, the standard model of an extensional object theory has one equation which is not definitional anymore: the interpretation of equality reflection. But we still get all the previous benefits from the other definitional equalities in the model.

Generally, if the equational theories on the object-level and the meta-level do not match exactly, shallow embedding is still usable.

If the metatheory has **too many** definitional equalities, then we can just modify the standard model in order to eliminate the extra equalities. For example, if the object theory does not have η for functions, we can introduce a wrapper type for functions, with η-equality turned off[1]:

```
record Π' {i}{j}(A : Set i)(B : A → Set j) : Set (i ⊔ j) where
  no-eta-equality
  constructor lam'
  field
    app' : ∀ x → B x
```

η can be still proven for `Π'` propositionally, however using the wrapping trick (Sect. 6) this equality won't be exported when using the syntax.

If the metatheory has **too few** definitional equalities, then shallow embedding might still be possible with some equations holding only propositionally. We saw such an example with the shallow embedding of equality reflection. However, if we can reflect some but not all equalities, that can be still very helpful in practical formalisations.

8.2 Recursors and Eliminators for the Embedded Syntax

Shallow embedding gave us a particular model with strict equalities. The question is: assuming that we only did morally correct constructions, is it consistent

[1] Or use an inductive type definition instead of a record.

to assume that the embedded syntax is really the syntax, i.e. it supports recursion and induction principles? For example, for our object theory, initiality (i.e. unique recursion) for the embedded syntax means that for any other model `M` containing `Conᴹ`, `Tyᴹ`, `Subᴹ` etc. components, there is a model morphism from the embedded syntax to `M` which includes the following functions:

```
⟦_⟧ : Con i → Conᴹ i
⟦_⟧ : Ty j Γ → Tyᴹ j ⟦ Γ ⟧
...
```

If "morally correct" means that all of our constructions can be in principle translated to constructions on deeply embedded syntax, then it is clearly consistent to rely on postulated initiality. We note here that the translation from shallow to deeply embedded syntax is an instance of translating from extensional type theory to intensional type theory [21,45], which introduces transports and invocations of function extensionality in order to make up for missing definitional equalities. However, in this paper we do not investigate moral correctness more formally.

If we do postulate initiality for the embedded syntax, we should be prepared that recursors and eliminators are unlikely to compute in any current proof assistant. In Agda, we attempted to use rewrite rules to make a postulated recursor compute on shallow syntax; this could be in principle possible, but the β-rules for the recursor seem to be illegal in Agda as rewrite rules. How great limitation the lack of computing recursion is? We argue that it is not as bad as it seems.

First, in the literature for semantics of type theory, it is rare that models of type theory make essential use of recursors of other models. The only example we know is in a previous work by two of the current authors and Altenkirch [29].

Second, many apparent uses of recursors in models are not essential, and can be avoided by reformulating models. We used such a technique in Sect. 7.3. Here we give a much simpler analogous example: writing a sorting function for lists of numbers, in two ways:

1. First, we write a sorting function, given by the recursor for a model of the theory of lists. Then, we prove by induction on lists that the function's output is really sorted. The latter step is given by a displayed model over the syntax of lists, which displayed model refers to the previous recursor.
2. We write a function which returns a Σ-type containing a list together with a proof that it is sorted.

In the latter case, we only use a single non-displayed model, and there is no need to refer to any recursor in the model.

8.3 Ergonomics

We consider here the experience of using shallowing embedding in proof assistants, in particular in Agda, where we have considerable experience as users of the technique. We focus on issues and annoyances, since the benefits of shallow embedding have been previously discussed.

Goal types and error messages are not the best, since they all talk about expressions in the wrapped standard model instead of the deeply embedded syntax. Hence, working with shallow embedding requires us to mentally translate between syntax and the standard model. It should be possible in principle to back-translate messages to deep syntax. In Agda, `DISPLAY` pragmas can be used to display expressions in user-defined way, but it seems too limited for our purpose.

Increased universe level of the embedded syntax. Let us assume an object type theory without a universe hierarchy. In this case the type of contexts can be given as `Con : Set₀` in an inductive `data` definition or a postulated quotient inductive definition. In contrast, the standard model *defines* `Con` as `Set`, hence `Con` has type `Set₁` in this case. In Agda, this increase in levels can cause additional boilerplate and usage of explicit level lifting. A way to remedy this is to define `Con` as a custom inductive-recursive universe, which can usually fit into `Set₀`, but in this case we get additional clutter in system messages arising from inductive-recursive decoding.

9 Conclusions

In this paper, we investigated the shallow embedding of a type theory into type theory. We motivated it as an effective technique to reflect definitional equalities of an object type theory. We showed that shallow embedding of a particular object theory is really an embedding, since it is injective in an external sense.

We do not suggest that shallow embedding can replace deep embedding in every use case. For example, when implementing a type checker or compiler, one has to use deep embeddings. We hope that future proof assistants will be robust and powerful enough to allow feasible direct formalisations and make shallow embeddings unnecessary.

A potential line of future work would be to try to use shallow embedding as presented here for other object theories and formalisations. Subjectively, shallow embedding made a huge difference when we formalised our case studies; a previous formalisation [3] of the parametricity translation took the current first author months to finish, while the current formalisation took less than a day, for a much larger object theory. Formalisations which were previously too tedious to undertake could be within reach now. Also, it could be explored in the future whether morally correct shallow embedding works for object theories which are not just syntactic fragments of the metatheory. For instance, structured categories other than CwFs, such as monoidal categories could be investigated for shallow embedding.

A The injectivity displayed model

We list the components of the displayed model for the injectivity proof described in Sect. 5. We don't write subscripts for metavariables and operators of the syntax, only for components of the displayed model ($_1$, $_2$, $_{12}$, $_{21}$ and $_=$).

$$
\begin{array}{ll}
\mathsf{Con}\, i\, \Gamma & := \Gamma \simeq (\bullet \rhd \underline{[\![\Gamma]\!]}) \\
\mathsf{Ty}\, j\, (\Gamma_1, \Gamma_2, \Gamma_{12}, \Gamma_{21})\, A & := A = \underline{\mathsf{app}\, [\![A]\!]}[\Gamma_1] \\
\mathsf{Sub}\, (\Gamma_1, \Gamma_2, \Gamma_{12}, \Gamma_{21})\, (\Delta_1, \Delta_2, \Delta_{12}, \Delta_{21})\, \sigma & := \sigma\ =\ \Delta_2 \circ (\epsilon, \mathsf{app}\, [\![\sigma]\!]) \circ \Gamma_1 \\
\mathsf{Tm}\, (\Gamma_1, \Gamma_2, \Gamma_{12}, \Gamma_{21})\, A_=\, t & := t\ =\ (\mathsf{app}\, [\![t]\!])[\Gamma_1]
\end{array}
$$

$$
\begin{array}{lll}
\mathsf{id}_= & : & \mathsf{id} = \\
& & \Gamma_2 \circ \Gamma_1 = \\
& & \Gamma_2 \circ (\epsilon, \mathsf{app}\, (\mathsf{lam}\, \mathsf{v}^0)) \circ \Gamma_1 = \\
& & \Gamma_2 \circ (\epsilon, \mathsf{app}\, [\![\mathsf{id}]\!]) \circ \Gamma_1 \\
\sigma_= \circ_= \delta_= & : & \sigma \circ \delta = \\
& & \Delta_2 \circ (\epsilon, \mathsf{app}\, [\![\sigma]\!]) \circ \Theta_1 \circ \Theta_2 \circ (\epsilon, \mathsf{app}\, [\![\delta]\!]) \circ \Gamma_1 = \\
& & \Delta_2 \circ (\epsilon, \mathsf{app}\, [\![\sigma]\!][\epsilon, \mathsf{app}\, [\![\delta]\!]]) \circ \Gamma_1 = \\
& & \Delta_2 \circ (\epsilon, ([\![\sigma]\!][\epsilon]\, \$([\![\delta]\!][\epsilon]\, \$\, \mathsf{v}^0))) \circ \Gamma_1 = \\
& & \Delta_2 \circ (\epsilon, \mathsf{app}\, [\![\sigma \circ \delta]\!]) \circ \Gamma_1 \\
A_=[\sigma_=]_= & : & A[\sigma] = \underline{\mathsf{app}[\![A]\!]}[\Delta_1][\Delta_2 \circ (\epsilon, \mathsf{app}\, [\![\sigma]\!]) \circ \Gamma_1] = \\
& & \underline{(\mathsf{app}\, [\![A]\!])[\epsilon, \mathsf{app}\, ([\![\sigma]\!][\epsilon])]}[\Gamma_1] = \underline{\mathsf{app}\, [\![A[\sigma]]\!]}[\Gamma_1] \\
t_=[\sigma_=]_= & : & t[\sigma] = (\mathsf{app}[\![t]\!])[\Delta_1][\Delta_2 \circ (\epsilon, \mathsf{app}\, [\![\sigma]\!]) \circ \Gamma_1] = \\
& & (\mathsf{app}\, [\![t]\!])[\epsilon, \mathsf{app}\, ([\![\sigma]\!][\epsilon])][\Gamma_1] = \mathsf{app}\, [\![t[\sigma]]\!][\Gamma_1] \\
\bullet_1 & := & (\epsilon, \mathsf{tt}) \\
\bullet_2 & := & \epsilon \\
\bullet_{12} & : & \bullet_1 \circ \bullet_2 = (\epsilon, \mathsf{tt}) \circ \epsilon = (\epsilon, \mathsf{tt}) = (\mathsf{p}, \mathsf{q}) = \mathsf{id} \\
\bullet_{21} & : & \bullet_2 \circ \bullet_1 = \epsilon \circ (\epsilon, \mathsf{tt}) = \epsilon = \mathsf{id} \\
\epsilon_= & : & \epsilon = \epsilon \circ \cdots = \bullet_2 \circ (\epsilon, \mathsf{app}\, [\![\sigma]\!]) \circ \Gamma_1 \\
(\Gamma_1, \dots) \rhd_1 A_= & := & (\epsilon, (\mathsf{v}^0[\Gamma_1 \circ \mathsf{p}], \mathsf{v}^0)) \\
(\Gamma_1, \Gamma_2, \dots) \rhd_2 A_= & := & (\Gamma_2 \circ (\epsilon, \mathsf{fst}\, \mathsf{v}^0), \mathsf{snd}\, \mathsf{v}^0)
\end{array}
$$

$$
\begin{aligned}
(\Gamma_1, \Gamma_2, \dots) \rhd_{12} A_= :\ & (\Gamma_1, \Gamma_2, \dots) \rhd_1 A_= \circ (\Gamma_1, \Gamma_2, \dots) \rhd_2 A_= \\
& (\epsilon, (\mathsf{v}^0[\Gamma_1 \circ \mathsf{p}], \mathsf{v}^0)) \circ (\Gamma_2 \circ (\epsilon, \mathsf{fst}\, \mathsf{v}^0), \mathsf{snd}\, \mathsf{v}^0) = \\
& (\epsilon, (\mathsf{v}^0[\Gamma_1 \circ \Gamma_2 \circ (\epsilon, \mathsf{fst}\, \mathsf{v}^0)], \mathsf{snd}\, \mathsf{v}^0)) = \\
& (\epsilon, (\mathsf{fst}\, \mathsf{v}^0, \mathsf{snd}\, \mathsf{v}^0)) = \\
& (\epsilon, \mathsf{v}^0) = \\
& (\mathsf{p}, \mathsf{q}) = \\
& \mathsf{id} \\
(\Gamma_1, \Gamma_2, \dots) \rhd_{21} A_= :\ & (\Gamma_1, \Gamma_2, \dots) \rhd_2 A_= \circ (\Gamma_1, \Gamma_2, \dots) \rhd_1 A_= \\
& (\Gamma_2 \circ (\epsilon, \mathsf{fst}\, \mathsf{v}^0), \mathsf{snd}\, \mathsf{v}^0) \circ (\epsilon, (\mathsf{v}^0[\Gamma_1 \circ \mathsf{p}], \mathsf{v}^0)) = \\
& (\Gamma_2 \circ (\epsilon, \mathsf{v}^0[\Gamma_1 \circ \mathsf{p}]), \mathsf{v}^0) = \\
& (\Gamma_2 \circ \Gamma_1 \circ \mathsf{p}, \mathsf{v}^0) = \\
& (\mathsf{p}, \mathsf{q}) = \\
& \mathsf{id} \\
\sigma_=,_= t_= \quad :\ & (\sigma, t) = \\
& (\Delta_2 \circ (\epsilon, \mathsf{app}\, [\![\sigma]\!]) \circ \Gamma_1, \mathsf{app}\, [\![t]\!][\Gamma_1]) = \\
& (\Delta_2 \circ (\epsilon, \mathsf{fst}\, \mathsf{v}^0), \mathsf{snd}\, \mathsf{v}^0) \circ (\epsilon, (\mathsf{app}\, [\![\sigma]\!], \mathsf{app}\, [\![t]\!])) \circ \Gamma_1 = \\
& (\Delta_1, \dots) \rhd_2 A_= \circ (\epsilon, \mathsf{app}\, [\![\sigma, t]\!]) \circ \Gamma_1 \\
\mathsf{p}_= \quad :\ & \mathsf{p} = \\
& \Gamma_2 \circ \Gamma_1 = \\
& \Gamma_2 \circ (\epsilon, \mathsf{fst}\, \mathsf{v}^0) \circ (\epsilon, (\mathsf{v}^0[\Gamma_1 \circ \mathsf{p}], \mathsf{v}^0)) = \\
& \Gamma_2 \circ (\epsilon, \mathsf{app}\, [\![\mathsf{p}]\!]) \circ (\Gamma_1, \dots) \rhd_1 A_= \\
\mathsf{q}_= \quad :\ & \mathsf{q} = \mathsf{v}^0 = \\
& \mathsf{lam}(\mathsf{snd}\, \mathsf{v}^0) = \\
& (\mathsf{snd}\, \mathsf{v}^0)[\epsilon, (\mathsf{v}^0[\Gamma_1 \circ \mathsf{p}], \mathsf{v}^0)] = \\
& \mathsf{app}\, [\![\mathsf{q}]\!][(\Gamma_1, \dots) \rhd_1 A_=] \\
\Pi_= A_= B_= \quad :\ & \Pi\, A\, B = \\
& \Pi\, \underline{\mathsf{app}\, [\![A]\!][\Gamma_1]}\ \underline{\mathsf{app}\, [\![B]\!][(\Gamma_1, \dots) \rhd_1 A_=]} = \\
& \Pi\, \underline{\mathsf{app}\, [\![A]\!][\Gamma_1]}\ \underline{\mathsf{app}\, [\![B]\!][\epsilon, (\mathsf{v}^1, \mathsf{v}^0)][\Gamma_1{}^{\uparrow}]} = \\
& \underline{\mathsf{app}\, [\![\Pi\, A\, B]\!]}[\Gamma_1]
\end{aligned}
$$

$$
\begin{aligned}
\mathsf{lam}_= \, t_= \quad &: \; \mathsf{lam}\, t = \\
&\quad \mathsf{lam}\,(\mathsf{app}\, [\![t]\!][(\Gamma_1, \dots) \rhd_1 A_=]) = \\
&\quad \mathsf{lam}\,(\mathsf{app}\, [\![t]\!][\epsilon, (\mathsf{v}^1, \mathsf{v}^0)][{\Gamma_1}^{\uparrow}]) = \\
&\quad \mathsf{lam}\,(\mathsf{app}\, [\![t]\!][\epsilon, (\mathsf{v}^1, \mathsf{v}^0)])[\Gamma_1] = \\
&\quad \mathsf{app}\,(\mathsf{lam}\,(\mathsf{lam}\,([\![t]\!][\epsilon]\,\$(\mathsf{v}^1, \mathsf{v}^0))))[\Gamma_1] = \\
&\quad \mathsf{app}\, [\![\mathsf{lam}\, t]\!][\Gamma_1] \\
\mathsf{app}_= \, t_= \quad &: \; \mathsf{app}\, t = \\
&\quad \mathsf{app}\,(\mathsf{app}\, [\![t]\!][\Gamma_1]) = \\
&\quad \mathsf{app}\,(\mathsf{app}\, [\![t]\!])[{\Gamma_1}^{\uparrow}] = \\
&\quad \mathsf{app}\,(\mathsf{app}\, [\![t]\!])[\epsilon, \mathsf{v}^1, \mathsf{v}^0][{\Gamma_1}^{\uparrow}] = \\
&\quad \mathsf{app}\,(\mathsf{app}\, [\![t]\!])[\epsilon, \mathsf{v}^0[\Gamma_1 \circ \mathsf{p}], \mathsf{v}^0] = \\
&\quad \mathsf{app}\,(\mathsf{app}\, [\![t]\!])[\epsilon, \mathsf{fst}\, \mathsf{v}^0, \mathsf{snd}\, \mathsf{v}^0][\epsilon, (\mathsf{v}^0[\Gamma_1 \circ \mathsf{p}], \mathsf{v}^0)] = \\
&\quad \mathsf{app}\,(\mathsf{app}\, [\![t]\!])[\epsilon, \mathsf{fst}\, \mathsf{v}^1, \mathsf{v}^0][\mathsf{id}, \mathsf{snd}\, \mathsf{v}^0][\epsilon, (\mathsf{v}^0[\Gamma_1 \circ \mathsf{p}], \mathsf{v}^0)] = \\
&\quad \mathsf{app}\,(\mathsf{app}\, [\![t]\!][\epsilon, \mathsf{fst}\, \mathsf{v}^0])[\mathsf{id}, \mathsf{snd}\, \mathsf{v}^0][(\Gamma_1, \dots) \rhd_1 A_=] = \\
&\quad ([\![t]\!][\epsilon]\,\$\,\mathsf{fst}\, \mathsf{v}^0\,\$\,\mathsf{snd}\, \mathsf{v}^0)[(\Gamma_1, \dots) \rhd_1 A_=] = \\
&\quad \mathsf{app}\,(\mathsf{lam}\,([\![t]\!][\epsilon]\,\$\,\mathsf{fst}\, \mathsf{v}^0\,\$\,\mathsf{snd}\, \mathsf{v}^0))[(\Gamma_1, \dots) \rhd_1 A_=] = \\
&\quad \mathsf{app}\, [\![\mathsf{app}\, t]\!][(\Gamma_1, \dots) \rhd_1 A_=] \\
\Sigma_= \, A_= \, B_= \quad &: \; \Sigma\, A\, B = \\
&\quad \Sigma\, \underline{\mathsf{app}\, [\![A]\!][\Gamma_1]}\; \underline{\mathsf{app}\, [\![B]\!][(\Gamma_1, \dots) \rhd_1 A_=]} = \\
&\quad \Sigma\, \underline{\mathsf{app}\, [\![A]\!][\Gamma_1]}\; \underline{\mathsf{app}\, [\![B]\!][\epsilon, (\mathsf{v}^1, \mathsf{v}^0)][{\Gamma_1}^{\uparrow}]} = \\
&\quad \underline{\mathsf{app}\, [\![\Sigma\, A\, B]\!]}[\Gamma_1] \\
u_= ,_= v_= \quad &: \; (u, v) = \\
&\quad (\mathsf{app}\, [\![u]\!][\Gamma_1], \mathsf{app}\, [\![v]\!][\Gamma_1]) = \\
&\quad (\mathsf{app}\, [\![u]\!], \mathsf{app}\, [\![v]\!])[\Gamma_1] = \\
&\quad \mathsf{app}\, [\![u, v]\!][\Gamma_1] \\
\mathsf{fst}_= \, t_= \quad &: \; \mathsf{fst}\, t = \\
&\quad \mathsf{fst}\,(\mathsf{app}\, [\![t]\!][\Gamma_1]) = \\
&\quad (\mathsf{fst}\,(\mathsf{app}\, [\![t]\!]))[\Gamma_1] = \\
&\quad \mathsf{app}\, [\![\mathsf{fst}\, t]\!][\Gamma_1] \\
\mathsf{snd}_= \, t_= \quad &: \; \mathsf{snd}\, t = \\
&\quad \mathsf{snd}\,(\mathsf{app}\, [\![t]\!][\Gamma_1]) = \\
&\quad (\mathsf{snd}\,(\mathsf{app}\, [\![t]\!]))[\Gamma_1] = \\
&\quad \mathsf{app}\, [\![\mathsf{snd}\, t]\!][\Gamma_1]
\end{aligned}
$$

$$
\begin{array}{lll}
\top_= & : & \top = \top[\Gamma_1] = \underline{\mathsf{app}\,(\mathsf{lam}\,(\mathsf{c}\,\top))[\Gamma_1]} = \underline{\mathsf{app}\,[\![\top]\!]}[\Gamma_1] \\
\mathsf{tt}_= & : & \mathsf{tt} = \mathsf{tt}[\Gamma_1] = \mathsf{app}\,(\mathsf{lam}\,\mathsf{tt})[\Gamma_1] = \mathsf{app}\,[\![\mathsf{tt}]\!][\Gamma_1] \\
\mathsf{U}_{=1} & : & \mathsf{U}\,i = \mathsf{U}\,i[\Gamma_1] = \underline{\mathsf{app}\,(\mathsf{lam}\,(\mathsf{c}\,(\mathsf{U}\,i)))}[\Gamma_1] = \underline{\mathsf{app}\,[\![\mathsf{U}\,i]\!]}[\Gamma_1] \\
\underline{a_{=}}_= & : & \underline{a} = \underline{\mathsf{app}\,[\![a]\!][\Gamma_1]} = \underline{\mathsf{app}\,[\![\underline{a}]\!]}[\Gamma_1] \\
\mathsf{c}_=\,A_= & : & A = \mathsf{app}\,[\![A]\!][\Gamma_1] = \mathsf{app}\,[\![\mathsf{c}\,A]\!][\Gamma_1] \\
\mathsf{Bool}_= & : & \mathsf{Bool} = \underline{\mathsf{c}\,\mathsf{Bool}}[\Gamma_1] = \underline{\mathsf{app}\,[\![\mathsf{Bool}]\!]}[\Gamma_1] \\
\mathsf{true}_= & : & \mathsf{true} = \mathsf{true}[\Gamma_1] = \mathsf{app}\,[\![\mathsf{true}]\!][\Gamma_1] \\
\mathsf{false}_= & : & \mathsf{false} = \mathsf{false}[\Gamma_1] = \mathsf{app}\,[\![\mathsf{false}]\!][\Gamma_1] \\
\mathsf{if}_=\,C_=\,u_=\,v_=\,t_= & : & \mathsf{if}\,C\,u\,v\,t = \\
 & & \mathsf{if}\,\underline{\mathsf{app}\,[\![C]\!][(\Gamma \rhd \mathsf{Bool})_1]}\,(\mathsf{app}\,[\![u]\!][\Gamma_1])\,(\mathsf{app}\,[\![v]\!][\Gamma_1]) \\
 & & \quad (\mathsf{app}\,[\![t]\!][\Gamma_1]) = \\
 & & \mathsf{if}\,\underline{\mathsf{app}\,[\![C]\!][\epsilon,(\mathsf{v}^1,\mathsf{v}^0)][{\Gamma_1}^{\uparrow}]}\,(\mathsf{app}\,[\![u]\!][\Gamma_1])\,(\mathsf{app}\,[\![v]\!][\Gamma_1]) \\
 & & \quad (\mathsf{app}\,[\![t]\!][\Gamma_1]) = \\
 & & \mathsf{if}\,\underline{[\![C]\!][\epsilon]\,\$(\mathsf{v}^1,\mathsf{v}^0)}\,(\mathsf{app}\,[\![u]\!])\,(\mathsf{app}\,[\![v]\!])\,(\mathsf{app}\,[\![t]\!])[\Gamma_1] = \\
 & & \mathsf{app}\,[\![\mathsf{if}\,C\,u\,v\,t]\!][\Gamma_1] \\
\mathsf{Id}_=\,A_=\,u_=\,v_= & : & \mathsf{Id}\,A\,u\,v = \\
 & & \mathsf{Id}\,\underline{\mathsf{app}\,[\![A]\!][\Gamma_1]}\,(\mathsf{app}\,[\![u]\!][\Gamma_1])\,(\mathsf{app}\,[\![v]\!][\Gamma_1]) \\
 & & \big(\mathsf{Id}\,\underline{\mathsf{app}\,[\![A]\!]}\,(\mathsf{app}\,[\![u]\!])\,(\mathsf{app}\,[\![v]\!])\big)\,[\Gamma_1] \\
 & & \underline{\mathsf{app}\,[\![\mathsf{Id}\,A\,u\,v]\!]}[\Gamma_1] \\
\mathsf{refl}_=\,u_= & : & \mathsf{refl}\,u = \\
 & & \mathsf{refl}\,(\mathsf{app}\,[\![u]\!][\Gamma_1]) = \\
 & & \mathsf{refl}\,(\mathsf{app}\,[\![u]\!])[\Gamma_1] = \\
 & & \mathsf{app}\,[\![\mathsf{refl}\,u]\!][\Gamma_1] \\
\mathsf{J}_=\,C_=\,w_=\,e_= & : & \mathsf{J}\,C\,w\,e = \\
 & & \mathsf{J}\,\underline{\mathsf{app}\,[\![C]\!][(\Gamma \rhd A \rhd \dots)_1]}\,(\mathsf{app}\,[\![w]\!][\Gamma_1])\,(\mathsf{app}\,[\![e]\!][\Gamma_1]) = \\
 & & \mathsf{J}\,\underline{\mathsf{app}\,[\![C]\!][\epsilon,(\mathsf{v}^2,\mathsf{v}^1,\mathsf{v}^0)][{\Gamma_1}^{\uparrow\uparrow}]}\,(\mathsf{app}\,[\![w]\!][\Gamma_1])\,(\mathsf{app}\,[\![e]\!][\Gamma_1]) = \\
 & & \big(\mathsf{J}\,\underline{\mathsf{app}\,[\![C]\!][\epsilon,(\mathsf{v}^2,\mathsf{v}^1,\mathsf{v}^0)]}\,(\mathsf{app}\,[\![w]\!])\,(\mathsf{app}\,[\![e]\!])\big)\,[\Gamma_1] = \\
 & & \mathsf{app}\,[\![\mathsf{J}\,C\,w\,e]\!][\Gamma_1]
\end{array}
$$

References

1. Abel, A., Öhman, J., Vezzosi, A.: Decidability of conversion for type theory in type theory. Proc. ACM Program. Lang. **2**(POPL), 23 (2017)
2. Altenkirch, T., Capriotti, P., Dijkstra, G., Kraus, N., Nordvall Forsberg, F.: Quotient inductive-inductive types. In: Baier, C., Dal Lago, U. (eds.) FoSSaCS 2018. LNCS, vol. 10803, pp. 293–310. Springer, Cham (2018). https://doi.org/10.1007/978-3-319-89366-2_16

3. Altenkirch, T., Kaposi, A.: Type theory in type theory using quotient inductive types. In: Bodik, R., Majumdar, R. (eds.) Proceedings of the 43rd Annual ACM SIGPLAN-SIGACT Symposium on Principles of Programming Languages, POPL 2016, 20–22 January 2016, St. Petersburg, FL, USA, pp. 18–29. ACM (2016). https://doi.org/10.1145/2837614.2837638
4. Altenkirch, T., Kaposi, A.: Normalisation by evaluation for type theory, in type theory. Logical Methods Comput. Sci. **13**(4) (2017). https://doi.org/10.23638/LMCS-13(4:1)2017
5. Anand, A., Boulier, S., Cohen, C., Sozeau, M., Tabareau, N.: Towards certified meta-programming with typed TEMPLATE-COQ. In: Avigad, J., Mahboubi, A. (eds.) ITP 2018. LNCS, vol. 10895, pp. 20–39. Springer, Cham (2018). https://doi.org/10.1007/978-3-319-94821-8_2
6. Bernardy, J.P., Jansson, P., Paterson, R.: Proofs for free – parametricity for dependent types. J. Funct. Program. **22**(02), 107–152 (2012). https://doi.org/10.1017/S0956796812000056
7. Birkedal, L., Mogelberg, R.E., Schwinghammer, J., Stovring, K.: First steps in synthetic guarded domain theory: step-indexing in the topos of trees. In: 2011 IEEE 26th Annual Symposium on Logic in Computer Science, pp. 55–64. IEEE (2011)
8. Boulier, S., Pédrot, P.M., Tabareau, N.: The next 700 syntactical models of type theory. In: Proceedings of the 6th ACM SIGPLAN Conference on Certified Programs and Proofs, CPP 2017, pp. 182–194. ACM, New York (2017). https://doi.org/10.1145/3018610.3018620
9. Brady, E.: Idris, a general-purpose dependently typed programming language: design and implementation. J. Funct. Program. **23**(5), 552–593 (2013)
10. Chapman, J.: Type theory should eat itself. Electron. Notes Theor. Comput. Sci. **228**, 21–36 (2009). https://doi.org/10.1016/j.entcs.2008.12.114
11. Chlipala, A.: Parametric higher-order abstract syntax for mechanized semantics. In: Proceedings of the 13th ACM SIGPLAN International Conference on Functional Programming, ICFP 2008, pp. 143–156. ACM, New York (2008). https://doi.org/10.1145/1411204.1411226
12. Cockx, J., Abel, A.: Sprinkles of extensionality for your vanilla type theory. In: TYPES 2016 (2016)
13. Cohen, C., Coquand, T., Huber, S., Mörtberg, A.: Cubical type theory: a constructive interpretation of the univalence axiom, December 2015
14. Coquand, T.: Canonicity and normalisation for dependent type theory. CoRR (2018). http://arxiv.org/abs/1810.09367
15. Coquand, T., Huber, S., Sattler, C.: Homotopy canonicity for cubical type theory. In: Geuvers, H. (ed.) Proceedings of the 4th International Conference on Formal Structures for Computation and Deduction (FSCD 2019) (2019)
16. Danielsson, N.A.: A formalisation of a dependently typed language as an inductive-recursive family. In: Altenkirch, T., McBride, C. (eds.) TYPES 2006. LNCS, vol. 4502, pp. 93–109. Springer, Heidelberg (2007). https://doi.org/10.1007/978-3-540-74464-1_7
17. Despeyroux, J., Felty, A., Hirschowitz, A.: Higher-Order Abstract Syntax in Coq. Technical Report RR-2556, INRIA, May 1995. https://hal.inria.fr/inria-00074124
18. Devriese, D., Piessens, F.: Typed syntactic meta-programming. In: Proceedings of the 2013 ACM SIGPLAN International Conference on Functional Programming (ICFP 2013). pp. 73–85. ACM, September 2013. https://doi.org/10.1145/2500365.2500575

19. Diehl, L.: Fully Generic Programming over Closed Universes of Inductive-Recursive Types. Ph.D. thesis, Portland State University (2017)
20. Dybjer, P.: Internal type theory. In: Berardi, S., Coppo, M. (eds.) TYPES 1995. LNCS, vol. 1158, pp. 120–134. Springer, Heidelberg (1996). https://doi.org/10.1007/3-540-61780-9_66
21. Hofmann, M.: Extensional concepts in intensional type theory. Thesis, University of Edinburgh, Department of Computer Science (1995)
22. Hofmann, M.: Syntax and semantics of dependent types. In: Semantics and Logics of Computation, pp. 79–130. Cambridge University Press (1997)
23. Hofmann, M.: Semantical analysis of higher-order abstract syntax. In: Proceedings of the 14th Annual IEEE Symposium on Logic in Computer Science, LICS 1999, p. 204. IEEE Computer Society, Washington (1999). http://dl.acm.org/citation.cfm?id=788021.788940
24. Hou (Favonia), K.B., Finster, E., Licata, D.R., Lumsdaine, P.L.: A mechanization of the Blakers-Massey connectivity theorem in homotopy type theory. In: Proceedings of the 31st Annual ACM/IEEE Symposium on Logic in Computer Science, LICS 2016, pp. 565–574. ACM, New York (2016). https://doi.org/10.1145/2933575.2934545
25. Huber, S.: Cubical Interpretations of Type Theory. Ph.D. thesis, University of Gothenburg (2016)
26. Jaber, G., Lewertowski, G., Pédrot, P.M., Sozeau, M., Tabareau, N.: The definitional side of the forcing. In: Logics in Computer Science, New York, United States, May 2016. https://doi.org/10.1145/2933575.2935320
27. Kaposi, A., Huber, S., Sattler, C.: Gluing for type theory. In: Geuvers, H. (ed.) Proceedings of the 4th International Conference on Formal Structures for Computation and Deduction (FSCD 2019) (2019)
28. Kaposi, A., Kovács, A.: A syntax for higher inductive-inductive types. In: Kirchner, H. (ed.) 3rd International Conference on Formal Structures for Computation and Deduction (FSCD 2018). Leibniz International Proceedings in Informatics (LIPIcs), vol. 108, pp. 20:1–20:18. Schloss Dagstuhl-Leibniz-Zentrum fuer Informatik, Dagstuhl, Germany (2018). https://doi.org/10.4230/LIPIcs.FSCD.2018.20
29. Kaposi, A., Kovács, A., Altenkirch, T.: Constructing quotient inductive-inductive types. Proc. ACM Program. Lang. **3**(POPL), 2 (2019)
30. Kaposi, A., Kovács, A., Kraus, N.: Formalisations in Agda using a morally correct shallow embedding, May 2019. https://bitbucket.org/akaposi/shallow/src/master/
31. Licata, D.: Running circles around (in) your proof assistant; or, quotients that compute (2011). http://homotopytypetheory.org/2011/04/23/running-circles-around-in-your-proof-assistant/
32. Martin-Löf, P.: An intuitionistic theory of types: predicative part. In: Rose, H., Shepherdson, J. (eds.) Logic Colloquium '73, Proceedings of the Logic Colloquium, Studies in Logic and the Foundations of Mathematics, North-Holland, vol. 80, pp. 73–118 (1975)
33. The Coq development team: The Coq proof assistant reference manual. LogiCal Project (2019). http://coq.inria.fr. version 8.9
34. McBride, C.: Outrageous but meaningful coincidences: dependent type-safe syntax and evaluation. In: Oliveira, B.C.d.S., Zalewski, M. (eds.) Proceedings of the ACM SIGPLAN Workshop on Generic Programming, pp. 1–12. ACM (2010). https://doi.org/10.1145/1863495.1863497
35. McBride, C., McKinna, J.: Functional pearl: I am not a number – I am a free variable. In: Proceedings of the 2004 ACM SIGPLAN Workshop on Haskell, Haskell

2004, pp. 1–9. ACM, New York (2004). https://doi.org/10.1145/1017472.1017477. http://doi.acm.org/10.1145/1017472.1017477
36. de Moura, L., Kong, S., Avigad, J., van Doorn, F., von Raumer, J.: The lean theorem prover (system description). In: Felty, A.P., Middeldorp, A. (eds.) CADE 2015. LNCS (LNAI), vol. 9195, pp. 378–388. Springer, Cham (2015). https://doi.org/10.1007/978-3-319-21401-6_26
37. Nordvall Forsberg, F.: Inductive-inductive definitions. Ph.D. thesis, Swansea University (2013)
38. Orton, I., Pitts, A.M.: Axioms for modelling cubical type theory in a topos. In: Talbot, J.M., Regnier, L. (eds.) 25th EACSL Annual Conference on Computer Science Logic (CSL 2016). Leibniz International Proceedings in Informatics (LIPIcs), vol. 62, pp. 24:1–24:19. Schloss Dagstuhl-Leibniz-Zentrum fuer Informatik, Dagstuhl, Germany (2016). https://doi.org/10.4230/LIPIcs.CSL.2016.24
39. Pfenning, F., Elliott, C.: Higher-order abstract syntax. SIGPLAN Not. **23**(7), 199–208 (1988). https://doi.org/10.1145/960116.54010
40. Pientka, B., Dunfield, J.: Beluga: a framework for programming and reasoning with deductive systems (system description). In: Giesl, J., Hähnle, R. (eds.) IJCAR 2010. LNCS (LNAI), vol. 6173, pp. 15–21. Springer, Heidelberg (2010). https://doi.org/10.1007/978-3-642-14203-1_2
41. Reynolds, J.C.: Types, abstraction and parametric polymorphism. In: Mason, R.E.A. (ed.) Information Processing 1983, Proceedings of the IFIP 9th World Computer Congress, Paris, 19–23 September 1983, pp. 513–523. Elsevier Science Publishers B. V. (North-Holland), Amsterdam (1983)
42. Tabareau, N., Tanter, É., Sozeau, M.: Equivalences for free. Proc. ACM Program. Lang. 1–29 (2018). https://hal.inria.fr/hal-01559073
43. The Agda development team: Agda (2015). http://wiki.portal.chalmers.se/agda
44. Wieczorek, P., Biernacki, D.: A Coq formalization of normalization by evaluation for Martin-Löf type theory. In: Proceedings of the 7th ACM SIGPLAN International Conference on Certified Programs and Proofs, CPP 2018, pp. 266–279. ACM, New York (2018). https://doi.org/10.1145/3167091
45. Winterhalter, T., Sozeau, M., Tabareau, N.: Eliminating reflection from type theory. In: Proceedings of the 8th ACM SIGPLAN International Conference on Certified Programs and Proofs, pp. 91–103. ACM (2019)

En Garde! Unguarded Iteration for Reversible Computation in the Delay Monad

Robin Kaarsgaard[1] and Niccolò Veltri[2(✉)]

[1] DIKU, Department of Computer Science, University of Copenhagen, Copenhagen, Denmark
robin@di.ku.dk
[2] Department of Computer Science, IT University of Copenhagen, Copenhagen, Denmark
nive@itu.dk

Abstract. Reversible computation studies computations which exhibit both forward and backward determinism. Among others, it has been studied for half a century for its applications in low-power computing, and forms the basis for quantum computing.

Though certified program equivalence is useful for a number of applications (e.g., certified compilation and optimization), little work on this topic has been carried out for reversible programming languages. As a notable exception, Carette and Sabry have studied the equivalences of the finitary fragment of Π°, a reversible combinator calculus, yielding a two-level calculus of type isomorphisms and equivalences between them. In this paper, we extend the two-level calculus of finitary Π° to one for full Π° (i.e., with both recursive types and iteration by means of a trace combinator) using the delay monad, which can be regarded as a "computability-aware" analogue of the usual maybe monad for partiality. This yields a calculus of iterative (and possibly non-terminating) reversible programs acting on user-defined dynamic data structures together with a calculus of certified program equivalences between these programs.

Keywords: Reversible computation · Iteration · Delay monad

1 Introduction

Reversible computation is an emerging computation paradigm encompassing computations that are not just deterministic when executed the *forward* direction, but also in the *backward* direction. While this may seem initially obscure, reversible computation forms the basis for quantum computing, and has seen applications in a number of different areas such as low-power computing [28],

Niccolò Veltri was supported by a research grant (13156) from VILLUM FONDEN.

G. Hutton (Ed.): MPC 2019, LNCS 11825, pp. 366–384, 2019.
https://doi.org/10.1007/978-3-030-33636-3_13

robotics [30], discrete event simulation [33], and the simultaneous construction of parser/pretty printer pairs [32]. Like classical computing, it has its own automata [4], circuit model [40], machine architectures [35], programming languages [20–22,34,41], semantic metalanguages [15,24,25], and so on.

Π° is a family of reversible combinator calculi comprising structural isomorphisms and combinators corresponding to those found in dagger-traced ω-continuous rig categories [26] (a kind of dagger category with a trace, monoidal sums $\oplus$ and products $\otimes$ such that they form a rig structure, and fixed points of the functors formed from the rig structure). Though superficially simple, Π° is expressive enough as a metalanguage to give semantics to the typed reversible functional programming language Theseus [22].

In [7], Carette and Sabry studied the equivalences of isomorphisms in the finitary fragment of Π° (i.e., without recursive types and iteration via the trace combinator), and showed that these equivalences could be adequately described by another combinator calculus of equivalences of isomorphisms, in sum yielding a two-level calculus of isomorphisms and equivalences of isomorphisms. In this paper, we build on this work to produce a (fully formalized) two-level calculus for full Π° (supporting both recursive types and iteration) via the delay monad, using insights gained from the study of its Kleisli category [8,37,39], as well as of join inverse categories in which reversible iteration may be modelled [25].

The full Π° calculus cannot be modelled in the same framework of [7], since Martin-Löf type theory is a total language which in particular disallows the specification of a trace operator on types. Consequently, it is necessary to move to a setting supporting the existence of partial maps, and in type theory this can be done by using monads, by considering partiality as an effect. Our choice fell on the coinductive delay monad, introduced by Capretta [6] as a way of representing general recursive functions in Martin-Löf type theory. The delay datatype has been employed in a large number of applications, ranging from operational semantics of functional languages [11] to formalization of domain theory in type theory [5] and normalization by evaluation [1]. Here it is used for giving denotational semantics to Π°. In particular, we show how to endow the delay datatype with a trace combinator, whose construction factors through the specification of a uniform iteration operator [16,17].

The uniform iteration operator introduces a notion of feedback, typically used to model control flow operations such as while loops. In the Kleisli category of the delay monad, this operation can be intuitively described as follows: We can apply a function $f : A \to B + A$ on an input $a : A$ and either produce an element $b : B$, or produce a new element $a' : A$ which can be fed back to f. This operation can be iterated, and it either terminates returning a value in B or it goes on forever without producing any output. This form of iteration is "unguarded" because it allows the possibility of divergence. The trace operator can then be seen as a particular form of iteration where, given a function $f : A + C \to B + C$, which can be decomposed as $f_{\mathsf{L}} : A \to B + C$ and $f_{\mathsf{R}} : C \to B + C$, we first apply f_{L} on an input $a : A$, and, if the latter operation produces a value $c : C$, we continue by iterating f_{R} on c. Notice that the notion of trace can be generally defined in monoidal categories where the monoidal structure is not necessarily given by coproducts, and it has been used to model other things besides iteration, such as partial traces in vector spaces [23], though this use falls outside of the scope of this paper.

Throughout the paper, we reason constructively in Martin-Löf type theory. Classically, the delay monad (quotiented by weak bisimilarity) is isomorphic to the maybe monad $\mathsf{Maybe}\, X = X + 1$, and thus just a complication of something that can be expressed much simpler. Constructively, however, they are very different. In particular, it is impossible to define a well-behaved trace combinator for the maybe monad without assuming classical principles such as the limited principle of omniscience.

We have fully formalized the development of the paper in the dependently typed programming language Agda [31]. The code is available online at https://github.com/niccoloveltri/pi0-agda. The formalization uses Agda 2.6.0.

Overview. In Sect. 2, we present the syntax of Π° as formalized in Agda, with particular emphasis on recursive types and the trace operator. In Sect. 3, we recall the definition of Capretta's delay datatype and weak bisimilarity. We discuss finite products and coproducts in the Kleisli category of the delay monad and we introduce the category of partial isomorphisms that serves as the denotational model of Π°. In Sect. 4, we build a complete Elgot monad structure on the delay datatype, that allows the encoding of a dagger trace operator in the category of partial isomorphisms. In Sect. 5, we formally describe the interpretation of Π° types, terms and terms equivalences. We conclude in Sect. 6 with some final remarks and discussion on future work.

The Type-Theoretical Framework. Our work is settled in Martin-Löf type theory with inductive and coinductive types. We write $(a : A) \to B\, a$ for dependent function spaces and $(a : A) \times B\, a$ for dependent products. We allow dependent functions to have implicit arguments and indicate implicit argument positions with curly brackets (as in Agda). We use the symbol $=$ for definitional equality of terms and $\equiv$ for propositional equality. Given $f : A \to C$ and $g : B \to C$, we write $[f, g] : A + B \to C$ for their copairing. The coproduct injections are denoted inl and inr. Given $h : C \to A$ and $k : C \to B$, we write $\langle h, k\rangle : C \to A \times B$ for their pairing. The product projections are denoted fst and snd. The empty type is 0 and the unit type is 1. We write Set for the category of types and functions between them. We also use Set to denote the universe of types. We define $A \leftrightarrow B = (A \to B) \times (B \to A)$.

We do not assume uniqueness of identity proofs (UIP), i.e. we do not consider two proofs of $x \equiv y$ necessarily equal. Agda natively supports UIP, so we have to manually switch it off using the `without-K` option.

In Sect. 3, we will need to quotient a certain type by an equivalence relation. Martin-Löf type theory does not support quotient types, but quotients can be simulated using setoids [3]. Alternatively, we can consider extensions of type theory with quotient types à la Hofmann [19], such as homotopy type theory [36]. Setoids and quotient types à la Hofmann are not generally equivalent approaches, but they are indeed equivalent for the constructions we develop in this work. Therefore, in the rest of the paper we assume the existence of quotient types and we refrain from technical discussions on their implementation.

2 Syntax of Π°

In this section, we present the syntax of Π°. The 1-structure of Π°, i.e. its types and terms, has originally been introduced by James and Sabry [21]. In particular, we include the presence of recursive types and a primitive trace combinator. Following Carette and Sabry's formalization of the finitary fragment of Π°, we consider a collection of equivalences between terms. Our list of axioms notably differs from theirs in that we do not require each term to be a total isomorphism, we ask only for the existence of a partial inverse.

Formally, the collection of types of Π° correspond to those naturally interpreted in dagger traced ω-continuous rig categories (see [26]).

2.1 Types

The types of Π° are given by the grammar:

$$A ::= \mathsf{Z} \mid A \oplus A \mid \mathsf{I} \mid A \otimes A \mid X \mid \mu X.A$$

where X ranges over a set of variables. In Agda, we use de Bruijn indexes to deal with type variables, so the grammar above is formally realized by the rules in Fig. 1. The type $\mathsf{Ty}\, n$ represents Π° types containing at most n free variables. Variables themselves are encoded as elements of $\mathsf{Fin}\, n$, the type of natural numbers strictly smaller then n. The type constructor μ binds a variable, which, for $A : \mathsf{Ty}\,(n+1)$, we consider to be $n+1$.

It is also necessary to define substitutions. In Agda, given types $A : \mathsf{Ty}\,(n+1)$ and $B : \mathsf{Ty}\, n$, we construct $\mathsf{sub}\, A\, B : \mathsf{Ty}\, n$ to represent the substituted type $A[B/X]$, where X corresponds to the $(n+1)$-th variable in context.

$$\frac{}{\mathsf{Z} : \mathsf{Ty}\, n} \qquad \frac{}{\mathsf{I} : \mathsf{Ty}\, n} \qquad \frac{i : \mathsf{Fin}\, n}{\mathsf{Var}\, i : \mathsf{Ty}\, n}$$

$$\frac{A : \mathsf{Ty}\, n \quad B : \mathsf{Ty}\, n}{A \oplus B : \mathsf{Ty}\, n} \qquad \frac{A : \mathsf{Ty}\, n \quad B : \mathsf{Ty}\, n}{A \otimes B : \mathsf{Ty}\, n} \qquad \frac{A : \mathsf{Ty}\,(n+1)}{\mu A : \mathsf{Ty}\, n}$$

Fig. 1. Types of Π°, as formalized in Agda

2.2 Terms

The terms of Π° are inductively generated by the rules in Fig. 2. They include the identity programs id and sequential composition of programs $\bullet$. $(\mathsf{Z}, \oplus)$ is a symmetric monoidal structure, with terms $\lambda_\oplus$, $\alpha_\oplus$ and $\sigma_\oplus$ as structural morphisms. Similarly for $(\mathsf{I}, \otimes)$. Moreover $\otimes$ distributes over Z and $\oplus$ from the right, as evidenced by κ and δ. Elements of $\mu X.A$ are built using the term constructor fold and destructed with unfold. Finally, we find the trace combinator.

Every Π° program is reversible. The (partial) inverse of a program is given by the function dagger : $(A \longleftrightarrow B) \to (B \longleftrightarrow A)$, recursively defined as follows:

$$
\begin{array}{llll}
\mathsf{dagger}\ \mathsf{id} & = \mathsf{id} & \mathsf{dagger}\ (g \bullet f) & = \mathsf{dagger}\ f \bullet \mathsf{dagger}\ g \\
\mathsf{dagger}\ (f \oplus g) & = \mathsf{dagger}\ f \oplus \mathsf{dagger}\ g & \mathsf{dagger}\ (f \otimes g) & = \mathsf{dagger}\ f \otimes \mathsf{dagger}\ g \\
\mathsf{dagger}\ \lambda_\oplus^{-1} & = \lambda_\oplus & \mathsf{dagger}\ \lambda_\oplus & = \lambda_\oplus^{-1} \\
\mathsf{dagger}\ \sigma_\oplus & = \sigma_\oplus & \mathsf{dagger}\ \alpha_\oplus & = \alpha_\oplus^{-1} \\
\mathsf{dagger}\ \alpha_\oplus^{-1} & = \alpha_\oplus & \mathsf{dagger}\ \lambda_\otimes & = \lambda_\otimes^{-1} \\
\mathsf{dagger}\ \lambda_\otimes^{-1} & = \lambda_\otimes & \mathsf{dagger}\ \sigma_\otimes & = \sigma_\otimes \\
\mathsf{dagger}\ \alpha_\otimes & = \alpha_\otimes^{-1} & \mathsf{dagger}\ \alpha_\otimes^{-1} & = \alpha_\otimes \\
\mathsf{dagger}\ \kappa & = \kappa^{-1} & \mathsf{dagger}\ \kappa^{-1} & = \kappa \\
\mathsf{dagger}\ \delta & = \delta^{-1} & \mathsf{dagger}\ \delta^{-1} & = \delta \\
\mathsf{dagger}\ \mathsf{fold} & = \mathsf{unfold} & \mathsf{dagger}\ \mathsf{unfold} & = \mathsf{fold} \\
\mathsf{dagger}\ (\mathsf{trace}\ f) & = \mathsf{trace}\ (\mathsf{dagger}\ f) & &
\end{array}
$$

The dagger operation is involutive. Notice that this property holds up to propositional equality. This is proved by induction on the term f.

$$\mathsf{daggerInvol} : (f : A \longleftrightarrow B) \to \mathsf{dagger}\ (\mathsf{dagger}\ f) \equiv f$$

The right unitor for $\oplus$ is given by $\rho_\oplus = \lambda_\oplus \bullet \sigma_\oplus : A \oplus \mathsf{Z} \longleftrightarrow A$, and $\rho_\otimes$ is defined similarly. Analogously, we can derive the left distributors $\kappa' : A \otimes \mathsf{Z} \longleftrightarrow A$ and $\delta' : A \otimes (B \oplus C) \longleftrightarrow (A \otimes B) \oplus (A \otimes C)$.

2.3 Term Equivalences

A selection of term equivalences of Π° is given in Fig. 3. We only include the equivalences that either differ or have not previously considered by Carette and Sabry in their formalization of the finite fragment of Π° [7]. In particular, we leave out the long list of Laplaza's coherence axioms expressing that types and terms of Π° form a rig category [29]. We also omit the equivalences stating that $\lambda_\oplus^{-1}$ is the total inverse of $\lambda_\oplus$, similarly for the other structural morphisms.

The list of term equivalences in Fig. 3 contains the trace axioms, displaying that the types of Π° form a traced monoidal category wrt. the additive monoidal structure $(\mathsf{Z}, \oplus)$ [23]. Next we ask for trace (dagger f) to be the partial inverse of trace f. Remember that we have defined dagger (trace f) to be trace (dagger f), so the axiom tracePIso is evidence that the trace combinator of Π° is a dagger trace.

$$\mathsf{id} : A \longleftrightarrow A \qquad \frac{g : B \longleftrightarrow C \quad f : A \longleftrightarrow B}{g \bullet f : A \longleftrightarrow C}$$

$$\frac{f : A \longleftrightarrow C \quad g : B \longleftrightarrow D}{f \oplus g : A \oplus B \longleftrightarrow C \oplus D} \qquad \frac{f : A \longleftrightarrow C \quad g : B \longleftrightarrow D}{f \otimes g : A \otimes B \longleftrightarrow C \otimes D}$$

$$\lambda_\oplus : \mathsf{Z} \oplus A \longleftrightarrow A \qquad \lambda_\oplus^{-1} : A \longleftrightarrow \mathsf{Z} \oplus A$$

$$\lambda_\otimes : \mathsf{I} \otimes A \longleftrightarrow A \qquad \lambda_\otimes^{-1} : A \longleftrightarrow \mathsf{I} \otimes A$$

$$\alpha_\oplus : (A \oplus B) \oplus C \longleftrightarrow A \oplus (B \oplus C) \qquad \alpha_\oplus^{-1} : A \oplus (B \oplus C) \longleftrightarrow (A \oplus B) \oplus C$$

$$\alpha_\otimes : (A \otimes B) \otimes C \longleftrightarrow A \otimes (B \otimes C) \qquad \alpha_\otimes^{-1} : A \otimes (B \otimes C) \longleftrightarrow (A \otimes B) \otimes C$$

$$\sigma_\oplus : A \oplus B \longleftrightarrow B \oplus A \qquad \sigma_\otimes : A \otimes B \longleftrightarrow B \otimes A$$

$$\kappa : \mathsf{Z} \otimes A \longleftrightarrow \mathsf{Z} \qquad \delta : (A \oplus B) \otimes C \longleftrightarrow (A \otimes C) \oplus (B \otimes C)$$

$$\kappa^{-1} : \mathsf{Z} \longleftrightarrow \mathsf{Z} \otimes A \qquad \delta^{-1} : (A \otimes C) \oplus (B \otimes C) \longleftrightarrow (A \oplus B) \otimes C$$

$$\mathsf{fold} : A[\mu X.A/X] \longleftrightarrow \mu X.A \qquad \mathsf{unfold} : \mu X.A \longleftrightarrow A[\mu X.A/X]$$

$$\frac{f : A \oplus C \longleftrightarrow B \oplus C}{\mathsf{trace}\, f : A \longleftrightarrow B}$$

Fig. 2. Terms of Π°

Afterwards we have two equivalences stating that unfold is the total inverse of fold.

It is possible to show that every term f has dagger f as its partial inverse. The notion of partial inverse used here comes from the study of *inverse categories* (see [27]) and amounts to saying that dagger f is the *unique* map that undoes everything which f does (unicity of partial inverses follows by the final equivalence of Fig. 3, see [27]). Note that this is different from requiring that f is an isomorphism in the usual sense, as dagger $f \bullet f$ is not going to be the identity when f is only partially defined, though it will behave as the identity on all points where f is defined.

The proof that every term has dagger f as its partial inverse proceeds by induction on f.

$$\mathsf{existsPIso} : (f : A \longleftrightarrow B) \rightarrow f \bullet \mathsf{dagger}\, f \bullet f \iff f$$

3 Delay Monad

The coinductive delay datatype was first introduce by Capretta for representing general recursive functions in Martin-Löf type theory [6]. Given a type A, elements of Delay A are possibly non-terminating "computations" returning a value

$$\mathsf{naturality_L} : f \bullet \mathsf{trace}\, g \iff \mathsf{trace}\,((f \oplus \mathsf{id}) \bullet g)$$

$$\mathsf{naturality_R} : \mathsf{trace}\, g \bullet f \iff \mathsf{trace}\,(g \bullet (f \oplus \mathsf{id}))$$

$$\mathsf{dinaturality} : \mathsf{trace}\,((\mathsf{id} \oplus f) \bullet g) \iff \mathsf{trace}\,(g \bullet (\mathsf{id} \oplus f))$$

$$\mathsf{superposing} : \mathsf{trace}\,(\alpha_\oplus^{-1} \bullet (\mathsf{id} \oplus f) \bullet \alpha_\oplus) \iff \mathsf{id} \oplus \mathsf{trace}\, f$$

$$\mathsf{vanishing}_\oplus : \mathsf{trace}\, f \iff \mathsf{trace}\,(\mathsf{trace}\,(\alpha_\oplus^{-1} \bullet f \bullet \alpha_\oplus))$$

$$\mathsf{vanishing_Z} : f \iff \rho_\oplus^{-1} \bullet \mathsf{trace}\, f \bullet \rho_\oplus \qquad \mathsf{yanking} : \mathsf{trace}\, \sigma_\oplus \iff \mathsf{id}$$

$$\mathsf{tracePIso} : \mathsf{trace}\, f \bullet \mathsf{trace}\,(\mathsf{dagger}\, f) \bullet \mathsf{trace}\, f \iff \mathsf{trace}\, f$$

$$\mathsf{foldIso} : \mathsf{fold} \bullet \mathsf{unfold} \iff \mathsf{id} \qquad \mathsf{unfoldIso} : \mathsf{unfold} \bullet \mathsf{fold} \iff \mathsf{id}$$

$$\mathsf{uniquePIso} : f \bullet \mathsf{dagger}\, f \bullet g \bullet \mathsf{dagger}\, g \iff g \bullet \mathsf{dagger}\, g \bullet f \bullet \mathsf{dagger}\, f$$

Fig. 3. Selection of term equivalences of Π°

of A whenever they terminate. Formally, $\mathsf{Delay}\, A$ is defined as a coinductive type with the following introduction rules:

$$\frac{a : A}{\mathsf{now}\, a : \mathsf{Delay}\, A} \qquad \frac{x : \mathsf{Delay}\, A}{\mathsf{later}\, x : \mathsf{Delay}\, A}$$

The constructor now embeds A into $\mathsf{Delay}\, A$, so $\mathsf{now}\, a$ represents the terminating computation returning the value a. The constructor later adds an additional unit of time delay to a computation. Double rule lines refer to a coinductive constructor, which can be employed an infinite number of times in the construction of a term of type $\mathsf{Delay}\, A$. E.g., the non-terminating computation never is corecursively defined as $\mathsf{never} = \mathsf{later}\, \mathsf{never}$.

The delay datatype is a monad. The unit is the constructor now, while the Kleisli extension bind is corecursively defined as follows:

$$\begin{aligned}
&\mathsf{bind} : (A \to \mathsf{Delay}\, B) \to \mathsf{Delay}\, A \to \mathsf{Delay}\, B \\
&\mathsf{bind}\, f\, (\mathsf{now}\, a) = f\, a \\
&\mathsf{bind}\, f\, (\mathsf{later}\, x) = \mathsf{later}\, (\mathsf{bind}\, f\, x)
\end{aligned}$$

The delay monad, like any other monad on Set, has a unique strength operation which we denote by $\mathsf{str} : A \times \mathsf{Delay}\, B \to \mathsf{Delay}\, (A \times B)$. Similarly, it has a unique costrength operation $\mathsf{costr} : (\mathsf{Delay}\, A) \times B \to \mathsf{Delay}\, (A \times B)$ definable using str. Moreover, the delay datatype is a commutative monad.

The Kleisli category of the delay monad, that we call $\mathbb{D}$, has types as objects and functions $f : A \to \mathsf{Delay}\, B$ as morphisms between A and B. In $\mathbb{D}$, the identity map on an object A is the constructor now, while the composition of morphisms $f : A \to \mathsf{Delay}\, B$ and $g : B \to \mathsf{Delay}\, C$ is given by $f \diamond g = \mathsf{bind}\, f \circ g$.

The delay datatype allows us to program with partial functions, but the introduced notion of partiality is intensional, in the sense that computations terminating with the same value in a different number of steps are considered different. To obtain an extensional notion of partiality, which in particular allows the specification of a well-behaved trace operator, we introduce the notion of (termination-sensitive) weak bisimilarity.

Weak bisimilarity is defined in terms of convergence. A computation x : $\mathsf{Delay}\,A$ converges to $a : A$ if it terminates in a finite number of steps returning the value a. When this happens, we write $x \downarrow a$. The relation $\downarrow$ is inductively defined by the rules:

$$\frac{}{\mathsf{now}\,a \downarrow a} \qquad \frac{x \downarrow a}{\mathsf{later}\,x \downarrow a}$$

Two computations in $\mathsf{Delay}\,A$ are weakly bisimilar if they differ by a finite number of applications of the constructor later. Alternatively, we can say that two computations x and y are weakly bisimilar if, whenever x terminates returning a value a, then y also terminates returning a, and vice versa. This informal statement can be formalized in several different but logically equivalent ways [8, 39]. Here we consider a coinductive formulation employed e.g. in [12].

$$\begin{array}{cc} \dfrac{}{\mathsf{now}_{\approx} : \mathsf{now}\,a \approx \mathsf{now}\,a} & \dfrac{p : x_1 \approx x_2}{\mathsf{later}_{\approx}\,p : \mathsf{later}\,x_1 \approx \mathsf{later}\,x_2} \\[2ex] \dfrac{p : x \approx \mathsf{now}\,a}{\mathsf{laterL}_{\approx}\,p : \mathsf{later}\,x \approx \mathsf{now}\,a} & \dfrac{p : \mathsf{now}\,a \approx x}{\mathsf{laterR}_{\approx}\,p : \mathsf{now}\,a \approx \mathsf{later}\,x} \end{array} \tag{1}$$

Notice that the constructor $\mathsf{later}_{\approx}$ is coinductive. This allows us to prove $\mathsf{never} \approx \mathsf{never}$. Weak bisimilarity is an equivalence relation and it is a congruence w.r.t. the later operation. For example, here is a proof that weak bisimilarity is reflexive.

$$\begin{aligned} &\mathsf{refl}_{\approx} : \{x : \mathsf{Delay}\,A\} \to x \approx x \\ &\mathsf{refl}_{\approx}\,\{\mathsf{now}\,a\} = \mathsf{now}_{\approx} \\ &\mathsf{refl}_{\approx}\,\{\mathsf{later}\,x\} = \mathsf{later}_{\approx}\,(\mathsf{refl}_{\approx}\,\{x\}) \end{aligned}$$

We call $\mathbb{D}_{\approx}$ the category $\mathbb{D}$ with homsets quotiented by pointwise weak bisimilarity. This means that in $\mathbb{D}_{\approx}$ two morphisms f and g are considered equal whenever $f\,a \approx g\,a$, for all inputs a. When this is the case, we also write $f \approx g$. The operation bind is compatible with weak bisimilarity, in the sense that $\mathsf{bind}\,f_1\,x_1 \approx \mathsf{bind}\,f_2\,x_2$ whenever $f_1 \approx f_2$ and $x_1 \approx x_2$.

As an alternative to quotienting the homsets of $\mathbb{D}$, we could have quotiented the delay datatype by weak bisimilarity: $\mathsf{Delay}_{\approx}\,A = \mathsf{Delay}\,A/{\approx}$. In previous work [8], we showed that this construction has problematic consequences if we employ Hofmann's approach to quotient types [19]. For example, it does not seem possible to lift the monad structure of Delay to $\mathsf{Delay}_{\approx}$ without postulating additional principles such as the axiom of countable choice. More fundamentally for this work, countable choice would be needed for modelling the trace operator of Π° in the Kleisli category of $\mathsf{Delay}_{\approx}$. Notice that, if the setoid approach to quotienting is employed, the latter constructions go through without the need for additional assumptions. In order to keep an agnostic perspective on quotient types and avoid the need for disputable semi-classical choice principles, we decided to quotient the homsets of $\mathbb{D}$ by (pointwise) weak bisimilarity instead of the objects of $\mathbb{D}$.

3.1 Finite Products and Coproducts

Colimits in $\mathbb{D}_{\approx}$ are inherited from Set. This means that 0 is also the initial object of $\mathbb{D}_{\approx}$, similarly $A+B$ is the binary coproduct of A and B in $\mathbb{D}_{\approx}$. Given $f : A \to \mathsf{Delay}\, C$ and $g : B \to \mathsf{Delay}\, C$, their copairing is $[f, g]_{\mathbb{D}} = [f, g] : A+B \to \mathsf{Delay}\, C$. The operation $[-,-]_{\mathbb{D}}$ is compatible with weak bisimilarity, in the sense that $[f_1, g_1]_{\mathbb{D}} \approx [f_2, g_2]_{\mathbb{D}}$ whenever $f_1 \approx f_2$ and $g_1 \approx g_2$. The coproduct injections are given by $\mathsf{inl}_{\mathbb{D}} = \mathsf{now} \circ \mathsf{inl} : A \to \mathsf{Delay}\,(A + B)$ and $\mathsf{inr}_{\mathbb{D}} = \mathsf{now} \circ \mathsf{inr} : B \to \mathsf{Delay}\,(A + B)$.

Just as limits in Set do not lift to limits in the category Par of sets and partial functions, they do not lift to $\mathbb{D}_{\approx}$ either. This is not an issue with these concrete formulations of partiality, but rather with the interaction of partiality (in the sense of restriction categories, a kind of categories of partial maps) and limits in general (see [10, Section 4.4]). In particular, 1 is not the terminal object and $A \times B$ is not the binary product of A and B in $\mathbb{D}_{\approx}$. In fact, 0 is (also) the terminal object, with $\lambda\_.\, \mathsf{never} : A \to \mathsf{Delay}\, 0$ as the terminal morphism. Nevertheless, it is possible to prove that 1 and $\times$ are partial terminal object and partial binary products respectively, in the sense of Cockett and Lack's restriction categories [9,10]. Here we refrain from making the latter statement formal. We only show the construction of the partial pairing operation, which we employ in the interpretation of Π°. Given $f : C \to \mathsf{Delay}\, A$ and $g : C \to \mathsf{Delay}\, B$, we define:

$$\begin{aligned} \langle f, g\rangle_{\mathbb{D}} &: C \to \mathsf{Delay}\,(A \times B) \\ \langle f, g\rangle_{\mathbb{D}} &= \mathsf{costr} \diamond (\mathsf{str} \circ \langle f, g\rangle) \end{aligned}$$

Since the delay monad is commutative, the function $\langle f, g\rangle_{\mathbb{D}}$ is equal to $\mathsf{str} \diamond (\mathsf{costr} \circ \langle f, g\rangle)$. The operation $\langle -,-\rangle_{\mathbb{D}}$ is compatible with weak bisimilarity.

3.2 Partial Isomorphisms

In order to model the reversible programs of Π°, we need to consider reversible computations in $\mathbb{D}_{\approx}$. Given a morphism $f : A \to \mathsf{Delay}\, B$, we say that it is a partial isomorphism if the following type is inhabited:

$$\mathsf{isPartialIso}\, f = (g : B \to \mathsf{Delay}\, A) \times ((a : A)(b : B) \to f\, a \downarrow b \leftrightarrow g\, b \downarrow a)$$

In other words, f is a partial isomorphism if there exists a morphism $g : B \to \mathsf{Delay}\, A$ such that, if $f\, a$ terminates returning a value b, then $g\, b$ terminates returning a, and vice versa. Given a partial isomorphism f, we denote its partial inverse by $\mathsf{dagger}_{\mathbb{D}}\, f$.

In $\mathbb{D}_{\approx}$, our definition of partial isomorphisms is equivalent to the standard categorical one [27] (see also [9]), which, translated in our type-theoretical setting, is

$$\mathsf{isPartialIsoCat}\, f = (g : B \to \mathsf{Delay}\, A) \times f \diamond g \diamond f \approx f \times g \diamond f \diamond g \approx g$$

We denote $A \simeq B$ the type of partial isomorphisms between A and B:

$$A \simeq B = (f : A \to \mathsf{Delay}\, B) \times \mathsf{isPartialIso}\, f$$

We call $\mathsf{Inv}\mathbb{D}_{\approx}$ the subcategory of $\mathbb{D}_{\approx}$ consisting of (equivalence classes of) partial isomorphisms. Note that $\mathsf{Inv}\mathbb{D}_{\approx}$ inherits neither partial products nor coproducts of $\mathbb{D}_{\approx}$, as the universal mapping property fails in both cases. However, it can be shown that in the category $\mathsf{Inv}\mathbb{D}_{\approx}$, 0 is a zero object, $A + B$ is the disjointness tensor product of A and B (in the sense of Giles [15]) with unit 0, and $A \times B$ a monoidal product of A and B with unit 1 (though it is *not* an inverse product in the sense of Giles [15], as that would imply decidable equality on all objects). In particular, we can derive the following operations, modelling the Π° term constructors $\oplus$ and $\otimes$:

$$\times_{\mathbb{D}\simeq} : A \simeq C \to B \simeq D \to A \times B \simeq C \times D$$
$$+_{\mathbb{D}\simeq} : A \simeq C \to B \simeq D \to A + B \simeq C + D$$

4 Elgot Iteration

A complete Elgot monad [16,17] is a monad T whose Kleisli category supports unguarded uniform iteration. More precisely[1], a monad T is Elgot if there exists an operation

$$\mathsf{iter}_{\mathsf{T}} : (A \to \mathsf{T}\,(B + A)) \to A \to \mathsf{T}\, B$$

satisfying the following axioms:

$$\mathsf{fixpoint} : \mathsf{iter}_{\mathsf{T}}\, f \equiv [\eta_{\mathsf{T}}, \mathsf{iter}_{\mathsf{T}}\, f] \diamond_{\mathsf{T}} f$$

$$\mathsf{naturality} : g \diamond_{\mathsf{T}} \mathsf{iter}_{\mathsf{T}}\, f \equiv \mathsf{iter}_{\mathsf{T}}\, ([\mathsf{T}\mathsf{inl} \circ g, \eta \circ \mathsf{inr}] \diamond_{\mathsf{T}} f)$$

$$\mathsf{codiagonal} : \mathsf{iter}_{\mathsf{T}}\, (\mathsf{iter}_{\mathsf{T}}\, g) \equiv \mathsf{iter}_{\mathsf{T}}\, (\mathsf{T}[\mathsf{id}, \mathsf{inr}] \circ g)$$

$$\frac{p : f \circ h \equiv \mathsf{T}(\mathsf{id} + h) \circ g}{\mathsf{uniformity}\, p : \mathsf{iter}_{\mathsf{T}}\, f \circ h \equiv \mathsf{iter}_{\mathsf{T}}\, g}$$

where η_{T} is the unit of T and $\diamond_{\mathsf{T}}$ denotes morphism composition in the Kleisli category of T. The standard definition of uniform iteration operator includes the

[1] Here we give the definition of complete Elgot monad on **Set**, but the definition of complete Elgot monad makes sense in any category with finite coproducts.

dinaturality axiom, which has recently been discovered to be derivable from the other laws [14,16].

The delay monad is a complete Elgot monad for which the axioms holds up to weak bisimilarity, not propositional equality. In other words, the category $\mathbb{D}_{\approx}$ can be endowed with a uniform iteration operator. The specification of the iteration operator relies on an auxiliary function $\mathsf{iter}'_{\mathbb{D}}$ corecursively defined as follows:

$$\begin{aligned}
&\mathsf{iter}'_{\mathbb{D}} : (A \to \mathsf{Delay}\,(B + A)) \to \mathsf{Delay}\,(B + A) \to \mathsf{Delay}\,B \\
&\mathsf{iter}'_{\mathbb{D}}\, f\,(\mathsf{now}\,(\mathsf{inl}\,b)) \quad = \mathsf{now}\,b \\
&\mathsf{iter}'_{\mathbb{D}}\, f\,(\mathsf{now}\,(\mathsf{inr}\,a)) = \mathsf{later}\,(\mathsf{iter}'_{\mathbb{D}}\, f\,(f\,a)) \\
&\mathsf{iter}'_{\mathbb{D}}\, f\,(\mathsf{later}\,x) \qquad\;\; = \mathsf{later}\,(\mathsf{iter}'_{\mathbb{D}}\, f\,x)
\end{aligned}$$

$$\begin{aligned}
&\mathsf{iter}_{\mathbb{D}} : (A \to \mathsf{Delay}\,(B + A)) \to A \to \mathsf{Delay}\,B \\
&\mathsf{iter}_{\mathbb{D}}\, f\,a = \mathsf{iter}'_{\mathbb{D}}\, f\,(f\,a)
\end{aligned}$$

The definition above can be given the following intuitive explanation. If $f\,a$ does not terminate, then $\mathsf{iter}_{\mathbb{D}}\, f\,a$ does not terminate either. If $f\,a$ terminates, there are two possibilities: either $f\,a$ converges to $\mathsf{inl}\,b$, in which case $\mathsf{iter}_{\mathbb{D}}\, f\,a$ terminates returning the value b; or $f\,a$ converges to $\mathsf{inr}\,a'$, in which case we repeat the procedure by replacing a with a'. Notice that in the latter case we also add one occurrence of later to the total computation time. This addition is necessary for ensuring the productivity of the corecursively defined function $\mathsf{iter}'_{\mathbb{D}}$. In fact, by changing the second line of its specification to $\mathsf{iter}'_{\mathbb{D}}\, f\,(\mathsf{now}\,(\mathsf{inr}\,a)) = \mathsf{iter}'_{\mathbb{D}}\, f\,(fa)$ and taking $f = \mathsf{inr}_{\mathbb{D}}$, we would have that $\mathsf{iter}'_{\mathbb{D}}\, f\,(\mathsf{now}\,(\mathsf{inr}\,a))$ unfolds indefinitely without producing any output. In Agda, such a definition would be rightfully rejected by the termination checker.

The operation $\mathsf{iter}_{\mathbb{D}}$ is compatible with weak bisimilarity, which means that $\mathsf{iter}_{\mathbb{D}}\, f_1 \approx \mathsf{iter}_{\mathbb{D}}\, f_2$ whenever $f_1 \approx f_2$.

As mentioned above, $\mathsf{iter}_{\mathbb{D}}$ satisfies the Elgot iteration axioms only up to weak bisimilarity. Here we show the proof of the fixpoint axiom, which in turns relies on an auxiliary proof $\mathsf{fixpoint}'_{\mathbb{D}}$.

$$\begin{aligned}
&\mathsf{fixpoint}'_{\mathbb{D}} : (f : A \to \mathsf{Delay}\,(B + A)) \to \mathsf{bind}\,[\mathsf{now}, \mathsf{iter}_{\mathbb{D}}\, f]_{\mathbb{D}} \approx \mathsf{iter}'_{\mathbb{D}}\, f \\
&\mathsf{fixpoint}'_{\mathbb{D}}\, f\,(\mathsf{now}\,(\mathsf{inl}\,b)) \quad = \mathsf{now}_{\approx} \\
&\mathsf{fixpoint}'_{\mathbb{D}}\, f\,(\mathsf{now}\,(\mathsf{inr}\,a)) = \mathsf{laterR}_{\approx}\,\mathsf{refl}_{\approx} \\
&\mathsf{fixpoint}'_{\mathbb{D}}\, f\,(\mathsf{later}\,x) \qquad\;\; = \mathsf{later}_{\approx}\,(\mathsf{fixpoint}'_{\mathbb{D}}\, f\,x)
\end{aligned}$$

$$\begin{aligned}
&\mathsf{fixpoint}_{\mathbb{D}} : (f : A \to \mathsf{Delay}\,(B + A)) \to [\mathsf{now}, \mathsf{iter}_{\mathbb{D}}\, f]_{\mathbb{D}} \diamond f \approx \mathsf{iter}_{\mathbb{D}}\, f \\
&\mathsf{fixpoint}_{\mathbb{D}}\, f\,x = \mathsf{fixpoint}'_{\mathbb{D}}\, f\,(f\,x)
\end{aligned}$$

4.1 Trace

From the Elgot iteration operator it is possible to derive a trace operator. First, given $f : A+B \to C$, we introduce $f_{\mathsf{L}} = f \circ \mathsf{inl} : A \to C$ and $f_{\mathsf{R}} = f \circ \mathsf{inr} : B \to C$, so that $f = [f_{\mathsf{L}}, f_{\mathsf{R}}]$. Graphically:

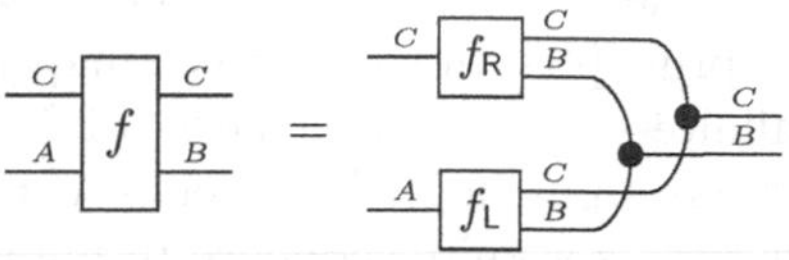

The trace operator in $\mathbb{D}_{\approx}$ is defined in terms of the iterator as follows:

$$\mathsf{trace}_{\mathbb{D}} : (A + C \to \mathsf{Delay}\,(B + C)) \to A \to \mathsf{Delay}\,B$$
$$\mathsf{trace}_{\mathbb{D}}\, f = [\mathsf{now}, \mathsf{iter}_{\mathbb{D}}\, f_{\mathsf{R}}]_{\mathbb{D}} \diamond f_{\mathsf{L}}$$

The operation $\mathsf{trace}_{\mathbb{D}}$ is compatible with weak bisimilarity. Graphically, we express the iterator on f as a wire looping back on the input, i.e., as

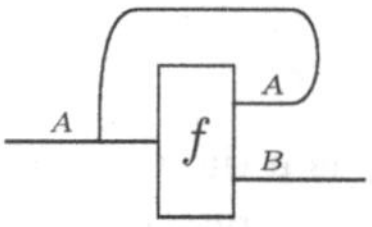

In this way, the definition of $\mathsf{trace}_{\mathbb{D}}$ may be expressed graphically as

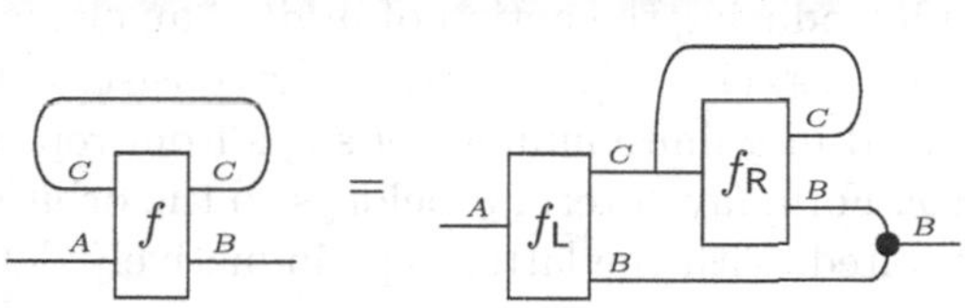

Intuitively, the function f_{L} initialises the loop. It either diverges, so that the trace of f diverges as well, or it terminates. It either terminates with an element $b : B$, in which case the loop ends immediately returning b, or it converges to a value $c : C$, and in this case we proceed by invoking the iteration of f_{R} on c.

It is well-known that a trace operator is obtainable from an iteration operator, as shown by Hasegawa [18]. His construction, instantiated to our setting, looks as follows:

$$\mathsf{traceH}_{\mathbb{D}} : (A + C \to \mathsf{Delay}\,(B + C)) \to A \to \mathsf{Delay}\,B$$
$$\mathsf{traceH}_{\mathbb{D}}\, f = \mathsf{iter}_{\mathbb{D}}(\mathsf{Delay}\,(\mathsf{id} + \mathsf{inr}) \circ f) \circ \mathsf{inl}$$

or graphically

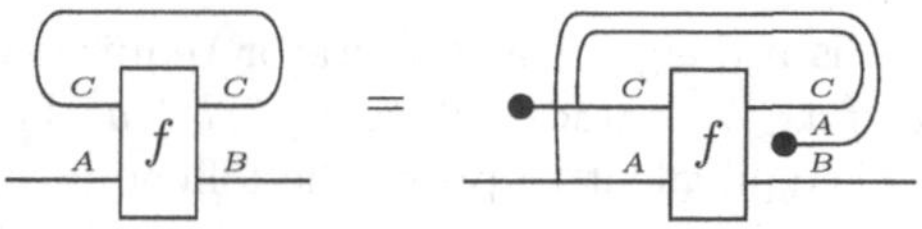

It is not difficult to prove that the two possible ways of defining a trace operator from Elgot iteration are equivalent, in the sense that $\mathsf{trace}_{\mathbb{D}}\, f \approx \mathsf{traceH}_{\mathbb{D}}\, f$ for all $f : A + C \to \mathsf{Delay}\,(B + C)$.

The trace axioms follow from the Elgot iteration axioms.

We conclude this section by remarking that the construction of a trace operator in the Kleisli category of the maybe monad is impossible without the assumption of additional classical principles. In fact, given a map $f : A{+}C \to B{+}C{+}1$, let xs be the possibly infinite sequence of elements of $B + C + 1$ produced by the iteration of f on a given input in A. In order to construct the trace of f, we need to decide whether xs is a finite sequence terminating with an element of $B + 1$, or xs is an infinite stream of elements of C. This decision requires the limited principle of omniscience, an instance of the law of excluded middle not provable in Martin-Löf type theory:

$$\mathsf{LPO} = (s : \mathbb{N} \to 2) \to ((n : \mathbb{N}) \times s\, n \equiv \mathsf{true}) + ((n : \mathbb{N}) \to s\, n \equiv \mathsf{false})$$

where 2 is the type of booleans, with true and false as only inhabitants.

4.2 Dagger Trace

We now move to show that $\mathsf{trace}_{\mathbb{D}}$ is a dagger trace operator, i.e. if f is a partial isomorphism, then $\mathsf{trace}_{\mathbb{D}}\, f$ is also a partial isomorphism with partial inverse $\mathsf{trace}_{\mathbb{D}}\,(\mathsf{dagger}_{\mathbb{D}}\, f)$.

This is proved by introducing the notion of *orbit* of an element $x : A{+}C$ wrt. a function $f : A{+}C \to \mathsf{Delay}\,(B{+}C)$. The orbit of x consists of the terms of type $B + C$ that are obtained in a finite number of steps from repeated applications of the function f on x. Formally, a term y belongs to the orbit of f wrt. x if the type $\mathsf{Orb}\, f\, x\, y$ is inhabited, with the latter type inductively defined as:

$$\frac{p : f\, x \downarrow y}{\mathsf{done}\, p : \mathsf{Orb}\, f\, x\, y} \qquad \frac{p : f\, x \downarrow \mathsf{inr}\, c \quad q : \mathsf{Orb}\, f\, (\mathsf{inr}\, c)\, y}{\mathsf{next}\, p\, q : \mathsf{Orb}\, f\, x\, y}$$

The notion of orbit can be used to state when the iteration of a function $f : A \to \mathsf{Delay}\,(B + A)$ on a input $a : A$ terminates with value $b : B$.

$$\mathsf{iter}_{\mathbb{D}}\, f\, a \downarrow b \leftrightarrow \mathsf{Orb}\, [\mathsf{inl}_{\mathbb{D}}, f]_{\mathbb{D}}\, (\mathsf{inr}\, a)\, (\mathsf{inl}\, b)$$

We refer the interested reader to our Agda formalization for a complete proof of this logical equivalence. Similarly, the orbit can be used to state when the trace of a function $f : A + C \to \mathsf{Delay}\,(B + C)$ on a input $a : A$ terminates with value $b : B$.

$$\mathsf{trace}_{\mathbb{D}}\, f\, a \downarrow b \leftrightarrow \mathsf{Orb}\, f\, (\mathsf{inl}\, a)\, (\mathsf{inl}\, b) \tag{2}$$

Showing that $\mathsf{trace}_{\mathbb{D}}$ is a dagger trace operator requires the construction of an inhabitant of $\mathsf{trace}_{\mathbb{D}}\, f\, a \downarrow b \leftrightarrow \mathsf{trace}_{\mathbb{D}}\,(\mathsf{dagger}_{\mathbb{D}}\, f)\, b \downarrow a$. Thanks to the logical equivalence in (2), this is equivalent to prove the following statement instead:

$$\mathsf{Orb}\, f\, (\mathsf{inl}\, a)\, (\mathsf{inl}\, b) \leftrightarrow \mathsf{Orb}\, (\mathsf{dagger}_{\mathbb{D}}\, f)\, (\mathsf{inl}\, b)\, (\mathsf{inl}\, a)$$

We give a detailed proof of the left-to-right direction, the other implication is derived in an analogous way. Notice that a term $p : \mathsf{Orb}\, f\, (\mathsf{inl}\, a)\, (\mathsf{inl}\, b)$ can be seen as a finite sequence of elements of C, precisely the intermediate values produced by $\mathsf{trace}_{\mathbb{D}}\, f\, a$ before converging to b. The orbit of b wrt. the partial inverse of f can therefore be computed by reversing the sequence of elements present in p. The construction of the reverse of an orbit is very similar to the way the reverse of a list is typically defined in a functional programming language like Haskell. We first consider an intermediate value $c : C$ and we assume to have already reversed the initial section of the orbit between $\mathsf{inl}\, a$ and $\mathsf{inr}\, c$, that is a term $p' : \mathsf{Orb}\, (\mathsf{dagger}_{\mathbb{D}}\, f)\, (\mathsf{inr}\, c)\, (\mathsf{inl}\, a)$.

$$
\begin{aligned}
&\mathsf{reverseOrb}' : (i : \{a : A\}\{b : B\} \to f\, a \downarrow b \to \mathsf{dagger}_{\mathbb{D}}\, f\, b \downarrow a) \to \\
&\qquad \{a : A\}\{b : B\}\{c : C\} \to \\
&\qquad \mathsf{Orb}\, f\, (\mathsf{inr}\, c)\, (\mathsf{inl}\, b) \to \mathsf{Orb}\, (\mathsf{dagger}_{\mathbb{D}}\, f)\, (\mathsf{inr}\, c)\, (\mathsf{inl}\, a) \to \\
&\qquad \mathsf{Orb}\, (\mathsf{dagger}_{\mathbb{D}}\, f)\, (\mathsf{inl}\, b)\, (\mathsf{inl}\, a) \\
&\mathsf{reverseOrb}'\, i\, (\mathsf{done}\, p) \quad p' = \mathsf{next}\, (i\, p)\, p' \\
&\mathsf{reverseOrb}'\, i\, (\mathsf{next}\, p\, q)\; p' = \mathsf{reverseOrb}'\, i\, q\, (\mathsf{next}\, (i\, p)\, p')
\end{aligned}
$$

The proof of $\mathsf{reverseOrb}'$ proceeds by structural induction on the final segment of the orbit between $\mathsf{inr}\, c$ and $\mathsf{inl}\, b$ that still needs to be reversed, which is the argument of type $\mathsf{Orb}\, f\, (\mathsf{inr}\, c)\, (\mathsf{inl}\, b)$. There are two possibilities.

- We have $p : f\, (\mathsf{inr}\, c) \downarrow \mathsf{inl}\, b$, in which case $i\, p : \mathsf{dagger}_{\mathbb{D}}\, f\, (\mathsf{inl}\, b) \downarrow \mathsf{inl}\, c$. Then we return $\mathsf{next}\, (i\, p)\, p'$.
- There exists another value $c' : C$ such that $p : f\, (\mathsf{inr}\, c) \downarrow \mathsf{inr}\, c'$ and $q : \mathsf{Orb}\, f\, (\mathsf{inr}\, c')\, (\mathsf{inl}\, b)$. Then we recursively invoke the function $\mathsf{reverseOrb}'\, i$ on arguments q and $\mathsf{next}\, (i\, p)\, p' : \mathsf{Orb}\, (\mathsf{dagger}_{\mathbb{D}}\, f)\, (\mathsf{inr}\, c')\, (\mathsf{inl}\, b)$.

The reverse of an orbit is derivable using the auxiliary function $\mathsf{reverseOrb}'$.

$$
\begin{aligned}
&\mathsf{reverseOrb} : (i : \{a : A\}\{b : B\} \to f\, a \downarrow b \to \mathsf{dagger}_{\mathbb{D}}\, f\, b \downarrow a) \to \\
&\qquad \{a : A\}\{b : B\} \to \\
&\qquad \mathsf{Orb}\, f\, (\mathsf{inl}\, a)\, (\mathsf{inl}\, b) \to \mathsf{Orb}\, (\mathsf{dagger}_{\mathbb{D}}\, f)\, (\mathsf{inl}\, b)\, (\mathsf{inl}\, a) \\
&\mathsf{reverseOrb}\, i\, (\mathsf{done}\, p) \quad = \mathsf{done}\, (i\, p) \\
&\mathsf{reverseOrb}\, i\, (\mathsf{next}\, p\, q) = \mathsf{reverseOrb}'\, i\, q\, (\mathsf{done}\, (i\, p))
\end{aligned}
$$

The proof of $\mathsf{reverseOrb}$ proceeds by structural induction on the orbit of type $\mathsf{Orb}\, f\, (\mathsf{inl}\, a)\, (\mathsf{inl}\, b)$. There are two possibilities.

- We have $p : f\, (\mathsf{inl}\, a) \downarrow \mathsf{inl}\, b$, in which case $i\, p : \mathsf{dagger}_{\mathbb{D}}\, f\, (\mathsf{inl}\, b) \downarrow \mathsf{inl}\, a$. Then we return $\mathsf{done}\, (i\, p)$.
- There exists a value $c : C$ such that $p : f\, (\mathsf{inl}\, a) \downarrow \mathsf{inr}\, c$ and $q : \mathsf{Orb}\, f\, (\mathsf{inr}\, c)\, (\mathsf{inl}\, b)$. We conclude by invoking the function $\mathsf{reverseOrb}'\, i$ on arguments q and $\mathsf{done}\, (i\, p) : \mathsf{Orb}\, (\mathsf{dagger}_{\mathbb{D}}\, f)\, (\mathsf{inr}\, c)\, (\mathsf{inl}\, a)$.

Summing up, in this section we have showed that the $\mathsf{trace}_{\mathbb{D}}$ operator can be restricted to act on partial isomorphisms. That is, the following type is inhabited:

$$\mathsf{trace}_{\mathbb{D}\simeq} : A + C \simeq B + C \to A \simeq B$$

5 Soundness

In this section, we provide some details on the interpretation of the syntax of Π°, presented in Sect. 2, into the category $\mathsf{Inv}\mathbb{D}_\approx$. Types of Π° are modelled as objects of $\mathsf{Inv}\mathbb{D}_\approx$, which are types of the metatheory. In Agda, the interpretation of types $[\![-]\!]_{\mathsf{Ty}}$ takes in input a Π° type $A : \mathsf{Ty}\,n$ and an environment $\rho : \mathsf{Fin}\,n \to \mathsf{Set}$ giving semantics to each variable in context. The interpretation is mutually inductively defined together with the operation $[\![-]\!]_\mu$ giving semantics to the μ type former. Remember that, given $A : \mathsf{Ty}\,(n+1)$ and $B : \mathsf{Ty}\,n$, we write $\mathsf{sub}\,A\,B$ for the substituted type $A[B/X]$, where X corresponds to the $(n+1)$-th variable in context.

$$
\begin{aligned}
&[\![\mathsf{Z}]\!]_{\mathsf{Ty}} && \rho = 0\\
&[\![A \oplus B]\!]_{\mathsf{Ty}} && \rho = [\![A]\!]_{\mathsf{Ty}}\,\rho + [\![B]\!]_{\mathsf{Ty}}\,\rho\\
&[\![\mathsf{I}]\!]_{\mathsf{Ty}} && \rho = 1\\
&[\![A \otimes B]\!]_{\mathsf{Ty}} && \rho = [\![A]\!]_{\mathsf{Ty}}\,\rho \times [\![B]\!]_{\mathsf{Ty}}\,\rho\\
&[\![\mathsf{Var}\,i]\!]_{\mathsf{Ty}} && \rho = \rho\,i\\
&[\![\mu A]\!]_{\mathsf{Ty}} && \rho = [\![A]\!]_\mu\,\rho
\end{aligned}
\qquad
\frac{x : [\![\mathsf{sub}\,A\,(\mu A)]\!]_{\mathsf{Ty}}\,\rho}{\mathsf{semFold}\,x : [\![A]\!]_\mu\,\rho}
$$

By abuse of notation, we use here $\times$ (respectively $+$) to refer to the product (respectively coproduct) in $\mathbb{D}_\approx$ even though it fails to be a product (respectively coproduct) in $\mathsf{Inv}\mathbb{D}_\approx$. However, both of these are symmetric monoidal products in $\mathsf{Inv}\mathbb{D}_\approx$, so their use as objects of $\mathsf{Inv}\mathbb{D}_\approx$ in the interpretation above is justified.

Terms of Π° are modelled as morphism of $\mathsf{Inv}\mathbb{D}_\approx$, i.e. partial isomorphisms. Here we only display the interpretation of a selection of programs, we refer the interested reader to our Agda formalization for a complete definition of the interpretation of terms.

$$
\begin{aligned}
&[\![-]\!]_{\longleftrightarrow} : (A \longleftrightarrow B) \to [\![A]\!]_{\mathsf{Ty}}\,\rho \simeq [\![B]\!]_{\mathsf{Ty}}\,\rho\\
&[\![f \oplus g]\!]_{\longleftrightarrow} \;= [\![f]\!]_{\longleftrightarrow} +_{\mathbb{D}\simeq} [\![g]\!]_{\longleftrightarrow}\\
&[\![f \otimes g]\!]_{\longleftrightarrow} \;= [\![f]\!]_{\longleftrightarrow} \times_{\mathbb{D}\simeq} [\![g]\!]_{\longleftrightarrow}\\
&[\![\mathsf{trace}\,f]\!]_{\longleftrightarrow} = \mathsf{trace}_{\mathbb{D}\simeq}\,[\![f]\!]_{\longleftrightarrow}
\end{aligned}
$$

Term equivalences of Π° are modelled as morphism equalities in $\mathsf{Inv}\mathbb{D}_\approx$, i.e. proofs of weak bisimilarity between two morphisms. Formally, we define an operation:

$$
[\![-]\!]_{\Longleftrightarrow} : (f \iff g) \to [\![f]\!]_{\longleftrightarrow} \approx [\![g]\!]_{\longleftrightarrow}
$$

Again we refer the interested reader to our Agda formalization for a complete definition of the interpretation of term equivalences.

6 Conclusions

In this paper, we have extended the work of Carette and Sabry [7] to a (fully formalized) two-level calculus of Π° programs and program equivalences. Key in this effort was the use of the Kleisli category of the delay monad on Set under weak bisimilarity, which turned out to support iteration via a trace that preserves all partial isomorphisms, in this way giving semantics to the dagger trace of Π°. Further, the work was formalized using Agda 2.6.0.

It is natural to wonder if our work can be ported to other monads of partiality in Martin-Löf type theory. As already discussed in the end of Sect. 4.1, the maybe monad is not suitable for modelling a well-behaved trace combinator without the assumption of classical principles such as LPO. The partial map classifier [13,38] $\mathsf{PMC}\,A = (P : \mathsf{Prop}) \to P \to A$, where Prop is the type of propositions (types with at most one inhabitant), supports the existence of a uniform iteration operator and therefore a trace. Nevertheless, the specification of iteration is more complicated then the one presented in Sect. 4 for the delay monad, which is a simple corecursive definition. The complete Elgot monad structure of PMC follows from its Kleisli category being a join restriction category, so iteration is defined in terms of least upper bounds of certain chains of morphisms. The subcategory of partial isomorphisms of the Kleisli category of PMC supports a dagger trace combinator, which can be proved following the general strategy in [25]. The exact same observations apply to the partiality monad in homotopy type theory [2,8], to which the quotiented delay monad $\mathsf{Delay}_\approx A = \mathsf{Delay}\,A/\!\approx$ is isomorphic under the assumption of countable choice.

Though the Kleisli category of the delay monad on Set is well studied, comparatively less is known about this monad on other categories. It could be interesting to study under which conditions its iterator exists – e.g., whether this is still the case when Set is replaced with an arbitrary topos. Another avenue concerns the study of *time invertible* programming languages: Though not immediately clear in the current presentation, the trace on $\mathsf{Inv}\mathbb{D}_\approx$ is not just reversible (in the sense that it preserves partial isomorphisms) but in fact time invertible, in the sense that the number of computation steps needed to perform $\mathsf{trace}_\mathbb{D}\,(\mathsf{dagger}_\mathbb{D}\,f)$ is *precisely* the same as what is needed to perform $\mathsf{trace}_\mathbb{D}\,f$ on any input. Since the delay monad conveniently allows the counting of computation steps, we conjecture that this is an ideal setting in which to study such intentional semantics of reversible programming languages.

References

1. Abel, A., Chapman, J.: Normalization by evaluation in the delay monad: a case study for coinduction via copatterns and sized types. In: Proceedings 5th Workshop on Mathematically Structured Functional Programming, MSFP@ETAPS 2014, Grenoble, France, 12 April 2014, pp. 51–67 (2014). https://doi.org/10.4204/EPTCS.153.4

2. Altenkirch, T., Danielsson, N.A., Kraus, N.: Partiality, revisited. In: Esparza, J., Murawski, A.S. (eds.) FoSSaCS 2017. LNCS, vol. 10203, pp. 534–549. Springer, Heidelberg (2017). https://doi.org/10.1007/978-3-662-54458-7_31
3. Barthe, G., Capretta, V., Pons, O.: Setoids in type theory. J. Funct. Program. **13**(2), 261–293 (2003). https://doi.org/10.1017/S0956796802004501
4. Bennett, C.H.: Logical reversibility of computation. IBM J. Res. Dev. **17**(6), 525–532 (1973)
5. Benton, N., Kennedy, A., Varming, C.: Some domain theory and denotational semantics in Coq. In: Berghofer, S., Nipkow, T., Urban, C., Wenzel, M. (eds.) TPHOLs 2009. LNCS, vol. 5674, pp. 115–130. Springer, Heidelberg (2009). https://doi.org/10.1007/978-3-642-03359-9_10
6. Capretta, V.: General recursion via coinductive types. Logical Methods Comput. Sci. **1**(2) (2005). https://doi.org/10.2168/LMCS-1(2:1)2005
7. Carette, J., Sabry, A.: Computing with semirings and weak rig groupoids. In: Thiemann, P. (ed.) ESOP 2016. LNCS, vol. 9632, pp. 123–148. Springer, Heidelberg (2016). https://doi.org/10.1007/978-3-662-49498-1_6
8. Chapman, J., Uustalu, T., Veltri, N.: Quotienting the delay monad by weak bisimilarity. Math. Struct. Comput. Sci. **29**(1), 67–92 (2019). https://doi.org/10.1017/S0960129517000184
9. Cockett, J.R.B., Lack, S.: Restriction categories I: categories of partial maps. Theoret. Comput. Sci. **270**(1–2), 223–259 (2002)
10. Cockett, J.R.B., Lack, S.: Restriction categories III: colimits, partial limits and extensivity. Math. Struct. Comput. Sci. **17**(4), 775–817 (2007). https://doi.org/10.1017/S0960129507006056
11. Danielsson, N.A.: Operational semantics using the partiality monad. In: ACM SIGPLAN International Conference on Functional Programming, ICFP 2012, Copenhagen, Denmark, 9–15 September 2012, pp. 127–138 (2012). https://doi.org/10.1145/2364527.2364546
12. Danielsson, N.A.: Up-to techniques using sized types. PACMPL **2**(POPL), 43:1–43:28 (2018). https://doi.org/10.1145/3158131
13. Escardó, M.H., Knapp, C.M.: Partial elements and recursion via dominances in univalent type theory. In: 26th EACSL Annual Conference on Computer Science Logic, CSL 2017, 20–24 August 2017, Stockholm, Sweden, pp. 21:1–21:16 (2017). https://doi.org/10.4230/LIPIcs.CSL.2017.21
14. Ésik, Z., Goncharov, S.: Some remarks on conway and iteration theories. CoRR abs/1603.00838 (2016). http://arxiv.org/abs/1603.00838
15. Giles, B.: An investigation of some theoretical aspects of reversible computing. Ph.D. thesis, University of Calgary (2014)
16. Goncharov, S., Milius, S., Rauch, C.: Complete elgot monads and coalgebraic resumptions. Electr. Notes Theor. Comput. Sci. **325**, 147–168 (2016). https://doi.org/10.1016/j.entcs.2016.09.036
17. Goncharov, S., Schröder, L., Rauch, C., Jakob, J.: Unguarded recursion on coinductive resumptions. Logical Methods Comput. Sci. **14**(3) (2018). https://doi.org/10.23638/LMCS-14(3:10)2018
18. Hasegawa, M.: Recursion from cyclic sharing: traced monoidal categories and models of cyclic lambda calculi. In: de Groote, P., Roger Hindley, J. (eds.) TLCA 1997. LNCS, vol. 1210, pp. 196–213. Springer, Heidelberg (1997). https://doi.org/10.1007/3-540-62688-3_37
19. Hofmann, M.: Extensional Constructs in Intensional Type Theory. CPHC/BCS Distinguished Dissertations. Springer, London (1997). https://doi.org/10.1007/978-1-4471-0963-1

20. Jacobsen, P.A.H., Kaarsgaard, R., Thomsen, M.K.: CoreFun: a typed functional reversible core language. In: Kari, J., Ulidowski, I. (eds.) RC 2018. LNCS, vol. 11106, pp. 304–321. Springer, Cham (2018). https://doi.org/10.1007/978-3-319-99498-7_21
21. James, R.P., Sabry, A.: Information effects. In: Proceedings of the 39th ACM SIGPLAN-SIGACT Symposium on Principles of Programming Languages, POPL 2012, Philadelphia, Pennsylvania, USA, 22–28 January 2012, pp. 73–84 (2012). https://doi.org/10.1145/2103656.2103667
22. James, R.P., Sabry, A.: Theseus: A high level language for reversible computing (2014). https://www.cs.indiana.edu/~sabry/papers/theseus.pdf. Work-in-progress report at RC 2014
23. Joyal, A., Street, R., Verity, D.: Traced monoidal categories. Math. Proc. Camb. Philos. Soc. **119**(3), 447–468 (1996). https://doi.org/10.1017/S0305004100074338
24. Kaarsgaard, R., Glück, R.: A categorical foundation for structured reversible flowchart languages: soundness and adequacy. Logical Methods Comput. Sci. **14**(3), 1–38 (2018)
25. Kaarsgaard, R., Axelsen, H.B., Glück, R.: Join inverse categories and reversible recursion. J. Logic Algebra Methods Program. **87**, 33–50 (2017). https://doi.org/10.1016/j.jlamp.2016.08.003
26. Karvonen, M.: The Way of the Dagger. Ph.D. thesis, School of Informatics, University of Edinburgh (2019)
27. Kastl, J.: Inverse categories. In: Hoehnke, H.J. (ed.) Algebraische Modelle, Kategorien und Gruppoide, Studien zur Algebra und ihre Anwendungen, vol. 7, pp. 51–60. Akademie-Verlag, Berlin (1979)
28. Landauer, R.: Irreversibility and heat generation in the computing process. IBM J. Res. Dev. **5**(3), 261–269 (1961)
29. Laplaza, M.L.: Coherence for distributivity. In: Kelly, G.M., Laplaza, M., Lewis, G., Mac Lane, S. (eds.) Coherence in Categories. LNM, vol. 281, pp. 29–65. Springer, Heidelberg (1972). https://doi.org/10.1007/BFb0059555
30. Laursen, J.S., Ellekilde, L.P., Schultz, U.P.: Modelling reversible execution of robotic assembly. Robotica **36**(5), 625–654 (2018)
31. Norell, U.: Dependently Typed Programming in Agda. In: Proceedings of TLDI 2009: 2009 ACM SIGPLAN International Workshop on Types in Languages Design and Implementation, Savannah, GA, USA, 24 January 2009, pp. 1–2 (2009)
32. Rendel, T., Ostermann, K.: Invertible syntax descriptions: unifying parsing and pretty printing. ACM SIGPLAN Not. **45**(11), 1–12 (2010)
33. Schordan, M., Jefferson, D., Barnes, P., Oppelstrup, T., Quinlan, D.: Reverse code generation for parallel discrete event simulation. In: Krivine, J., Stefani, J.-B. (eds.) RC 2015. LNCS, vol. 9138, pp. 95–110. Springer, Cham (2015). https://doi.org/10.1007/978-3-319-20860-2_6
34. Schultz, U.P.: Reversible object-oriented programming with region-based memory management. In: Kari, J., Ulidowski, I. (eds.) RC 2018. LNCS, vol. 11106, pp. 322–328. Springer, Cham (2018). https://doi.org/10.1007/978-3-319-99498-7_22
35. Thomsen, M.K., Axelsen, H.B., Glück, R.: A reversible processor architecture and its reversible logic design. In: De Vos, A., Wille, R. (eds.) RC 2011. LNCS, vol. 7165, pp. 30–42. Springer, Heidelberg (2012). https://doi.org/10.1007/978-3-642-29517-1_3
36. Univalent Foundations Program: Homotopy Type Theory: Univalent Foundations of Mathematics, Institute for Advanced Study (2013). https://homotopytypetheory.org/book

37. Uustalu, T., Veltri, N.: The delay monad and restriction categories. In: Hung, D., Kapur, D. (eds.) Theoretical Aspects of Computing - ICTAC 2017. LNCS, vol. 10580, pp. 32–50. Springer, Cham (2017). https://doi.org/10.1007/978-3-319-67729-3_3
38. Uustalu, T., Veltri, N.: Partiality and container monads. In: Chang, B.-Y.E. (ed.) APLAS 2017. LNCS, vol. 10695, pp. 406–425. Springer, Cham (2017). https://doi.org/10.1007/978-3-319-71237-6_20
39. Veltri, N.: A Type-Theoretical Study of Nontermination. Ph.D. thesis, Tallinn University of Technology (2017). https://digi.lib.ttu.ee/i/?7631
40. de Vos, A.: Reversible Computing: Fundamentals, Quantum Computing, and Applications. Wiley, Weinheim (2010)
41. Yokoyama, T., Glück, R.: A reversible programming language and its invertible self-interpreter. In: Proceedings of Partial Evaluation and Program Manipulation, pp. 144–153. ACM (2007)

Completeness and Incompleteness of Synchronous Kleene Algebra

Jana Wagemaker[1(✉)], Marcello Bonsangue[2], Tobias Kappé[1], Jurriaan Rot[1,3], and Alexandra Silva[1]

[1] University College London, London, UK
j.wagemaker@ucl.ac.uk
[2] Leiden University, Leiden, The Netherlands
[3] Radboud University, Nijmegen, The Netherlands

Abstract. *Synchronous Kleene algebra* (*SKA*), an extension of Kleene algebra (KA), was proposed by Prisacariu as a tool for reasoning about programs that may execute synchronously, i.e., in lock-step. We provide a countermodel witnessing that the axioms of SKA are incomplete w.r.t. its language semantics, by exploiting a lack of interaction between the *synchronous product* operator and the Kleene star. We then propose an alternative set of axioms for SKA, based on Salomaa's axiomatisation of regular languages, and show that these provide a sound and complete characterisation w.r.t. the original language semantics.

1 Introduction

Kleene algebra (*KA*) is applied in various contexts, such as relational algebra and automata theory. An important use of KA is as a logic of programs. This is because the axioms of KA correspond well to properties expected of sequential program composition, and hence they provide a logic for reasoning about control flow of sequential programs presented as Kleene algebra expressions. Regular languages then provide a canonical semantics for programs expressed in Kleene algebra, due to a tight connection between regular languages and the axioms of KA: an equation is provable using the Kleene algebra axioms if and only if the corresponding regular languages coincide [5,18,21].

In [24], Prisacariu proposes an extension of Kleene algebra, called *synchronous Kleene algebra* (*SKA*). The aim was to introduce an algebra useful for studying not only sequential programs but also *synchronous* concurrent programs. Here, synchrony is understood as in Milner's SCCS [23], i.e., each program executes a single action instantaneously at each discrete time step. Hence, the synchrony paradigm assumes that basic actions execute in one unit of time and that at each time step, all components capable of acting will do so. This model permits a *synchronous product* operator, which yields a program that, at each

This work was partially supported by ERC Starting Grant ProFoundNet (679127), a Leverhulme Prize (PLP–2016–129) and a Marie Curie Fellowship (795119). The first author conducted part of this work at Centrum Wiskunde & Informatica, Amsterdam.

G. Hutton (Ed.): MPC 2019, LNCS 11825, pp. 385–413, 2019.
https://doi.org/10.1007/978-3-030-33636-3_14

time step, executes some combination of the actions put forth by the operand programs.

This new operator is governed by various expected axioms such as associativity and commutativity. Another axiom describes the interaction between the synchronous product and the sequential product, capturing the intended lock-step behaviour. Crucially, the axioms do not entail certain equations that relate the Kleene star (used to describe loops) and the synchronous product.

The contributions of this paper are twofold. First, we show that the lack of connection between the Kleene star and the synchronous product is problematic. In particular, we exploit this fact to devise a countermodel that violates a semantically valid equation, thus showing that the SKA axioms are incomplete w.r.t. the language semantics. This invalidates the completeness result in [24].

The second and main contribution of this paper is a sound and complete characterisation of the equational theory of SKA in terms of a generalisation of regular languages. The key difference with [24] is the shift from *least* fixpoint axioms in the style of Kozen [18] to a *unique* fixpoint axiom in the style of Salomaa [26]. In the completeness proof, we give a reduction to the completeness result of Salomaa via a normal form for SKA expressions. As a by-product, we get a proof of the correctness of the partial derivatives for SKA provided in [7].

This paper is organised as follows. In Sect. 2 we discuss the necessary preliminaries. In Sect. 3 we discuss SKA as presented in [24]. Next, in Sect. 4, we demonstrate why SKA is incomplete, and in Sect. 5 go on to provide a new set of axioms, which we call SF_1. The latter section also includes basic results about the partial derivatives for SKA from [7]. In Sect. 6 we provide an algebraic characterisation of SF_1-terms; this characterisation is used in Sect. 7, where we prove completeness of SF_1 w.r.t. to its language model. In Sect. 8 we consider related work and conclude by discussing directions for future work in Sect. 9. For the sake of readability, some of the proofs appear in the appendix.

2 Preliminaries

Throughout this paper, we write 2 for the two-element set $\{0, 1\}$.

Languages. Throughout the paper we fix a finite alphabet Σ. A *word* formed over Σ is a finite sequence of symbols from Σ. The *empty word* is denoted by ε. We write Σ^* for the set of all words over Σ. *Concatenation* of words $u, v \in \Sigma^*$ is denoted by $uv \in \Sigma^*$. A *language* is a set of words. For $K, L \subseteq \Sigma^*$, we define

$$K \cdot L = \{uv : u \in K, v \in L\} \qquad K + L = K \cup L \qquad K^* = \bigcup_{n \in \mathbb{N}} K^n,$$

where $K^0 = \{\varepsilon\}$ and $K^{n+1} = K \cdot K^n$.

Kleene Algebra. We define a *Kleene algebra* [18] as a tuple $(A, +, \cdot, ^*, 0, 1)$ where A is a set, * is a unary operator, $+$ and $\cdot$ are binary operators and 0 and 1 are constants. Moreover, for all $e, f, g \in A$ the following axioms are satisfied:

$$e+(f+g)=(e+f)+g \quad e+f=f+e \quad e+0=e \quad e+e=e$$
$$e\cdot 1=e=1\cdot e \quad e\cdot 0=0=0\cdot e \quad e\cdot(f\cdot g)=(e\cdot f)\cdot g$$
$$e^*=1+e\cdot e^*=1+e^*\cdot e \quad (e+f)\cdot g=e\cdot g+f\cdot g \quad e\cdot(f+g)=e\cdot f+e\cdot g$$

Additionally, we write $e \leq f$ as a shorthand for $e+f=f$, and require that the *least fixpoint axioms* [18] hold, which stipulate that for $e, f, g \in A$ we have

$$e+f\cdot g\leq g \implies f^*\cdot e\leq g \qquad e+f\cdot g\leq f \implies e\cdot g^*\leq f$$

The set of *regular expressions*, denoted $\mathcal{T}_{\mathsf{KA}}$, is described by the grammar:

$$\mathcal{T}_{\mathsf{KA}} \ni e, f ::= 0 \mid 1 \mid a\in\Sigma \mid e+f \mid e\cdot f \mid e^*$$

Regular expressions can be interpreted in terms of languages. This is done by defining $[\![-]\!]_{\mathsf{KA}} : \mathcal{T}_{\mathsf{KA}} \to \mathcal{P}(\Sigma^*)$ inductively, as follows.

$$[\![0]\!]_{\mathsf{KA}}=\emptyset \qquad [\![a]\!]_{\mathsf{KA}}=\{a\} \qquad [\![e\cdot f]\!]_{\mathsf{KA}}=[\![e]\!]_{\mathsf{KA}}\cdot[\![f]\!]_{\mathsf{KA}}$$
$$[\![1]\!]_{\mathsf{KA}}=\{\varepsilon\} \qquad [\![e+f]\!]_{\mathsf{KA}}=[\![e]\!]_{\mathsf{KA}}+[\![f]\!]_{\mathsf{KA}} \qquad [\![e^*]\!]_{\mathsf{KA}}=[\![e]\!]_{\mathsf{KA}}^*$$

A language L is called *regular* if and only if $L=[\![e]\!]_{\mathsf{KA}}$ for some $e\in\mathcal{T}_{\mathsf{KA}}$.

We write $\equiv_{\mathsf{KA}}$ for the smallest congruence on $\mathcal{T}_{\mathsf{KA}}$ induced by the Kleene algebra axioms—e.g., for all $e\in\mathcal{T}_{\mathsf{KA}}$, we have $1+e\cdot e^* \equiv_{\mathsf{KA}} e^*$. Intuitively, $e\equiv_{\mathsf{KA}} f$ means that the regular expressions e and f can be proved equivalent according to the axioms of Kleene algebra. A pivotal result in the study of Kleene algebras tells us that $[\![-]\!]_{\mathsf{KA}}$ characterises $\equiv_{\mathsf{KA}}$, in the following sense:

Theorem 2.1 (Soundness and Completeness of KA [18]**).** *For all $e, f\in\mathcal{T}_{\mathsf{KA}}$, we have that $e\equiv_{\mathsf{KA}} f$ if and only if $[\![e]\!]_{\mathsf{KA}}=[\![f]\!]_{\mathsf{KA}}$.*

Remark 2.2. The above can be generalised, as follows. Let $\mathcal{K}=(A,+,\cdot,^*,0,1)$ be a KA, and let $\sigma:\Sigma\to A$. Then for all $e,f\in\mathcal{T}_{\mathsf{KA}}$ such that $e\equiv_{\mathsf{KA}} f$, interpreting e and f according to σ in $\mathcal{K}$ yields the same result. For instance, since $(a^*)^*\equiv_{\mathsf{KA}} a^*$, we know that for *any* element e of *any* KA $\mathcal{K}$, we have that $(e^*)^*=e$.

Linear Systems. Let Q be a finite set. A *Q-vector* is a function $x: Q\to\mathcal{T}_{\mathsf{KA}}$. A *$Q$-matrix* is a function $M: Q\times Q\to\mathcal{T}_{\mathsf{KA}}$. Let x and y be Q-vectors. Addition is defined pointwise, setting $(x+y)(q)=x(q)+y(q)$. Multiplication by a Q-matrix M is given by

$$(M\cdot x)(q)=\sum_{e\in Q} M(q,e)\cdot x(e)$$

When $x(q)\equiv_{\mathsf{KA}} y(q)$ for all $q\in Q$, we write $x\equiv_{\mathsf{KA}} y$.

Definition 2.3. *A Q-*linear system *is a pair (M,x) with M a Q-matrix and x a Q-vector. A solution to (M,x) in KA is a Q-vector y such that $M\cdot y+x\equiv_{\mathsf{KA}} y$.*

Non-deterministic Finite Automata. A *non-deterministic automaton (NDA)* over an alphabet Σ is a triple (X, o, d) where $o\colon X \to 2$ is called the *termination function* and $d\colon X \times \Sigma \to X$ called the *continuation function.* If X is finite, (X, o, d) is referred to as a *non-deterministic finite automaton (NFA).*

The semantics of an NDA (X, o, d) can be characterised recursively as the unique map $\ell : X \to \mathcal{P}(\Sigma^*)$ such that

$$\ell(x) = \{\varepsilon : o(x) = 1\} \cup \bigcup_{x' \in d(x,a)} \{a\} \cdot \ell(x') \tag{1}$$

This coincides with the standard definition of language acceptance.

3 Synchronous Kleene Algebra

Synchronous Kleene algebra extends Kleene algebra with an additional operator denoted $\times$, which we refer to as the synchronous product [24].

Definition 3.1 (Synchronous Kleene Algebra). *A* synchronous KA (SKA) *is a tuple* $(A, S, +, \cdot, ^*, \times, 0, 1)$ *such that* $(A, +, \cdot, ^*, 0, 1)$ *is a Kleene algebra and* $\times$ *is a binary operator on* A*, with* $S \subseteq A$ *closed under* $\times$ *and* $(S, \times)$ *a semilattice. Furthermore, the following hold for all* $e, f, g \in A$ *and* $\alpha, \beta \in S$*:*

$$e \times (f + g) = e \times f + e \times g \qquad e \times (f \times g) = (e \times f) \times g \qquad e \times 0 = 0$$
$$(\alpha \cdot e) \times (\beta \cdot f) = (\alpha \times \beta) \cdot (e \times f) \qquad e \times f = f \times e \qquad e \times 1 = e$$

Note that 0 and 1 need not be elements of S. The *semilattice terms*, denoted $\mathcal{T}_{\mathsf{SL}}$, are given by the following grammar.

$$\mathcal{T}_{\mathsf{SL}} \ni e, f ::= a \in \Sigma \mid e \times f$$

The *synchronous regular terms*, denoted $\mathcal{T}_{\mathsf{SKA}}$, are given by the grammar:

$$\mathcal{T}_{\mathsf{SKA}} \ni e, f ::= 0 \mid 1 \mid a \in \mathcal{T}_{\mathsf{SL}} \mid e + f \mid e \cdot f \mid e \times f \mid e^*$$

Thus we have $\mathcal{T}_{\mathsf{SL}} \subseteq \mathcal{T}_{\mathsf{SKA}}$. We then define $\equiv_{\mathsf{SKA}}$ as the smallest congruence on $\mathcal{T}_{\mathsf{SKA}}$ satisfying the axioms of SKA Here, $\mathcal{T}_{\mathsf{SL}}$ plays the role of the semilattice; for instance, for $a \in \mathcal{T}_{\mathsf{SL}}$ we have that $a \times a \equiv_{\mathsf{SKA}} a$.

Remark 3.2. In [24], $\times$ is declared to be idempotent on the *generators* of the semilattice, whereas in our definition it holds for semilattice elements in general. This does not change anything, as the axiom $a \times a = a$ for generators together with commutativity and associativity results in idempotence on the semilattice. We present SKA as in Definition 3.1 to prevent a meta-definition of a third sort (namely the semilattice generated by Σ) present in the signature of the algebra. We have also left out the second distributivity and unit axioms that follow immediately from the ones presented and commutativity.

3.1 A Language Model for SKA

Similar to Kleene algebra, there is a language model for SKA [24].

Words over $\mathcal{P}(\Sigma) \setminus \{\emptyset\} = \mathcal{P}_n(\Sigma)$ are called *synchronous strings*, and sets of synchronous strings are called *synchronous languages*. The standard language operations (sum, concatenation, Kleene closure) are also defined on synchronous languages. The synchronous product of synchronous languages K, L is given by:

$$K \times L = \{u \times v : u \in K, v \in L\}$$

where we define $\times$ inductively for $u, v \in (\mathcal{P}_n(\Sigma))^*$ and $x, y \in \mathcal{P}_n(\Sigma)$, as follows:

$$u \times \varepsilon = u = \varepsilon \times u \qquad \text{and} \qquad (x \cdot u) \times (y \cdot v) = (x \cup y) \cdot (u \times v)$$

To define the language semantics for all elements in $\mathcal{T}_{\mathsf{SKA}}$, we first give an interpretation of elements in $\mathcal{T}_{\mathsf{SL}}$ in terms of non-empty finite subsets of Σ.

Definition 3.3. *For $a \in \Sigma$ and $e, f \in \mathcal{T}_{\mathsf{SL}}$, define $[\![-]\!]_{\mathsf{SL}} : \mathcal{T}_{\mathsf{SL}} \to \mathcal{P}_n(\Sigma)$ by*

$$[\![a]\!]_{\mathsf{SL}} = \{a\} \qquad\qquad [\![e \times f]\!]_{\mathsf{SL}} = [\![e]\!]_{\mathsf{SL}} \cup [\![f]\!]_{\mathsf{SL}}$$

Denote the smallest congruence on $\mathcal{T}_{\mathsf{SL}}$ with respect to idempotence, associativity and commutativity of $\times$ with $\equiv_{\mathsf{SL}}$. It is not hard to show that $[\![-]\!]_{\mathsf{SL}}$ characterises $\equiv_{\mathsf{SL}}$, in the following sense.

Lemma 3.4 (Soundness and Completeness of SL). *For all $e, f \in \mathcal{T}_{\mathsf{SL}}$, we have $[\![e]\!]_{\mathsf{SL}} = [\![f]\!]_{\mathsf{SL}}$ if and only if $e \equiv_{\mathsf{SL}} f$.*

The semantics of synchronous regular terms is given in terms of a mapping to synchronous languages: $[\![-]\!]_{\mathsf{SKA}} : \mathcal{T}_{\mathsf{SKA}} \to \mathcal{P}((\mathcal{P}_n(\Sigma))^*)$. We have:

$$\begin{array}{llll} [\![0]\!]_{\mathsf{SKA}} = \emptyset \quad [\![1]\!]_{\mathsf{SKA}} = \{\varepsilon\} & [\![a]\!]_{\mathsf{SKA}} = \{[\![a]\!]_{\mathsf{SL}}\} \quad \forall a \in \mathcal{T}_{\mathsf{SL}} & [\![e^*]\!]_{\mathsf{SKA}} = [\![e]\!]_{\mathsf{SKA}}^* \\ [\![e \cdot f]\!]_{\mathsf{SKA}} = [\![e]\!]_{\mathsf{SKA}} \cdot [\![f]\!]_{\mathsf{SKA}} & [\![e + f]\!]_{\mathsf{SKA}} = [\![e]\!]_{\mathsf{SKA}} + [\![f]\!]_{\mathsf{SKA}} & [\![e \times f]\!]_{\mathsf{SKA}} = [\![e]\!]_{\mathsf{SKA}} \times [\![f]\!]_{\mathsf{SKA}} \end{array} \qquad (2)$$

A synchronous language L is called *regular* when $L = [\![e]\!]_{\mathsf{SKA}}$ for some $e \in \mathcal{T}_{\mathsf{SKA}}$.

Let $S = \{\{x\} : x \in \mathcal{P}_n(\Sigma)\}$, that is to say, S is the set of synchronous languages consisting of a single word, whose single letter is in turn a subset of Σ. Furthermore, let $\mathcal{L}_\Sigma$ denote the set of synchronous languages over Σ. It is straightforward to prove that $\mathcal{L}_\Sigma$ together with S is closed under the SKA operations and satisfies the SKA axioms [24]; more precisely, we have:

Lemma 3.5. *The structure $(\mathcal{L}_\Sigma, S, +, \cdot, ^*, \times, \emptyset, \{\varepsilon\})$ is an SKA, that is, synchronous languages over Σ form an SKA.*

As a consequence of Lemma 3.5, we obtain soundness of the SKA axioms with respect to the language model based on synchronous regular languages:

Lemma 3.6 (Soundness of SKA). *For all $e, f \in \mathcal{T}_{\mathsf{SKA}}$, we have that $e \equiv_{\mathsf{SKA}} f$ implies $[\![e]\!]_{\mathsf{SKA}} = [\![f]\!]_{\mathsf{SKA}}$.*

Remark 3.7. The above generalises almost analogously to Remark 2.2. Let $\mathcal{M}$ be an SKA with semilattice S, and let $\sigma : \Sigma \to S$ be a function. Then for all $e, f \in \mathcal{T}_{\mathsf{SKA}}$ such that $e \equiv_{\mathsf{SKA}} f$, if we interpret e in $\mathcal{M}$ according to σ, then we should get the same result as when we interpret f in $\mathcal{M}$ according to σ.

In other words, when $e \equiv_{\mathsf{SKA}} f$ holds, it follows that $e = f$ is a valid equation in *every* SKA, provided that the symbols from Σ are interpreted as elements of the semilattice. It is not hard to show that this claim does not hold when symbols from Σ can be interpreted as elements of the carrier at large.

4 Incompleteness of SKA

We now prove incompleteness of the SKA axioms as presented in [24]. Fix alphabet $\mathcal{A} = \{a\}$. First, note that the language model of SKA has the following property.

Lemma 4.1. *For $\alpha \in \mathcal{T}_{\mathsf{SL}}$, we have $[\![\alpha^* \times \alpha^*]\!]_{\mathsf{SKA}} = [\![\alpha^*]\!]_{\mathsf{SKA}}$.*

If $\equiv_{\mathsf{SKA}}$ were complete w.r.t. $[\![-]\!]_{\mathsf{SKA}}$, then the above implies that $a^* \times a^* \equiv_{\mathsf{SKA}} a^*$ holds. In this section, we present a countermodel where all the axioms of SKA are true, but $\alpha^* \times \alpha^* = \alpha^*$ does not hold for any $\alpha \in S$. This shows that $a^* \times a^* \not\equiv_{\mathsf{SKA}} a^*$; consequently, $\equiv_{\mathsf{SKA}}$ cannot be complete w.r.t. $[\![-]\!]_{\mathsf{SKA}}$.

Countermodel for SKA

We define our countermodel as follows. For the semilattice, let $S = \{\{\{s\}\}\}$, the set containing the synchronous language $\{\{s\}\}$. We denote the set of all synchronous languages over alphabet $\{s\}$ with $\mathcal{L}_s$; the carrier of our model is formed by $\mathcal{L}_s \cup \{\dagger\}$, where $\dagger$ is a symbol not found in $\mathcal{L}_s$. The symbol $\dagger$ exists only in the model, and not in the algebraic theory. It remains to define the SKA operators on this carrier, which we do as follows.

Definition 4.2. *An element of $\mathcal{L}_s \cup \{\dagger\}$ is said to be infinite when it is an infinite language. For $K, L \in \mathcal{L}_s \cup \{\dagger\}$, define the SKA operators as follows:*

$$K + L = \begin{cases} \dagger & K = \dagger \vee L = \dagger \\ K \cup L & \textit{otherwise} \end{cases}$$

$$K \cdot L = \begin{cases} \emptyset & K = \emptyset \vee L = \emptyset \\ \dagger & K = \dagger \vee L = \dagger \\ \{u \cdot v : u \in K, v \in L\} & \textit{otherwise} \end{cases}$$

$$K \times L = \begin{cases} \emptyset & K = \emptyset \vee L = \emptyset \\ \dagger & K = \dagger \vee L = \dagger \vee K, L \textit{ infinite} \\ \{u \times v : u \in K, v \in L\} & \textit{otherwise} \end{cases}$$

$$K^* = \begin{cases} \dagger & K = \dagger \\ \bigcup_{n \in \mathbb{N}} K^n & \textit{otherwise} \end{cases}$$

where $u \times v$ for $u \in K$ and $v \in L$ and K^n is as defined in Sect. 3. Here, the cases are given in order of priority—e.g., if $K = \emptyset$ and $L = \dagger$, then $K \cdot L = \emptyset$.

The intuition behind this model is that SKA has no axioms that relate to the synchronous execution of starred expressions, such as in $\alpha^* \times \alpha^*$, nor can such a relation be derived from the axioms, meaning that a model has some leeway in defining the outcome in such cases. Since the language of a starred expression is generally infinite, we choose $\times$ such that it diverges to the extra element $\dagger$ when given infinite languages as input; for the rest of the operators, the behaviour on $\dagger$ is chosen to comply with the axioms.

First, we verify that our operators satisfy the SKA axioms.

Lemma 4.3. $\mathcal{M} = (\mathcal{L}_s \cup \{\dagger\}, \{\{\{s\}\}\}, +, \cdot, ^*, \times, \emptyset, \{\varepsilon\})$ *with the operators as defined in Definition 4.2 forms an SKA.*

Proof. For the sake of brevity, we validate one of the least fixpoint axioms and the synchrony axiom; the other axioms are treated in the appendix.

Let $K, L, J \in \mathcal{L}_s \cup \{\dagger\}$. We verify that $K + L \cdot J \leq J \implies L^* \cdot K \leq J$. Assume that $K + L \cdot J \leq J$. If $J = \dagger$, then the result follows by definition of $\leq$ and our choice of $+$. Otherwise, if $J \in \mathcal{L}_s$, we distinguish two cases. If $L = \dagger$, then J must be $\emptyset$ (otherwise $J = \dagger$); hence $K = \emptyset$, and the claim holds. Lastly, if $L \in \mathcal{L}_s$, then $K \in \mathcal{L}_s$. In this case, all of the operands are languages, and thus the proof goes through as it does for KA.

For the synchrony axiom, we need only check

$$(A \cdot K) \times (A \cdot L) = (A \times A) \cdot (K \times L)$$

for $A = \{\{s\}\}$ as that is the only element in S. Let $K, L \in \mathcal{L}_s \cup \{\dagger\}$. If either K or L is $\emptyset$, both sides of the equation reduce to $\emptyset$. Otherwise, if K or L is $\dagger$, then both sides of the equation reduce to $\dagger$. If K and L are both infinite then $A \cdot K$ and $A \cdot L$ are infinite and the claim follows. In all the remaining cases where K and L are elements of $\mathcal{L}_s$ and at most one of them is infinite, the proof goes through as it does for synchronous regular languages (Lemma 3.6). □

This leads us to the following theorem:

Theorem 4.4. *The axioms of SKA presented in Definition 3.1 are incomplete. That is, there exist* $e, f \in \mathcal{T}_{\mathsf{SKA}}$ *such that* $[\![e]\!]_{\mathsf{SKA}} = [\![f]\!]_{\mathsf{SKA}}$ *but* $e \not\equiv_{\mathsf{SKA}} f$.

Proof. Take $a \in \mathcal{A}$. We know from Lemma 4.1 that $[\![a^* \times a^*]\!]_{\mathsf{SKA}} = [\![a^*]\!]_{\mathsf{SKA}}$. Now suppose $a^* \times a^* \equiv_{\mathsf{SKA}} a^*$. As our countermodel is an SKA that means in particular that $\{\{s\}\}^* \times \{\{s\}\}^* = \{\{s\}\}^*$ should hold (c.f. Remark 3.7). However, in this model we can calculate that $\{\{s\}\}^* \times \{\{s\}\}^* = \dagger \neq \{\{s\}\}^*$. Hence, we have a contradiction. Thus $a^* \times a^* \not\equiv_{\mathsf{SKA}} a^*$, rendering SKA incomplete. □

5 A New Axiomatisation

We now create an alternative algebraic formalism, which we call SF_1, and prove that its axioms are sound and complete w.r.t the model of synchronous regular languages. Whereas the definition of SKA relies on Kleene algebras (with *least fixpoint axioms*) as presented by Kozen [18], the definition of SF_1 builds on F_1-algebras (with a *unique fixpoint axiom*) as presented by Salomaa [26]. The axioms of Salomaa are strictly stronger than Kozen's [10], and we will see that the unique fixpoint axiom allows us to derive a connection between the synchronous product and the Kleene star, even though this connection is not represented in an axiom directly (see Remark 5.8).

Definition 5.1. *An* F_1-algebra *[26] is a tuple* $(A, +, \cdot, {}^*, 0, 1, H)$ *where* A *is a set,* * *is a unary operator,* $+$ *and* $\cdot$ *are binary operators and* 0 *and* 1 *are constants, and such that for all* $e, f, g \in A$ *the following axioms are satisfied:*

$$e + (f + g) = (e + f) + g \qquad e + f = f + e \qquad e + 0 = e \qquad e + e = e$$
$$e \cdot 1 = e = 1 \cdot e \qquad e \cdot 0 = 0 = 0 \cdot e \qquad e \cdot (f \cdot g) = (e \cdot f) \cdot g$$
$$e^* = 1 + e \cdot e^* = 1 + e^* \cdot e \qquad (e + f) \cdot g = e \cdot g + f \cdot g \qquad e \cdot (f + g) = e \cdot f + e \cdot g$$

Additionally, the loop tightening *and* unique fixpoint axiom *hold:*

$$(e + 1)^* = e^* \qquad\qquad H(f) = 0 \wedge e + f \cdot g = g \implies f^* \cdot e = g$$

Lastly, we have the following axioms for H*:*

$$H(0) = 0 \qquad H(e + f) = H(e) + H(f) \qquad H(e^*) = (H(e))^*$$
$$H(1) = 1 \qquad H(e \cdot f) = H(e) \cdot H(f)$$

In [26], an $e \in A$ with $H(e) = 1$ is said to have the *empty word property*, which will be reflected in the semantic interpretation of $H(e)$ stated below.

The set of F_1-*expressions*, denoted $\mathcal{T}_{\mathsf{F}_1}$, is described by:

$$\mathcal{T}_{\mathsf{F}_1} \ni e, f ::= 0 \mid 1 \mid a \in \Sigma \mid e + f \mid e \cdot f \mid e^* \mid H(e)$$

We can interpret F_1-expressions in terms of languages through $[\![-]\!]_{\mathsf{F}_1} : \mathcal{T}_{\mathsf{F}_1} \to \mathcal{P}(\Sigma^*)$, defined analogously to $[\![-]\!]_{\mathsf{KA}}$, where furthermore for $e \in \mathcal{T}_{\mathsf{F}_1}$ we have

$$[\![H(e)]\!]_{\mathsf{F}_1} = [\![e]\!]_{\mathsf{F}_1} \cap \{\varepsilon\}$$

We write $\equiv_{\mathsf{F}_1}$ for the smallest congruence on $\mathcal{T}_{\mathsf{F}_1}$ induced by the F_1-axioms. Additionally, we require that for $a \in \Sigma$, we have $H(a) \equiv_{\mathsf{F}_1} 0$. A characterisation similar to Theorem 2.1 can then be established as follows[1]:

[1] Unlike [26], we include H in the syntax; one can prove that for any $e \in \mathcal{T}_{\mathsf{F}_1}$ it holds that $H(e) \equiv 0$ or $H(e) \equiv 1$, and hence any occurence of H can be removed from e. This is what allows us to apply the completeness result from op. cit. here.

Theorem 5.2 (Soundness and Completeness of F_1 [26]). *For all $e, f \in \mathcal{T}_{\mathsf{F}_1}$, we have that $e \equiv_{\mathsf{F}_1} f$ if and only if $[\![e]\!]_{\mathsf{F}_1} = [\![f]\!]_{\mathsf{F}_1}$.*

Remark 5.3. Kozen [18] noted that the above does not generalise along the same lines as in Remark 2.2. In particular, the axiom $H(a) \equiv_{\mathsf{SKA}} 0$ is not stable under substitution; for instance, if we interpret $H(a)$ according to the valuation $a \mapsto \{\epsilon\}$ in the F_1-algebra of languages, then we obtain $\{\epsilon\}$, whereas 0 is interpreted as $\emptyset$.

Definition 5.4. *A* synchronous F_1-algebra *(SF_1-algebra for short) is a tuple $(A, S, +, \cdot, ^*, 0, 1, H)$, such that $(A, +, \cdot, ^*, 0, 1, H)$ is an F_1-algebra and $\times$ is a binary operator on A, with $S \subseteq A$ closed under $\times$ and $(S, \times)$ a semilattice. Furthermore, the following hold for all $e, f, g \in A$ and $\alpha, \beta \in S$:*

$$e \times (f + g) = e \times f + e \times g \qquad e \times (f \times g) = (e \times f) \times g \qquad e \times 0 = 0$$
$$(\alpha \cdot e) \times (\beta \cdot f) = (\alpha \times \beta) \cdot (e \times f) \qquad e \times f = f \times e \qquad e \times 1 = e$$

Moreover, H is compatible with $\times$ as well, i.e., for $e, f \in A$ we have that $H(e \times f) = H(e) \times H(f)$. Lastly, for $\alpha \in S$ we require that $H(\alpha) = 0$.

Remark 5.5. The countermodel from Sect. 4 cannot be extended to a model of SF_1. To see this, note that we have $H(\{\{s\}\}) = 0$ and $\emptyset + \{\{s\}\} \cdot \dagger = \dagger$, but $\{\{s\}\}^* \cdot \emptyset \neq \dagger$—contradicting the unique fixpoint axiom.

The set of SF_1-*expressions* over Σ, denoted $\mathcal{T}_{\mathsf{SF}_1}$, is described by:

$$\mathcal{T}_{\mathsf{SF}_1} \ni e, f ::= 0 \mid 1 \mid a \in \mathcal{T}_{\mathsf{SL}} \mid e + f \mid e \cdot f \mid e \times f \mid e^* \mid H(e)$$

We interpret $\mathcal{T}_{\mathsf{SF}_1}$ in terms of languages via $[\![-]\!]_{\mathsf{SF}_1} : \mathcal{T}_{\mathsf{SF}_1} \to \mathcal{L}_\Sigma$, defined analogously to $[\![-]\!]_{\mathsf{SKA}}$, where furthermore for $e \in \mathcal{T}_{\mathsf{SF}_1}$ we have

$$[\![H(e)]\!]_{\mathsf{SF}_1} = [\![e]\!]_{\mathsf{SF}_1} \cap \{\varepsilon\}$$

Note that when $e \in \mathcal{T}_{\mathsf{SKA}}$, then $e \in \mathcal{T}_{\mathsf{SF}_1}$ and $[\![e]\!]_{\mathsf{SKA}} = [\![e]\!]_{\mathsf{SF}_1}$.

Define $\equiv_{\mathsf{SF}_1}$ as the smallest congruence on $\mathcal{T}_{\mathsf{SF}_1}$ induced by the axioms of SF_1, where $\mathcal{T}_{\mathsf{SL}}$ fulfills the role of the semilattice—e.g., if $a \in \mathcal{T}_{\mathsf{SL}}$, then $a \times a \equiv_{\mathsf{SF}_1} a$. This axiomatisation is sound with respect to the language model.[2]

Lemma 5.6. *Let $e, f \in \mathcal{T}_{\mathsf{SF}_1}$. If $e \equiv_{\mathsf{SF}_1} f$ then $[\![e]\!]_{\mathsf{SF}_1} = [\![f]\!]_{\mathsf{SF}_1}$.*

Remark 5.7. The caveat from Remark 5.3 can be transposed to this setting. However, the condition that for $\alpha \in S$ we have that $H(\alpha) = 0$ allows one to strengthen the above along the same lines as Remark 3.7, that is, if $e \equiv_{\mathsf{SF}_1} f$, then interpreting e and f in some SKA according to some valuation of Σ in terms of semilattice elements will produce the same outcome.

[2] Note that for the synchronous language model we know the least fixpoint axioms are sound as well (Lemma 3.6). However, there might be other SF_1-models where the least fixpoint axioms are not valid.

Remark 5.8. To demonstrate the use of the new axioms, we give an algebraic proof of $\alpha^* \times \alpha^* \equiv_{\mathsf{SF}_1} \alpha^*$ for $\alpha \in \mathcal{T}_{\mathsf{SL}}$:

$$\begin{aligned}\alpha^* \times \alpha^* &\equiv_{\mathsf{SF}_1} (1 + \alpha \cdot \alpha^*) \times (1 + \alpha \cdot \alpha^*) \equiv_{\mathsf{SF}_1} 1 + \alpha \cdot \alpha^* + (\alpha \cdot \alpha^*) \times (\alpha \cdot \alpha^*) \\ &\equiv_{\mathsf{SF}_1} 1 + \alpha \cdot \alpha^* + (\alpha \times \alpha) \cdot (\alpha^* \times \alpha^*) \equiv_{\mathsf{SF}_1} \alpha^* + \alpha \cdot (\alpha^* \times \alpha^*)\end{aligned}$$

Since $H(\alpha) = 0$, we can apply the unique fixpoint axiom to find $\alpha^* \cdot \alpha^* \equiv_{\mathsf{SF}_1} \alpha^* \times \alpha^*$. In SF_1, it is not hard to show that $\alpha^* \cdot \alpha^* \equiv_{\mathsf{F}_1} \alpha^*$; hence, we find $\alpha^* \times \alpha^* \equiv_{\mathsf{SF}_1} \alpha^*$.

Remark 5.9. Adding $\alpha^* \times \alpha^* = \alpha^*$ for $\alpha \in \mathcal{T}_{\mathsf{SL}}$ as an axiom to the old axiomatisation of SKA would not have been sufficient; one can easily find another semantical truth that does not hold in our countermodel, such as $[\![(\alpha \cdot \beta)^* \times (\alpha \cdot \beta)^*]\!]_{\mathsf{SKA}} = [\![(\alpha \cdot \beta)^*]\!]_{\mathsf{SKA}}$. Adding $e^* \times e^* = e^*$ as an axiom is also not an option, as this is not sound; for instance, take $e = a + b$ for $a, b \in \Sigma$. In order to keep the axiomatisation finitary, a unique fixpoint axiom provided an outcome.

5.1 Partial Derivatives

In this section we develop the theory of SKA and set up the necessary machinery for Sect. 6 and the completeness proof in Sect. 7. We start by presenting partial derivatives, which provide a termination and continuation map on $\mathcal{T}_{\mathsf{SF}_1}$. These derivatives allow us to turn the set of synchronous regular terms into a non-deterministic automaton structure, such that the language accepted by $e \in \mathcal{T}_{\mathsf{SF}_1}$ as a state in this automaton is the same as the semantics of e. Furthermore, partial derivatives turn out to provide a way to algebraically characterise a term by means of acceptance and reachable terms, which is useful in the completeness proof of SF_1.

The termination and continuation map for SF_1-expressions presented below are a trivial extension of the ones from [7]. Intuitively, the termination map is 1 if an expression can immediately terminate, and 0 otherwise; the continuation map of a term w.r.t. A gives us the set of terms reachable with an A-step.

Definition 5.10 (Termination map). *For $a \in \Sigma$, we define $o : \mathcal{T}_{\mathsf{SF}_1} \to 2$ inductively, as follows:*

$$\begin{array}{llll} o(0) = 0 & o(e^*) = 1 & o(e + f) = \max(o(e), o(f)) & o(e \times f) = \min(o(e), o(f)) \\ o(1) = 1 & o(a) = 0 & o(e \cdot f) = \min(o(e), o(f)) & o(H(e)) = o(e) \end{array}$$

Definition 5.11 (Continuation map). *For $a \in \Sigma$, we inductively define $\delta : \mathcal{T}_{\mathsf{SF}_1} \times \mathcal{P}_n(\Sigma) \to \mathcal{P}(\mathcal{T}_{\mathsf{SF}_1})$ as follows:*

$$\begin{array}{rlrl} \delta(0, A) = \delta(1, A) &= \emptyset & \delta(e \times f, A) &= \Delta(e, f, A) \cup \Delta(f, e, A) \\ \delta(H(e), A) &= \emptyset & & \cup \{e' \times f' : e' \in \delta(e, B_1), \\ \delta(a, A) &= \{1 : A = \{a\}\} & & f' \in \delta(f, B_2), B_1 \cup B_2 = A\} \\ \delta(e^*, A) &= \{e' \cdot e^* : e' \in \delta(e, A)\} & \delta(e \cdot f, A) &= \{e' \cdot f : e' \in \delta(e, A)\} \\ \delta(e + f, A) &= \delta(e, A) \cup \delta(f, A) & & \cup \Delta(f, e, A) \end{array}$$

where $\Delta(e, f, A)$ is defined to be $\delta(e, A)$ when $o(f) = 1$, and $\emptyset$ otherwise.

Definition 5.12 (Syntactic Automaton). *We call the NDA* $(\mathcal{T}_{\mathsf{SF}_1}, o, \delta)$ *the* syntactic automaton *of* SF_1*-expressions.*

In Sect. 6 we give a proof of correctness of partial derivatives: for $e \in \mathcal{T}_{\mathsf{SF}_1}$ the semantics of e is equivalent to the language accepted by e as a state in the syntactic automaton. An analogous property holds for (partial) derivatives in Kleene algebras [1,8], which makes derivatives a powerful tool for reasoning about language models and deciding equivalences of terms [6].

In the next two sections, we want to use terms reachable from e, that is to say, terms that are a result of repeatedly applying the continuation map on e. To this end, we define the following function:

Definition 5.13. *For* $e, f \in \mathcal{T}_{\mathsf{SF}_1}$ *and* $a \in \Sigma$, *we inductively define the* reach *function* $\rho : \mathcal{T}_{\mathsf{SF}_1} \to \mathcal{P}(\mathcal{T}_{\mathsf{SF}_1})$ *as follows:*

$$\begin{aligned}
&\rho(e + f) = \rho(e) \cup \rho(f) && \rho(0) = \emptyset \\
&\rho(e \cdot f) = \{e' \cdot f : e' \in \rho(e)\} \cup \rho(f) && \rho(1) = \{1\} \\
&\rho(e^*) = \{1\} \cup \{e' \cdot e^* : e' \in \rho(e)\} && \rho(a) = \{1, a\} \\
&\rho(e \times f) = \{e' \times f' : e' \in \rho(e), f' \in \rho(f)\} \cup \rho(e) \cup \rho(f) && \rho(H(e)) = \{1\}
\end{aligned}$$

Using a straightforward inductive argument, one can prove that for all $e \in \mathcal{T}_{\mathsf{SF}_1}$, $\rho(e)$ is finite. Note that e is not always a member of $\rho(e)$. To see that $\rho(e)$ indeed contains all terms reachable from e, we record the following.

Lemma 5.14. *For all* $e \in \mathcal{T}_{\mathsf{SF}_1}$ *and* $A \in \mathcal{P}_n(\Sigma)$, *we have* $\delta(e, A) \subseteq \rho(e)$. *Also, if* $e' \in \rho(e)$, *then* $\delta(e', A) \subseteq \rho(e)$.

5.2 Normal Form

In this section we develop a *normal form* for expressions in $\mathcal{T}_{\mathsf{SL}}$, which we will use in the completeness proof for SF_1. As $[\![-]\!]_{\mathsf{SL}}$ is a surjective function it has at least one right inverse. Let us pick one and denote it by $(-)^{\Pi}$. We thus have $(-)^{\Pi} : \mathcal{P}_n(\Sigma) \to \mathcal{T}_{\mathsf{SL}}$ such that $[\![-]\!]_{\mathsf{SL}} \circ (-)^{\Pi}$ is the identity on $\mathcal{P}_n(\Sigma)$.

The normal form for expressions in $\mathcal{T}_{\mathsf{SL}}$ is defined as follows:

Definition 5.15 (Normal form). *For all* $e \in \mathcal{T}_{\mathsf{SL}}$ *the* normal form *of* e, *denoted as* $\overline{e}$, *is defined as* $([\![e]\!]_{\mathsf{SL}})^{\Pi}$. *Let* $\overline{\mathcal{T}_{\mathsf{SL}}} = \{\overline{e} : e \in \mathcal{T}_{\mathsf{SL}}\}$.

Intuitively, an expression in normal form is standardised with respect to idempotence, associativity and commutativity. For instance, for a term $(a \times a) \times (c \times b)$ with $a, b, c \in \Sigma$, the chosen normal form, dictated by the chosen right inverse, could be $(a \times b) \times c$, and all terms provably equivalent to $(a \times a) \times (c \times b)$ will have this same normal form. Using Lemma 3.4, we can formalise this in the following two results:

Lemma 5.16. *For all* $e \in \mathcal{T}_{\mathsf{SL}}$, *we have that* e *is provably equivalent to its normal form:* $e \equiv_{\mathsf{SL}} \overline{e}$. *Moreover, if two expressions* $e, f \in \mathcal{T}_{\mathsf{SL}}$ *are provably equivalent, they have the same normal form: if* $e \equiv_{\mathsf{SL}} f$, *then* $\overline{e} = \overline{f}$.

Proof. As $(-)^{\Pi}$ is a right inverse of $[\![-]\!]_{\mathsf{SL}}$, we can derive the following:

$$[\![\overline{e}]\!]_{\mathsf{SL}} = [\![([\![e]\!]_{\mathsf{SL}})^{\Pi}]\!]_{\mathsf{SL}} = [\![e]\!]_{\mathsf{SL}}$$

From completeness we get $e \equiv_{\mathsf{SL}} \overline{e}$. For the second part of the statement we obtain via soundness that $[\![e]\!]_{\mathsf{SL}} = [\![f]\!]_{\mathsf{SL}}$ and subsequently that $\overline{e} = \overline{f}$. □

Normalising normalised terms does not change anything.

Lemma 5.17. *For all $e \in \overline{\mathcal{T}_{\mathsf{SL}}}$ we have that $\overline{e} = e$.*

We extend $(-)^{\Pi}$ from synchronous strings of length one to words and synchronous languages in the obvious way. For a synchronous string aw with $a \in \mathcal{P}_n(\Sigma)$ and $w \in (\mathcal{P}_n(\Sigma))^*$, and synchronous language $L \in \mathcal{L}_\Sigma$ we define:

$$\varepsilon^{\Pi} = \varepsilon \qquad (aw)^{\Pi} = a^{\Pi} \cdot (w^{\Pi}) \qquad L^{\Pi} = \{w^{\Pi} : w \in L\}$$

We abuse notation and assume the type of $(-)^{\Pi}$ is clear from the argument.

Since $(-)^{\Pi}$ is a homomorphism of languages, we have the following.

Lemma 5.18. *For synchronous languages L and K, all of the following hold: (i) $(L \cup K)^{\Pi} = L^{\Pi} \cup K^{\Pi}$, (i) $(L \cdot K)^{\Pi} = L^{\Pi} \cdot K^{\Pi}$, and (iii) $(L^*)^{\Pi} = (L^{\Pi})^*$.*

6 A Fundamental Theorem for SF_1

In this section we shall algebraically capture an expression in terms of its partial derivatives. This characterisation of an SF_1-term will be useful later on in proving completeness but also provides us with a straightforward method to prove correctness of the partial derivatives. Following [25,27], we call this characterisation a *fundamental theorem* for SF_1. Before we state and prove the fundamental theorem, we prove an intermediary lemma:

Lemma 6.1. *For all $e, f \in \mathcal{T}_{\mathsf{SF}_1}$, we have*

$$\sum_{e' \in \delta(e,A)} (A^{\Pi} \cdot e') \times \sum_{e' \in \delta(f,A)} (A^{\Pi} \cdot e') \equiv_{\mathsf{SF}_1} \sum_{\substack{e' \in \delta(e,A) \\ e'' \in \delta(f,B)}} (A \cup B)^{\Pi} \cdot (e' \times e'')$$

Proof. First note the following derivation for $A, B \in \mathcal{P}_n(\Sigma)$, using Lemma 5.16, the fact that all axioms of $\equiv_{\mathsf{SL}}$ are included in $\equiv_{\mathsf{SF}_1}$, and that $(-)^{\Pi}$ is a right inverse of $[\![-]\!]_{\mathsf{SL}}$:

$$\begin{aligned} A^{\Pi} \times B^{\Pi} &\equiv_{\mathsf{SF}_1} \overline{A^{\Pi} \times B^{\Pi}} = ([\![A^{\Pi} \times B^{\Pi}]\!]_{\mathsf{SL}})^{\Pi} \\ &= ([\![A^{\Pi}]\!]_{\mathsf{SL}} \cup [\![B^{\Pi}]\!]_{\mathsf{SL}})^{\Pi} = (A \cup B)^{\Pi} \end{aligned}$$

Using distributivity, the synchrony axiom and the equation above, we can derive:

$$\sum_{e' \in \delta(e,A)} (A^{\Pi} \cdot e') \times \sum_{e' \in \delta(f,A)} (A^{\Pi} \cdot e') \equiv_{\mathsf{SF}_1} \sum_{\substack{e' \in \delta(e,A) \\ e'' \in \delta(f,B)}} (A^{\Pi} \cdot e') \times (B^{\Pi} \cdot e'')$$

$$\equiv_{\mathsf{SF}_1} \sum_{\substack{e' \in \delta(e,A) \\ e'' \in \delta(f,B)}} (A^{\Pi} \times B^{\Pi}) \cdot (e' \times e'') \equiv_{\mathsf{SF}_1} \sum_{\substack{e' \in \delta(e,A) \\ e'' \in \delta(f,B)}} (A \cup B)^{\Pi} \cdot (e' \times e'')$$

The synchrony axiom can be applied because $A^{\Pi}, B^{\Pi} \in \mathcal{T}_{\mathsf{SL}}$. □

Theorem 6.2 (Fundamental Theorem). *For all $e \in \mathcal{T}_{\mathsf{SF}_1}$, we have*

$$e \equiv_{\mathsf{SF}_1} o(e) + \sum_{c' \in \delta(e,A)} A^{\Pi} \cdot e'.$$

Proof. This proof is mostly analogous to the proof of the fundamental theorem for regular expressions, such as the one that can be found in [27].

We proceed by induction on e. In the base, we have three cases to consider: $e \in \{0,1\}$ or $e = a$ for $a \in \Sigma$. For $e \in \{0,1\}$, the result follows immediately. For $e = a$, the only non-empty derivative is $\delta(a, \{a\})$ and the result follows:

$$o(a) + \sum_{e' \in \delta(a,A)} A^{\Pi} \cdot e' \equiv_{\mathsf{SF}_1} o(a) + \overline{a} \cdot 1 \equiv_{\mathsf{SF}_1} \overline{a} \equiv_{\mathsf{SF}_1} a$$

In the inductive step, wc treat only the case for synchronous composition; the others can be found in the appendix. If $e = e_0 \times e_1$, derive as follows:

$$\begin{aligned}
& e_0 \times e_1 \\
& \equiv_{\mathsf{SF}_1} \Big(o(e_0) + \sum_{e' \in \delta(e_0,A)} A^{\Pi} \cdot e'\Big) \times \Big(o(e_1) + \sum_{e' \in \delta(e_1,A)} A^{\Pi} \cdot e'\Big) && \text{(Ind. hyp.)} \\
& \equiv_{\mathsf{SF}_1} o(e_0) \times o(e_1) + \sum_{e' \in \delta(e_0,A)} (A^{\Pi} \cdot e') \times o(e_1) + o(e_0) \times \sum_{e' \in \delta(e_1,A)} A^{\Pi} \cdot e' \\
& \qquad + \sum_{e' \in \delta(e_0,A)} (A^{\Pi} \cdot e') \times \sum_{e' \in \delta(e_1,A)} (A^{\Pi} \cdot e') && \text{(Distributivity)} \\
& \equiv_{\mathsf{SF}_1} o(e_0 \times e_1) + \sum_{e' \in \delta(e_0,A)} (A^{\Pi} \cdot e') \times o(e_1) + o(e_0) \times \sum_{e' \in \delta(e_1,A)} A^{\Pi} \cdot e' \\
& \qquad + \sum_{\substack{e' \in \delta(e_0,A) \\ e'' \in \delta(e_1,B)}} (A \cup B)^{\Pi} \cdot (e' \times e'') && \text{(Def. } o\text{, Lemma 6.1)} \\
& \equiv_{\mathsf{SF}_1} o(e_0 \times e_1) + \sum_{e' \in \Delta(e_0,e_1,A)} A^{\Pi} \cdot e' + \sum_{e' \in \Delta(e_1,e_0,A)} A^{\Pi} \cdot e' + \sum_{\substack{e' \in \{e_0' \times e_1' : e_0' \in \delta(e_0,A), \\ e_1' \in \delta(e_1,B), C = A \cup B\}}} C^{\Pi} \cdot e' \\
& \equiv_{\mathsf{SF}_1} o(e_0 \times e_1) + \sum_{e' \in \delta(e_0 \times e_1,A)} A^{\Pi} \cdot e' && \text{(Def. } \delta\text{)} \quad \square
\end{aligned}$$

Correctness of Partial Derivatives for SF_1
We now relate the partial derivatives for SF_1 to their semantics. Let $\ell : \mathcal{T}_{\mathsf{SF}_1} \to \mathcal{L}_\Sigma$ be the semantics of the syntactic automaton $(\mathcal{T}_{\mathsf{SF}_1}, o, \delta)$ (Definition 5.12), uniquely defined by Eq. 1:

$$\ell(e) = \{\varepsilon : o(e) = 1\} \cup \bigcup_{e' \in \delta(e,A)} \{A\} \cdot \ell(e') \tag{2}$$

To prove correctness of derivatives for SF_1, we prove that the language semantics of the syntactic automaton and the SF_1-expression coincide:

Theorem 6.3 (Soundness of derivatives). *For all $e \in \mathcal{T}_{\mathsf{SF}_1}$ we have:*

$$\ell(e) = [\![e]\!]_{\mathsf{SF}_1}$$

Proof The claim follows almost immediately from the fundamental theorem. From Lemma 3.6 and Theorem 6.2, we obtain

$$[\![e]\!]_{\mathsf{SF}_1} = \{\varepsilon : o(e) = 1\} \cup \bigcup_{e' \in \delta(e,A)} \{A\} \cdot [\![e']\!]_{\mathsf{SF}_1}$$

Note that $[\![A^\Pi]\!]_{\mathsf{SF}_1} = \{[\![A^\Pi]\!]_{\mathsf{SL}}\} = \{A\}$ by definition of the SF_1 semantics of a term in $\mathcal{T}_{\mathsf{SL}}$ and the fact that $(-)^\Pi$ is a right inverse. Because ℓ is the only function satisfying Eq. 2, we obtain the desired equality between $[\![e]\!]_{\mathsf{SF}_1}$ and the language $\ell(e)$ accepted by e as a state of the automaton $(\mathcal{T}_{\mathsf{SF}_1}, o, \delta)$. □

7 Completeness of SF_1

In this section we prove completeness of the SF_1-axioms with respect to the synchronous language model: we prove that for $e, f \in \mathcal{T}_{\mathsf{SF}_1}$, if $[\![e]\!]_{\mathsf{SF}_1} = [\![f]\!]_{\mathsf{SF}_1}$, then $e \equiv_{\mathsf{SF}_1} f$. We first prove completeness of SF_1 for a subset of SF_1-expressions, relying on the completeness result of F_1 (Lemma 7.3). Then we demonstrate that for every SF_1-expression we can find an equivalent SF_1-expression in this specific subset (Theorem 7.6). This subset is formed as follows.

Definition 7.1. *The set of SF_1-expressions in* normal form, $\mathcal{T}_{\mathsf{NSF}}$, *is described by the grammar*

$$\mathcal{T}_{\mathsf{NSF}} \ni e, f ::= 0 \mid 1 \mid a \in \overline{\mathcal{T}_{\mathsf{SL}}} \mid e + f \mid e \cdot f \mid e^*$$

where $\overline{\mathcal{T}_{\mathsf{SL}}}$ is as defined in Definition 5.15.

From this description it is immediate that an SF_1-term $e \in \mathcal{T}_{\mathsf{NSF}}$ is formed from terms of $\overline{\mathcal{T}_{\mathsf{SL}}}$ connected via the regular F_1-algebra operators. Hence, F_1-expressions formed over the alphabet $\overline{\mathcal{T}_{\mathsf{SL}}}$ are the same set of terms as $\mathcal{T}_{\mathsf{NSF}}$. We shall use this observation to prove completeness for $\mathcal{T}_{\mathsf{NSF}}$ with respect to the language model by leveraging completeness of F_1.

We use the function $(-)^\Pi$ to give a translation between the SF_1 semantics of a term in $\mathcal{T}_{\mathsf{NSF}}$ and the F_1 semantics of that same term:

Lemma 7.2. *For all $e \in \mathcal{T}_{\mathsf{NSF}}$, we have $(\llbracket e \rrbracket_{\mathsf{SF}_1})^{\Pi} = \llbracket e \rrbracket_{\mathsf{F}_1}$.*

Proof. We proceed by induction on the construction of e. In the base, there are three cases to consider. If $e = 0$, then $\llbracket e \rrbracket_{\mathsf{SF}_1} = \emptyset = \llbracket e \rrbracket_{\mathsf{F}_1}$, and we are done. If $e = 1$, then $(\llbracket e \rrbracket_{\mathsf{SF}_1})^{\Pi} = (\{\varepsilon\})^{\Pi} = \{\varepsilon\} = \llbracket 1 \rrbracket_{\mathsf{F}_1}$, and the claim follows. If $e = a$ for $a \in \overline{\mathcal{T}_{\mathsf{SL}}}$, we use Lemma 5.17 to obtain $\overline{a} = a$. As $a \in \overline{\mathcal{T}_{\mathsf{SL}}} \subseteq \mathcal{T}_{\mathsf{SL}}$, we know that $(\llbracket a \rrbracket_{\mathsf{SF}_1})^{\Pi} = (\{\llbracket a \rrbracket_{\mathsf{SL}}\})^{\Pi} = \{(\llbracket a \rrbracket_{\mathsf{SL}})^{\Pi}\} = \{\overline{a}\} = \{a\} = \llbracket a \rrbracket_{\mathsf{F}_1}$, and the claim follows.

For the inductive step, first consider $e = H(e_0)$. $(\llbracket H(e_0) \rrbracket_{\mathsf{SF}_1})^{\Pi} = \{\varepsilon\}$ if $\varepsilon \in \llbracket e_0 \rrbracket_{\mathsf{SF}_1}$ and $\emptyset$ otherwise. We also have $\llbracket H(e_0) \rrbracket_{\mathsf{F}_1} = \{\varepsilon\}$ if $\varepsilon \in \llbracket e_0 \rrbracket_{\mathsf{F}_1}$ and $\emptyset$ otherwise. The induction hypothesis states that $(\llbracket e_0 \rrbracket_{\mathsf{SF}_1})^{\Pi} = \llbracket e_0 \rrbracket_{\mathsf{F}_1}$, from which we obtain that $\varepsilon \in \llbracket e_0 \rrbracket_{\mathsf{SF}_1} \Leftrightarrow \varepsilon \in \llbracket e_0 \rrbracket_{\mathsf{F}_1}$. Hence we can conclude that $(\llbracket H(e_0) \rrbracket_{\mathsf{SF}_1})^{\Pi} = \llbracket H(e_0) \rrbracket_{\mathsf{F}_1}$. All other inductive cases follow immediately from Lemma 5.18. The details can be found in the appendix. □

We are now ready to prove completeness of SF_1 for terms in normal form.

Lemma 7.3. *Let $e, f \in \mathcal{T}_{\mathsf{NSF}}$. If $\llbracket e \rrbracket_{\mathsf{SF}_1} = \llbracket f \rrbracket_{\mathsf{SF}_1}$, then $e \equiv_{\mathsf{SF}_1} f$.*

Proof. By the premise, we have that $(\llbracket e \rrbracket_{\mathsf{SF}_1})^{\Pi} = (\llbracket f \rrbracket_{\mathsf{SF}_1})^{\Pi}$. From Lemma 7.2 we get $(\llbracket e \rrbracket_{\mathsf{SF}_1})^{\Pi} = \llbracket e \rrbracket_{\mathsf{F}_1}$ and $(\llbracket f \rrbracket_{\mathsf{SF}_1})^{\Pi} = \llbracket f \rrbracket_{\mathsf{F}_1}$, which results in $\llbracket e \rrbracket_{\mathsf{F}_1} = \llbracket f \rrbracket_{\mathsf{F}_1}$. From Theorem 5.2 we know that this entails that $e \equiv_{\mathsf{F}_1} f$. As SF_1 contains all the axioms of F_1, we may then conclude that $e \equiv_{\mathsf{SF}_1} f$ and the claim follows. □

In order to prove completeness with respect to the language model for all $e \in \mathcal{T}_{\mathsf{SF}_1}$, we prove that for every $e \in \mathcal{T}_{\mathsf{SF}_1}$ there exists a term $\hat{e} \in \mathcal{T}_{\mathsf{NSF}}$ in normal form such that $e \equiv_{\mathsf{SF}_1} \hat{e}$. To see this is indeed enough to establish completeness of SF_1, imagine we have such a procedure to transform e into $\hat{e}$ in normal form. We can then conclude that $\llbracket e \rrbracket_{\mathsf{SF}_1} = \llbracket f \rrbracket_{\mathsf{SF}_1}$ implies $\llbracket \hat{e} \rrbracket_{\mathsf{SF}_1} = \llbracket \hat{f} \rrbracket_{\mathsf{SF}_1}$, which by Lemma 7.3 implies $\hat{e} \equiv_{\mathsf{SF}_1} \hat{f}$, and consequently that $e \equiv_{\mathsf{SF}_1} f$.

To obtain $\hat{e}$, we will make use of the "unfolding" of an SF_1-expression e in terms of partial derivatives, given by the fundamental theorem, which will give rise to a linear system. We will then show that this linear system has a unique solution that has the properties we require from $\hat{e}$. Since e is also a solution to this linear system, we can conclude that they are provably equivalent.

Let us start with the following property of linear systems over SF_1. A Q-vector is a function $x : Q \to \mathcal{T}_{\mathsf{SF}_1}$ and a Q-matrix is a function $M : Q \times Q \to \mathcal{T}_{\mathsf{SF}_1}$. We call a matrix M *guarded* if $H(M(i,j)) = 0$ for all $i, j \in Q$. We say a vector p and matrix M are in normal form if $p(i) \in \mathcal{T}_{\mathsf{NSF}}$ for all $i \in Q$ and $M(i,j) \in \mathcal{T}_{\mathsf{NSF}}$ for all $i, j \in Q$. The following lemma is a variation of [26, Lemma 2] and the proof is a direct adaptation of the proof found in [17, Lemma 3.12].

Lemma 7.4. *Let (M, p) be a Q-linear system such that M and p are guarded. We can construct Q-vector x that is the unique (up to SF_1-equivalence) solution to (M, p) in SF_1. Moreover, if M and p are in normal form, then so is x.*

We now define the linear system associated to an SF_1-expression e. This linear system makes use of the partial derivatives for SF_1, and essentially represents an NFA that accepts the language described by e.

Definition 7.5. *Let $e \in \mathcal{T}_{\mathsf{SF}_1}$, and choose $Q_e = \rho(e) \cup \{e\}$, where ρ is the reach function from Definition 5.13. Define the Q_e-vector x_e and the Q_e-matrix M_e by*

$$x_e(e') = o(e') \qquad\qquad M_e(e', e'') = \sum_{e'' \in \delta(e', A)} A^{\Pi}$$

We can now use Lemma 7.4 to obtain the desired normal form $\hat{e}$:

Theorem 7.6. *For all $e \in \mathcal{T}_{\mathsf{SF}_1}$, there exists an $\hat{e} \in \mathcal{T}_{\mathsf{NSF}}$ such that $\hat{e} \equiv_{\mathsf{SF}_1} e$.*

Proof. It is clear from their definition that x_e and M_e are both in normal form and that M_e is guarded. From Lemma 7.4 we then get that there exists a unique solution s_e to (M_e, x_e), and s_e is a Q_e-vector in normal form. Now consider the Q_e-vector y such that $y(q) = q$ for all $q \in Q_e$. Using Lemma 5.14 and Theorem 6.2, we can derive the following:

$$\begin{aligned} x_e(q) + M_e \cdot y(q) &\equiv_{\mathsf{SF}_1} x_e(q) + \sum_{q' \in Q_e} M_e(q, q') \cdot y(q') \\ &\equiv_{\mathsf{SF}_1} o(q) + \sum_{q' \in Q_e} \sum_{q' \in \delta(q, A)} A^{\Pi} \cdot q' \\ &\equiv_{\mathsf{SF}_1} o(q) + \sum_{q' \in \delta(q, A)} A^{\Pi} \cdot q' \equiv_{\mathsf{SF}_1} q = y(q) \end{aligned}$$

This demonstrates that y is also a solution to (M_e, x_e). As we know from Lemma 7.4 that s_e is unique, we get that $y \equiv_{\mathsf{SF}_1} s_e$. This means that $e = y(e) \equiv_{\mathsf{SF}_1} s_e(e)$. As s_e is in normal form we get that $s_e(e) \in \mathcal{T}_{\mathsf{NSF}}$. Thus, if we take $s_e(e) = \hat{e}$, then we have obtained the desired result. □

Combining Theorem 7.6 and Lemma 7.3 gives the main result of this section:

Theorem 7.7 (Soundness and Completeness). *For all $e, f \in \mathcal{T}_{\mathsf{SF}_1}$, we have*

$$e \equiv_{\mathsf{SF}_1} f \Leftrightarrow [\![e]\!]_{\mathsf{SF}_1} = [\![f]\!]_{\mathsf{SF}_1}$$

As a corollary of Theorem 6.3 and Theorem 7.7 we know that SF_1 is decidable by deciding language equivalence in the syntactic automaton.

8 Related Work

Synchonous cooperation among processes has been extensively studied in the context of process calculi such as ASP [4] and SCCS [23]. SKA bears a strong resemblance to SCCS, with the most notable differences being the equivalence axiomatised (bisimulation vs. language equivalence), and the use of Kleene star (unbounded finite recursion) instead of fixpoint (possibly infinite recursion). Contrary to ASP, but similar to SCCS, SKA cannot express incompatibility of action synchronisation.

In the context of Kleene algebra based frameworks for concurrent reasoning, a synchronous product is just one possible interpretation of concurrency. An interleaving-based approach with a concurrent operator (a parallel operator denoted with $\|$) is explored in Concurrent Kleene Algebra [14,15,17,22].

We have proved that $\equiv_{\mathsf{SF}_1}$ is sound and complete with respect to the synchronous language model by making use of the completeness of F_1 [26]. The strategy of transforming an expression e to an equivalent expression $\hat{e}$ with a particular property is often used in literature [16,17,20,22]. In particular, we adopted the use of linear systems as a representation of automata, which was first done by Conway [9] and Backhouse [2]. The machinery that we used to solve linear systems in F_1 is based on Salomaa [26] and can also be found in [17] and [19]. The idea of the syntactic automaton originally comes from Brzozowski, who did this for regular expressions [8]. He worked with derivatives which turn a Kleene algebra expression into a deterministic automaton. We worked with partial derivatives instead, resulting in a non-deterministic finite automaton for each SF_1-expression. Partial derivatives were first proposed by Antimirov [1].

Other related work is that of Hayes et al. [12]. They explore an algebra of synchronous atomic steps that interprets the synchrony model SKA is based on (Milner's SCCS calculus). However, their algebra is not based on a Kleene algebra—they use concurrent refinement algebra [11] instead. Later, Hayes et al. presented an abstract algebra for reasoning about concurrent programs with an abstract synchronisation operator [13], of which their earlier algebra of atomic steps is an instance. A key difference seems to be that Hayes et al. use different units for synchronous and sequential composition. It would be interesting to compare expressive powers of the two algebras more extensively.

A decision procedure for equivalence between SKA terms is given by Broda et al. [7]. They defined partial derivatives for SKA that we also used in the proof of completeness, and used those to construct an NFA that accepts the semantics of a given SKA expression. Deciding language equivalence of two automata then leads to a decision procedure for semantic equivalence of SKA expressions.

9 Conclusions and Further Work

We have presented a complete axiomatisation with respect to the model of synchronous regular languages. We have first proved incompleteness of SKA via a countermodel, exploiting the fact that SKA did not have any axioms relating the synchronous product to the Kleene star. We then provided a set of axioms based on the F_1-axioms from Salomaa [26] and the axioms governing the synchronous product familiar from SKA. This was shown to be a sound and complete axiomatisation with respect to the synchronous language model.

In the original SKA paper there is a presentation of *synchronous Kleene algebra with tests* including a wrongful claim of completeness. An obvious next step would be to see if we can prove completeness of SF_1 with tests. We conjecture SF_1 with tests is indeed complete and that this is straightforward to prove via a reduction to SF_1 in a style similar to the completeness proof of KAT [20].

Another generalisation is to add a unit to the semilattice, making it a bounded semilattice. This will probably lead to a type of delay operation [23].

Our original motivation to study SKA was to use it as an axiomatisation of Reo, a modular language of connectors combining synchronous data flow with an asynchronous one [3]. The semantics of Reo is based on an automata model very similar to that of SKAT, in which transitions are labelled by sets of ports (representing a synchronous data flow) and constraints (the tests of SKAT). Interestingly, automata are combined using an operation analogous to the synchronous product of SKAT expressions. We aim to study the application of SKA or SKAT to Reo in future work.

Acknowledgements. The first author is grateful for discussions with Hans-Dieter Hiep and Benjamin Lion.

A Appendix

Lemma 3.5. *The structure* $(\mathcal{L}_\Sigma, S, +, \cdot, ^*, \times, \emptyset, \{\varepsilon\})$ *is an SKA, that is, synchronous languages over* Σ *form an SKA.*

Proof. The carrier $\mathcal{L}_\Sigma$ is obviously closed under the operations of synchronous Kleene algebra. We need only argue that each of the SKA axioms is valid on synchronous languages.

The proof for the Kleene algebra axioms follows from the observation that synchronous languages over the alphabet Σ are simply languages over the alphabet $\mathcal{P}_n(\Sigma)$. Thus we know that the Kleene algebra axioms are satisfied, as languages over alphabet $\mathcal{P}_n(\Sigma)$ with $1 = \{\varepsilon\}$ and $0 = \emptyset$ form a Kleene algebra.

For the semilattice axioms, note that S is isomorphic to $\mathcal{P}_n(\Sigma)$ (by sending a singleton set in S to its sole element), and that the latter forms a semilattice when equipped with $\cup$. Since the isomorphism between S and $\mathcal{P}_n(\Sigma)$ respects these operators, it follows that $(S, \times)$ is also a semilattice.

The first SKA axiom that we check is commutativity. We prove that $\times$ on synchronous strings is commutative by induction on the paired length of the strings. Consider synchronous strings u and v. For the base, where u and v equal ε, the result is immediate. In the induction step, we take $u = xu'$ with $x \in \mathcal{P}_n(\Sigma)$. If $v = \varepsilon$ we are done immediately. Now for the case $v = yv'$ with $y \in \mathcal{P}_n(\Sigma)$. We have $u \times v = (xu') \times (yv') = (x \cup y) \cdot (u' \times v')$. From the induction hypothesis we know that $u' \times v' = v' \times u'$. Combining this with commutativity of union we have $u \times v = (x \cup y) \cdot (v' \times u') = v \times u$. Take synchronous languages K and L. Now consider $w \in K \times L$. This means that $w = u \times v$ for $u \in K$ and $v \in L$. From commutativity of synchronous strings we know that $w = u \times v = v \times u$. And thus we have $w \in L \times K$. The other inclusion is analogous.

It is obvious that the axioms $K \times \emptyset = \emptyset$ and $K \times \{\varepsilon\} = K$ are satisfied.

For associativity we again first argue that $\times$ on synchronous strings is associative. Take synchronous strings u, v and w. We will show by induction on the paired length of u, v and w that $u \times (v \times w) = (u \times v) \times w$. If $u, v, w = \varepsilon$ the result is immediate. Now consider $u = xu'$ for $x \in \mathcal{P}_n(\Sigma)$. If v or w equals

ε the result is again immediate. Hence we consider the case where $v = yv'$ and $w = zw'$ for $y, z \in \mathcal{P}_n(\Sigma)$. From the induction hypothesis we know that $u' \times (v' \times w') = (u' \times v') \times w'$. We can therefore derive

$$\begin{aligned} u \times (v \times w) &= (xu') \times (yv' \times zw') = (xu') \times ((y \cup z) \cdot (v' \times w')) \\ &= (x \cup (y \cup z)) \cdot (u' \times (v' \times w')) = (x \cup (y \cup z)) \cdot ((u' \times v') \times w') \end{aligned}$$

From associativity of union, we then know that $(x \cup (y \cup z)) \cdot ((u' \times v') \times w') = (u \times v) \times w$. Now consider $t \in K \times (L \times J)$ for K, L and J synchronous languages. Thus $t = u \times (v \times w)$ for $u \in K$, $v \in L$ and $w \in J$. From associativity of synchronous strings we know that $t = u \times (v \times w) = (u \times v) \times w$, and thus we have $t \in (K \times L) \times J$. The other inclusion is analogous.

For distributivity consider $w \in K \times (L + J)$ for K, L, J synchronous languages. This means that $w = u \times v$ for $u \in K$ and $v \in L + J$. Thus we know $v \in L$ or $v \in J$. We immediately get that $u \times v \in K \times L$ or $u \times v \in K \times J$ and therefore that $w \in K \times L + K \times J$. The other direction is analogous.

For the synchrony axiom we take synchronous languages K, L and $A, B \in S$. Suppose $A = \{x\}$ and $B = \{y\}$ for $x, y \in \mathcal{P}_n(\Sigma)$. Take $w \in (A \cdot K) \times (B \cdot L)$. This means that $w = u \times v$ for $u \in A \cdot K$ and $v \in B \cdot L$. Thus we know that $u = xu'$ with $u' \in K$ and $v = yv'$ with $v' \in L$. From this we conclude $w = u \times v = (xu') \times (yv') = (x \cup y) \cdot (u' \times v')$. As $u' \in K$ and $v' \in L$ and $x \cup y = x \times y$ with $x \in A$ and $y \in B$, we have that $w \in (A \times B) \cdot (K \times L)$. For the other direction, consider $w \in (A \times B) \cdot (K \times L)$. This entails $w = t \cdot v$ for $t \in A \times B$ and $v \in K \times L$. As $A \times B = \{x \cup y\}$ we have $t = x \cup y$. And $v = u \times s$ for $u \in K$ and $s \in L$. Thus $t \cdot v = (x \cup y) \cdot (u \times s) = (xu) \times (ys)$ for $u \in K$, $s \in L$, $x \in A$ and $y \in B$. Hence $w \in (A \cdot K) \times (B \cdot L)$. □

Lemma 3.6 (Soundness of SKA). *For all $e, f \in \mathcal{T}_{\mathsf{SKA}}$, we have that $e \equiv_{\mathsf{SKA}} f$ implies $[\![e]\!]_{\mathsf{SKA}} = [\![f]\!]_{\mathsf{SKA}}$.*

Proof. This is proved by induction on the construction of $\equiv_{\mathsf{SKA}}$. In the base case we need to check all the axioms generating $\equiv_{\mathsf{SKA}}$, which we have already done for Lemma 3.5. For the inductive step, we need to check the closure rules for congruence preserve soundness. This is all immediate from the definition of the semantics of SKA and the induction hypothesis. For instance, if $e = e_0 + e_1$, $f = f_0 + f_1$, $e_0 \equiv_{\mathsf{SKA}} f_0$ and $e_1 \equiv_{\mathsf{SKA}} f_1$, then $[\![e]\!]_{\mathsf{SKA}} = [\![e_0]\!]_{\mathsf{SKA}} + [\![e_1]\!]_{\mathsf{SKA}} = [\![f_0]\!]_{\mathsf{SKA}} + [\![f_1]\!]_{\mathsf{SKA}} = [\![f]\!]_{\mathsf{SKA}}$, where use that $[\![e_0]\!]_{\mathsf{SKA}} = [\![f_0]\!]_{\mathsf{SKA}}$ and $[\![e_1]\!]_{\mathsf{SKA}} = [\![f_1]\!]_{\mathsf{SKA}}$ as a consequence of the induction hypothesis.

Lemma 4.1. *For $\alpha \in \mathcal{T}_{\mathsf{SL}}$, we have $[\![\alpha^* \times \alpha^*]\!]_{\mathsf{SKA}} = [\![\alpha^*]\!]_{\mathsf{SKA}}$.*

Proof. For the first inclusion, take $w \in [\![a^* \times a^*]\!]_{\mathsf{SKA}} = [\![a^*]\!]_{\mathsf{SKA}} \times [\![a^*]\!]_{\mathsf{SKA}}$. Thus we have $w = u \times v$ for $u, v \in [\![a^*]\!]_{\mathsf{SKA}}$. Hence $u = x_1 \cdots x_n$ for $x_i \in [\![a]\!]_{\mathsf{SKA}}$ and $v = y_1 \cdots y_m$ for $y_i \in [\![a]\!]_{\mathsf{SKA}}$. As $[\![a]\!]_{\mathsf{SKA}} = \{[\![a]\!]_{\mathsf{SL}}\}$ with $[\![a]\!]_{\mathsf{SL}} \in \mathcal{P}_n(\Sigma)$, we know that $x_i = [\![a]\!]_{\mathsf{SL}}$ and $y_i = [\![a]\!]_{\mathsf{SL}}$. Assume that $n \le m$ without loss of generality. We then know that $v = u \cdot [\![a]\!]_{\mathsf{SL}}^{m-n}$, where synchronous string e^n indicates n copies of string e concatenated. Unrolling the definition of $\times$ on words, we find $u \times v = u \times (u \cdot [\![a]\!]^k_{\mathsf{SL}}) = (u \times u) \cdot [\![a]\!]^k_{\mathsf{SL}} = u \cdot [\![a]\!]^k_{\mathsf{SL}} = v$, and hence $w = u \times v = v \in [\![a^*]\!]_{\mathsf{SKA}}$. For the other inclusion, take $w \in [\![a^*]\!]_{\mathsf{SKA}}$. As $\varepsilon \in [\![a^*]\!]_{\mathsf{SKA}}$ and $w \times \varepsilon = w$, we immediately have $w \in [\![a^*]\!]_{\mathsf{SKA}} \times [\![a^*]\!]_{\mathsf{SKA}}$. □

Lemma A.1. *For $K, L \in \mathcal{L}_s$, K a non-empty finite language and L an infinite language, $K \times L$ is an infinite language.*

Proof. Suppose that $K \times L$ is a finite language. Hence we have an upper bound on the length of words in $K \times L$. Since the length of the synchronous product of two words is obviously the maximum of the length of the operands, this means we also have an upper bound on the length of words in L, and as we have finite words over a finite alphabet in L this means that L is finite. Hence we get a contradiction, thus $K \times L$ is infinite.

Lemma 4.3. $\mathcal{M} = (\mathcal{L}_s \cup \{\dagger\}, \{\{\{s\}\}\}, +, \cdot, ^*, \times, \emptyset, \{\varepsilon\})$ *with the operators as defined in Definition* 4.2 *forms an SKA.*

Proof. In the main text we treated one of the least fixpoint axioms and the synchrony axiom, and here we will treat all the remaining cases. For the sake of brevity, for each axiom we omit the cases where we can appeal to the proof for (synchronous) regular languages.

The proof that $(S, \times)$ is a semilattice is the same as in Lemma 3.5. Next, we take a look at the Kleene algebra axioms. If $K \in \mathcal{L}_s$, then $K + \emptyset = \emptyset$ holds by definition of union of sets. If $K = \dagger$, we get $\dagger + \emptyset = \dagger$, and the axiom also holds.

For $K \in \mathcal{L}_s \cup \{\dagger\}$, the axiom $K + K = K$ also easily holds by definition of the plus operator. Same for $K \cdot \{\varepsilon\} = K = K \cdot \{\varepsilon\}$ and $K \cdot \emptyset = \emptyset = \emptyset \cdot K$ by definition of the operator for sequential composition.

It is easy to see the axioms $1 + e \cdot e^* \equiv_{\mathsf{SKA}} e^*$ and $1 + e^* \cdot e \equiv_{\mathsf{SKA}} e^*$ hold for $K \in \mathcal{L}_s$. In case $K = \dagger$, for $1 + e \cdot e^* \equiv_{\mathsf{SKA}} e^*$ we have

$$1 + \dagger \cdot \dagger^* = 1 + \dagger \cdot \dagger = 1 + \dagger = \dagger = \dagger^*$$

and a similar derivation for $1 + e^* \cdot e \equiv_{\mathsf{SKA}} e^*$.

For the commutativity of $+$ we take $K, L \in \mathcal{L}_s \cup \{\dagger\}$. If $K = \dagger$ or $L = \dagger$, we have $K + L = \dagger = L + K$.

For associativity of the plus operator we take $K, L, J \in \mathcal{L}_s \cup \{\dagger\}$. If any of K, L or J is $\dagger$, it is easy to see the axiom holds.

For associativity of the sequential composition operator, consider $K, L, J \in \mathcal{L}_s \cup \{\dagger\}$. We first can observe that if one of K, L or J is empty, then the equality holds trivially. Otherwise, if one of K, L and J is $\dagger$, then $(K \cdot L) \cdot J = \dagger = K \cdot (L \cdot J)$.

Next, we verify distributivity of concatenation over $+$. We will show a detailed proof for left-distributivity only; right-distributivity can be proved similarly. Let $K, L, J \in \mathcal{L}_s \cup \{\dagger\}$. If one of K, L, or J is empty, then the claim holds immediately (the derivation is slightly different for K versus L or J). Otherwise, if one of K, L or J is $\dagger$, then $K \cdot (L + J) = \dagger = K \cdot L + K \cdot J$.

For the remaining least fixpoint axiom, let $K, L, J \in \mathcal{L}_s \cup \{\dagger\}$. Assume that $K + L \cdot J \leq L$. We need to prove that $K \cdot J^* \leq L$. If $L = \dagger$, then the claim holds immediately. If $L \in \mathcal{L}_s$ and $J = \dagger$, then L must be empty, hence K is empty, and the claim holds. If $L, J \in \mathcal{L}_s$, then also $K \in \mathcal{L}_s$ and the proof goes through as it does for KA.

We now get to the axioms for the $\times$-operator. The commutativity axiom is obvious from the commutative definition of $\times$ (as we already know that $\times$ is commutative on synchronous strings). The axiom $K \times \emptyset = \emptyset$ is also satisfied by definition. The same holds for the axiom $K \times \{\varepsilon\} = K$ as $\{\varepsilon\}$ is finite.

For associativity of the synchronous product, consider $K, L, J \in \mathcal{L}_s \cup \{\dagger\}$. If one of K, L or J is empty, then both sides of the equation evaluate to $\emptyset$. Otherwise, if one of K, J, or L is $\dagger$, then both sides of the equation evaluate to $\dagger$. If K, J and L are all languages, and at most one of them is finite, then either $K \times L = \dagger$, in which case the left-hand side evaluates to $\dagger$, or $K \times L$ is infinite (by Lemma A.1) and $J = \dagger$, in which case the right-hand side evaluates to $\dagger$ again. The right-hand side can be shown to evaluate to $\dagger$ by a similar argument. In the remaining cases (at least two out of K, J and L are finite languages and none of them is $\dagger$ or $\emptyset$), the proof of associativity for the language model applies.

For distributivity of synchronous product over $+$, let $K, L, J \in \mathcal{L}_s \cup \{\dagger\}$. If one of K, L or J is $\emptyset$, then the proof is straightforward. Otherwise, if one of K, L or J is $\dagger$, then both sides evaluate to $\dagger$. If K and $L + J$ are infinite, then the outcome is again $\dagger$ on both sides (note that $L + J$ being infinite implies that either L or J is infinite). In the remaining cases, K, L and J are languages and either K or $L + J$ (hence L and J) is finite. In either case the proof for synchronous regular languages goes through. □

Lemma 5.6. *Let $e, f \in \mathcal{T}_{\mathsf{SF}_1}$. If $e \equiv_{\mathsf{SF}_1} f$ then $[\![e]\!]_{\mathsf{SF}_1} = [\![f]\!]_{\mathsf{SF}_1}$.*

Proof. We need to verify each of the axioms of SF_1. The proof for the axioms of F_1 is immediate via the observation that synchronous languages over the alphabet Σ are simply languages over the alphabet $\mathcal{P}_n(\Sigma)$. Thus we know that the F_1-axioms are satisfied, as languages over alphabet $\mathcal{P}_n(\Sigma)$ with $1 = \{\varepsilon\}$ and $0 = \emptyset$ form an F_1-algebra. The additional axioms are the same as the ones that were added to KA for SKA, and we know they are sound from Lemma 3.6. □

Lemma 5.14. *For all $e \in \mathcal{T}_{\mathsf{SF}_1}$ and $A \in \mathcal{P}_n(\Sigma)$, we have $\delta(e, A) \subseteq \rho(e)$. Also, if $e' \in \rho(e)$, then $\delta(e', A) \subseteq \rho(e)$.*

Proof. We prove the first statement by induction on the structure of e. In the base, if we have $e \in \{0, 1\}$, the claim holds vacuously. If we have $a \in \Sigma$, then $\rho(a) = \{1, a\}$ and $\delta(a, A) = \{1 : A = \{a\}\}$, so the claim follows. For the inductive step, there are five cases to consider.

- If $e = H(e_0)$, then immediately $\delta(H(e_0), A) = \emptyset$ so the claim holds vacuously.
- If $e = e_0 + e_1$, then by induction we have $\delta(e_0, A) \subseteq \rho(e_0)$ and $\delta(e_1, A) \subseteq \rho(e_1)$. Hence, we find that $\delta(e, A) = \delta(e_0, A) \cup \delta(e_1, A) \subseteq \rho(e_0) \cup \rho(e_1) = \rho(e)$.

- If $e = e_0 \cdot e_1$, then by induction we have $\delta(e_0, A) \subseteq \rho(e_0)$ and $\delta(e_1, A) \subseteq \rho(e_1)$. Hence, we can calculate that

$$\begin{aligned}\delta(e, A) &= \{e_0' \cdot e_1 : e_0' \in \delta(e_0, A)\} \cup \Delta(e_1, e_0, A) \\ &\subseteq \{e_0' \cdot e_1 : e_0' \in \rho(e_0)\} \cup \rho(e_1) = \rho(e)\end{aligned}$$

- If $e = e_0 \times e_1$, then by induction we have $\delta(e_0, A) \subseteq \rho(e_0)$ and $\delta(e_1, A) \subseteq \rho(e_1)$ for all $A \in \mathcal{P}_n(\Sigma)$. Hence, we can calculate that

$$\begin{aligned}\delta(e, A) &= \{e_0' \times e_1' : e_0' \in \delta(e_0, B_1), e_1' \in \delta(e_1, B_2), B_1 \cup B_2 = A\} \\ &\quad \cup \Delta(e_0, e_1, A) \cup \Delta(e_1, e_0, A) \\ &\subseteq \{e_0' \times e_1' : e_0' \in \rho(e_0), e_1' \in \rho(e_1)\} \cup \rho(e_0) \cup \rho(e_1) = \rho(e)\end{aligned}$$

- If $e = e_0^*$, then by induction we have $\delta(e_0, A) \subseteq \rho(e_0)$. Hence, we find that

$$\delta(e, A) = \{e_0' \cdot e_0^* : e_0' \in \delta(e_0, A)\} \subseteq \{e_0' \cdot e_0^* : e_0' \in \rho(e_0)\} \subseteq \rho(e)$$

For the second statement, we prove that if $e' \in \rho(e)$, then $\rho(e') \subseteq \rho(e)$. The result of the first part tells us that $\delta(e', A) \subseteq \rho(e')$, which together with $\rho(e') \subseteq \rho(e)$ proves the claim. We proceed by induction on e. In the base, there are two cases to consider. First, if $e = 0$, then the claim holds vacuously. If $e = 1$, then the only $e' \in \rho(e)$ is $e' = 1$, so the claim holds. If $e = a$ for $a \in \Sigma$, we have $\rho(e) = \{1, a\}$. It trivially holds that $\rho(e') \subseteq \rho(e)$ for $e' \in \rho(e)$.

For the inductive step, there are four cases to consider.

- If $e = H(e_0)$, then $\rho(e) = \{1\}$, and the proof is as in the case where $e = 1$.
- If $e = e_0 + e_1$, assume w.l.o.g. that $e' \in \rho(e_0)$. By induction, we derive that

$$\rho(e') \subseteq \rho(e_0) \subseteq \rho(e)$$

- If $e = e_0 \cdot e_1$ then there are two cases to consider.
  - If $e' = e_0' \cdot e_1$ where $e_0' \in \rho(e_0)$, then we calculate

$$\begin{aligned}\rho(e') &= \{e_0'' \cdot e_1 : e_0'' \in \rho(e_0')\} \cup \rho(e_1) \\ &\subseteq \{e_0'' \cdot e_1 : e_0'' \in \rho(e_0)\} \cup \rho(e_1) = \rho(e)\end{aligned}$$

  - If $e' \in \rho(e_1)$, then by induction we have $\rho(e') \subseteq \rho(e_1) \subseteq \rho(e)$.
- If $e = e_0 \times e_1$ then there are three cases to consider.
  - The first case is $e' = e_0' \times e_1'$ where $e_0' \in \rho(e_0)$ and $e_1' \in \rho(e_1)$, we get $\rho(e_0') \subseteq \rho(e_0)$ and $\rho(e_1') \subseteq \rho(e_1)$ by induction. We calculate

$$\begin{aligned}\rho(e') &= \{e_0'' \times e_1'' : e_0'' \in \rho(e_0'), e_1'' \in \rho(e_1')\} \cup \rho(e_0') \cup \rho(e_1') \\ &\subseteq \{e_0'' \cdot e_1'' : e_0'' \in \rho(e_0), e_1'' \in \rho(e_1)\} \cup \rho(e_0) \cup \rho(e_1) \\ &= \rho(e)\end{aligned}$$

  - For $e' \in \rho(e_0)$, then by induction we have $\rho(e') \subseteq \rho(e_0) \subseteq \rho(e)$.
  - For $e' \in \rho(e_1)$, the argument is similar to the previous case.

– If $e = e_0^*$, then either $e' = 1$ or $e' = e_0' \cdot e_0^*$ for some $e_0' \in \rho(e_0)$. In the former case, $\rho(e') = \{1\} \subseteq \rho(e)$. In the latter case, we find by induction that

$$\begin{aligned}\rho(e') &= \{e_0'' \cdot e_0^* : e_0'' \in \rho(e_0')\} \cup \rho(e_0^*) \\ &\subseteq \{e_0'' \cdot e_0^* : e_0'' \in \rho(e_0)\} \cup \rho(e_0^*) \subseteq \rho(e_0^*)\end{aligned}$$

□

Lemma 5.17. *For all* $e \in \overline{\mathcal{T}_{\mathsf{SL}}}$ *we have that* $\overline{e} = e$.

Proof. As $e \in \overline{\mathcal{T}_{\mathsf{SL}}}$ we have that $e = \overline{e_0}$ for some $e_0 \in \mathcal{T}_{\mathsf{SL}}$. From Lemma 5.16 we know that $\overline{e_0} \equiv_{\mathsf{SF}_1} e_0$. So we get $e \equiv_{\mathsf{SF}_1} e_0$. Again from Lemma 5.16 we then know that $\overline{e} = \overline{e_0} = e$. □

Lemma A.2. *For* $x, y \in (\mathcal{P}_n(\Sigma))^*$, *we have* $(x \cdot y)^{\Pi} = x^{\Pi} \cdot y^{\Pi}$.

Proof. We proceed by induction on the length of xy. In the base, we have $xy = \varepsilon$. Thus $x = \varepsilon$ and $y = \varepsilon$. We have $\varepsilon^{\Pi} = \varepsilon$ so the result follows immediately. In the inductive step we consider $xy = aw$ for $a \in \mathcal{P}_n(\Sigma)$. We have to consider two cases. In the first case we have $x = ax'$. The induction hypothesis gives us that $(x' \cdot y)^{\Pi} = x'^{\Pi} \cdot y^{\Pi}$. We then have $(x \cdot y)^{\Pi} = (ax' \cdot y)^{\Pi} = a^{\Pi} \cdot (x' \cdot y)^{\Pi} = a^{\Pi} \cdot x'^{\Pi} \cdot y^{\Pi} = x^{\Pi} \cdot y^{\Pi}$. In the second case we have $x = \varepsilon$ and $y = aw$. We then conclude that $(x \cdot y)^{\Pi} = y^{\Pi} = x^{\Pi} \cdot y^{\Pi}$. □

Lemma 5.18. *For synchronous languages* L *and* K, *all of the following hold: (i)* $(L \cup K)^{\Pi} = L^{\Pi} \cup K^{\Pi}$, *(ii)* $(L \cdot K)^{\Pi} = L^{\Pi} \cdot K^{\Pi}$, *and(iii)* $(L^*)^{\Pi} = (L^{\Pi})^*$.

Proof. (i) First, suppose $w \in (L \cup K)^{\Pi}$. Thus we have $w = v^{\Pi}$ for $v \in L \cup K$. This gives us $v \in L$ or $v \in K$. We assume the former without loss of generality. Thus we know $w = v^{\Pi} \in L^{\Pi}$. Hence we know $w \in L^{\Pi} \cup K^{\Pi}$. The other direction can be proved analogously.

(ii) First, suppose $w \in (L \cdot K)^{\Pi}$. Thus we have $w = v^{\Pi}$ for some $v \in L \cdot K$. This gives us $v = v_1 \cdot v_2$ for some $v_1 \in L$ and some $v_2 \in K$. By definition of $(-)^{\Pi}$ we know that $v_1^{\Pi} \in L^{\Pi}$ and $v_2^{\Pi} \in K^{\Pi}$. Thus we have $v_1^{\Pi} \cdot v_2^{\Pi} \in L^{\Pi} \cdot K^{\Pi}$. From Lemma A.2 we know that $w = v^{\Pi} = (v_1 \cdot v_2)^{\Pi} = v_1^{\Pi} \cdot v_2^{\Pi}$, which gives us the desired result of $w \in L^{\Pi} \cdot K^{\Pi}$. The other direction can be proved analogously.

(iii) Take $w \in (L^*)^{\Pi}$. Thus we have $w = v^{\Pi}$ for some $v \in L^*$. By definition of the star of a synchronous language we know that $v = u_1 \cdots u_n$ for $u_i \in L$. As $u_i \in L$, we have $u_i^{\Pi} \in L^{\Pi}$ and $u_1^{\Pi} \cdots u_n^{\Pi} \in (L^{\Pi})^*$. By Lemma A.2, we know that $w = v^{\Pi} = (u_1 \cdots u_n)^{\Pi} = u_1^{\Pi} \cdots u_n^{\Pi}$. Thus we have $w \in (L^{\Pi})^*$, which is the desired result. The other direction can be proved analogously. □

Theorem 6.2 (Fundamental Theorem). *For all* $e \in \mathcal{T}_{\mathsf{SF}_1}$, *we have*

$$e \equiv_{\mathsf{SF}_1} o(e) + \sum_{e' \in \delta(e,A)} A^{\Pi} \cdot e'.$$

Proof. Here we treat the inductive cases not displayed in the main proof, where we treated only the synchronous case.

– If $e = H(e_0)$, derive:

$$\begin{aligned}
H(e_0) &\equiv_{\mathsf{SF}_1} H(o(e_0)) + \sum_{e' \in \delta(e_0, A)} H(A^{\Pi}) \cdot H(e') && \text{(IH, compatibility of } H\text{)} \\
&\equiv_{\mathsf{SF}_1} H(o(e_0)) && (H(A^{\Pi}) = 0) \\
&\equiv_{\mathsf{SF}_1} o(H(e_0)) && (o(H(e_0)) \in 2) \\
&\equiv_{\mathsf{SF}_1} o(H(e_0)) + \sum_{e' \in \delta(H(e_0), A)} A^{\Pi} \cdot e' && \text{(Def. } \delta\text{)}
\end{aligned}$$

– If $e = e_0 + e_1$, derive:

$$\begin{aligned}
e_0 + e_1 &\equiv_{\mathsf{SF}_1} o(e_0) + \sum_{e' \in \delta(e_0, A)} A^{\Pi} \cdot e' + o(e_1) + \sum_{e' \in \delta(e_1, A)} A^{\Pi} \cdot e' && \text{(IH)} \\
&\equiv_{\mathsf{SF}_1} o(e_0 + e_1) + \sum_{e' \in \delta(e_0, A) \cup \delta(e_1, A)} A^{\Pi} \cdot e' && \text{(Def. } o\text{, merge sums)} \\
&\equiv_{\mathsf{SF}_1} o(e_0 + e_1) + \sum_{e' \in \delta(e_0 + e_1, A)} A^{\Pi} \cdot e' && \text{(Def. } \delta\text{)}
\end{aligned}$$

– If $e = e_0 \cdot e_1$, derive:

$$\begin{aligned}
e_0 \cdot e_1 &\equiv_{\mathsf{SF}_1} \Big(o(e_0) + \sum_{e' \in \delta(e_0, A)} A^{\Pi} \cdot e'\Big) \cdot e_1 && \text{(IH)} \\
&\equiv_{\mathsf{SF}_1} o(e_0) \cdot e_1 + \sum_{e' \in \delta(e_0, A)} (A^{\Pi} \cdot e' \cdot e_1) && \text{(Distributivity)} \\
&\equiv_{\mathsf{SF}_1} o(e_0) \cdot \Big(o(e_1) + \sum_{e' \in \delta(e_1, A)} A^{\Pi} \cdot e'\Big) + \sum_{e' \in \delta(e_0, A)} (A^{\Pi} \cdot e' \cdot e_1) && \text{(IH)} \\
&\equiv_{\mathsf{SF}_1} o(e_0 \cdot e_1) + o(e_0) \cdot \sum_{e' \in \delta(e_1, A)} A^{\Pi} \cdot e' + \sum_{e' \in \delta(e_0, A)} (A^{\Pi} \cdot e' \cdot e_1) \\
& && \text{(Def. } o\text{, distributivity)} \\
&\equiv_{\mathsf{SF}_1} o(e_0 \cdot e_1) + \sum_{e' \in \Delta(e_1, e_0, A)} A^{\Pi} \cdot e' + \sum_{e' \in \{e'_0 \cdot e_1 : e'_0 \in \delta(e_0, A)\}} A^{\Pi} \cdot e' \\
&\equiv_{\mathsf{SF}_1} o(e_0 \cdot e_1) + \sum_{e' \in \delta(e_0 \cdot e_1, A)} A^{\Pi} \cdot e' && \text{(Def. } \delta\text{)}
\end{aligned}$$

– If $e = e_0^*$, we derive:

$$\begin{aligned}
e_0^* &\equiv_{\mathsf{SF}_1} \Big(o(e_0) + \sum_{e' \in \delta(e_0, A)} A^{\Pi} \cdot e'\Big)^* && \text{(Induction hypothesis)} \\
&\equiv_{\mathsf{SF}_1} \Big(\sum_{e' \in \delta(e_0, A)} A^{\Pi} \cdot e'\Big)^* && (o(e_0) \in 2 \text{ and loop tightening}) \\
&\equiv_{\mathsf{SF}_1} 1 + \Big(\sum_{e' \in \delta(e_0, A)} A^{\Pi} \cdot e'\Big) \cdot \Big(\sum_{e' \in \delta(e_0, A)} A^{\Pi} \cdot e'\Big)^* && \text{(star axiom of } \mathsf{SF}_1) \\
&\equiv_{\mathsf{SF}_1} 1 + \Big(\sum_{e' \in \delta(e_0, A)} A^{\Pi} \cdot e'\Big) \cdot e_0^* && \text{(first two steps)} \\
&\equiv_{\mathsf{SF}_1} 1 + \sum_{e' \in \delta(e_0, A)} (A^{\Pi} \cdot e' \cdot e_0^*) && \text{(Distributivity)} \\
&\equiv_{\mathsf{SF}_1} o(e_0^*) + \sum_{e' \in \delta(e_0^*, A)} A^{\Pi} \cdot e' && \text{(Def. } o, \text{ def. } \delta) \ \square
\end{aligned}$$

Lemma 7.2. *For all* $e \in \mathcal{T}_{\mathsf{NSF}}$*, we have* $(\llbracket e \rrbracket_{\mathsf{SF}_1})^{\Pi} = \llbracket e \rrbracket_{\mathsf{F}_1}$.

Proof. In the main text we have treated the base cases. The inductive cases work as follows. There are three cases to consider. If $e = e_0 + e_1$, then $(\llbracket e \rrbracket_{\mathsf{SKA}})^{\Pi} = (\llbracket e_0 \rrbracket_{\mathsf{SKA}} \cup \llbracket e_1 \rrbracket_{\mathsf{SKA}})^{\Pi} = (\llbracket e_0 \rrbracket_{\mathsf{SKA}})^{\Pi} \cup (\llbracket e_1 \rrbracket_{\mathsf{SKA}})^{\Pi}$ (Lemma 5.18). From the induction hypothesis we obtain $(\llbracket e_0 \rrbracket_{\mathsf{SKA}})^{\Pi} = \llbracket e_0 \rrbracket_{\mathsf{KA}}$ and $(\llbracket e_1 \rrbracket_{\mathsf{SKA}})^{\Pi} = \llbracket e_1 \rrbracket_{\mathsf{KA}}$. Combining these results we get $(\llbracket e \rrbracket_{\mathsf{SKA}})^{\Pi} = \llbracket e_0 \rrbracket_{\mathsf{KA}} \cup \llbracket e_1 \rrbracket_{\mathsf{KA}} = \llbracket e_0 \rrbracket_{\mathsf{KA}} + \llbracket e_1 \rrbracket_{\mathsf{KA}} = \llbracket e_0 + e_1 \rrbracket_{\mathsf{KA}}$, so the claim follows. Secondly, if $e = e_0 \cdot e_1$, then $(\llbracket e \rrbracket_{\mathsf{SKA}})^{\Pi} = (\llbracket e_0 \rrbracket_{\mathsf{SKA}} \cdot \llbracket e_1 \rrbracket_{\mathsf{SKA}})^{\Pi} = (\llbracket e_0 \rrbracket_{\mathsf{SKA}})^{\Pi} \cdot (\llbracket e_1 \rrbracket_{\mathsf{SKA}})^{\Pi}$ (Lemma 5.18). From the induction hypothesis we obtain $(\llbracket e_0 \rrbracket_{\mathsf{SKA}})^{\Pi} = \llbracket e_0 \rrbracket_{\mathsf{KA}}$ and $(\llbracket e_1 \rrbracket_{\mathsf{SKA}})^{\Pi} = \llbracket e_1 \rrbracket_{\mathsf{KA}}$. We can then conclude that $(\llbracket e \rrbracket_{\mathsf{SKA}})^{\Pi} = \llbracket e_0 \rrbracket_{\mathsf{KA}} \cdot \llbracket e_1 \rrbracket_{\mathsf{KA}} = \llbracket e_0 \cdot e_1 \rrbracket_{\mathsf{KA}}$. Lastly, if $e = e_0^*$, we get $(\llbracket e_0^* \rrbracket_{\mathsf{SKA}})^{\Pi} = ((\llbracket e_0 \rrbracket_{\mathsf{SKA}})^*)^{\Pi} = ((\llbracket e_0 \rrbracket_{\mathsf{SKA}})^{\Pi})^*$ (Lemma 5.18). From the induction hypothesis we obtain $(\llbracket e_0 \rrbracket_{\mathsf{SKA}})^{\Pi} = \llbracket e_0 \rrbracket_{\mathsf{KA}}$. Thus we have $(\llbracket e \rrbracket_{\mathsf{SKA}})^{\Pi} = \llbracket e_0 \rrbracket_{\mathsf{KA}}^* = \llbracket e_0^* \rrbracket_{\mathsf{KA}}$ and the claim follows. □

Lemma 7.4. *Let* (M, p) *be a* Q*-linear system such that* M *and* p *are guarded. We can construct* Q*-vector* x *that is the unique (up to* SF_1*-equivalence) solution to* (M, p) *in* SF_1*. Moreover, if* M *and* p *are in normal form, then so is* x*.*

Proof. We will construct x by induction on the size of Q. In the base, let $Q = \emptyset$. In this case the unique Q-vector is a solution. In the inductive step, take $k \in Q$ and let $Q' = Q \setminus \{k\}$. Then construct the Q'-linear system (M', p') as follows:

$$\begin{aligned}
M'(i, j) &= M(i, k) \cdot M(k, k)^* \cdot M(k, j) + M(i, j) \\
p'(i) &= p(i) + M(i, k) \cdot M(k, k)^* \cdot p(k)
\end{aligned}$$

As Q' is a strictly smaller set than Q and M' is guarded, we can apply our induction hypothesis to (M', p'). So we know by induction that (M', p') has a unique solution x'. Moreover, if M' and p' are in normal form, so is x'; note that if M and p are in normal form, then so are M' and p'.

We use x' to construct the Q-vector x:

$$x(i) = \begin{cases} x'(i) & i \neq k \\ M(k,k)^* \cdot \left(p(k) + \sum_{j \in Q'} M(k,j) \cdot x'(j)\right) & i = k \end{cases}$$

The first thing to show now is that x is indeed a solution of (M, p). To this end, we need to show that $M \cdot x + p \equiv_{\mathsf{SF}_1} x$. We have two cases. For $i \in Q'$ we derive:

$$\begin{aligned}
x(i) &= x'(i) && \text{(Def. } x\text{)} \\
&\equiv_{\mathsf{SF}_1} p'(i) + \sum_{j \in Q'} M'(i,j) \cdot x'(j) && (x' \text{ solution of } (M', p')) \\
&\equiv_{\mathsf{SF}_1} p(i) + M(i,k) \cdot M(k,k)^* \cdot p(k) \\
&\quad + \sum_{j \in Q'} (M(i,k) \cdot M(k,k)^* \cdot M(k,j) + M(i,j)) \cdot x'(j) && \text{(Def. } (M', p')\text{)} \\
&\equiv_{\mathsf{SF}_1} p(i) + \sum_{j \in Q'} M(i,j) \cdot x'(j) \\
&\quad + M(i,k) \cdot M(k,k)^* \cdot \left(p(k) + \sum_{j \in Q'} M(k,j) \cdot x'(j)\right) && \text{(Distributivity)} \\
&\equiv_{\mathsf{SF}_1} p(i) + \sum_{j \in Q'} M(i,j) \cdot x(j) + M(i,k) \cdot x(k) && \text{(Def. } x\text{)} \\
&\equiv_{\mathsf{SF}_1} p(i) + \sum_{j \in Q} M(i,j) \cdot x(j) && \text{(Merge sum)}
\end{aligned}$$

For $i = k$, we derive:

$$\begin{aligned}
x(k) &= M(k,k)^* \cdot \left(p(k) + \sum_{j \in Q'} M(k,j) \cdot x'(j)\right) && \text{(Def. } x\text{)} \\
&\equiv_{\mathsf{SF}_1} (1 + M(k,k) \cdot M(k,k)^*) \cdot \left(p(k) + \sum_{j \in Q'} M(k,j) \cdot x'(j)\right) && \text{(star axiom)} \\
&\equiv_{\mathsf{SF}_1} p(k) + \sum_{j \in Q'} M(k,j) \cdot x'(j) \\
&\quad + M(k,k) \cdot M(k,k)^* \cdot \left(p(k) + \sum_{j \in Q'} M(k,j) \cdot x'(j)\right) && \text{(Distributivity)} \\
&\equiv_{\mathsf{SF}_1} p(k) + \sum_{j \in Q'} M(k,j) \cdot x(j) + M(k,k) \cdot x(k) && \text{(Def. } x\text{)} \\
&\equiv_{\mathsf{SF}_1} p(k) + \sum_{j \in Q} M(k,j) \cdot x(j) && \text{(Merge sum)}
\end{aligned}$$

We now know that x is a solution to (M, p) because $M \cdot x + p \equiv_{\mathsf{SF}_1} x$. Furthermore, if M and p are in normal form, then so is x', and thus x is in normal form by construction.

Next we claim that x is unique. Let y be any solution of (M, p). We choose the Q'-vector y' by taking $y'(i) = y(i)$. To see that y' is a solution to (M', p'), we first claim that the following holds:

$$y(k) \equiv_{\mathsf{SF}_1} M(k,k)^* \cdot \Big(p(k) + \sum_{j \in Q'} M(k,j) \cdot y(j)\Big) \tag{3}$$

To see that this is true, derive

$$\begin{aligned}
y(k) &\equiv_{\mathsf{SF}_1} p(k) + \sum_{j \in Q} M(k,j) \cdot y(j) && (y \text{ solution of } (M,p)) \\
&\equiv_{\mathsf{SF}_1} p(k) + M(k,k) \cdot y(k) + \sum_{j \in Q'} M(k,j) \cdot y(j) && \text{(Split sum)} \\
&\equiv_{\mathsf{SF}_1} M(k,k)^* \cdot \Big(p(k) + \sum_{j \in Q'} M(k,j) \cdot y(j)\Big) && \text{(Unique fixpoint axiom)}
\end{aligned}$$

Note that we can apply the unique fixpoint axiom because we know that M is guarded and thus that $H(M(k,k)) = 0$.

Now we can derive the following:

$$\begin{aligned}
y'(i) &= y(i) && (\text{Def. } y) \\
&\equiv_{\mathsf{SF}_1} p(i) + \sum_{j \in Q} M(i,j) \cdot y(j) && (y \text{ solution of } (M,p)) \\
&\equiv_{\mathsf{SF}_1} p(i) + M(i,k) \cdot y(k) + \sum_{j \in Q'} M(i,j) \cdot y(j) && \text{(Split sum)} \\
&\equiv_{\mathsf{SF}_1} p(i) + \sum_{j \in Q'} M(i,j) \cdot y(j) \\
&\qquad + M(i,k) \cdot M(k,k)^* \cdot \Big(p(k) + \sum_{j \in Q'} M(k,j) \cdot y(j)\Big) && \text{(Equation 3)} \\
&\equiv_{\mathsf{SF}_1} p(i) + M(i,k) \cdot M(k,k)^* \cdot p(k) \\
&\qquad + \sum_{j \in Q'} \big(M(i,k) \cdot M(k,k)^* \cdot M(k,j) + M(i,j)\big) \cdot y(j) && \text{(Distributivity)} \\
&\equiv_{\mathsf{SF}_1} p'(i) + \sum_{j \in Q'} M'(i,j) \cdot y(j) && (\text{Def. } (M', p'))
\end{aligned}$$

Thus y' is a solution to (M', p'). As x' is the unique solution to (M', p'), we know that $y' \equiv_{\mathsf{SF}_1} x'$.

For $i \neq k$ we know that $x(i) = x'(i) \equiv_{\mathsf{SF}_1} y'(i) = y(i)$. For $i = k$ we can derive:

$$\begin{aligned}
y(k) &\equiv_{\mathsf{SF}_1} M(k,k)^* \cdot \Big(p(k) + \sum_{j \in Q'} M(k,j) \cdot y(j)\Big) && \text{(Equation 3)} \\
&\equiv_{\mathsf{SF}_1} M(k,k)^* \cdot \Big(p(k) + \sum_{j \in Q'} M(k,j) \cdot y'(j)\Big) && \text{(Def. } y'\text{)} \\
&\equiv_{\mathsf{SF}_1} M(k,k)^* \cdot \Big(p(k) + \sum_{j \in Q'} M(k,j) \cdot x'(j)\Big) && (x' \equiv_{\mathsf{SF}_1} y') \\
&\equiv_{\mathsf{SF}_1} M(k,k)^* \cdot \Big(p(k) + \sum_{j \in Q'} M(k,j) \cdot x(j)\Big) && \text{(Def. } x'\text{)} \\
&\equiv_{\mathsf{SF}_1} x(k) && \text{(Def. } x\text{)}
\end{aligned}$$

Thus, $y \equiv_{\mathsf{SF}_1} x$, thereby proving that x is the unique solution to (M, p). □

References

1. Antimirov, V.M.: Partial derivatives of regular expressions and finite automaton constructions. Theor. Comput. Sci. **155**(2), 291–319 (1996). https://doi.org/10.1016/0304-3975(95)00182-4
2. Backhouse, R.: Closure algorithms and the star-height problem of regular languages. PhD thesis, University of London (1975)
3. Baier, C., Sirjani, M., Arbab, F., Rutten, J.: Modeling component connectors in reo by constraint automata. Sci. Comput. Program. **61**(2), 75–113 (2006). https://doi.org/10.1016/j.scico.2005.10.008
4. Bergstra, J.A., Klop, J.W.: Process algebra for synchronous communication. Inf. Control **60**(1–3), 109–137 (1984). https://doi.org/10.1016/S0019-9958(84)80025-X
5. Boffa, M.: Une remarque sur les systèmes complets d'identités rationnelles. ITA **24**, 419–428 (1990)
6. Bonchi, F., Pous, D.: Checking NFA equivalence with bisimulations up to congruence. In: Proceedings of Principles of Programming Languages (POPL), pp. 457–468 (2013). https://doi.org/10.1145/2429069.2429124
7. Broda, S., Cavadas, S., Ferreira, M., Moreira, N.: Deciding synchronous Kleene algebra with derivatives. In: Drewes, F. (ed.) CIAA 2015. LNCS, vol. 9223, pp. 49–62. Springer, Cham (2015). https://doi.org/10.1007/978-3-319-22360-5_5
8. Brzozowski, J.A.: Derivatives of regular expressions. J. ACM **11**(4), 481–494 (1964). https://doi.org/10.1145/321239.321249
9. John Horton Conway: Regular Algebra and Finite Machines. Chapman and Hall Ltd., London (1971)
10. Foster, S., Struth, G.: On the fine-structure of regular algebra. J. Autom. Reason. **54**(2), 165–197 (2015). https://doi.org/10.1007/s10817-014-9318-9
11. Hayes, I.J.: Generalised rely-guarantee concurrency: an algebraic foundation. Formal Asp. Comput. **28**(6), 1057–1078 (2016). https://doi.org/10.1007/s00165-016-0384-0

12. Hayes, I.J., Colvin, R.J., Meinicke, L.A., Winter, K., Velykis, A.: An algebra of synchronous atomic steps. In: Fitzgerald, J., Heitmeyer, C., Gnesi, S., Philippou, A. (eds.) FM 2016. LNCS, vol. 9995, pp. 352–369. Springer, Cham (2016). https://doi.org/10.1007/978-3-319-48989-6_22
13. Hayes, I.J., Meinicke, L.A., Winter, K., Colvin, R.J.: A synchronous program algebra: a basis for reasoning about shared-memory and event-based concurrency. Formal Asp. Comput. **31**(2), 133–163 (2019). https://doi.org/10.1007/s00165-018-0464-4
14. Kozen, D.: Myhill-Nerode relations on automatic systems and the completeness of Kleene algebra. In: Ferreira, A., Reichel, H. (eds.) STACS 2001. LNCS, vol. 2010, pp. 27–38. Springer, Heidelberg (2001). https://doi.org/10.1007/3-540-44693-1_3
15. Hoare, T., van Staden, S., Möller, B., Struth, G., Zhu, H.: Developments in concurrent Kleene algebra. J. Log. Algebr. Meth. Program. **85**(4), 617–636 (2016). https://doi.org/10.1016/j.jlamp.2015.09.012
16. Kappé, T., Brunet, P., Rot, J., Silva, A., Wagemaker, J., Zanasi, F.: Kleene algebra with observations. arXiv:1811.10401
17. Kappé, T., Brunet, P., Silva, A., Zanasi, F.: Concurrent Kleene algebra: free model and completeness. In: Proceedings of European Symposium on Programming (ESOP), pp. 856–882 (2018). https://doi.org/10.1007/978-3-319-89884-1_30
18. Kozen, D.: A completeness theorem for Kleene algebras and the algebra of regular events. Inf. Comput. **110**(2), 366–390 (1994). https://doi.org/10.1006/inco.1994.1037
19. Kozen, D.: Myhill-Nerode relations on automatic systems and the completeness of Kleene algebra. In: Proceedings of Symposium on Theoretical Aspects of Computer Science (STACS), pp. 27–38 (2001). https://doi.org/10.1007/3-540-44693-1_3
20. Kozen, D., Smith, F.: Kleene algebra with tests: completeness and decidability. In: van Dalen, D., Bezem, M. (eds.) CSL 1996. LNCS, vol. 1258, pp. 244–259. Springer, Heidelberg (1997). https://doi.org/10.1007/3-540-63172-0_43
21. Krob, D.: Complete systems of B-rational identities. Theor. Comput. Sci. **89**(2), 207–343 (1991). https://doi.org/10.1016/0304-3975(91)90395-I
22. Laurence, M.R., Struth, G.: Completeness theorems for pomset languages and concurrent Kleene algebras. arXiv:1705.05896
23. Milner, R.: Calculi for synchrony and asynchrony. Theor. Comput. Sci. **25**, 267–310 (1983). https://doi.org/10.1016/0304-3975(83)90114-7
24. Prisacariu, C.: Synchronous Kleene algebra. J. Log. Algebr. Program. **79**(7), 608–635 (2010). https://doi.org/10.1016/j.jlap.2010.07.009
25. Rutten, J.J.M.M.: Behavioural differential equations: a coinductive calculus of streams, automata, and power series. Theor. Comput. Sci. **308**(1–3), 1–53 (2003). https://doi.org/10.1016/S0304-3975(02)00895-2
26. Salomaa, A.: Two complete axiom systems for the algebra of regular events. J. ACM **13**(1), 158–169 (1966). https://doi.org/10.1145/321312.321326
27. Silva, A.: Kleene Coalgebra. PhD thesis, Radboud Universiteit Nijmegen (2010)

Unraveling Recursion: Compiling an IR with Recursion to System F

Michael Peyton Jones[1(✉)], Vasilis Gkoumas[1], Roman Kireev[1], Kenneth MacKenzie[1], Chad Nester[2], and Philip Wadler[2]

[1] IOHK, Hong Kong, China
{michael.peyton-jones,vasilis.gkoumas,roman.kireev,
kenneth.mackenzie}@iohk.io
[2] University of Edinburgh, Edinburgh, UK
{cnester,wadler}@inf.ed.ac.uk

Abstract. Lambda calculi are often used as intermediate representations for compilers. However, they require extensions to handle higher-level features of programming languages. In this paper we show how to construct an IR based on System F^{μ}_{ω} which supports recursive functions and datatypes, and describe how to compile it to System F^{μ}_{ω}. Our IR was developed for commercial use at the IOHK company, where it is used as part of a compilation pipeline for smart contracts running on a blockchain.

1 Introduction

Many compilers make use of *intermediate representations* (IRs) as stepping stones between their source language and their eventual target language. Lambda calculi are tempting choices as IRs for functional programming languages. They are simple, well-studied, and easy to analyze.

However, lambda calculi also have several features that make them poor IRs.

- They are hard to read and write. Although they are mostly read and written by computers, this complicates writing compilers and debugging their output.
- They can be hard to optimize. Some optimizations are much easier to write on a higher-level language. For example, dead-binding elimination is much easier with explicit let-bindings.
- They make the initial compilation step "too big". Compiling all the way from a high-level surface language to a lambda calculus can involve many complex transformations, and it is often advantageous from an engineering standpoint to break it into smaller steps.

Hence it is common to design an IR by extending a lambda calculus with additional features which make the IR more legible, easier to optimize, or closer to the source language (e.g. GHC Core [26], Henk [25], Idris' TT [4], and OCaml's Lambda [20]). However, given that such IRs are desirable, there is little material on implementing or compiling them.

G. Hutton (Ed.): MPC 2019, LNCS 11825, pp. 414–443, 2019.
https://doi.org/10.1007/978-3-030-33636-3_15

In this paper we construct an IR suitable for a powerful functional programming language like Haskell. We take as our lambda calculus System F_ω^μ (System F_ω with indexed fixpoints: see [27, Chapter 30], formalized recently in [8]), which allows us to talk about higher-kinded recursive types, and extend it to an IR called FIR which adds the following features:

- Let-binding of non-recursive terms, types, and datatypes.
- Let-binding of recursive terms and datatypes.

This is a small, but common, subset of the higher-level features that functional programming languages usually have, so this provides a reusable IR for compiler writers targeting System F_ω^μ.

Moreover, all of the compilation passes that we provide are *local* in the sense that they do not access more than one level of the syntax tree, and they do not require any type information that is not present in type annotations. So while we provide typing rules for FIR, it is not necessary to perform type synthesis in order to compile it.

Encoding recursive terms has traditionally been done with fixpoint combinators. However, the textbook accounts typically do not cover mutual recursion, and where it *is* handled it is often assumed that the calculus is non-strict. We construct a generalized, polyvariadic fixpoint combinator that works in both strict and non-strict base calculi, which we use to compile recursive terms.

In order to compile datatypes, we need to encode them and their accompanying constructors and destructors using the limited set of types and terms we have available in our base calculus. The Church encoding [27, Chapter 5.2, Chapter 23.4] is a well-known method of doing this in plain System F. With it, we can encode even recursive datatypes, so long as the recursion occurs only in positive positions.

However, some aspects of the Church encoding are not ideal, for example, it requires time proportional to the size of a list to extract its tail. We use a different encoding, the Scott encoding [1], which can encode any recursive datatype, but requires adding a fixpoint operator to System F in order to handle arbitrary type-level recursion.

To handle mutually recursive datatypes we borrow some techniques from the generic programming community, in particular indexed fixpoints, and the use of type-level tags to combine a family of mutually recursive datatypes into a single recursive datatype. While this technique is well-known (see e.g. [32]), the details of our approach are different, and we face some additional constraints because we are targeting System F_ω^μ rather than a full dependently-typed calculus.

We have used FIR as an IR in developing Plutus [16], a platform for developing smart contracts targeting the Cardano blockchain. Users write programs in Haskell, which are compiled by a GHC compiler plugin into Plutus Core, a small functional programming language. Plutus Core is an extension of System F_ω^μ, so in order to easily compile Haskell's high-level language features we developed FIR as an IR above Plutus Core. We have used this compiler to write substantial programs in Haskell and compile them to Plutus Core, showing that the techniques in this paper are usable in practice. The compiler is available for public use at [17].

Contributions. We make the following contributions.

- We give syntax and typing rules for FIR, a typed IR extending System F_ω^μ.
- We define a series of local compilation passes which collectively compile FIR into System F_ω^μ.
- We provide a reference implementation of the syntax, type system, and several of the compilation passes in Agda [24], a powerful dependently typed programming language.
- We have written a complete compiler implementation in Haskell as part of a production system for the Plutus platform.

Our techniques for encoding datatypes are not novel [21,32]. However, we know of no complete presentation that handles mutual recursion and parameterized datatypes, and targets a calculus as small as System F_ω^μ.

We believe our techniques for encoding mutually recursive functions are novel.

While the Agda compiler implementation is incomplete, and does not include soundness proofs, we believe that the very difficulty of doing this makes our partial implementation valuable. We discuss the difficulties further in Sect. 5.

Note on the Use of Agda. Although System F_ω^μ is a complete programming language in its own right, it is are somewhat verbose and clumsy to use for the *exposition* of the techniques we are presenting.

Consequently we will use:

- Agda code, typeset colourfully, for exposition.
- System F_ω^μ code, typeset plainly, for the formal descriptions.

We have chosen to use $*$ for the kind of types, whereas Agda normally uses Set. To avoid confusion we have aliased Set to $*$ in our Agda code. Readers should recall that Agda uses $\rightarrow$ following binders rather than a . character.

The Agda code in this paper and the Agda compiler code are available in the Plutus repository.

Notational Conventions. We will omit kind signatures in System F_ω^μ when they are $*$, and any other signatures when they are obvious from context or repetition.

We will be working with a number of constructs that have sequences of elements. We will adopt the metalanguage conventions suggested by Guy Steele [29], in particular:

- $t[x := v]$ is a substitution of v for x in t.
- $\overline{t}$ is expanded to any number of (optionally separated) copies of t. Any underlined portions of t must be expanded the same way in each copy. Where we require access to the index, the overline is superscripted with the index. For example:

  - $\overline{x : T}$ is expanded to $x_1 : T_1 \ldots x_n : T_n$
  - $\overline{\Gamma \vdash J}$ is expanded to $\Gamma \vdash J_1 \ldots \Gamma : J_n$
  - $\overline{x_j : T_{j+1}}^j$ is expanded to $x_1 : T_2 \ldots x_n : T_{n+1}$
- $\overline{t} \rightarrow u$ is expanded to $t_1 \rightarrow \ \ldots \ \rightarrow t_n \rightarrow u$, similarly for $\Rightarrow$.

2 Datatype Encodings

The Scott encoding represents a datatype as the type of the pattern-matching functions on it. For example, the type of booleans, Bool, is encoded as

$$\forall R.R \rightarrow R \rightarrow R$$

That is, for any output type R you like, if you provide an R for the case where the value is false and an R for the case where the value is true, then you have given a method to construct an R from all possible booleans, thus performing a sort of pattern-matching. In general the arguments to the encoded datatype value are functions which transform the arguments of each constructor into an R.

The type of naturals, Nat, is encoded as

$$\forall R.R \rightarrow (\mathsf{Nat} \rightarrow R) \rightarrow R$$

Here we see an occurrence of Nat in the definition, which corresponds to recursive use in the "successor" constructor. We will need type-level recursion to deal with recursive references.

The Church encoding of Bool is the same as the Scott encoding. This is true for all non-recursive datatypes, but not for recursive datatypes. The Church encoding of Nat is:

$$\forall R.R \rightarrow (R \rightarrow R) \rightarrow R$$

Here the recursive occurrence of Nat has disappeared, replaced by an R. This is because while the Scott encoding corresponds to a pattern-match on a type, the Church encoding corresponds to a *fold*, so recursive occurrences have already been folded into the output type.

This highlights the tradeoffs between the two encodings (see [19] for further discussion):

- To operate on a Church encoded value we must perform a fold on the entire structure, which is frequently inefficient. For a Scott encoded value, we only have to inspect the surface level of the term, which is inexpensive.
- Since recursive occurrences of the type are already "folded" in the Church encoding, there is no need for a type-level recursion operator. Contrast this with the situation with the Scott encoding, in which additional type-level machinery (fixed points) is needed to define type-level recursion.

In this paper we will use the Scott encoding to encode datatypes.

3 Syntax and Type System of System F_ω^μ and FIR

FIR is an extension of System F_ω^μ, which is itself an extension of the well-known System F_ω. In the following figures we give

- Syntax (Fig. 1)
- Kinding (Fig. 2)
- Well-formedness of constructors and bindings (Fig. 4)
- Type equivalence (Fig. 5)
- Type synthesis (Fig. 6)

for full FIR. Cases without highlighting are for System F_ω, while we highlight additions for System F_ω^μ and FIR .

There are a number of auxiliary definitions in Fig. 3 for dealing with datatypes and bindings. These define kinds and types for the various bindings produced by datatype bindings. We will go through examples of how they work in Sect. 4.3.

3.1 Recursive Types

System F_ω is very powerful, but does not allow us to define (non-positive) recursive types. Adding a type-level fixed point operator enables us to do this (see e.g. [27, Chapter 20]). However, we must make a number of choices about the precise nature of our type-level fixed points.

Isorecursive and Equirecursive Types. The first choice we have is between two approaches to exposing the fixpoint property of our recursive types. Systems with *equirecursive* types identify $(\texttt{fix}\, f)$ and $f(\texttt{fix}\, f)$; whereas systems with *isorecursive* types provide an isomorphism between the two, using a term `unwrap` to convert the first into the second, and a term `wrap` for the other direction.

The tradeoff is that equirecursive types add no additional terms to the language, but have a more complicated metatheory. Indeed, typechecking System F_ω^μ with equirecursive types is not known to be decidable in general [7,11]. Isorecursive types, on the other hand, have a simpler metatheory, but require additional terms. It is not too important for an IR to be easy to program by hand, so we opt for isorecursive types, with our witness terms being `wrap` and `unwrap`.

Choosing an Appropriate Fixpoint Operator. We also have a number of options for *which* fixpoint operator to add. The most obvious choice is a fixpoint operator `fix` which takes fixpoints of type-level endofunctions at any kind K (i.e. it has signature $\texttt{fix} : (K \Rightarrow K) \Rightarrow K$). In contrast, our language System F_ω^μ has a fixpoint operator `ifix` ("indexed fix") which allows us to take fixpoints only at kinds $K \Rightarrow *$.

The key advantage of `ifix` over `fix` is that it is much easier to give fully-synthesizing type rules for `ifix`. To see this, suppose we had a `fix` operator

terms	t, u	$::= x$	variable
		$\lambda x : T.t$	lambda abstraction
		$t\ t$	function application
		$\Lambda X :: K.t$	type abstraction
		$t\ \{T\}$	type application
		$\texttt{wrap}\ T\ U\ t$	wrap
		$\texttt{unwrap}\ t$	unwrap
		$\texttt{let}\ [\texttt{rec}]\ \overline{b}\ \texttt{in}\ t$	let
bindings	b	$::= x : T = t$	term binding
		$X :: K = T$	type binding
		$\texttt{data}\ X\ (\overline{Y :: K}) = \overline{c}\ \texttt{with}\ x$	datatype binding
constructors	c	$::= x\ (\overline{T})$	
values	v	$::= \lambda x : T.t$	lambda abstraction
		$\Lambda X :: K.t$	type abstraction
		$\texttt{wrap}\ T\ U\ v$	wrap
types	T, U	$::= X$	type variable
		$T \rightarrow U$	arrow type
		$\forall X :: K.T$	universal type
		$\lambda X :: K.T$	function type
		$T\ U$	function application
		$\texttt{ifix}\ T\ U$	fixpoint type
contexts	Γ	$::= \varnothing$	empty
		$\Gamma, x : T$	term variable binding
		$\Gamma, X :: K$	type variable binding
kind	K	$::= *$	type kind
		$K \Rightarrow K$	arrow kind

Fig. 1. Syntax of FIR

$$\text{K-TVar}\ \frac{X :: K \in \Gamma}{\Gamma \vdash X :: K} \qquad \text{K-Abs}\ \frac{\Gamma, X :: K_1 \vdash T :: K_2}{\Gamma \vdash (\lambda X :: K_1.T) :: K_1 \Rightarrow K_2}$$

$$\text{K-App}\ \frac{\Gamma \vdash T_1 :: K_1 \Rightarrow K_2 \quad \Gamma \vdash T_2 :: K_1}{\Gamma \vdash (T_1\ T_2) :: K_2} \qquad \text{K-Arrow}\ \frac{\Gamma \vdash T_1 :: * \quad \Gamma \vdash T_2 :: *}{\Gamma \vdash (T_1 \rightarrow T_2) :: *}$$

$$\text{K-All}\ \frac{\Gamma, X :: K \vdash T :: *}{\Gamma \vdash (\forall X :: K.T) :: *} \qquad \text{K-Ifix}\ \frac{\Gamma \vdash T :: K \quad \Gamma \vdash F :: (K \Rightarrow *) \Rightarrow (K \Rightarrow *)}{\Gamma \vdash (\texttt{ifix}\ F\ T) :: *}$$

Fig. 2. Kinding for FIR

Throughout this figure when d or c is an argument
$d = \mathtt{data}\ X\ (\overline{Y :: K}) = (\overline{c})\ \mathtt{with}\ x$
$c = x(\overline{T})$

AUXILIARY FUNCTIONS

$$
\begin{array}{ll}
\mathsf{branchTy}(c, R) & = \overline{T \rightarrow} R \\
\mathsf{dataTy}(d) & = \lambda(\overline{Y :: K}).\forall R.(\overline{\mathsf{branchTy}(c, \underline{R})}) \rightarrow R \\
\mathsf{dataKind}(d) & = \overline{K \Rightarrow} * \\
\mathsf{constrTy}(d, c) & = \forall(\overline{Y :: K}).\overline{T \rightarrow} X\ \overline{Y} \\
\mathsf{matchTy}(d) & = \forall(\overline{Y :: K}).(X\ \overline{Y}) \rightarrow (\mathsf{dataTy}(d)\ \overline{Y})
\end{array}
$$

BINDER FUNCTIONS

$$
\begin{array}{ll}
\mathsf{dataBind}(d) & = X :: \mathsf{dataKind}(d) \\
\mathsf{constrBind}(d, c) & = c : \mathsf{constrTy}(d, c) \\
\mathsf{constrBinds}(d) & = \overline{\mathsf{constrBind}(\underline{d}, c)} \\
\mathsf{matchBind}(d) & = x : \mathsf{matchTy}(d) \\
\mathsf{binds}(x : T = t) & = x : T \\
\mathsf{binds}(X : K = T) & = X : K \\
\mathsf{binds}(d) & = \mathsf{dataBind}(d), \mathsf{constrBinds}(d), \mathsf{matchBind}(d), x : \mathsf{matchTy}(d)
\end{array}
$$

Fig. 3. Auxiliary definitions

$$
\text{W-Con}\ \frac{c = x(\overline{T}) \qquad \overline{\underline{\Gamma} \vdash T :: *}}{\Gamma \vdash_{\mathsf{ok}} c}
$$

$$
\text{W-Term}\ \frac{\Gamma \vdash T :: * \qquad \Gamma \vdash t : T}{\Gamma \vdash_{\mathsf{ok}} x : T = t} \qquad \text{W-Type}\ \frac{\Gamma \vdash T :: K}{\Gamma \vdash_{\mathsf{ok}} X : K = T}
$$

$$
\text{W-Data}\ \frac{\begin{array}{c} d = \mathtt{data}\ X\ (\overline{Y :: K}) = (\overline{c})\ \mathtt{with}\ x \\ \Gamma' = \Gamma, \overline{Y :: K} \qquad \overline{\underline{\Gamma'} \vdash_{\mathsf{ok}} c} \end{array}}{\Gamma \vdash_{\mathsf{ok}} d}
$$

Fig. 4. Well-formedness of constructors and bindings

$$
\text{Q-Refl}\ \frac{}{T \equiv T} \qquad \text{Q-Symm}\ \frac{T \equiv S}{S \equiv T}
$$

$$
\text{Q-Trans}\ \frac{S \equiv U \qquad U \equiv T}{S \equiv T} \qquad \text{Q-Arrow}\ \frac{S_1 \equiv S_2 \qquad T_1 \equiv T_2}{(S_1 \rightarrow T_1) \equiv (S_2 \rightarrow T_2)}
$$

$$
\text{Q-All}\ \frac{S \equiv T}{(\forall X :: K.S) \equiv (\forall X :: K.T)} \qquad \text{Q-Abs}\ \frac{S \equiv T}{(\lambda X :: K.S) \equiv (\lambda X :: K.T)}
$$

$$
\text{Q-App}\ \frac{S_1 \equiv S_2 \qquad T_1 \equiv T_2}{S_1 T_1 \equiv S_2 T_2} \qquad \text{Q-Beta}\ \frac{}{(\lambda X :: K.T_1)T_2 \equiv T_1[X := T_2]}
$$

Fig. 5. Type equivalence for FIR

$$\text{T-Var}\ \frac{x : T \in \Gamma}{\Gamma \vdash x : T} \qquad \text{T-Abs}\ \frac{\Gamma, x : T_1 \vdash t : T_2 \qquad \Gamma \vdash T_1 :: *}{\Gamma \vdash (\lambda x : T_1 . t) : T_1 \to T_2}$$

$$\text{T-App}\ \frac{\Gamma \vdash t_1 : T_1 \to T_2 \qquad \Gamma \vdash t_2 : T_1}{\Gamma \vdash (t_1\ t_2) : T_2} \qquad \text{T-TAbs}\ \frac{\Gamma, X :: K \vdash t : T}{\Gamma \vdash (\Lambda X :: K . t) : (\forall X :: K . T)}$$

$$\text{T-TApp}\ \frac{\Gamma \vdash t_1 : \forall X :: K_2 . T_1 \qquad \Gamma \vdash T_2 :: K_2}{\Gamma \vdash (t_1\ \{T_2\}) : T_1[X := T_2]} \qquad \text{T-Eq}\ \frac{\Gamma \vdash t : S \qquad S \equiv T}{\Gamma \vdash t : T}$$

$$\text{T-Wrap}\ \frac{\Gamma \vdash M : (F\ (\lambda(X :: K).\, \mathtt{ifix}\, F\ X))\ T \qquad \Gamma \vdash T :: K \qquad \Gamma \vdash F :: (K \Rightarrow *) \Rightarrow (K \Rightarrow *)}{\Gamma \vdash \mathtt{wrap}\ F\ T\ M : \mathtt{ifix}\, F\ T}$$

$$\text{T-Unwrap}\ \frac{\Gamma \vdash M : \mathtt{ifix}\, F\ T \qquad \Gamma \vdash T :: K}{\Gamma \vdash \mathtt{unwrap}\, M : (F\ (\lambda(X :: K).\, \mathtt{ifix}\, F\ X))\ T}$$

$$\text{T-Let}\ \frac{\Gamma \vdash T :: * \qquad \overline{\Gamma \vdash_{\mathsf{ok}} b} \qquad \Gamma, \overline{\mathsf{binds}(b)} \vdash t : T}{\Gamma \vdash (\mathtt{let}\, \overline{b}\, \mathtt{in}\, t) : T}$$

$$\text{T-LetRec}\ \frac{\Gamma \vdash T :: * \qquad \Gamma' = \Gamma, \overline{\mathsf{binds}(b)} \qquad \overline{\Gamma' \vdash_{\mathsf{ok}} b} \qquad \Gamma' \vdash t : T}{\Gamma \vdash (\mathtt{let\,rec}\, \overline{b}\, \mathtt{in}\, t) : T}$$

Fig. 6. Type synthesis for FIR

in our language, with corresponding `wrap` and `unwrap` terms. We now want to write typing rules for `wrap`. However, `fix` allows us to take fixpoints at *arbitrary* kinds, whereas `wrap` and `unwrap` are terms, which always have types of kind $*$. Thus, the best we can hope for is to use `wrap` and `unwrap` with *fully applied* fixed points, i.e.:

$$\begin{array}{lll}
\mathtt{wrap}_0\ f_0 & t : \mathtt{fix}\, f_0 & \text{where } t : f_0\ (\mathtt{fix}\, f_0) \\
\mathtt{wrap}_1\ f_1\ a1 & t : \mathtt{fix}\, f_1\ a1 & \text{where } t : f_1\ (\mathtt{fix}\, f_1)\ a1 \\
\mathtt{wrap}_2\ f_2\ a1\ a2 & t : \mathtt{fix}\, f_2\ a1\ a2 & \text{where } t : f_2\ (\mathtt{fix}\, f_2)\ a1\ a2 \\
\ldots & &
\end{array}$$

This must be accounted for in our typing rules for fixed points.

It is possible to give typing rules for `wrap` that will do the right thing regardless of how the fixpoint type is applied. One approach is to use *elimination contexts*, which represent the context in which a type will be eliminated (i.e. applied). This is the approach taken in [10]. However, this incurs a cost, since we cannot *guess* the elimination context (since type synthesis is bottom-up), so we must attach elimination contexts to our terms somehow.

An alternative approach is to pick a more restricted fixpoint operator. Using `ifix` avoids the problems of `fix`: it always produces fixpoints at kind $K \Rightarrow *$, which must be applied to precisely one argument of kind K before producing a type of kind $*$. This means we can give relatively straightforward typing rules as shown in Fig. 6.

Adequacy of `ifix`. Perhaps surprisingly, `ifix` is powerful enough to give us fixpoints at any kind K. We give a semantic argument here, but the idea is simply stated: we can "CPS-transform" a kind K into $(K \Rightarrow *) \Rightarrow *$, which then has the correct shape for `ifix`.

Definition 1. *Let J and K be kinds. Then J is a* retract *of K if there exist functions $\phi : J \Rightarrow K$ and $\psi : K \Rightarrow J$ such that $\psi \circ \phi = id$.*

Proposition 1. *Suppose J is a retract of K and there is a fixpoint operator $\mathtt{fix}_K$ on K. Then there is fixpoint operator $\mathtt{fix}_J$ on J.*

Proof. Take $\mathtt{fix}_J(f) = \psi(\mathtt{fix}_K(\phi \circ f \circ \psi))$.

Proposition 2. *Let K be a kind in System F^{μ}_{ω}. Then there is a unique (possibly empty) sequence of kinds $(K_0, \ldots, K_n)$ such that $K = \overline{K \Rightarrow}\ *$.*

Proof. Simple structural induction.

Proposition 3. *For any kind K in System F^{μ}_{ω}, K is a retract of $(K \Rightarrow *) \Rightarrow *$.*

Proof. Let $K = \overline{K \Rightarrow}\ *$ (by Proposition 2), and take

$$\begin{aligned}
&\phi : K \Rightarrow (K \Rightarrow *) \Rightarrow * \\
&\phi = \lambda(x :: K).\lambda(f :: K \Rightarrow *).f\ x \\
&\psi : ((K \Rightarrow *) \Rightarrow *) \Rightarrow K \\
&\psi = \lambda(w :: (K \Rightarrow *) \Rightarrow *).\lambda(\overline{a :: K}).w(\lambda(o :: K).o\ \overline{a})
\end{aligned}$$

Corollary 1. *If there is a fixpoint operator at kind $(K \Rightarrow *) \Rightarrow *$ then there is a fixpoint operator at any kind K.*

We can instantiate `ifix` with $K \Rightarrow *$ to get fixpoints at $(K \Rightarrow *) \Rightarrow *$, so `ifix` is sufficient to get fixpoints at any kind.

Note that since our proof relies on Proposition 2, it will not go through for arbitrary kinds when there are additional kind forms beyond $*$ and $\Rightarrow$. However, it will still be true for all kinds of the structure shown in Proposition 2.

The fact that retractions preserve the fixed point property is well-known in the context of algebraic topology: see [12, Exercise 4.7] or [5, Proposition 23.9] for example. While retractions between datatypes are a common tool in theoretical computer science (see e.g. [30]), we have been unable to find a version of Proposition 1 in the computer science literature. Nonetheless, we suspect this to be widely known.

3.2 Datatypes

FIR includes *datatypes*. A FIR datatype defines a type with a kind, parameterized by several type variables. The right-hand side declares of a list of constructors with type arguments, and the name of a matching function.[1] They thus are similar to the familiar style of defining datatypes in languages such as Haskell.

For example,

$$\texttt{data}\ \mathsf{Maybe}\ (A :: *) = (\mathsf{Nothing}(), \mathsf{Just}(A))\ \texttt{with}\ \mathsf{matchMaybe}$$

defines the familiar Maybe datatype, with constructors Nothing and Just, and matching function `matchMaybe`.

The type of `matchMaybe` is $\mathsf{Maybe}\, A \to \forall R.R \to (A \to R) \to R$. This acts as a pattern-matching function on Maybe—exactly as we saw the Scott encoding behave in Sect. 2. The matcher converts the abstract datatype into the raw, Scott-encoded type which can be used as a pattern matcher. We will see the full details in Sect. 4.3, and the type is given by $\mathsf{matchTy}(\mathsf{Maybe})$ as defined in Fig. 10.

Since FIR includes recursive datatypes, we could have removed `ifix`, `wrap` and `unwrap` from FIR. However, in practice it is useful for the target language (System F_ω^μ) to be a true subset of the source language (FIR), as this allows us to implement compilation as a series of FIR-to-FIR passes.

3.3 Let

FIR also features let terms. These have a number of bindings in them which bind additional names which are in scope inside the body of the let, and inside the right-hand-sides of the bindings in the case of a recursive let.

FIR supports let-binding terms, (opaque) types, and datatypes.

The typing rules for let are somewhat complex, but are crucially responsible for managing the scopes of the bindings defined in the let. In particular:

- The bindings defined in the let are *not* in scope when checking the right-hand sides of the bindings if the let is non-recursive, but *are* in scope if it is recursive.
- The bindings defined in the let are *not* in scope when checking the type of the entire binding.[2]

The behaviour of type-let is also worth explaining. Type-lets are more like *polymorphism* than type aliases in a language like Haskell. That is, they are opaque inside the body of the let, whereas a type alias would be transparent.

[1] Why declare a matching function explicitly, rather than using case expressions? The answer is that we want to be *local*: matching functions can be defined and put into scope when processing the datatype binding, whereas case expressions require additional program analysis to mach up the expression with the corresponding datatype.

[2] This is the same device usually employed when giving typing rules for existential types to ensure that the inner type does not escape.

This may make them seem like a useless feature, but this is not so. Term-lets are useful for binding sub-expressions of term-level computations to reusable names; type-lets are similarly useful for binding sub-expressions of *type-level* computations.

4 Compilation

We will show how to compile FIR by defining a compilation scheme for each feature in FIR:

- Non-recursive bindings of terms ($\mathbb{C}_{\text{term}}$, Sect. 4.1) and types ($\mathbb{C}_{\text{type}}$, Sect. 4.1)
- Recursive bindings of terms ($\mathbb{C}_{\text{termrec}}$, Sect. 4.2)
- Non-recursive bindings of datatypes ($\mathbb{C}_{\text{data}}$, Sect. 4.3)
- Recursive bindings of datatypes ($\mathbb{C}_{\text{datarec}}$, Sect. 4.4)

We do not consider recursive bindings of types, since the case of recursive datatypes is much more interesting and subsumes it.

Although our goal is to compile to System F_ω^μ, since it is a subset of FIR we can treat each pass as targeting FIR, by eliminating one feature from the language until we are left with precisely the subset that corresponds to System F_ω^μ. This has the advantage that we can continue to features of FIR until the point in the pipeline in which they are eliminated.[3]

In particular, we will use non-recursive let-bindings in $\mathbb{C}_{\text{termrec}}$ and $\mathbb{C}_{\text{datarec}}$, which imposes some ordering constraints on our passes.

Homogeneous Let-Bindings. We have said that we are going to compile e.g. term and type bindings separately, but our syntax (and typing rules) allows for let terms with many bindings of both sorts. While this is technically true, it is an easy problem to avoid.

Non-recursive bindings do not interfere with each other, since the newly-defined variables cannot occur in the right-hand sides of other bindings. That means that we can always decompose a single term with n bindings into n separate terms, one for each binding. Hence we can consider each sort of binding (and indeed, each individual binding) in isolation.

The same is not true for recursive bindings. To simplify the presentation we add a restriction to the programs that we compile: we require recursive lets to be *homogeneous*, in that they must only contain one sort of binding (term, type, or datatype). This means that we can similarly consider each sort of binding in isolation, although we will of course need to consider *multiple* bindings of the same sort.

This restriction is not too serious in practice. Given a recursive let term with arbitrary bindings:

[3] An elegant extension of this approach would be to define an indexed *family* of languages with gradually fewer features. However, this would be a distraction from the main point of this paper, so we have not adopted it.

- Types cannot depend on terms, so there are no dependencies from types or datatypes to terms.
- We do not support recursive type bindings, so there are no dependencies from types or datatypes to types.

So we can always pull out the term and type bindings into separate (recursive) let terms. The situation would be more complicated if we wanted to support recursive types or dependent types.

$$\begin{aligned}\mathbb{C}_{\mathsf{term}}(\texttt{let}\ x : t = b\ \texttt{in}\ v) &= (\lambda(x : t).v)\ b\\ \mathbb{C}_{\mathsf{type}}(\texttt{let}\ t :: k = b\ \texttt{in}\ v) &= (\Lambda(t :: k).v)\ \{b\}\end{aligned}$$

Fig. 7. Compilation of non-recursive term and type bindings

4.1 Non-recursive Term and Type Bindings

Non-recursive term and type bindings are easy to compile. They are encoded as immediately-applied lambda- and type-abstractions, respectively. We define the compilation scheme in Fig. 7.

4.2 Recursive Term Bindings

Self-reference and Standard Combinators. It is well-known that we cannot encode the Y combinator in the polymorphic lambda calculus, but that we *can* encode it if we have recursive types [14, Section 20.3].[4] We need the following types:

```
Fix0 : (∗ → ∗) → ∗
Fix0 = IFix (λ Rec F → F (Rec F))

- Type for values which can be applied to themselves
Self : ∗ → ∗
Self a = Fix0 (λ Rec → Rec → a)

self : ∀ {A} → (Self A → A) → Self A
self f = wrap f

unself : ∀ {A} → Self A → Self A → A
unself s = unwrap s

selfApply : ∀ {A} → Self A → A
selfApply s = unself s s
```

[4] We here mean *arbitrary* recursive types, not merely strictly positive types. We cannot encode the Y combinator in Agda, for example, without disabling the positivity check.

The first thing we defined was $\mathsf{Fix}_0 : (* \Rightarrow *) \Rightarrow *$, which is a fixpoint operator that only works at kind $*$. We won't need the full power of `ifix` for this section, so the techniques here should be applicable for other recursive variants of System F_ω, provided they are able to define Fix_0.

Now we can define the Y combinator and its η-expanded version, the Z combinator.

$$\begin{aligned}
&\mathsf{y} : \forall \{A\} \to (A \to A) \to A\\
&\mathsf{y}\, f = (\lambda\, z \to f\,(\mathsf{selfApply}\; z))\\
&\qquad\;(\mathsf{self}\;(\lambda\, z \to f\,(\mathsf{selfApply}\; z)))
\end{aligned}$$

$$\begin{aligned}
&\mathsf{z} : \forall \{A\; B : *\} \to ((A \to B) \to (A \to B)) \to (A \to B)\\
&\mathsf{z}\, f = (\lambda\, z \to f\,(\lambda\, a \to (\mathsf{selfApply}\; z)\; a))\\
&\qquad\;(\mathsf{self}\;(\lambda\, z \to f\,(\lambda\, a \to (\mathsf{selfApply}\; z)\; a)))
\end{aligned}$$

In strict lambda calculi the Y combinator does not terminate, and we need to use the Z combinator, which has a more restricted type (it only allows us to take the fixpoint of things of type $A \to B$).

Mutual Recursion. The Y and Z combinators allow us to define singly recursive functions, but we also want to define *mutually* recursive functions.

This is easy in a non-strict lambda calculus: we have the Y combinator, and we know how to encode tuples, so we can simply define a recursive *tuple* of functions. However, this is still easy to get wrong, as we must be careful not to force the recursive tuple too soon.

Moreover, this approach does not work with the Z combinator, since a tuple is not a function (the Scott-encoded version is a function, but a *polymorphic* function).

We can instead construct a more generic fixpoint combinator which will be usable in both a non-strict and strict setting. We will present the steps using recursive definitions for clarity, but all of these can be implemented with the Z combinator.

Let us start with the function fix_2 which takes the fixpoint of a function of 2-tuples.

$$\begin{aligned}
&\mathsf{fix}_2 : \forall \{A\; B\} \to (A \times B \to A \times B) \to A \times B\\
&\mathsf{fix}_2\; f = f\,(\mathsf{fix}_2\; f)
\end{aligned}$$

We can transform this as follows: first we curry f.

$$\begin{aligned}
&\mathsf{fix}_2\text{-}\mathsf{uncurry} : \forall \{A\; B\} \to (A \to B \to A \times B) \to A \times B\\
&\mathsf{fix}_2\text{-}\mathsf{uncurry}\; f = (\mathsf{uncurry}\; f)\;(\mathsf{fix}_2\text{-}\mathsf{uncurry}\; f)
\end{aligned}$$

Now, we replace both the remaining tuple types with Scott-encoded versions, using the corresponding version of uncurry for Scott-encoded 2-tuples.

$$
\begin{array}{l}
\mathsf{uncurry_2\text{-}scott} \\
\quad : \forall \{A\ B\ R : *\} \\
\quad \to (A \to B \to R) \\
\quad \to ((\forall \{Q\} \to (A \to B \to Q) \to Q) \to R) \\
\mathsf{uncurry_2\text{-}scott}\ f\ g = g\ f
\end{array}
$$

$$
\begin{array}{l}
\mathsf{fix_2\text{-}scott} \\
\quad : \forall \{A\ B\} \\
\quad \to (A \to B \to \forall \{Q\} \to (A \to B \to Q) \to Q) \\
\quad \to \forall \{Q\} \to (A \to B \to Q) \to Q \\
\mathsf{fix_2\text{-}scott}\ f = (\mathsf{uncurry_2\text{-}scott}\ f)\ (\mathsf{fix_2\text{-}scott}\ f)
\end{array}
$$

Finally, we reorder the arguments to f to make it look as regular as possible.

$$
\begin{array}{l}
\mathsf{fix_2\text{-}rearrange} \\
\quad : \forall \{A\ B\} \\
\quad \to (\forall \{Q\} \to (A \to B \to Q) \to A \to B \to Q) \\
\quad \to \forall \{Q\} \to (A \to B \to Q) \to Q \\
\mathsf{fix_2\text{-}rearrange}\ f\ k = (\mathsf{uncurry_2\text{-}scott}\ (f\ k))\ (\mathsf{fix_2\text{-}rearrange}\ f)
\end{array}
$$

This gives us a fixpoint function pairs of mutually recursive values, but we want to handle *arbitrary* sets of recursive values. At this point, however, we notice that all we need to do to handle, say, triples, is to replace $A \to B$ with $A \to B \to C$ and the binary uncurry with the ternary uncurry. And we can abstract over this pattern.

$$
\begin{array}{l}
\mathsf{fixBy} \\
\quad : \forall \{F : * \to *\} \\
\quad \to ((\forall \{Q\} \to F\ Q \to Q) \to \forall \{Q\} \to F\ Q \to Q) \\
\quad \to (\forall \{Q\} \to F\ Q \to F\ Q) \to \forall \{Q\} \to F\ Q \to Q \\
\mathsf{fixBy}\ by\ f = by\ (\mathsf{fixBy}\ by\ f) \circ f
\end{array}
$$

To get the behaviour we had before, we instantiate by appropriately:

$$
\begin{array}{l}
\mathsf{by_2} \\
\quad : \forall \{A\ B\} \\
\quad \to (\forall \{Q\} \to (A \to B \to Q) \to Q) \\
\quad \to \forall \{Q\} \to (A \to B \to Q) \to Q \\
\mathsf{by_2}\ r\ k = (\mathsf{uncurry_2\text{-}scott}\ k)\ r \\
\mathsf{fixBy_2} \\
\quad : \forall \{A\ B\} \\
\quad \to (\forall \{Q\} \to (A \to B \to Q) \to A \to B \to Q) \\
\quad \to \forall \{Q\} \to (A \to B \to Q) \to Q \\
\mathsf{fixBy_2} = \mathsf{fixBy}\ \mathsf{by_2}
\end{array}
$$

How do we interpret by? Inlining uncurry into our definition of $\mathsf{by_2}$ we find that it is in fact the identity function! However, by choosing the exact definition we can tweak the termination properties of our fixpoint combinator. Indeed, our current definition does not terminate even in a non-strict language like Agda, since it evaluates the components of the recursive tuple before feeding them into

f. However, we can avoid this by "repacking" the tuple so that accessing one of its components will no longer force the other.[5]

$$
\begin{array}{l}
\texttt{- Repacking tuples.} \\
\mathsf{repack_2} : \forall \{A\ B\} \rightarrow A \times B \rightarrow A \times B \\
\mathsf{repack_2}\ tup = (\mathsf{proj_1}\ tup\ ,\ \mathsf{proj_2}\ tup)
\end{array}
$$

$$
\begin{array}{l}
\texttt{- Repacking Scott-encoded tuples.} \\
\mathsf{by_2\text{-}repack} \\
\quad : \forall \{A\ B : *\} \\
\quad \rightarrow (\forall \{Q : *\} \rightarrow (A \rightarrow B \rightarrow Q) \rightarrow Q) \\
\quad \rightarrow \forall \{Q : *\} \rightarrow (A \rightarrow B \rightarrow Q) \rightarrow Q \\
\mathsf{by_2\text{-}repack}\ r\ k = k\ (r\ (\lambda\ x\ y \rightarrow x))\ (r\ (\lambda\ x\ y \rightarrow y))
\end{array}
$$

Passing `by₂-repack` to `fixBy` gives us a fixpoint combinator that terminates in a non-strict language like Agda or Haskell.

Can we write one that terminates in a strict language? We can, but we incur the same restriction that we have when using the Z combinator: the recursive values must all be functions. This is because we use exactly the same trick, namely η-expanding the values.

$$
\begin{array}{l}
\texttt{- with tuples} \\
\mathsf{repack_2\text{-}strict} \\
\quad : \forall \{A_1\ B_1\ A_2\ B_2 : *\} \\
\quad \rightarrow (A_1 \rightarrow B_1) \times (A_2 \rightarrow B_2) \\
\quad \rightarrow (A_1 \rightarrow B_1) \times (A_2 \rightarrow B_2) \\
\mathsf{repack_2\text{-}strict}\ tup = ((\lambda\ x \rightarrow \mathsf{proj_1}\ tup\ x)\ ,\ (\lambda\ x \rightarrow \mathsf{proj_2}\ tup\ x))
\end{array}
$$

$$
\begin{array}{l}
\texttt{- with Scott-encoded tuples} \\
\mathsf{by_2\text{-}strict} \\
\quad : \forall \{A_1\ B_1\ A_2\ B_2 : *\} \\
\quad \rightarrow (\forall \{Q : *\} \rightarrow ((A_1 \rightarrow B_1) \rightarrow (A_2 \rightarrow B_2) \rightarrow Q) \rightarrow Q) \\
\quad \rightarrow \forall \{Q : *\} \rightarrow ((A_1 \rightarrow B_1) \rightarrow (A_2 \rightarrow B_2) \rightarrow Q) \rightarrow Q \\
\mathsf{by_2\text{-}strict}\ r\ k = k\ (\lambda\ x \rightarrow r\ (\lambda\ f_1\ f_2 \rightarrow f_1\ x))\ (\lambda\ x \rightarrow r\ (\lambda\ f_1\ f_2 \rightarrow f_2\ x))
\end{array}
$$

This gives us general, n-ary fixpoint combinators in System F_ω^μ.

Formal Encoding of Recursive Let-Bindings. We define the compilation scheme for recursive term bindings in Fig. 8, along with a number of auxiliary functions.

The definitions of fixBy, by, and fix are as in our Agda presentation. The function sel_k is what we pass to a Scott-encoded tuple to select the kth element. The Z combinator is defined as in the previous section (we do not repeat the definition here). We have given the lazy version of by, but it is straightforward

[5] We have defined × as a simple datatype, rather than using the more sophisticated version in the Agda standard library. The standard library version has different strictness properties—indeed, for that version `repack₂` is precisely the identity.

AUXILIARY FUNCTIONS

$$
\begin{aligned}
\mathsf{Id} &= \lambda(X :: *).X \\
F \rightsquigarrow G &= \forall Q.F\ Q \to G\ Q \\
\mathsf{fixBy} &= \Lambda(F :: * \Rightarrow *).\lambda(by : (F \rightsquigarrow \mathsf{Id}) \to (F \rightsquigarrow \mathsf{Id})). \\
&\qquad z\ (\lambda(r : (F \rightsquigarrow F) \to (F \rightsquigarrow \mathsf{Id})).\lambda(f : F \rightsquigarrow F).by(\Lambda Q.\lambda(k : F\ Q).r\ f\ \{Q\}\ (f\ \{Q\}\ k))) \\
\mathsf{sel}_k(\overline{T}) &= \lambda(\overline{x : T}).x_k \\
\mathsf{by}(\overline{T}) &= \lambda(r : \forall Q.(\overline{T \to Q}) \to Q).\Lambda Q.\lambda(k : \overline{T \to Q}).k\ \overline{\underline{r}\ \{\underline{Q}\}\ (\mathsf{sel}_j\ \underline{(\overline{T})})}^{j} \\
\mathsf{fix}(\overline{T}) &= \mathsf{fixBy}\ \{\lambda Q.\overline{T \to Q}\}\ \mathsf{by}(\overline{T})
\end{aligned}
$$

COMPILATION FUNCTION

$$
\begin{aligned}
&\mathbb{C}_{\mathrm{termrec}}(\mathtt{let\ rec}\ \overline{x : T = t}\ \mathtt{in}\ u) \\
&\qquad = \mathtt{let}\ r = \mathsf{fix}(\overline{T})\ \Lambda Q.\lambda(k : \overline{T \to Q})\ (\overline{x : T}).k\ \overline{t} \\
&\qquad\quad \mathtt{in\ let}\ \overline{x = \underline{r}\ \{T\}\ (\mathsf{sel}_j\ \underline{(\overline{T})})}^{j}\ \mathtt{in}\ u
\end{aligned}
$$

Fig. 8. Compilation of recursive let-bindings

to define the strict version, in exchange for the corresponding restriction on the types of the recursive bindings.

The compilation function is a little indirect: we create a recursive tuple of values, then we let-bind each component of the tuple again! Why not just pass a single function to the tuple that consumes all the components and produces t? The answer is that in order to use the Scott-encoded tuple we need to give it the *type* of the value that we are producing, which in this case would be the type of t. But we do not know this type without doing type inference on FIR. This way we instead extract each of the components, whose types we *do* know, since they are in the original let-binding.

Polymorphic Recursion with the Z Combinator. Neither the simple Z combinator nor our strict fixBy allow us to define recursive values which are not of function type. This might not seem too onerous, but this also forbids defining *polymorphic* values, such as polymorphic functions. For example, we cannot define a polymorphic map function this way.

Sometimes we can get around this problem by floating the type abstraction out of the recursion. This will work in many cases, but fails in any instance of polymorphic recursion, which includes most recursive functions over irregular datatypes.

However, we can work around this restriction if we are willing to transform our program. The *thunking* transformation is a variant of the well-known transformation for simulating call-by-name evaluation in a call-by-value language [9,28]. Conveniently, this also has the property that it transforms the "thunked" parameters into values of *function* type, thus making them computable with the Z combinator.

The thunking transformation takes a set of recursive definitions $f_i : T_i = b_i$ and transforms it by:

- Defining the Unit datatype with a single, no-argument constructor ().

- Creating new (recursive) definitions $f'_i : \mathsf{Unit} \to T_i = \lambda(u : \mathsf{Unit}).b_i$.
- Replacing all uses of f_i in the b_is with f'_i (),
- Creating new (non-recursive) definitions $f_i : T_i = f'_i$ () to replace the originals.

Now our recursive value is truly of function type, rather than universal type, so we can compile it as normal.

An example is given in Fig. 9 of transforming a polymorphic map function.

$$
\begin{array}{l}
\texttt{let rec}\ \mathit{map} : \forall A\ B.(A \to B) \to (\mathsf{List}\,A \to \mathsf{List}\,B) = \\
\qquad \Lambda A\ B.\ \lambda(f : A \to B)\ (l : \mathsf{List}\,A) \\
\qquad \mathsf{matchList}\,\{A\}\ l\ \{\mathsf{List}\,B\} \\
\qquad (\mathsf{Nil}\ \{B\}) \\
\qquad (\lambda(h : A)(t : \mathsf{List}\,A).\,\mathsf{Cons}\ \{B\}\ (f\ h)\ (\mathit{map}\ \{A\}\ \{B\}\ f\ t)) \\
\texttt{in}\ t \\
\\
\Rightarrow \\
\\
\texttt{let rec}\ \mathit{map'} : \mathsf{Unit} \to \forall A\ B.(A \to B) \to (\mathsf{List}\,A \to \mathsf{List}\,B) = \\
\qquad \lambda(u : \mathsf{Unit}).\ \Lambda A\ B.\ \lambda(f : A \to B)\ (l : \mathsf{List}\,A) \\
\qquad \mathsf{matchList}\ \{A\}\ l\ \{\mathsf{List}\,B\} \\
\qquad (\mathsf{Nil}\ \{B\}) \\
\qquad (\lambda(h : A)(t : \mathsf{List}\,A).\,\mathsf{Cons}\ \{B\}\ (f\ h)\ (\mathit{map'}\ ()\ \{A\}\ \{B\}\ f\ t)) \\
\texttt{in let}\ \ \mathit{map} : \forall A\ B.(A \to B) \to (\mathsf{List}\,A \to \mathsf{List}\,B) = \mathit{map'}\ () \\
\texttt{in}\ t
\end{array}
$$

Fig. 9. Example of transforming polymorphic recursion

4.3 Non-recursive Datatype Bindings

Non-recursive datatypes are fairly easy to compile. We will generalize the Scott-encoding approach described in Sect. 2.

We define the compilation scheme for non-recursive datatype bindings in Fig. 10, along with a number of auxiliary functions in addition to those in Fig. 3.

Let's go through the auxiliary functions in turn (both those in Figs. 3 and 10), using the Maybe datatype as an example.

$$d := \texttt{data}\ \ \mathsf{Maybe}\ \ A = (\mathsf{Nothing}(), \mathsf{Just}(A))\ \texttt{with}\ \mathsf{match}$$

- $\mathsf{branchTy}(c, R)$ computes the type of a function which consumes all the arguments of the given constructor, producing a value of type R.

$$
\begin{aligned}
\mathsf{branchTy}(\mathsf{Nothing}(), R) &= R \\
\mathsf{branchTy}(\mathsf{Just}\,A, R) &= A \to R
\end{aligned}
$$

- $\mathsf{dataKind}(d)$ computes the kind of the datatype type. This is a kind arrow from the kinds of all the type arguments to $*$.

$$\mathsf{dataKind}(\mathsf{Maybe}) = * \Rightarrow *$$

Throughout this figure when d or c is an argument
$d = \texttt{data}\ X\ (\overline{Y :: K}) = (\overline{c})\ \texttt{with}\ x$
$c = x(\overline{T})$

AUXILIARY FUNCTIONS
$\mathsf{unveil}(d, t) \quad = t[X := \mathsf{dataTy}(d)]$
$\mathsf{constr}_k(d, c) = \mathsf{unveil}(d, \Lambda(\overline{Y :: K}).\lambda(\overline{a : T}).\Lambda R.\lambda(\overline{b : \mathsf{branchTy}(c, \underline{R})})\ b_k\ \overline{a})$
$\mathsf{constrs}(d) \quad = \overline{\mathsf{constr}_j(\underline{d}, c_j)}^j$
$\mathsf{match}(d) \quad = \Lambda(\overline{Y :: K}).\lambda(x : (\mathsf{dataTy}(d)\ \overline{Y}).x$

COMPILATION FUNCTION
$\mathbb{C}_{\mathtt{data}}(\texttt{let}\ d\ \texttt{in}\ t)$
$= (\Lambda(\mathsf{dataBind}(d)).\lambda(\mathsf{constrBinds}(d)).\lambda(\mathsf{matchBind}(d)).t)$
$\{\mathsf{dataTy}(d)\}$
$\mathsf{constrs}(d)$
$\mathsf{match}(d)$

Fig. 10. Compilation of non-recursive datatype bindings

- $\mathsf{dataTy}(d)$ computes the Scott-encoded datatype. This binds the type variables with a lambda and then constructs the pattern matching function type using the branch types.

$$\mathsf{dataTy}(d) = \lambda A.\forall R.R \to (A \to R) \to R$$

- $\mathsf{constrTy}(c, T)$ computes the type of a constructor of the datatype d.

$$\mathsf{constrTy}(\mathsf{Nothing}(), \mathsf{Maybe}) = \forall A.\ \mathsf{Maybe}\ A$$
$$\mathsf{constrTy}(\mathsf{Just}\ A, \mathsf{Maybe}) = \forall A.A \to \mathsf{Maybe}\ A$$

- $\mathsf{unveil}(d, t)$ "unveils" the datatype inside a type or term, replacing the abstract definition with the concrete, Scott-encoded one. We apply this to the *definition* of the constructors, for a reason we will see shortly. This makes no difference for non-recursive datatypes, but will matter for recursive ones.

$$\mathsf{unveil}(d, t) = t[\mathsf{Maybe} := \lambda A.\forall R.R \to (A \to R) \to R]$$

- $\mathsf{constr}_k(d, c)$ computes the definition of the kth constructor of a datatype. To match the signature of the constructor, this is type abstracted over the type variables and takes arguments corresponding to each of the constructor arguments. Then it constructs a pattern matching function which takes branches for each alternative and uses the kth branch on the constructor arguments.

$$\mathsf{constr}_1(d, \mathsf{Nothing}()) = \Lambda A.\Lambda R.\lambda(b_1 : R)(b_2 : A \to R).b_1$$
$$\mathsf{constr}_2(d, \mathsf{Just}(A)) = \Lambda A.\lambda(v : A).\Lambda R.\lambda(b_1 : R)(b_2 : A \to R).b_2\ v$$

- $\mathsf{matchTy}(d)$ computes the type of the datatype matcher, which converts from the abstract datatype to a pattern-matching function—that is, precisely the Scott-encoded type.

$$\mathsf{matchTy}(d) = \forall A.\ \mathsf{Maybe}\ A \to (\forall R.R \to (A \to R) \to R)$$

- $\mathsf{match}(d)$ computes the definition of the matcher of the datatype, which is the identity.

$$\mathsf{match}(d) = \Lambda A.\lambda(v : \mathsf{Maybe}\, A).v$$

The basic idea of the compilation scheme itself is straightforward: use type abstraction and lambda abstraction to bind names for the type itself, its constructors, and its match function.

There is one quirk: usually when encoding let-bindings we create an *immediately* applied type- or lambda-abstraction, but here they are *interleaved*. The reason for this is that the datatype must be abstract inside the *signature* of the constructors and the match function, since otherwise any uses of those functions inside the body will not typecheck. But inside the *definitions* the datatype must be concrete, since the definitions make use of the concrete structure of the type. This explains why we needed to use $\mathsf{unveil}(d, t)$ on the definitions of the constructors, since they appear outside the scope in which we define the abstract type. Note that this means we really must perform a substitution rather than creating a let-binding, since that would simply create another abstract type.[6]

Consider the following example:

$$
\begin{array}{ll}
\mathbb{C}_{\text{data}}(\texttt{let data}\ \mathsf{Maybe}\ A = (\mathsf{Nothing}(), \mathsf{Just}(A))\ \texttt{with}\ \mathsf{match} & \\
\quad\quad \texttt{in}\ \mathsf{match}\ \{\mathsf{Int}\}\ (\mathsf{Just}\{\mathsf{Int}\}1)\ 0\ (\lambda x : \mathsf{Int}.x + 1)) & \\
= (\Lambda(\mathsf{Maybe} :: * \Rightarrow *). & \text{(signature of Maybe)} \\
\quad \lambda(\mathsf{Nothing} : \forall A.\, \mathsf{Maybe}\, A). & \text{(signature of Nothing)} \\
\quad \lambda(\mathsf{Just} : \forall A.A \to \mathsf{Maybe}\, A). & \text{(siganture of Just)} \\
\quad \lambda(\mathsf{match} : \forall A.\, \mathsf{Maybe}\, A \to \forall R.R \to (A \to R) \to R). & \text{(signature of match)} \\
\quad \mathsf{match}\ \{\mathsf{Int}\}\ (\mathsf{Just}\{\mathsf{Int}\}1)\ 0\ (\lambda x : \mathsf{Int}.x + 1)) & \text{(body of the let)} \\
\quad (\lambda A.\forall R.R \to (A \to R) \to R) & \text{(definition of Maybe)} \\
\quad (\Lambda A.\Lambda R.\lambda(b_1 : R)\ (b_2 : A \to R).b_1) & \text{(definition of Nothing)} \\
\quad (\Lambda A.\lambda(v_1 : A).\Lambda R.\lambda(b_1 : R)\ (b_2 : A \to R).b_2\ v_1) & \text{(definition of Just)} \\
\quad (\Lambda A.\lambda(v : \forall R.R \to (A \to R) \to R).v) & \text{(definition of match)}
\end{array}
$$

Here we can see that:

- Just needs to produce the abstract type inside the body of the let, otherwise the application of match will be ill-typed.
- The definition of Just produces the Scott-encoded type.
- match maps from the abstract type to the Scott-encoded type inside the body of the let.
- The definition of match is the identity on the Scott-encoded type.

[6] It is well-known that abstract datatypes can be encoded with existential types [22]. The presentation we give here is equivalent to using a value of existential type which is immediately unpacked, and where existential types are given the typical encoding using universal types.

4.4 Recursive Datatype Bindings

Adding singly recursive types is comparatively straightforward. We can write our datatype as a type-level function (often called a "pattern functor" [3]) with a parameter for the recursive use of the type, and then use our fixpoint operator to produce the final datatype.[7]

$$
\begin{array}{l}
\mathsf{ListF} : (* \to *) \to (* \to *) \\
\mathsf{ListF}\ \mathit{List}\ A = \\
\quad \texttt{- This is the normal Scott-encoding, using the} \\
\quad \texttt{- recursive `List' provided by the pattern functor.} \\
\quad \forall\ \{R : *\} \to R \to (A \to \mathit{List}\ A \to R) \to R \\
\\
\mathsf{List} : * \to * \\
\mathsf{List}\ A = \mathsf{IFix}\ \mathsf{ListF}\ A
\end{array}
$$

However, it is not immediately apparent how to use this to define *mutually* recursive datatypes. The type of `ifix` is quite restrictive: we can only produce something of kind $k \Rightarrow *$.

If we had kind-level products and an appropriate fixpoint operator, then we could do this relatively easily by defining a singly recursive product of our datatypes. However, we do not have products in our kind system.

But we can *encode* type-level products. In [32] the authors use the fact that an n-tuple can be encoded as a function from an index to a value, and thus type-level naturals can be used as the index of a type-level function to encode a tuple of types. We take a similar approach except that we will not use a natural to index our type, but rather a richer datatype. This will prove fruitful when encoding parameterized types.

Let's consider an example: the mutually recursive types of trees and forests.

$$
\begin{array}{l}
\mathsf{mutual} \\
\quad \mathsf{data}\ \mathsf{Tree}_0\ (A : *) : *\ \mathsf{where} \\
\quad\quad \mathsf{node}_0 : A \to \mathsf{Forest}_0\ A \to \mathsf{Tree}_0\ A \\
\\
\quad \mathsf{data}\ \mathsf{Forest}_0\ (A : *) : *\ \mathsf{where} \\
\quad\quad \mathsf{nil}_0 : \mathsf{Forest}_0\ A \\
\quad\quad \mathsf{cons}_0 : \mathsf{Tree}_0\ A \to \mathsf{Forest}_0\ A \to \mathsf{Forest}_0\ A
\end{array}
$$

First of all, we can rewrite this with a "tag" datatype indicating which of the two cases in our datatype we want to use. That allows us to use a single data

[7] This is where the Scott encoding really departs from the Church encoding: the recursive datatype itself appears in our encoding, since we are only doing a "one-level" fold whereas the Church encoding gives us a full recursive fold over the entire datastructure.

declaration to cover both of the types. Moreover, the tag can include the type parameters of the datatype, which is important in the case that they differ between the different datatypes.

```
data TreeForest^t : * where
  - 'Tree^t A' tags the type 'Tree A'
  Tree^t : * → TreeForest^t
  - 'Forest^t A' tags the type 'Forest A'
  Forest^t : * → TreeForest^t

module Single where
  - This mutual recursion is not strictly necessary,
  - and is only there so we can define the 'Tree'
  - and 'Forest' aliases for legibility.
  mutual
    - Type alias for the application of the main
    - datatype to the 'Tree' tag
    Tree : * → *
    Tree A = TreeForest (Tree^t A)

    - Type alias for the application of the main
    - datatype to the 'Forest' tag
    Forest : * → *
    Forest A = TreeForest (Forest^t A)

    data TreeForest : TreeForest^t → * where
  node : ∀ {A} → A → Tree A → Tree A
  nil : ∀ {A} → Forest A
  cons : ∀ {A} → Tree A → Forest A → Forest A
```

That has eliminated the mutual recursion, but we still have a number of problems:

- We are relying on Agda's data declarations to handle recursion, rather than our fixpoint combinator.
- We are using inductive families, which we don't have a way to encode.
- $\texttt{TreeForest}^t$ is being used at the kind level, but we don't have a way to encode datatypes at the kind level.

Fortunately, we can get past all of these problems. Firstly we need to make our handling of the different constructors more uniform by encoding them as sums of products.

```
module Constructors where
  mutual
    Tree : * → *
    Tree A = TreeForest (Tree^t A)

    Forest : * → *
    Forest A = TreeForest (Forest^t A)

    - This chooses the type of the constructor
    - given the tag
    TreeForestF : TreeForest^t → *
    - The 'Tree' constructor takes a pair of
    - an 'A' and a 'Forest A'
    TreeForestF (Tree^t A) = A × Forest A
    - The 'Forest' constructor takes either nothing,
    - or a pair of a 'Tree A' and a 'Forest A'
    TreeForestF (Forest^t A) = ⊤ ⊎ Tree A × Forest A

    {-# NO_POSITIVITY_CHECK #-}
    data TreeForest (tag : TreeForest^t) : * where
      treeForest : TreeForestF tag → TreeForest tag
```

If we now rewrite `TreeForestF` to take the recursive type as a parameter instead of using it directly, we can write this with `ifix`.

```
module IFixed where
  TreeForestF : (TreeForest^t → *) → (TreeForest^t → *)
  TreeForestF Rec (Tree^t A) = A × Rec (Forest^t A)
  TreeForestF Rec (Forest^t A) = ⊤ ⊎ Rec (Tree^t A) × Rec (Forest^t A)

  TreeForest : TreeForest^t → *
  TreeForest = IFix TreeForestF
```

Finally, we need to encode the remaining datatypes that we have used. The sums and products in the right-hand-side of `TreeForestF` should be Scott-encoded as usual, since they represent the constructors of the datatype.

The tag type is more problematic. The Scott encoding of the tag type we have been using would be:

$$\text{Scott-tag} = \forall\, \{R : *\} \to (* \to R) \to (* \to R) \to R$$

However, we do not have polymorphism at the kind level! But if we look at how we use the tag we see that we only ever match on it to produce something of kind $*$, and so we can get away with immediately instantiating this to $*$.

```
module Encoded where
  - Tag type instantiated to '*'
  TreeForest^e : *
  TreeForest^e = (* → *) → (* → *) → *

  - Encoded 'Tree^t' tag
  Tree^e : * → TreeForest^e
  Tree^e A = λ T F → T A

  - Encoded 'Forest^t' tag
  Forest^e : * → TreeForest^e
  Forest^e A = λ T F → F A

  TreeForestF : (TreeForest^e → *) → (TreeForest^e → *)
  TreeForestF Rec tag =
    - Pattern matching has been replaced by application
    tag
      - The encoded 'Tree' constructor
      (λ A → ∀ {R} → (A → Rec (Forest^e A) → R) → R)
      - The encoded 'Forest' constructor
      (λ A → ∀ {R} → R → (Rec (Tree^e A) → Rec (Forest^e A) → R) → R)

  TreeForest : TreeForest^e → *
  TreeForest = IFix TreeForestF
```

This, finally, gives us a completely System F^{μ}_{ω}-compatible encoding of our mutually recursive datatypes.

Formal Encoding of Recursive Datatypes. We define the compilation scheme for recursive datatype bindings in Fig. 11, along with a number of auxiliary functions. We will reuse some of the functions from Fig. 10, but many of them need variants for the recursive case, which are denoted with a `rec` superscript.

Let's go through the functions again, this time using Tree and Forest as examples:

$$d_1 := \texttt{data}\ \mathsf{Tree}\ A = (\mathsf{Node}(A, \mathsf{Forest}\, A))\ \texttt{with}\ \mathsf{matchTree}$$
$$d_2 := \texttt{data}\ \mathsf{Forest}\ A = (\mathsf{Nil}(), \mathsf{Cons}(\mathsf{Tree}\, A, \mathsf{Forest}\, A))\ \texttt{with}\ \mathsf{matchForest}$$

- $\mathsf{tagKind}(l)$ defines the kind of the type-level tags for our datatype family, which is a Scott-encoded tuple of types.

$$\mathsf{tagKind}(l) = (* \Rightarrow *) \Rightarrow (* \Rightarrow *) \Rightarrow *$$

- $\mathsf{tag}_k(l, d)$ defines the tag type for the datatype d in the family.

$$\mathsf{tag}_1(l, \mathsf{Tree}) = \lambda A.\lambda(v_1 :: * \Rightarrow *)(v_2 :: * \Rightarrow *).v_1\ A$$
$$\mathsf{tag}_2(l, \mathsf{Forest}) = \lambda A.\lambda(v_1 :: * \Rightarrow *)(v_2 :: * \Rightarrow *).v_2\ A$$

Throughout this figure when l, d, or c is an argument
$l = \texttt{let rec}\ \overline{d}\ \texttt{in}\ t$
$d = \texttt{data}\ X\ (\overline{Y :: K}) = (\overline{c})\ \texttt{with}\ x$
$c = x(\overline{T})$

AUXILIARY FUNCTIONS

$$
\begin{array}{ll}
\mathsf{tagKind}(l) & = \overline{\mathsf{dataKind}(d) \Rightarrow} * \\
\mathsf{tag}_k(l,d) & = \lambda(\overline{Y :: K}).\lambda(\overline{X :: \mathsf{dataKind}(d)}).X_k\ \overline{Y} \\
\mathsf{inst}_k(f,l,d) & = \lambda(\overline{Y :: K}).f\ (\mathsf{tag}_k(l,d)\ \overline{Y}) \\
\mathsf{family}(l) & = \lambda(r :: \overline{\mathsf{dataKind}(d) \Rightarrow} *)\ .\lambda(t :: \mathsf{tagKind}(l)).\,\texttt{let}\ \overline{X = \mathsf{inst}_j(\underline{r}, \underline{l}, d_j)}^j\ \texttt{in}\ t\ \overline{\mathsf{dataTy}(d)} \\
\mathsf{instFamily}_k(l,d) & = \lambda(\overline{Y :: K}).\,\texttt{ifix}\ \mathsf{family}(l)\ (\mathsf{tag}_k(l,d)\ \overline{Y}) \\
\mathsf{unveil}^{\texttt{rec}}(l,t) & = t\overline{[X := \mathsf{instFamily}_j(\underline{l}, d_j)]}^j \\
\mathsf{constr}^{\texttt{rec}}_{k,m}(l,d,c) & = \Lambda(\overline{Y :: K}).\lambda(\overline{a : T}).\,\texttt{wrap}\ \mathsf{family}(l)\ (\mathsf{tag}_k(l,d)\ \overline{Y})\ (\Lambda R.\lambda(\overline{b : \mathsf{branchTy}(c, \underline{R})}).\ b_m\ \overline{a}) \\
\mathsf{constrs}^{\texttt{rec}}_k(l,d) & = \overline{\mathsf{constr}^{\texttt{rec}}_{k,j}(\underline{l}, \underline{d}, c_j)}^j \\
\mathsf{match}^{\texttt{rec}}_k(l,d) & = \Lambda(\overline{Y :: K}).\lambda(x : \mathsf{instFamily}_k(l,d)\ \overline{Y}).\,\texttt{unwrap}\ x
\end{array}
$$

COMPILATION FUNCTION

$$
\begin{array}{ll}
\mathbb{C}_{\mathsf{datarec}}(l) & = (\Lambda(\overline{\mathsf{dataBind}(d)}).\lambda(\overline{\mathsf{constrBinds}(d)}).\lambda(\overline{\mathsf{matchBind}(d)}).t) \\
& \quad \{\overline{\mathsf{instFamily}_j(\underline{l}, d_j)}^j\} \\
& \quad \overline{\mathsf{constrs}^{\texttt{rec}}_j(\underline{l}, d_j)}^j \\
& \quad \overline{\mathsf{match}^{\texttt{rec}}_j(\underline{l}, d_j)}^j
\end{array}
$$

Fig. 11. Compilation of recursive datatype bindings

- $\mathsf{inst}_k(f,l,d)$ instantiates the family type f for the datatype d in the family by applying it to the datatype tag.

$$
\begin{aligned}
\mathsf{inst}_1(f, l, \mathsf{Tree}) &= \lambda A.f\ (\mathsf{tag}_1(l, \mathsf{Tree})\ A) \\
\mathsf{inst}_2(f, l, \mathsf{Forest}) &= \lambda A.f\ (\mathsf{tag}_2(l, \mathsf{Forest})\ A)
\end{aligned}
$$

- $\mathsf{family}(l)$ defines the datatype family itself. This takes a recursive argument and a tag argument, and applies the tag to the Scott-encoded types of the datatype components, where the types themselves are instantiated using the recursive argument.

$$
\begin{aligned}
\mathsf{family}(l) &= \lambda r\ t.\,\texttt{let} \\
&\qquad\qquad \mathsf{Tree} = \mathsf{inst}_1(r, l, \mathsf{Tree}) \\
&\qquad\qquad \mathsf{Forest} = \mathsf{inst}_2(r, l, \mathsf{Forest}) \\
&\qquad\quad \texttt{in}\ t\ \mathsf{dataTy}(d_1)\ \mathsf{dataTy}(d_2) \\
\mathsf{dataTy}(d_1) &= \lambda A.\forall R.(A \to \mathsf{Forest}\ A \to R) \to R \\
\mathsf{dataTy}(d_2) &= \lambda A.\forall R.R \to (\mathsf{Tree}\ A \to \mathsf{Forest}\ A \to R) \to R
\end{aligned}
$$

- $\mathsf{instFamily}_k(l,d)$ is the full recursive datatype family instantiated for the datatype d, much like $\mathsf{inst}_k(f,l,d)$, but with the full datatype family.

$$
\mathsf{instFamily}_1(l, \mathsf{Tree}) = \lambda A.\,\texttt{ifix}\ (\mathsf{family}(l))\ (\mathsf{tag}_1(l, \mathsf{Tree})\ A)
$$

- $\mathsf{unveil}^{\mathsf{rec}}(l,t)$ "unveils" the datatypes as before, but unveils all the datatypes and replaces them with the full recursive definition instead of just the Scott-encoded type.

- $\mathsf{constr}^{\texttt{rec}}_{k,m}(l, d, c)$ defines the constructor c of the datatype d in the family. It is similar to before, but includes a use of `wrap`.

$$\begin{aligned}
\mathsf{constr}^{\texttt{rec}}_{1,1}(l, \mathsf{Tree}, \mathsf{Node}) &= \Lambda A.\lambda(v_1 : A)(v_2 : \mathsf{Forest}\, A). \\
&\quad \texttt{wrap}\ (\mathsf{instFamily}_1(l, \mathsf{Tree}))\ A \\
&\quad (\Lambda R.\lambda(b_1 : A \to \mathsf{Forest}\, A \to R).b_1\ v_1\ v_2) \\
\mathsf{constr}^{\texttt{rec}}_{2,1}(l, \mathsf{Forest}, \mathsf{Nil}) &= \Lambda A. \\
&\quad \texttt{wrap}\ (\mathsf{instFamily}_2(l, \mathsf{Forest}))\ A \\
&\quad (\Lambda R.\lambda(b_1 : R)(b_2 : \mathsf{Tree}\, A \to \mathsf{Forest}\, A \to R).b_1) \\
\mathsf{constr}^{\texttt{rec}}_{2,2}(l, \mathsf{Forest}, \mathsf{Cons}) &= \Lambda A.\lambda(v_1 : \mathsf{Tree}\, A)(v_2 : \mathsf{Forest}\, A). \\
&\quad \texttt{wrap}\ (\mathsf{instFamily}_2(l, \mathsf{Forest}))\ A \\
&\quad (\Lambda R.\lambda(b_1 : R)(b_2 : \mathsf{Tree}\, A \to \mathsf{Forest}\, A \to R).b_2\ v_1\, v_2)
\end{aligned}$$

- $\mathsf{match}^{\texttt{rec}}_{k}(l, d)$ defines the matcher of the datatype d as before, but includes a use of `unwrap`.

$$\begin{aligned}
\mathsf{match}^{\texttt{rec}}_{1}(l, \mathsf{Tree}) &= \Lambda A.\lambda(v : \mathsf{Tree}\, A).\, \texttt{unwrap}\ v \\
\mathsf{match}^{\texttt{rec}}_{2}(l, \mathsf{Forest}) &= \Lambda A.\lambda(v : \mathsf{Forest}\, A).\, \texttt{unwrap}\ v
\end{aligned}$$

5 Compiler Implementation in Agda

As a supplement to the presentation in this paper, we have written a formalisation of a FIR compiler in Agda.[8] The compiler includes the syntax, the type system (the syntax is intrinsically typed, so there is no need for a typechecker), and implementations of several of the passes. In particular, we have implemented:

- Type-level compilation of mutually recursive datatypes into System F^{μ}_{ω} types.
- Term-level compilation of mutually recursive terms into System F^{μ}_{ω} terms.

The Agda presentation uses an intrinsically-typed syntax, where terms are identified with their typing derivations [2]. This means that the compilation process is provably kind- and type-preserving.

However, the implementation is incomplete. The formalization is quite involved since the term-level parts of datatypes (constructors) must exactly line up with the type-level parts. Moreover, we have not proved any soundness results beyond type preservation. The complexity of the encodings makes it very hard to prove soundness. The artifact contains some further notes on the difficulties in the implementation.

6 Optimization

FIR has the virtue that it is significantly easier to optimize than System F^{μ}_{ω}. Here are two examples.

[8] The complete source can be found in the Plutus repository.

6.1 Dead Binding Elimination

Languages with let terms admit a simple form of dead code elimination: any bindings in let terms which are unused can be removed. A dead binding in a FIR term can be easily identified by constructing a dependency graph over the variables in the term, and eliminating any bindings for unreachable variables.

We can certainly do something with the compiled form of simple, non-recursive let bindings in System F_ω^μ. These are compiled to immediately-applied lambda abstractions, which is an easy pattern to identify, and it is also easy to work out whether the bound variable is used.

Recursive let bindings are much trickier. Here the compiled structure is obscured by the fixpoint combinator and the construction and deconstruction of the encoded tuple, which makes the pattern much harder to spot. Datatype bindings are similarly complex.

The upshot is that it is much easier to perform transformations based on the structure of variable bindings when those bindings are still present in their original form.

6.2 Case-of-Known-Constructor

The case-of-known-constructor optimization is very important for functional programming languages with datatypes (see e.g. [26, section 5]). When we perform a pattern-match on a term which we know is precisely a constructor invocation, we can collapse the immediate construction and deconstruction.

For example, we should be able to perform the following transformation:

$$\mathsf{match}\ \{\mathsf{Int}\}\ (\mathsf{Just}\ \{\mathsf{Int}\}\ 1)\ 0\ (\lambda x.x+1) \Longrightarrow (\lambda x.x+1)\ 1$$

This is easy to implement in FIR, since we still have the knowledge of which constructors and destructors belong to the same datatype. But once we have compiled to System F_ω^μ we lose this information. A destructor-constructor pair is just an inner redex of the term, which happens to reduce nicely. But reducing arbitrary redexes is risky (since we have no guarantee that it will not grow the program), and we do not know of a general approach which would identify these redexes as worth reducing.

7 Why Not Support These Features Natively?

The techniques in this paper cause a significant amount of runtime overhead. The combinator-based approach to defining recursive functions requires many more reductions than a direct implementation which could implement recursive calls by jumping directly to the code pointer for the recursive function.

Similarly, representing datatype values as functions is much less efficient than representing them as tagged data.

However, there are tradeoffs here for the language designer. If the language is intended to be a competitive general-purpose programming language like

Haskell, then these performance losses may be unacceptable. On the other hand, if we care less about performance and more about correctness, then the benefits of having a minimal, well-studied core may dominate.

Moreover, even if a language has a final target language which provides these features natively, a naive but higher-assurance backend can provide a useful alternative code generator to test against.

Of course, the proof is in the pudding, and we have practical experience using these techniques in the Plutus platform [16]. Experience shows that the overhead proves not to be prohibitive: the compiler is able to compile and run substantial real-world Haskell programs, and is available for public use at [17].

8 Related Work

8.1 Encoding Recursive Datatypes

Different approaches to encoding datatypes are compared in [19]. The authors provide a schematic formal description of Scott encoding, but ours is more thorough and includes complete handling of recursive types.

Indexed fixpoints are used in [32] to encode regular and mutually recursive datatypes as fixpoints of pattern functors. We use the same fixpoint operator—they call it "hfix", while we call it "ifix". They also use the trick of encoding products with a tag, but they use the natural numbers as an index, and they do not handle parameterized types. Later work in [21] does handle parameterized types, but our technique of putting the parameters into the tag type appears to be novel. Neither paper handles non-regular datatypes.

There are other implementations of System F_ω with recursive types. Brown and Palsberg [6] use isorecursive types, and includes an indexed fixpoint operator as well as a typecase operator. However, the index for the fixpoint must be of kind $*$, whereas ours may be of any kind. Cai et al. [7] differ from this paper both in using equirecursive types and in that their fixpoint operator only works at kind $*$. Moreover, algebraic datatypes are supported directly, rather than via encoding.

8.2 Encoding Recursive Terms

There is very little existing material on compiling multiple mutually recursive functions, especially in a strict language. Some literature targets lower-level or specialized languages [15,23,31], whereas ours is a much more standard calculus. There are some examples which use fixpoint combinators (such as [18], extending [13] for typed languages) which use different fixpoint combinators.

8.3 Intermediate Representations

GHC Haskell is well-known for using a fairly small lambda-calculus-based IR ("Core") for almost all of its intermediary work [26]. FIR is very inspired by GHC

Core, but supports far fewer features and is aimed at eliminating constructs like datatypes and recursion, whereas they are native features of GHC Core.

A more dependently-typed IR is described in [25]. We have not yet found the need to generalize our base calculus to a dependently-typed one like Henk, but all the techniques in this paper should still apply in such a setting. Extensions to Henk that handle let-binding and datatypes are discussed, but it appears that these are intended as additional native features rather than being compiled away into a base calculus.

9 Conclusion

We have presented FIR, a reusable, typed IR which provides several typical functional programming language features. We have shown how to compile it into System F_ω^μ via a series of local compilation passes, and given a reference implementation for the compiler.

There is more work to do on the theory and formalisation of FIR. We have not given a direct semantics, in terms of reduction rules or otherwise. We would also like to prove our compilation correct, in that it commutes with reduction. A presentation of a complete compiler written in Agda with accompanying proofs would be desirable.

We could also remove some of the restrictions present in this paper: in particular the lack of mutually recursive type bindings, and the requirement that recursive let terms be homogeneous.

Acknowledgments. The authors would like to thank Mario Alvarez-Picallo and Manuel Chakravarty for their comments on the manuscript, as well as IOHK for funding the research.

References

1. Abadi, M., Cardelli, L., Plotkin, G.: Types for the Scott numerals (1993)
2. Altenkirch, T., Reus, B.: Monadic presentations of lambda terms using generalized inductive types. In: Flum, J., Rodriguez-Artalejo, M. (eds.) CSL 1999. LNCS, vol. 1683, pp. 453–468. Springer, Heidelberg (1999). https://doi.org/10.1007/3-540-48168-0_32
3. Backhouse, R., Jansson, P., Jeuring, J., Meertens, L.: Generic programming — an introduction. In: Swierstra, S.D., Oliveira, J.N., Henriques, P.R. (eds.) AFP 1998. LNCS, vol. 1608, pp. 28–115. Springer, Heidelberg (1999). https://doi.org/10.1007/10704973_2
4. Brady, E.: Idris, a general-purpose dependently typed programming language: design and implementation. J. Funct. Program. **23**(5), 552–593 (2013). https://doi.org/10.1017/S095679681300018X
5. Bredon, G.E.: Topology and Geometry. Graduate Texts in Mathematics, vol. 139. Springer, New York (1993). https://doi.org/10.1007/978-1-4757-6848-0

6. Brown, M., Palsberg, J.: Typed self-evaluation via intensional type functions. In: Proceedings of the 44th ACM SIGPLAN Symposium on Principles of Programming Languages, POPL 2017, pp. 415–428. ACM, New York (2017). https://doi.org/10.1145/3009837.3009853. http://doi.acm.org/10.1145/3009837.3009853
7. Cai, Y., Giarrusso, P.G., Ostermann, K.: System F-omega with equirecursive types for datatype-generic programming. In: Proceedings of the 43rd Annual ACM SIGPLAN-SIGACT Symposium on Principles of Programming Languages, POPL 2016, pp. 30–43. ACM, New York (2016). https://doi.org/10.1145/2837614.2837660. http://doi.acm.org/10.1145/2837614.2837660
8. Chapman, J., Kireev, R., Nester, C., Wadler, P.: System F in Agda, for fun and profit. In: Hutton, G. (ed.) MPC 2019. LNCS, vol. 11825, pp. 255–297. Springer, Cham (2019)
9. Danvy, O., Hatcliff, J.: Thunks (continued). In: Proceedings of Actes WSA 1992 Workshop on Static Analysis, Bordeaux, France, Laboratoire Bordelais de Recherche en Informatique (LaBRI), pp. 3–11, September 1992
10. Dreyer, D.: Understanding and evolving the ML module system. Ph.D. thesis, Carnegie Mellon University, Pittsburgh, PA, USA (2005). aAI3166274
11. Dreyer, D.: A type system for recursive modules. In: Proceedings of the 12th ACM SIGPLAN International Conference on Functional Programming, ICFP 2007, pp. 289–302. ACM, New York (2007). https://doi.org/10.1145/1291151.1291196. http://doi.acm.org/10.1145/1291151.1291196
12. Fulton, W.: Algebraic Topology: A First Course. Graduate Texts in Mathematics, vol. 153. Springer, New York (1995). https://doi.org/10.1007/978-1-4612-4180-5
13. Goldberg, M.: A variadic extension of curry's fixed-point combinator. High. Order Symbol. Comput. **18**(3–4), 371–388 (2005). https://doi.org/10.1007/s10990-005-4881-8
14. Harper, R.: Practical Foundations for Programming Languages. Cambridge University Press, New York (2012)
15. Hirschowitz, T., Leroy, X., Wells, J.B.: Compilation of extended recursion in call-by-value functional languages. In: Proceedings of the 5th ACM SIGPLAN International Conference on Principles and Practice of Declaritive Programming, PPDP 2003, pp. 160–171. ACM, New York (2003). https://doi.org/10.1145/888251.888267. http://doi.acm.org/10.1145/888251.888267
16. IOHK: Plutus, May 2019. https://github.com/input-output-hk/plutus. Accessed 01 May 2019
17. IOHK: Plutus playground, May 2019. https://prod.playground.plutus.iohkdev.io/. Accessed 01 May 2019
18. Kiselyov, O.: Many faces of the fixed-point combinator, August 2013. http://okmij.org/ftp/Computation/fixed-point-combinators.html. Accessed 21 Feb 2019
19. Koopman, P., Plasmeijer, R., Jansen, J.M.: Church encoding of data types considered harmful for implementations: functional pearl. In: Proceedings of the 26th 2014 International Symposium on Implementation and Application of Functional Languages, IFL 2014, pp. 4:1–4:12. ACM, New York (2014). https://doi.org/10.1145/2746325.2746330. http://doi.acm.org/10.1145/2746325.2746330
20. Leroy, X.: The ZINC experiment: an economical implementation of the ML language. Ph.D. thesis, INRIA (1990)
21. Löh, A., Magalhães, J.P.: Generic programming with indexed functors. In: Proceedings of the Seventh ACM SIGPLAN Workshop on Generic Programming, WGP 2011, pp. 1–12. ACM, New York (2011). https://doi.org/10.1145/2036918.2036920. http://doi.acm.org/10.1145/2036918.2036920

22. Mitchell, J.C., Plotkin, G.D.: Abstract types have existential types. In: Proceedings of the 12th ACM SIGACT-SIGPLAN Symposium on Principles of Programming Languages, POPL 1985, pp. 37–51. ACM, New York (1985). https://doi.org/10.1145/318593.318606. http://doi.acm.org/10.1145/318593.318606
23. Nordlander, J., Carlsson, M., Gill, A.J.: Unrestricted pure call-by-value recursion. In: Proceedings of the 2008 ACM SIGPLAN Workshop on ML, ML 2008, pp. 23–34. ACM, New York (2008). https://doi.org/10.1145/1411304.1411309. http://doi.acm.org/10.1145/1411304.1411309
24. Norell, U.: Towards a practical programming language based on dependent type theory. Ph.D. thesis, Chalmers University of Technology (2007)
25. Peyton Jones, S., Meijer, E.: Henk: A typed intermediate language. In: Proceedings of the First International Workshop on Types in Compilation (1997)
26. Peyton, S.L., Santos, A.L.M.: A transformation-based optimiser for haskell. Sci. Comput. Program. **32**(1–3), 3–47 (1998). https://doi.org/10.1016/S0167-6423(97)00029-4
27. Pierce, B.C.: Types and Programming Languages. MIT Press, Cambridge (2002)
28. Steele, G.L., Sussman, G.J.: Lambda: the ultimate imperative. Technical report, Cambridge, MA, USA (1976)
29. Steele Jr., G.L.: It's time for a new old language. In: PPOPP, p. 1 (2017)
30. Stirling, C.: Proof systems for retracts in simply typed lambda calculus. In: Fomin, F.V., Freivalds, R., Kwiatkowska, M., Peleg, D. (eds.) ICALP 2013, Part II. LNCS, vol. 7966, pp. 398–409. Springer, Heidelberg (2013). https://doi.org/10.1007/978-3-642-39212-2_36
31. Syme, D.: Initializing mutually referential abstract objects: the value recursion challenge. Electron. Notes Theor. Comput. Sci. **148**(2), 3–25 (2006). https://doi.org/10.1016/j.entcs.2005.11.038
32. Yakushev, A.R., Holdermans, S., Löh, A., Jeuring, J.: Generic programming with fixed points for mutually recursive datatypes. SIGPLAN Not **44**(9), 233–244 (2009). https://doi.org/10.1145/1631687.1596585. http://doi.acm.org/10.1145/1631687.1596585

Coding with Asymmetric Numeral Systems

Jeremy Gibbons(✉)

University of Oxford, Oxford, UK
jeremy.gibbons@cs.ox.ac.uk

Abstract. Asymmetric Numeral Systems (ANS) are an entropy-based encoding method introduced by Jarek Duda, combining the Shannon-optimal compression *effectiveness* of arithmetic coding with the execution *efficiency* of Huffman coding. Existing presentations of the ANS encoding and decoding algorithms are somewhat obscured by the lack of suitable presentation techniques; we present here an equational derivation, calculational where it can be, and highlighting the creative leaps where it cannot.

1 Introduction

Entropy encoding techniques compress symbols according to a model of their expected frequencies, with common symbols being represented by fewer bits than rare ones. The best known entropy encoding technique is *Huffman coding* (HC) [18], taught in every undergraduate course on algorithms and data structures: a classic greedy algorithm uses the symbol frequencies to construct a trie, from which an optimal *prefix-free binary code* can be read off. For example, suppose an alphabet of $n = 3$ symbols $s_0 = $ `'a'`, $s_1 = $ `'b'`, $s_2 = $ `'c'` with respective expected relative frequencies $c_0 = 2, c_1 = 3, c_2 = 5$ (that is, `'a'` is expected $^2/_{2+3+5} = 20\%$ of the time, and so on); then HC might construct the trie and prefix-free code shown in Fig. 1. A text is then encoded as the concatenation of its symbol codes; thus, the text `"cbcacbcacb"` encodes to 1 01 1 00 1 01 1 00 1 01. This is optimal, in the sense that no prefix-free binary code yields a shorter encoding of any text that matches the expected symbol frequencies.

But HC is only 'optimal' among encodings that use a whole number of bits per symbol; if that constraint is relaxed, more effective encoding becomes possible. Note that the two symbols `'a'` and `'b'` were given equal-length codes 00 and 01 by HC, despite having unequal frequencies—indeed, any expected frequencies in the same order $c_0 < c_1 < c_2$ will give the same code. More starkly, if the alphabet has only two symbols, HC can do no better than to code each symbol as a single bit, whatever their expected frequencies; that might be acceptable when the frequencies are similar, but is unacceptable when they are not.

G. Hutton (Ed.): MPC 2019, LNCS 11825, pp. 444–465, 2019.
https://doi.org/10.1007/978-3-030-33636-3_16

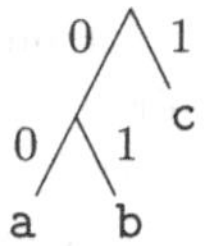

'a' $\mapsto$ 00
'b' $\mapsto$ 01
'c' $\mapsto$ 1

Fig. 1. A Huffman trie and the corresponding prefix-free code

Arithmetic coding (AC) [23,26] is an entropy encoding technique that allows for a fractional number of bits per symbol. In a nutshell, a text is encoded as a half-open subinterval of the unit interval; the model assigns disjoint subintervals of the unit interval to each symbol, in proportion to their expected frequencies (as illustrated on the left of Fig. 2); encoding starts with the unit interval, and narrows this interval by the model subinterval for each symbol in turn (the narrowing operation is illustrated on the right of Fig. 2). The encoding is the shortest binary fraction in the final interval, without its final '1'. For example, with the model illustrated in Fig. 2, the text "abc" gets encoded via the narrowing sequence of intervals

$$[0,1) \xrightarrow{\texttt{'a'}} [0, {}^1\!/\!_5) \xrightarrow{\texttt{'b'}} [{}^1\!/\!_{25}, {}^1\!/\!_{10}) \xrightarrow{\texttt{'c'}} [{}^7\!/\!_{100}, {}^1\!/\!_{10})$$

from which we pick the binary fraction ${}^3\!/\!_{32}$ (since ${}^7\!/\!_{100} \leqslant {}^3\!/\!_{32} < {}^1\!/\!_{10}$) and output the bit sequence 0001. We formalize this sketched algorithm in Sect. 3.

This doesn't look like much saving: this particular example is only one bit shorter than with HC; and similarly, the arithmetic coding of the text "cbcacbcacb" is 14 bits, where HC uses 15 bits. But AC can do much better; for example, it encodes the permutation "cabbacbccc" of that text in 7 bits, whereas of course HC uses the same 15 bits as before.

In fact, AC is *Shannon-optimal*: the number of bits used tends asymptotically to the Shannon entropy of the message—the sum $\sum_i -\log_2 p_i$ of the negative logarithms of the symbol probabilities. Moreover, AC can be readily made *adaptive*, whereby the model evolves as the text is read, whereas HC entails separate modelling and encoding phases.

'a' $\mapsto [0, {}^1\!/\!_5)$
'b' $\mapsto [{}^1\!/\!_5, {}^1\!/\!_2)$
'c' $\mapsto [{}^1\!/\!_2, 1)$

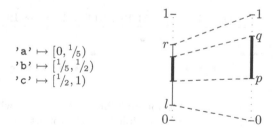

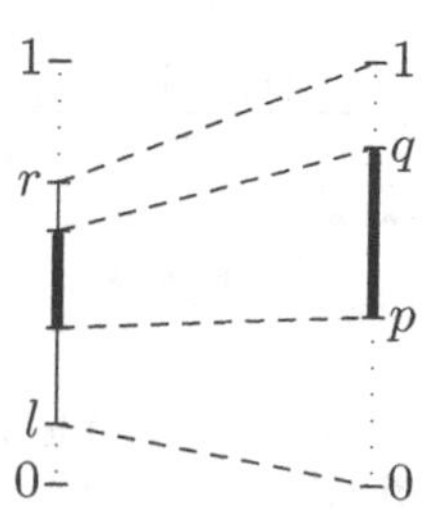

Fig. 2. A text model in interval form. Narrowing interval $[l, r)$ by interval $[p, q)$ yields the interval marked in bold on the left, which stands in relation to $[l, r)$ as $[p, q)$ does to $[0, 1)$.

However, AC does have some problems. One problem is a historical accident: specific applications of the technique became mired in software patents in the 1980s, and although those patents have now mostly expired, the consequences are still being felt (for example, Seward's `bzip` compressor [24] switched in 1996 from AC to HC because of patents, and has not switched back since). A more fundamental problem is that AC involves a lot of arithmetic, and even after slightly degrading coding effectiveness in order to use only single-word fixed-precision rather than arbitrary-precision arithmetic, state-of-the-art implementations are still complicated and relatively slow.

A recent development that addresses both of these problems has been Jarek Duda's introduction of *asymmetric numeral systems* (ANS) [10,11,13]. This is another entropy encoding technique; in a nutshell, rather than encoding longer and longer messages as narrower and narrower subintervals of the unit interval, they are simply encoded as larger and larger integers. Concretely, with the same frequency counts $c_0 = 2, c_1 = 3, c_2 = 5$ as before, and cumulative totals $t_0 = 0, t_1 = t_0 + c_0 = 2, t_2 = t_1 + c_1 = 5, t = t_2 + c_2 = 10$, encoding starts with an accumulator at 0, and for each symbol s_i (traditionally from right to left in the text) maps the current accumulator x to $(x \operatorname{div} c_i) \times t + t_i + (x \bmod c_i)$, as illustrated in Fig. 3. Thus, the text `"abc"` gets encoded via the increasing (read from right to left) sequence of integers:

$$70 \xleftarrow{\texttt{'a'}} 14 \xleftarrow{\texttt{'b'}} 5 \xleftarrow{\texttt{'c'}} 0$$

It is evident even from this brief sketch that the encoding process is quite simple, with just a single division and multiplication per symbol; it turns out that decoding is just as simple. The encoding seems quite mysterious, but it is very cleverly constructed, and again achieves Shannon-optimal encoding; ANS combines the *effectiveness* of AC with the *efficiency* of Huffman coding, addressing the more fundamental concern with AC. The purpose of this paper is to motivate and justify the development, using calculational techniques where possible.

As it happens, Duda is also fighting to keep ANS in the public domain, despite corporate opposition [20], thereby addressing the more accidental concern too. These benefits have seen ANS recently adopted by large companies for

	0	1	2	3	4	5	6	7	8	9	10	11	12	13	14	15	16	17	18	19	20	21	···
'a'	•	•									•	•									•	•	
'b'			•	•	•								•	•	•								
'c'						•	•	•	•	•						•	•	•	•	•			

Fig. 3. The start of the coding table for alphabet `'a'`, `'b'`, `'c'` with counts 2, 3, 5. The indices 0.. are distributed across the alphabet, in proportion to the counts: two for `'a'`, three for `'b'`, and so on. Encoding symbol s with current accumulator x yields the index of the xth blob in row s as the new accumulator. For example, with $x = 4$ and next symbol $s = \texttt{'b'} = s_i$ with $i = 1$, we have $c_i = 3, t_i = 2, t = 10$ so $x' = (x \operatorname{div} c_i) \times t + t_i + (x \bmod c_i) = 13$, and indeed the 4th blob in row `'b'` (counting from zero) is in column 13.

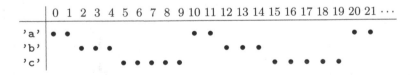

products such as Facebook Zstd [8], Apple LZFSE [9], Google Draco [6], and Dropbox DivANS [22], and ANS is expected [1] to be featured in the forthcoming JPEG XL standard [19].

One disadvantage of ANS is that, whereas AC acts in a first-in first-out manner, ANS acts last-in first-out, in the sense that the decoded text comes out in the reverse order to which it went in. Our development will make clear where this happens. This reversal makes ANS unsuitable for encoding a communications channel, and also makes it difficult to employ adaptive text models. (DivANS [22] processes the input forwards for statistical modelling, and then uses this information backwards to encode the text; one could alternatively batch process the text in fixed-size blocks. In some contexts, such as encoding the video stream of a movie for distribution to set-top boxes, it is worth expending more effort in offline encoding in order to benefit online decoding.)

The remainder of this paper is structured as follows. Section 2 recaps various well-known properties of folds and unfolds on lists. Section 3 presents the relevant basics of AC, and Sect. 4 a proof of correctness of this basic algorithm. Section 5 presents the key step from AC to ANS, namely the switch from accumulating fractions to accumulating integers. Section 6 shows how to modify this naive ANS algorithm to work in bounded precision, and Sect. 7 shows how to make the resulting program 'stream' (to start generating output before consuming all the input). Section 8 discusses related work and concludes.

We use Haskell [21] as an algorithmic notation. Note that function application binds tightest of all binary operators, so that for example $f\ x + y$ means $(f\ x) + y$; apart from that, we trust that the notation is self-explanatory. We give definitions of functions from the Haskell standard library as we encounter them. The code from the paper is available online [15], as is a longer version [16] of the paper including proofs and other supporting material.

2 Origami Programming

In this section, we recap some well-studied laws of folds

$$
\begin{aligned}
&\mathit{foldr} :: (a \to b \to b) \to b \to [a] \to b\\
&\mathit{foldr}\ f\ e\ [\,] \qquad\quad = e\\
&\mathit{foldr}\ f\ e\ (x : xs) = f\ x\ (\mathit{foldr}\ f\ e\ xs)
\end{aligned}
$$

$$
\begin{aligned}
&\mathit{foldl} :: (b \to a \to b) \to b \to [a] \to b\\
&\mathit{foldl}\ f\ e\ [\,] \qquad\quad = e\\
&\mathit{foldl}\ f\ e\ (x : xs) = \mathit{foldl}\ f\ (f\ e\ x)\ xs
\end{aligned}
$$

and unfolds

$$
\begin{aligned}
&\mathit{unfoldr} :: (b \to \mathit{Maybe}\ (a, b)) \to b \to [a]\\
&\mathit{unfoldr}\ f\ y = \textbf{case}\ f\ y\ \textbf{of}\ \mathit{Nothing} \quad\ \to [\,]\\
&\qquad\qquad\qquad\qquad\qquad\quad\ \mathit{Just}\ (x, y') \to x : \mathit{unfoldr}\ f\ y'
\end{aligned}
$$

on lists.

Folds. The *First Duality Theorem* of *foldl* and *foldr* [4, §3.5.1] states that

$$\mathit{foldr}\ f\ e = \mathit{foldl}\ f\ e$$

when f and e form a monoid. The *Third Duality Theorem*, from the same source, says:

$$\mathit{foldr}\ f\ e \cdot \mathit{reverse} = \mathit{foldl}\ (\mathit{flip}\ f)\ e$$

where $\mathit{flip}\ f\ a\ b = f\ b\ a$ swaps the arguments of a binary function. (The published version [4, §3.5.1] has the *reverse* on the other side, and holds only for finite lists.)

We will also use the *Fusion Law* for *foldr* [3, §4.6.2]:

$$h \cdot \mathit{foldr}\ f\ e = \mathit{foldr}\ f'\ e' \quad \Leftarrow \quad h\ e = e' \wedge h\ (f\ x\ y) = f'\ x\ (h\ y)$$

(at least, on finite lists), and its corollaries the *Map Fusion* laws:

$$\begin{aligned} \mathit{foldr}\ f\ e \cdot \mathit{map}\ g &= \mathit{foldr}\ (\lambda x\ y \rightarrow f\ (g\ x)\ y)\ e \\ \mathit{foldl}\ f\ e \cdot \mathit{map}\ g &= \mathit{foldl}\ (\lambda x\ y \rightarrow f\ x\ (g\ y))\ e \end{aligned}$$

Unfolds. The sole law of *unfoldr* stated in the Haskell 98 standard [21, §17.4] gives conditions under which it inverts a *foldr*: if

$$\begin{aligned} g\ (f\ x\ z) &= \mathit{Just}\ (x, z) \\ g\ e &= \mathit{Nothing} \end{aligned}$$

for all x and z, then

$$\mathit{unfoldr}\ g\ (\mathit{foldr}\ f\ e\ xs) = xs$$

for all finite lists xs. We call this the *Unfoldr–Foldr Theorem*. (The proof is a straightforward induction on xs.)

We make two generalisations to this theorem. The first, the *Unfoldr–Foldr Theorem with Junk*, allows the unfold to continue after reconstructing the original list: if only

$$g\ (f\ x\ z) = \mathit{Just}\ (x, z)$$

holds, for all x and z, then

$$\exists ys\ .\ \mathit{unfoldr}\ g\ (\mathit{foldr}\ f\ e\ xs) = xs \mathbin{+\!\!+} ys$$

for all finite lists xs—that is, the *unfoldr* inverts the *foldr* except for appending some (possibly infinite) junk ys to the output. This too can be proved by induction on xs.

The second generalisation is the *Unfoldr–Foldr Theorem with Invariant*. We say that predicate p is an invariant of $\mathit{foldr}\ f\ e$ and $\mathit{unfoldr}\ g$ if

$$\begin{aligned} p\ (f\ x\ z) &\Leftarrow p\ z \\ p\ z' &\Leftarrow p\ z \wedge g\ z = \mathit{Just}\ (x, z') \end{aligned}$$

for all x, z, z'. The theorem states that if p is such an invariant, and the conditions

$$\begin{array}{llll} g\,(f\,x\,z) & = Just\,(x,z) & \Leftarrow & p\,z \\ g\,e & = Nothing & \Leftarrow & p\,e \end{array}$$

hold for all x and z, then

$$unfoldr\;g\;(foldr\;f\;e\;xs) = xs \quad \Leftarrow \quad p\;e$$

for all finite lists xs. Again, there is a straightforward proof by induction. And of course, there is an *Unfoldr–Foldr Theorem with Junk and Invariant*, incorporating both generalisations; this is the version we will actually use.

3 Arithmetic Coding

We start from a simplified version of arithmetic coding: we use a fixed rather than adaptive model, and rather than picking the shortest binary fraction within the final interval, we simply pick the lower bound of the interval.

Intervals and Symbols. We represent intervals as pairs of rationals,

$$\mathbf{type}\;Interval = (Rational, Rational)$$

so the unit interval is $unit = (0, 1)$ and the lower bound is obtained by fst. We suppose a symbol table

$$counts :: [(Symbol, Integer)]$$

that records a positive count for every symbol in the alphabet; in the interests of brevity, we omit the straightforward definitions of functions

$$\begin{array}{l} encodeSym :: Symbol \rightarrow Interval \\ decodeSym :: Rational \rightarrow Symbol \end{array}$$

that work on this fixed global model, satisfying the central property: for $x \in unit$,

$$decodeSym\;x = s \quad \Leftrightarrow \quad x \in encodeSym\;s$$

For example, with the same alphabet of three symbols `'a'`, `'b'`, `'c'` and counts 2, 3, and 5 as before, we have $encodeSym$ `'b'` $= (^1/_5, ^1/_2)$ and $decodeSym\,(^1/_3) =$ `'b'`.

We have operations on intervals:

$$\begin{array}{ll} weight, scale :: Interval \rightarrow Rational \rightarrow Rational & \\ weight\,(l,r)\,x & = l + (r-l) \times x \\ scale\,(l,r)\,y & = (y-l)\;/\;(r-l) \end{array}$$

$$\begin{array}{ll} narrow & :: Interval \rightarrow Interval \rightarrow Interval \\ narrow\;i\;(p,q) & = (weight\;i\;p, weight\;i\;q) \end{array}$$

that satisfy

$$\begin{array}{lll} \mathit{weight}\; i\; x \in i & \Leftrightarrow & x \in \mathit{unit} \\ \mathit{weight}\; i\; x = y & \Leftrightarrow & \mathit{scale}\; i\; y = x \end{array}$$

Informally, $\mathit{weight}\,(l, r)\, x$ is 'fraction x of the way between l and r', and conversely $\mathit{scale}\,(l, r)\, y$ is 'the fraction of the way y is between l and r'; and *narrow* is illustrated in Fig. 2.

Encoding and Decoding. Now we can specify arithmetic encoding and decoding by:

$$\begin{array}{l} \mathit{encode}_1 :: [\mathit{Symbol}] \to \mathit{Rational} \\ \mathit{encode}_1 = \mathit{fst} \cdot \mathit{foldl}\; \mathit{estep}_1\; \mathit{unit} \end{array}$$

$$\begin{array}{l} \mathit{estep}_1 :: \mathit{Interval} \to \mathit{Symbol} \to \mathit{Interval} \\ \mathit{estep}_1\; i\; s = \mathit{narrow}\; i\, (\mathit{encodeSym}\; s) \end{array}$$

$$\begin{array}{l} \mathit{decode}_1 :: \mathit{Rational} \to [\mathit{Symbol}] \\ \mathit{decode}_1 = \mathit{unfoldr}\; \mathit{dstep}_1 \end{array}$$

$$\begin{array}{l} \mathit{dstep}_1 :: \mathit{Rational} \to \mathit{Maybe}\, (\mathit{Symbol}, \mathit{Rational}) \\ \mathit{dstep}_1\; x = \mathbf{let}\; s = \mathit{decodeSym}\; x\; \mathbf{in}\; \mathit{Just}\, (s, \mathit{scale}\, (\mathit{encodeSym}\; s)\; x) \end{array}$$

For example, with the same alphabet and counts, the input text `"abc"` gets encoded symbol by symbol, from left to right (because of the *foldl*), starting with the unit interval $(0, 1)$, via the narrowing sequence of intervals

$$\begin{array}{ll} \mathit{estep}_1\, (0, 1)\; \texttt{'a'} & = (0, {}^{1}\!/_{5}) \\ \mathit{estep}_1\, (0, {}^{1}\!/_{5})\; \texttt{'b'} & = ({}^{1}\!/_{25}, {}^{1}\!/_{10}) \\ \mathit{estep}_1\, ({}^{1}\!/_{25}, {}^{1}\!/_{10})\; \texttt{'c'} & = ({}^{7}\!/_{100}, {}^{1}\!/_{10}) \end{array}$$

from which we select the lower bound ${}^{7}\!/_{100}$ of the final interval. Conversely, decoding starts with ${}^{7}\!/_{100}$, and proceeds as follows:

$$\begin{array}{ll} \mathit{dstep}_1\, ({}^{7}\!/_{100}) & = \mathit{Just}\, (\texttt{'a'}, {}^{7}\!/_{20}) \\ \mathit{dstep}_1\, ({}^{7}\!/_{20}) & = \mathit{Just}\, (\texttt{'b'}, {}^{1}\!/_{2}) \\ \mathit{dstep}_1\, ({}^{1}\!/_{2}) & = \mathit{Just}\, (\texttt{'c'}, 0) \\ \mathit{dstep}_1\, 0 & = \mathit{Just}\, (\texttt{'a'}, 0) \\ \mathit{dstep}_1\, 0 & = \mathit{Just}\, (\texttt{'a'}, 0) \\ \ldots & \end{array}$$

Note that decoding runs forever; but the finite encoded text is a prefix of the decoded output—for any input text xs, there is an infinite sequence of junk ys such that

$$\mathit{decode}_1\, (\mathit{encode}_1\; xs) = xs +\!\!+\; ys$$

(indeed, $ys = \mathit{repeat}\; \texttt{'a'}$ when we pick the *fst* of an interval).

4 Correctness of Arithmetic Coding

Using the laws of folds, we can 'fission' the symbol encoding out of $encode_1$, turn the *foldl* into a *foldr* (because *narrow* and *unit* form a monoid), fuse the *fst* with the *foldr*, and then re-fuse the symbol encoding with the fold:

$$
\begin{aligned}
& encode_1 \\
= \quad & \{ \text{ definition } \} \\
& fst \cdot foldl\ estep_1\ unit \\
= \quad & \{ \text{ map fusion for } foldl\text{, backwards } \} \\
& fst \cdot foldl\ narrow\ unit \cdot map\ encodeSym \\
= \quad & \{ \text{ duality: } narrow \text{ and } unit \text{ form a monoid } \} \\
& fst \cdot foldr\ narrow\ unit \cdot map\ encodeSym \\
= \quad & \{ \text{ fusion for } foldr \text{ (see below) } \} \\
& foldr\ weight\ 0 \cdot map\ encodeSym \\
= \quad & \{ \text{ map fusion; let } estep_2\ s\ x = weight\ (encodeSym\ s)\ x \ \} \\
& foldr\ estep_2\ 0
\end{aligned}
$$

For the fusion step, we have

$$
\begin{aligned}
& fst\ unit &&= 0 \\
& fst\ (narrow\ i\ (p, q)) &&= weight\ i\ (fst\ (p, q))
\end{aligned}
$$

as required. So we have calculated $encode_1 = encode_2$, where

$$
\begin{aligned}
& encode_2 :: [Symbol] \rightarrow Rational \\
& encode_2 = foldr\ estep_2\ 0 \\
& estep_2 :: Symbol \rightarrow Rational \rightarrow Rational \\
& estep_2\ s\ x = weight\ (encodeSym\ s)\ x
\end{aligned}
$$

Now encoding is a simple *foldr*, which means that it is easier to manipulate.

Inverting Encoding. Let us turn now to decoding, and specifically the question of whether it faithfully decodes the encoded text. We use the Unfoldr–Foldr Theorem. Of course, we have to accept junk, because our decoder runs indefinitely. We check the inductive condition:

$$
\begin{aligned}
& dstep_1\ (estep_2\ s\ x) \\
= \quad & \{ \ estep_2\text{; let } x' = weight\ (encodeSym\ s)\ x \ \} \\
& dstep_1\ x' \\
= \quad & \{ \ dstep_1\text{; let } s' = decodeSym\ x' \ \} \\
& Just\ (s', scale\ (encodeSym\ s')\ x')
\end{aligned}
$$

Now, we hope to recover the first symbol; that is, we require $s' = s$:

$$
\begin{aligned}
& s' = s \\
\Leftrightarrow \quad & \{ \ s' = decodeSym\ x'\text{; central property } \} \\
& x' \in encodeSym\ s
\end{aligned}
$$

$$\begin{array}{l} \Leftrightarrow \quad \{\ \text{definition of } x' \ \} \\ \quad weight\,(encodeSym\ s)\ x \in encodeSym\ s \\ \Leftrightarrow \quad \{\ \text{property of } weight \ \} \\ \quad x \in unit \end{array}$$

Fortunately, it is an invariant of the computation that the state x is in the unit interval, as is easy to check; so indeed $s' = s$. Continuing:

$$\begin{array}{l} \quad dstep_1\,(estep_2\ s\ x) \\ = \quad \{\ \text{above}\ \} \\ \quad Just\,(s, scale\,(encodeSym\ s)\,(weight\,(encodeSym\ s)\ x)) \\ = \quad \{\ scale\ i \cdot weight\ i = id\ \} \\ \quad Just\,(s, x) \end{array}$$

as required. Therefore, by the Unfoldr–Foldr Theorem with Junk and Invariant, decoding inverts encoding, up to junk: for all finite xs,

$$\exists ys\ .\ decode_1\,(encode_2\ xs) = xs + ys$$

But we can discard the junk, by pruning to the desired length:

$$take\,(length\ xs)\,(decode_1\,(encode_2\ xs)) = xs$$

for all finite xs. Alternatively, we can use an 'end of text' marker ω that is distinct from all proper symbols:

$$takeWhile\,(\neq \omega)\,(decode_1\,(encode_2\,(xs + [\omega]))) = xs \quad \Leftarrow \quad \omega \notin xs$$

for all finite xs. Either way, arithmetic decoding does indeed faithfully invert arithmetic coding.

5 From Fractions to Integers

We now make the key step from AC to ANS. Whereas AC encodes longer and longer messages as more and more precise fractions, ANS encodes them as larger and larger integers. Given the symbol table *counts* as before, we can easily derive definitions of the following functions (for example, by tabulating *cumul* using a scan)—we again omit the definitions for brevity:

$$\begin{array}{l} count :: Symbol \rightarrow Integer \\ cumul :: Symbol \rightarrow Integer \\ find \quad :: Integer \rightarrow Symbol \end{array}$$

such that $count\ s$ gives the count of symbol s, $cumul\ s$ gives the cumulative counts of all symbols preceding s in the symbol table, and $find\ x$ looks up an integer $0 \leqslant x < t$:

$$find\ x = s \quad \Leftrightarrow \quad cumul\ s \leqslant x < cumul\ s + count\ s$$

(where $t = sum\,(map\ snd\ counts)$ is the total count).

The Integer Encoding Step. We encode a text as an integer x, containing $\log_2 x$ bits of information. The next symbol s to encode has probability $p = \mathit{count}\ s\ /\ t$, and so requires an additional $\log_2 ({}^1\!/_p)$ bits; in total, that makes $\log_2 x + \log_2 ({}^1\!/_p) = \log_2 ({}^x\!/_p) = \log_2 (x \times t\ /\ \mathit{count}\ s)$ bits. So entropy considerations tell us that, roughly speaking, to incorporate symbol s into state x we want to map x to $x' \simeq x \times t\ /\ \mathit{count}\ s$. Of course, in order to decode, we need to be able to invert this transformation, to extract s and x from x'; this suggests that we should do the division by *count s* first:

$$x' = (x \operatorname{div} \mathit{count}\ s) \times t \quad \text{-- not final}$$

so that the multiplication by the known value t can be undone first:

$$x \operatorname{div} \mathit{count}\ s = x' \operatorname{div} t$$

(we will refine this definition shortly). How do we reconstruct s? Well, there is enough headroom in x' to add any value u with $0 \leqslant u < t$ without affecting the division; in particular, we can add *cumul s* to x', and then we can use *find* on the remainder:

$$x' = (x \operatorname{div} \mathit{count}\ s) \times t + \mathit{cumul}\ s \quad \text{-- still not final}$$

so that

$$\begin{aligned} x \operatorname{div} \mathit{count}\ s &= x' \operatorname{div} t \\ \mathit{cumul}\ s &= x' \operatorname{mod} t \\ s &= \mathit{find}\,(\mathit{cumul}\ s) = \mathit{find}\,(x' \operatorname{mod} t) \end{aligned}$$

(this version still needs to be refined further). We are still missing some information from the lower end of x through the division, namely $x \operatorname{mod} \mathit{count}\ s$; so we can't yet reconstruct x. Happily,

$$\mathit{find}\,(\mathit{cumul}\ s) = \mathit{find}\,(\mathit{cumul}\ s + r)$$

for any r with $0 \leqslant r < \mathit{count}\ s$; of course, $x \operatorname{mod} \mathit{count}\ s$ is in this range, so there is still precisely enough headroom in x' to add this lost information too, without affecting the *find*, allowing us also to reconstruct x:

$$x' = (x \operatorname{div} \mathit{count}\ s) \times t + \mathit{cumul}\ s + x \operatorname{mod} \mathit{count}\ s \quad \text{-- final}$$

so that

$$\begin{aligned} x \operatorname{div} \mathit{count}\ s &= x' \operatorname{div} t \\ s &= \mathit{find}\,(\mathit{cumul}\ s + x \operatorname{mod} \mathit{count}\ s) \\ &= \mathit{find}\,(x' \operatorname{mod} t) \\ x &= \mathit{count}\ s \times (x \operatorname{div} \mathit{count}\ s) + x \operatorname{mod} \mathit{count}\ s \\ &= \mathit{count}\ s \times (x' \operatorname{div} t) + x' \operatorname{mod} t - \mathit{cumul}\ s \end{aligned}$$

This is finally the transformation we will use for encoding one more symbol.

Integer ANS. We define

$$
\begin{aligned}
&\mathit{encode}_3 :: [\mathit{Symbol}] \to \mathit{Integer} \\
&\mathit{encode}_3 = \mathit{foldr}\ \mathit{estep}_3\ 0
\end{aligned}
$$

$$
\begin{aligned}
&\mathit{estep}_3 :: \mathit{Symbol} \to \mathit{Integer} \to \mathit{Integer} \\
&\mathit{estep}_3\ s\ x = \mathbf{let}\ (q, r) = x\ \mathrm{divMod}\ \mathit{count}\ s\ \mathbf{in}\ q \times t + \mathit{cumul}\ s + r
\end{aligned}
$$

$$
\begin{aligned}
&\mathit{decode}_3 :: \mathit{Integer} \to [\mathit{Symbol}] \\
&\mathit{decode}_3 = \mathit{unfoldr}\ \mathit{dstep}_3
\end{aligned}
$$

$$
\begin{aligned}
&\mathit{dstep}_3 :: \mathit{Integer} \to \mathit{Maybe}\ (\mathit{Symbol}, \mathit{Integer}) \\
&\mathit{dstep}_3\ x = \mathbf{let}\ (q, r) = x\ \mathrm{divMod}\ t \\
&\qquad\qquad\qquad s = \mathit{find}\ r \\
&\qquad\qquad \mathbf{in}\ \mathit{Just}\ (s, \mathit{count}\ s \times q + r - \mathit{cumul}\ s)
\end{aligned}
$$

Correctness of Integer ANS. Using similar reasoning as for AC, we can show that a decoding step inverts an encoding step:

$$
\begin{aligned}
&\quad \mathit{dstep}_3\ (\mathit{estep}_3\ s\ x) \\
=&\quad \{\ \mathit{estep}_3;\ \text{let}\ (q, r) = x\ \mathrm{divMod}\ \mathit{count}\ s,\ x' = q \times t + \mathit{cumul}\ s + r\ \} \\
&\quad \mathit{dstep}_3\ x' \\
=&\quad \{\ \mathit{dstep}_3;\ \text{let}\ (q', r') = x'\ \mathrm{divMod}\ t,\ s' = \mathit{find}\ r'\ \} \\
&\quad \mathit{Just}\ (s', \mathit{count}\ s' \times q' + r' - \mathit{cumul}\ s') \\
=&\quad \{\ r' = \mathit{cumul}\ s + r, 0 \leqslant r < \mathit{count}\ s,\ \text{so}\ s' = \mathit{find}\ r' = s\ \} \\
&\quad \mathit{Just}\ (s, \mathit{count}\ s \times q' + r' - \mathit{cumul}\ s) \\
=&\quad \{\ r' - \mathit{cumul}\ s = r, q' = x'\ \mathrm{div}\ t = q\ \} \\
&\quad \mathit{Just}\ (s, \mathit{count}\ s \times q + r) \\
=&\quad \{\ (q, r) = x\ \mathrm{divMod}\ \mathit{count}\ s\ \} \\
&\quad \mathit{Just}\ (s, x)
\end{aligned}
$$

Therefore decoding inverts encoding, modulo pruning, by the Unfoldr–Foldr Theorem with Junk:

$$\mathit{take}\ (\mathit{length}\ \mathit{xs})\ (\mathit{decode}_3\ (\mathit{encode}_3\ \mathit{xs})) = \mathit{xs}$$

for all finite xs. For example, with the same alphabet and symbol counts as before, encoding the text `"abc"` proceeds (now from right to left, because of the *foldr* in encode_3) as follows:

$$
\begin{aligned}
&\mathit{estep}_3\ \texttt{'c'}\ 0\ \ = 5 \\
&\mathit{estep}_3\ \texttt{'b'}\ 5\ \ = 14 \\
&\mathit{estep}_3\ \texttt{'a'}\ 14 = 70
\end{aligned}
$$

and the result is 70. Decoding inverts this:

$$
\begin{aligned}
&\mathit{dstep}_3\ 70 = \mathit{Just}\ (\texttt{'a'}, 14) \\
&\mathit{dstep}_3\ 14 = \mathit{Just}\ (\texttt{'b'}, 5) \\
&\mathit{dstep}_3\ 5\ \ = \mathit{Just}\ (\texttt{'c'}, 0) \\
&\mathit{dstep}_3\ 0\ \ = \mathit{Just}\ (\texttt{'a'}, 0) \\
&\ldots
\end{aligned}
$$

Huffman as an Instance of ANS. Incidentally, we can see here that ANS is in fact a generalisation of HC. If the symbol counts and their sum are all powers of two, then the arithmetic in $estep_3$ amounts to simple manipulation of bit vectors by shifting and insertion. For example, with an alphabet of four symbols `'a'`, `'b'`, `'c'`, `'d'` with counts $4, 2, 1, 1$, encoding operates on a state x with binary expansion $\cdots x_3\, x_2\, x_1\, x_0$ (written most significant bit first) as follows:

$$\begin{aligned}
estep_3\ \texttt{'a'}\ (\cdots x_3\, x_2\, x_1\, x_0) &= \cdots x_3\, x_2\, 0\, x_1\, x_0 \\
estep_3\ \texttt{'b'}\ (\cdots x_3\, x_2\, x_1\, x_0) &= \cdots x_3\, x_2\, x_1\, 1\, 0\, x_0 \\
estep_3\ \texttt{'c'}\ (\cdots x_3\, x_2\, x_1\, x_0) &= \cdots x_3\, x_2\, x_1\, x_0\, 1\, 1\, 0 \\
estep_3\ \texttt{'d'}\ (\cdots x_3\, x_2\, x_1\, x_0) &= \cdots x_3\, x_2\, x_1\, x_0\, 1\, 1\, 1
\end{aligned}$$

That is, the symbol codes $0, 10, 110, 111$ are inserted into rather than appended onto the state so far; the binary expansion of the ANS encoding of a text yields some permutation of the HC encoding of that text.

A Different Starting Point. As it happens, the inversion property of $encode_3$ and $decode_3$ holds, whatever value we use to start encoding with (since this value is not used in the proof); in Sect. 6, we start encoding with a certain lower bound l rather than 0. Moreover, $estep_3$ is strictly increasing on states strictly greater than zero, and $dstep_3$ strictly decreasing; which means that the decoding process can stop when it returns to the lower bound. That is, if we pick some $l > 0$ and define

$$\begin{aligned}
&encode_4 :: [Symbol] \rightarrow Integer \\
&encode_4 = foldr\ estep_3\ l \\
&decode_4 :: Integer \rightarrow [Symbol] \\
&decode_4 = unfoldr\ dstep_4 \\
&dstep_4 :: Integer \rightarrow Maybe\,(Symbol, Integer) \\
&dstep_4\ x = \mathbf{if}\ x == l\ \mathbf{then}\ Nothing\ \mathbf{else}\ dstep_3\ x
\end{aligned}$$

then the stronger version of the Unfoldr–Foldr Theorem (without junk) holds, and we have

$$decode_4\,(encode_4\ xs) = xs$$

for all finite xs.

6 Bounded Precision

The previous versions all used arbitrary-precision arithmetic, which is expensive. We now change the approach slightly to use only bounded-precision arithmetic. As usual, there is a trade-off between effectiveness (a bigger bound on the numbers involved means more accurate approximations to ideal entropies) and efficiency (a smaller bound generally means faster operations). Fortunately, the reasoning does not depend much on the actual bounds. We will pick a base b

and a lower bound l, and represent the integer accumulator x as a pair (w, ys) which we call a *Number*:

$$\mathbf{type}\ \mathit{Number} = (\mathit{Int}, [\mathit{Int}])$$

such that ys is a list of digits in base b, and w is an integer in the range $l \leqslant w < u$ (where we define $u = l \times b$ for the upper bound), under the abstraction relation $x = \mathit{abstract}\,(w, ys)$ induced by

$$\mathit{abstract}\,(w, ys) = \mathit{foldl}\ \mathit{inject}\ w\ ys$$

where

$$\mathit{inject}\ w\ y = w \times b + y$$

We call w the 'window' and ys the 'remainder'. For example, with $b = 10$ and $l = 100$, the pair $(123, [4, 5, 6])$ represents the value 123456.

Properties of the Window. Specifying a range of the form $l \leqslant w < l \times b$ induces nice properties. If we introduce an operation

$$\mathit{extract}\ w = w\ \mathrm{divMod}\ b$$

as an inverse to *inject*, then we have

$$\begin{array}{lll} \mathit{inject}\ w\ y < u & \Leftrightarrow & w < l \\ l \leqslant \mathit{fst}\,(\mathit{extract}\ w) & \Leftrightarrow & u \leqslant w \end{array}$$

(we omit the straightforward proofs, using the universal property

$$u < v \times w \quad \Leftrightarrow \quad u\ \mathrm{div}\ w < v$$

of integer division). That is, given an in-range window value w, injecting another digit will take it outside (above) the range; but if w is initially below the range, injecting another digit will keep it below the upper bound. So starting below the range and repeatedly injecting digits will eventually land within the range (it cannot hop right over), and injecting another digit would take it outside the range again. Conversely, given an in-range window value w, extracting a digit will take it outside (below) the range; but if w is initially above the range, extracting a digit will keep it at least the lower bound. So starting above the range and repeatedly extracting digits will also eventually land within the range (it cannot hop right over), and extracting another digit would take it outside the range again. This is illustrated in Fig. 4. In particular, for any $x \geqslant l$ there is a unique representation of x under *abstract* that has an in-range window.

For fast execution, b should be a power of two, so that multiplication and division by b can be performed by bit shifts; and arithmetic on values up to u should fit within a single machine word. It is also beneficial for t to divide evenly into l, as we shall see shortly.

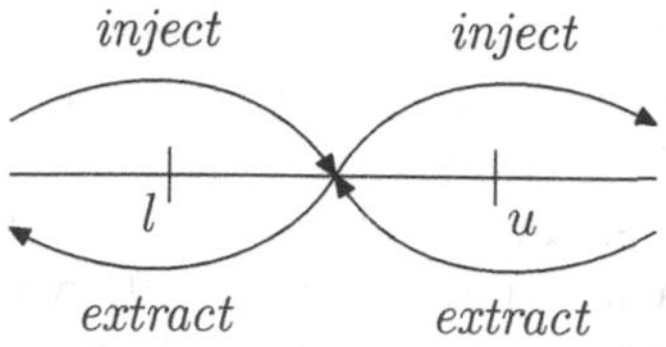

Fig. 4. 'Can't miss it' properties of the range: injecting an extra digit can only land within the range $[l, u)$ when starting below it, and will land above the range when starting within it; and conversely, extracting a digit can only land within the range when starting above it, and will land below the range when starting within it.

Encoding with Bounded Arithmetic. The encoding step acts on the window in the accumulator using estep_3, which risks making it overflow the range; we therefore renormalize with enorm_5 by shifting digits from the window to the remainder until this overflow would no longer happen, before consuming the symbol.

$$\begin{array}{l}
\mathit{econsume}_5 :: [\mathit{Symbol}] \rightarrow \mathit{Number} \\
\mathit{econsume}_5 = \mathit{foldr}\ \mathit{estep}_5\ (l, [\,])
\end{array}$$

$$\begin{array}{l}
\mathit{estep}_5 :: \mathit{Symbol} \rightarrow \mathit{Number} \rightarrow \mathit{Number} \\
\mathit{estep}_5\ s\ (w, ys) = \mathbf{let}\ (w', ys') = \mathit{enorm}_5\ s\ (w, ys)\ \mathbf{in}\ (\mathit{estep}_3\ s\ w', ys')
\end{array}$$

$$\begin{array}{l}
\mathit{enorm}_5 :: \mathit{Symbol} \rightarrow \mathit{Number} \rightarrow \mathit{Number} \\
\mathit{enorm}_5\ s\ (w, ys) = \mathbf{if}\quad \mathit{estep}_3\ s\ w < u \\
\qquad\qquad\qquad\quad \mathbf{then}\ (w, ys) \\
\qquad\qquad\qquad\quad \mathbf{else}\ \ \mathbf{let}\ (q, r) = \mathit{extract}\ w\ \mathbf{in}\ \mathit{enorm}_5\ s\ (q, r : ys)
\end{array}$$

That is, enorm_5 preserves the abstract value of a *Number*:

$$\mathit{abstract} \cdot \mathit{enorm}_5 = \mathit{abstract}$$

and leaves the window safe for estep_3 to incorporate the next symbol.

Note that if t divides l, then we can rearrange the guard in enorm_5:

$$\begin{array}{ll}
& \mathit{estep}_3\ s\ w < u \\
\Leftrightarrow & \{\ \mathit{estep}_3;\ \text{let}\ (q, r) = w\ \text{divMod}\ \mathit{count}\ s\ \} \\
& q \times t + \mathit{cumul}\ s + r < u \\
\Leftrightarrow & \{\ t\ \text{divides}\ l,\ \text{so}\ u = (u\ \text{div}\ t) \times t\ \} \\
& q \times t + \mathit{cumul}\ s + r < (u\ \text{div}\ t) \times t \\
\Leftrightarrow & \{\ \text{universal property of division: } u < v \times w \Leftrightarrow u\ \text{div}\ w < v\ \} \\
& (q \times t + \mathit{cumul}\ s + r)\ \text{div}\ t < u\ \text{div}\ t \\
\Leftrightarrow & \{\ 0 \leqslant r < \mathit{count}\ s,\ \text{so}\ 0 \leqslant \mathit{cumul}\ s + r < t\ \} \\
& q < u\ \text{div}\ t \\
\Leftrightarrow & \{\ q = w\ \text{div}\ \mathit{count}\ s\ \} \\
& w\ \text{div}\ \mathit{count}\ s < u\ \text{div}\ t \\
\Leftrightarrow & \{\ \text{universal property of } \mathit{div} \text{ again}\ \} \\
& w < (u\ \text{div}\ t) \times \mathit{count}\ s \\
\Leftrightarrow & \{\ u = l \times b,\ t\ \text{divides}\ l\ \} \\
& w < b \times (l\ \text{div}\ t) \times \mathit{count}\ s
\end{array}$$

This is worthwhile because $b \times (l \operatorname{div} t)$ is a constant, independent of s, so the comparison can be done with a single multiplication, whereas the definition of $estep_3$ involves a division by $count\ s$.

For example, consider again encoding the text `"abc"`, with $b = 10$ and $l = 100$. The process is again from right to left, with the accumulator starting at $(100, [\,])$. Consuming the `'c'` then the `'b'` proceeds as before, because the window does not overflow u:

$$\begin{aligned}&estep_5\ \texttt{'c'}\ (100,[\,]) = (205,[\,])\\&estep_5\ \texttt{'b'}\ (205,[\,]) = (683,[\,])\end{aligned}$$

Now directly consuming the `'a'` would make the window overflow, because $estep_3\ \texttt{'a'}\ 683 = 3411 \geqslant u$; so we must renormalize to $(68, [3])$ before consuming the `'a'`, leading to the final state $(340, [3])$:

$$\begin{aligned}&enorm_5\ \texttt{'a'}\ (683,[\,]) = (68,[3])\\&estep_5\ \texttt{'a'}\ (683,[\,]) \;= (estep_3\ \texttt{'a'}\ 68,[3]) = (340,[3])\end{aligned}$$

Note that the move from arbitrary to fixed precision is not just a data refinement—it is not the case that $econsume_5\ xs$ computes some representation of $encode_4\ xs$. For example, $encode_4$ `"abc"` $= 3411$, whereas $econsume_5$ `"abc"` $= (340, [3])$, which is not a representation of 3411. We have really sacrificed some effectiveness in encoding in return for the increased efficiency of fixed precision arithmetic.

Decoding with Bounded Arithmetic. Decoding is an unfold using the accumulator as state. We repeatedly output a symbol from the window; this may make the window underflow the range, in which case we renormalize if possible by injecting digits from the remainder (and if this is not possible, because there are no more digits to inject, it means that we have decoded the entire text).

$$\begin{aligned}&dproduce_5 :: Number \to [Symbol]\\&dproduce_5 = unfoldr\ dstep_5\end{aligned}$$

$$\begin{aligned}&dstep_5 :: Number \to Maybe\,(Symbol, Number)\\&dstep_5\,(w, ys) = \ \mathbf{let}\ Just\,(s, w') \;= dstep_3\ w\\&\qquad\qquad\qquad\qquad\ (w'', ys'') \;= dnorm_5\,(w', ys)\\&\qquad\qquad\qquad\ \mathbf{in\ if}\ w'' \geqslant l\ \mathbf{then}\ Just\,(s,(w'', ys''))\ \mathbf{else}\ Nothing\end{aligned}$$

$$\begin{aligned}&dnorm_5 :: Number \to Number\\&dnorm_5\,(w, y : ys) = \mathbf{if}\ w < l\ \mathbf{then}\ dnorm_5\,(inject\ w\ y, ys)\ \mathbf{else}\ (w, y : ys)\\&dnorm_5\,(w,[\,]) \quad\;\; = (w,[\,])\end{aligned}$$

Note that decoding is of course symmetric to encoding; in particular, when encoding we renormalize before consuming a symbol; therefore when decoding we renormalize *after* emitting a symbol. For example, decoding the final encoding $(340, [3])$ starts by computing $dstep_3\ 340 = Just\,(\texttt{'a'}, 68)$; the window value 68 has underflowed, so renormalization consumes the remaining digit 3, restoring the accumulator to $(683, [\,])$; then decoding proceeds to extract the `'b'` and `'c'` in turn, returning the accumulator to $(100, [\,])$ via precisely the same states as for encoding, only in reverse order.

$$\begin{array}{lll} dstep_5\,(340,[3]) & = & Just\,(\texttt{'a'},(683,[\,])) \\ dstep_5\,(683,[\,]) & = & Just\,(\texttt{'b'},(205,[\,])) \\ dstep_5\,(205,[\,]) & = & Just\,(\texttt{'c'},(100,[\,])) \\ dstep_5\,(100,[\,]) & = & Nothing \end{array}$$

Correctness of Decoding. We can prove that decoding inverts encoding, again using the Unfoldr–Foldr Theorem with Invariant. Here, the invariant p is that the window w is in range $(l \leqslant w < u)$, which is indeed maintained by the consumer $estep_5$ and producer $dstep_5$. As for the conditions of the theorem: in the base case, $dstep_3\,l = Just\,(s, w')$ with $w' < l$, and $dnorm_5\,(w', [\,]) = (w', [\,])$, so indeed

$$dstep_5\,(l, [\,]) = Nothing$$

For the inductive step, suppose that $l \leqslant w < u$; then we have:

$$\begin{array}{ll} & dstep_5\,(estep_5\,s\,(w, ys)) \\ = & \{\ estep_5;\ \text{let } (w', ys') = enorm_5\,s\,(w, ys)\ \} \\ & dstep_5\,(estep_3\,s\,w', ys') \\ = & \{\ dstep_5, dstep_3;\ \text{let } (w'', ys'') = dnorm_5\,(w', ys')\ \} \\ & \mathbf{if}\ w'' \geqslant l\ \mathbf{then}\ Just\,(s, (w'', ys''))\ \mathbf{else}\ Nothing \\ = & \{\ \text{see below: } dnorm_5 \text{ inverts } enorm_5\,s, \text{ so } (w'', ys'') = (w, ys)\ \} \\ & \mathbf{if}\ w \geqslant l\ \mathbf{then}\ Just\,(s, (w, ys))\ \mathbf{else}\ Nothing \\ = & \{\ \text{invariant holds, so in particular } w \geqslant l\ \} \\ & Just\,(s, (w, ys)) \end{array}$$

The remaining proof obligation is to show that

$$dnorm_5\,(enorm_5\,s\,(w, ys)) = (w, ys)$$

when $l \leqslant w < u$. We prove this in several steps. First, note that $dnorm_5$ is idempotent:

$$dnorm_5 \cdot dnorm_5 = dnorm_5$$

Second, when $l \leqslant w$ holds,

$$dnorm_5\,(w, ys) = (w, ys)$$

Finally, the key lemma is that, for $w < u$ (but not necessarily $w \geqslant l$), $dnorm_5$ is invariant under $enorm_5$:

$$dnorm_5\,(enorm_5\,s\,(w, ys)) = dnorm_5\,(w, ys)$$

When additionally $w \geqslant l$, the second property allows us to conclude that $dnorm_5$ inverts $enorm_5$:

$$dnorm_5\,(enorm_5\,s\,(w, ys)) = (w, ys)$$

The 'key lemma' is proved by induction on w. For $w = 0$, we clearly have

$$\begin{array}{ll} & \mathit{dnorm}_5\,(\mathit{enorm}_5\,s\,(w, \mathit{ys})) \\ = & \{\ \mathit{estep}_3\,s\,0 = \mathit{cumul}\,s \leqslant t \leqslant l, \text{ so } \mathit{enorm}_5\,s\,(w, \mathit{ys}) = (w, \mathit{ys})\ \} \\ & \mathit{dnorm}_5\,(w, \mathit{ys}) \end{array}$$

For the inductive step, we suppose that the result holds for all $q{<}w$, and consider two cases for w itself. In case $\mathit{estep}_3\,s\,w < u$, we have:

$$\begin{array}{ll} & \mathit{dnorm}_5\,(\mathit{enorm}_5\,s\,(w, \mathit{ys})) \\ = & \{\ \text{assumption; } \mathit{enorm}_5\ \} \\ & \mathit{dnorm}_5\,(w, \mathit{ys}) \end{array}$$

as required. And in case $\mathit{estep}_3\,s\,w \geqslant u$, we have:

$$\begin{array}{ll} & \mathit{dnorm}_5\,(\mathit{enorm}_5\,s\,(w, \mathit{ys})) \\ = & \{\ \text{assumption; } \mathit{enorm}_5\text{; let } (q, r) = \mathit{extract}\,w\ \} \\ & \mathit{dnorm}_5\,(\mathit{enorm}_5\,s\,(q, r : \mathit{ys})) \\ = & \{\ q < w\text{; induction}\ \} \\ & \mathit{dnorm}_5\,(q, r : \mathit{ys}) \\ = & \{\ w < u, \text{ so } q = w \operatorname{div} b < l;\ \mathit{dnorm}_5\ \} \\ & \mathit{dnorm}_5\,(\mathit{inject}\,q\,r, \mathit{ys}) \\ = & \{\ q, r\ \} \\ & \mathit{dnorm}_5\,(w, \mathit{ys}) \end{array}$$

Note that we made essential use of the limits of the range: $w < u \Rightarrow w \operatorname{div} b < l$.

Therefore decoding inverts encoding:

$$\mathit{dproduce}_5\,(\mathit{econsume}_5\,\mathit{xs}) = \mathit{xs}$$

for all finite xs.

7 Streaming

The version of encoding in the previous section yields a *Number*, that is, a pair consisting of an integer window and a digit-sequence remainder. It would be more conventional for encoding to take a sequence of symbols to a sequence of digits alone, and decoding to take the sequence of digits back to a sequence of symbols. For encoding, we have to flush the remaining digits out of the window at the end of the process, reducing the window to zero:

$$\begin{array}{l} \mathit{eflush}_5 :: \mathit{Number} \rightarrow [\mathit{Int}] \\ \mathit{eflush}_5\,(0, \mathit{ys}) \ = \mathit{ys} \\ \mathit{eflush}_5\,(w, \mathit{ys}) = \textbf{let}\,(w', y) = \mathit{extract}\,w\,\textbf{in}\,\mathit{eflush}_5\,(w', y : \mathit{ys}) \end{array}$$

For example, $\mathit{eflush}_5\,(340,[3]) = [3,4,0,3]$. Then we can define

$$\begin{aligned}&\mathit{encode}_5 :: [\mathit{Symbol}] \to [\mathit{Int}]\\&\mathit{encode}_5 = \mathit{eflush}_5 \cdot \mathit{econsume}_5\end{aligned}$$

Correspondingly, decoding should start by populating an initially-zero window from the sequence of digits:

$$\begin{aligned}&\mathit{dstart}_5 :: [\mathit{Int}] \to \mathit{Number}\\&\mathit{dstart}_5\; ys = \mathit{dnorm}_5\,(0, ys)\end{aligned}$$

For example, $\mathit{dstart}_5\,[3,4,0,3] = (340,[3])$. Then we can define

$$\begin{aligned}&\mathit{decode}_5 :: [\mathit{Int}] \to [\mathit{Symbol}]\\&\mathit{decode}_5 = \mathit{dproduce}_5 \cdot \mathit{dstart}_5\end{aligned}$$

One can show that dstart_5 inverts eflush_5 on in-range values:

$$\mathit{dstart}_5\,(\mathit{eflush}_5\,(w, ys)) = (w, ys) \quad \Leftarrow \quad l \leqslant w < u$$

and therefore again decoding inverts encoding:

$$\begin{aligned}&\quad \mathit{decode}_5\,(\mathit{encode}_5\; xs)\\&= \quad \{\; \mathit{decode}_5, \mathit{encode}_5 \;\}\\&\quad \mathit{dproduce}_5\,(\mathit{dstart}_5\,(\mathit{eflush}_5\,(\mathit{econsume}_5\; xs)))\\&= \quad \{\; \mathit{econsume}_5 \text{ yields in-range, on which } \mathit{dstart}_5 \text{ inverts } \mathit{eflush}_5 \;\}\\&\quad \mathit{dproduce}_5\,(\mathit{econsume}_5\; xs)\\&= \quad \{\; \mathit{dproduce}_5 \text{ inverts } \mathit{econsume}_5 \;\}\\&\quad xs\end{aligned}$$

for all finite xs.

Introducing Streaming. We would now like to *stream* the encoding and decoding processes, so that each can start generating output before having consumed all its input. With some effort, it is possible to persuade the definitions of encode_5 and decode_5 into *metamorphism* form [14]; however, that turns out to be rather complicated. Here, we take a more direct route instead.

For encoding, we have

$$\mathit{encode}_5 = \mathit{eflush}_5 \cdot \mathit{foldr}\; \mathit{estep}_5\; (l, [\,])$$

A first step for streaming is to make as much of this as possible tail-recursive. The best we can do is to apply the Third Duality Theorem to transform the *foldr* into a *foldl*:

$$\mathit{encode}_5 = \mathit{eflush}_5 \cdot \mathit{foldl}\,(\mathit{flip}\; \mathit{estep}_5)\,(l, [\,]) \cdot \mathit{reverse}$$

Now we note that the remainder component of the *Number* behaves like a queue, in the sense that already-enqueued digits simply pass through without being further examined:

$$
\begin{array}{ll}
\mathit{eflush}_5\,(w, ys \mathbin{+\!\!+} zs) & = \mathit{eflush}_5\,(w, ys) \mathbin{+\!\!+} zs \\
\mathit{enorm}_5\,s\,(w, ys \mathbin{+\!\!+} zs) & = \mathbf{let}\,(w', ys') = \mathit{enorm}_5\,s\,(w, ys)\,\mathbf{in}\,(w', ys' \mathbin{+\!\!+} zs) \\
\mathit{estep}_5\,s\,(w, ys \mathbin{+\!\!+} zs) & = \mathbf{let}\,(w', ys') = \mathit{estep}_5\,s\,(w, ys)\;\;\mathbf{in}\,(w', ys' \mathbin{+\!\!+} zs)
\end{array}
$$

If we then introduce the auxilliary functions e_1, e_2 specified by

$$
\begin{array}{ll}
\mathit{reverse}\,(e_1\,w\,ss) \mathbin{+\!\!+} ys & = \mathit{eflush}_5\,(\mathit{foldl}\,(\mathit{flip}\;\mathit{estep}_5)\,(w, ys)\,ss) \\
\mathit{reverse}\,(e_2\,w) \mathbin{+\!\!+} ys & = \mathit{eflush}_5\,(w, ys)
\end{array}
$$

and unfold definitions, exploiting the queueing properties, we can synthesize $\mathit{encode}_5 = \mathit{encode}_6$, where:

$$
\begin{array}{l}
\mathit{encode}_6 :: [\mathit{Symbol}] \rightarrow [\mathit{Int}] \\
\mathit{encode}_6 = \mathit{reverse} \cdot e_1\,l \cdot \mathit{reverse}\;\mathbf{where} \\
\quad e_1\,w\,(s : ss) = \mathbf{let}\,(q, r) = w\,\mathrm{divMod}\,\mathit{count}\,s \\
\qquad\qquad\qquad\qquad\quad\; w' \quad\;\, = q \times t + \mathit{cumul}\,s + r\,\mathbf{in} \\
\qquad\qquad\qquad\quad\;\; \mathbf{if}\;w' < u\;\mathbf{then}\;e_1\,w'\,ss \\
\qquad\qquad\qquad\qquad\qquad\quad\; \mathbf{else}\;\;\mathbf{let}\,(q', r') = w\,\mathrm{divMod}\,b\,\mathbf{in}\,r' : e_1\,q'\,(s : ss) \\
\quad e_1\,w\,[\,] = e_2\,w \\
\quad e_2\,w = \mathbf{if}\;w == 0\;\mathbf{then}\,[\,]\,\mathbf{else}\,\mathbf{let}\,(w', y) = w\,\mathrm{divMod}\,b\,\mathbf{in}\,y : e_2\,w'
\end{array}
$$

In this version, the accumulator w simply maintains the window, and digits in the remainder are output as soon as they are generated. Note that the two *reverses* mean that encoding effectively reads its input and writes its output from right to left; that seems to be inherent to ANS.

Streaming Decoding. Decoding is easier, because dnorm_5 is already tail-recursive. Similarly specifying functions d_0, d_1, d_2 by

$$
\begin{array}{ll}
d_0\,w\,ys & = \mathit{dproduce}_5\,(\mathit{dnorm}_5\,(w, ys)) \\
d_1\,w\,ys & = \mathit{dproduce}_5\,(w, ys) \\
d_2\,s\,w'\,ys & = \mathbf{let}\,(w'', ys'') = \mathit{dnorm}_5\,(w', ys)\,\mathbf{in} \\
 & \quad\;\mathbf{if}\;w'' \geqslant l\;\mathbf{then}\;s : d_1\,w''\,ys''\;\mathbf{else}\,[\,]
\end{array}
$$

and unfolding definitions allows us to synthesize directly that $\mathit{decode}_5 = \mathit{decode}_6$, where:

$$
\begin{array}{lll}
\multicolumn{3}{l}{\mathit{decode}_6 :: [\mathit{Int}] \rightarrow [\mathit{Symbol}]} \\
\multicolumn{3}{l}{\mathit{decode}_6 = d_0\,0\;\mathbf{where}} \\
\quad d_0\,w\,(y : ys) & \mid w < l & = d_0\,(w \times b + y)\,ys \\
\quad d_0\,w\,ys & & = d_1\,w\,ys \\
\quad d_1\,w\,ys & & = \mathbf{let}\,(q, r) = w\,\mathrm{divMod}\,t \\
 & & \qquad\quad\; s \quad\;\; = \mathit{find}\,r \\
 & & \qquad\quad\; w' \quad\; = \mathit{count}\,s \times q + r - \mathit{cumul}\,s \\
 & & \quad\;\mathbf{in}\;d_2\,s\,w'\,ys \\
\quad d_2\,s\,w\,(y : ys) & \mid w < l & = d_2\,s\,(w \times b + y)\,ys \\
\quad d_2\,s\,w\,ys & \mid w \geqslant l & = s : d_1\,w\,ys \\
\quad d_2\,s\,w\,[\,] & & = [\,]
\end{array}
$$

Ignoring additions and subtractions, encoding involves one division by *count s* and one multiplication by t for each input symbol s, plus one division by b for each output digit. Conversely, decoding involves one multiplication by b for each input digit, plus one division by t and one multiplication by *count s* for each output symbol s. Encoding and decoding are both tail-recursive. The arithmetic in base b can be simplified to bit shifts by choosing b to be a power of two. They therefore correspond rather directly to simple imperative implementations [17].

8 Conclusion

We have presented a development using the techniques of constructive functional programming of the encoding and decoding algorithms involved in asymmetric numeral systems, including the step from arbitrary- to fixed-precision arithmetic and then to streaming processes. The calculational techniques depend on the theory of folds and unfolds for lists, especially the duality between *foldr* and *foldl*, fusion, and the Unfoldr–Foldr Theorem. We started out with the hypothesis that the theory of streaming developed by the author together with Richard Bird for arithmetic coding [2,14] would be a helpful tool; but although it can be applied, it seems here to be more trouble than it is worth.

To be precise, what we have described is the *range* variant (rANS) of ANS. There is also a *tabled* variant (tANS), used by Zstd [8] and LZFSE [9], which essentially tabulates the functions $estep_5$ and $dstep_5$; for encoding this involves a table of size $n \times (u - l)$, the product of the alphabet size and the window width, and for decoding two tables of size $u - l$. Tabulation makes even more explicit that HC is a special case of ANS, with the precomputed table corresponding to the Huffman trie. Tabulation also allows more flexibility in the precise allocation of codes, which slightly improves effectiveness [13]. For example, the coding table in Fig. 3 corresponds to the particular arrangement `"aabbbccc"` of the three symbols in proportion to their counts, and lends itself to implementation via arithmetic; but any permutation of this arrangement would still work, and a permutation such as `"cbcacbcacb"` which distributes the symbols more evenly turns out to be slightly more effective and no more difficult to tabulate.

One nice feature of AC is that the switch from arbitrary-precision to fixed-precision arithmetic can be expressed in terms of a carefully chosen adaptive model, which slightly degrades the ideal distribution in order to land on convenient rational endpoints [25]. We do not have that luxury with ANS, because of the awkwardness of incorporating adaptive coding; consequently, it is not clear that there is any simple relationship between the arbitrary-precision and fixed-precision versions. But even with AC, that nice feature only applies to encoding; the approximate arithmetic seems to preclude a correctness argument in terms of the Unfoldr–Foldr Theorem, and therefore a completely different (and more complicated) approach is required for decoding [2].

The ANS algorithms themselves are of course not novel here; they are due to Duda [12,13]. Our development in Sect. 5 of the key bit of arithmetic in ANS encoding was informed by a very helpful commentary on Duda's paper published in a series of twelve blog posts [5] by Charles Bloom. The illustration in Fig. 3 derives from Duda, and was also used by Roman Cheplyaka [7] as the basis of a (clear but very inefficient) prototype Haskell implementation.

Acknowledgements. I am grateful to participants of IFIP WG2.1 Meeting 78 in Xitou and IFIP WG2.11 Meeting 19 in Salem, who gave helpful feedback on earlier presentations of this work, and to Jarek Duda and Richard Bird who gave much helpful advice on content and presentation. I thank the Programming Research Laboratory at the National Institute of Informatics in Tokyo, for providing a Visiting Professorship during which some of this research was done; in particular, Zhixuan Yang and Josh Ko of NII explained to me the significance of t dividing evenly into l, as exploited in Sect. 6.

References

1. Alakuijala, J.: Google's compression projects, May 2019. https://encode.ru/threads/3108-Google-s-compression-projects#post60072
2. Bird, R., Gibbons, J.: Arithmetic coding with folds and unfolds. In: Jeuring, J., Peyton Jones, S. (eds.) AFP 2002. LNCS, vol. 2638, pp. 1–26. Springer, Heidelberg (2003). https://doi.org/10.1007/978-3-540-44833-4_1
3. Bird, R.S.: Introduction to Functional Programming Using Haskell. Prentice-Hall, Upper Saddle River (1998)
4. Bird, R.S., Wadler, P.L.: An Introduction to Functional Programming. Prentice-Hall, Upper Saddle River (1988)
5. Bloom, C.: Understanding ANS, January 2014. http://cbloomrants.blogspot.com/2014/01/1-30-14-understanding-ans-1.html
6. Brettle, J., Galligan, F.: Introducing Draco: compression for 3D graphics, January 2017. https://opensource.googleblog.com/2017/01/introducing-draco-compression-for-3d.html
7. Cheplyaka, R.: Understanding asymmetric numeral systems, August 2017. https://ro-che.info/articles/2017-08-20-understanding-ans
8. Collet, Y., Turner, C.: Smaller and faster data compression with Zstandard, August 2016. https://code.fb.com/core-data/smaller-and-faster-data-compression-with-zstandard/
9. De Simone, S.: Apple open-sources its new compression algorithm LZFSE. InfoQ, July 2016. https://www.infoq.com/news/2016/07/apple-lzfse-lossless-opensource
10. Duda, J.: Kodowanie i generowanie układówstatystycznych za pomocą algorytmów probabilistycznych (Coding and generation of statistical systems using probabilistic algorithms). Master's thesis, Faculty of Physics, Astronomy and Applied Computer Science, Jagiellonian University (2006)
11. Duda, J.: Optimal encoding on discrete lattice with translational invariant constrains using statistical algorithms. CoRR, abs/0710.3861 (2007). English translation of [11]
12. Duda, J.: Asymmetric numeral systems. CoRR, 0902.0271v5 (2009)
13. Duda, J.: Asymmetric numeral systems: entropy coding combining speed of Huffman coding with compression rate of arithmetic coding. CoRR, 1311.2540v2 (2014)
14. Gibbons, J.: Metamorphisms: streaming representation-changers. Sci. Comput. Program. **65**(2), 108–139 (2007)

15. Gibbons, J.: Coding with asymmetric numeral systems (Haskell code), July 2019. http://www.cs.ox.ac.uk/jeremy.gibbons/publications/asymm.hs
16. Gibbons, J.: Coding with asymmetric numeral systems (long version), July 2019. http://www.cs.ox.ac.uk/jeremy.gibbons/publications/asymm-long.pdf
17. Giesen, F.: Simple rANS encoder/decoder, February 2014. https://github.com/rygorous/ryg_rans
18. Huffman, D.A.: A method for the construction of minimum-redundancy codes. Proc. IRE **40**(9), 1098–1101 (1952)
19. Joint photographic experts group. Overview of JPEG XL. https://jpeg.org/jpegxl/
20. Lee, T.B.: Inventor says Google is patenting work he put in the public domain. Ars Technica, October 2018. https://arstechnica.com/tech-policy/2018/06/inventor-says-google-is-patenting-work-he-put-in-the-public-domain/
21. Peyton Jones, S.: The Haskell 98 Language and Libraries: The Revised Report. Cambridge University Press, Cambridge (2003)
22. Reiter Horn, D., Baek, J.: Building better compression together with DivANS, June 2018. https://blogs.dropbox.com/tech/2018/06/building-better-compression-together-with-divans/
23. Rissanen, J.J., Langdon, G.G.: Arithmetic coding. IBM J. Res. Dev. **23**(2), 149–162 (1979)
24. Seward, J.: `bzip2` (1996). https://en.wikipedia.org/wiki/Bzip2
25. Stratford, B.: A formal treatment of lossless data compression algorithms. D. Phil. thesis, University of Oxford (2005)
26. Witten, I.H., Neal, R.M., Cleary, J.G.: Arithmetic coding for data compression. Commun. ACM **30**(6), 520–540 (1987)

Author Index

Printed in the United States
By Bookmasters